Gels Horizons: From Science to Smart Materials

Series Editor

Vijay Kumar Thakur, School of Aerospace, Cranfield University, Cranfield, Bedfordshire, UK

This series aims at providing a comprehensive collection of works on the recent advances and developments in the domain of *Gels*, particularly as applied to the various research fields of sciences and engineering disciplines. It covers a broad range of topics related to *Gels* ranging from *Polymer Gels, Protein Gels, Self-Healing Gels, Colloidal Gels, Composites/Nanocomposites Gels, Organogels, Aerogels, Metallogels & Hydrogels* to *Micro/Nano gels*. The series provides timely and detailed information on advanced synthesis methods, characterization and their application in a broad range of interrelated fields such as chemistry, physics, polymer science & engineering, biomedical & biochemical engineering, chemical engineering, molecular biology, mechanical engineering and materials science & engineering.

This Series accepts both edited and authored works, including textbooks, monographs, reference works, and professional books. The books in this series will provide a deep insight into the state-of-art of *Gels* and serve researchers and professionals, practitioners, and students alike.

More information about this series at http://www.springer.com/series/15205

Anuj Tripathi · Jose Savio Melo
Editors

Immobilization Strategies

Biomedical, Bioengineering and Environmental Applications

Springer

Editors
Anuj Tripathi
Nuclear Agriculture and Biotechnology Division
Bhabha Atomic Research Centre
Mumbai, Maharashtra, India

Jose Savio Melo
Nuclear Agriculture and Biotechnology Division
Bhabha Atomic Research Centre
Mumbai, Maharashtra, India

ISSN 2367-0061 ISSN 2367-007X (electronic)
Gels Horizons: From Science to Smart Materials
ISBN 978-981-15-7997-4 ISBN 978-981-15-7998-1 (eBook)
https://doi.org/10.1007/978-981-15-7998-1

This Springer imprint is published by the registered company Springer Nature Singapore Pte Ltd.
The registered company address is: 152 Beach Road, #21-01/04 Gateway East, Singapore 189721, Singapore

Preface

Immobilization is a universal phenomenon that does occur naturally and can attribute its occurrence to the origin of life. The process of arresting the movement of functional molecules like cells and enzymes by tethering or impregnating them in the stationary phase during process development is referred to as *immobilization*.

Considerable research in recent times has focused on the aspect of immobilization in multidisciplinary areas. Biocatalysis is a remarkable discovery in biological sciences with outstanding advantages to the bioprocess technologies but has inadequate scaling-up potential for societal benefits. The process of immobilization thus paves the way to utilize biocatalysts so as to overcome its limitations. In the initial phase, various active enzymes and cells have been successfully immobilized on support matrices for advancing the performance like improved stability and protection to labile biomolecules, biocatalysis in batch as well as continuous mode, repetitive application of biomolecules, evasion of product contamination by system components and easy separation. Further, the immobilization practices have been expanded in new dimensions so as not only to advance the biological process but also to develop smart materials and their performance too. New era of immobilization even broadens its application and deals with immobilization of non-biological substances, for example, smart nanoparticles for accelerating non-biological catalysis. On the other hand, few limitations usually do come along with immobilization practices, for example, deactivation of function of biomolecule during immobilization, abrasion of the immobilization matrix, diffusion limitation of the immobilization matrix and, most importantly, economic viability of the immobilization that depend on the application. However, the ultimate aim of immobilization is to significantly increase the performance of a complex system under the most adverse environmental conditions from in vitro to in situ in terms of industrial and clinical developments.

Advantages of immobilization in biological sciences and bioengineering have compelled scientists since several decades to rummage around for the development of advanced methods and to overcome the existing problems of bioprocess at the same time. In this book, the introductory chapter has dealt with the journey of immobilization, its fundamental principle followed by multidisciplinary research potential. Further, the chapters are arranged based on three major disciplines, i.e., biomedical, bioengineering and environmental applications, wherein direct or indirect immobilization strategies are applied to develop an improved and advanced system, for example, delivery vehicles, cell immobilized porous matrix as functional tissue substitute, development of biosensors for forensic and pesticides detection, development of functional materials using cold plasma, super-cooled liquids, biofilms, as well as recovery and remediation of scarce and toxic molecules. Each chapter highlights the principal factors involved in the development of immobilization-mediated advance application, which include the selection of molecule and its support structure used for immobilization, conditions and methods with respect to activity and stability of the immobilized molecule. We speculate that this book will stimulate further interest and research in immobilization.

Arduous task of compiling recent development in 'immobilization strategies: biomedical, bioengineering and environmental applications' and succinct pretension about their future entails greater leisure and extensive reading. In this book, we have sought-after scientist's expertise involved in the fundamental and applied areas of biological sciences and materials sciences, and performing immobilization-based advanced research around the globe, with the aim of providing an overview on contemporary preparation and diversified applications of functional molecules and materials. Although each chapter is a stand-alone reference, covering a given facet of immobilization strategies written by an expert in his/her zone of expertise, the editorial process has ensured that this edition is an excellent introductory reference book with broad field spectrum to academicians and scientists. This book is appropriate for advanced graduates, introductory-level researchers and also interdisciplinary and multidisciplinary scientists. It should also serve as a source book for statuary authorities for approving clinical and industrial applications. We once again thank all the contributory authors for their quality contribution and consistent cooperation that has facilitated us to make this book possible.

Mumbai, India

Dr. Anuj Tripathi
Dr. Jose Savio Melo

Contents

About the Editors

Dr. Anuj Tripathi is a scientist in the Bio-Science Group at Bhabha Atomic Research Center (BARC) and faculty at Homi Bhabha National Institute (HBNI), Mumbai, India. He completed his M.Sc. and Ph.D. in the field of Biotechnology. His doctoral study involved research on designing and applications of porous biomaterials in biomedicine and bioengineering. He has been a visiting research fellow at Newcastle University (UK), Lund University (Sweden), Kyushu University (Japan) and Protista Biotechnology AB, (Sweden). He received research fellowships/grants from different Government of India funding organizations like DBT, CSIR, DST, and from International Atomic Energy Agency (IAEA). He has published high-impact peer-reviewed research articles, book chapters, and patents in his area of research. He has edited a book with Springer Nature on the subjects Advances in Biomaterials and bulletin on Nano-biomaterials for theragnostic applications. He is an editorial board member and reviewer for several biomaterials and bioscience journals. He is a founding member of the society for interdisciplinary research in materials and biology (SIRMB) and also honored as a life member of several international science societies. His research is focused in development of novel nano-biomaterials and superabsorbent sponges by chemical and radiation chemistry for advancing the field of bioengineering, sustainable agriculture and environmental biotechnology. He has contributed significantly to cryostructurization of polymer systems and translated technologies of societal benefits.

Dr. Jose Savio Melo career spans over three decades in the field of research and development. He received his doctoral degree in Biochemistry from Mumbai University. He is currently the head of the "Enzyme and Microbial Technology Section" in the Bio-Science Group at the Bhabha Atomic Research Centre. He has contributed to science education as invited guest faculty at Institute of Chemical Technology, Mumbai and as resource person at National Workshops and Refresher Courses. He is also a Professor at Homi Bhabha National Institute and associated with Mumbai University and Savitribai Phule Pune University as a Ph.D. guide. He has published over a hundred papers in peer reviewed international journals, symposiums, workshops proceedings and book chapters and is credited with an overall citation score of 3500 and h-index of 31. In the field of enzyme and microbial technology, he has developed a number of novel heterogeneous biocatalyst systems that can be used in bioprocess development, waste treatment and biosensors. Dr. Melo's current field of interest is in interdisciplinary research and nanobiotechnology for better understanding and broadening the scope of ongoing programs. Dr. Melo has been honored as a Fellow of the Maharashtra Academy of Sciences and Society of Applied Biotechnology.

Immobilization: Then and Now

Jose Savio Melo, Anuj Tripathi, Jitendra Kumar, Archana Mishra, Bhanu Prakash Sandaka, and Kuber C. Bhainsa

Abstract Immobilization of biological molecules such as cells, enzymes, antibodies and other catalytically active cellular components greatly enhances their application in biocatalysis. Therefore, immobilized biomolecules have extensively been investigated for successful technological advancement. This chapter provides a comprehensive overview on the evolution of immobilization that took place over the years with the fundamental understanding on immobilization and importance of immobilization support structures called "scaffolds" followed by practical applicability reported in multidisciplinary fields like industrial bioprocessing, biomedicine, biosensing and biorecognition, as well as environmental biotechnology but partly restricted to activities in authors laboratory. In last section of the chapter, latest computational practices like biosimulation techniques have been discussed which helps in predicting the mode of steady execution and realization of biomolecule's immobilization for viable bioprocess developments.

Keywords Immobilization · Biocatalysis · Scaffold · Biosensor · Invertase · Bioremediation · Biosimulation

Abbreviations

1D	One-dimensional
2D	Two-dimensional
3D	Three-dimensional

J. S. Melo (✉) · A. Tripathi (✉) · J. Kumar · A. Mishra · B. P. Sandaka · K. C. Bhainsa
Nuclear Agriculture and Biotechnology Division, Bhabha Atomic Research Centre, Mumbai 400085, India
e-mail: jsmelo@barc.gov.in

A. Tripathi
e-mail: anujtri@barc.gov.in; chianuj@gmail.com

Homi Bhabha National Institute, Mumbai 400094, India

A. Tripathi and J. S. Melo (eds.), *Immobilization Strategies*,
Gels Horizons: From Science to Smart Materials,
https://doi.org/10.1007/978-981-15-7998-1_1

6-APA	6-Aminopenicillanic acid
AA	Agarose–alginate
AA	Ascorbic acid
AAM	Agarose–alginate–magnetite
AM	Additive manufacturing
AMP	Adenosine monophosphate
APTES	(3-Aminopropyl)triethoxysilane
AuNPNTs	Gold nanoparticle nanotubes
BMP-2	Bone morphogenetic protein-2
BSA	Bovine serum albumin
CAT	Catalase
ChEt	Cholesterol esterase
ChOx	Cholesterol oxidase
CLE	Cross-linked dissolved enzymes
CLEA	Cross-linked enzyme aggregates
CLEC	Cross-linked enzyme crystals
CNTs	Carbon nantubes
CPS	Controlled-pore silica
CSE	Centre of science and environment
CV	Cyclic voltammetry
DA	Dopamine
DNA	Deoxyribonucleic acid
DO	Dissolved oxygen
DOP	Degree of polymerization
DPV	Differential pulse voltammetry
ECM	Extra cellular matrix
EDC	1-Ethyl-3-(3-dimethylaminopropyl)-carbodiimide
EDTA	Ethylenediamine tetracaetic acid
EPA	Environmental protection agency
f-MWCNTs	Functionalized multiwalled carbon nanotubes
FG	Functionalized graphene
FGF	Fibroblast growth factor
GC	Gas chromatography
GM-CSF	Granulocyte macrophage-colony-stimulating factor
GO-SH	Thiol-functionalized graphene oxidase
GOD	Glucose oxidase
GQD	Graphene quantum dots
Gr	Graphite
HFCS	High fructose corn syrups
HPLC	High-performance liquid chromatography
HRP	Horseradish peroxidase
hUCMSCs	Human umbilical cord-derived mesenchymal stem cells
IGF-1	Insulin-like growth factor-1
IUPAC	International Union of Pure and Applied Chemistry
LBL	Layer by layer

MAAS	Melanin–agarose–alginate–silica
MAC	Melanin-functionalized agarose–chitosan
MD	Molecular dynamics
MM	Molecular mechanics
MP	Methyl parathion
MWCNTs	Multi-walled carbon nanotubes
NHS	N-hydroxysuccinimide
NP	Nanoparticles
OPH	Organophosphorus hydrolase
PAAM	Polyacrylamide
PC	Polycarbonates
PCG	Polyacryonitrile-chitosan-graphene oxide
PDDA	Poly(diallyldimethylammonium chloride)
PDLA	Poly(d-lactic acid)
PDMS	Poly(dimethylsiloxane)
PDS	Polydioxanone
PE	Polyethylene
PEEK	Polyether ether ketone
PEG	Polyethylene glycol
PEI	Polyethylene imine
PES	Polyether sulfone
PET	Poly(ethylene terephthalate)
PGA	Poly(glycolic acid)
PH3B	Poly(3-hydroxybutyrate)
PHA	Polyhydroxyalkanoates
PHBV	Poly(3-hydroxybutyrate-co-3-hydroxyvalerate)
pHEMA	Poly (hydroxyethyl methacrylate)
PHV	Polyhydroxyvalerate
PLLA	Poly(l-lactic acid)
PMMA	Poly (methyl methacrylates)
PMMA	Poly-methyl methacrylate
PP	Polypropylene
PSS	Polystyrene sulfonate
PTFE	Poly(tetrafluoroethylene)
PTh	Polymerized thiophene
PVA	Polyvinyl alcohol
PVC	Poly(vinyl chloride)
Q-AuNPs	Quercetin stabilized gold nanoparticles
QD	Quantum dots
QM	Quantum mechanics
Rd	Rhoeo discolour
RDF	Radial distribution function
RMSD	Root mean square deviation
RMSF	Root mean square fluctuation
RNA	Ribonucleic acid

ROS	Reactive oxygen species
SASA	Solvent accessible surface area
SCPL	Solvent casting and particulate leaching
SPCE	Screen printed carbon electrode
SWCNTs	Single-walled carbon nanotubes
TGF- β	Transforming growth factor β
TIPS	Thermally induce phase separation
TSA	Transition state analogue
UA	Uric acid
VEGF	Vascular endothelial growth factor
WHO	World Health Organization
ZnO	Zinc oxide

1 Introduction

Immobilization means "attachment of a molecule with a suitable support with objective to improve/retain their catalytic activities which can be used repeatedly and continuously". Immobilization has been widely explored for immobilizing a wide variety of biomolecule as well as inorganic metals/pollutant too. Immobilized molecules especially enzymes/micro-organism are very crucial for industrial processes and have gained lot of importance in comparison to their free counterparts. Using immobilization techniques, various biotechnology products which have application in bioprocessing, diagnostics, chromatography and biosensors have been developed.

During the initial phase of immobilization, the focus was on immobilization of single enzymes usually with adsorption technique. However, with time immobilization has evolved and after 1970s, two enzyme system and living cells were successfully immobilized. The technique of immobilization also plays a very important role. A biomolecule can be attached to the support either by reversible physical adsorption or irreversible stable covalent bonds. The choice of the suitable immobilization technique depends on the nature of the biomolecule and the support. Immobilization helps to improve/retain catalytic as well as thermal stability of immobilized biomolecule. Thus, it has been widely explored as a technique to improve operational stability of various biomolecule.

Although the development/progress of biomolecule immobilization techniques has been discussed for years (Cao 2005), yet there is still a need to discuss important milestones in the history of immobilization of enzymes and whole cells from way back 1916 till today in order to understand the immobilization techniques and its future applications.

2 History of Immobilization

The history of biomolecule immobilization can be broadly divided into three phases (Fig. 1):

- Pre-mature phase (1916–1950),
- Maturation phase (1960–1980) and
- Advanced designing phase (1990–present).

In this classification, the focus will be on the major developments in immobilization techniques and a brief discussion regarding the evolution of immobilization in the last 100 years.

2.1 *The Pre-mature Phase (1916–1950)*

In 1916, Nelson and Griffin discovered that invertase immobilized on charcoal was catalytically active (Nelson and Griffin 1916). This was the first scientific observation that led to the discovery of immobilization of enzyme and widely recognized as the milestone of the various enzyme immobilization techniques currently available.

During this phase, the application of immobilization techniques was mainly used to prepare adsorbents for isolation of proteins using adsorption on inorganic carriers such as glass (Harkins et al. 1940), alumina (Gale and Epps 1944) or hydrophobic compound-coated glass (Langmuir and Schaefer 1938). However, in some studies, enzyme was also immobilized irreversibly by covalent attachment (Micheel and Evers 1949).

Till 1950s, the method of immobilization was dominated by physical methods, i.e. non-specific physical adsorption of enzymes on solid carriers, for example, α-amylase adsorbed on activated carbon, bentonite or clay (Stone 1955), AMP deaminase on silica (Dickey 1955) and chymotrypsin on kaolinite (Mclaren 1957). During this time, the method of adsorption was shifted from simple physical adsorption to

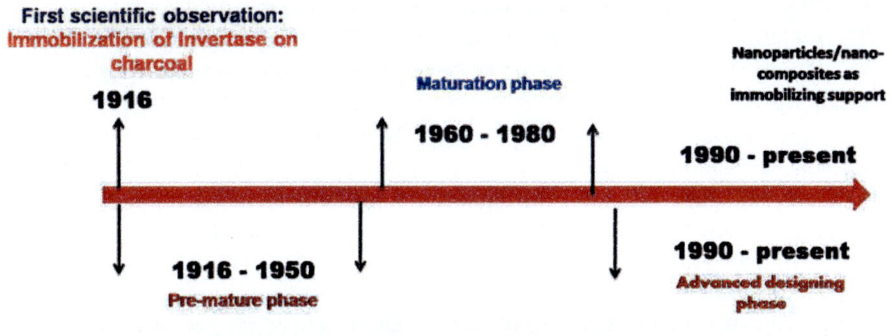

Fig. 1 Evolution of immobilization (year-wise progression)

ionic adsorption which was successfully demonstrated by immobilization of ribonuclease on the anionic exchanger Dowex-2 and the cationic exchanger Dowex-50, respectively (Barnet and Bull 1959).

Herein, adsorption was widely explored for immobilization along with covalent immobilization (Grubhofer and Schleith 1954). However, supports developed in this phase were highly hydrophobic in nature and not suitable for covalent enzyme immobilization, because of poor retention of activity (2–20% of the native activity) (Brandenberg 1955). For the first time, Dickey showed that enzyme adenosine monophosphate (AMP) deaminase entrapped in the sol–gel inorganic matrix formed by silicic acid derived glasses retained reasonable biological activity (Dickey 1955). However, the importance of this report was not fully explored at that time (Reetz et al. 2000).

2.2 *Maturation Phase (1960–1980)*

In 1960s

Herein, the focus was on covalent methods of immobilization, yet the non-covalent process of immobilization like adsorption (Tosa et al. 1967) and entrapment (Hicks and Updike 1966; Mosbach and Mosbach 1966) was also explored. Along with this, encapsulation of enzymes in semi-permeable spherical membranes was first proposed by Chang (1964). Enzyme entrapment techniques were also further explored by the use of PAAM (Leuschner 1966), PVA (Brown et al. 1968), starch (Pennington et al. 1968) and silicon elastomers for the sol–gel process (Quiocho and Richards 1964; Haynes and Walsh 1969). Other methods of enzyme immobilization like adsorptive cross-linking of enzymes on membranes and films or beads for the formation of enzyme envelopes were also developed (Quiocho and Richards 1966).

There were reports which showed that insoluble carrier-free immobilized enzymes were synthesized by cross-linking of crystalline enzymes (Axen et al. 1967) and dissolved enzymes (Kay and Crook 1967) by use of a glutaraldehyde. Yet, the potential of cross-linking of enzyme crystals was not recognized at that time, intensive studies were carried out for preparation of carrier-free immobilized enzymes, especially CLE (cross-linked dissolved enzymes), as immobilized enzymes.

The scope of immobilization was greatly explored in order to use more hydrophilic insoluble carriers with defined geometric shapes such as cross-linked dextran, agarose and cellulose beads. Also, new methods were developed for activation of support such as cyanogen bromide (Manecke and Gunzel 1967) and triazine for polysaccharides (Patel et al. 1967), isothiocyanate for coupling amino groups (Bernfeld et al. 1969) and Woodward reagents (Mosbach 1976a) for activation of carboxyl groups. One of the important findings is that not only the soluble enzyme but also the enzyme crystals could be entrapped in gel matrix and their activity was retained (Buchholz and Kasche 1997). By the end of 1960s, the first industrial application of an immobilized enzyme for production of L-amino acids from racemic amino acid derivatives was developed (Tosa et al. 1967). This work showed the industrial applicability of immobilized enzymes and also inspired several new research interests in the field of immobilization (Hipwell et al. 1974; Kennedy and Epton 1973).

In 1970s

In this period, earlier developed immobilizing techniques were used for immobilization of other industrially important enzymes such as invertase, α-amylase, penicillin G-acylase, acylase, etc. Along with four main methods (adsorption, covalent, entrapment and encapsulation), many new subgroups of immobilization techniques, for example affinity binding, coordination binding, etc., were developed. New techniques were developed in order to improve the performance of the immobilized enzymes. It was observed that immobilization of the enzymes through affinity adsorption using spacer can improve the enzyme activity (Martinek et al. 1977). Coordination immobilization which deals with combined immobilization and regeneration of the carrier was also explored during this phase (Wykes et al. 1971). Enzymes were also immobilized on soluble supports (Monsan et al. 1971) and complimentary multi-point attachment was used with the motive to improve enzyme stability (Hueper 1974).

Another observation was that through chemical modification also, the characteristics of enzymes could be improved. Thus, chemically modified enzymes with improved properties (like enhanced stability) have been further immobilized (Bartling and Brown 1973). Another important observation was that enzyme immobilization could be performed in organic solvents also (Hartdegen and Swann 1976; Manecke and Polakowaski 1981). This method had great impact for modulation of enzyme conformation; however, it was not fully explored at the time.

In this period, the focus was to gain insight on the effect of various factors like the microenvironment of the carrier (Zaborsky 1972), the effect of the spacer or arm (Barker et al. 1970), different modes of binding (Ollis 1972), enzyme loading (Bernath and Vieth 1972), changes in the conformation of the enzyme, diffusion barriers (Miwa and Ohtomo 1975; Mosbach 1976a) and orientation of the enzyme (Messing 1975) on the activity of the immobilized enzymes.

Many new strategies were developed to improve the performance of the immobilized enzymes and published (Messing and Filbert 1975; Mosbach 1976b). Because of this extensive work, the potential of enzyme immobilization techniques in commercial processes got recognized and many commercial processes with use of immobilized enzymes have been developed, like use of immobilized penicillin G-acylase for production of 6-aminopenicillanic acid (6-APA) and the application of immobilized glucose isomerase for production of fructose syrup from glucose. Immobilized enzyme was further explored in other areas like controlled release protein drugs and biomedical application.

By the end of 1970s, enzyme immobilization techniques had matured to such extent that every enzyme could be immobilized by selecting a suitable method of immobilization (entrapment, encapsulation, covalent attachment and adsorption) as shown in Fig. 2, or a suitable support and immobilization conditions (aqueous, organic solvents, pH, temperature, etc.). Herein, it comes in the picture that the main problem was not immobilization of the enzymes on the support but how to obtain the desired performance from the immobilized enzyme.

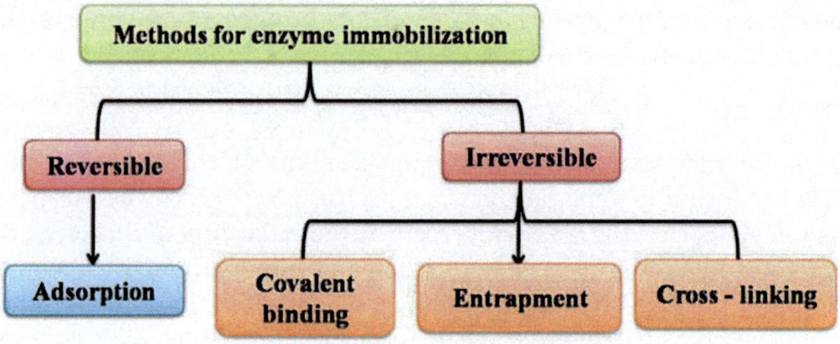

Fig. 2 Different methods of immobilization

In 1980s

During this period, work was carried out to design robust immobilized enzymes in order to increase their applicability in manufacturing of pharmaceuticals and agrochemicals which would help to overcome strict environmental regulations, lower energy consumption and less time consumption compared with conventional chemical processes.

It was observed that many enzymes are also active and stable in organic solvents (Zaks and Kllbanov 1985; Lee et al. 1986). It was proposed that immobilizing an enzyme by a combination of covalent, layer-by-layer(LBL) and cross-linking techniques could improve enzyme loading and enzyme stability. Also, cross-linking of crystalline enzymes can be used to create stable biocatalysts for biotransformation, especially in organic solvents because of their high stability in these solvents. Various methods and concepts were developed to make the enzymes more active in organic solvents. The concept of the water activity of the reaction medium was proposed (1986).

Extensive work was carried out to understand the effects of support and immobilization methods on the catalytic property of the immobilized enzymes by encaging the enzyme (Tor et al. 1989) or multipoint attachment to the support (Martinek et al. 1977; Blanco et al. 1988). Many earlier developed techniques were further explored to improve enzyme performance.

2.3 *Advanced Designing Phase (1990–Present)*

In the 1990s, there was an important transition in the development of immobilized enzymes. Approaches used for the design of immobilized enzymes became more rational in order to overcome the limitations of previously developed single immobilization approaches. Site-specific enzyme immobilization gained lot of attention because of improved features (Persson et al. 1990). Along with the free enzymes, whole cells expressing enzymes were also successfully used for different applications in non-aqueous/aqueous media (Niiolova and Ward 1993; Ramkrishna and Prakasham 1999).

During this phase, the major concern of enzyme immobilization was developing robust enzymes that are stable and selective in organic solvents. Although in the period from the 1970s to the 1980s, it was recognized that many enzymes are active and stable in organic solvents under appropriate conditions; however, the used enzymes are usually less active or stable in organic solvents than in conventional aqueous media (Klibanov 1997). Due to this, development of robust immobilized enzymes which can work in non-aqueous media became area of interest in this period (Kawakami et al. 1992). Studies were carried out for the development of cross-linked enzyme crystals (CLEC) suitable for biotransformation in non-aqueous media or in organic-water mixtures, because of the higher stability of the enzymes (St Clair and Navia 1992; Margolin 1996). However, a deep investigation of the performance of CLEC and comparison with conventional support-bound immobilized enzymes is still missing. There are reports which show that the turnover frequency of cross-linked enzymes in organic solvents is lower than that of carrier-bound immobilized enzymes which suggests that confinement of the enzyme molecules in the compact crystal lattice or diffusion limitation could be major factors responsible for the lower activity (Secundo et al. 1999). Many efforts have also been devoted to developing novel strategies for improving the performance of immobilized enzymes (Fernandex-Lafuente et al. 1995; Partridge et al. 1998; Furukawa et al. 2002). Reetz et al. performed entrapment of lipases in hydrophobic sol–gel materials and reported efficient heterogeneous biocatalysts in aqueous medium (Reetz et al. 2000).

In the 1990s, it was also observed that the performance of the immobilized enzymes can be also improved by post-immobilization techniques like strengthening the multipoint attachment and consecutive treatment of immobilized enzymes by chemical or physical modification or other stabilization techniques (Guisan et al. 1993; Rocha et al. 1998).

The molecular imprinting techniques, which were proposed in the 1970s, were further investigated and it was observed that even the stable immobilized enzymes can be imprinted (Costantino et al. 1997). In another study, it was found that stable form of *Candida rugosa* lipase which was covalently immobilized on silanized controlled-pore silica (CPS) previously activated with glutaraldehyde in the presence of PEG-1500 was increased five-fold compared with the immobilized enzyme without addition of PEG-1500 (Soares et al. 2001). This shows that the enzyme conformation induced by the effectors (or additives) was imprinted.

In the end of 1990s, it was discovered that not only enzyme crystals but also physical enzyme aggregates could be cross-linked to form catalytically active insoluble immobilized enzymes which nowadays are known as cross-linked enzyme aggregates (CLEA) (Cao et al. 2000; Cao and Elzinga 2003). By adding the advantages of carrier-bound and carrier-free immobilized enzymes, CLEA can be tailor-made in order to have both non-catalytic and catalytic function which has impact in industrial applications.

In general, the techniques currently used for creation of robust immobilized enzymes, which meet both catalytic requirements (desired activity, selectivity and stability) and non-catalytic requirements (desired geometric properties such as shape, size and length) expected for a given process, are all characterized and used to solve problems that are unsolvable by the straightforward method.

In this duration, also the focus was shifted on application of nanoparticles as immobilizing support due to various unique and advantageous properties of nanoparticles. These nanoparticles have far fetching impact in the field of bioprocessing, bioremediation, environmental monitoring, immunosensing and biomedical. There are large number of publications where enzymes were immobilized on scaffolds like spheres, fibres, tubes and monoliths (Mitchell et al. 2002; Martin and Kohli 2003; Yim et al. 2003; Kim et al. 2006; Tripathi and Kumar 2012).

The motive behind using nanoscale structures for immobilization is to reduce diffusional limitations and maximize the functional surface area to increase enzyme loading (Xie et al. 2009). Also improved thermal stability, increased surface area and irradiation resistance are the other advantages (Moghaddam et al. 2009). Liu et al. (2009a) have reported covalent attachment of NAD(H) to silica nanoparticles and found successful coordination of particle-immobilized enzymes which enabled multistep biotransformation. Immobilization of glucose oxidase on the surface of silica nanoparticles was performed and on entrapment leads to decrease in reaction rate and an increase in apparent K_m of immobilized glucose oxidase (Jang et al. 2010). In another study, Husain et al. have performed comparative stability studies of β galactosidase immobilized on native ZnO and ZnO-NP and results showed that enzyme immobilized on ZnO-NP retains higher acitivity (Husain et al. 2011). Gold nanoparticles were also functionalized and used for immobilization of enzymes like pepsin and xanthine oxidase (Zhao et al. 1996; Gole et al. 2001; Jun et al. 2001). Crespilho et al. (2009) developed silver nanoparticles impregnated electrode for immobilization of urease enzyme and results showed a fast increase in cathodic current on addition of urea. Due to its inherent feature of magnetic separation, various magnetic supports have been used for immobilization successfully (Koneracka et al. 1999; Dyal et al. 2003; Saiyed et al. 2003; Kouassi et al. 2005).

Organic porous scaffolds and nanoparticles were also synthesized and used as immobilizing support for enzymes due to its biocompatibility and functionality aspect. Crude form of tyrosinase (extracted from *A. Campanulatus*) was immobilized on the porous cryogel scaffolds for biogenic transformation of L-DOPA to melanin (a natural pigment), which showed high free-radical scavangening activity and electro-catalytic property and advocating its potential drug delivery applications (Saini et al. 2015). Ho et al. (2008) have developed one-step method for preparing cellulase-immobilized on nanoparticles that consisted of well-defined poly-methyl methacrylate (PMMA) cores and cellulase shells. Immobilized cellulase showed improved thermostability and retained significantly very high activity at broader pH range as compared to soluble enzyme. Miletic et al. (2010) have immobilized lipase on polystyrene nanoparticles and found that the activity of immobilized enzyme was remarkably improved as compared to the free enzyme.

In further studies, quantum dots (QDs) were explored as immobilizing support. QDs are nanometer-scale semiconductor crystals with unique quantum confinement effects. They work on the principle of fluorescence transduction due to direct or indirect interaction of analyte with the QD surface, either through photoluminescent activation or through quenching. These are used for a wide variety of applications ranging from detection of pH and ions to quantification of organic derivatives and biomolecules (DNA, RNA, enzymes, proteins, amino acids and drugs). A list of cellular components and proteins labelled with quantum dots can be found in the review by Mintz et al. (2019). Quantum dot can also function as nanoscaffold for proteins (Gupta et al. 2011). Graphene quantum dots (GQD) impregnated electrodes were also synthesized and successfully used for immobilization of glucose oxidase enzyme (Razmi and Mohammad-Rezaei 2013).However, the applications of QDs are limited because of their toxicity and limited reusability. There is need to develop suitable conjugation methods for the immobilization of protein/enzymes with quantum dots which will find application in biocatalysis/biosensor in near future.

In this period, a wide range of nanoparticles has been applied as immobilizing support. However, there are some issues which are associated with application of nanoparticles like separation of nanoparticles and resistance towards substrate diffusion. To address this, efforts have been made to assemble the nanoparticles in higher length scale and synthesize fibrous/dendritic-shaped structures. Polshettiwar et al. (2010) have used chemical method for synthesis of fibrous silica nanospheres which are of different sizes. Hierachically structured hollow silica spheres were also synthesized by Cao et al (2013) and results showed high enzyme immobilization efficiency. Recently, our group has developed a simple and efficient route for synthesis of fibrous silica-based biohybrid material wherein for the first time a seed (*Ocimum basilicum*) has been explored as template to assemble silica nanoparticles. The study led to the synthesis of two different morphological biohybrid materials (Silica@seed and Silica@PEI-seed). Immobilized invertase enzyme showed improved catalytic performance (Mishra et al. 2020).

In the span of 104 years (1916–2020), various supports (natural/organic/inorganic) have been applied successfully for immobilization of enzymes as well as whole cells, and with time, immobilizing techniques have also evolved a lot. The milestones achieved in the field of immobilization during these 100 years are listed in Table 1.

3 Scaffolds: Significance in Immobilization

3.1 Fundamentals of Biomedical Scaffolds

The immobilization surfaces are widely explored in order to provide a suitable microenvironment to the biologically active molecules that catalyzes the complex chemical reaction under the benign biological conditions. Selection of precursor materials with suitable properties (Table 2) and their fabrication to achieve an ideal

Table 1 List of different immobilizing supports, immobilizing methods and achievements in the field of immobilization

Duration (Years)	Immobilizing supports	Progression of immobilizing methods and achievements of immobilization
1916–1950	**Natural polymers** Cellulose **Synthetic polymers** Amberlite **Inorganic supports** Activated Carbon Silica Glass Alumina Clay Kaolinite	First scientific observation of immobilization of invertase (Nelson and Griffin 1916) Physical adsorption Adsorption on inorganic carriers such as glass (Harkins et al. 1940), alumina (Gale and Epps 1944) Covalent attachment of enzymes (Micheel and Evers 1949) AMP deaminase immobilization on silica (Dickey 1955) Ionic adsorption (Barnet and Bull 1959)
In 1960s	**Synthetic polymers** Polyacrylamide PAAM PVA Nylon Polystyrene **Natural polymers** Starch Agarose Destran DEAE-cellulose **Inorganic supports** Carbon Clay Silica gel Silicon elastomers	Entrapment of enzyme and cells (Hicks and Updike 1966; Mosbach and Mosbach 1966) First time, encapsulation of enzymes in semi-permeable spherical membranes (Chang 1964) Adsorptive cross-linking of enzymes to develop enzyme envelopes (Quiocho and Richards 1966) Synthesis of insoluble carrier-free immobilized enzymes (Axen et al. 1967) by cross-linking of crystalline enzymes (Axen et al. 1967) and dissolved enzymes (Kay and Crook 1967) First industrial application of an immobilized enzyme for production of L-amino acids from racemic amino acid derivatives (Tosa et al. 1967)
In 1970s	**Synthetic supports** Halogen, Epoxy ring Acylazide, Carbonate **Polymers used for entrapment** PVA-SbQ, PEG-DMA PEG-CA, ENTP **Inorganic support** Silica **Natural polymers** Alginate Gelatin Agarose	Affinity immobilization (Martinek et al. 1977) Coordination immobilization (Wykes et al. 1971) Immobilization on soluble supports (Monsan et al. 1971) Complimentary multipoint attachment (Hueper 1974) Enzyme immobilization in organic solvents (Hartdegen and Swann 1976; Manecke and Polakowaski 1981) Application of immobilized penicillin G-acylase and glucose isomerase in industry

(continued)

Table 1 (continued)

Duration (Years)	Immobilizing supports	Progression of immobilizing methods and achievements of immobilization
In 1980s	**Synthetic porous supports** Commercialization of supports like Eupergit C **Other types of support groups** Various other polymeric supports have been made commercially available	Observed that enzymes are active and stable in organic solvents (Zaks and Kllbanov 1985; Lee et al. 1986) Introduction of concept of the water activity of the reaction medium (1986) Encaging of the enzyme (Tor et al. 1989) Introduction of orientation groups to the carrier-bound multipoint attachment (Martinek et al. 1977; Blanco et al. 1989)
1990-present	***Biomaterials/nanocomposites*** Silica nanoparticles Zinc oxide nanoparticles Gold nanoparticles, Silver nanoparticles Iron nanoparticles Mesoporous nanomaterials Mesoporous silica Quantum dots Porous cryobeads Dendritic/fibrous support	Site-specific enzyme immobilization (Persson et al. 1990) Immobilization of whole cells expressing enzymes in organic solvents (Niiolova and Ward 1993) Development of CLEC for biotransformation in non-aqueous media or in organic-water mixture (St Clair and Navia 1992; Margolin 1996) Heterogeneous biocatalysts: entrapment of lipases in hydrophobic sol–gel materials (Reetz et al. 2000) Introduction of post-immobilization techniques (Guisan et al. 1993; Rocha et al. 1998) Enzyme imprinting (Costantino et al. 1997) Development of catalytically active insoluble immobilized enzymes CLEA (Cao et al. 2000; Cao and Elzinga 2003) Enzyme immobilization on nanosized scaffolds like spheres, fibres and tubes (Mitchell et al. 2002; Martin and Kohli 2003; Yim et al. 2003; Kim et al. 2006) Application of nanoscale supports to reduce diffusional limitations and maximize surface area to improve enzyme loading (Xie et al. 2009) Immobilization of whole cells in cryobeads (Tripathi et al. 2010) Quantum dots as immobilizing support (Mintz et al. 2019)

matrix plays a vital role in the development of a successful technology. These immobilization surfaces have been evaluated by immobilizing of biomolecules in many phases that include one-dimensional (1D), two-dimensional (2D) and three-dimensional (3D) (Tripathi and Melo 2017). This section is focused on the fundamental aspect of matrix designing, principal parameters and applications in biomedicine by emphasizing the tissue engineering and drug delivery fields.

As the name implies, process of immobilization is merely arresting or entrapping of biomolecule within a matrix that may be constructed de novo around the biomolecules. A wide range of immobilization mediums from biological and non-biological origins are available to develop porous polymeric scaffolds (Table 3).

Table 2 Classification of hydrogels based on their properties and types

<table>
<tr><th>Properties</th><th>Types</th></tr>
<tr><td>Polymer charges</td><td>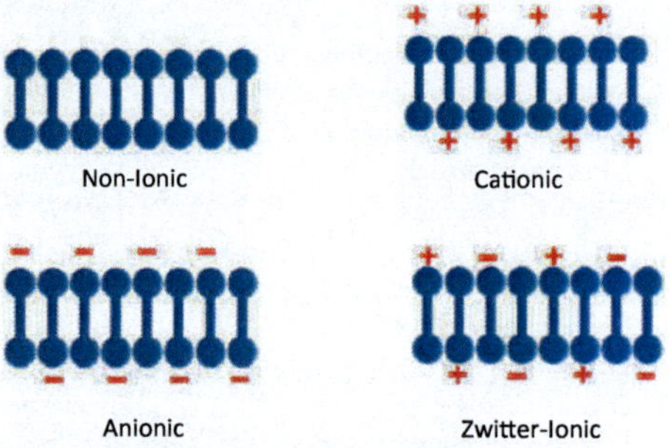
</td></tr>
<tr><td>Polymer networks</td><td>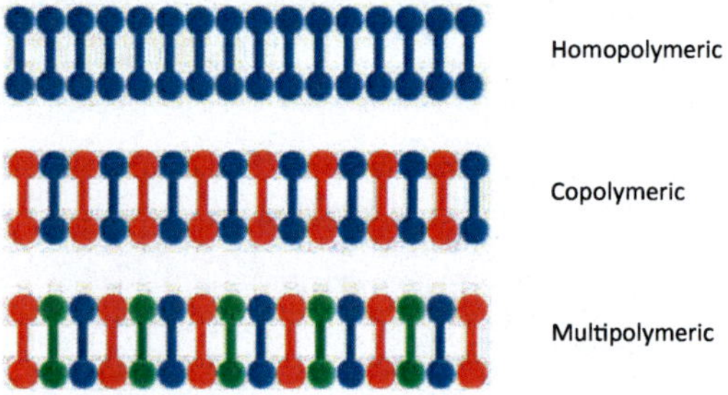
</td></tr>
<tr><td>Cross-linking</td><td>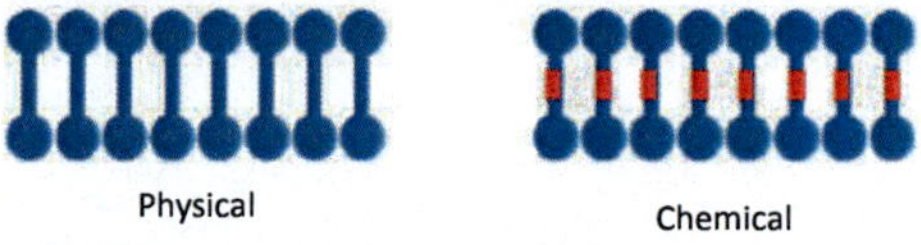
</td></tr>
<tr><td>Molecular arrangements</td><td>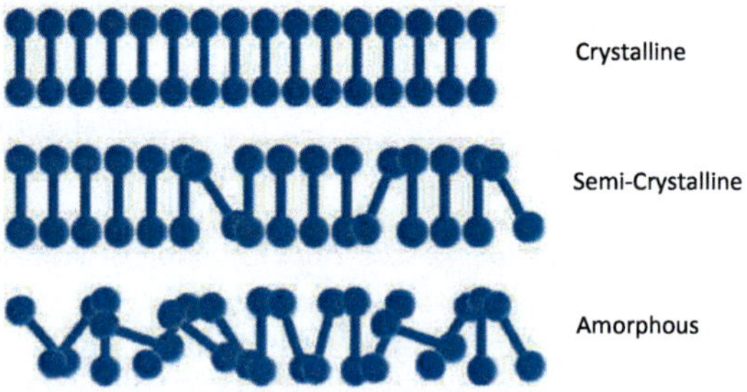
</td></tr>
</table>

(continued)

Table 2 (continued)

Properties	Types
Structure	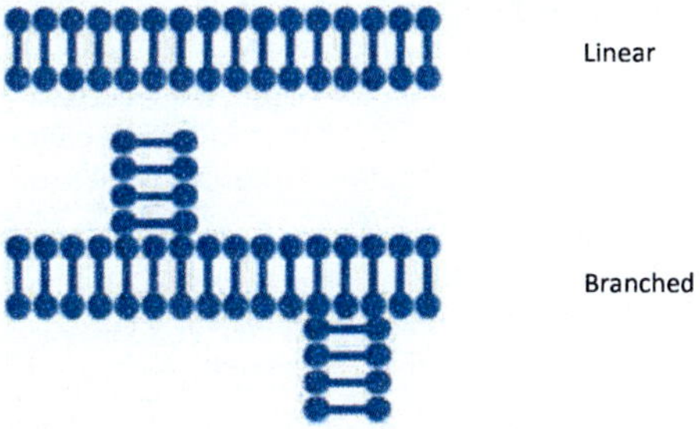
Porosity	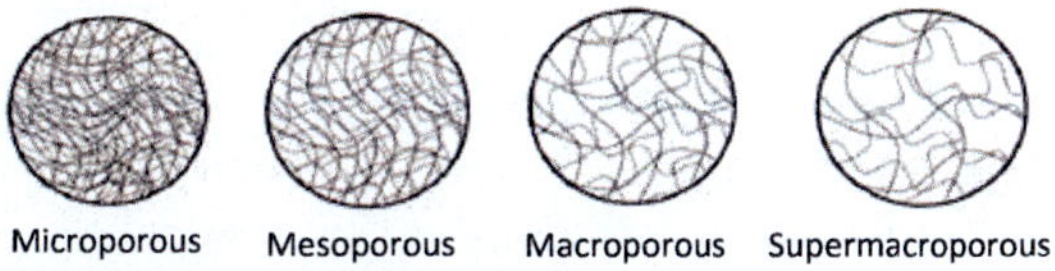
Composition	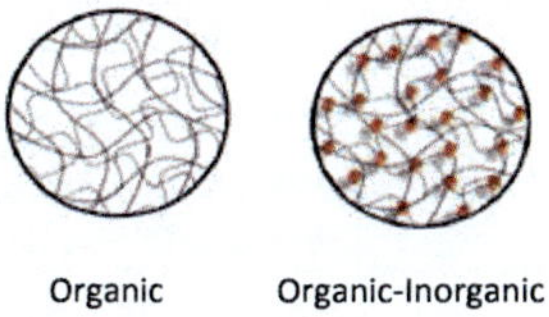

However, necessity of biomedical biomaterials has to endure through time in an intimate contact with cells and tissues. Hydrogels are considered as one of the most promising categories of biomaterials in biomedicine due to their many unique properties. Many synthetic polymers are listed in Table 3 that have demonstrated their vital role in the development of novel polymer materials for soft tissue replacement and also in the development of miniaturized devices, prosthesis or biomimetic tissues that can be used to replace arteries and heart valves. Besides, biodegradable polymers have truly modernized the field of controlled drug delivery developments and biomaterials fabrication for regenerative medicine. In addition to the benefits of synthetic polymers, naturally derived biopolymers have several addional advantages wherein polymeric materials may act as a blueprint of natural ECM with analogous molecular structure and morphology. Biopolymers work not only at macroscopic level but also on cellular level when implanted inside the body and degrade by simple enzymatic action of body's enzymes. Several biopolymers and their modified forms are being continuously examined by researchers for potential clinical developments in regenerative medicine. Cell to material interaction is essential information which aids to the design of new biocompatible and biodegradable polymeric materials in a three-dimensional format for specific tissue engineering application. These structures are also useful for encapsulation of single or multiple growth factors and maintain them biologically active for prolonged period of time, thus significantly enhancing the

Table 3 Classification of polymers based on their origin and source for their use in biomedicine

Origin	Class	Polymers
Natural	Polysaccharides	Alginate, agarose, amylase, arabinogalactan, carrageenan, cellulose, chitosan, chondroitin sulphate, dextran, hyaluronic acid, inulin, pectin, pullulan, glucomannan, guar gum, xantham gum, starch, tragacanth
	Polypeptides	Collagen, gelatin, silk fibroin, sericin, elastin, albumin, soy protein, zein, wheat gluten
	Polynucleotides	deoxyribonucleic acid (DNA) and ribonucleic acid (RNA)
	Polyhydroxyalkanoates (PHA)	poly(3-hydroxybutyrate) (PH3B), polyhydroxyvalerate (PHV) and poly(3-hydroxybutyrate-co-3-hydroxyvalerate) (PHBV)
Synthetic	Polyolefins	polyethylene (PE) and polypropylene (PP)
	Poly(tetrafluoroethylene) (PTFE)	Gore-Tex® made of ePTFE
	Poly(vinyl chloride) (PVC)	poly(vinyl chloride) (PVC)
	Silicone	poly(dimethylsiloxane) (PDMS)
	Polyacrylates	poly (methyl methacrylates) (PMMA) and poly (hydroxyethyl methacrylate) (pHEMA)
	Polyesters	polycarbonates (PC), poly(ethylene terephthalate) (PET, dacron), poly(glycolic acid) (PGA), poly(l-lactic acid) (PLLA), poly(d-lactic acid) (PDLA) and polydioxanone (PDS)
	Polyethers	polyether ether ketone (PEEK) and polyether sulfone (PES)
	Polyurethane	polyurethane
	Polyamide	Nylon-6

proliferation and differentiation of encapsulated cells and thus improve confined tissue regeneration.

Over the past decades, comprehensive research on novel biomaterials provides the flexibility to produce permanent or reversible matrices. Soft and reversible gel matrices may have poor stability and functional performance while the permanent matrix may involve toxic steps during its preparation. All these gel/matrix preparation are standardized based on the system requirement, for example, mild synthesis procedures (biologically less toxic) are used to develop immobilization matrix for retaining the high cellular viability besides it can be adequate where the matrix durability for long performance is essential. Also, the adequate number of cells can propagate by supplying suitable nutrients and growth parameters, thus process intensification. Though, this may lead to the abrasion and disruption of immobilization matrix, it results in leakage of cells in case of milder matrix application.

In the field of tissue reconstruction and biomolecules delivery, immobilization is a key event that leads a well-organized integration of various cellular and molecular events (like cell adhesion, proliferation, migration and differentiation) which ultimately direct a re-forging of 3D functional tissue. Although, the start of cell culture in two-dimensional began in early 1900s and gained wide acceptance during 1940–1950. In 2D culture systems, cells grow on the flat surfaces typically made of surface-coated plastic to provide suitable adherence and spreading. However, the 2D systems do not provide biomimetic microenvironment to the cells; thus, it lacks predictive nature that increases the trial cost of system as well as chances of failure of clinical drug discovery and functional tissue genesis. Despite of these inherent flaws, 2D culture systems are still the first preference for cell-expansion and drug testing analysis because it is well established and economically less-expensive. Cell adhesion and growth in 2D can be easily monitored under the microscope and measurements are not influenced by the inert support matrix. Expansion of an established cell line is also quiet easy as it can sub-culture by easy detachment from 2D surfaces. Due to this, researchers have performed several comparative studies using different cell types and culture conditions which are available as published literature to understand the fundamentals. Few studies have shown well-augmented 3D tissue genesis by immobilizing various cell types including stem cells and vascularization using magnetic nanoparticles under optimized magnetic field on 2D surfaces (Tripathi et al. 2013c; Horie et al. 2017). Still these properties delimit the advancement of regenerative medicine.

On the other hand, cell culturing in three-dimensional systems has been around for longer than what we might think. The first established in vitro cell culture of frog embryo nerve fibres entrapped in hanging drop system was reported by Ross Granville Harrison in the year 1907. Revolution in cell and tissue culture in term of functional tissue engineering was established by Lanza et al. (2000) and Atala and Lanza (2002). Later, Atala and his group have demonstrated 3D tissue and organ bioprinting techniques (Murphy and Atala 2014).

The process of tissue engineering utilizes coactions of cells, porous scaffold and growth factors in an integrated system which eventually leads to the formation of an analogous functional 3D tissue structure resembling identical to the native tissue produced by same cell or cells in the living human body (Fig. 3). Therefore, designing of a complex information-coded biomimetic 3D scaffold can have greater influence in interaction of biomolecules like cells to the surface followed by its fate. During the tissue regeneration, polymeric hydrogels act like an extra cellular matrix (ECM) that serve as a 3D microenvironment for growing the cells in three dimensions. This synthetic ECM is providing amiable porosity for governing the desired biological functions similar to native tissue. Thus, fabrication of synthetic ECM that can biomimic the desired tissue or organ functions is of prime importance in many biomedical applications. Disciplines like cell and molecular biology, materials sciences, chemical sciences, computational biology, bioreactors and bioprocessing, and clinical operations are some of the essential fields for successful reconstructive surgery and targeted drug delivery.

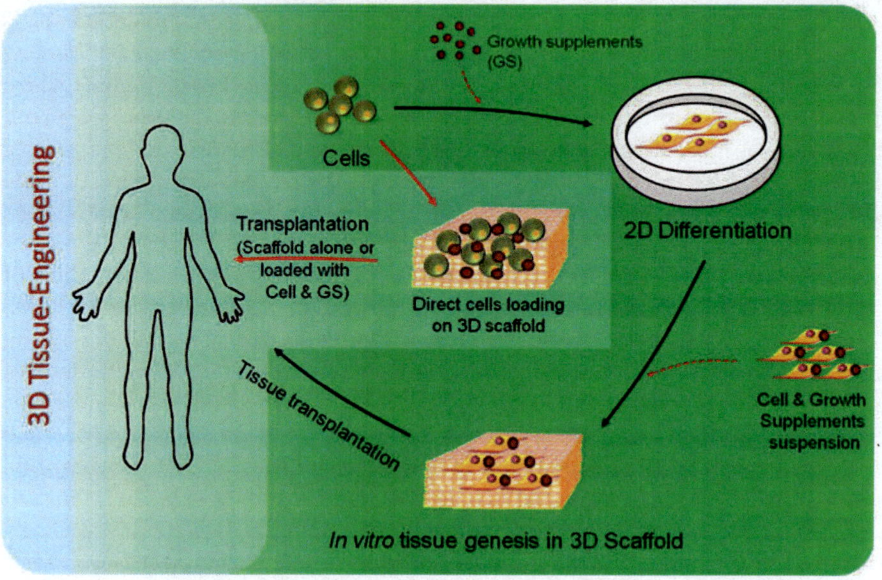

Fig. 3 Schematic representation of 3D tissue engineering approach which involves two ways; **a** 2D cell differentitation, adhesion and growth of cell in 3D porous scaffold followed by in vitro tissue genesis in presence of suitable growth supplements and finally tissue transplantation, and **b** transplantation of scaffold alone or loaded with cells and growth supplements for in situ tissue regeneration

Especially in bioengineering, scaffolds are defined as a temporary three-dimensional porous architecture which supports the immobilization of biomolecules for the genesis, delivery and production of other industrial important molecules. Porosity is one of the important parameters in scaffold fabrication. IUPAC standard nomenclature defines different porosity range in materials, i.e. micro (0.2–2 nm), meso (2–50 nm) and macro (50–1000 nm) (IUPAC 2019). Supermacroporous (pore size is $\geq$10 μm) is another informally known class of porous materials with superior immobilization potential and versatile applications (Tripathi and Melo 2019). In tissue engineering, macro and supermacro size porosity is considered to be ideal for allowing migration of cells from one end to the other for uniform tissue development. Additionally, macropores allow the exchange of nutrients and gasses to give the cells a 3D microenviroment they need to thrive (Tripathi et al. 2013a). This renders them crucial for the development of 3D tissues, drug delivery systems and other type of research. Some of the most commonly used growth factors that have been entrapped in hydrogel include fibroblast growth factor (FGF), vascular endothelial growth factor (VEGF), bone morphogenetic protein-2 (BMP-2), insulin-like growth factor-1 (IGF-1) and transforming growth factor β (TGF-β) (Burdick and Anseth 2002; Peters et al. 1998; Elisseeff et. al. 2001; Tabata et al. 1999; Lutolf et al. 2003).

3.2 *Methods of Fabrication and Properties*

Biomaterial fabrication methods have been evolved since many decades to improve their property and performance in various biomedical applications. Some of the well-evaluated methods are solvent casting and particulate leaching (SCPL), emulsification, microfluidics, freeze-drying (lyophilization), gas foaming, thermally induced phase separation (TIPS), spray drying, nano-fibre self-assembly, supercritical fluid technology, sol–gel, electrospinning,additive manufacturing (AM) technologies and cryogelation. Among them, salt leaching, gas foaming and lyophilization techniques are conventional techniques that have been used to produce porous scaffolds with tunable pore parameters while designing the scaffolds (Kim et al. 2017). With advancements in technology and the onset of bioinspired design principles, some innovative methods such as electrospinning, AM technologies (Unnikrishnan et al. 2020), cryogelation (Tripathi and Melo 2019) and self-assembly have been widely applied to produce novel biomaterial scaffolds for tissue engineering. Before choosing a fabrication method for scaffold synthesis, it is important to understand that an ideal scaffold for tissue engineering (cell delivery) and drug delivery should posses the following important properties:

For tissue engineering (cell delivery):

1. Significant mechanical support for new tissue growth and also to shield cells from tensile forces without inhibiting biomechanical cues
2. Flexibility to fabricate in desired shape and volume
3. Acceptable biocompatibility
4. A highly porous and well-interconnected open pore structure that allow high cell seeding density and tissue in growth
5. Guided tissue function by enriching a specific cellular response while constraining others
6. Augmentation of cell adhesion and subsequent cellular activation (e.g. cell adhesion and proliferation, facilitating cell–cell contact, cell migration and production of ECM)
7. Bioadsorption and bioresorption should take place under predetermined time period
8. Biocompatible chemical compositions and their degradation products, causing minimal immune or inflammatory responses.

For drug delivery:

1. Homogenous drug dispersion throughout the scaffold
2. Ability to release the drug at a predetermined rate
3. Drug binding affinity that is sufficiently low to allow the drug released to be stable when incorporated in the scaffold at a physiological temperature
4. Stable physical dimension, chemical structure and biological activity over a prolonged period of time.

There is a significant challenge in the designing and manufacturing of a scaffold that possess all of the above requirements and the ability to control the release kinetics of drug or growth factors over the period of treatment or tissue regeneration (Tripathi and Kumar 2011; Ho-Shui-Ling et al. 2018).

3.3 Biomedical Applications

Immobilization by encapsulation or entrapment into a porous scaffold to retain, support and control its release on demand is the fundamental prerequisite to utilize a biologically active molecule more effectively for therapeutics and economic standpoint. Beside two main advantages of scaffold, i.e. long-term transplantation of biomolecule to restore of improve tissue/organ function and long-term secretion of therapeutics, it also circumvents immune rejection. A well-optimized encapsulation system may prevent entering and destroying of encapsulated system from immune cells and antibodies, thereby reducing the chronic administration of immunosuppressant, thus facilitating an improved clinical alternative (Sarnowska et al. 2013).

Bisceglie (1933) has developed a polymer structure consisting of tumour cells and transplanted in the abdominal cavity of pig model as one of the documented preliminary endeavours in transplantation of encapsulated system. Later in 1960s, the concept of "artificial cell" or "minimal cell" was invoked which consists of biologically active molecules entrapped in a semi-permeable polymer membrane and mimic one or more functions of biological cell (Chang 1964). Number of structures such as liposomes, nanoparticles, microcapsules are referred as artificial cell that may carry various biomolecules like cells, enzyme, microorganism, vaccines, haemoglobin, genes, etc. Some of the notable advantages of artificial cell are increased solubility of cargo, improved imitation in the host and vanishing of immune response. One of its successful clinical applications is in hemoperfusion (Gebelein 1984). However, its first clinical use in hemoperfusion by the encapsulation of activated charcoal was reported by Chang (1996). Since then, tremendous efforts have been made all around the world for advancing the nano/micro-assisted strategies and potential application in regenerative medicine. Nano/micro-materials including particles, composites and surfaces provide a wide range of advanced immobilization systems. Based on the different spatial scales of biomaterial structures, they can be divided into nanoscale ($\leq$100 nm), sub-micronscale (100 nm–1 μm) and micronscale ($\geq$1 μm) (Chen et al. 2018). Besides several advantages, encapsulation of cells also shows limitation in certain cases, for example, cells which have proliferation potential in immobilization medium may increase its population beyond the limit of polymer structure resulting in compromising with long-term viability of cells and diminishing the property of therapeutics diffusion.

One of the interesting studies of cell immobilization was demonstrated by Lim and his colleague by immobilization of xenograft islet cells in the composite polymer matrix of aginate-poly-L-lysin and transplanted in animal model for controlling the glucose level in diabetic condition (Lim and Sun 1980). With this proof of concept, allogenic transplantation and drug delivery have successfully demonstrated in diseased animal models of haemophilia (Hortelano et al. 1996), cancer (Xu et al. 2002) and renal failure (Prakash and Chang 1996). Importance of cell immobilization to achieve a functional bioartificial liver (BAL) system is a urgent need for the patients suffering with chronic liver disease (Kumar et al. 2011) and maintaining the liver-like finction in 3D scaffolds could be a beneficial diagnostic tool for pharmacokinetics studies which ultimately reduce the burden of animal trials of drug molecules (Tripathi and Melo 2015).

Due to the shortage of organ donors, allogeneic and xenogeneic cells and tissues have been evaluated as a potential clinical application. However, xenografts are of major concern due to the possible source of transmission of infection. To address this issue, encapsulation of cells could serve as a prospective therapeutics delivery system wherein immobilized cells can act as a reactor for the in vivo secretion of desired biomolecules. For example, immobilization of granulocyte macrophage-colony-stimulating factor (GM-CSF) also known as colony stimulating factor-2 (CSF-2) (functions as a cytokine) secreting cells (like macrophages, T cells, mast cells, natural killer cells, endothelial cells and fibroblasts) has been investigated on human and animal models as a potential adjuvant or immunemodulator (Hong 2016). The pharmaceutical analogues (developed by recombinant DNA technology) of naturally occurring GM-CSF are known as molgramostim (made in *E. coli*) and sargramostim (made in yeast). A study suggests that the delivery of adequate amount of therapeutic gene by consistent and long-term administration of vector can be replaced by the use of immobilized cells producing retroviral vector for in vivo gene transfer (Saller et al. 2002).

Sakurai et al (2003) have demonstrated the immobilization of angiogenic factor secreting cells or bFGF or VEGF in the gelatin microsphere that can advance the process of angiogenesis and neovascularization. Similarly, scaffold-based cell delivery has dramatically improved the treatment of many diseases (Langer 2000). On account of biopolymer-based tissue-engineered scaffolds applications, few studies have reported the drawbacks of natural polymers, like immunogenicity, infection risk and also uncontrolled rate of degradation (Porter et al. 2009). However, based on the published literature, higher success rate of tissue regeneration has been recorded by the use of naturally derived polymeric scaffolds.

Encapsulation of osteoblast cells in the alginate scaffold was found to promote the bone mineralization (Alsberg et al. 2001). However, to mimic the bone strength and microenvironment, hydroxyapetite (a key inorganic component of natural bone)-based composite scaffolds have been fabricated and implantation studies were performed. Observations for twelve successive months suggested suitable biocompatibility of scaffold without any chronic inflammatory response. Another study evaluated the impregnation of nanoparticles (10%) in polymer which improved the mechanical properties of the composite scaffold and encouraged bone formation in animal model (Cao et al. 2014). A bone-like hierarchical nanostructure with mineralized collagen was used for the fabrication of a biomimetic 3D scaffold with high porosity and interconnected pores favouring the cell homing, migration, multidifferentiation, vascularization and formation of new bone (Liu et al. 2019).

Based on the nature of soft tissues (both structural and molecular arrangement) like cartilage, skin, neural, cardiac, corneal, various polymeric compositions have been developed to mimic the natural tissue features (Kumar and Tripathi 2012). It is worth mentioning that polymer surface should not affect the phenotypic characteristics of cells in 3D microenvironment but provide suitable adhesion to cells (Tripathi and Kumar 2011). For example, agarose and alginate are soft polymer materials that are suitable for the entrapment of chondrocyte cells and do not affect cell physiology due to their inert surface chemistry (Hauselmann et al. 1994; Tripathi and Kumar 2011). However, these matrices in porous 3D form did not find their suitability for cell adhesion. Incorporation of cell attractive moieties like gelatin (mimics RGD-like sequences) into inert polymer provides a cell attractive porous scaffold with suitable mechanical strength (Tripathi et al. 2009). There are several studies of surface functionalization of polymeric scaffolds for increasing their biological activity and cell-liking surface augmentation using atmospheric plasma or "cold-plasma technology", which is brilliantly elaborated by Trimukhe and his collegues (Trimukhe et al. 2017). Including cold-plasma surface modification, various nanoparticles have also been utilized for successful clinical application of antimicrobial polymeric biomaterials (Agnihotri et al. 2018).

Engineered tissue implants are pivotal in large tissue defects that cannot re-grow without any bridging medium. For example, the neural signal processing in nervous system takes place at nano-size gap called synaptic cleft, and dissemination of cell signals is not suitable for proper function of central as well as peripheral nervous systems. In addition, development of scar in the large defect region could be a life threatening situation. Therefore, delivery of immobilized neural cells like Schwann cells using non-immunogenic polymer medium as a robust implant or 3D engineered tissue could be suitable strategy to fill the gap of damaged tissue (Phillips et al. 2005). Studies have demonstrated successful application of 3D scaffold in developing functional neural tissue, differentiation of human umbilical cord-derived mesenchymal stem cells (hUCMSCs) to nerve cells by co-culturing and delivery into brain in the in vitro, in vivo and in situ conditions (Jurga et al. 2011; Sarnowska et al. 2013).This study suggests that scaffold-mediated delivery of hUCMSCs owing not only a strong immune modulation property and commitment for neuro-epithelialization, but also ought to be considered a double-edged sword in case of brain injuries where it helps in bringing neural tissue and attenuating inflammation.

A collagen scaffold used along with undifferentiated embryonic stem cells and implanted in the intra-myocardial region of rat heart showed successive integration of implant with myocardial tissue function (Kofidis et al. 2005). In another study, collagen type I-based scaffold without any cell was sutured in the left ventricle of rat's heart which showed well tissue-integration, cell migration and formation of vascularized ECM (van Amerongen et al. 2006). Similarly, various clinically tested scaffold alone and cell-loaded scaffolds are available in market for skin tissue regeneration (Kumar and Tripathi 2012).

3.4 *Immune Response and Future Outlook*

The application of polymeric biomaterials in biomedical field has shown tremendous potential alone or in combination with one or more organic and inorganic materials with improved and tuneable physico-chemical and mechanical properties. However, plausible induction of immunological response by the use of biopolymers is also widespread concern and a topic of debate (Mariani et al. 2019).

Biopolymers are inflammation active entities and could lower the physiological pH of tissue because of carrying of inherent functional groups. Several studies have been performed to understand the host immune response to various polymers for identification of responsible components of ideal immunological reactions during the transport of foreign polymeric materials with improved host resistance. Polymeric scaffolds made of proteins, glycoproteins, peptides, lipopolysaccharides are some of the potent candidate of immunomoldulation. Many polysaccharides have long been used in the development of conjugated vaccine development (Tzianabos 2000). Although, they serve as a weak antigen in the conjugated form and boost the immune system by activation of T-cell followed by development of memory against same antigen. Experimental evidences advocate that the drawbacks associated with naive biopolymers may be reduced or vanished if they are cross-linked to produce new composite and hybrid scaffolds with cell-liking bioactive molecules for regulating cell behaviour (Ren et al. 2009). Moreover, the formation of small degraded polymeric entities (responsible for high immune reactions) will automatically reduce after controlling the degree of cross-linking and help to control immune responses. Despite of some challenges in developing immunologically ideal scaffold, biopolymer presents an advantage of swift and successful body clearance by hydrolytic degradation and enzymatic metabolism. Overall, the distinctive properties of biopolymers and overcoming of their disadvantages by using chemically modified analogues could be a potential strategy for successful application of biopolymeric scaffold in biomedicine. With the advent of novel manufacturing practices and cellular advancements, myriad of future prospective scaffolding materials is anticipated for advanced therapeutics and other biomedical applications.

4 Immobilization in Biosensor Development

4.1 *Concept of Biosensors*

Biosensors are analytical devices that convert a biological response or signal into an electrical signal which can be read out by transducers for the detection of analytes. The term "biosensor" is a short form for "biological sensor". The term "biosensor" was coined by Cammann, and its definition was introduced by IUPAC. According to IUPAC recommendations 1999, a biosensor is an independently integrated receptor transducer device, which is capable of providing selective quantitative or semi-quantitative analytical information using a biological recognition element (Thevenot et al. 1999, 2001a; b; Turner et al. 1987). Although definition states that biocomponent is required for detection, recently many biosensors have been developed for the detection of biological analytes without the use of biocomponent as a detection component and still included in biosensor category (Manjunatha et al. 2010; Mallesha et al. 2011).

So basically biosensor is a marriage between biological receptors and electronic transducers. In biosensors, analytes diffuse from the solution to the surface of the biosensors where analytes specifically and efficiently react with the biocomponents integrated on transducers. The interaction and reaction of the analytes with biocomponents generates some physiochemical changes as a signal which is converted into optical or electrical signal by transducer and that is proportional to the concentration of analytes (Fig. 4).

Biosensors consist of three main parts.

Biocomponents or biological detection system: Biocomponents may be an enzyme, an antibody or similar binding molecule, DNA probe, a living cell and organelles.

Transducers: It transforms the signal resulting from the interaction of the analyte with the biocomponent into a measurable electrical signal.

A signal processing system: It converts the measured signal into workable and displayed form.

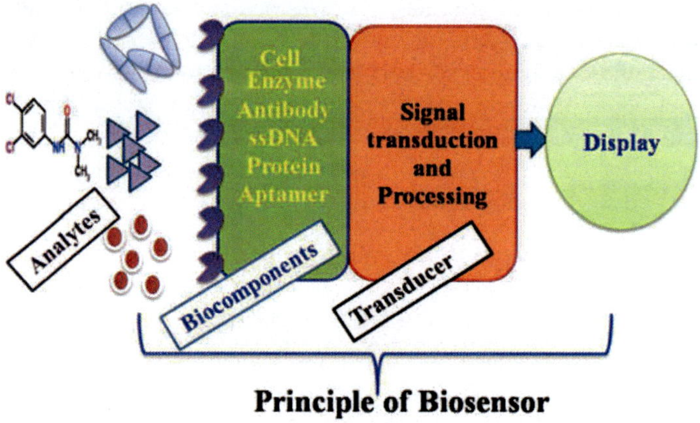

Fig. 4 Schematic diagram of principle of biosensor

4.2 Advantages of Biosensor Over Conventional Technique

Many conventional and traditional analytical methods like GC and HPLC have been widely used for analysis and monitoring of analytes, but they require not only expensive equipments but also highly trained technicians. Also, these conventional and traditional techniques are time consuming and laborious because it requires pre-sample preparation before analysis. Over a course of time, researchers have put efforts to develop promising alternatives for the detection of analytes, which can be used for easy, online and prompt detection with comparable accuracy and sensitivity. Also approach is such that the sample preparation can be avoided and minimized. The key benefits of biosensors include, rapid and continuous measurement, high specificity, very less usage of reagents required for calibration, fast response time and ability to measure non-polar molecules that cannot be estimated by other conventional devices.

Characteristics feature of biosensors

Biosensors are characterized with following parameters during the development:

a. Selectivity: Selectivity is the ability of the biosensor to respond only to the target analyte. Selectivity usually comes from biocomponents and its interaction with analyte. It defines the lack of response to other interfering element present in the solution.
b. Sensitivity: Sensitivity is the response of the sensor to unit change in analyte concentration. Sensitivity usually comes from the transducer part of the biosensor.
c. Detection range: It is the concentration range over which the sensitivity and reproducibility of the biosensor is good.
d. Detection limit: It is the lowest concentration of the analyte to which there is a measurable response from the biosensor.
e. Reproducibility: It is the accuracy with which the biosensor's output can be obtained.
f. Response time: It is the time required by the biosensor to produce response.
g. Reusability: It is the number over which accuracy of same biosensors can be used for many samples.
h. Storage stability: It is the time period over which the sensors can be used without significant deterioration in performance characteristics. It also characterizes the change in its baseline or sensitivity over a fixed period of time.

4.3 Immobilization Strategy in Developing Biosensors

In the progressing field of developments of biosensors, most prominent footstep is to immobilize biocomponents onto the transducer to read out the signal generated due to interaction between the analytes and biocomponents. Immobilization step is playing an extremely crucial role in the development of biosensors. Immobilization of biological elements onto a surface must be stable, permit diffusion of substrates and products and allow excellent electron transfer. The benefits provided by an effective and optimized immobilization method always extend the utilization of the biosensor and predict the increase storage stability. Choice of right and optimum immobilization methods gives rise to an efficient, simple and cheap biosensor which can be commercialized for societal benefit and for large number of populations. As discussed earlier in this chapter, there are different methods of immobilization such as adsorption, covalent bonding, cross-linking and encapsulation that can be used as single or combined methods. There are many review papers which report the immobilization technique and its role in development of biosensor (D'Souza 2001a, b; Sassolas et al. 2012; Bhardwaj 2014; Putzbach and Ronkainen 2013; Nguyen et al. 2019). Many biosensors were developed for monitoring of analytes in the field of health, environment, agriculture and food. In the very early periods, simple and direct methods of immobilizations were used for development of biosensors but in the course of time with advancement in immobilization methods, led to integrating nanotechnology and other related techniques.

4.3.1 For Health Applications

The emphasis of this section is on the immobilization strategy used for developing biosensors for detection of analytes related to health applications. In our laboratory, various strategies had been used for developing biosensor related to health monitoring such as glucose, cholesterol, hydrogen peroxide, urea, dopamine, acetaminophen and anticancer molecules as mentioned in Table 4.

The glucose biosensor has been widely used as a clinical indicator of diabetes. Development of the blood glucose biosensor is one of the greatest contributions of the biosensor field to improve human health, allowing home monitoring of glucose levels by those suffering from diabetes, greatly aiding them in the management of their condition and improving their quality of life. Our group had developed a dissolved oxygen-based biosensor where glucose oxidase (GOD) enzyme was used as biocomponent and entrapped on PVA membrane, developed using PVA with low and high degree of polymerization (DOP), acetone as a mixture of solvent, benzoic acid (BA) as sensitizer and cross-linked using UV treatment (Kumar and D'Souza 2008). In another study, inner epidermal membrane of the onion bulb scales was used as a natural polymer support for immobilization of the glucose oxidase (GOD) enzyme (Kumar and D'Souza 2009). These two biosensors were developed by simple immobilization strategy and used glucose oxidase as biocomponent like first biosensor developed by Leland Clark. In another study, a non-enzymatic method of graphite electrode modified with functionalized graphene was used for the determination of β, D (+)-glucose (Mallesha et al. 2012). Accordingly with the advancement in development, enzyme-based glucose biosensors were categorized to first-, second- and third-generation biosensors.

Serum cholesterol level plays an important role in the diagnosis and treatment of various diseases and is considered a biomarker for cardiovascular diseases, heart

Table 4 Biosensors for monitoring of health-related analytes

Health-related analytes	Biocomponents	Transducers	Immobilization strategy	Detection Range	References
Gluocse	Glucose oxidase GOD	Dissolved oxygen (DO) probe	GOD immobilized in PVA membrane prepared using PVA with low and high degree of polymerization	0.9–225 mg/dl	Kumar and D'Souza (2008)
Glucose	Glucose oxidase	DO probe	Inner epidermal membrane of the onion bulb scales	4.5–900 mg/dl	Kumar and D'Souza (2009)
β, D (+)-glucose	Non- enzymatic	Electrochemical	Graphite electrode modified with functionalized graphene	0.5 × 10–3–7.5 × 10 -3 M	Mallesha et al. (2012)
Cholesterol	Cholesterol oxidase (ChOx)	Electrochemical	Surface of graphite electrode by immobilizing positively charged ChOx and negatively charged multiwalled carbon nanotubes	0.2–1 mM	Manjunatha et al. (2011a)
Cholesterol	Cholesterol oxidase (ChOx) and cholesterol esterase (ChEt)	Cyclic Voltammetry, electrochemical impedance spectroscopy	Cholesterol oxidase (ChOx) and cholesterol esterase (ChEt) covalently immobilized onto functionalized graphene (FG) modified graphite electrode	50–300 μM cholesterol 0.5–7 mM hydrogen peroxide	Manjunatha et al. (2012)
Cholesterol	Cholesterol oxidase	Electrochemical	Cholesterol oxidase (ChOx) immobilized on AuNs appended on the graphite (Gr) electrode via chemisorption onto thiol-functionalized graphene oxide (GO-SH)	0.05–11.45 mM	Nandini et al. (2016)
Hydrogen peroxide	Horseradish peroxidase	Electrochemical	Co-deposition of palladium, horseradish peroxidase on functionalized graphene modified graphite electrode as composite	25 μM–3.5 mM	Nandini et al. (2013)
Hydrogen Peroxide	*Pichia pastoris* catalase	Electrochemical	*Pichia pastoris* catalase immobilized on gold nanoparticle nanotubes and polythiophene hybrid	0.05–18.5 mM	Nandini et al. (2014a)

(continued)

Table 4 (continued)

Health-related analytes	Biocomponents	Transducers	Immobilization strategy	Detection Range	References
Hydrogen peroxide	Catalase	Electrochemical	Metal ion co-ordination assembly based multilayer of one-dimensional gold nanostructures and catalase	0.05–19.35 mM	Seetharamaiah et al. (2017)
Hydrogen peroxide and cholesterol	Horseradish peroxidase (HRP) or cholesterol oxidase (ChOx)	Electrochemical	Functionalized grapheme (straggled sheets) and horseradish peroxidase (HRP) or cholesterol oxidase (ChOx), on graphite electrode	25 μM–19.35 mM for hydrogen peroxide and 0.1–4.5 mM for cholesterol	Nalini et al. (2014)
Glucose and hydrogen peroxide	Glucose oxidase (GOx) and horseradish peroxidase (HRP)	Electrochemical	Glucose oxidase on Rhoeo discolour (Rd) leaf extract with glutaraldehyde (GLD) on functionalized multiwalled carbon nanotubes (f-MWCNTs) modified graphite (Gr) electrode	0.5–28.5 mM Glucose 0.2–6.8 mM hydrogen peroxide	Nandini et al. (2014b)
Dopa and Dopamine	Tyrosinase from Amorphophallus companulatus	Electrochemical	Novel composite of two biopolymers: agarose and guar gum	2–10 μM	Tembe et al. (2006)
l-Dopa	Tyrosinase	Optical	Tyrosinase enzyme extracted from Amorphophallus campanulatus and immobilized on the surface of the microplate wells by adsorption followed by cross-linking	10–1000 μM	Saini et al. (2014)
Dopamine	Non-enzymatic	Electrochemical	Functionalized graphene modified graphite electrode for the selective determination of dopamine in presence of uric acid and ascorbic acid	3.0–60 μM	Mallesha et al. (2011)

(continued)

Table 4 (continued)

Health-related analytes	Biocomponents	Transducers	Immobilization strategy	Detection Range	References
Simultaneous determination of ascorbic acid, dopamine and uric acid	Non-enzymatic	Electrochemical	Layer-by-layer (LBL) technique on graphite electrode, by positively charged poly(diallyldimethylammonium chloride) (PDDA) and negatively charged multiwalled carbon nanotubes (MWCNTs) wrapped with polystyrene sulfonate (PSS) through electrostatic interaction	1–150 μM dopamine 1–120 μM Uric acid 0.5–2.5 mM ascorbic acid	Manjunatha et al. (2010)
Urea	Urease	Potentiometric	Egg shell emembrane	0.5–10 mM	D'Souza et al. (2013)
Urea	Urease	Colorimetric	Jackfruit membrane	0.5–10 mM	Kumar and Melo (2016)
Urea	urease from Arthrobacter creatinolyticus	Potentiometric	PAN [poly(acrylonitrile-methylmethacrylate-sodium vinylsulfonate)] membrane	1 to 100 mM	Ramesh et al. (2015)
Anticancer	Agrobacterium tumefaciens (At) Clitoria ternatea (Ct)	Electrochemical techniques such as CV and EIS	Au nanostructures, quercetin stabilized gold nanoparticles (Q-AuNPs) were synthesized using onion peel	–	Nalini et al. (2016)
Acetaminophen	Non-enzymatic	Electrochemical	Layer-by-layer (LBL) method using both positively and negatively charged multiwalled carbon nanotubes (MWCNTs) on poly (diallyldimetheylammonium chloride) (PDDA)/poly styrene sulfonate (PSS) modified graphite electrode	25–400 μM	Manjunatha et al. (2011b)

attack, strokes, peripheral arterial disease, type 2 diabetes and high blood pressure. Various strategies had been used for developing cholesterol oxidase enzyme-based electrochemical biosensors. Our group had also developed an electrochemical biosensor using different methods of immobilization. In one method, surface of graphite electrode was modified by immobilizing positively charged cholesterol oxidase and negatively charged multiwalled carbon nanotubes through electrostatic interaction using layer-by-layer technique (Manjunatha et al. 2011a). Beside immobilizing single enzyme cholesterol oxidase, advancement was also made where two enzymes cholesterol oxidase (ChOx) and cholesterol esterase (ChEt) were covalently immobilized onto functionalized graphene (FG) modified graphite electrode. For the free cholesterol determination, ChOx-FG/Gr electrode exhibits a sensitive response from 50 to 350 μM with a detection limit of 5 μM. For total cholesterol determination, co-immobilization of ChEt and ChOx on modified electrode, i.e. (ChEt/ChOx)-FG/Gr electrode, showed linear range from 50 to 300 μM with a detection limit of 15 μM (Manjunatha et al. 2012). In another study, nanoparticles were integrated with the aim to develop a highly sensitive cholesterol biosensor, cholesterol oxidase (ChOx) was immobilized on AuNs which were appended on the graphite (Gr) electrode via chemisorption onto thiol-functionalized graphene oxide (GO-SH) and Gr/GO-SH/AuNs/ChOx biosensor was characterized using cyclic voltammetry (CV), electrochemical impedance spectroscopy and chronoamperometry. The sensitivity determined for this biosensor was found to be 273 $mA/mM/cm^2$ and detection limit 0.2 nM (Nandini et al. 2016).

Hydrogen peroxide determination has been an important analyte in very wide fields, such as clinical, pharmaceutical, environmental and industrial applications. It is a reactive oxygen species (ROS) that is present throughout the body, playing various roles in physiological processes, including cellular signalling, where it regulates cell growth, immune activation and apoptosis. However, at high levels, H_2O_2 can be detrimental to the body, causing cell damage, inflammatory disease and cancer. Horseradish peroxidase, a redox active enzyme, is being used in development of amperometric biosensors for determining H_2O_2, organic peroxides, alcohol and also for sensing certain biomolecule such as glucose and amino acid by co-immobilization of corresponding oxidase on the same electrode surface. Researchers have been developing sensors using different strategies of immobilization of enzyme to detect and quantify hydrogen peroxide under various conditions to understand biological as well as the health status. In one strategy, a sensitive and noble amperometric biosensor was developed by the co-deposition of palladium and horseradish peroxidase (HRP) on functionalized graphene (f-graphene) modified graphite electrode. This biosensor showed linearity in detection with increase of the H_2O_2 concentration in the range of 25 μM–3.5 mM (Nandini et al. 2013). In the other approach, a simple and innovative electrochemical hydrogen peroxide biosensor was studied using catalase (CATpp) derived from *Pichia pastoris* as bioelectrocatalyst. Biocomponent was immobilized on gold nanoparticle nanotubes (AuNPNTs) and polythiophene composite using 1-ethyl-3-(3-dimethylaminopropyl)-carbodiimide and N-hydroxysuccinimide (EDC–NHS) coupling reagent. The assembly of AuNPNTs onto the graphite (Gr) electrode was achieved via S–Au chemisorption. The latter was pre-coated with electropolymerized thiophene (PTh) to enable S groups to bind AuNPNTs. The combination of AuNPNTs–PTh, i.e. an inorganic–organic hybrid, provides a stable enzyme immobilization platform. The fabricated bioelectrode Gr/PTh/AuNPNTs/EDC–NHS/CATpp exhibited a wide linear range from 0.05 to 18.5 mM of H_2O_2 (Nandini et al. 2014a). In another strategy, first time an electrochemical biosensor was developed for the quantification of hydrogen peroxide (H_2O_2) based on one-dimensional gold nanostructures (1D-AuNs) and catalase (CAT) multilayer fabricated on graphite electrode (GE) through metal ion coordination assembly technique and integrated with CV, DPV and chronoamperometry. Biosensor showed a wide linear range (0.05–19.35 mM) of detection and low detection limit (0.98 nM) (Seetharamaiah et al. 2017).

Advancement in strategy for immobilization of enzyme has also been carried out in such a way that more than one analyte can be detected simultaneously. In this direction, an amperometric hydrogen peroxide and cholesterol biosensors were designed by using hierarchical curtailed silver flowers functionalized graphene and horseradish peroxidase (HRP) or cholesterol oxidase (ChOx) enzymes deposits and the resulting biosensors named Nf/(HRP-f-graphene-Ag)/Gr and Nf/(ChOx-f-graphene-Ag)/Gr were evaluated for electrochemical activity using cyclic voltammetry (CV), differential pulse voltammetry (DPV) and chronoamperometry. It demonstrated a good linear range of 25 μM to 19.35 mM with detection limit of 5 μM for hydrogen peroxide and a linear range of 0.1–4.5 mM with detection limit of 0.514 mM for cholesterol (Nalini et al. 2014). In a similar strategy, two enzymes, glucose oxidase (GOx) and horseradish peroxidase (HRP), were immobilized using *Rhoeo discolour* (Rd) leaf extract with 2.5% glutaraldehyde (GLD) on functionalized multiwalled carbon nanotubes (f-MWCNTs) modified graphite (Gr) electrode. The Gr/f-MWCNTs/(Rd-GLD)/GOx and Gr/f-MWCNTs/(Rd-GLD)/HRP biosensors showed excellent electrocatalytic activity concerning the detection of glucose (0.5–28.5 mM) and hydrogen peroxide (0.2–6.8 mM) (Nandini et al. 2014b).

Phenolic compounds such as L-Dopa and dopamine have relevant significance in health care and pollution monitoring. Dopamine (DA) is naturally produced and widely distributed in the central nervous system of mammals. Abnormal levels of DA in body fluids are the indications of many serious diseases like Schizophrenia, Huntington's disease and Parkinson's disease. For developing biosensor for detection of these phenolic compounds, tyrosinase enzyme was immobilized by various strategies. Earlier simple strategy was used for the fabrication of electrochemical tyrosinase-based biosensor where tyrosinase, extracted from a plant source *Amorphophallus companulatus* was immobilized in a novel composite of two biopolymers: agarose and guar gum. This composite matrix-containing enzyme forms a self-adhering layer on the active surface of glassy carbon electrode used for detection of dopamine with a linear detection range of 2–10 μM (Tembe et al. 2006). In another study, tyrosinase enzyme extracted from *Amorphophallus campanulatus* was also immobilized on the surface of the microplate wells, integrated with an optical transducer for L-Dopa detection and showed a linear range of detection from 10–1000 μM and detection limit 3 μM (Saini et al. 2014). DA coexists with ascorbic acid and uric acid in the extra cellular fluids of central nervous system and serum in mammals. The concentrations of ascorbic acid and uric acid are much higher (100–1000 times) than that of DA in body fluids and offer greater interference during the determination of one in presence of the other. There is need to detect dopamine in presence of ascorbic acid (AA) and uric acid (UA). Sometimes, there is need of simultaneous determination of AA, DA and UA although all three electroactive compounds may have very similar electrochemical properties and they will be oxidized at nearly same potential; our group has also developed a promising electrochemical biosensor fabricated using layer-by-layer (LBL) technique on graphite electrode, by positively charged poly(diallyldimethylammonium chloride) (PDDA) and negatively charged multiwalled carbon nanotubes (MWCNTs) wrapped with polystyrene sulfonate (PSS) through electrostatic interaction, for the simultaneous determination of ascorbic acid (AA), dopamine (DA) and uric acid (UA). The modified electrode exhibits superior electrocatalytic activity towards AA, DA and UA than the bare graphite electrode. The three separated anodic peaks were obtained at 192, 123 and 315 mV between AA-DA, DA-UA and AA-UA, respectively, in CV and corresponding separated anodic peaks were 210, 119 and 329 mV in DPV, respectively (Manjunatha et al. 2010). In the above biosensor, there is improvement in the separation of oxidation peak of the analytes due to integration of nanotechnology in biosensor. In a further modification to this sensor, MWCNTs were replaced with functionalized grapheme sheet. Since grapheme shows 60 times more conductivity than MWCNTs, only a single layer of functionalized grapheme was deposited on the surface of electrode instead of the multilayers of CNTs above. The functionalized graphene modified graphite electrode-based electrochemical biosensor was able to determine dopamine with a linear range of 1.75–90 μM in presence of UA and AA (Mallesha et al. 2011).

4.3.2 For Pesticide Detection

Although extensive use of pesticides has improved in securing enough crops, these pesticides are equally toxic or harmful to non-target organisms like mammals, birds, etc. and thus their presence even in small amounts can cause serious health and environmental problems. Pesticides have thus become environmental pollutants, and they are often found in soil, water and atmosphere products. Thus, monitoring of these pesticides and its residues become extremely important (Kumar and Melo 2017).

Organophosphates pesticides are a class of insecticides, several of which are highly toxic. Earlier, they were among the most widely used insecticides; however, in the past decade, several notable organophosphates pesticides have been discontinued for use, including parathion, which is no longer registered for any use. Methyl parathion (MP) is an organophosphate pesticide, which was used as non-systemic insecticide in agriculture to protect the crops from insects. MP was used as insecticide in agriculture to protect the crops from insects. The IUPAC chemical name of MP is *O,O*-dimethyl *O*-4-nitrophenyl phosphorothioate (Fig. 5) (Kumar et al. 2006; Kumar and Melo 2020).

Fig. 5 Chemical structure of methyl parathion

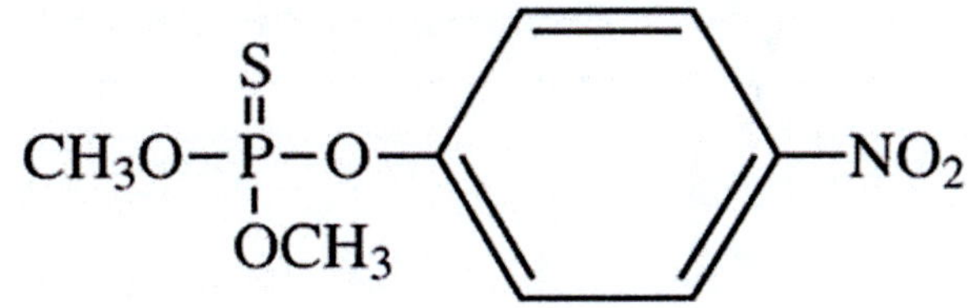

MP kills pests by acting as a stomach poison and acts as potent irreversible acetylcholinesterase inhibitor. It was used to control a variety of insects and mites, including thrips, weevils, aphids and leafhoppers, in a very wide range of crops including cereals, fruit, nuts, vines, vegetables, ornamentals, cotton and field crops (Kumar and Melo 2015, 2017, 2020; Kumar et al. 2018). MP causes inhibition of acetylcholinesterase and that leads to excess accumulation of acetylcholine and causes overstimulation of muscles and nerve fibres, uncontrollable twitching, convulsions, difficulty in breathing and death. It was classified by the World Health Organization (WHO) as a Category Ia (extremely toxic) and by the United States EPA (US EPA) as a Toxicity Category I (most toxic) insecticide (US EPA 2003). As per statistical report by Directorate of PPQS, India and Centre of Science and Environment (CSE), consumption of MP in India was 5286 MT during 2010–2016. Recently, some of the pesticides including MP were completely banned by Government of India with effect from August 2018 because of the high toxicity concern (Kumar and Melo 2020).

Pesticide biosensors have been developed using different immobilization strategies depending on the biocomponent used and its interaction with analyte (Kumar and Melo 2015, 2017, 2020; Kumar et al. 2018). In general, biocomponents were either immobilized on the surface of transducer or on membrane and then the immobilized membrane was integrated with surface of the transducer to read out the signal. A lot of work has been carried out to develop the biosensor for detecting pesticides. Different types of biocomponents such as enzyme, microbial cells and antibody were used to read out the signal. Table 5 lists number of publications available describing the detection of methyl parathion pesticide.

Initially, researchers have used simple methods for immobilization of biocomponents, and later on, advance strategy such as nanoparticles was integrated for sensitive detection of methyl parathion. Mulchandani et al. 1999 had developed a biosensor for detecting methyl parathion pesticide by using basic element of pH electrode modified with an immobilized organophosphorus hydrolase (OPH) layer formed by cross-linking OPH with bovine serum albumin (BSA) and glutaraldehyde. Later in 2001, same group had developed an amperometric biosensor based on a carbon paste electrode containing genetically engineered cells expressing OPH on the cell surface. This biosensor was used to measure as low as 1 μM methyl parathion (Mulchandani et al. 2001a). In 2006, our group had also developed an optical biosensor where whole cells of *Flavobacterium* sp. were immobilized by trapping in glass fibre filter and were used as biocomponent along with optic fibre system. The biocomponent of this biosensor was disposable and showed a detection range between 4–80 μM (Kumar et al. 2006). Later on, our group developed a microplate-based optical biosensor

Table 5 Biosensors for monitoring of pesticide (Methyl parathion)

Pesticides	Biocomponents	Transducers	Immobilization strategy	Detection range	References
Methyl parathion	Organophosphorus hydrolase (OPH)	Electrochemical flow-through detector	Covalently immobilized on activated aminopropyl controlled-pore glass beads	Upto 120 μM	Mulchandani et al. (2001a)
Methyl parathion	*Moraxella* sp. express INPNC-OPH on the cell surface	Electrochemical	Whole cell was mixed with carbon paste electrode	Upto 175 μM	Mulchandani et al. (2001b)
Methyl parathion	*Flavobacterium* sp. whole cells	Optical	Flavobacterium sp. whole cells adsorbed on glass fibre filters	4–80 μM	Kumar et al. (2006)
Methyl parathion	Whole cells of Sphingomonas	Optical	Whole cells immobilized directly onto the surface of the wells of polystyrene microplates (96 wells) using glutaraldehyde as the cross-linker	4–80 μM	Kumar and D'Souza (2010)
Methyl parathion	Recombinant *E.coli* having opd gene	Electrochemical CV	Whole cells immobilized on screen printed carbon electrode (SPCE) using glutaraldehyde	2–80 μM	Kumar and D'Souza (2011b)
Methyl parathion	Whole cells of Sphingomonas	Optical	Whole cells immobilized on inner epidermis of onion bulb scale by adsorption followed by cross-linking methods	4–80 μM	Kumar and D'Souza (2011a)
Methyl parathion	Organophosphorus hydrolase	Amperometric	Bilayer approach with the OPH layer atop of the CNT film	Upto 2 μM	Deo et al. (2005)

(continued)

Table 5 (continued)

Pesticides	Biocomponents	Transducers	Immobilization strategy	Detection range	References
Methyl parathion	Methyl parathion hydrolase	Amperometric	Covalent attachment on AuNP modified GC substrate	0.2–100 ng mL^{-1}	Liu et al. (2014)
Methyl parathion	Sphingomonas	Microplate-based Optical	Biohybrid of cells-silica nanoparticles	0.1–1.0 ppm	Mishra et al. (2017)
Methyl parathion	–	Eloectrochemical	Magnetic molecularly imprinted polymers	0.0–526.0 mg L^{-1}	Hassan et al. (2018)

where whole cells of Sphingomonas bacteria were immobilized directly onto the surface of the wells of polystyrene microplates (96 wells) using glutaraldehyde as the cross-linker. The microplate-based biosensor is having advantages as it has 96 reaction vessels, and therefore, it provides a convenient system for detecting multiple numbers of samples in a single platform (Kumar and D'Souza 2010). In the other strategy, whole cells of *Sphingomonas* sp. were immobilized on inner epidermis of onion bulb scale by adsorption followed by cross-linking methods. Cells immobilized on onion membrane were directly placed in the wells of the microplate and associated with the optical transducer (Kumar and D'Souza 2011a)But both biosensors had shown the same detection range as the previous one. In the same period, whole cells of recombinant *Escherichia coli*, having high periplasmic expression of organophosphorus hydrolase enzyme, were immobilized on the screen printed carbon electrode (SPCE) using glutaraldehyde and associated with cyclic voltammetry and cyclic voltammograms to increase the sensitivity. This electrochemical biosensor showed little improvement in detection range (2–80 μM) (Kumar and D'Souza 2011b).

Advancement in immobilization strategy led to integrating the biocomponents with nanomaterials. Deo et al. (2005) developed an amperometric biosensor for oganophosphorus (OP) pesticides based on a carbon nanotube (CNT)-modified transducer and an organophosphorus hydrolase (OPH) biocatalyst. A bilayer approach with the OPH layer atop of the CNT film was used for preparing the CNT/OPH biosensor. The CNT layer leads to a greatly improved anodic detection of the enzymatically generated *p*-nitrophenol product, including higher sensitivity and stability. In 2014, Liu and his colleagues have developed a biosensor based on AuNP modified GC electrodes for direct detection of methyl parathion. AuNP can be introduced to mixed monolayers of aryldiazonium salt modified GC electrodes by Au–C bonding through aryldiazonium salt chemistry, which provides a stable interface showing efficient electron transfer between biomolecules and electrodes. This biosensor showed a detection range 0.2–100 ppb methyl parathion. In 2017, our group has also integrated the silica nanoparticles with Sphingomonas cells and immobilized on wells of microplate and showed a very sensitive detection (0.1–1.0 ppm) and stable biosensors (Mishra et al. 2017). In 2018, Hassan et al. developed a biosensor based on electrochemical detection of methyl parathion in fish by preconcentrating the pesticide on magnetic molecularly imprinted polymer and further readout on magneto-actuated electrode by square wave voltammetry. The magneto-actuated electrochemical sensor showed outstanding analytical performance for the detection of methyl parathion in fish.

5 Immobilization in Bioprocessing

Immobilization when integrated with bioprocessing makes the process economically more viable through reuseability and continuous operations. Besides biocatalyts like isolated enzymes or the whole cells after immobilization onto the support display improved stability to harsh conditions including temperature, pH and organic solvents. We have thus seen the use of immobilization technology becoming a reality for a number of bioprocesses pertaining to the food and pharmaceutical industry. In this chapter, we will however limit our discussion to the bioprocess involved in preparation of invert sugar syrups which has been extensively investigated in the author's laboratory.

5.1 Bioprocess for Invert Sugar Syrups

Sweeteners are the most important component of confections as they contribute a broad range of functionalities (Dziezak 1989). Wild honey is the oldest sweetener known to man. However, it is believed that cane sugar dates back to at least 8000 years to the South Pacific. Sucrose manufactured from cane sugar or sugar beet is very much abundant than honey and as such became the sweetener of choice. Although artificial sweeteners have helped fulfil demand of health conscious consumers, there is an increasing trend towards use of natural ingredients.

To-date only a few processes using immobilized enzymes are operative on an industrial scale. By far one of the most important applications of immobilized enzyme in the food industry today is in the conversion of glucose to fructose for producing high fructose corn syrups (HFCS) using glucose isomerase (Chen 1980). The traditional method for manufacture of fructose was by hydrolysis of inulin from Jerusalem artichoke or dahlias using inulinase (Nicol 1977). HFCS improve properties in relation to sucrose and helps in formulating products having a high quality with good temporal characteristics, as it lends the product a better moisture retention capacity, thereby increasing its shelf life as well as preventing undesirable crystal formation. Compared to sucrose, HFCS have a higher osmotic pressure. This helps to keep down microbial growth in the product and helps in rapid penetration of cell membrane as in sweet pickle manufacture (Fruin and Scallet 1975) HFCS are differentiated on the basis of their fructose content into products containing 42, 55 and 90% fructose (Coker and Venkatasubramanien 1985). Fructose is reported to have 1.5–2.0 times the sweetness of sucrose. Fructose has a higher solubility, is less viscous and less carcinogenic than sucrose and low levels can be metabolized by diabetics without the need for insulin (Fleming and Groot-Wassink 1979). Fructose can be separated and concentrated from HFCS on ion exchange resins. HFCS has thus resulted in an important impact on the sugar industry.

Similar sugar mixture can be obtained from the hydrolysis of sucrose (Kotwal and Shankar 2009). It is perhaps a better alternative especially in tropical countries like India where growth of cane sugar is abundant. Sucrose inversion can be effected either in the presence of heat and acid or enzymatically using the enzyme invertase (EC 3.2.1.26) or B-D-fructofuranosidase fructohydrolase. In spite of its wide distribution in nature, only the invertase formed in yeast has been extensively used. Although acid hydrolysis is a simple process but the practical limit of conversion is only about 65–70%. It also results in undesirable side reactions and excessive colour formation in the product stream. Enzymatic hydrolysis on the other hand leads to very high conversion (above 95%) without the above problems.

5.2 *Immobilization of Invertase*

Immobilization of invertase has been investigated by many workers. Michael's and Ehrenreich (1908) were the first to report that yeast invertase could be partially adsorbed on charcoal and aluminium hydroxide gel. A few years later, Nelson and Griffin (1916) showed that invertase retains its activity after adsorption on charcoal and certain biocolloids. A number of reports then followed based on the selection of a suitable technique of immobilization or a suitable support and immobilization condition (Table 6). The approach to assist in the rational selection of support has been discussed by Kolot (1981). The supports used for immobilization range from elegantly produced spheroids to naturally available materials. Particle size, pore size and surface area can affect the performance of the immobilized preparation. Large beads encounter less pressure drop and good flow properties. However, a greater diffusion path has to be followed by the substrate to reach all the active sites. Thus, enzyme activity has been shown often to decrease with increase in bead diameter (Deshpande et al. 1987; Nilsson et al. 1980). This in turn has led to the development of methods to retain biocatalysts at the outer edge of particles using pellicular supports which are chemically synthesized (Horvath and Engasser 1973). To-date most of such supports are expensive and tedious to prepare. Another important point of concern is whether these matrices can be easily made in case it needs to be scaled up.

5.2.1 *Ocimum basilicum* Seeds as a Pellicular Support for Immobilization of Invertase

In our studies, a natural pellicular support viz. *Ocimum basillicum* seeds also called sweet basil or "sabja" were investigated for the first time to immobilize invertase enzyme, thus overcoming the step of synthesizing pellicular matrix (Melo and D'Souza 1992). Seeds when stepped in water swell to form an outer pectinous matrix with an inner hard core. The swelled seeds are uniform in size (3 mm). The mucilaginous outer layer consists of thread like microfibrillar structure with a large surface area as observed under microscope. Three different techniques were developed to covalently link invertase enzyme to this support. For all the three techniques, the first step involved epoxy activation with 1,4 butanediol diglycidyl ether. In the next step, two of the above were further activated with ethylenediamine, and in the last step, one of the two was further treated with glutaraldehyde. To the three preparations, invertase enzyme was then added except for the activated support where post-ethylenedianmine treatment enzyme oxidized with periodate was added. Of the various methods investigated for binding enzyme, periodate oxidation of the carbohydrate moiety of the glycoprotein enzyme prior to its immobilization to the

Table 6 Immobilization of invertase enzyme

Matrices	Applications	References
Immobilized on charcoal and aluminium hydroxide	Adsorption of Invertase	Michaelis and Ehrenreich (1908)
Charcoal, serum and egg albumin	Adsorption of invertase	Nelson and Griffin (1916)
Porous glass	Continuous inversion of sucrose	Mason and Weetall (1972)
Adsorption on amberlite ion exchange resin	Continuous sucrose hydrolysis in a tubular reactor	Boudrant and Cheftel (1975)
Radiocopolymerization with synthetic monomers	Preperation of membranous immobilized invertase	Kawashima and Umeda (1976)
Entrapment in photo-cross-linkable oligomer	Immobilization and characterization	Tanaka et al. (1977)
Immobilized on collagen membrane	Continuous inversion of sucrose in semi pilot scale reactor	Goldstein et al. (1977)
Covalently on porous glass and ionically on ion exchange resin	Kinetic characteristics for sucrose hydrolysis	Ooshima et al. (1980)
Immobilized on corn grits	Batch and continuous sucrose hydrolysis. Scaled up to 17.6 L pilot reactor	Monson and Combes (1984)
Covalently coupled to Con A Sepharose	Continuous hydrolysis studied for 60 days	Iqbal and Saleemudin (1985)
Immobilized on PEI-coated cotton thread	Batch and continuous studies.PBR of 1.5L gave 97% hydrolysis of 80% sucrose at 50 °C	Godbole et al. (1990)
Covalently bound to *O. basilicum* seeds through protein or carbohydrate moiety	Biochemical characterization	Melo and D'Souza (1992)
Immobilized on graft copolymer poly(ethylene-g-acrylic acid)	Studies on Kinetic parameters	Querioz et al. (1996)
Immobilization in alginate capsules	Biochemical characterization	Tanriseven and Dogan (2001)
Covalently bound to nylon-6 microbeads	Biochemical characterization and continuous sucrose hydrolysis in tubular fixed bed reactor at different sugar concentrations and flow rates	Amaya-Delgado et al. (2006)
Adsorbed on polystyrene beads of DOWEX	Comparison of continuous sucrose hydrolysis in UF and MF membrane bioreactor	Tomotani and Vitolo (2007)
Covalently immobilized on PMTM-PTAA matrix	Biochemical characterization and batch hydrolysis	Dizge et al. (2008)

(continued)

Table 6 (continued)

Matrices	Applications	References
Review article	Immobilization techniques and biotechnological applications	Kotwal and Shankar (2009)
Covalently bound to porous silicon layer	Biochemical characterization	Azodi et al. (2011)
Invertase –Silica hybrid particles	Biochemical studies	Rai et al. (2012)
Covalently linked with glutaraldehyde on chitosan nanoparticles	Biochemical studies	Valerio et al. (2012)
Covalent attachment to monolithic silica roads and mesoporous cellular foam	Biochemical characterization and sucrose hydrolysis in continuous mode	Szymanska et al. (2013)
Beidellite nanoclay modified with surfactanct, pillaring with Al/Fe and acid modification	Biochemical characterization and batch hydrolysis	Andjelkovic et al. (2015)
Covalently immobilized on chitosan spheres using glutaraldehyde	Comparison of sucrose hydrolysis in packed bed and fluidized bed reactors	Lorenzoni et al. (2015)
Covalently bound to silane-coated silica carriers	Continuous hydrolysis in a sandwich microreactor	Carvalho and Fernandes (2015)
Covalently bound to magnetic nanoparticles via p(GMA) and hexamethylene diamine	Biochemical characterization and batch hydrolysis	Bayramoglu et al. (2017a, b)
Covalently bound to magnetic diatomaceous earth nanoparticles via APTES and glutaraldehyde	Statistical studies for enzyme binding and batch hydrolysis of sucrose	Cabrera et al. (2017)
Immobilized using soy proteins to form CLEA's	Biochemical characterization and sucrose hydrolysis in feed batch process	Mafra et al. (2018)
Immobilized on a biohybrid of silica nanoparticles and *O. basilicum* seeds	Biochemical characterization and batch hydrolysis	Mishra et al. (2020)

activated matrix showed the maximum activity. About 10^4 I.U. of the enzyme could be bound to 10 g dry weight seeds. The preparation where enzyme was directly bound to epoxy activated matrix had the lowest activity. Loss in activity may be attributed to either active site amino acids being involved in bond formation or to conformational changes besides possible inactivation due to the alkaline pH required to bind to epoxy activated matrix. It has been reported that the active form of invertase requires an unprotonated COO^- group for substrate binding and a protonated imidazole residue involved in substrate binding. It has also been suggested that a SH group is involved in catalytic action. Deactivation of invertase due to immobilization at higher pH cannot be ruled out. The alternate method of binding invertase to epoxy seeds via ethylenediamine glutaraldehyde arm did not improve retention yield of enzyme activity. Glutaraldehyde is known to bind to amino or sulphydryl groups which may be present in the active site of enzyme and cause the inactivation of the same. This study has not only provided a useful method for the immobilization of invertase through its carbohydrate moiety, but has brought out the use of *O. basilicum* seeds as a natural, non-toxic, inexpensive, pellicular polysaccharide support.

5.2.2 Surface Adhesion on Glass of *Saccharomyces cerevisiae* Cells Expressing Invertase Activity

Invertase can also be immobilized by using whole cells. This obviates the need for extraction of the enzyme and risk of enzyme inactivation during isolation, purification and immobilization process. The whole cells expressing invertase activity and immobilized by different authors are summarized in Table 7. Development of a suitable immobilized invertase system for commercial application depends on the availability of the enzyme source in bulk. Invertase even though is present in a number of plants, animals, insects and microbes, but only certain strains of yeast with GRASS status have been found to be a potent source. Even though methods are available to grow yeast cells for induction of invertase, this is capital intensive proposal. Most of the bakers and brewer's yeast contain large amount of invertase and use of this yeast which is cheap can bring down the cost of the process. In view of this, studies were undertaken to identify a readily available baker's yeast strain with maximum invertase activity.

Entrapment has been the most commonly used technique for immobilizing cells. However, mass transfer problems associated with this immobilization technique as mentioned earlier are of concern. In order to minimize these problems, a simple and novel technique for the surface immobilization of yeast cells on glass through adhesion was investigated. Bacterial adhesion on to solid surfaces is being extensively studied in different areas such as marine fouling, soil and plant ecology, dental plaque formation, biomaterial associated infections, fermentation and in waste treatment. Unlike naturally adhered biofilms wherein it takes time to build a microbial film sufficient to affect a bioprocess, there are also a number of industrially useful organisms that do not normally adhere. This has attracted attention in developing techniques to induce adhesion. Under normal pH conditions, the surfaces of most organisms including yeast have a net negative charge (Thonart et al 1982). It was shown that yeast cells could be imparted polycationic characteristics by treating them with polyethyleimine (PEI). The PEI-treated cells were found to adhere to glass surface uniformly (D'Souza et al. 1986). The control cells which were not coated with PEI did not show any adhesion. Viability of cells was not found to be affected. The operational stability was studied in a specially designed slide reactor which could be reused for over ten batches for sucrose (10%) inversion without appreciable loss in activity. The PEI-coated cells were found to adhere very strongly on glass surface. High ionic concentrations or extreme pH conditions which normally disrupt the ionic interactions failed to desorb the cells. One of the notable observations was the ability

Table 7 Immobilization of *Saccharomyces cerevisiae* cells

Matrices	Applications	References
Entrapment in acryalamide polymerized by gamma rays	Biochemical characterization and continuous inversion in packed bed reactor	D'Souza and Nadkarni (1980)
Entrapment in gelatin	Comparison study with cells entrapped in alginate and k-carrageenan	Dhulster et al. (1983)
Adhesion on glass using PEI	Biochemical characterization and batch sucrose hydrolysis	D'Souza et al. (1986)
Entrapment in alginate	Studies on enzyme activity based on cell loading, gelation method, gel concentration and composition	Johansen and Flink (1986)
Entrapment in chemically polymerized acryalamide	Biochemical characterization and batch hydrolysis	Ayukt et al. (1988)
Adhesion to PEI-coated waste cotton thread	Biochemical studies and continuous hydrolysis in PBR	Melo et al. (1992)
Entrapped in films of poly(2-hydroxymethyl methacrylate)	Membrane reactor studies	Cantarella et al. (1992)
Yeast cells covalently bonded with PEI and glutaraledhyde	Biochemical characterization and batch sucrose hydrolysis	Hasal et al. (1992)
Immobilized in a SiO2 sol layer coated on glass	Kinetic studies and cell discharge studies from matrix	Inama et al. (1993)
Entrapment in gel matrix	Growth rate studies of entrapped cells using different media and correlating enzyme activity	Parascandola et al. (1993)
Entrapped in calcium alginate beads	Long-term metabolic activity of cells stored under feasting and starving conditions	Melzoch et al. (1994)
Immobilized in sintered glass rings	Batch and continuous glycerol synthesis	Gonzales Benito et al. (1994)
Entrapment in alumina particles with binder through spray drying	Ethanol conversion	Isono et al. (1995)
Direct immobilization in PBR packed with PEI-treated cotton threads	Continuous hydrolysis of sucrose	Melo and D'Souza (1999)
Immobilization on PEI-treated jute fabric	Continuous sucrose hydrolysis in a annular reactor	D'Souza and Melo (2001)
Cells modified with ferrofluid	Batch sucrose hydrolysis	Safarikova et al. (2009)

(continued)

Table 7 (continued)

Matrices	Applications	References
Cells entrapped in magnetically responsive alginate microbeads	Sucrose conversion in batch mode	Safarik et al. (2009)
Cells entrapped in magnetic particles of alginate	Used in magnetically stabilized fluidized bed reactor for ethanol fermentation	Liu et al. (2009b)
Cells immobilized in luffa sponge disc	Ethanol production from mahula flowers under submerged fermentation	Behera et al. (2011)
Cells immobilized on thin shell silk cocoon	Comparison studies with free cells for ethanol fermentation	Eiadpum et al. (2012)
Immobilized inside monolithic macroporous hydrogel of acryalamide	Batch fermentation of ethanol	Mulko et al. (2016)

of PEI-coated cells to bind also onto PEI-coated glass surface. Based on these observations, it was postulated that in addition to ionic interactions between positively charged PEI cells and negatively charged glass surface, hydrophobic interactions and other weak forces may play a role in strong adhesion. This accidental observation of cells binding to glass instantly, uniformly and efficiently, paved the way for many more studies in our laboratory and became well accepted and acknowledged by other groups also.

5.2.3 Immobilization of Baker's Yeast on Waste Cotton Thread

The major limitation of glass slides is its limited surface area. Even though highly porous glass support are currently available, high cost of such supports limits its application for large-scale use. Besides from an economic point of view industrial scale, enzymatic hydrolysis must be achieved using concentrated sucrose syrups to minimize energy intensive post-concentrations steps. Application of immobilized invertase for hydrolysis of concentrated sugar syrups has been hampered by the viscous nature of substrate and the substrate product inhibition exhibited by the enzyme with product inhibition being more significant. Thus, a packed bed reactor is more useful for such a conversion. Thus, this necessitates the search for a more feasible support which is both inexpensive and in addition to providing large surface area is also amenable for being packed into columns.

In view of this, the modification of the technique of adhesion to immobilize cells on cotton thread waste which fulfils some of the above characteristics was worked upon. Like yeast cells, cotton threads could also be imparted anion exchanger property by coating it with PEI. The yeast cells were found to adhere strongly on the PEI threads (Melo et al. 1992). About 4.4 g wet weight of cells could be bound to 10 g cotton threads. The binding efficiency was not altered even at temperature of 60°C. Inversion of sucrose was carried out in a jacketed packed bed column (2.6 × 40 cm). Cotton threads (equivalent to 30 g dry weight) containing adhered cells were packed in the column. Commercial sucrose 60% (w/v) adjusted to the required pH with benzoic acid was continuously pumped through the column using a peristaltic pump. Studies on mass transfer at same cell loading and under identical operational conditions indicated that adhered cells could hydrolyze 75% of the sucrose as compared to calcium alginate entrapped cells which could hydrolyze only 37% of the original sucrose. The effect of temperature on inversion was compared both in terms of productivity and operational stability. High temperature is normally preferred for the inversion of sucrose both in terms of efficiency as well as reduced microbial contamination. Operational stability however depends on the temperature. Majority of the studies on immobilized invertase have considered only the effect of temperature on reaction efficiency for short-term studies and half life has been predicted based on limited operational studies. Although a temperature of 60 °C showed a better initial efficiency, the system showed a drastic reduction in activity with a half life of four days. Constant conversion rate of 75% could however be obtained with a column run at 45 °C for over a month. Optimal rate of hydrolysis was seen at pH 4.5 and at a flow rate of 5L/day with a productivity of 2.25 kg sucrose hydrolyzed/day.

5.2.4 Immobilization of Cells and Sucrose Hydrolysis in a Column Reactor

One of the limitations of adhering cells on to cotton threads and then packing them into the column was the leaching of cells from the matrix when loading the matrix into the column. One of the practical solutions would be to adhere the cells directly in the column packed with the cotton threads. This prompted us to explore the possibility of using PEI-coated threads in a packed bed reactor as a novel microbial filter. Adsorption capacity, affinity and selectivity are the major criteria for successful application of such material. Yeast cells were found to adhere to PEI-coated cotton threads even from a flowing suspension (Melo and D'Souza 1999). To study the efficiency of adhesion from a flowing suspension, the cells were perfused through the column using a peristaltic pump. Effluent from the column was monitored for presence of cells. Pumping of cells through the column was stopped when the column was saturated with the cells, which was indicated by turbidity in the effluent. Amount of cells bound to the matrix was calculated from the total volume of suspension perfused through the column. Moreover, optimization of the effect of PEI concentration on the efficiency of binding yeast cells showed that the biomass adhesion capacity of the threads could be increased threefolds by coating the threads with 2.5% PEI as compared with threads coated with 0.2% PEI as in the previous study. Ten grams of 2.5% PEI-coated threads packed in column (2.6 × 40 cm) showed a thousand fold reduction in cell count for over 2L of yeast cell suspension (10^7 cells/mL) perfused at the rate of 210 mL/h. Following the binding of cells, the column was used for the continuous inversion of sucrose. One of the notable features was the capability of this column which was loaded with just one third the amount of cotton thread loaded in the previous study, being able to perform equally well as the above packed column. These studies thus indicate that prior loading of PEI-coated threads into column can serve as a direct method to immobilize cells in the column from a flowing suspension and also avoid the limitations of the previous method. In addition to the high capacity of the PEI threads, other important features include its simplicity of preparation and low cost when used in packed bed reactors.

5.2.5 An Annular Column Reactor for Sucrose Hydrolysis

Selection of materials as suitable adsorbent is still made on a rather empirical basis. Adsorption capacity, affinity, selectivity and good flow properties are the major criteria. The above studies have shown that waste cotton threads have been found to be a useful adsorbent of microbes from flowing suspension. However, waste cotton threads packed in column have limitations of bed compaction/pressure drop when perfusing concentrated sucrose syrups at high flow rate through packed bed column. These problems can be obviated using other geometries of the same matrix like cloth. Since cotton cloth was used in another study for binding ureolytic cells, in this study, jute fabric which is also a lignocellulosic support was investigated for immobilizing yeast cells.

Jute fabric (3.5×30.5 cm^2) treated with 2% PEI (after optimizing for cell binding) was set on a stainless steel mesh of similar dimension and rolled into an annular column with an outer diameter of 4.8 cm. As seen from the top of the cylinder, it appeared as three rolls of jute fabric separated by the stainless steel mesh. This cylindrical module was then placed in a glass cylinder containing yeast cells (13.6 g in 680 ml water). After cells got bound to the matrix, the module was rinsed extensively with water to remove any unbound cells and then mounted in a temperature controlled column reactor with an inner diameter of 5 cm. The temperature was maintained at 45 °C by circulating water through the outer jacket using a thermoregulated water circulation system. The column was operated for 45 days without loss in efficiency. When 60% sucrose syrup was perfused through the column, 70% conversion was achieved at flow rate of 2.4L/day. However, when 80% sucrose syrup was passed through column at flow rate of 1.2 L/day, 60% conversion was achieved. At higher temperatures, conversion rates could be increased but operational stability decreased drastically. Such spiral wound annular tubular bioreactors provide excellent performance over packed bed reactors as no change in flow characteristics was observed when perusing viscous or particulate materials in substrate stream and thus makes it amenable for scale up (D'Souza and Melo 2001).

5.2.6 Nanomaterials for the Immobilization of Invertase

A more recent breakthrough features the concept of nanotechnology wherein various nanostructured materials such as nanoparticles, nanofibres, mesoporous silica and nanoparticles prepared through sol–gel encapsulation are being explored for various applications including enzyme stabilization. Interestingly, some of these materials have also been investigated using invertase enzyme.

Valerio et al. (2012) immobilized invertase on organic nanoparticles of chitosan. They achieved good immobilization yields and efficiencies. However, there was no uniformity in the nature of matrix prepared as besides the nanoform aggregated chitosan was observed under TEM because of the presence of bifunctional reagent used. Most of the studies using nanoparticles have shown improved enzyme activity and loading, but the hurdle remains in the separation from the reaction medium. In this context, magnetic separation is an attractive alternative.

Cabrera et al. (2017) immobilized invertase on magnetic diatomaceous earth nanoparticles functionalized with (3-Aminopropyl)triethoxysilane (APTES). They used design of experiment approach to optimize the best conditions for binding of enzyme and hydrolysis of sucrose. They also compared the specific activity with other studies and reported that they had achieved the highest specific activity. Here also, the morphology of the DE particles was irregular and showed a sheet-like structure.

In a further modification, magnetic particles functionalized with APTES were grafted with polymer glycidyl methacrylate through surface initiated atom transfer radial polymerization and also hexamethylene diamine spacer arm. Both the matrices were activated with bifunctional reagent glutaraldehyde for immobilizing invertase enzyme. The amount of invertase immobilized was 36 and 33 mg/g, respectively (Bayramoglu et al. 2017a, b).

Szymanska et al (2013) performed a study to compare invertase immobilized on mesoporous silica and monolithic silica rods with native enzyme. They observed that the apparent Km for the mesoporous silica was marginally higher than for native enzyme indicating that the affinities for sucrose were similar corroborating the elimination of mass transfer. But when monolithic rod-shaped silica having hierarchial pores were compared with native enzyme, the apparent Km was lower which correlated with the fold increase in apparent Vmax. Whereas mesoporous silica can develop backpressure in a fixed bed reactor the monolithic silica rods could operate at flow rate up to 20 mL/min due to their bimodal pore structure.

5.2.7 A Biohybrid of *O. basilicum* Seeds and Silica Nanoparticles for Immobilizing Invertase

As seen in the above study, although effective enzyme loading may be achieved on nanostructured materials, the pore size is a matter of concern as it restricts the accessibility of the substrate to the enzyme. To overcome these, Polshettiwar et al. (2010) have proposed the use of dendritic silica structures which they prepared through a chemical route. To achieve similar structures at a larger length scale which are amenable for easy retrieval from a reaction mixture, we proposed a green route for synthesizing fibrous silica macrostructures through template-based assembly of silica nanoparticles on the fibrillar structure of *O. basilicum* seeds which has featured earlier. To achieve maximum enzyme loading, two different methods were followed for loading silica viz. silica nanoparticles were directly contacted with seeds so as to entrap the silica during the swelling of the outer pellicular surface (Silica@seeds), and in the other methods, the seeds were first treated with polyethylenimine so that the fibres which are a pectinous matrix and negatively charged bind to the positively charged PEI, and next, when silica nanoparticles were added, it got adhered to the PEI-coated fibres(Silica@PEI-seeds) (Mishra et al 2020). Surprisingly, when the amount of silica bound was quantified gravimetrically, it was found that silica@seeds had twice the amount of silica when compared under identical conditions. Various techniques such as SEM, SAXS, BET, FTIR and synchrotron radiaton-based X-ray micro-computed tomography were used to characterize and put forth a mechanism for silica binding on the seeds. We observed a more ordered binding of silica on silica@PEI-seeds unlike the bulk loading in silica@seeds explaining for higher loading in the later. When invertase was loaded on these matrices, the immobilization yield was highest (99%) in the case of silica@PEI-seeds followed by silica@seeds (94%) and least (79%) in plain seeds at the lowest initial enzyme concentrations of 4000 IU and decreased to 52, 32 and 23%, respectively, when starting enzyme concentration was increased to 48,500 IU. We have also shown that silica nanoparticles can be used for entrapping microbial cells using spray dryer through evaporation induced self-assembly process (Melo et al 2013; Mishra et al 2014). In future, we thus propose to use silica nanoparticles to entrap *Sacharomyces cerevisiae* cells and use it for sucrose hydrolysis.

6 Immobilization in Bioremediation

6.1 Bioremediation

Pollution of environment is a big environmental challenge. Metallic pollutants are non-degradable and persist in nature, and hence, they have more scope of entering into the food chain. Heavy metals, such as Cd, Co, Cu, Ni, Pb, Hg, etc., can cause serious toxic effect once they enter into the organism. Unlike these metals, long-lived radionuclides such as U, Th, Pu, etc. can be harmful even without entering into the organism. World Health Organization (WHO) considers uranium as a human carcinogen (Han et al. 2007). Toxicity in this case is contributed both by the radioactivity as well as the chemical nature of the heavy metal (Volesky 1990; Hu et al. 1996; Bhainsa and D'Souza 1999). Uranium toxicity depends on various factors such as concentration of the ions and its solubility and types of species available. For example, uranyl ions binds to –SH groups of amino acids in a protein affecting its function (DNA repair enzymes, membrane proteins, antioxidant enzymes, etc.). Further, binding of uranyl ion to the biologically important anions can cause uranium toxicity. Other important factors of toxicity include route and time of exposure. Thus, environmental pollution of heavy metals and radionuclides is adversely affecting human health and environment. To mitigate the metal pollution, many conventional methods like adsorption, precipitation, filtration, evaporation, coagulation, ion exchange, cementation, electro-dialysis, electro-winning, membrane separation, electro-coagulation and reverse osmosis have been developed (Wilde and Benemann 1993; He and Chen 2014). However, these methods suffer from some deficiency such as high cost, secondary waste, decreased efficiency at lower concentration, etc. In view of this, there has been an increased interest in developing green, eco-friendly biobased treatment methods commonly known as bioremediation. Bioremediation is often considered as a cost effective environmental-friendly method and is gradually making inroads for environmental clean-up application (Volesky 1990). In bioremediation, the natural ability of the organisms for binding and uptake of metal is utilized. Both live and dead biomass (bacteria, fungi, plant, biomolecules, etc.) are applied for removal of metal from the contaminated medium. These biomass posses various metal-binding functional groups such as carboxyl, phosphate, sulphate, hydroxyl and amine. The amount of these functional groups, affinity and their accessibility determines the metal-binding capacity of the biomass. This binding of metal ions onto the biomass can be considered as a natural immobilization process which could involve single or combination of mechanism adsorption/sorption, covalent binding, micro-precipitation, etc. (Hu et al. 1996; Lylod and Renshaw 2005).

Bioremediation by dead biomass is usually referred to as biosorption. Biosorption by microbial and plant biomass has been studied extensively for removal of metal ions from aqueous solution. Various parameters affecting the biosorption process include effect of pH, contact time, metal ion concentration, biomass dose, cations and anions, etc. which have been investigated. Mechanism of metal biosorption with respect to the kinetics and isotherms has been investigated in detail (Michalak et al. 2013). These parameters help to optimize the performances and understand the mechanism of metal ion interaction with the biomass. It also plays important role in scale up studies. Maximum loading capacity of the biomass is used for screening and selection of the various biosorbents. A biosorbent with a maximum loading capacity of more than 15% is considered economically viable (Hu et al. 1996). Therefore, biosorbents with less uptake capacity are modified to improve their performance. Often, chemical modification and/or functionalization of the biomass yields better result. Biosorption by nonliving/dead biomass allows desorption of the bound metal and its recovery without damaging the biosorbent significantly. However, free biomass has its own limitation such as difficulties in separation. Often, the biomass needs immobilization onto a support to improve its stability, regeneration and reusablility, suitability for scale up, etc. (Tripathi et al. 2010, 2013d). In this regard, various support materials with desirable property have been investigated and they have been applied for removal of radionuclides such as uranium. This chapter aims at presenting a progressive development of newer approaches regarding how the biomass/biomolecule using immobilization process could be utilized for uranium removal.

6.2 *Uranium Bioremediation Using Natural Biosorbent*

6.2.1 Microbes

Bacteria are ubiquitous in nature and they can be grown on variety of substrates. They grow very fast and can be multiplied in large scale. Further they posses natural ability to bind to heavy metals and exhibit tolerance to metal ions. Due to these properties, many bacterial biomass such as *Citrobacter*, *Pseudomonas aeruginosa* etc., have been investigated for removal of harmful metal ions including radionuclides (Hu et al. 1996; Lloyd and Renshaw 2005). Sar and D'Souza (2001) investigated a *Pseudomonas* sp. which exhibited excellent uranium uptake properties. The maximum loading capacity of the lyophilized and live *Pseudomonas* biomass was found to be 541 and 410 mg U g^{-1} respectively at pH 5.0 (Table 8). Uranium binding was rapid and achieved >90% sorption efficiency within 10 min of contact while the equilibrium was attained after 1 h contact time. Factors such as culture age, the presence of an energy source or metabolic inhibitor affected the uptake process. The results fitted to both Freundlich and Langmuir isotherm models indicating monolayered uranium binding by the test biomass.

Despite having good biosorption capacity, the bacterial biomass faces certain disadvantages such as difficulty in separation from the liquid medium during both upstream and downstream step due to its small (micron) size. Other microbial biomass such as filamentous fungal biomass and algae with metal-binding affinity can be used to overcome the limitations. Fungal biomass can be grown easily on variety of substrates including waste material. They can be harvested easily and processed for application (Kapoor and Viraraghavan 1995; Bhainsa and D'Souza 2009). In this respect, study by our group (Bhainsa and D'Souza 1999) involving *Aspergillus fumigatus,* the cellulose degrading fungal soil isolate showed very good uranium binding property. The uptake was time and pH dependent and rapid. It reached equilibrium within 1 h of contact of biomass with the aqueous metal solution. pH 5.0 was found to be optimum for biosorption of the metal ions. Biosorption result fitted to Langmuir model of isotherm indicating monolayer binding of metal ions onto the biomass. The maximum loading capacity of 423 mg U g^{-1} dry weight obtained was better than many fungal biomass *Penicilliumctrinium, Russulasanguine* etc., (Table 8). It is important to note here that the uranium uptake capacity exhibited by this biomass is comparable to our earlier study using bacterial biomass, *Pseudomonas* indicating they can be good candidate for further application.

6.2.2 Algae

Algae are autotrophs and possess very good metal uptake capacity (Yang and Volesky 1999; Lee et al. 2014; Khani 2011; Gok et al. 2017). Also, they are cheap and available in plenty in several habitats including the sea water (He and Chen 2014). An algal biomass *Catenella repens* (a red alga) was collected from the Konkan sea coast and

Table 8 Immobilization of uranium by free and immobilized biosorbents

Biomaterials		pH	Loading capacity (mg g^{-1})	References
Natural biosorbents				
Microbes	*Aspergillus fumigatus*	5.0	423	Bhainsa and D'Souza (1999)
	Pseudomonas aeruginosa sp.	5.0	541 (lyophilized) 410 (live)	Sar and D'Souza (2001)
	Penicillium citrinum	6.0	103.1	Pang et al. (2011)
	Aspergillus niger	7.0	4.314	Li et al. (2015)
	Russulasan guinea	5.0	174.3	Bagda et al. (2018)
Algae	*Sargassum fluitans*	4.0	378.4	Yang and Volesky (1999)
	Catenella repenes	4.5	303	Bhat et al. (2008)
	Laminaria japonica	4.0	96.4	Lee et al. (2014)
	Cystoseira	3.0	468.01	Gok et al. (2017)
Plants	*Eichhornia crassipes*	6.0	371	Bhainsa and D'Souza. (2001)
	Hydrilla verticillata (live)	5.0	78	Srivastava et al. (2010)
	Hydrilla verticillata (live)	6.6	0.426	Srivastava and Bhainsa (2016)
	Rice husk	3.0	15.14	Xia et al. (2017)
	Lemna	4.0	162.078	Vieira et al. (2019)
	Pistia	4.0	6.807	Vieira et al. (2019)
Modified biomass	Chitin (shrimp shells)	3.6	25.77	Ahmed et al. (2014)
	Alkali-treated hair	4.5	62.5	Saini and Melo (2015)

(continued)

Table 8 (continued)

Biomaterials		pH	Loading capacity (mg g^{-1})	References
	Amidoxane modified *Aspergillus niger*	5.0	621	Li et al. (2015)
	Nitrilotriacetate modified *Bangia atropurpurea*	5.5	328.8	Bayramoglu et al. (2017a, b)
	EDTA modified sugarcane bagasse	5	1394.1	Su et al. (2018)
	Alkali-treated wheat straw	3.0	19.23	Wang et al. (2011)
	Alkali-treated rice Stem	4.0	18.975	Xiao-teng et al. (2019)
Immobilized Bio/Nanomaterial	*Pseudomonas* immobilized onto polyacrylamide beads	5.0	202	D'Souza et al. (2006)
	Yeast cell entrapped in calcium-alginate-PVA-GO-cross-linked	5.0	142.1	Chen et al. (2019)
	Streptococcus lactis cells in SiNP	5.0	169.5	Mishra et al. (2014)
	Magnetic *Arachis hypogaea* leaves powder in chitosan	5.0	232.4	Yuvaraja et al. (2020)
	Chitosan (oxo-2- glutaric acid grafting)	6.0	450	Guibal et al. (1994)
	Alginate	4.0	400	Gok and Aytas (2009)
	Melanin	5	588.24	Saini and Melo (2013)
	Alginate–agarose–magnetite cryobeads	5.0	120.5	Tripathi et al. (2013d)
	Melanin -functionalized agarose–chitosan (MAC)	5.5	435	Tripathi and Melo (2016)
	Melanin–agarose–alginate–silica (MAAS)	5.5	232	Tripathi and Melo (2019)
	Chitosan composite hydrogel-containing amidoxime polyacryonitrile-chitosan-graphene oxide (PCG)	6.0	247	Lu et al. (2020)

used for uranium uptake studies (Bhat et al. 2008). This biomass exhibited very good uranium binding capacity in the range of 1.5–7.5. pH 4.5 was found to be the optimum pH for the uptake of uranium(VI). At a low pH of 2.5, where most of the biomasses show either no or less metal uptake, this biomass showed a good (>15%) metal loading capacity of 25%. This was unique observation related to this algal biomass as none other biomass tested so far showed appreciable uptake capacity at low (acidic) pH. Generally reduction in particle size increases the surface area and improves biosorption performance. However, in this study, this biomass did not show significant change in the biosorption capacity upon reduction in particle size. The metal removal was rapid, with more than 90% of total biosorption taking place in 30 min, while equilibrium occurred after 45 min of contact time. The maximum metal loading capacity of the algae was 303 mg U g^{-1} which was comparable to other algal biomass *S. fluitans, L. japonica,* etc. reported for uranium uptake (Table 8).

6.2.3 Plants

Notwithstanding the advantage of algae to be used for biosorption of metal ions, plant offers wider choice and availability. Hence, studies using easily available natural occurring plant have been investigated by many workers. Like algae, plants too are autotroph. They can be grown easily to prepare the biomass. Plant residual (waste) biomass after use offers a good choice to be utilized for biosorption of metal ions. The other cheap alternative is aquatic plant with naturally hairy root system providing wide surface area for metal ions uptake. *Eichhornia craasipes* biomass collected from a local pond was used as a biosorbent for uranium uptake (Bhainsa and D'Souza 2001). Uranium uptake by these dried roots was rapid and the biomass could remove 54% of the initial uranium present within 4 min of contact time. The process was favoured at pH 5–6 and was least influenced by temperature. Biosorption data fitted to both Langmuir and Freundlich isotherm indicating favourable binding of the metal ions onto the biomass. The maximum loading capacity obtained was 371 mg U g^{-1} dry biomass, much superior in comparison to the uptake capacity shown by other plants biomass such as *Lemna* and *Pistia* and wheat straw (Vieira et al. 2019; Wang et al, 2011) (Table 8).

6.2.4 Modified Biomass

Many of the naturally available biomasses are cheap source of biosorbent for uptake of uranium. However, often these biosorbents suffer from low uptake capacity and becomes less attractive for practical application. This problem could be addressed by modifying the biomass to improve the loading capacity (Table 8). All kind of biomasses, irrespective of their origin (microbial, plant and animal) can be modified using suitable chemical treatment. Some of the easily available and cheap animal and plant-based biomass includes chitin, human hair, sugarcane bagasse, wheat straw, rice stem, etc.

Chitin, the long chain polymer of *N*-acetyl-*D*-glucosamine, is the second most abundant resource in nature next to cellulose. It is biodegradable and ecofriendly. Chitin forms the major components of the exoskeletons of arthropods such as crustaceans. Chitin extracted from fresh shrimp shells was found to have good uranium loading capacity 25.77 mg U g^{-1} biomass at pH 3.6. (Ahmed et al. 2014).

Similarly, hair is a naturally available low cost waste material which contains melanin, the pigment with many hydroxyl groups. These hydroxyl groups play important role in binding to the metal ions. A study in our laboratory has shown increased uranium uptake capacity with alkali-treated hair (Saini and Melo 2013). The uranium binding capacity of alkali-treated hair was found to be 62.5 mg U g^{-1} biomass at pH 4.5. This uptake was better than the other animal waste biomass such as shrimp shells reported by Ahmed et al. 2014 (Table.8). Investigation into the mechanism of binding of the metal ions onto the biomass using FT-IR spectroscopy study showed involvement of amine, hydroxyl and carboxyl group, etc. Further, more than sixty percent of the bound metal could also be desorbed using 1 M nitirc acid.

Amidoxane modification of *Aspergillus niger* biomass exhibited very good uptake capacity of 621 mg U g^{-1}. Use of EDTA for modification of sugarcane bagasse surpassed this value and recorded uranium binding capacity of 1394.1 mg U g^{-1}. Other reported modification using nitrilotriacetate for *Bangia atropurpurea* also showed significant improvement in its uranium loading capacity. Alkali treatments of cellulosic plant biomass such as wheat straw and rice stem were marginally above the consideration of economically viable biosorbents, i.e. 15 mg metal ions g^{-1} biosorbents.

6.3 *Uranium Bioremediation Using Immobilized Biomaterials*

While good loading capacity is a primary requirement for immobilization of metal ions onto the matrix, consideration should be given to easy solid liquid separation and large-scale application for developing continuous process. This happens due to smaller scale biosorbents in micrometre range. To increase the length scale, the biomass (whole or components) can be immobilized suitably to increase the size for easy separation and application in continuous study. Cross-linking and entrapment of the biomass is considered as a suitable technique of immobilization for whole cells. It helps improvement in mechanical strength, easy separation, desorption and recovery and developing continuous process. *Pseudomonas,* a bacterium with very good binding potential of uranium (430 mg U/g), was immobilized in radiation-induced polyacrylamide matrix for its application in radionuclide containing wastewater treatment (D'Souza et al. 2006). The cross-linked beads with bacterial biomass exhibited a high U sorption of 202 mg g^{-1} dry wt. with its optimum at pH 5.0 which was less than the free cells. This could be due to the bead size (2 mm diameter) creating additional diffusion barrier and less accessibility to binding sites. The uptake result fitted well to Freundlich model suggesting multilayered uranium binding with an affinity distribution among multiple metal-binding sites. SEM study showed accumulation of uranium by the immobilized biomass without any physical damage to the cells. Ninety percent of the bound uranium could be successfully desorbed using sodium bicarbonate. Biomass was also found to retain resorption capacity for multiple sorption–desorption cycles. Study was further extended to continuous up-flow packed bed column reactor which showed its effectiveness in removing uranium from low concentration (50 mg U L^{-1}) followed by its recovery resulting in a 4–5 fold waste volume reduction, thus proving its advantage.

The other commonly used method for cell immobilization is performed using calcium alginate beads. In a recent report, yeast cell was entrapped in calcium alginate-PV-graphene oxide-cross-linked gel beads which showed good uranium immobilization capacity, 142.1 mg U g^{-1} (Chen et al. 2019). In this study, the yeast cell entrapped bead size was 4 mm in diameter which could adversely affect the performance.

To address the problem of diffusion barrier, effort was made to take the help of nanotechnology for immobilization of the cells. Wherein, *Streptococcus lactis* bacterial cells (2–4 micron) were immobilized using silica nanoparticles (app. 15 nm) by spray drying method. The resultant spray dried doughnut-shaped microstructures with diameter of 10–20 micron have bacterial cells exposed on the surface as well as inside. This not only improved the surface area but also the porosity in comparison to the above methods. The uranium loading capacity of the cell bound matrix was found to be 169.5 mg U g^{-1} at pH 5.0. By modulating the parameter of the cell suspension and spray dryer operation, other microstructures such as spherical shape also can be obtained. This establishes a proof of concept for immobilization of cells using silica nanoparticles by spray drying method for many such applications (Mishra et al 2014).

Another approach of easy solid liquid separation is by using the principle of magnetism. Yuvaraja et al (2020) in a recent investigation used chitosan modified magnetic *Arachis hypogaea* leaves powder for uranium uptake. The leave powder immobilized 232.4 mg U g^{-1} at optimum pH of 5.0. Together with its property of magnetic separation, this system may prove useful in large-scale operation.

Although whole biomass is often used for immobilization of toxic metal ions including uranium, only few components (biomolecules/functional group) of the biomass are found to be responsible for the binding of meal ions as discussed previously in this chapter. Therefore, it makes sense if those biocomponents are used specifically for the purpose of metal binding. Some of the metal-binding biomolecules, such as chitosan, alginate and melanin, etc., have been investigated either singly (Guibal et al. 1994; Gok and Aytas 2009; Saini and Melo 2013) or as composite matrix (Tripathi et al. 2013d; Tripathi and Melo 2016,2019; Yuvaraja et al. 2020; Lu et al. 2020) (Table 8).

Uranium sorption using chitosan grafted with oxo-2-gutaric acid was reported by Guibal et al. (1994). The maximum binding capacity of the matrix was found to be

450 mg U g^{-1} at pH 6.0. Similarly, calcium alginate bead was found to have good uranium uptake capacity of 400 mg U g^{-1} matrix at pH 4.0 (Gok and Aytas 2009). Whereas, tyrosinase-based biosynthesized melanin showed excellent uptake capacity (588.24 mg U g^{-1}) at pH 5.0. Binding of uranium to the functional groups of melanin could be proved using FTIR and Energy Dispersive Spectroscopy (EDS) study. These studies indicate that the specific metal-binding capacity of the biomolecules can play decisive role in developing remediation process.

As can be seen alginate, melanin and chitosan showed their pH optimum at pH 4, 5 and 6, respectively. Instead of using one type of molecule at a time, if they are used together in various combinations, their metal-binding properties along with others may be combined yielding desirable outcome. Therefore, many naturally derived polymers in different composition, size with variable densities have been developed and investigated in environmental bioprocessing and bioremediation. Besides the other environmental applications of agarose–alginate composite cryomatrix like whole cell immobilization for biobutanol production (Tripathi et al. 2010) and immobilization of spore-protein crystal complex for mosquitocidal formulation development (Tripathi et al. 2013b), these soft natural polymers have also been developed with a novel floating feature in macroporus alginate–agarose–magnetic cryobeads and used for recovery of uranium successfully by Tripathi et al (2013d) (Fig. 6). Uranium loading upto 120.5 mg U g^{-1} matrix was observed at pH 5.0. Further, the uranium loaded floating beads could be separated using magnet. Melanin-functionalized agarose–chitosan (MAC) beads were prepared by the process of cryotropic polymerization and used for uranium recovery (Tripathi and Melo 2016). Maximum loading capacity of this matrix was found to be 435 mg U g^{-1} at pH 5.5 with multiple sorption and desorption feasibility. Similarly, composite matrix was prepared using self-assembled biogenic melanin, agarose, alginate and colloidal silica nanoparticles (MAACS) to combine the property of organic biomolecules and inorganic, stable, biocompatible silica nanoparticles. The loading capacity for uranium was found to be 232 mg U g^{-1} matrix at pH 5.5 (Tripathi and Melo 2019). Further, in a recent investigation amidoximepolyacrylonitrile, chitosan and graphene oxide (PCG) was used to synthesize chitosan composite hydrogel for selective adsorption of uranium with multiple sorption and desorption cycles (Lu et al. 2020). These studies have proved that synthesis of composite materials in various combinations will be very much helpful to achieve the desired result of immobilizing uranium onto it and scale up the process.

6.4 *Outlook*

Biological materials such as microbes, plants, animals and biomolecules with metal-binding property have been used for removal of toxic metallic pollutant like uranium. The metal immobilization potential of the materials is due to its chemical composition, i.e. type, abundance, form and orientation of the functional groups. Often,

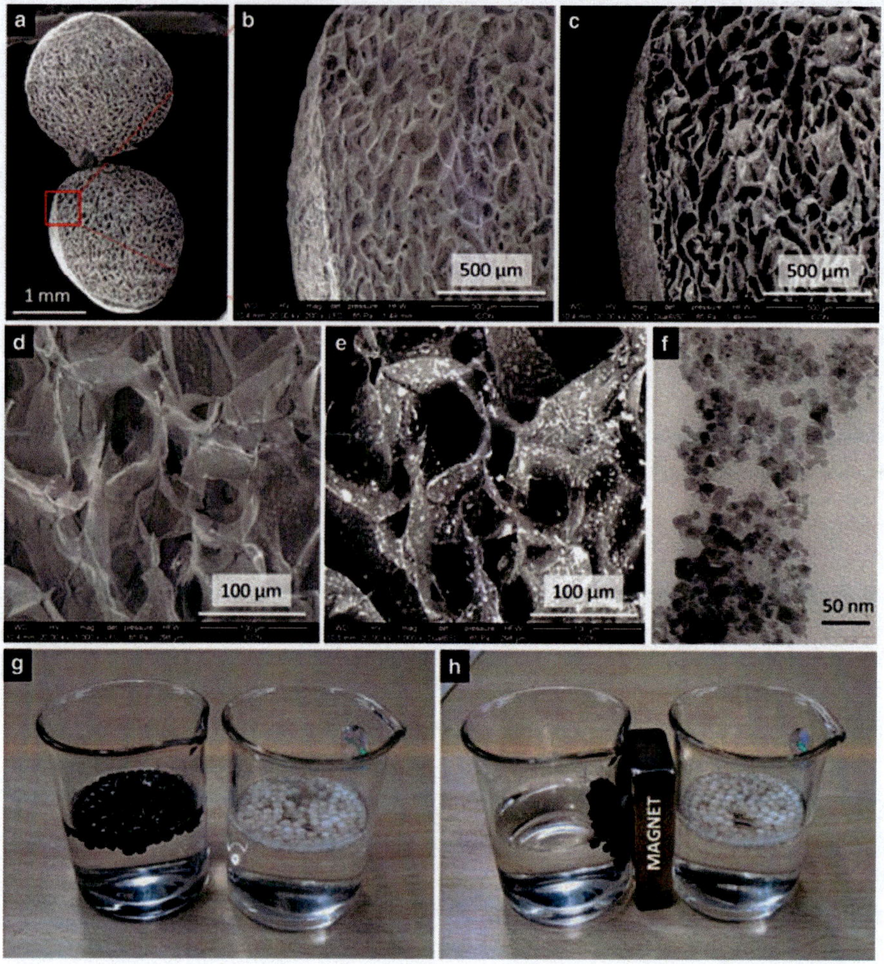

Fig. 6 Scanning electron micrograph shows porous morphology (**a**) and uniform pores distribution throughout the cryobeads (**b**, **c**). Images **d** and **e** show the interconnected pores at higher magnification (1000×). Images **c** (magnification 200×) and **e** (magnification 1000×) are the bi-scattered images show the presence of magnetite (white spots) in the pore walls of AAM cryobeads. The image **f** is the transmission electron micrograph of the magnetite nanoparticles. Digital image shows the floating behaviour of AAM and AA cryobeads (**g**), which showed magnetic suseptibility under external magnetic field (**h**). (Reproduced with permission from Tripathi et al. 2013d; © 2012 Elsevier)

these properties could be improved by modification of the biomass by suitable treatment. Further, several biocompatible and ecofriendly biomolecules have been used to improve the efficiency and specificity of the binding. However, the biomaterials showing good binding property typically operates at optimal physico-chemical condition limiting wide range application. This problem is compounded further due to its poor solid liquid separation, which is a significant impediment for large-scale studies. To overcome these problems, both suitable biomaterials and immobilization matrix would be required. This could be compensated by using materials at nanoscale, i.e. specific biomolecule such as alginate, melanin, etc., either singly or in combination and immobilization matrix such as silica, gelatin, etc. In addition, the immobilization matrices such as magnetic nanomaterial have been shown to offer enhanced operational properties for better separation which would be of significant advantage. Going forward, multiple types of biomolecules together with suitable nanomaterials immobilized at larger length scale would offer greater choice and flexibility for successful development of bioremediation process.

7 Biosimulation for Advancing Immobilization

7.1 Computational Tools in Understanding and Improving Enzyme Immobilization

As discussed hitherto, many of the bound enzymes show better stability, higher catalytic efficiency, manifold reusability and longer shelf life (Basso and Simona 2019). Therefore, the qualitative and quantitative studies of bound enzymes have advanced from lab scale experiments to large-scale plants (Chapman et al. 2018). Furthermore, the matrices for enzyme immobilization have evolved from natural polymers (cloth, alginate, chitin, etc.) (Bilal and Iqbal 2019), membranes (egg-shell membranes, onion epidermis and artificial membranes like PVA) (Kumar and Melo 2016) to the use of advanced matrices viz nanomaterials (quantum dots, core–shell, yolk-shell particles, virus like organo-silica nanoparticles, biohybrids) (Mishra et al. 2017), (functionalized) graphene sheets (Soozanipour and Taheri-kafrani 2018), carbon nanotubes (SWCNTs, MWCNTs, functionalized CNTs, etc.) (Singh and Chauhan 2020) and mesoporous and macroporous matrices (Saini et al. 2015; Shivudu et al. 2019). This could be accomplished because of imbibing the advances in other fields into enzymology and immobilization techniques. One such field, which has grown vastly in recent days and has potential applications in enzyme studies, is computation (Lodola and Mulholland 2013). By means of the availability of advanced algorithms with better reliability, rapidly growing computing resources and softwares, many computational techniques have found their way into enzyme studies. These tools have helped in understanding of enzymatic reactions, their interaction with substrates, surfaces (immobilization matrices) and process optimization in an unprecedented manner.

The process of immobilization, as discussed up till now, is a complex phenomenon mediated by surface molecules where the functional groups of the enzyme and the matrix interact with each other either by weak non-covalent interactions or by strong covalent bond(s) (Fig. 7) (Garcia-Galan et al. 2011). Knowledge of the enzyme structure, its surface residues and functional groups of the matrix would help understand

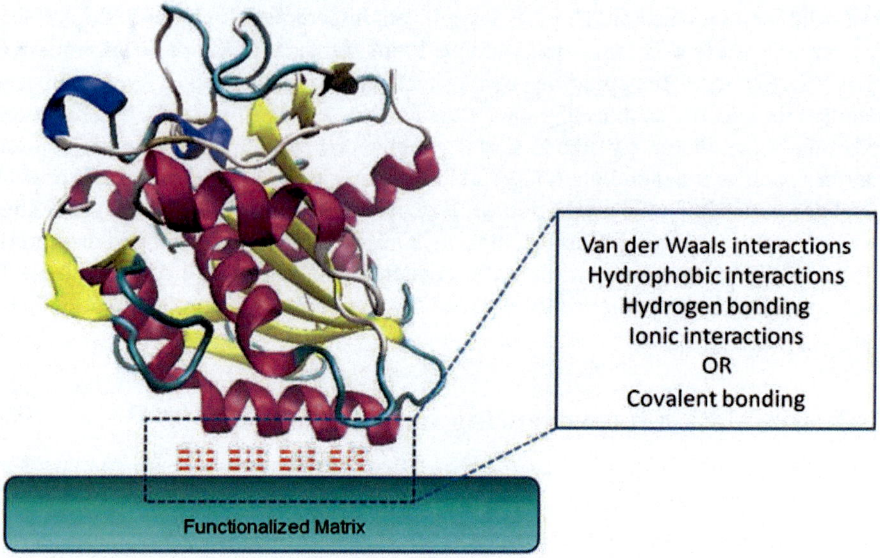

Fig. 7 Possible interaction between enzyme and matrix

as well as forecast the possible interactions between the enzyme and the matrix resulting in immobilization. Regardless of the interaction between the enzyme and the matrix, the process of immobilization is always a random phenomenon, without any control over the orientation of enzyme with respect to the matrix. If the enzyme is improperly oriented onto the matrix, it might reduce the ease of the substrate access towards the active site, leading to subjacent materials (e.g. nitrate reductase exhibits different activities, if it is immobilized parallel and perpendicular to its axis (Schroeder et al. 2017), lipase exhibits different activity if immobilized on C terminus than N-terminus) (Raeeszadeh-Sarmazdeh et al. 2015). With orientation control and proper placing of the enzymes onto the matrix, the immobilization efficiency and therefore catalytic output of the immobilized enzymes can possibly be improved.

This can be achieved by functionalizing the matrices by functional groups viz acids (–COOH), alcohols (–OH), amines ($-NH_2$), imines (–NH) for improved ionic, hydrogen bond interactions, benzenes (C_6H_6), phenols (C_6H_7O) and other aromatic groups for π-π stacking between aromatic residues of the enzyme and maximize the attachment of enzyme with the matrix. Capping agents of nanomaterials, spacer arms, edges of graphene and CNTs are also functionalized with these groups for better interactions with the enzymes (Fig. 8). Functionalizing the matrix may provide better binding with the enzyme, but orientation control is still an issue.

The other alternative, which has improved orientation control, is changing the surface residues of the enzyme for better binding with the matrix. But mutating the residues often alter stability of the enzyme. For attaching maximum number of properly oriented, catalytically active enzymes on the matrix, we need to understand

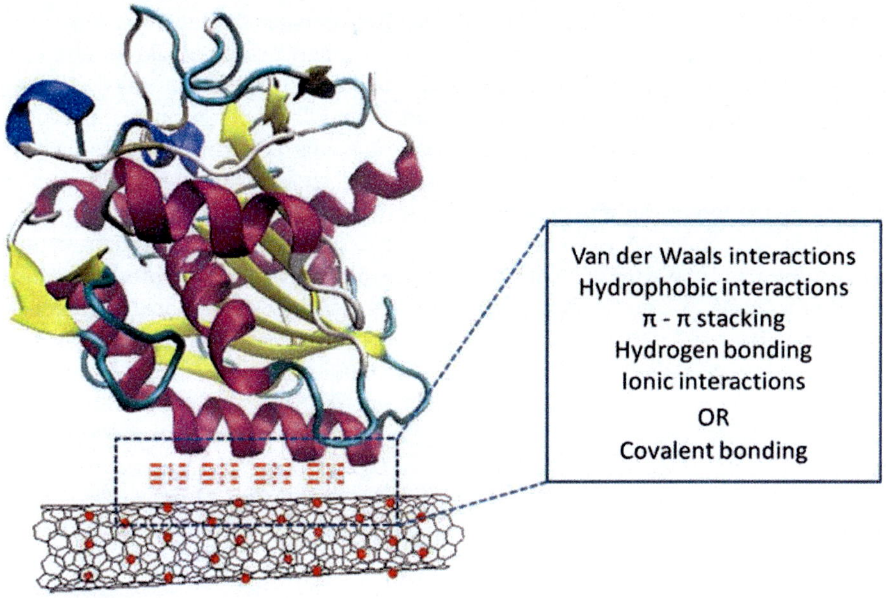

Fig. 8 *Thermomyces lanuginosus* lipase immobilized on functionalized CNT

the molecular interactions between the enzyme and the matrix that lead to immobilization, study the behaviour of the enzyme after altering the surface residues, estimate the stability of the mutant and assess the integrity of the catalytic site after the desired mutation. At the same time, understanding enzyme interactions with the functionalized matrix is also essential.

Advanced spectroscopic techniques, TEM, SEM, AFM, SPR, time resolved techniques can give an insight into the molecular interactions between the enzyme and the matrix (Mohamad et al. 2015; Hoarau et al. 2017; Bolivar and Nidetzky 2019). At the same time, complementing the experimental studies with computational techniques have become an essential part of understanding any molecular interaction per se, as the level of molecular detailing provided by computational techniques is unparalleled. Advancements in the computational tool repertoire empowers us with better algorithms for atomistic molecular mechanics (MM) simulations that can help computationally determine the interactions between enzyme-matrix, the structural stability of the mutant and substrate access to the active site. Largely, MM techniques constitute molecular dynamics (MD) and Monte Carlo (MC) simulations can help determine the non-covalent interactions between the enzyme and matrix. Among these, MD is time-dependent simulation where as MC is time-independent, in general (Daan and Berend 2001). The selection of method among these two depends on the subject under investigation.

Molecular, non-covalent interactions between the enzyme and the matrix, time-dependent changes in these interactions, structural and thermodynamic stability of the mutant can be studied by MD simulations and its variants (Hollingsworth and Dror 2018). Apart from the non-covalent interactions, the bond making/breaking, catalysis can also be studied by hybrid quantum mechanics/molecular mechanics (QM/MM) studies, wherein a part of the enzyme structure is simulated by quantum mechanics and rest of the system is studied by classical molecular mechanics (Ahmadi et al. 2018). MD simulations are therefore the method of choice when we are interested in studying the molecular interactions, as closely as observing hydrogen vibrations. Additional advantage of MD simulation is the ease of altering the milieu of the system. One can choose the solvent system as per requirement and therefore study the effect of aqueous or non-aqueous effects of immobilization conditions. Commonly used softwares for MD simulations are AMBER, CHARMM, Gromacs, NAMD, MOE, Discovery Studio, DL-POLY, etc. Understanding MD simulations of immobilization would give us an extensive insight on what happens on the surface of the matrix before, during and after immobilization. Therefore, MD is commonly regarded as molecular microscope (Dror et al. 2012).

7.2 *Molecular Dynamics Simulation of Enzyme Immobilization*

7.2.1 Basics of Molecular Dynamics

Assuming the atoms of the enzymes and the support matrix as hard spheres, with charges, polarization (as allowed on amino acids) and solving the Newton's equations of motion for N-particle system, assisted by suitable force field would help understand the non-covalent molecular interactions between the components of the system (Fig. 9) (Schlick 2010). Structures of the protein and the matrix are used as input for MD simulation. Protein structures can be obtained by RCSB protein data bank (Berman et al. 2000) or can be modelled. The matrix structures such as graphene, CNTs, nanoparticle surfaces, polymer surfaces, gels (silica, hydrogels), hybrids, etc. can be designed by computational tools viz Materials Studio, Schrodinger, CNT Builder, Samson-Connect (www.samson-connect.com), Molden, Avagadro, VMD and others (Schrodinger 2020: https://www.schrodinger.com/maestro), (Nanotube Modeller: https://jcrystal.com/products/wincnt/index.htm), (Schaftenaar et al. 2017; Hanwell et al. 2017; Humphrey et al. 1996).

Trajectory, the output file of MD simulation containing information of the time-dependent position, velocities and forces of all atoms in the simulation is analyzed to extract the structural and thermodynamic information of the system. Immobilization often leads to formation of new ionic, hydrophobic, vdW interactions and hydrogen bonds between the enzyme and the matrix. Changes in these weak interactions that keep the protein in native state could lead to structural changes, which can be observed by MD simulations.

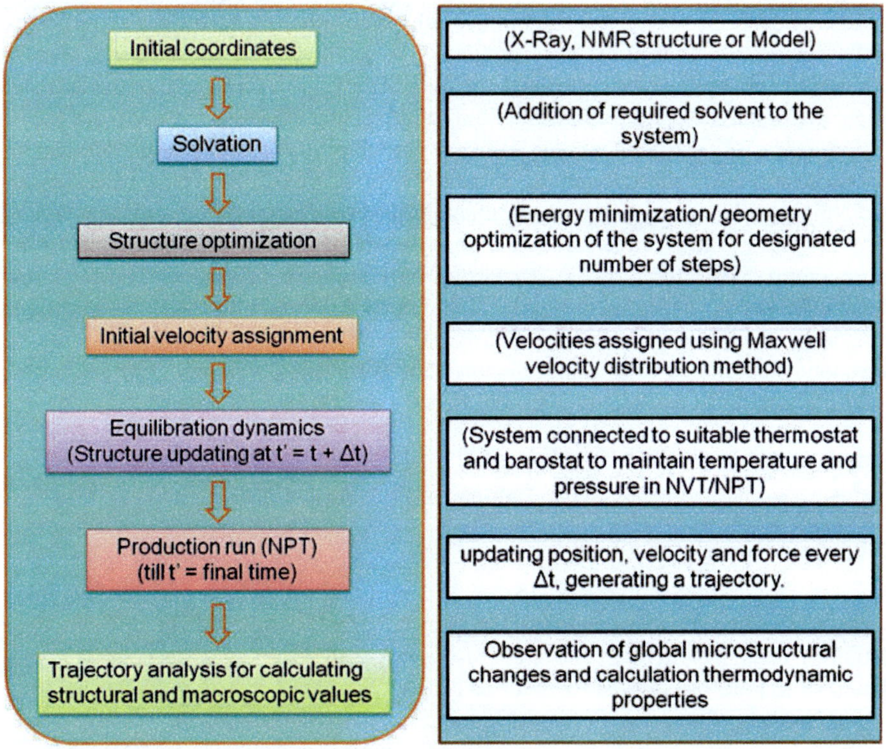

Fig. 9 Flow chart of molecular dynamics simulations

7.2.2 Analysis of Enzyme-Matrix Interactions

Structural Changes in the Enzyme

Immobilization involving multiple ionic/hydrogen bond/van der Waals/π-π stacking hydrophobic interactions often leads to changes in these interactions, resulting in global structural deviation of the enzyme (or the matrix). These deviations, if plotted against the native protein with respect to time, will help understand the effect of adsorption on the overall structure of the enzyme (or matrix). Besides, comparative root mean square deviation (RMSD) analysis between native and mutants (and non-functionalized vs functionalized matrix) will also help understand the impact of a mutation(s) on the structural stability of the protein/matrix.

Immobilization often results in movement of residues and in some cases, domains of the enzyme. Root mean square fluctuation (RMSF) analysis of the trajectory would indicate the mobile atoms/residues/domains and help understanding the areas with greatest mobility. Observing the RMSF in addition with solvent accessible surface area of surface residues will suggest about the extent to which particular residue is involved in interaction with the matrix and mark the hotspots. The comparative

RMSF analysis of native and mutant proteins would help whether the mutation is inducing additional mobility in the enzyme.

Not just movement of the residues, the overall secondary structures of the enzyme might also change during the course of attraction towards the matrix and binding to it. The secondary structural changes in the enzyme during/after immobilization can be studied using DSSP (Dictionary of Secondary Structure of Proteins) tool (Kabasch and Sander 1983), which observes the changes in secondary structures with respect to time and yields information about the time at which the change in secondary structure has occured. Most often, the *N*-, *C*-termini move quite rapidly towards/away from the matrix. In case of globular proteins, the outer α-helices, folds, bends and random coils are susceptible to alterations. Domain dynamics hinge like motions if any and the changes in them due to immobilization could be found out comparing the initial and final structures using DynDom tool (Lee et al. 2003).

Solvent Accessible Surface Area, Volume and Packing Studies

In the field of enzyme immobilization, solvent accessible surface area (SASA) is an essential parameter predictable from MD simulations. SASA is an indicator of strength of adsorption/desorption of the enzyme onto the matrix. The difference of {(area of protein + area of the matrix)—area of protein-matrix complex} would represent the area of interaction between the enzyme and matrix. Lower SASA of enzyme-matrix complex indicates stronger adsorption of enzyme, whereas a reduced no of atoms would indicate desorption of the enzyme from the matrix. By voronoi polyhedral method, the packed volume, free volume of the enzyme and the changes in enzyme volume before, during and after immobilization, can be assessed.

Hydrogen Bond Dynamics

The number and dynamics of inter, intramolecular hydrogen bonds between the enzyme and the matrix indicate the strength of adsorption between the protein and enzyme. MD simulations would furnish time-dependent information about the hydrogen bond dynamics (changes in the number, angles, distances, hydrogen bond life time of hydrogen bonds and changes in donors/acceptors with respect to time, etc.) between enzyme and matrix, which would furnish information about the strength of enzyme-matrix attachment. Increasing number of hydrogen bonds between the enzyme and the matrix would indicate stronger affinity between the two leading to adsorption, whereas the reducing number between them indicates non-favourable interactions between them leading to desorption. Also, comparing these parameters of native and mutant enzymes and their interactions with normal and functionalized matrices would help identifying the ideal choice of mutations/functionalization.

Ionic Interactions

Similar to hydrogen bonds, ionic interactions between charged residues that keep the native structure of the protein intact might as well get affected due to immobilization. Interactions between charged surface residues of the enzyme and the oppositely charged functional groups of the matrices would lead to stronger attachment of enzyme to the matrix but also may result in changes in protein structure. Besides the charge on the residues/matrix, strength of ionic interactions (inter and intramolecular) depends on the distance between the charged species and the dielectric constant of the medium. Considering these points, MD simulations would give the time-dependent position of the charged groups and thereby the interactions between them, resulting in altered intramolecular/intermolecular ionic interactions leading to adsorption/desorption and at times enzyme denaturation. Some solvent molecules diffuse into the intermolecular space between the enzyme and the matrix and affect the interaction between them. This is more important in case of aqueous media, due to higher dielectric constant of water, compared to other commonly used solvents.

Radial Distribution/density Distribution Function

Additional important parameter that could be derived from MD simulation is radial distribution function (RDF, also known as pair correlation function), which is dependent on the distribution of one type of atoms/molecules around the other. If there is an interaction between them, the density distribution of one type of particle around the other is more and their ratio would be more than 1. If there is no/neutral interactions, the ratio would be less than 1. The RDF between two types of particles A & B can be defined as:

$$g_{AB}(r) = \langle \rho_B(r) \rangle / \langle \rho_B \rangle_{\text{local}}$$

where $\langle \rho_B(r) \rangle$ is the ensemble average of B type particles, around A type particles at a distance r and $\langle \rho_B \rangle_{\text{local}}$ is the particle density of type B averaged over all spheres around particles A with radius r_{max}, which is usually half the simulation box length. RDF can also be an indicator for density-based structural properties of the system and can report the interactions between the given enzyme and matrix.

Stronger attachment of enzyme to the matrix would result in higher RDF values, whereas weaker attachment/desorption would lead to lower RDF value. Therefore, radial distribution functions between the interacting atoms of protein surface and that of the matrix would give imperative structural information that will help deciding which mutations stabilize the protein and which are structurally detrimental to the protein.

Integrity of the Active Site

The extent of fluctuations of catalytic residues and the change in the distances between them would signify the integrity of the catalytic site. Immobilization may lead to many structural changes in the enzyme but the change in the active site conformation is most undesirable one. Binding of enzyme or mutant to the functionalized matrix should "not" result in an enzyme that is strongly bound to the matrix, in right orientation, structurally stable but unable to catalyze due to loss of active site integrity.

The active site residues are maintained in catalytic conformation, by optimum conformation of the protein as well as the hydrogen bonds between the catalytic residues. Calculating the number, dynamics and lifetime of hydrogen bonds between them as well as the distance between the active site residues throughout the simulation time would provide information about the integrity of the active site. Not only the overall structural stability but the integrity of the active site is also important for efficient catalysis. Therefore, integrity of active site is one of the prime parameters to be analyzed in all immobilization simulations.

Thermodynamics Studies

Besides the information about the structural changes and their implications, MD simulations can also help deduce the thermodynamic studies of binding between the enzyme and the matrix. Calculating the binding free energy between the enzyme and matrix can help deciding the affinity of the protein towards the matrix and help choosing the ideal matrix for immobilization or the necessary mutations/functionalization to maximize the interactions. Identifying residues with highest binding energy would help understanding the possible interactions between enzyme and the matrix (Zhang et al. 2014). Identifying the components of binding energy would help understanding the major contributing residues for binding energy and mutations/matrix functionalization can be planned accordingly. Potential of mean force is another study giving information about binding-free energy between enzyme and matrix in various milieus. Enthalpy, entropy, potential, kinetic energies, Lennard Jones potential values can all be extracted from the trajectory.

MD Simulations as a Tool to Understand Immobilization

Domain-specific changes in RMSD, RMSF, reduced SASA, rapid/sharp rise in number of contact atoms (between the protein and the matrix), number of hydrogen bonds, ionic interactions between the enzyme and the matrix would indicate a quick and strong binding between them. Integral catalytic site, higher binding energy (overall and per residue) between enzyme and matrix would result in stronger attachment of the enzyme onto the matrix. Rationally planned mutations/matrix functionalizations would keep the enzyme in relevant orientation, allowing the enzyme not only to stay strongly footed onto the matrix but also allow easy substrate access to the active site, maintaining high catalytic activity.

7.3 Selection of Ideal Residues for Mutation and Designer Enzymes:

7.3.1 Protein Engineering for Better Attachment to Matrix and Catalysis:

As discussed above, for better binding, mutating the surface residues is one of the options, altering the functional groups of the matrix being the other. Mutating residues would sometimes lead to enzymes with altered folding and thereby possible altered functions, called designer enzymes. Extensively modifying the biocatalysts as per our requirement especially to make designer enzymes constitutes the third wave of biocatalysis (Bornscheuer et al. 2012). Designer enzymes have the desirable properties that we aspire the enzyme should possess as well as the structural features, not present hitherto, and therefore, they can quite possibly catalyze a novel reaction. However, redesigning the native enzymes to produce new enzymes with desirable features is rarely successful in a single attempt. Enzyme engineering with multiple iterative cycles of mutations assisted with computational tools produce novel catalysts.

7.3.2 Enzyme Engineering to Produce Designer Enzymes

Significant changes in the nature of the enzyme usually require changes in protein structure assisted by multiple mutations. Random mutagenesis (by error prone PCR or chemical/radiation mutagenesis) would lead to multiple mutants, which need to be screened further for selection of mutants with better, desirable properties (Fig. 10). Often many subsequent rounds of selections are required. By an in vitro version of Darwinian evolution, termed as directed evolution, the evolution of designer enzymes can be accelerated to generate mutants iteratively to get best possible protein structure with all desired features viz better binding to matrices, increased catalysis, altered active site space (enzymes would then be capable of accepting either bulky or smaller substrates than they are destined to. Eg: synthesis of cholesterol reducing atorvastatin, antihistamine Montelukast, antidiabetic sitagliptin). If the structure of the enzyme is known, the mutagenesis can be targeted to specific regions.

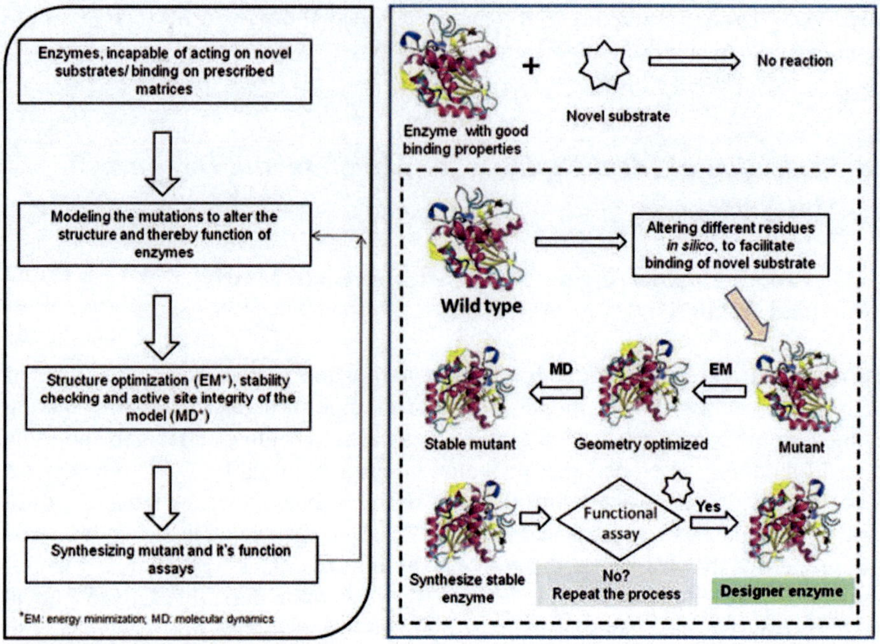

Fig. 10 Developing designer enzymes

Targeted Mutagenesis

Unlike random mutagenesis/directed evolution, if the structural details of the enzyme are known, an educated mutation to the desired location can be planned (Fig. 11). In targeted mutagenesis, often single/double mutations are sufficient to achieve the desired output. Besides, it might change the nature of catalysis as well, if the mutation is in the active site. (e.g. Leu142Arg substitution of N-acetylneuraminatelyase changes its catalytic activity to that of Dihydrodipicolate synthase). Unlike screening multiple mutants, minimum numbers of educated mutations are often sufficient to improve the binding properties of enzyme with the matrix. Essential residues contributing to majority of binding interactions are selected as hot-spots and are accommodated in the locations likely to augment binding interactions. The remaining portion of the interface may then be filled-in with residues that support and compound the packing on all sides of the hotspots, and also provide high binding affinity. Directed evolution-based iterations can help further improve the binding constants and strength of the binding between the enzyme and matrix. The computational designs can be ameliorated by considering the behaviour of solvent phases, electrostatic models and the associated protein dynamics.

PiSQRD tool (Aleksiev et al. 2009) can also be used for selection of ideal residues to mutate, for better binding to matrix. It is used to catalogue the rigid, solvent accessible domains of the protein surface, identify 1–2 residues with highest solvent

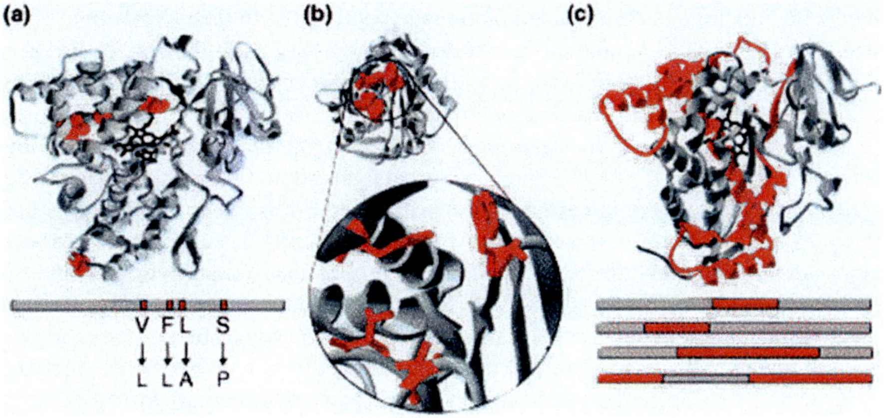

Fig. 11 Laboratory evolution of proteins: enzyme designing for improved properties. **a** Site directed mutagenesis, **b** active site mutations/multiple mutations, and **c** recombination of multiple sequences (Reproduced with permission from Bloom et al. 2005; © 2005 Elsevier)

accessible surface area (except Gly and pro and reactive residues viz Asp, Glu, Lys, Asnetc) in each domain. These residues are mutated in silico (commonly into Cys, so as to facilitate site-specific immobilization by thiol vs maleimide/EDC reaction, occasionally into Trp as well, to facilitate π-π stacking hydrophobic interactions between the aromatic rings of the enzyme and matrix, to hold the enzyme linked onto the surface of the matrix). Caution must be exercised, not to amend any residues close to the active site. In some enzymes like lipases, the moment of domains is important for catalysis. In those enzymes, these domains, though highly mobile, solvent exposed and away from the catalytic site, should not be mutated.

Assessing stability of mutant and choosing the ideal one: Once mutant is generated in silico, its comprehensive structural acceptability analysis (sterical acceptability can be checked with Ramachandran plots (Ramachandran et al. 1963), and rotamer analysis can be ensured by ProCheck/What if) has to be done. Structural stability of the mutants (with single or multiple mutations), their flexibility compared to wild type can be studied by molecular mechanics or Monte Carlo simulations. These techniques can also be used to understand the interactions between the enzyme vs altered matrix with different functional groups. Computational approach can be the preliminary approach to study the stability of the mutants/functionalized matrices generated, before commissioning the tedious site directed mutagenesis experiments and NMR studies to assess the structural stability in solvents.

Ab Initio Enzyme Design

Enzymes can be "created" de novo, to catalyze the reactions known to us and hitherto unknown as well. Enzymes with excellent catalytic and binding abilities can be made by adding residue by residue to make peptides, with their geometries optimized by quantum mechanics and side chain rotamers well packed. Peptides are then joined together to make desired folds, optimized to reduce steric clashes, keeping the desired residues on the surface and maximize packing. Active site mimics/analogues are added to it at suitable location, thus completing the design of enzyme. If the designed protein is in the required geometry, catalysis should be probable. However, de novo synthesis is computationally intensive, requiring massive computational resources and is often crowd sourced. At every step, dead end elimination, Monte Carlo sampling, iterative flexible backbone minimization and ab initio packing of side chains (rotamers) in their minimum energy conformation are carried out using tools like Rosetta Match, Rosetta Design and online tools viz Robetta. The resulting stable, rotamer adjusted enzyme is expected to get maximum binding (to the matrix) and catalysis. Top 7 is the de novo designed, crowd-sourced protein with α/β topology. Retro-aldolases, Kemp eliminases (KE07, KE59 and KE70) are some known designer enzymes with industrial applications.

Catalytic Antibodies

Another designer enzyme with great orientation control is catalytic antibody (abzymes or catalytic monoclonal antibodies, CatMAbs) (Padiolleau-Lefevre et al. 2014). Antibodies inherently possess great binding abilities to antigens/haptens. If a hapten, similar to the transition state of a reaction (transition state analogue, TSA), is used to generate the antibodies, they should bind to the real substrate in its transition state and there is a possibility of a catalytic reaction, if the residues are in catalytic geometry. As of today, abzymes can accelerate catalysis at 10^4–10^6 times the uncatalyzed reaction. However, the catalytic prowesses of catalytic antibodies are less, compared to enzymes which are perfected over billions of years. Improving the rate of abzyme catalysis, either by better hapten selection or by other methods viz active site redesign, random/targeted mutations, directed evolution could be the futuristic work in this aspect.

7.4 Outlook

Understanding the molecular intricacies of enzyme immobilization, identifying the changes needed for improvising the process and executing the changes it need to maximize the attachment of enzyme to the matrix with orientation control is possible with the aid of computational studies. This can therefore be the most rational approach of immobilization, giving us the best opportunity to ease out diffusion of substrate to the active site and maximizing the catalytically active enzymes on the matrix. Further, the effects of any changes in the structures of the enzyme or the matrix, on the overall process of immobilization can be predicted before commencing experiments. Computational studies can therefore be a valuable tool in understanding immobilization and many molecular processes.

References

Agnihotri S, Dhiman NK, Tripathi A (2018) Antimicrobial surface modification of polymeric biomaterials. In: Tiwari A (ed) Handbook of antimicrobial coatings, Elsevier pp 435–486. ISBN: 9780128119822. https://dx.doi.org/10.1016/B978-0-12-811982-2.00020-2

Ahmadi S, Herrera LB, Chehelamirani M, Hostas J, Jalife S, Salahurb DR (2018) Miltiscale modelling of enzymes: QM-Cluster, QM/MM and QM/MM/MD—a tutorial. Int J Quantum Chem 118(9):e25558. https://doi.org/10.1002/qua.25558

Ahmed SH, Shiekh EM, Morsy AMA (2014) Potentiality of uranium biosorption from nitric acid solutions using shrimp shells. J Environ Radioact 134:120–127

Aleksiev T, Potestio R, Pontiggia F, Cozzini S, Micheletti C (2009) PiSQRD: a web server for decomposing protein into quasi-rigid dynamical domains. Bioinformatics 25(20):2743–2744

Alsberg E, Anderson KW, Albeiruti A, Fransceschi RT, Mooney DJ (2001) Cell interactive alginate hydrogels for bone tissue engineering. J Dent Res 80:2025–2029. https://doi.org/10.1177/00220345010800111501

Amaya-Delgado L, Hidalgo-Lara ME, Montes-Horcasitas MC (2006) Hydrolysis of sucrose by invertase immobilized on nylon-6 microbeads. Food Chem 99:299–304

Andjelković U, Milutinović-Nikolić A, Jović-Jovičić N, Banković P, Jovanović D (2015) Efficient stabilization of *Saccharomyces cerevisiae* external invertase by immobilisation on modified beidellite nanoclays. Food Chem 1681:262–269

Atala A, Lanza RP (eds) (2002) Methods of tissue engineering. Academic Press, San Diego

Axen R, Porath J, Ernback S (1967) Chemical coupling of peptides and proteins to poly-saccharides by means of cyanogens halides. Nature 214:1302–1304. https://doi.org/10.1038/2141302a0

Ayukt G, Hasirci VN, Alaeddinoglu G (1988) Immobilization of yeast cells in acrylamide gel matrix. Biomaterials 9:168–172

Azodi M, Falamaki C, Mohsenifar A (2011) Sucrose hydrolysis by invertase immobilized on functionalized porous silicon. J Mol Catal B: Enzym 69:154–160

Bagda E, Sari A, Tuzen M (2018) Effective uranium biosorption by macrofungus (*Russulasanguinea*) from aqueous solutions: equilibrium, thermodynamic and kinetic studies. J Radioanal Nucl Chem 317:1387–1397

Barker SA, Somers PJ, Epton R, McLaren JV (1970) Cross-linked polyacrylamide derivatives as water insoluble carriers of amylolytic enzymes. Carbohydr Res 14(3):287–296. https://doi.org/10.1016/S0008-6215(00)80001-4

Barnet LB, Bull HB (1959) The optimum pH of adsorbed ribonuclease. Biochim Biophys Acta 36:244–246. https://doi.org/10.1016/0006-3002(59)90090-3

Bartling GJ, Brown HD, Chattopadhyay SK (1973) Synthesis of matrix-supported enzyme in non-aqueous conditions. Nature 243:342–344. https://doi.org/10.1038/243342b0

Basso A, Simona S (2019) Industrial applications of immobilized enzymes. Mol Catal 479:110607. https://doi.org/10.1016/j.mcat.2019.110607

Bayramoglu G, Akbulut Y, AcIkogz-Erkaya I, Arica MY (2017a) Uranium sorption by native and nitriloacetate-modified Bangiaatropurpurea biomass: kinetic and thermodynamics. J Appl Phycol 30:649–661

Bayramoglu G, Doz T, Ozlap C, Arica MY (2017b) Improvement stability and performance of invertase via immobilization on to silanized and polymer brush grafted magnetic nanoparticles. Food Chem 221:1442–1450

Behera S, Mohanty RC, Ray RC (2011) Ethanol production from mahula (*Madhuca latifolia* L.) flowers with immobilized cells of *Saccharomyces cerevisiae* in Luffa cylindrica L. sponge discs. Appl Energy 88:212–215

Berman HM, Westbrook J, Feng Z, Gililand G, Bhat TN, Weissig H, Shindyalov IN, Bourne PE (2000) The protein databank. Nucleic Acid Res 28:235–242

Bernath FR, Vieth WR (1972) Lysozyme activity in the presence of nonionic detergent micelles. Biotechnol Bioeng 14:737–752. https://doi.org/10.1002/bit.260140505

Bernfeld P, Bieber RE, Watson DM (1969) Kinetics of water-insoluble phosphoglycerate mutase. Biochim Biophys Acta Enzymol1 91(3):570–578. doi.org/10.1016/0005-2744(69)90350-7

Bhainsa KC, D'Souza SF (1999) Biosorption of uranium (VI) by *Aspergillus fumigatus*. Biotechnol Tech 13:695–699

Bhainsa KC, D'Souza SF (2001) Uranium (VI) biosorption by dried roots of *Eichhorniacrassipes* (water hyacinth). J Environ Sci Health 36(9):1621–1631

Bhainsa KC, D'Souza SF (2009) Thorium biosorption by *Aspergillus fumigatus*, a filamentous fungal biomass. J Hazard Mater 165:670–676

Bhardwaj T (2014) A review on immobilization techniques of biosensors. IntJEng TechRes 3(5):294–298

Bhat SV, Melo JS, Chaugule BB, D'Souza SF (2008) Biosorption characteristics of uranium(VI) from aqueous medium onto *Catenellarepens*, a red alga. J Hazard Mater 158:628–635

Bilal M, Iqbal HMN (2019) Naturally derived biopolymers: potential platforms for enzyme immobilization. Int J Biol Macromol 130:462–482

Bisceglie V (1933) Uber die antineoplastische immunitat; heterologe Einpflnzung von Tumoren in Huhner-embryonen. Ztschr Krebsforsch 40:122–140

Blanco RM, Calvete JJ, Guisan JM (1989) Immobilisation-stabilisation of enzymes: variables that control the intensity of the trypsin (amine) agarose (aldehyde) multi-point covalent attachment. Enzyme MicrobTechnol11(6):353–359. doi.org/10.1016/0141-0229(89)90019-7

Bloom JD, Meyer MM, Meinhold P, Otey CR, MacMillan D, Arnold FH (2005) Evolving strategies for enzyme engineering. Curr Opin Struct Biol 15(4):447–452. https://doi.org/10.1016/j.sbi.2005.06.004

Bolivar JM, Nidetzky B (2019) The microenvironment in immobilized enzymes: Methods of characterization and its role in determining enzyme performance. Molecules 24:3460. https://doi.org/10.3390/molecules24193460

Bornscheuer UT, Huisman GW, Kazalauskas RJ, Lutz S, Moore JC, Robins K (2012) Engineering the third wave of Biocatalysis. Nature 485:185–194

Boudrant J, Cheftel C (1975) Continuous hydrolysis of sucrose by invertase adsorbed in a tubular reactor. Biotechnol Bioeng 17:827–844

Brandenberg H (1955) Methods for linking enzymes to insoluble carriers.AngewChem 67:661–661

Brown HD, Patel AB, Chattopadhyay SK (1968) Enzyme entrapment within hydrophobic and hydrophilic matrices. J Biomed Mater Res 2(2):231–235. https://doi.org/10.1002/jbm.820020206

Buchholz K, Kasche V (1997) Biokatalysatoren und Enzym-technologie. Wiley-VCH 7–11

Burdick JA, Anseth KS (2002) Photoencapsulation of osteoblastsin injectable RGD-modified PEG hydrogels for bone tissue engineering. Biomaterials 23(22):4315–4323

Cabrera MP, Assis CRD, Neri DFM Pereira CF, Soria F, Carvalho Jr LB (2017) High sucrolytic activity by invertase immobilized onto magnetic diatomaceous earth nanoparticles. Biotechnol Rep 14:38-46

Cantarella M, Cantarella L, Gallifuoco A, Alfani F (1992) Invertase activity of *Saccharomyces cerevisiae* cells entrapped in poly (2-hydroxyethyl methacrylate) gels: kinetic and thermostability study in membrane reactors. J Biotechnol 24:159–168

Cao L (2005) Carrier-bound immobilized enzymes: principles, applications and design. Wiley-VCH 1–52.ISBN: 978-3-527-31232-0

Cao L, Elzinga J (2003) Cross-linked enzyme aggregates and crosslinker agents therefore.WO 03/066:850

Cao L, Van RF, Sheldon RA (2000) Cross-linked enzyme aggregates: a simple and effective method for the immobilization of penicillin acylase. Org Lett 2(10):1361–1364. https://doi.org/10.1021/ol005593x

Cao L, Werkmeister JA, Wang J, Glattauer V, McLean KM, Liu C (2014) Bone regeneration using photocrosslinked hydrogel incorporating rhBMP-2 loaded 2-N, 6-O-sulfated chitosan nanoparticles. Biomaterials 35(9):2730–2742. https://doi.org/10.1016/j.biomaterials.2013.12.028

Cao S, Fang L, Zhao Z, Ge Y, Piletsky S, Turner APF (2013) Hierachically structured hollow silica spheres for high efficiency immobilization of enzymes. Adv Funct Mater 23:2162–2167. https://doi.org/10.1002/adfm.201201793

Carvalho F, Fernandes P (2015) Packed bed enzyme microreactor: application in sucrose hydrolysis as proof-of-concept. Biochem Eng J 104:74–81

Chang T (1996) Editorial: past, present and future perspectives on the 40th anniversary of hemoglobin based red blood cell substitutes. Artif Cells Blood Substit Immobil Biotechnol 24: ixxxvi

Chang TMS (1964) Semipermeable microcapsules. Science 146(3643):524–525. https://doi.org/10.1126/science.146.3643.524

Chapman J, Ismail AI, Dince CZ (2018) Industrial applications of enzymes: recent advances, techniques, and outlooks. Catalysts 8(6):238. https://doi.org/10.3390/catal8060238

Chen C, Hu J, Wang J (2019) Uranium biosorption by immobilized active yeast cells entrapped in calcium-algintae-PVA-GO-crosslinked gel beads. Radiochim Acta.https://doi.org/10.1515/ract-2019-3150

Chen WP (1980) Glucose isomerase (a review). Process Biochem 15:30

Chen X, Fan H, Deng X, Wu L, Yi T, Gu L, Zhou C, Fan Y, Zhang X (2018) Scaffold structural microenvironmental cues to guide tissue regeneration in bone tissue applications. Nanomaterials (Basel) 8(11):960. https://doi.org/10.3390/nano8110960

Cooker LE, Venkatasubramanien K (1985) The practice of biotechnology: current commodity product. In: Blanch HW, Drew S, Wand DIC (eds) Comprehensive biotechnology, vol 3, Permagon Press, Oxford, p 777

Costantino HR, Griebenow K, Langer R, Klibanov AM (1997) On the pH memory of lyophilized compounds containing protein functional groups. Biotechnol Bioeng 53(3):345–348. https://doi.org/10.1002/(SICI)1097-0290(19970205)53:3%3c345::AID-BIT14%3e3.0.CO;2-J

Crespilho FN, Iost RM, Travain SA, Zucolotto V (2009) Enzyme immobilization on Ag nanoparticles/polyaniline nanocomposites. Biosens Bioelectron 24:3073–3077. https://doi.org/10.1016/j.bios.2009.03.026

D'Souza SF (2001a) Immobilization and stabilization of biomaterials for biosensor applications. Appl Biochem Biotechnol 96:225–238. https://doi.org/10.1385/ABAB:96:1-3:225

D'Souza SF (2001b) Microbial biosensors. Biosens Bioelectron 16(6):337–353. https://doi.org/10.1016/S0956-5663(01)00125-7

D'Souza SF, Nadkarni GB (1980) Continuous inversion of sucrose by gel-entrapped yeast cells. Enzyme Microb Technol 2:217–222

D'Souza SF, Melo JS (2001) Immobilization of bakers yeast on jute fabric through adhesion using polyethylenimine: Application in an annular column reactor for the inversion of sucrose. Process Biochem 36:677–681

D'Souza SF, Melo JS, Deshpande A, Nadkarni GB (1986) Immobilization of yeast cells by adhesion to glass surface using polyethylenimine. Biotechnol Lett 8:643–648

D'Souza SF, Kumar J, Jha SK, Kubal BS (2013) Immobilization of the urease on eggshell membrane and its application in biosensor. Mater Sci Eng C 33(2):850–854. https://doi.org/10.1016/j.msec.2012.11.010

D'Souza SF, Sar P, Kazy S, Kubal BS (2006) Uranium sorption by Pseudomonas biomass immobilized in radiation polymerized polyacrylamide bio-beads. J Environ Sci Health 41(3):487–500

Daan F, Berend S (2001) Understanding molecular simulation. Academic Press, San Diego. ISBN: 9780122673511

Deo RP, Wang J, Block I, Mulchandani A, Joshi KA, Trojanowicz M, Scholz F, Chen W, Lin Y (2005) Determination of organophosphate pesticides at a carbon nanotube/organophosphorus hydrolase electrochemical biosensor. Anal Chim Acta 530(2):185–189. https://doi.org/10.1016/j.aca.2004.09.072

Deshpande A, D'Souza SF, Nadkarni GB (1987) Co-immobilization of D-amino acid oxidase and catalase by entrapment of Trignopsis variabilis in radiation polymerized polyacrylamide beads. J Bioscience 11:137–144

Dhulster P, Parascandola P, Scardi V (1983) Improved method for immobilizing invertase-active whole cells of Saccharomyces cerevisiae in gelatin. Enzyme Microb Technol 5:65–69

Dickey FH (1955) Specific adsorption. J Phys Chem 59(8):695–707. doi.org/10.1021/j150530a006

Dizge N, Gunaydin O, Yilmaz F, Tanriseven A (2008) Immobilization of invertase onto poly(3-methylthienyl methacrylate)/poly(3-thiopheneacetic acid) matrix. Biochem Eng J 40:64–71

Dror RO, Dirks RM, Grossman JP, Xu H, Shaw DE (2012) Biomolecular Simulation: a computational microscope for molecular biology. Annu Rev Biophys 41:429–452

Dyal A, Loos K, Noto M, Chang SW, Spagnoli C, Shafi KVPM et al (2003) Activity of Candida rugosa lipase immobilized on Fe_2O_3 magnetic nanoparticles. J Am ChemSoc 125(7):1684–1985. https://doi.org/10.1021/ja021223n

Dziezak JD (1989) Ingredients for sweet success. Food Technol 43:94–116

Eiadpum A, Limtong S, Phisalaphong M (2012) High-temperature ethanol fermentation by immobilized coculture of Kluyveromyces marxianus and Saccharomyces cerevisiae. J Biosci Bioeng 114:325–329

Elisseeff J, McIntosh W, Fu K, Blunk BT, Langer R (2001) Controlled-release of IGF-I and TGF-beta1 in a photopolymerizing hydrogel for cartilage tissue engineering. J Orthop Res 19(6):1098–1104. https://doi.org/10.1016/S0736-0266(01)00054-7

Fernandex-Lafuente R, Cowan DA, Wood ANP (1995) Hyperstabilization of a thermophilic esterase by multipoint covalent attachment.Enzyme MicrobTechnol17(4):366–372 doi.org/10.1016/0141–0229(94)00089–1

Fleming SE, Groot Wassink JWD (1979) Preperation of high fructose syrups from the tubers of Jerusalem artichoke. CRC Crit Rev Food Sci Nutr 12:1–28

Fruin JC, Scallet BL (1975) Isomerized corn syrups in food products. Food Technol 29:40

Furukawa S, Ono T, Ijima H, Kawakami K (2002) Effect of imprinting sol–gel immobilized lipase with chiral template substrates in esterification of (*R*)-(+)- and (*S*)-(–)-glycidol. J MolCatal B: Enzymatic 17(1):23–28. https://doi.org/10.1016/S1381-1177(01)00076-5

Gale EF, Epps MR (1944) Studies on bacterial amino-acid decarboxylases 1. 1-(+)-lysine decarboxylase. Biochem J 38(3):232–242 doi:https://doi.org/10.1042/bj0380232

Garcia-Galan C, Berenguer-Murcia A, Fernandez-Lafuente R, Rodriguez RC (2011) Potential of different enzyme Immobilization strategies to improve enzyme performance. Adv Synth Catal 353(16):2885–2904

Gebelein CG (1984) Polymeric materials and artificial organs . ACS Symp Ser Am Chem Soc 256:1–11. https://doi.org/10.1021/bk-1984-0256.ch001

Godbole SS, Kubal BS, D'Souza SF (1990) Hydrolysis of concentrated sucrose syrups by invertase immobilized on anion exchanger waste cotton thread. Enzyme Microb Technol 12:214–217

Gok C, Aytas S (2009) Biosorption of uranium(VI) from aqueous solution using calcium alginate beads. J Hazard Mater 168:368–375

Gok C, Aytas S, Sezer H (2017) Modeling uranium biosorption by *Cystoseira* sp. and application studies. Sep Sci Technol 52(5): 792–803

Goldstein H, Barry PW, Rizzuto AB, Venkatasubramanian K, Vieth WR (1977) Continuous enzymatic production of invert sugar. J Ferment Technol 55:516–524

Gole A, Dash C, Ramakrishnan V, Sainkar SR, Mandale AB, Rao M, Sastry M (2001) Pepsin–gold colloid conjugates: preparation, characterization and enzymatic activity. Langmuir 17(5):1674–1679. https://doi.org/10.1021/la001164w

González Benito G, Ozores M, Peña M (1994) Continuous glycerol production in a packed-bed bioreactor with immobilized cells of *Saccharomyces cerevisiae.* Bioresour Technol 49:209–212

Grubhofer N, Schleith L (1954) Coupling of proteins on diazotized polyaminostyrene. Hoppe-Seylers Z Physiol Chem 297(2):108–112 PMID:13221271

Guibal E, Saucedo I, Jansson-charrier M, Delanghe B, Le Chloirec P (1994) Uranium and vanadium sorption by chitosan and its derivatives. Water Sci Technol 30(9):183–190

Guisan JM, Alvaro G, Fernandez-Lafuente R, Rosell CM, Garcia JL, Tagliani A (1993) Stabilization of heterodimeric enzyme by multipoint covalent immobilization: penicillin G acylase from *Kluyveracitrophila*. Biotechnol Bioeng 42(4):455–464. https://doi.org/10.1002/bit.260420408

Gupta MN, Kaloti M, Kapoor M, Solanki K (2011) Nanomaterials as Matrices for Enzyme Immobilization. J Artif Cells Blood Substitutes Biotechnol 39(2):98–109. https://doi.org/10.3109/10731199.2010.516259

Hahn-Hagerdal B (1986) Water activity: a possible external regulator in biotechnical processes. Enzyme MicrobTechnol 8(6):322–327. doi.org/10.1016/0141-0229(86)90129-8

Han RP, Zou WH, Wang Y, Zhu L (2007) Removal of uranium (VI) from aqueous solutions by manganese oxide coated zeolite: discussion of adsorption isotherm and pH effect. J Environ Radioact 93(3):127–143

Hanwell MD, Curtis DE, Lonie DC, Vandermeersch T, Zurech E, Hutchinson GR (2017) Avogadro: an advanced semantic chemical editor, visualisation and analysis platform. J Cheminformatics 4:17. https://doi.org/10.1186/1758-2946-4-17

Harkins WD, Fourt L, Fourt PC (1940) Immunochemistry of catalase II. Activity in multilayers. J Biol Chem 132:111–118

Hartdegen FJ, Swann WE (1976) Polyurethane foams containing bound immobilized proteins. Ger Offen DE 2(612):138

Hasal P, Vojtíšek V, Čejková A, Kleczek P, Kofroňová O (1992) An immobilized whole yeast cell biocatalyst for enzymatic sucrose hydrolysis. Enzyme Microb Technol 14:221–229

Hassan AHA, Moura SL, Ali FHM, Moselhy WA, Taboada Sotomayor MDP, Pividori MI (2018) Electrochemical sensing of methyl parathion on magnetic molecularly imprinted polymer. Biosens Bioelectron 118:181–187. https://doi.org/10.1016/j.bios.2018.06.052

Häuselmann HJ, Fernandes RJ, Mok SS, Schmid TM, BlockJA AMB, Kuettner KE, Thonar EJMA (1994) Pheno-typic stability of bovine articular chondrocytes after long-termculture in alginate beads. J Cell Sci 107:17–27

Haynes R, Walsh KA (1969) Enzyme envelopes on colloidal particles. Biochem Biophys Res Commun 36(2):235–242. https://doi.org/10.1016/0006-291X(69)90320-9

He J, Chen JP (2014) A comprehensive review on biosorption of heavy metals by algal biomass: materials, performance, chemistry and modelling simulation tools. Bioresour Technol 160:67–78

Hicks GP, Updike SJ (1966) The preparation and characterisation of lyophilized polyacrylamide enzyme gels for chemical analysis. Anal Chem 38(6):726–730. https://doi.org/10.1021/ac60238a014

Hipwell MC, Harvey MJ, Dean PDG (1974) Affinity chromatography on a homologous series of immobilized N6-amega-aminoalkyl AMP. Effect of ligand-matrix spacer length on ligand-enzyme interaction. FEBS Lett 42:355–359

Ho KM, Mao X, Gu L, Li P (2008) Facile route to enzyme immobilization: core–shell nanoenzyme particles consisting of well-defined poly (methyl methacrylate) cores and cellulase shells. Langmuir 24(19):11036–11042. https://doi.org/10.1021/la8016529

Hoarau M, Badiean S, Marsh ENG (2017) Immobilised enzymes: understanding enzyme-substrate interactions at the molecular level. Org Biomol Chem 15:9539–9551

Hollingsworth SA, Dror RO (2018) Molecular dynamics simulations for all. Neuron 99:1129–1143

Hong IS (2016) Stimulatory versus suppressive effects of GM-CSF on tumor progression in multiple cancer types. Exp Mol Med 48(7):e242. https://doi.org/10.1038/emm.2016.64

Horie M, Tripathi A, Ito A, Kawabe Y, Kamihira M (2017) Magnetic nanoparticles:functionalization and manufacturing of pluripotent stem cells. In: Tripathi A, Melo JS (eds) Advances in biomaterials for biomedical applications. Springer-Nature, Singapore, pp 363–384 ISBN 978-981-10-3327-8

Hortelano G, Al-Hendy A, Ofosu FA, Chang PL (1996) Delivery of human factor IX in mice by encapsulated recombinant myoblasts: a novel approach towards allogeneic gene therapy of hemophilia B. Blood 87:5095–5103

Horvath C, Engasser JM (1973) Pellicular heterogeneous catalaysts; A theoretical study of advantages of shell structured immobilized enzyme particle. Ind Eng Chem Fundam 12:229–235

Ho-Shui-Ling A, Bolander J, Rustom LE, Johnson AW, Luyten FP, Picart C (2018) Bone regeneration strategies: Engineered scaffolds, bioactive molecules and stem cells current stage and future perspectives. Biomaterials 180:143–162. https://doi.org/10.1016/j.biomaterials.2018.07.017

Hu MZ, Norman MJ, Faison BD, Reeves ME (1996) Biosorption of uranium by Peudomonas*aeruginosa* strain CSU: characterization and comparison studies. Biotechnol Bioeng 51(2):232–247

Hueper F (1974) Water soluble, polymeric substrate covalently bound penicillin acylase for preparing 6-aminopenicillanic acid. DE 2,312,824

Humphrey W, Dalke A, Schulten K (1996) VMD: Visual Molecular Dynamics. J Mol Graph 14(1):33–38

Husain Q, Ansari SA, Fahad M, Azam A (2011) Immobilization of Aspergillus oryzae β galactosidase on zinc oxide nanoparticles via simple adsorption mechanism. Int J Biol Macromol 49:37–43. https://doi.org/10.1016/j.ijbiomac.2011.03.011

Inama L, Diré S, Carturan G, Cavazza A (1993) Entrapment of viable microorganisms by SiO2 sol-gel layers on glass surfaces: Trapping, catalytic performance and immobilization durability of *Saccharomyces cerevisiae.* J Biotechnol 30:197–210

Iqbal J, Saleemuddin M (1985) Sucrose hydrolysis using invertase immobilized on concanavalin A-sepharose. Enzyme Microb Technol 7:175–178

Isono Y, Araya G, Hoshino A (1995) Immobilization of *Saccharomyces cerevisiae* for ethanol fermentation on γ-alumina particles using a spray-dryer. Process Biochem 30:743–746

IUPAC (1997) Compendium of chemical terminology, 2nd edn. (the "Gold Book"). Compiled by McNaught AD, Wilkinson A. Blackwell Scientific Publications, Oxford. Online version (2019) created by SJ Chalk. ISBN 0–9678550–9–8. https://doi.org/10.1351/goldbook

Jang E, Park S, Park S, Lee Y, Kim D, Kim B, Koh W (2010) Fabrication of polyethylene glycolbased hydrogels entrapping enzyme-immobilized silica nanoparticles. Polym Adv Technol 21:476–82. https://doi.org/10.1002/pat.1455

Johansen A, Flink JM (1986) Influence of alginate properties on sucrose inversion by immobilized whole cell invertase. Enzyme Microb Technol 8:485–490

Jun L, Zhou W, Kumbhar J, Wiemann J, Fang J, Carpentier EE, O'Connor CJ (2001) Gold-coated iron (Fe@Au) nanoparticles: synthesis, characterization and magnetic field induced self-assembly. Solid State Chem 159:26–33. https://doi.org/10.1006/jssc.2001.9117

Jurga M, Dainiak MB, Sarnowska A, Tripathi A, Plieva FM, Strojek L, Jungvid H, Kumar A, Firraz N, McGuckin C (2011) The performance of laminin-containing cryogel scaffolds in neural tissue regeneration. Biomaterials 32(13):3423–3434. https://doi.org/10.1016/j.biomaterials.2011.01.049

Kabasch W, Sander C (1983) Dictionary of protein secondary structures: pattern recognition of hydrogen bonded geometrical features. Biopolymers 22(12):2577–2637. https://doi.org/10.1002/bip.360221211

Kapoor A, Viraraghavan T (1995) Fungal biosorption-an alternative treatment option for heavy metal bearing waste water-a review. Bioresour Technol 53:195–206

Kawakami K, Abe T, Yoshida T (1992) Silicon immobilised biocatalyst effective for bioconversions in non-aqueous media. Enzyme Microb Technol 14(5):371–375. doi:https://doi.org/10.1016/0141-0229(92)90005-9

Kawashima K, Umeda K (1976) Preparation of membranous immobilized invertase and it's characterisics. Agri Biol Chem 40:1151–1157

Kay G, Crook EM (1967) Coupling of enzymes to cellulose using chloro-*s* -triazines. Nature 216(5114):514–515. https://doi.org/10.1038/216514a0

Kennedy JF, Epton J (1973) Poly (N-acryloyl-4- and -5-aminosalicylic acids). III. Uses as their titanium complexes for the insolubilisation of enzymes. Carbohyd Res 27:11–20

Khani MH (2011) Uranium biosorption by *Padina* sp. algae biomass: kinetics and thermodynamics. Environ Sci Pollut Res Int 18:1593. https://doi.org/10.1007/s11356-011-0518-0

Kim HD, Amirthalingam S, Kim SL, Lee SS, Rangasamy J, Hwang NS (2017) Biomimetic materials and fabrication approaches for bone tissue engineering. Adv Healthc Mater 6(23):10.1002/adhm.201700612. doi: https://doi.org/10.1002/adhm.20170061

Kim J, Grate JW, Wang P (2006) Nanostructures for enzyme stabilization. Chem Eng Sci 61:1017–1026. doi:https://doi.org/10.1016/j.ces.2005.05.067

Klibanov AM (1997) Why are enzymes less active in organic solvents than in water? Trends Biotechnol 15(3):97–101. https://doi.org/10.1016/S0167-7799(97)01013-5

Kofidis T, de Bruin JL, Hoyt G, Ho Y, Tanaka M, Yamane T, Lebl DR, Swijnenburg RJ, Chang CP, Quertermous T, Robbins C (2005) Myocardial restoration with embryonic stem cell bioartificial tissue transplantation. J Heart Lung Transplant 24(6):737–744. https://doi.org/10.1016/j.healun.2004.03.023

Kolot FB (1981) Microbial carriers—strategy for selection 2. Process Biochem 16:30–33

Koneracka M, Kopcansky P, Antalmk M, Timko M, Ramchand CN, Lobo D, Mehta RV, Upadhyay RV (1999) Immobilization of proteins and enzymes to the magnetic particles. J Magnet Magn Mater 201(1–3):427–430. https://doi.org/10.1016/S0304-8853(99)00005-0

Kotwal SM, Shankar V (2009) Immobilized invertase. Biotechnol Adv 27:311–322

Kouassi GK, Irudayaraj J, McCarty G (2005) Examination of cholesterol oxidase attachment to magnetic nanoparticles. J Nanobiotechnol 3:1–9. https://doi.org/10.1186/1477-3155-3-1

Kumar J, D'Souza SF (2008) Preparation of PVA membrane for immobilization of GOD for glucose biosensor. Talanta 75(1):183–188. https://doi.org/10.1016/j.talanta.2007.10.048

Kumar J, D'Souza SF (2009) Inner epidermis of onion bulb scale: as natural support for immobilization of glucose oxidase and its application in dissolved oxygen based biosensor. Biosens Bioelectron 24(6):1792–1795. https://doi.org/10.1016/j.bios.2008.08.022

Kumar J, D'Souza SF (2010) An optical microbial biosensor for detection of methyl parathion using *Sphingomonas* sp. immobilized on microplate as a reusable biocomponent. Biosens Bioelectron 26(4):1292–1296.doi:https://doi.org/10.1016/j.bios.2010.07.016

Kumar J, D'Souza SF (2011a) Immobilization of microbial cells on inner epidermis of onion bulb scale for biosensor application. Biosens Bioelectron 26(11):4399–4404. https://doi.org/10.1016/j.bios.2011.04.049

Kumar J, D'Souza SF (2011b) Microbial biosensor for detection of methyl parathion using screen printed carbon electrode and cyclic voltammetry. Biosens Bioelectron 26(11):4289–4293. https://doi.org/10.1016/j.bios.2011.04.027

Kumar A, Tripathi A (2012) Biopolymeric scaffolds for Tissue Engineering. In: Tiwari A, Srivastava RB (ed) Biotechnology in biopolymers developments, applications and challenging areas, i-Smithers Repra Publication Ltd. United Kingdom, pp 233–285

Kumar J, Jha SK, D'Souza SF (2006) Optical microbial biosensor for detection of methyl parathion pesticide using *Flavobacterium* sp. whole cells adsorbed on glass fiber filters as disposable biocomponent. Biosens Bioelectron 21(11):2100–2105. doi:https://doi.org/10.1016/j.bios.2005.10.012

Kumar J, Melo JS (2015) Nova Science Publishers, Inc. Microbial biosensors for methyl parathion: From single to multiple samples analysis. Adv Biosens Res 89–111. ISBN: 978-1-63463-652-0.

Kumar J, Melo JS (2016) Jackfruit membrane as new biomaterial for biocatalytic application in detection of urea. J Biosens Bioelectron 7(4):1000226. https://doi.org/10.4172/2155-6210.1000226

Kumar J, Melo JS (2017) Overview on biosensors for detection of organophosphate pesticides. Cur Trend Biomed Eng Bios 5(3):555663. https://doi.org/10.19080/CTBEB.2017.05.555663

Kumar J, Melo JS (2020) Methyl parathion biosensor: From lab to field. IANCAS Bulletin 14(1):59–64

Kumar A, Tripathi A, Jain S (2011) Extracorporeal bioartificial liver (BAL) for treating acute liver diseases. Am J Extracorp Technol 43(4):195–206

Kumar J, Mishra A, Melo JS (2018) Biodegradation of methyl parathion and its application in biosensors. Austin J Environ Toxicol 4(1):1024. ISSN: 2472–372X.

Langer R (2000) Biomaterials in drug delivery and tissue engineering: one laboratory's experience. Acc Chem Res 33(2):94–101. https://doi.org/10.1021/ar9800993

Langmuir I, Schaefer VJ (1938) Activities of urease and monolayers. J Am Chem Soc 60:1351–1360

Lanza RP, Langer R, Vacanti J (eds) (2000) Principles of tissue engineering. Academic Press, San Diego, USA

Lee KM, Blaghen M, Samama JP, Biellmann JF (1986) Cross-linked crystalline horse liver alcohol dehydrogenase as a redox catalyst: activity and stability towards organic solvent. Bioorg Chem 14(2):202–210. doi.org/10.1016/0045-2068(86)90031-3

Lee RA, Razaz M, Hayward S (2003) The dyndom database of protein domain motions. Bioinformatics 19(10):1290–1291

Lee KY, Kim W, Baek YJ, Chung DY, Lee EH, Lee SY, Moon JK (2014) Biosorption of uranium(VI) from aqueous solution by biomass of brown algae *Laminaria japonica.* Water Sci Technol 70(1):136–143

Leuschner F (1966) Shaped structures for biological processes. German Patent 1(227):855

Li L, Hu N, Ding D, Xin X, Wang Y, Xue J, Zhang H, Tan Y (2015) Adsorption and recovery of U(VI) from low concentration uranium solution by amidoxime modified *Aspergillus niger*. RSC Adv 5(81):65827–65839

Lim F, Sun AM (1980) Microencapsulated islets as bioartificial endocrine pancreas. Science 210:908–909

Liu CZ, Wang F, Ou-Yang F (2009a) Ethanol fermentation in a magnetically fluidized bed reactor with immobilized *Saccharomyces cerevisiae* in magnetic particles. Bioresour Technol 100:878–882

Liu W, Zhang S, Wang P (2009b) Nanoparticle-supported multi-enzyme biocatalysis with in situ cofactor regeneration. J Biotechnol 139:102–107. https://doi.org/10.1016/j.jbiotec.2008.09.015

Liu G, Guo W, Yin Z (2014) Covalent fabrication of methyl parathion hydrolase on gold nanoparticles modified carbon substrates for designing a methyl parathion biosensor. Biosens Bioelectron 53:440–446. https://doi.org/10.1016/j.bios.2013.10.025

Liu Y, Luo D, Yu M, Wang Y, Jin S, Li Z, Cui S, He D, Zhang T, Wang T, Zhou Y (2019) Thermodynamically controlled self-assembly of hierarchically staggered architecture as an osteoinductive alternative to bone autografts. Adv Funct Mater 29(10):1806445

Lloyd JR, Renshaw JC (2005) Bioremediation of radioactive waste: radionuclide-microbe interactions in laboratory and filed-scale studies. Curr Opin Biotech 16:254–260

Lodola A, Mulholland AJ (2013) Computational enzymology. In: Monticelli L, Salonen E (eds) Biomolecular simulations: methods and protocols. Springer Science+ Business Media, New York, pp 67–89

Lorenzoni ASG, Aydos LF, Klein MP, Ayub MAZ, Hertz PF (2015) Continuous production of fructooligosaccharides and invert sugar by chitosan immobilized enzymes: comparison between in fluidized and packed bed reactors. J Mol Catal B: Enzymatic 111:51–55

Lu W, Dai Z, Li L, Liu J, Wang S, Yang H, Cao C, Liu L, Chen T, Zhu B, Sun L, Chen L, Li H, Zhang P (2020) Preparation of composite hydrogel (PCG) and its adsorption performance for uranium(VI). J Mol Liq 303:112604

Lutolf MP, Weber FE, Schmoekel HG, Schense JC, Kohler T, Muller R, Hubbell JA (2003) Repair of bone defects using synthetic mimetics of collagenous extracellular matrices. Nat Biotechnol 21(5):513–518. https://doi.org/10.1038/nbt818

Mafra ACO, Beltrame MB, Ulrich LG, Giordano de Lima R, Giordano C, Tardioli PW (2018) Combined CLEAs of invertase and soy protein for economically feasible conversion of sucrose in a fed-batch reactor. Food Bioprod Process 110:145–157

Mallesha M, Manjunatha R, Nethravathi C, Suresh GS, Rajamathi M, Melo JS, Venkatesha TV (2011) Functionalized-graphene modified graphite electrode for the selective determination of

dopamine in presence of uric acid and ascorbic acid. Bioelectrochemistry 81(2):104–108. https://doi.org/10.1016/j.bioelechem.2011.03.004
Mallesha M, Manjunatha R, Suresh GS, Melo JS, D'Souza SF, Venkatesha TV (2012) Direct electrochemical non-enzymatic assay of glucose using functionalized grapheme. J Solid State Electrochem 16(8):2675–2681. https://doi.org/10.1007/s10008-012-1674-y
Manecke G, Gunzel G (1967) Polymere Isothiocyanate zur Darstellunghochwirksamer Enzymharze. Naturwissenschaften 54:531–533. https://doi.org/10.1007/BF00627209
Manecke G, Polakowaski D (1981) Some carriers for the immobilisation of enzymes based on copolymers of derivatizedpoly(vinyl alcohol) and on copolymers of methacrylates with different spacer lengths. J Chromatogr 215:13–24. https://doi.org/10.1016/S0021-9673(00)81381-3
Manjunatha R, Nagaraju DH, Suresh GS, Melo JS, D'Souza SF, Venkatesha TV (2011a) Direct electrochemistry of cholesterol oxidase on MWCNTs. J Electroanal Chem 651(1):24–29. https://doi.org/10.1016/j.jelechem.2010.11.009
Manjunatha R, Nagaraju DH, Suresh GS, Melo JS, D'Souza SF, Venkatesha TV (2011b) Electrochemical detection of acetaminophen on the functionalized MWCNTs modified electrode using layer-by-layer technique. Electrochim Acta 56(19):6619–6627. https://doi.org/10.1016/j.electacta.2011.05.018
Manjunatha R, Suresh GS, Melo JS, D'Souza SF, Venkatarangaiah VT (2012) An amperometric bienzymatic cholesterol biosensor based on functionalized graphene modified electrode and its electrocatalytic activity towards total cholesterol determination. Talanta 99: 302–309. doi:https://doi.org/10.1016/j.talanta.2012.05.056
Manjunatha R, Suresh GS, Melo JS, D'Souza SF, Venkatesha TV (2010) Simultaneous determination of ascorbic acid, dopamine and uric acid using polystyrene sulfonate wrapped multiwalled carbon nanotubes bound to graphite electrode through layer-by-layer technique. Sens Actuators B Chem 145(2):643–650. https://doi.org/10.1016/j.snb.2010.01.011
Margolin AL (1996) Novel crystalline catalysts.Trends Biotechnol 14(7):223–230. https://doi.org/10.1016/0167-7799(96)10031-7
Mariani E, Lisignoli G, Borzì RM, Pulsatelli L (2019) Biomaterials: foreign bodies or tuners for the immune response? Int J Mol Sci 20(3):636. https://doi.org/10.3390/ijms20030636
Martin CR, Kohli P (2003) The emerging field of nanotube biotechnology. Nat Rev Drug Discov 2:29–37. https://doi.org/10.1038/nrd988
Martinek K, Klibanov AM, Goldmacher VS, Berezin IV (1977) The principles of enzyme stabilization: 1. Increase in thermostability of enzymes covalently bound to a complimentary surface of a polymer support in a multipoint fashion. Biochim Biophys Acta 485(1):1–12. https://doi.org/10.1016/0005-2744(77)90188-7
Mason RD, Weetall HH (1972) Invertase covalently coupled to porous glass: preparation and characterization. Biotechnol Bioeng 14:637–645
Mclaren AD (1957) Concerning the pH dependence of enzyme reactions on cells, particulates and in solution. Science 125(3250):697–697. https://doi.org/10.1126/science.125.3250.697
Melo JS, D'souza SF (1992) Immobilization of invertase through its carbohydrate moiety on *Ocimum basilicum* seed. Appl Biochem Biotechnol 32:159–170
Melo JS, D'Souza SF (1999) Simultaneous filtration and immobilization of cells from a flowing suspension using a bioreactor containing polyethylenimine coated cotton threads: application in the continuous inversion of concentrated sucrose syrups. World J Microb Biot 15:23–27
Melo JS, Kubal BS, D'Souza SF (1992) Production of inverted sucrose syrup using yeast cells adhered to polyethylenimine treated cotton threads. Food Biotechnol 6(2):175–186
Melo JS, Sen D, Mazumder S, D'Souza SF (2013) Spray drying as a novel technique for obtaining microbial imprinted microspheres and its application in filtration. Soft Matter 9:805–810
Melzoch K, Rychtera M, Hábová V (1994) Effect of immobilization upon the properties and behaviour of *Saccharomyces cerevisiae* cells. J Biotechnol 32:59–65
Messing RA (ed) (1975) Immobilized enzymes for industrial reactors. Academic Press, New York, pp 63–77

Messing RA, Filbert AM (1975) Immobilized glucose isomerase for the continuous conversation of glucose to fructose. J Agric Food Chem 23(5):920–923. https://doi.org/10.1021/jf60201a018

Michaelis L, Ehrenreich M (1908) Die adsorptionsanalyse der fermente. Biochem Z 10:283–290

Michalak I, Chojnacka K, Witek-Krowiak A (2013) State of the art for the biosorption process—a review. Appl Biochem Biotechnol 170:1389–1416

Micheel F, Evers J (1949) Synthesis of cellulose-bound proteins. Macromol Chem 3:200–209

Miletic N, Abetz V, Ebert K, Loos K (2010) Immobilization of Candida antarctica lipase B on polystyrene nanoparticles. Macromol Rapid Commun 31:71–74. https://doi.org/10.1002/marc.200900497

Mintz KJ, Zhou Y, Leblanc RM (2019) Recent development of carbon quantum dots regarding their optical properties, photoluminescence mechanism and core structure. Nanoscale 11:4634–4652. https://doi.org/10.1039/C8NR10059D

Mishra A, Kumar J, Melo JS (2017) An optical microplate biosensor for the detection of methyl parathion pesticide using a biohybrid of Sphingomonas sp. cells-silica nanoparticles. Biosens Bioelectron 87:332–338. https://doi.org/10.1016/j.bios.2016.08.048

Mishra A, Melo JS, Agrawal A, Kashyap Y, Sen D (2020) Preparation and application of silica nanoparticles-*Ocimumbasilicum* seeds bio-hybrid for the efficient immobilization of invertase enzyme. Colloids Surf B Biointerfaces 188:110796. https://doi.org/10.1016/j.colsurfb.2020.110796

Mishra A, Melo JS, Sen D, D'Souza SF (2014) Evaporation induced self assembled microstructures of silica nanoparticles and *Streptococcus lactis* cells as sorbent for uranium (VI). J Colloid Interf Sci 414:33–40

Mitchell DT, Lee SB, Trofin L, Li N, Nevanen TK, Soderlund H, Martin CR (2002) Smart nanotubes for bioseparations and biocatalysis. J Am ChemSoc 124:11864–11865. https://doi.org/10.1021/ja027247b

Miwa N, Ohtomo K (1975) Enzyme immobilization in the presence of substrates and inhibitors. Jpn Kokai Tokkyo Koho JP 56(045):591

Moghaddam AB, TayebeNazari T, Badraghi J, Kazemzad M (2009) Synthesis of ZnO nanoparticles and electrodeposition of polypyrrole/ZnO nanocomposite film. Int J ElectrochemSci 4:247–257

Mohamad NR, Marzuki NHC, Buang NA, Fuyop S, Wahab RA (2015) An overview of technologies for Immobilization of enzymes and surface analysis techniques for immobilized enzymes. Biotechnol Biotechnol Equip 29(2):205–220

Monsan et al (1971) Nouvelle method de preparation d'enzymes fixes sur des supports mineraux. C R AcadSci Paris T 273:33–36

Monson P, Combes D (1984) Application of immobilized invertase to continuous hydrolysis of concentrated sucrose solutions. Biotechnol Bieng 26:347–351

Mosbach K (1976a) Immobilised enzymes. FEBS Lett 62:80–95

Mosbach K (ed) (1976b) Immobilized enzymes. Methods enzymol. Academic Press, New York, vol 44, pp 3–999

Mosbach K, Mosbach R (1966) Entrapment of enzymes and microorganisms in synthetic cross-linked polymers and their applications in column techniques. Acta Chem Scand 20(10):2807–2810. https://doi.org/10.3891/acta.chem.scand.20-2807

Mulchandani A, Chen W, Mulchandani P, Wang J, Rogers KR (2001) Biosensors for direct determination of organophosphate pesticides. Biosens Bioelectron 16(4–5):225–230. https://doi.org/10.1016/S0956-5663(01)00126-9

Mulchandani P, Chen W, Mulchandani A (2001) Flow injection amperometric enzyme biosensor for direct determination of organophosphate nerve agents. Environ Sci Technol 35(12):2562–2565. https://doi.org/10.1021/es001773q

Mulchandani P, Mulchandani A, Kaneva I, Chen W (1999) Biosensor for direct determination of organophosphate nerve agents. 1. Potentiometric enzyme electrode. Biosens Bioelectron 14(1):77–85. doi. https://doi.org/10.1016/S0956-5663(98)00096-7

Mulko L, Rivarola CR, Barbero CA, Acevedo DF (2016) Bioethanol production by reusable *Saccharomyces cerevisiae* immobilized in a macroporous monolithic hydrogel matrices. J Biotechnol 23310:56–65

Murphy SV, Atala A (2014) 3D bioprinting of tissues and organs. Nat Biotechnol 32(8):773–785. https://doi.org/10.1038/nbt.2958

Nalini S, Nandini S, Shanmugam S, Neelagund SE, Melo JS, Suresh GS (2014) Amperometric hydrogen peroxide and cholesterol biosensors designed by using hierarchical curtailed silver flowers functionalized graphene and enzymes deposits. J Solid State Electrochem 18(3):685–701. https://doi.org/10.1007/s10008-013-2305-y

Nalini S, Nandini S, Madhusudana Reddy MB, Suresh GS, Melo JS, Neelagund SE, Naveenkumar HN, Shanmugam S (2016) A novel bioassay based gold nanoribbon biosensor to aid the preclinical evaluation of anticancer properties. RSC Adv 6(65):60693–60703. https://doi.org/10.1039/C6RA07501K

Nandini S, Nalini S, Manjunatha R, Shanmugam S, Melo JS, Suresh GS (2013) Electrochemical biosensor for the selective determination of hydrogen peroxide based on the co-deposition of palladium, horseradish peroxidase on functionalized-graphene modified graphite electrode as composite. J Electroanal Chem 689:233–242. https://doi.org/10.1016/j.jelechem.2012.11.004

Nandini S, Nalini S, Sanetuntikul J, Shanmugam S, Niranjana P, Melo JS, Suresh GS (2014a) Development of a simple bioelectrode for the electrochemical detection of hydrogen peroxide using Pichia pastoris catalase immobilized on gold nanoparticle nanotubes and polythiophene hybrid. Analyst 139(22):5800–5812. https://doi.org/10.1039/C4AN01262C

Nandini S, Nalini S, Shanmugam S, Niranjana P, Melo JS, Suresh GS (2014b) Rhoeo discolor leaf extract as a novel immobilizing matrix for the fabrication of an electrochemical glucose and hydrogen peroxide biosensor. Anal Methods 6(3):863–877. https://doi.org/10.1039/C3AY41795F

Nandini S, Nalini S, Reddy MBM, Suresh GS, Melo JS, Niranjana P, Sanetuntikul J, Shanmugam S (2016) Synthesis of one-dimensional gold nanostructures and the electrochemical application of the nanohybrid containing functionalized graphene oxide for cholesterol biosensing. Bioelectrochemistry 110:79–90. https://doi.org/10.1016/j.bioelechem.2016.03.006

Nanotube Modeler. https://jcrystal.com/products/wincnt/index.htm. Accessed on 15 march 2020

Nelson JM, Griffin EG (1916) Adsorption of invertase. J Am Chem Soc 38(5):1109–1115. https://doi.org/10.1021/ja02262a018

Nguyen HH, Lee SH, Lee UJ, Fermin CD, Kim M (2019) Immobilized enzymes in biosensor applications. Materials 12:121. https://doi.org/10.3390/ma12010121

Nicol WM (1977) Sucrose and food technology in sugar, science and technology. In: Birch GG, Parker KJ (eds), London Applied Science, pp 211–230

Niiolova P, Ward OP (1993) Whole cell biocatalysis in nonconventional media. J Ind Microbiol 12:76–86. doi.org/10.1007/BF01569905

Nilsson I, Ohlson S, Haggstrom L, Molin N, Mosbach K (1980) Denitrification of water using immobilized *Pseudomonas denitirificans* cells. Eur J Appl Microbiol 10:261–274

Ollis DF (1972) Diffusion influence in denaturableinsolubilised enzyme catalysts. Bio technol Bioeng 14:871–884

Ooshima H, Saimoto M, Harano Y (1980) Characteristics of immobilized invertase. Biotechnol Bioeng 22:2155–2167

Padiolleau-Lefevre S, Naya RB, Shahsavarian MA, Friboulet A, Avalle B (2014) Catalytic antibodies and their applications in biotechnology: state of the art. Biotechnol Lett 36:1369–1379

Pang C, Lu Y, Cao X, Li M, Huang G, Hua R, Wang C, Liu Y, An X (2011) Biosorption of uranium (VI) from aqueous solution by dead fungal biomass of *Penicillium citrinum.* Chem Eng J 171(1):1–6

Parascandola P, de Alteriis E, Scardi V (1993) Invertase and acid phosphatase in free and gel-immobilized cells of *Saccharomyces cerevisiae* grown under different cultural conditions. Enzyme Microb Technol 15:42–49

Partridge J, Halling PJ, Moore BD (1998) Practical route to high activity enzyme preparations for synthesis in organic media. J Chem Soc Chem Commun 7:841–842 doi:https://doi.org/10.1039/A800408K

Patel RP,Lopiekes DV, Brown SR, Price S (1967) Derivatives of proteins II. Coupling of α-chymotrypsin to carboxyl containing polymers by use of *N*-ethyl-5- phenylisoxazolium-3′-sulfonate.Biopolymers 5(6):577–582. doi.org/10.1002/bip.1967.360050611

Pennington SN, Brown HD, Patel AB, Chattopadhyay SK (1968) Silastic entrapment of glucose oxidase-peroxidase and acetylcholine esterase. J Biomed Mater Res 2(4):443–446. https://doi.org/10.1002/jbm.820020404

Persson M, Bulow L, Mosbach K (1990) Purification and site-specific immobilization of genetically engineered glucose dehydrogenase on thiopropyl-Sepharose. FEBS Lett 270(1–2):41–44. doi: https://doi.org/10.1016/0014-5793(90)81230-1

Peters MC, Isenberg BC, Rowley JA, Mooney DJ (1998) Release from alginate enhances the biological activity of vascular endothelial growth factor. J Biomater Sci Polym Ed 9(12):1267–1278. https://doi.org/10.1163/156856298x00389

Phillips JB, Bunting SC, Hall SM, Brown RA (2005) Neural tissue engineering: a self-organizing collagen guidance conduit. Tissue Eng 11(9–10):1611–1617. https://doi.org/10.1089/ten.2005.11.1611

Polshettiwar V, Cha D, Zhang X, Basset JM (2010) High-surface-area silica nanospheres (KCC-1) with a fibrous morphology. Angew Chem Int Ed 49(50):9652–9656. doi: https://doi.org/10.1002/anie.201003451

Porter JR, Ruckh TT, Popat KC (2009) Bone tissue engineering: a review in bone biomimetics and drug delivery strategies. Biotechnol Prog 25(6):1539–1560. https://doi.org/10.1002/btpr.246

Prakash S, Chang TMS (1996) Microencapsulated genetically engineered live E. Coli DH5 cells administered orally to maintain normal plasma urea level in uremic rats. Nat Med 2:883–887

Putzbach W, Ronkainen NJ (2013) Immobilization techniques in the fabrication of nanomaterial-based electrochemical biosensors: a review. Sensors 13:4811–4840. https://doi.org/10.3390/s130404811

Queiroz AAA, Vitolo M, de Oliveira RC, Higa OZ (1996) Invertase immobilization onto radiation-induced graft copolymerized polyethylene pellets. Radiat Phys Chem 47:873–880

Quiocho FA, Richards FM (1964) Intermolecular cross-linking of a protein in the crystalline state: carboxypeptidase. A Proc Natl Acad Sci USA 52(3):833–839. doi: https://doi.org/10.1073/pnas.52.3.833

Quiocho FA, Richards FM (1966) Enzyme behavior of carboxypeptidase-A in the solid state. Biochemistry 5(12):4062–4076. https://doi.org/10.1021/bi00876a041

Raeeszadeh-Sarmazdeh M, Parthasarathy R, Boder ET (2015) Site-specific immobilization of protein layers on gold surfaces via orthogonal sortases. Colloids Surf B Biointerfaces 128:457–463

Rai A, Prabhune A, Perry CC (2012) Entrapment of commercially important invertase in silica particles at physiological pH and the effect of pH and temperature on enzyme activity. Mat Sci Eng C 32:785–789

Ramachandran GN, Ramakrishnan C, Sasisekaran V (1963) Stereochemistry of polypeptide chain configuration. J Mol Bio 7:95–99

Ramesh R, Puhazhendi P, Kumar J, Gowthaman MK, D'Souza SF, Kamini NR (2015) Potentiometric biosensor for determination of urea in milk using immobilized Arthrobacter creatinolyticus urease. Mater Sci Eng C 49:786–792. https://doi.org/10.1016/j.msec.2015.01.048

Ramkrishna SV, Prakasham RS (1999) Microbial Fermentation with Immobilized cells. CurrSci 77(1):87–100

Razmi H, Mohammad-Rezaei R (2013) Graphene quantum dots as a new substrate for immobilization and direct electrochemistry of glucose oxidase: application to sensitive glucose determination. Biosens Bioelectron 41:498–504. https://doi.org/10.1016/j.bios.2012.09.009

Reetz MT, Wenkel R, Avnir D (2000) Entrapment of lipases in hydrophobic sol–gel materials: efficient heterogeneous biocatalysts in aqueous medium. Synthesis 6:781–783. https://doi.org/10.1055/s-2000-6276

Ren YJ, Zhang H, Huang H, Wang X-M, Zhou Z-Y, Cui F-Z, An Y-H (2009) In vitro behavior of neural stem cells in response to different chemical functional groups. Biomaterials 30(6):1036–1044. https://doi.org/10.1016/j.biomaterials.2008.10.028

Rocha JMS, Gil MH, Garcia FAP (1998) Effects of additives on the activity of a covalently immobilised lipase in organic media. J Biotechnol 66(1):61–67. https://doi.org/10.1016/S0168-1656(98)00157-6

Safarik I, Sabatkova Z, Safarikova M (2009) Invert sugar formation with *Saccharomyces cerevisiae* cells encapsulated in magnetically responsive alginate microparticles. J Magn Magn Mater 321:1478–1481

Safarikova M, Maderova Z, Safarik I (2009) Ferrofluid modified *Saccharomyces cerevisiae* cells for biocatalysis. Food Res Int 42:521–524

Saini AS, Kumar J, Melo JS (2014) Microplate based optical biosensor for l-Dopa using tyrosinase from Amorphophallus campanulatus. Anal ChimActa 849:50–56. https://doi.org/10.1016/j.aca.2014.08.016

Saini AS, Melo JS (2013) Biosorption of uranium by melanin: kinetic, equilibrium and thermodynamic studies. Bioresour Technol 149:155–162

Saini AS, Melo JS (2015) Biosorption of uranium by human black hair. J Environ Radioact 142:29–35

Saini AS, Tripathi A, Melo JS (2015) On-column enzymatic synthesis of melanin nanoparticles using cryogenic poly(AAM-co-AGE) monolith and its free radical scavenging and electro-catalytic properties. RSC Adv 5:87206–87215

Saiyed ZM, Telang SD, Ramchand CN (2003) Application of magnetic techniques in the field of drug discovery and biomedicine. Biomag Res Technol 1(1):2–8. https://doi.org/10.1186/1477-044X-1-2

Sakurai T, Satake A, Nagata N, Gu Y, Hiura A, Doo-Hoon K, Hori H, Tabata Y, Sumi S, Inoue K (2003) The development of new immunoisolatory devices possessing the ability to induce neovascularization. Cell Transplant 12(5):527–535. https://doi.org/10.3727/000000003108746984

Saller RM, Indraccolo S, Coppola V, Esposito G, Stange J, Mitzner S, Amadori A, Salmons B, Günzburg WH (2002) Encapsulated cells producing retroviral vectors for in vivo gene transfer. J Gene Med 4(2):150–160

Sar P, D'Souza SF (2001) Biosorptive uranium uptake by a *Pseudomonas* strain: characterization and equilibrium studies. J Chem Technol Biotech 76(12):1286–1294

Sarnowska A, Jablonska A, Jurga M, Dainiak M, Strojek L, Drela K, Wright K, Tripathi A, Kumar A, Jungvid H, Lukomska B, Forraz N, McGuckin C, Domanska-Janik K (2013) Encapsulation of mesenchymal stem cells by bio-scaffolds protects cells survival and attenuates neuroinflammatory reaction in injured brain tissue after transplantation. Cell Transplant 22:S67–S82. https://doi.org/10.3727/096368913X672172

Sassolas A, Blum LJ, Leca-Bouvier BD (2012) Immobilization strategies to develop enzymatic biosensors. Biotechnology Adv 30:489–511. https://doi.org/10.1016/j.biotechadv.2011.09.003

Schaftenaar G, Vlieg E, Vriend G (2017) Molden 2.0: quantum chemistry meets proteins. J Comput Aided Mol Des 31:789–800

Schlick T (2010) Molecular dynamics: basics. In: Schlick T (ed) Molecular modeling and simulation: an interdisciplinary guide. Springer-Nature, pp 425–461. ISBN 978-1-4419-6351-2

Schrodinger Maestro (2020) Schrodinger LLC, New York. https://www.schrodinger.com/maestro. Accessed on 15 Mar 2020

Schroeder MM, Wang Q, Badieyan S, Chen Z, Marsh ENG (2017) Effect of surface crowding and surface hydrophilicity on the activity, stability and molecular orientation of a covalently tethered enzyme. Langmuir 33:7152–7159

Secundo F, Spadaro S, Carrea G, Overbeek PLA (1999) Optimisation of *Pseudomonascepacia*lipase preparations for catalysis in organic solvents. Biotechnol Bioeng 62(5):554–561. https://doi.org/10.1002/(sici)1097-0290(19990305)62:5%3c554::aid-bit7%3e3.0.co;2-2

Seetharamaiah N, Seetharamaiah N, Pathappa N, Melo JS, Gurukar SS (2017) Metal-ion co-ordination assembly based multilayer of one dimensional gold nanostructures and catalase as electrochemical sensor for the analysis of hydrogen peroxide. Sens Actuators B Chem 245:726–740. https://doi.org/10.1016/j.snb.2017.02.003

Shivudu G, Khan S, Chandraraj K, Selvan P (2019) Immobilization of Endo 1,4-beta xylanase on ordered mesoporous matrices for xylooligosachharides production. ChemistrySelect 4(38):11214–11221

Singh RS, Chauhan K (2020) Functionalization of multiwalled carbon nanotubes for enzyme immobilization. Methods Enzymol 630:25–38. https://doi.org/10.1016/bs.mie.2019.10.014

Soares CMF, de Castro HF, Santana MHA, Zanin GM (2001) Selection of stabilizing additive for lipase immobilization on controlled pore silica by factorial design. Appl Biochem Biotechnol 91(93):703–718. https://doi.org/10.1385/abab:91-93:1-9:703

Soozanipour A, Taheri-kafrani A (2018) Enzyme immobilization on functionalized graphene oxide nanosheets: efficient and robust biocatalysis. Methods Enzymol 609:371–403

Srivastava S, Bhainsa KC (2016) Evaluation of uranium removal by Hydrillaverticillata (L.f.) Royle from low level nuclear waste under laboratory conditions. J Environ Manage 167:124–129

Srivastava S, Bhainsa KC, D'Souza SF (2010) Investigation of uranium accumulation potential and biochemical responses of an aquatic weed *Hydrillaverticillata* (L.f.) Royle. Bioresour Technol 101:2573–2579

St Clair NL, Navia MA (1992) Cross-linked enzyme crystals as robust biocatalysts. J Am Chem Soc 114(18):7314–7316. doi.org/10.1021/ja00044a064

Stone I (1955) Method of making dextrose using starch glucogenase. US patent 2,717,852

Su S, Liu Q, Liu J, Zhang H, Li R, Jing X, Wang J (2018) Functionalized sugarcane bagasse for U(VI) adsorption from acid and alkaline conditions. Sci Rep 8:793. https://doi.org/10.1038/s41598-017-18698-9

Szymanska K, Pudio W, Mroowiec-Bialon J, Czarddybon A, Kocurek J, Jarzebski AB (2013) Immobilization of invertase on silica monolits with hierarchial pore structure to obtain continuous flow enzymatic microreactors of high performance. Microporous Mesoporous Mater 170:75–82

Tabata Y, Nagano A, Ikada Y (1999) Biodegradation of hydrogel carrier incorporating fibroblast growthfactor. Tissue Eng 5(2):127–138

Tanaka A, Yasuhara S, Fukui S, Iida T, Hasegawa E (1977) Immobilization of invertase by the use of photocrosslinkable resin oligomers and properties of the immobilized enzyme. J Ferment Technol 55:71–75

Tanriseven A, Doğan S (2001) Immobilization of invertase within calcium alginate gel capsules. Process Biochem 36:1081–1083

Tembe S, Karve M, Inamdar S, Haram S, Melo J, D'Souza SF (2006) Development of electrochemical biosensor based on tyrosinase immobilized in composite biopolymeric film. Anal Biochem 349(1):72–77. https://doi.org/10.1016/j.ab.2005.11.016

Thévenot DR, Toth K, Durst RA, Wilson GS (2001a) Electrochemical biosensors: recommended definitions and classification. Biosens Bioelectron 16(1–2):121–131. https://doi.org/10.1016/S0956-5663(01)00115-4

Thévenot DR, Toth K, Durst RA, Wilson GS (2001b) Electrochemical biosensors: recommended definitions and classification. Anal Lett 34(5):635–659. https://doi.org/10.1081/AL-100103209

Thonart P, Custinne M, Paquot M (1982) Zeta potential of yeast cells: application in cell immobilization. Enzyme Microb Technol 4:191–194

Thvenot DR, Toth K, Durst RA, Wilson GS (1999) Electrochemical biosensors: recommended definitions and classification (technical report). Pure Appl Chem 71(12):2333–2348

Tomotani EJ, Vitolo M (2007) Production of high-fructose syrup using immobilized invertase in a membrane reactor. J Food Eng 80:662–667

Tor R, Dror Y, Freeman A (1989) Enzyme stabilization by bilayer "encagement". Enzyme MicrobTechnol 11(5):306–312. doi.org/10.1016/0141-0229(89)90047-1

Tosa T, Mori T, Fuse N, Chibata I (1967) Studies on continuous enzyme reactions. IV. Preparation of a DEAE-sephadex–aminoacylase column and continuous optical resolution of acyl-DL-amino acids. Biotechnol Bioeng 9(4):603–615. https://doi.org/10.1002/bit.260090413

Trimukhe AM, Pandiyaraj KN, Tripathi A, Melo JS, Deshmukh RR (2017) Plasma surface modification of biomaterials for biomedical applications. In: Tripathi A, Melo JS (eds) Advances in biomaterials for biomedical applications. Springer-Nature, Singapore, pp 95–166

Tripathi A, Kumar A (2011) Multi-featured macroporous agarose-alginate cryogel: synthesis and characterization for bioengineering applications. Macromol Biosci 11(1):22–35

Tripathi A, Kumar A (2012) Integrated approach for β-glucosidase purification from non-clarified crude homogenate using macroporous cryogel matrix. Sep Sci Technol 48(16):2410–2417

Tripathi A, Melo JS (2015) Preparation of sponge-like biocomposite agarose-chitosan scaffold with primary hepatocytes for establishing an in-vitro 3D liver tissue model. RSC Adv 5:30701–30710

Tripathi A, Melo JS (2016) Synthesis of low-density biopolymeric chitosan-agarose cryomatrix and its surface functionalization with bio-transformed melanin for the enhanced recovery of uranium(VI) from aqueous subsurfaces. RSC Adv 6:37067–37078

Tripathi A, Melo JS (eds) (2017) Advances in biomaterials for biomedical applications. Springer-Nature, Singapore

Tripathi A, Melo JS (2019a) Cryostructurization of polymeric systems for developing macroporous cryogel as a foundational framework in bioengineering applications. J Chem Sci (2019) 131:92. https://doi.org/10.1007/s12039-019-1670-1

Tripathi A, Melo JS (2019b) Self-assembled biogenic melanin modulated surface-chemistry of biopolymers-colloidal silica composite porous matrix for the recovery of uranium. J Appl Polym Sci 136(5):46937. https://doi.org/10.1002/app.46937

Tripathi A, Kathuria N, Kumar A (2009) Elastic and macroporous agarose-gelatin cryogels with isotropic and anisotropic porosity for tissue engineering. J Biomed Mater Res 90(3):680–694. https://doi.org/10.1002/jbm.a.32127

Tripathi A, Sami H, Jain SR, Viloria-cols M, Zhuravleva N, Nilsson G, Jungvid H, Kumar A (2010) Improved bio-catalytic conversion by novel immobilization process using cryogel beads to increase solvent production. Enzyme Microb Technol 47:44–51. https://doi.org/10.1016/j.enzmictec.2010.03.009

Tripathi A, Hadapad AB, Hire RS, Melo JS, D'Souza SF (2013a) Polymeric macroporous formulations for the control release of mosquitocidal Bacillus sphaericus ISPC-8. Enzyme Microb Technol 53(6–7):398–405. https://doi.org/10.1016/j.enzmictec.2013.08.006

Tripathi A, Vishnoi T, Singh D, Kumar A (2013b) Modulated crosslinking of macroporous polymeric cryogel affects in vitro cell adhesion and growth. Macromol Biosci 13(7):838–850. doi: 10.1002/mabi.201200398

Tripathi A, Melo JS, D'Souza SF (2013c) Magnetic nanoparticles in tissue regeneration. In: Tiwari A, Tiwari A (eds) Nanomaterials in drug delivery, imaging, and tissue engineering. Wiley-Scrivener Publisher, USA, pp 443–492. ISBN 9781118290323

Tripathi A, Melo JS, D'Souza SF (2013d) Uranium(VI) recovery from aqueous medium using novel floating macroporous alginate-agarose-magnetite cryobeads. J Hazard Mater 246–247:87–95

Turner APF, Karube I, Wilson GS (eds) (1987) Biosensors, fundamentals and applications. Oxford University Press, Oxford

Tzianabos AO (2000) Polysaccharide immunomodulators as therapeutic agents: structural aspects and biologic function. Clin Microbiol Rev 13(4):523–533. https://doi.org/10.1128/CMR.13.4.523

Unnikrishnan BS, Preethi GU, Joseph MM, Melo JS, TT Sreelekha, Tripathi A (2020) 3D printing in Dental Implants. In: Thomas DJ, Singh D (eds) 3D printing in medicine and surgery, Elsevier-Woodhead Publishing, USA. ISBN: 9780081025420

US EPA (2003) Interim reregistration eligibility decision for methyl parathion. Case No. 0153. United States Environmental Protection Agency

Valerio SG, Alves JS, Klein MP, Rodrigues RC, Hertz PF (2012) High operational stability of invertase from *Saccharomyces cerevisiae*. Carbohyr Polym 92:462–468

van Amerongen MJ, Harmsen MC, Petersen AH, Kors G, van Luyn MJ (2006) The enzymatic degradation of scaffolds and their replacement by vascularized extracellular matrix in the murine myocardium. Biomaterials 27(10):2247–2257. https://doi.org/10.1016/j.biomaterials.2005.11.002

Vieira LC, Araujo LG, Ferreira RVP, Silva EA, Canevesi RLS, Marumo JT (2019) Uranium biosorption by *Lemna* sp. and *Pistia stratiotes* . J Environ Radioact 203:179–186

Volesky B (1990) Biosorption of heavy metals. CRC Press, USA

Wang X, Xia L, Tan K and Zheng W (2011) Studies on adsorption of Uraniu (VI) from aqueous solution by wheat straw. Environ Prog Sustain Energy. doi 10.1002/ep

Wildey EW, Benemann JR (1993) Bioremoval of heavy metals by the use of microalgae. Biotechnol Adv 11:781–782

www.samson-connect.com. Accessed on 15 march 2020

Wykes JR, Dunnill P, Lilly MD (1971) Immobilisation of α-amylase by attachment to soluble support materials. Biochim Biophys Acta 250(3):522–529. https://doi.org/10.1016/0005-2744(71)90252-X

Xia L, Li R, Xiao Y, Zheng W, Tan K (2017) Removal of uranium from aqueous solutions by rice husk. Environ Prot Eng 43(4):41–52

Xiao-teng Z, Dong-mei J, Yi-qun X, Jun-chang C, Shuai H, Liang-shu X (2019) Adsorption of uranium(VI) from aqueous solution by modified rice stem. J Chem 2019: Article ID 6409504. https://doi.org/10.1155/2019/6409504

Xie T, Wang A, Huang L, Li H, Chen Z, Wang Q, Yin X (2009) Recent advances in the support and technology used in enzyme immobilization. Afr J Biotechnol 8(19):4724–4733

Xu W, Liu L, Charles IG (2002) Microencapsulated iNOS-expressing cells cause tumor suppression in mice. FASEB J 16:213–215

Yang J, Volesky B (1999) Biosorption of uranium on *Sargassusm* biomass. Water Res 33:3357–3363

Yim TJ, Kim DY, Karajanagi SS, Lu TM, Kane R, Dordick JS (2003) Silicon nanocolumns as novel nanostructured supports for enzyme immobilization. J Nanosci Nanotechnol 3(6):479–482. https://doi.org/10.1166/jnn.2003.240

Yuvaraja G, Zheng N, Pang y, Su M, Chen D, Cong L, Mehmood S, Subbaiah MV, Wen J (2020) Removal of U(VI) from aqueous and polluted water solutions using magnetic Arachishypogaea leaves powder impregnated into chitosan macromolecule. Int J Biol Macromol 148:887–897

Zaborsky OR (1972) Alteration of enzymic properties prior to immobilization. Biotechnol Bioeng Symp 3:211–217

Zaks A, Kllbanov AM (1985) Enzyme-catalyzed processes in organic solvents. Proc Natl Acad Sci USA 82(10):3192–3196. https://doi.org/10.1073/pnas.82.10.3192

Zhang L, Xiao X, Yuan Y, Guo Y, Li M, Pu X (2014) Probing Immobilization mechanism of alfa-chymotrypsin onto carbon nanotube in organic media by molecular dynamics simulations. Sci Rep 5:9297. https://doi.org/10.1038/srep09297

Zhao J, O'Daly JP, Henkens RW, Stonehuerner J, Crumbliss AL (1996) A xanthine oxidase/colloidal gold enzyme electrode for amperometric biosensor applications. Biosens Bioelectron 11(5):493–502. https://doi.org/10.1016/0956-5663(96)86786-8

Cell Immobilization Strategies for Tissue Engineering: Recent Trends and Future Perspectives

Pallavi Kulkarni, Rohit Parkale, Surbhi Khare, Prasoon Kumar, and Neha Arya

Abstract Cell immobilization is the process of localizing intact cells onto specific regions in a device or material without the loss of requisite biological function. Immobilization of cells can generally be performed through physical adsorption, encapsulation, entrapment and self-aggregation. Various types of cells, including microbial, plant, mammalian and insect cells, have been immobilized on materials for improving the bio-synthesis, bioanalytics, as well as cell therapy applications. Recently, cell immobilization on biomaterials-based 3-D scaffolds has gained attention in the area of regenerative medicine and tissue engineering wherein the transplantation of immobilized cells has been utilized in repair, restoration or improvement of tissue function. This chapter will majorly focus on different mammalian cell immobilization techniques, physio-chemical properties of matrix materials employed for cell immobilization and their applications in tissue engineering.

Keywords Microencapsulation · Regenerative medicine · Cell immobilization · Cell entrapment · 3D scaffolds · Hydrogels

Abbreviations

ALR	Airlift bioreactor
ATP	Adenosine triphosphate
BAL	Bioartificial liver
CPBA	4- Carboxyphenylboronic acid

P. Kulkarni · S. Khare · N. Arya (✉)
Department of Biochemistry, All India Institute of Medical Sciences, Bhopal, Saket Nagar, Bhopal 462020, Madhya Pradesh, India
e-mail: neha.arya@gmail.com

R. Parkale · P. Kumar
Department of Medical Devices, National Institute of Pharmaceutical Education and Research-Ahmedabad (NIPER-A), Opposite Air force Station, Palaj, Gandhinagar 382355, Gujarat, India

A. Tripathi and J. S. Melo (eds.), *Immobilization Strategies*,
Gels Horizons: From Science to Smart Materials,
https://doi.org/10.1007/978-981-15-7998-1_2

CSTR	Continuous stirred tank reactor
DEAE cellulose	Diethylaminoethyl cellulose
DEP	Dielectrophoresis
DM	Degree of methylation
DMFBR	Diversion-type microcapsule-suspension FBR
DO	Dissolved oxygen
ECM	Extracellular matrix
ESC	Embryonic stem cells
FBR	Fluidized bed reactor
FSK	Forskolin
(b-FGF)	Basic-fibroblast growth factor
GMA	Glycidyl methacrylate
GMSCs	Gingival mesenchymal stem cells
GT	Gum tragacanth
hAMSCs	Human amniotic mesenchymal stromal cells
(HACs)	Human articular chondrocytes
HBMSCs	Human bone marrow mesenchymal stem cells
HFM	Hollow fiber membrane
hGBM	Human glioblastoma
HMEC	Human microvascular endothelial cell line
HPTC	Human primary renal proximal tubule cells
IL	Interleukines
iPSCs	Induced pluripotent stem cells
MBR	Membrane bioreactor
MHC	Major histocompatibility complex
MeBIO	(6-Bromo-1-methylindirubin-3′-oxime
MSC	Mesenchymal stem cells
NK cells	Natural killer cells
NSPC	Neural stem/progenitor cells
NWF	Non-woven fibers
OCR	Oxygen consumption rate
PAAm	Polyacrylamide
PBR	Packed bed reactor
PDMS	Polydimethylsiloxane
PEG	Polyethylene glycol
PEG–PCL-PMs	Poly(ethyleneglycol)-*b*-poly(ε-caprolactone) polymersomes
PEMA	Pectin methacrylate
PET	Positron emission tomography
PLA	Polylactic acid
PLLA	Poly (l-lactic acid)
PMBVS	Phenylboronic acid-*co*–*N*-succinimidyloxycarbonyl tetra(ethylene glycol)methacrylate)
PPG	Poly (propylene glycol)
PUF	Polyurethane foam
PUM	Polyurethane membrane

PVA	Polyvinyl alcohol
PVF	Polyvinyl fluoride
SCP	Single cell protein
sFvIL-2	Interleukin-2 fusion protein
STB	Stirred tank bioreactor
STR	Stirred tank reactor
VPBR	Volume packed bed reactor
κ-CA	κ-Carrageenan

1 Introduction

Tissue engineering combines the principles of biology, medicine and engineering and aims at regenerating as well as maintaining tissue function (Langer and Vacanti 1993). Towards this, tissue engineering encompasses three factors namely, (i) cells, for the formation of functional tissue, (ii) 3-D transient structures called "scaffolds", for supporting cell attachment and (iii) bio-reactive molecules or growth factors for growth and development of the cells into the desired tissue (De Isla et al. 2010). In order to successfully generate a tissue-engineered graft, one of the crucial steps is the seeding of cells on 3-D scaffolds. Cells can either be seeded directly onto the pre-fabricated scaffolds or can be immobilized by bio-encapsulation within the polymeric matrices (Wu and Elisseeff 2014). These matrices are usually based on biocompatible materials with potential cell-binding domains. Cell immobilization in a matrix has been defined by Karel et al. as "the physical confinement or localization of intact cells to a certain defined region of space with preservation of some desired catalytic activity" (Karel et al. 1985). Towards the development of a successful tissue-engineered graft, immobilized cells offer many advantages over freely suspended cells, such as enhanced biological stability, high biomass concentration, improved mass transfer, high cell density per volume of the reactor, increased productivity, increased substrate utilization and reduction in the lag phase of cell growth (Dervakos and Webb 1991; Duarte et al. 2013).

Looking at the success of enzyme immobilization techniques in recent years, the concept of cell immobilization has been borrowed from enzyme immobilization (Kierstan and Bucke 1977). Mammalian cells need a substrate to latch upon and proliferate while performing their role as a molecular factory. As a molecular factory, various biochemicals are synthesized by mammalian cells and are used in conjunction with ATP for the growth, development and maintenance of the cell. So, as molecular factories, mammalian cells are used for the production of monoclonal antibodies, therapeutic and diagnostic proteins (Khan 2013; Moussavou et al. 2015; Lalonde and Durocher 2017). Although both cells and enzymes have similar strategies of immobilization such as encapsulation, entrapment and adsorption (Ge et al. 2016), the hallmark of a cell immobilization system is the trapping of cells in a confined space of a substrate or matrix. The criteria for choosing a technique for cell immobilization depends on various parameters; the trapped cells should be capable of communicating among themselves, latching upon the substrate material, should be able to perform transport functions such as gas/nutrients/waste exchange, maintain high density of cells and enable the recovery of these cells after downstream processing. In this regard, 3-D matrices based on different materials and architecture such as 3-D printed scaffolds, microfluidics devices, matrigel (gelatinous protein mixture), hydrogels, aerogels and cryogels have been used as either stand-alone biocompatible materials or coupled with RGD peptides, proteins and other biomolecules for cell immobilization (Thakar et al. 2019; Tripathi and Melo 2019). Additionally, the morphological cues on the matrix material also guide the differentiation of mammalian cells (Akhmanova et al. 2015). Although these 3-D matrices have demonstrated success in cell immobilization for production of biomolecules/metabolites as well as in tissue engineering applications, the choice of materials to be used largely relies on the housing of cell types towards their growth and development.

This chapter provides a review on fundamentals and advantages of cell immobilization, types of cell immobilization systems, bioreactors used for cell immobilization and their application in tissue engineering.

2 Fundamentals of Cell Immobilization

Cell immobilization techniques were introduced to address the challenges associated with immobilization and usage of multiple enzymes simultaneously for biocatalysis (Chang and Wang 2011). Success in immobilization techniques led to the immobilization of prokaryotic cells to achieve multiple biocatalysis activity in a single reactor. However, immobilization of prokaryotic cells against their mobile nature brings the challenges associated with the design of substrate to facilitate immobilization without compromising on the function of prokaryotic cells. Therefore, porous substrates with interconnected pores were used (Zhao et al. 2006). These substrates ensured that the prokaryotic cells were physically entrapped along with media in their growth phase towards the production of desired biochemicals. Therefore, efforts have been taken to design substrates for prokaryotic cell entrapment based on the desired biochemical activity (Lee et al. 1994).

However, the complexity increases further during immobilization of mammalian cells. This is because mammalian cells require cell–cell communication and exhibit contact inhibition in a crowded space while residing within the substrate. The substrate should exhibit various characteristics such as adequate surface topology and chemistry to facilitate cell adherence and migration, better transport of nutrients and waste products toward the survival of cells for a long period of time, easy recovery of biomolecules and maintenance of appropriate microenvironment for the cells (Huang et al. 2012). Therefore, similar to prokaryotic cells, material selection and substrate design play a vital role in governing mammalian cell immobilization as well.

Towards this, one of the important characteristics of the substrate is that it should not inhibit cell growth or cause cell lysis. There are a number of polymeric materials that can be used as 3-D matrices. Few examples include alginate, agar, gelatin, κ-carrageenan, chitosan, pectin, polyacrylamide, epoxy resin, and silica sol, and porous or nonporous preformed support materials such as wood chips, stainless steel, cotton cloth, porous glass particles, DEAE cellulose, porous silica, porous ceramics and diatomaceous earth (Klein and Kressdorf 1989; Salter and Kell 1991). These materials have been used to immobilize cells through entrapment, attachment or encapsulation. However, as mentioned previously, depending upon the application and type of cells, the immobilization method needs to be adapted. As a result, there are different techniques that support cell immobilization suited to different application types. Considering the variety of mammalian cells and their associated microenvironments for appropriate phenotypic expression, they require adequate physical, chemical and physiological cues from the substrate.

Additionally, substrate design is another essential characteristic of efficient and appropriate mammalian cell immobilization. As an example, the substrate selected for mammalian cell immobilization should have high porosity and pore connectivity so as to support high mammalian cell density within the matrix (Mota et al. 2001). The substrate also needs to be mechanically, chemically and thermally stable during the process of immobilization towards the generation of a tissue-engineered graft. Further, the substrate should facilitate storage at ambient conditions, eliminating the need for cold storage. For example, lyopreservation of human amniotic membrane encapsulating cells has been demonstrated to have viable cells until 21 days without cold storage (Mao et al. 2019). It is demanded that the substrate should provide a large surface area for maximum cell adherence and differentiation into requisite phenotype. For this, the surfaces need to be functionalized accordingly for efficient cell binding (Tallawi et al. 2015). The surface topology also needs to be tailored such that it provides adequate hydrophobicity for protein binding and eventually appropriate cellular adhesion (Rahmany and Van Dyke 2013; Ferrari et al. 2019). Additionally, cell number is an important criterion for mammalian cells as they show contact inhibition when immobilized in high numbers and also face viability issues due to inadequate cell-to-cell communication when seeded at a low density (Abercrombie 1970). Hence, cell immobilization density should be optimized for a specific application and as per the surface area available on the substrate.

In addition to these factors, considering the immobilization of mammalian cells for tissue development using a bioreactor (which will be discussed in the later section), there is an increased probability of cell exposure to shear stress. It has been reported in the literature that shear stress tends to alter the cellular behavior (Huber et al. 2018). In response to shear stress, the mechano-transduction pathways are usually triggered, wherein the cells produce chemicals or signaling molecules that may either change the cellular microenvironment or result in a deleterious physiological response (De Meester et al. 1998; Kandow et al. 2007; Polacheck et al. 2011). Hence, optimization of shear stress on cells immobilized within a bioreactor needs to be optimized according to the tissue type. As an example, in a continuous bioreactor, the flow of fluid (both media and cellular waste) should be such that there is an efficient waste and nutrient exchange with the fluid media. The design of bioreactor should also enable proper utilization of cell culture media and easy low-cost separation of biochemicals and cells for different applications and reusability, respectively.

Further, the immobilized cell growth needs to be maintained in a logarithmic phase, ensuring minimal death due to poor bio-transport of media/waste (Kovárová-Kovar and Egli 1998). As a result of the continuous exchange of waste and nutrients, the pH and ionic concentration are usually altered in cell culture media. Hence, measures have to be taken for continuous monitoring of media composition, followed by tailoring the composition according to the type of the tissue. This will also ensure effective maintenance of the external microenvironment of cells for desired phenotypic expression and continuous flow of media would also take care of any end product inhibition in the immobilized cells. As a result, enabling a conducive environment for the growth and development of mammalian cells will also lead to increased genetic stability (Ramasubramanyan and Venkatasubramanian 1990). In addition to these features, the substrate design should not only enable mammalian cell culture but should also possess antibacterial activity. It has been reported that substrate microtopography can lead to the generation of bactericidal surfaces (Glinel et al. 2012). Further, impregnation of porous substrates with lipids leads to inhibition of bacterial colonies (Prabhawathi et al. 2014). Thus, substrate designing or use of antibacterial compounds/materials can reduce the risk of microbial contamination.

There are several cell immobilization techniques reported in the literature and will be explained in the upcoming section. The first technique will discuss the immobilization through support materials followed by self-immobilization strategies. The type of technique chosen for the immobilization of mammalian cells is usually guided by the type of application and mammalian cells. For instance, an application demanding production of primary metabolites through cells that are in the logarithmic phase cannot be performed in any gel matrix, whereas the bulk production of low-value products such as biofuels may be best achieved by self-immobilization of cells (Gungormusler-Yilmaz et al. 2016). Therefore, understanding the application of cell immobilization will serve as a guide to select the cell immobilization technique and substrate type for immobilization.

3 Types of Cell Immobilization

There are various types of immobilized cell systems, which can be selected, according to the intended application and the properties of the immobilization matrix. To start with, the system should be non-toxic and biocompatible, should have easy access to the nutrients, should be economically viable and should have a high surface-to-volume ratio for maximum cell immobilization. In addition, the system should facilitate mass transfer as well as promote easy separation of the cells, in case of both suspended and anchorage-dependent cells (Willaert 2011). Based on the types of support materials and cell types, immobilization systems can be classified into different categories. Figure 1 and Table 1 show various types of cell immobilization

Table 1 Advantages and disadvantages of various types of immobilization techniques

S. No.	Immobilization technique	Advantage	Disadvantage
1	Physical adsorption	Simple and cheap, high catalytic activity, no need to use reagent	Low stability, possible loss of biomolecule, weak bond formation
2	Encapsulation	Protection of biocatalyst, prevent leakage	Limitation of mass transfer
3	Entrapment	Allows transport of low molecular weight compounds, enables continuous operation due to maintained cell density	Low cell loading and leakage
4	Cross-linking	Allows controlled release of product prevents leakage	Diffusion limitation
5	Covalent bonding	Strong binding high heat stability	Less effective for cell immobilization

with their advantages and disadvantages, respectively. The upcoming section will discuss the support materials that have been used for cell immobilization in various studies.

3.1 Immobilization with Support Material

3.1.1 Surface Attachment

Surface immobilization of cells on a support material is dependent on surface properties of the material, such as hydrophilicity, roughness, presence of functional entities, surface charge, surface chemistry, pore size as well as the shape of the material.

Hydrophilicity can be defined as the factor that governs cell adhesion. It has been reported that positive charge and hydrophilicity enhance cellular adhesion as compared to negative charge and hydrophobicity. The negative charge and presence of polar head-group of lipid by-layer membrane of any cell enhance their interaction with surfaces of substrates having positive charge with hydrophilic nature. This enhanced interaction leads to better adhesion and spreading of cells for minimization of surface energy (Tripathi et al. 2013; Ferrari et al. 2019). In a study, Guo et al. demonstrated that 3T3 fibroblast cell adhesion was highest on the surfaces having high hydrophilicity and decorated with positive charge moieties (Guo et al. 2016).

Surface roughness also plays a vital role in cell immobilization and may be used for cell entrapment or directing cellular behavior (Rao et al. 2013). Kim et al. showed that human mammary epithelial cells grown on a substrate grafted with dendrimers of different sizes exhibited different cellular adhesion. While the naked dendrimer surface demonstrated F-actin rich cells as compared to non-modified plain surfaces, the dendrimers with mean roughness greater than 4 nm inhibited the stretching of cells, resulting in round-shaped cells (Kim et al. 2009). Apart from this, patterned roughness with aligned nanofiber mat of PCL/gelatin composite or parallel microchannel grooves in collagen-chitosan scaffolds have been reported to direct the orientation of cardiomyocytes in the direction of the major axis of the channel/fibers (Kai et al. 2011; Zhang et al. 2012).

Further, the presence of various functional groups on the material surface can modulate protein adsorption via integrin binding and hence can facilitate cellular adhesion (Arima and Iwata 2007). Towards this, a study demonstrated that immature osteoblast-like cell line MC3T3-E1 cultured on surfaces grafted with amine and hydroxyl group led to enhanced alkaline phosphatase activity, bio-mineralization as well as gene expression specific to osteoblast lineage as compared to surfaces grafted with alkyl or carboxylic groups (Keselowsky et al. 2003). Another study utilized nanotextured polydimethylsiloxane (PDMS) substrate to increase the effective surface area for immobilization of various functional groups (Wan et al. 2012). Wan et al. immobilized anti-EGFR aptamers on the nanotextured PDMS surfaces and showed that the nanotextured surfaces captured higher amounts of human glioblastoma (hGBM) cells as compared to the conventional smooth substrates based on glass, thereby demonstrating specific cell capture application.

Additionally, the functional groups grafted on a substrate influence the presence of positive or negative charge on the surface, which can also affect cell immobilization. Mammalian cells are negatively charged; their attachment to negatively charged substrates causes electrostatic repulsion thereby hindering cell immobilization. On the other hand, the positive charge assists in cell binding to substrates and modulates cellular behavior (Chang and Wang 2011). In one study, the positively charged surface of hydrogels based on 2-hydroxyethyl methacrylate demonstrated better attachment of osteoblast and fibroblast cells as compared to negative or neutral surfaces (Schneider et al. 2004; Kim et al. 2009). In a few other studies, polylysine has been coated onto the substrate material to enhance the cellular attachment to the substrate. The positive charge moieties such as NH_2 groups demonstrated better cell adhesion, growth and spreading rate as compared to COOH and CH_2OH groups (Lee et al. 1994). Although positive charge assists in cellular attachment and proliferation, the negative charge of oligo (poly(ethylene glycol) fumarate) hydrogels has also demonstrated increased propensity to differentiate chondrocytes and hence the expression of collagen type II and glycosaminoglycan (Dadsetan et al. 2011). Hence, the choice of surface charge on any material is dependent on the type of cells to be immobilized and the intended application. In summary, the aforementioned factors when combined together can play a crucial role in altering protein adsorption, cellular adhesion and hence cell fate processes toward specific tissue engineering application.

3.1.2 Immobilization Within 3-D Polymeric Substrates

Entrapment in Gel Matrices: Hydrogels

Among various cell entrapment methods, the entrapment within porous matrices is the most preferred method since it supports high cell viability and offers low stress on cells (Mazzei et al. 2010). Cell entrapment within hydrogel is achieved through the formation of matrices around the cells to be immobilized. Various natural and synthetic polymers have been used in this regard. Polymers from natural origin such as alginate, carrageenan, agar and agarose, chitin and chitosan, protein, polysaccharide and synthetic polymers such as alumina pellets, kaolinite, polyester foam, polystyrene have also been employed in cell entrapment (Willaert and Baron 1996; Gasperini et al. 2014). These hydrogels are prepared when natural or synthetic polymers are cross-linked by chemical bonds through a cross-linker, leading to ionic, physical as well as hydrophobic interactions among polymer chains. Certain polysaccharides, synthetic polymers and proteins can form hydrogels with hydrophilic matrices under mild conditions, thereby resulting in cell entrapment with low cell death and high cell mass loading (Ferrari et al. 2019). Hydrogels absorb water and readily swell in aqueous solution while maintaining their integrity (Willaert and Baron 1996) and also protect the entrapped cells within their 3-D porous network.

Cell entrapment within hydrogels requires that the hydrogel is conducive to cell viability and also allows uniform cell distribution throughout the matrix. Further, the pores and pore size should enable the efficient mass transfer of nutrients and waste products. Hydrogels can take up different shapes and sizes. This property of hydrogels has been exploited for the various applications; in a study, this was used for stem cell encapsulation through high-throughput droplet microfluidics method (Choe et al. 2018). The size of the hydrogels spherical beads can reach up to 2.5 mm diameter or 3D bio-printed complex structures/scaffolds spanning a couple of centimeters (Jen et al. 2000; Belgrano et al. 2018; Jasińska et al. 2018). The upcoming section will discuss about the various natural and synthetic polymers that have been used towards the generation of hydrogels for cell immobilization studies.

Cell Entrapment in Hydrogels Based on Natural Polymers

Alginic acid is a polysaccharide-based on poly-anionic linear chain and contains G and M block, i.e., guluronic acid and mannuronic acid, produced from bacteria *(Azotobacter and Pseudomonas spp)* as well as seaweed. Calcium alginate is one of the most widely used alginates for cell immobilization. Immobilization is performed when calcium ion is introduced into an alginate solution with dispersed cells; calcium ion binds to the G block and forms an interchain bridge resulting in a sol–gel system. This sol–gel system gains rigidity and mechanical strength depending on the ratio of G to M block and also on the percentage of calcium ions introduced to the solution. The cells thus entrapped within these sol–gel matrices are guarded against the external environment. Figure 2 shows the procedure of whole cell entrapment into alginate polymer. Generally, alginate-cell suspension forms a spherical ball of cross-linked alginate when it interacts with calcium ion in the form of a droplet. Further, the diffusion rate of the hardening ion and rate of cross-linking governs the pore size distribution in an alginate bead; pore size is typically in the range of 5–200 μm (Andersen et al. 2015).

Based on its ability to gel under mild conditions, retain water and demonstrate biocompatibility, alginic acid finds application in tissue engineering and cell immobilization (Szekalska et al. 2016). Alginate beads have been extensively studied to immobilize pancreatic islets as a means to protect allogeneic and xenogeneic transplanted pancreatic islets from immune rejection and have also shown to improve the insulin production of islet progenitor cells (Hoesli et al. 2011; Suarez-Arnedo et al. 2018). Hoesli et al. optimized the generation of alginate beads using emulsion and internal gelation process to increase insulin expression in MIN6 mouse cells. The authors then extended the emulsion process to primary pancreatic exocrine cell immobilization, demonstrating 67 ± 32 fold enhanced insulin expression following 10 days of culture (Hoesli et al. 2011).

Although alginic acid demonstrates the potential for cell immobilization and various tissue engineering applications, it possesses poor cell adhesion properties; as a result, it is modified with various cell adhesion peptides (Ansari et al. 2016), inorganic materials and other polymers to improve its biological properties. One such study developed an injectable RGD-based alginate hydrogel loaded with multiple growth factors namely, Forskolin (FSK), MeBIO (6-Bromo-1-methylindirubin-3′-oxime) and basic-fibroblast growth factor (b-FGF) (Ansari et al. 2016). The authors then encapsulated gingival mesenchymal stem cells (GMSCs) and human bone marrow mesenchymal stem cells (HBMSCs) within the hydrogels and evaluated their ability to differentiate into the myogenic lineage. The results of the study demonstrated higher expression of markers related to muscle regeneration in encapsulated GMSCs as compared to HBMSCs, both in vitro and in vivo, thereby demonstrating the potential of GMSCs encapsulated RGD-based hydrogels for muscle tissue engineering. Another study reported the blending of gum tragacanth (GT), an anionic, heterogeneous and highly branched polysaccharide obtained from the exudates of several plant species of the *Astragalus* genus, with alginate (Kulanthaivel et al. 2017). It was observed that the presence of GT improved the viability, differentiation and proliferation of bone cells in alginate matrix and also demonstrated to be pro-angiogenic as compared to native alginate hydrogels.

Although there have been multiple reports of cell immobilization using the alginate system, its enhanced sensitivity toward chelating agents such as phosphate, lactate, citrate and metal ions may limit its application. To overcome this, alginate beads are placed in a medium having free calcium ions and with low sodium to calcium ion ratio (Martinsen et al. 1989).

Carrageenan is a polymer isolated from red seaweed and possesses a backbone of α(1→3)D-galactose and β(1→4)D-galactose. There are three different types of carrageenan available, namely lambda (λ), kappa (κ) and iota (ι). These are defined based on the location of the ester and sulfate groups on the sugar moiety. Among all the forms of carrageenan, κ form is mostly considered for cell immobilization (Chibata et al. 1987). κ form is mainly obtained from the seaweed, *Eucheuma cottoni*.

It can form gels of various shapes and sizes, generated by the dripping method. Gelation can be achieved by cooling or treating the polymer solution with gel inducing reagents such as K^+, NH^{4+}, Ca^{2+}, Cu^{2+}, Mg^{2+}, Fe^{3+}, amines as well as water-miscible organic solvents. The mechanical strength of κ form can be enhanced by treating with amine and glutaraldehyde that act as a hardening agent. Carrageenan can also form a hydrocolloid gel due to double helix formation (Willaert 2011). In a study, κ-carrageenan (κ-CA) microfibers were developed with chitosan coating through the two-step process, ionotropic gelation of κ-CA followed by polyelectrolyte complexation with chitosan for enhanced fiber stability and utilized them for vascularized bone tissue engineering application. These microfibers demonstrated suitability to carry and deliver microvascular-like endothelial cells without loss in their viability and phenotypic characteristics. These cell immobilized fibers were then tested for their integration within a 3-D hydrogel containing osteoblast-like cells towards the development of complex tissues (Mihaila et al. 2014).

Another natural material, ***agarose***, has been used immobilizing cells for various tissue engineering applications. It is a preparation of agar and an extract from marine sources such as algae. The type of agar differs from their gelling temperature and presence or absence of methyl and ether groups. Agarose is a linear polymer comprising β-D-galactose and 3,6-anhydro-α-L-galactose. The pentagonal pore is an essential feature of an agarose gel that provides easy access to the proteins of high molecular weight. Agarose gels exhibit thermo-responsive reversible sol–gel property. Typically, the polymer solution is cooled to the temperature at 10 °C above the gel-forming temperature and thereafter mixed with the cells to be immobilized with the hydrogel. During the formation of agarose gel, the transition of the random coil in solution to double helix allows its further transformation into bundles of helices (Zarrintaj et al. 2018). Agarose-based hydrogels have been employed in cell immobilization for tissue engineering applications (Tripathi and Kumar 2011; Jen et al. 2000). In one such study, Jutila et al. encapsulated chondrocytes in agarose hydrogels of varying stiffness corresponding to pathological and physiological conditions. They demonstrated a relationship between agarose stiffness and chondrocyte mechanotransduction as a result of distinct metabolic profiles in various small molecules affecting cell physiology (Jutila et al. 2015). In another study, Arya et al. encapsulated human articular chondrocytes (HACs) in carboxylated agarose hydrogels of varying stiffness and demonstrated a relationship between the hydrogel stiffness and integrin-binding peptide sequence conjugated to the hydrogels on the differentiation potential of HACs (Arya et al. 2019). Agarose/gelatin hydrogels have also been used for immobilizing small tissue or embryo. In a study, McClelland et al. provided a protocol for oriented paraffin embedding and sectioning following immobilizing small tissue or embryo ($\leq$5 mm) in agarose/gelatin cubes. This protocol of tissue immobilization in agarose/gelatin cubes followed by conventional embedding and sectioning demonstrated defined orientation of the tissue of interest as well as easy visualization (McClelland et al. 2016).

Another polymer, ***chitosan*** is obtained from chitin deacetylation; chitin is a polysaccharide obtained from crab shells and is the second-largest polysaccharide available from marine sources. Chitosan is a linear polymer comprising of β-linked

D-glucosamine and N-acetyl-D-glucosamine. In an acidic solution, chitosan gets protonated to become a poly-cationic entity that can form a complex gel with any poly-anionic group. In one study, Liouni et al. developed chitosan/alginate polyelectrolyte complex coated beads for yeast entrapment (Liouni et al. 2008). Additionally, cross-linking of chitosan with low molecular weight ions such as ferrocyanide, ferricyanide and polyphosphates results in globule formation, wherein the cells of choice can be entrapped (Górecka and Jastrzębska 2011; Pandey et al. 2017). In one such study, animal cells were encapsulated in the chitosan microbeads via emulsification method using thermosensitive gel formulation. Microbeads were formed using water-in-oil emulsion, wherein cells (L929 fibroblast or human mesenchymal stromal cells), chitosan solution and gelling agents (sodium hydrogen carbonate and phosphate buffer or beta-glycerophosphate) served as the aqueous phase. The report concluded that mesenchymal stromal cells encapsulated within chitosan microbeads released higher vascular epithelial growth factor as compared to manually casted macrogels, thereby demonstrating the potential of emulsion generated chitosan microbeads in cell therapy applications (Alinejad et al. 2020).

Another polymer, **pectin**, is present in the plant cell walls and is a linear polymer chain with 1,4-linked α-D-galacturonate as a backbone and randomly branched with 1,2 linked L-rhamnose. Upon interaction with calcium ions, pectin forms a gel; due to the versatility in its chemical structure, these gels can be formed by various methods. Polygalacturonic acid, the principal constituent of these gels, is partly esterified with methoxyl groups, while the free acid groups may be partly or fully neutralized with monovalent ions (i.e., Na^+, K^+, NH^{4+}). Solubility and gelation properties of pectin majorly depend upon the degree of methylation (DM); pectins with a DM lower than 5% are called pectic acids, while those with a higher DM are termed as pectinic acids (Gawkowska et al. 2018). Pectins gels also demonstrated potential as cell immobilizing matrices for tissue engineering applications. A recent report suggests the combination of human amniotic mesenchymal stromal cells (hAMSCs) with pectin gels for osteochondral regeneration and maintenance of immunological characteristics. Following encapsulation of hAMSCs in pectin/hydroxyapatite gel bio-composite, the authors found the hAMSCs differentiation toward osteogenic lineage with no induction of immune response; however, the cells retained their ability to reduce T-cell proliferation (Silini et al. 2018). In another study, Mehrali M and colleagues developed mechanically strong and durable pectin by generating UV-cross-linkable pectin methacrylate (PEMA) polymer. The system was then incorporated with thiolated gelatin to improve the hydrogel matrix for cell differentiation and spreading property and further fine-tuned to achieve compressive modulus between ~ 0.5 to ~24 kPa. It was demonstrated that the PEGMA-gelatin system supported the growth of encapsulated cells including muscle progenitor (C2C12), neural progenitor (PC12) and human mesenchymal stem cells (hMSCs). PEMA-gelatin-encapsulated hMSCs cells facilitated bone-like structure after 5 weeks of incubation thereby demonstrating the potential of pectin-based hydrogels in tissue engineering applications (Mehrali et al. 2019).

Collagen is a fibrillar protein that forms a major part of the extracellular matrix of most organs. It mainly consists of glycine, proline and hydroxyproline and has a triple-stranded helical structure. Collagen hydrogels have been widely used in cell immobilization due to the self-assembly property of collagen; tropocollagen molecules self assemble via electrostatic interaction and is stabilized by hydrogen and hydrophobic interactions. Cell laden hydrogels require mixing of collagen and cell suspension at low temperatures followed by gel formation at 37 °C. A study by Yuan et al. led to the generation of injectable hydrogels based on collagen type I and II with modulated mechanical properties. Chondrocytes embedded within this hydrogel produced cartilage-specific ECM, thereby demonstrating the potential in cartilage tissue engineering (Yuan et al. 2016). In another study, injectable hydrogels based on collagen type 2 were fabricated, and encapsulated chondrocytes were tested for their potential to regenerate rabbit full-thickness cartilage defects without a periosteal graft (Funayama et al. 2008); 8 weeks following implantation showed hyaline cartilage regeneration with appropriate chondrocyte morphology. Collagen hydrogels along with other polymers such as hyaluronic acid have also been used for cartilage tissue engineering applications (Kontturi et al. 2014). In yet another study, Kuo et al. demonstrated the potential of murine collagen-phenolic hydroxyl hydrogels in vascular network formation both under in vitro and in vivo conditions (Kuo et al. 2015).

A derivative of collagen, **gelatin**, has been used for various cell immobilization for tissue engineering applications (Singh et al. 2011; Sarem et al. 2018) since it mimics important features of the ECM such as cell binding and protease-responsive peptides. Unlike collagen, gelatin is non-immunogenic, can be easily functionalized thereby demonstrating potential in tissue engineering applications. Although gelatin can form physically cross-linked hydrogels based on its thermo-responsive properties, gelatin hydrogels of tunable mechanical properties can be obtained as a result of its covalent cross-linking. In this regard, Blocki and colleagues generated a gelatin-based cell carrier micro-spherical hydrogels. Through photo-polymerization, fibroblasts were encapsulated in glycidyl methacrylate (GMA)—functionalized gelatin and stable microspheres were synthesized by adjusting the cross-linking density of hydrogel by varying the time of irradiation (0.5–2 min) or average gelatin-chain length. Evaluation of cell viability showed that microgels with the same polymer weight content and with lower storage and elastic moduli as a result of decreased cross-linking density showed better potential as cell encapsulating materials. The results of the study demonstrated the potential of GMA-gelatin-based hydrogels with elastic moduli comparable to the native ECM as suitable material for various biomedical applications (Blocki et al. 2017).

Another natural polymer, ***hyaluronic acid***, is a linear anionic polysaccharide present in connective tissue and comprising D-glucuronic acid and D-N-acetylglucosamine that are linked together through alternating β-1,4 and β -1,3 glycosidic bonds. It is a hydrophilic polymer that is viscous even at low concentrations as a result of chain entanglement. For generating hydrogels, it requires basic modification in the carboxyl and the hydroxyl group by esterification and cross-linking with glutaraldehyde and divinyl sulfone (Collins and Birkinshaw 2007). Being a naturally occurring ECM component, it is mostly preferred for encapsulation of mammalian cells whose native matrix contains high glycosaminoglycans and hyaluronic acid.

In a study, Vilariño-Feltrer et al. used hyaluronic acid to design tubular scaffold for immobilization of Schwann cells toward cell transplantation therapies in the nervous system. They fabricated a novel, single-channel tubular conduits based on hyaluronic acid with or without poly-l-lactic acid (PLLA) fibers within their lumen followed by seeding of rat Schwann cells. Conduits comprised of a three-layered porous structure, which allowed cell invasion from the exterior surface and nutrients for cell survival, thereby resulting in the generation of Schwann cells tubes with 0.55 mm diameter and variable lengths. These conduits would also demonstrate potential in cell sustenance and protection for transplantation (Vilariño-Feltrer et al. 2016).

In summary, cell entrapment within natural hydrogels has demonstrated potential in various tissue engineering applications. Although natural polymers are superior in terms of biocompatibility and possess biological information that supports cell attachment, however, they demonstrate batch to batch variability, immunogenicity, possible risk of disease transfer as well as lack requisite mechanical strength. Hence, synthetic polymers have been used due to their excellent mechanical properties and tailorability. The next section will discuss few examples of synthetic hydrogels utilized for tissue engineering applications.

Cell Entrapment in Hydrogels Based on Synthetic Polymers

As mentioned in the previous section, synthetic polymers are being extensively explored for cell immobilization due to their suitable methodology of fabrication as compared to natural polymers. Their properties such as porosity, hydrophilicity or hydrophobicity, mechanical properties can be designed artificially according to the desired application. This section will discuss the various synthetic polymers that have been used for cell immobilization studies.

Polyacrylamide (PAAm) is one of the first synthetic materials that were used for packing cells into the polymer matrix. Due to its ease of fabrication, PAAm gels have been widely used for studying the role of mechanical properties of the substrate on cell fate processes (Kandow et al. 2007). The hydrogels of PAAm are transparent and non-fluorescent in nature and offer an advantage in terms of observing cellular activity through optical microscopy (Trelles and Rivero 2013). Cell entrapment within the PAAm matrix is achieved by polymerization of acrylamide that is toxic to cells as against PAAm. Therefore, is it crucial to optimize polymerization time and temperature; 100% viability of some of the cell types have been reported (Park et al. 2014). Further, the mechanical properties and porosity of the gel can be varied by using different concentrations of cross-linker, bisacrylamide. PAAm hydrogels have been used in immobilization of various types of cells (Plieva et al. 2008; Guo et al. 2014; Rana et al. 2016). In one of the reports, PAAm-alginate beads have been employed to encapsulate human bone marrow-derived mesenchymal stem cells (hBMSCs), wherein the authors demonstrated viable cells with round morphology encapsulated within PAAm-alginate beads in comparison with the spindle cell morphology when seeded on a flat 2-D substrate. The authors demonstrated cell encapsulation within PAAm-alginate beads as an opportunity to immobilize the stem cells at the site of injury during regeneration therapy (Rana, Tabasum and Ramalingam 2016).

A vinyl polymer, ***polyvinyl alcohol (PVA)***, is a less expensive polymer that is non-toxic to the cells. Generally, PVA can be used for the entrapment of living cells during the gelation process that is achieved by freezing followed by thawing; the strength of gel can be enhanced by repeated freezing and thawing process. PVA can produce a rubber-like, elastic hydrogel without using any external chemical agent in the above process of during gelation. However, sometimes, repeated freezing and thawing can compromise the cellular activity of the entrapped cells. However, it can be overcome by the addition of cryoprotectants such as glycerol. Further, elastic PVA gel with high strength and durability can be formed by cross-linking with the boric acid solution without compromise on the cellular activity. Nevertheless, owing to the highly acidic nature of the cross-linking agent, boric acid, the cell viability sometimes gets affected. Another challenge associated with PVA is its highly viscous nature that results in agglomeration of PVA beads and affects the performance of fluidized bed reactors. This is solved by using a mixture of PVA and boric acid with a small amount of calcium alginate for hardening the mixture of PVA. Sodium alginate with a mixed solution of boric acid and calcium chloride has been reported for the entrapment of *Pseudomonas* cells in PVA gels (Ting and Sun 2000). In another study, the hydrogel based on PVA and poly(2-methacryloyloxyethyl phosphorylcholine-*co*–*n*-butyl methacrylate-*co*-*p*-vinyl phenylboronic acid-*co*–*N*-succinimidyloxycarbonyl tetra(ethylene glycol)methacrylate) (PMBVS) were used for cell immobilization. Further, stimulation of encapsulated mouse fibroblast L929 cells with basic-fibroblast growth factor (b-FGF) conjugated to PMBVS resulted in ECM infiltration followed by generation of a hybrid hydrogel based on ECM and encapsulated cells (Zhang et al. 2019). In a recent study by Zhao et al., PVA hydrogels were cross-linked with 4-carboxyphenylboronic acid (CPBA) to generate a borate bond as well as ionic interaction in the presence of multivalent cations. Hydrogels of tunable mechanical strengths were generated and demonstrated potential in cartilage repair (Zhao et al. 2018). ***Polyurethanes*** consist of isocyanate group at both the terminals of a linear chain and are synthesized at a high temperature through polyaddition reaction of organic compounds with reactive hydroxyl groups, such as di/polyisocyanate with a di/polyols (of aromatic and aliphatic nature). Cell entrapment within polyurethane gels can be achieved by mixing the water-miscible pre-polymers with aqueous cell suspension; here, the terminally located isocyanate groups react to form urea linkages resulting in gel formation. Cell entrapment in the pores of polyurethanes can also be obtained by polycondensation of polyisocyanates. In one such study, the polyurethane scaffolds seeded with mesenchymal stromal cells were investigated for the repair of a critical defect in avascular meniscal lesions. In vivo studies demonstrated the integration of scaffolds and stable healing of approximately 7 mm broad meniscus lesions (Koch et al. 2018). The study further demonstrated polyurethane-mediated vascular ingrowth and faster healing of meniscus lesions in the presence of MSCs seeded polyurethane scaffolds.

Furthermore, hydrogels based on ***methacrylates*** have been widely applied in cell immobilization studies. Here, the gel-forming agent, the pre-polymer, is cross-linked with the help of illumination, UV light in most cases. The method of entrapment of cells varies with the selection of suitable resin with near UV light initiating free radical polymerization of the pre-polymer, and the complete gel is formed within 3–5 min. Different types of pre-polymers having a photosensitive functional group have been developed (eg., polyethylene glycol dimethacrylate, gelatin methacrylate). Entrapment of cells within the network of gel can be achieved by irradiating the mixture containing photosensitizer (Benzoin ethyl ether), the pre-polymer and cell suspension (Rosiak and Ulański 1999).

In a study, the photo-encapsulation of swine auricular chondrocytes in a polyethylene glycol dimethacrylate hydrogels led to neocartilage generation, its integration with the native cartilage and also demonstrated potential in in vivo chondrogenesis (Papadopoulos et al. 2011). In another study, Krishnamoorthy et al. studied the effect of encapsulated 3T3 fibroblasts on the physicomechanical properties as well as microstructure of gelatin methacrylate hydrogels (Krishnamoorthy et al. 2019). Further, Sawyer et al. generated electrically conductive hydrogels based on gelatin methacrylate and poly(aniline) with osteoid-like soft mechanical properties. Further encapsulation of human osteogenic cells resulted in bone mineralization activity; however, the mineral content was lower than gelatin methacrylate hydrogels. The study demonstrated the potential of printing these hydrogels into complex 3-D structures by digital projection stereolithography (Sawyer et al. 2018).

Cell Entrapment in Preformed Scaffolds

The previous section discussed cell encapsulation within hydrogels that were generated post cell immobilization. This section will discuss the cell immobilization in preformed support materials. There are certain prerequisites of an ideal preformed support material suitable for cell immobilization. In short, the support material should be porous in order to allow cell entrapment. Further, the porous carrier should have the desired pore size for cell seeding. The preformed carrier for immobilization involves passive natural immobilization usually performed in a bioreactor or culture medium (Mavituna 2004). In this, the cells are usually added to the sterile medium containing empty preformed carrier; the cells get entrapped via shear forces within the bioreactor. In the second type of immobilization, cells are seeded onto the pores in an adsorption method, and during this, the cells migrate through the tortuous pores of the porous substrate. The nutrient and mass transport to the cells are achieved by diffusion and convection process (Rouwkema et al. 2009). On porous matrices, cell immobilization methods are simple and usually have a high degree of cell viability. The cell packing density in a porous matrix increases due to surface attachment and self-aggregation of the cell. Various forms of preformed matrices have been used for cell immobilization for tissue engineering applications. On the basis of the geometry of the tissue need to be engineered, the preformed scaffolds could be designed from different methods. There are several types of scaffolds such as *sponge or foam porous scaffolds, solid free-form scaffolds, microspheres/microparticles and fibrous scaffolds* (Nikolova and Chavali 2019). These scaffolds being porous in nature as well as high surface area facilitate cell immobilization by virtue of surface properties in the pore volume. Preformed scaffolds have been explored for several tissue engineering applications. Few of the methods that are commonly used to fabricate scaffolds are electrospinning, salt leaching, 3-D printing by fused deposition molding, freeze drying and cryo-gelation. Cell seeded scaffolds have been used in orthopedic, dental, connective and soft tissue engineering (Nikolova and Chavali 2019). For instance, the fibrous scaffolds of PLGA-silk fibroin blend generated by electrospinning and decorated with HAp demonstrated the potential to support adhesion, proliferation and differentiation of MSCs to osteoblastic lineage (Gao et al. 2018). In another study, hMSCs were extruded with biomaterials like gelatin, alginate and HAp; the scaffold supported the cell viability till 3 days (Wüst et al. 2014). Further, macroporous 3-D scaffolds based on biomaterials such as agarose-chitosan composite, agarose-gelatin and gelatin-HA-alginate composite scaffolds, were used for various tissue engineering applications (Tripathi et al. 2009; Singh et al. 2014; Tripathi and Melo 2015). To add to this, similar to hydrogels, the preformed matrices can also be made up of various natural and synthetic polymers (Loh and Choong 2013).

3.1.3 Containment Behind a Barrier

Containment of cells behind a barrier is usually achieved by attachment of cells on a preformed membrane, cell entrapment within a microcapsule or through cell immobilization on the interface between two immiscible liquids. This type of immobilization is preferred when cell needs to be separated from its effluent and for production of cell-free specific product. These are also beneficial in applications where the cells and their products are separated through a physical barrier, such as during product inhibition and water-insoluble substrates. Different existing polymeric barriers such as membrane system or hollow fiber system and capsule have been reported (Storm et al. 2016). For application of membrane systems in micro and ultrafiltration, synthetics membranes are used. Mass transfer through these membranes largely depends on the pore size, charge as well as the hydrophobicity/hydrophilicity. Further, the transport is usually through diffusion or as a result of pressure difference-induced flow. Additionally, cell entrapment through these membranes does not involve harsh chemical agents. An advantage of using cell entrapment on membranes is the possible recycling of cell-containing phase, since the cell-containing phase and its products are separated through a barrier. This might be difficult with other immobilization strategies.

In case of hollow fiber systems, mass transfer is as a result of convective flow, however, there are possibilities of pore clogging of the barrier thereby leading to a reduction in the efficiency of cell product and waste removal at the outlet. Further, the interaction between binary fluids used in a cross-flow system across the barrier can cause unintended reactions (Kourkoutas et al. 2004). In case of hollow fiber systems, the cells are either immobilized onto the membrane support layer, wherein the medium flows in the lumen space or the cells can be present within the lumen space. Hollow fiber membrane bioreactors have been explained in the later sections.

Microencapsulation

During microencapsulation, cells are immobilized in the hydrogel and can be protected from the harsh environment such as extreme pH, temperature, toxic compound or solvents. A membrane can then be deposited on this hydrogel. Further, the hydrogel can also be liquified eventually resulting in membrane and encapsulated cells. The polymer constituting the membrane can have porosity based on the size of the molecule to be permeabilized. Further, resistance by the membrane should not affect the mass transfer activities and should be made up of biocompatible materials. Semipermeable membranes for microencapsulation can be generated using coacervation, interfacial polymerization, pre-gel dissolving and liquid droplet forming (Park and Chang 2000).

Microencapsulation of human cells has been used for many applications. In particular, it is extremely useful during transplant, wherein the membrane can create a barrier between the host cells and transplanted cells. Such kind of immune-isolation not only prevents the rejection of the transplanted cells but can also prevent the use of immunosuppressive drugs.

Cell Immobilization Using Membranes

Cell immobilization on membrane bioreactors is advantageous since it provides combinatorial effects including bioconversion and product separation and is well suited for high-value biological products. However, for the production of low-value biological products, conventional bioreactors can be used because of the complexity and expense associated with the membrane bioreactors. Membrane bioreactors can be either used as a flat-sheet or as hollow fiber modules. While hollow fiber modules provide a high surface area to volume ratio for cell immobilization, the configuration of flat-sheet modules is simple due to control over distance between the membranes.

Cell immobilization using membranes can either immobilize the cells within the membrane, immobilization of the cells on the membrane and immobilization in the cell compartment. In a study, an artificial artery was fabricated by Janke et al., wherein bioreactor made of hollow fiber membranes was used. These membranes were immobilized with endothelial cells (HUVECs) on the inside and smooth muscle cells (HUASMCs) on the outside. The authors demonstrated that arterial functional characteristics such as flow conditions (0.1 and 3 N/m^2), metabolic exchange as well as the cross-talk between HUVECs and HUASMCs through hollow fibers recapitulated the vascular system, similar to in vivo physiological conditions. This model demonstrates potential as a screening model for various small molecules (Janke et al. 2013).

3.2 *Self-aggregation of Cells*

Various cells can naturally aggregate and form a pellet/flocculate; this is also termed as a type of cell immobilization process. Although there have been various industrial applications of self-aggregation using culture of fungal, microbial or yeast aggregates, self-aggregation has also been used in animal cell culture. Numerous medicinally important products such as antibiotics are produced during the secondary metabolism of the aggregated fungal cells. The yeast cells aggregate and flocculate in the last step of fermentation during many industrial applications including wine making and brewing (Willaert and Baron 1996). Aggregation occurs because of many biological (secretion, genetics, growth rate, nutrition, strain type), physical (ionic, temperature, interfacial forces, hydrodynamic properties) and chemical (chelating agents, enzymes, trace metals, carbon-to-nitrogen ratio) factors influencing the cell wall/membrane region of the cells. Further, the process of cell aggregation can either be natural or artificially induced.

Self-aggregation has been well reported in case of cartilage tissue engineering. In particular, there have been studies for micromass culture or pellet culture of prechondrocytes or mesenchymal stem cells to induce chondrogenesis resulting in the formation of cartilage-like tissue (Zhang et al. 2010). These culture systems provide three-dimensionality to the cells, which is crucial for cell–cell and cell–matrix interaction during chondrogenesis.

In conclusion, this section summarizes various types of cell immobilization techniques by various support materials and self-aggregation. Apart from the techniques mentioned in this section, mammalian cell immobilization can also be achieved through bioreactors for large-scale applications. For long-term culture of immobilized mammalian cells, bioreactors are a better option, as the regulation of diverse growth conditions for different cells can be achieved. The next section will describe various types of bioreactors for mammalian cell immobilization.

4 Bioreactors for Cell Immobilization

Bioreactors are vessels used for culturing microbial/plant/mammalian cells and are mostly used for the production of various biochemicals (Bhatia and Bera 2015). Bioreactors are of various types ranging from batch reactors that allow static culture of cells in a vessel or a continuous flow reactors where the media flows through a bed of immobilized microbial cells/enzymes/plant cells in a perfusion mode (Antolli and Liu 2012). Apart from the production of various biochemicals, bioreactors also find application in tissue engineering. The cell survival, tissue structure, tissue organization as well as tissue function are addressed through immobilization of cells in a matrix within a bioreactor (Blose et al. 2014). In general, mammalian cell immobilization not only entails cell trapping in pores of the substrate but also in providing effective surface area for attachment, migration, cell-to-cell communication, proliferation as well as differentiation. Thus, the design of a bioreactor for mammalian cell culture should be conducive to the aforementioned downstream processes. Several parameters such as pore size, perfusion system, thermal, chemical and mechanical properties as well as the surface topology of the matrix have been shown to play a role in these processes. As an example, in order to pass the culture medium forcefully through the pores of solid porous 3-D scaffolds, perfusion systems are used, which enhance the nutrient transport and also provide the mechanical stimuli to the cells, not only at the periphery of the 3-D scaffolds but also within the internal pores (Yates et al. 2012; Salehi-Nik et al. 2013). Mechanical properties such as shear stress have been shown to have a dominant impact on cell function and viability inside a bioreactor. There are different reported values for the maximal sustainable shear stress for different types of cells (Zoro et al. 2008; Bayati et al. 2011). Further, during the cell growth in a bioreactor, the mechanical interaction between cells, water and scaffold material determines whether cells would form cell aggregates or will be dispersed throughout the scaffold (Lemon et al. 2006; Vatsa et al. 2007; Bacabac et al. 2008).

Therefore, the choice of the bioreactor is determined by four basic considerations: (a) types of cells to be immobilized for requisite downstream application, (b) technique used for cell immobilization process (c) behavior and properties of the cell aggregates (d) hydrodynamics and mass transfer of the bioreactors (Bai et al. 2011). Figure 3 and Table 2 describe various types of bioreactors along with their advantages and disadvantages. This section will discuss various bioreactors that have been used for cell immobilization for tissue engineering applications.

4.1 Stirred Tank Reactor (STRs)

Stirred tank reactors (STRs) are the most common type of bioreactors that are used for culturing immobilized cells (Gao 2012). They are typically of three types, namely, continuous (CSTRs), batch and fed-batch (Ge et al. 2016). While in a batch mode-bioreactor, the substrate concentration may increase over time leading to inhibition of cellular activities, fed-batch and continuous bioreactors do not show substrate inhibition on cells (Kanojia et al. 2017). Connecting CSTRs in series leads to proper mixing in each reactor, complete utilization of the volume for reaction and avoid any dead space (Denbigh and Turner 1984). As a result, effective seeding of cells on the scaffold is the major advantage of this type of bioreactor; this is achieved using spinnerets (Ellis et al. 2005).

In general, STRs have a small height-to-diameter ratio as compared to other bioreactors, and the tank diameter is approximately 10 m for industrial applications. The core component of the STR is an agitator/impeller, which is mainly responsible for heat and mass transfer, aeration and homogenization as well as for maintenance of fluid dynamics. Conventionally, there are two types of agitators/impellers, radial and axial (Wang and Zhong 2007). The choice of agitator depends on the reactor

Table 2 Classification of bioreactors, their advantages and disadvantages (Margaritis and Wallace 1984; Kargi 1992; Antolli and Liu 2012; Sen et al. 2017; Khan et al. 2019)

Classification of bioreactors for immobilized cell systems			
Reactor Type	Advantages	Disadvantages	Applications
Mixed, suspended particles			
Stirred tank bioreactor	• Flexible • Variable mixing intensity • Suitable for high viscosity • Temperature control with exothermic reactions • Sterile conditions can be maintained	• Costly • High power consumption • Shear damage to the matrices • High weight of the bioreactor • Need of shaft seals and bearings	• Biomass production • Mostly used for processing of antibiotic, citric acid, exo-polysaccharides, cellulose, preferred in solid state fermentation etc. • Pharmaceutical production and large-scale production
Airlift bioreactor	• No moving parts • Low shear and energy demand, simple in construction and easy to handle • Easy sterilization • Good temperature control, easily adapts to two phase runs • Low operating (labor) cost	• Low mixing intensity • High amount of air and high pressure needed • No breaking of bubble and foam due to absence of shaft and blade, etc	• For animal cell culture • Production of single cell proteins (SCP), Enzyme, secondary metabolite, bio surfactant

(continued)

Table 2 (continued)

Classification of bioreactors for immobilized cell systems

Reactor Type	Advantages	Disadvantages	Applications
Fluidized bed	• No moving parts • Simple to handle • Good heat transfer • Variable mixing characteristics for liquids and solids. Uniform particle mixing due to fluid like behavior inside the reactor • Easy to scale up and reactor can operate in continuous state • Uniform temperature gradients	• Difficult matching of feed and fluidization rate • Requirements on particle density (dense support) • Good local mixing intensity • Only for low viscosity • High cost-due to increased vessel size • Requires high maintenance • associated with loss of fluidization pressure leading to runaway reaction	• Tissue engineering (for example Hybridoma technology and mammalian cell culture) • Production of ethanol, secondary metabolite and waste water treatment • Oil decontamination of sand and radioactive waste solidification • Coal gasification and nuclear power plants

(continued)

Table 2 (continued)

Classification of bioreactors for immobilized cell systems

Reactor Type	Advantages	Disadvantages	Applications
Fixed particle or large surfaces			
Packed bed reactors	• Simple, low cost • Immobilized enzymes can be reused • Continuous mode of operations, low substrate and product inhibition, high yield of desired product • Ease of operation • High mass transfer and reaction rate • Constant product recovery	• Plugging by solids at low flow rates • PBR has high temperatures due to exothermic reaction which affect sensitivity of the reactor • Simultaneous diffusion of substrate and products at the time of reaction • Difficulties in growing some cells at high density • Limitation related to developing complex bioartificial organ in ideal conditions	• PBR as bioartificial organ (bioartificial liver) and tissues • Widely applied in the dairy, food, beverage, and agricultural industries to manufacture nutraceuticals as well as wastewater treatment • For ethanol and biodiesel production • Bio energy generation
Membrane reactor	• High cell densities • High productivities • Perfusion operation possible • Simultaneous product separation • Separate feed of Gas and Liquid • Low shear • Continuous removal of toxic metabolite and addition of nutrient • Gentle hydrodynamic condition	• Sterilization problem • Microbial damage (perforation of membrane) • Low capacity • High Cost	• For tissue engineering • Pharmaceutical production • Fermentation process • Waste water treatment • Bioconversion of agricultural residue

dimension, viscosity, volume, density and nature of media to be dispersed within the bioreactor (Jagani et al. 2010).

Stirred tank reactors find application in large-scale stem cell production and immobilized anaerobic cell cultures for production of chemicals and fuels (Zhu 2007; Petry et al. 2018). Olmer et al. expanded the static culture of human pluripotent stem cells and human embryonic stem cells to suspension culture in stirred tank bioreactor for scaling up the application (Olmer et al. 2012). In another study, Lock et al. worked on pseudo islets in stirred tank bioreactor. As a result, the formation of beta cells pseudo islets and insulin secretion was found to be enhanced as compared to static culture due to an increase in mass transfer (Lock et al. 2011). In yet another study, Papas et al. developed a stirred micro-chamber for rapidly measuring the oxygen consumption rate (OCR) in pancreatic islets with high precision and demonstrated that measured OCR was directly proportional to cell viability. Hence, development of such STR-based micro-chambers would monitor the oxygen consumption and would suggest the quality of tissue before transplantation (Papas et al. 2007).

In conclusion, STRs have shown to demonstrate more substrate contact, pH and temperature control, removal of toxic by-products, uniform distribution of cells in the bioreactor and reduction in particle size due to shear stress applied via agitation and blending (Kadic and Heindel 2010). However, this type of bioreactors exhibits damage to the matrix due to shear stress and is associated with high cost.

4.2 *Fluidized Bed Reactor (FBR)*

Fluidized bed gas generator was developed by Fritz Winkler in Germany in 1920 and was initially used in the petroleum industry (Tavoulareas 1991; Gonzlez et al. 2012). This is a continuous type of bioreactor with multiphase reactions and is typically used for bioprocesses that require short reactor residence time. FBR system consists of a cylindrical vessel through which nutrient flows from the bottom through the carrier matrix; carrier possesses specific gravity sufficient enough to remain suspended during the flow. The cell system usually used for fluidized bed reactor (FBR) is immobilized bacteria, yeast and other fungi. Herein, small granular biocatalyst particles are used for fluidization that remains highly stable even during large shear forces for long time periods. As a result, FBRs are preferred where there is a high demand for liquid mixing for application with high cell density.

FBRs have been used for scaling up the mammalian cell culture and hybridoma culture for by-products and bioartificial organs. In a study, Detzel et al. developed a centrifugal bioreactor with a fluidized bed for culturing mammalian cells with density up to 0.1 billion cells/ml. This type of bioreactor modification provided optimal conditions for cell growth and proliferation (Detzel et al. 2010). In another study, Ray et al. cultured mouse hybridoma cells and Chinese hamster ovary cells in collagen microsphere in an FBR to determine the effect of dilution rates on immunoglobulin production, glucose consumption and ammonia production. It was seen that increasing dilution rate increased the bioreactor productivity of the bioreactor (Ray et al. 1990). Recently, a modified FBR with encapsulated hepatic cells was developed by Lu et al. as a potential alternative to hollow fiber membrane-based bioreactors with improved mass transfer (hollow fiber membrane bioreactor will be explained later in the chapter). Herein, the authors modified the reactor as diversion-type microcapsule-suspension fluidized bed bioreactor (DMFBR) to overcome the damage to cell-containing microcapsules and large void volume of usual FBR. Alginate/chitosan microcapsules of C3A hepatic cells were found to be more viable in DMFBR compared to FBR, which could be an alternative option for bioartificial liver (Lu et al. 2016).

Although FBRs have no agitators, they exhibit moving parts and have good heat transfer and uniform particle mixing; however, the reactors are high maintenance, and it is difficult to match the feed and the fluidization rate.

4.3 Airlift Bioreactor (ALR)

Airlift bioreactors (ALRs) are relatively new type of fermenters and can be used for aerobic fermentation, especially when the reactants or final products are in a gaseous state. In airlift bioreactors, the mixing of culture broth is achieved with the help of compressed air, which is provided via airlift pump situated at the base of the reactor. The compressed air then mixes with the media, and this mixture (gas/liquid) that is less dense than the remaining media, gets transferred upward via a discharge tube. Since ALR has no stirrer to achieve mixing, mixing is achieved by the air bubbles generated through air sparging by an air compressor. These generated air bubbles help in turbulent liquid mixing as well as mass transfer. ALRs can be designed on the basis of two different configurations, external and internal loop of baffles or tubes. For the internal loop ALR, flow is controlled by the executing barrier (tube) which forms the channel for circulation for better liberation of gases in liquid, while in case of an external loop reactor, circulation takes place by distinct connections.

Airlift fermenter was successfully scaled up by ICI England, wherein they used methanol as a substrate (*US4910228A—Methanol—Google Patents*, 1990). For tissue engineering applications, Bugarski et al. worked on mouse hybridoma cells encapsulated in alginate-PLL microcapsules for production of a monoclonal antibody, IgG. These microcapsules contained million hybridoma cells/ml which were loaded into an ALR and the authors demonstrated that microencapsulation in ALR provided retention of cells and IgG with diffusion of nutrients, metabolites and oxygen (Bugarski et al. 1999). Another inventor Philip C. Familletti patented an apparatus, which is a modified version of ALR in which the cell immobilization was performed by the microbeads comprising alginate and gelatin. This modified system improved the cell density and increased gas absorption from liquid growth media into the matrix within the bioreactor (*US5073491A—Immobilization of cells in alginate beads containing cavities for growth of cells in airlift bioreactors—Google Patents*, 1991). ALRs with modified air sparging meet the requirements for those animal cells whose culture is limited with shear sensitivity in suspension. However, this type of bioreactor has low mixing intensity and therefore requires high air pressure.

4.4 Packed Bed Reactor (PBR)

Packed bed reactors (PBRs) have a relatively simple design consisting of a column packed with biocatalyst and continuously perfusing liquid phase. Herein, the cells are immobilized on large particles, which are not affected by the liquid phase. PBRs are usually preferred when long resident time is required to generate less biomass (Scott 1987). The design consideration of PBR must include the optimization of bed structure, packing density, carrier type based on the mass transfer, heat transfer and biochemical reactions involved in the reactor. In a few cases, the pressure drop also needs to be adjusted based on the length of the bed in a reactor. The pressure drop can be reduced by using larger catalyst particles, but this causes lower intraparticle diffusion, resulting in the slow progress of the reactions (Bey and Eigenberger 1997). Further, the design of PBR can be also modified by using horizontal packed beds having a dry or gas-phase system or multiple columns in sequence. In this regard, gas sparging in a packed bed has been tested to remove the fermentation product (*US4032407A—Tapered bed bioreactor—Google Patents*, 1977). The cell systems used that have been typically used in these types of bioreactors are immobilized bacteria, yeast, fungi and plant cells. These types of bioreactors are also employed for bioengineering applications such as culturing mammalian cells apart from the production of medicinal products, enzymes and ethanol and have demonstrated high productivity in a compact size bioreactor (Meuwly et al. 2007). These types of reactors are mostly used for studying cellular processes during immobilization and commercial production of pharmaceutical products. The cells are immobilized onto appropriate matrices such as macroporous microcarriers, porous ceramic beads and porous glass beads that act as a bed in a packed bed bioreactor (Warnock et al. 2005). Higher porosities of bead sphere enhance the surface-to-volume ratio to support the proliferation of cells to increase cell density.

In one example, Osmokrovic et al. cultivated murine BMSCs in the form of alginate beads for cartilage tissue engineering using a PBR. Here, the alginate beads were prepared by electrostatic droplet generation, and the medium perfusion flow was maintained at around 100 μm/s, which is physiological to cartilage tissue. This study indicated that PBRs with immobilized alginate beads can be a potential platform for maintaining the cellular microenvironment for cartilage tissue engineering (Osmoković et al. 2006). Another study was performed by Highfill et al. toward the expansion of haematopoietic cells in PBRs. The authors tried to set up a large-scale production for murine bone marrow cells, where the stromal cell-conditioned medium was used in a bioreactor, which led to haematopoiesis. This process of perfusion culture led to the production of 3.6×10^8 cells in a span of 11 weeks (Highfill et al. 2000).

Ahrens et al. developed a perfusion system based in-house bioreactor for studying the efficacy of biotinylated poly(ethylene terephthalate) meshes to capture avidin-coated microspheres; this could be extrapolated to mammalian cell capture (Ahrens et al. 2019). PBRs have also been adapted for developing bioartificial livers. One such study was done by Batch et al. in which authors cultured primary rat hepatocytes in PBR with internal silicon tubing. Here, the authors studied the cellular functionality over a period of 48–148 h. The result showed that in 48 h, the cell viability, urea and albumin production increased, while there was a decrease in the values when cultured for 148 h (Bratch and Al-Rubeai 2001).

PBRs are a very promising tool for cell immobilization applications in tissue engineering. Due to the immobilization of cells in the porous matrices of PBRs, various types of cells can be grown for a long duration with less shear stress on the cells. However, there are some limitations to this type of culture setup, such as non-uniform fluid flow across the reactor and limited mass transfer. PBRs are also associated with issues of cell harvesting during ex vivo cell expansions (Nielsen 1999). The limitations of these bioreactors were overcome by a hollow fiber membrane reactor with increased productivity and scaling up the culture (Warnock et al. 2005).

4.5 Membrane Bioreactor (MBR)

Membrane bioreactor (MBR) is an advanced type of bioreactor, wherein different configurations are employed for different applications (Krzeminski et al. 2017). However, this section will focus on the multi-tube type fibrous MBR. These fibrous membrane bioreactors were introduced in the 1960s for micro- and ultrafiltration applications. Primarily, they comprise of a flat-sheet type membrane made up of cellulose acetate and polysulfone with a pore size of 0.03–0.001 μm (Belfort 1989). Although hollow fiber type MBRs have been evaluated for plant cells, bacteria, yeast and enzyme culture (Wung et al. 2014), they have also been used for immobilizing mammalian cells for tissue engineering (Wang and Zhong 2007). The fiber type MBR has a high surface area to volume ratio; as a result, they enable enhanced cell attachment. They also enable homogeneous cell viability and distribution; fibrous membrane mimics the uniform extracellular matrix (ECM) that allows cells to produce their own ECM thereby enhancing cell proliferation as well as differentiation. As an example, the hollow fiber MBR has been used in culturing metabolic cells such as hepatocytes, which require high mass transfer capabilities as compared to other cell types (Attanasio and Netti 2017). Roberts et al. have also used a hollow fiber MBR for a pilot study to demonstrate the commercialization of embryonic stem cells for cell therapy. In this regard, the authors cultured 60 million stem cells on MBR and obtained 708 million stem cells after a period of 5 days (Roberts et al. 2012). MBRs have also been used to culture lymphocytes and osteogenic cell lines, and the results have demonstrated better understanding of tissue physiology, cellular metabolism and fluid dynamics in vitro (Gramer and Poeschl 2000; Gloeckner and Lemke 2001; Ellis and Chaudhuri 2007).

MBRs have also been used as bioartificial organs. In one such example, Oo et al. developed a hollow fiber membrane (HFM) bioreactor-based bioartificial kidney for extensive characterization of the human primary renal cells. The bioreactor comprised of single hollow fiber membranes with a skin layer as outer surface, made up of polyethersulfone/polyvinylpyrrolidone and suggested that these are more suitable as compared to membranes with inner skin layer. Further, they demonstrated successful generation of human primary renal proximal tubule cells (HPTC)-based functional epithelium in double-coated HFM bioreactors; however, differentiation of HPTCs needs to be addressed further (Oo et al. 2011). Further, in another study, improved gas exchange in the bioartificial lung was obtained following immobilization of carbonic anhydrase on conventional HFM; this increased the diffusion of CO_2 into fiber due to catalysis of bicarbonate ions in the blood by carbonic anhydrase. Authors demonstrated that this type of modification in fibers of HFM might lead to development of biocompatible and compact bioartificial lung (Oh et al. 2010).

MBRs are advantageous over the other bioreactors because of continuous removal of toxic metabolites and presence of gentle hydrodynamic conditions (Kargi 1992). However, the only disadvantage with this bioreactor is high investment and membrane cost and the regular replacement of the membrane (Kootenaei 2014).

5 Applications of Immobilized Cells in Tissue Engineering

This section will discuss about few applications of cell immobilization in tissue engineering (Ma et al. 2005; Cui et al. 2006; Ito 2007; Shachar et al. 2011).

5.1 *Recombinant Protein Production*

Cell immobilization has been used for the production of enzymes, alcohols, organic acids and recombinant proteins, which is dependent on their cell growth. Recombinant protein production has been developed as an alternative to the extraction of proteins from natural sources. It is mainly useful in biopharmaceuticals and enzymes production. Mammalian and insect cell cultures have been used for time to time to study the expression of heterologous proteins, such as hormones, enzymes, cytokines and antibodies, for research or human therapy (Wurm 2004; Drugmand et al. 2012). Mammalian/insect cells are preferred systems for the production of complex human proteins because these cells synthesize and secrete a wide variety of functional proteins and human-like complex glycan structures (Zhu 2012). Immobilization of mammalian/insect cells also protects the cells from shear damage occurring during the mechanical agitation and gas sparing in bioreactors. Various studies have been conducted on the immobilization of mammalian and insect cells for therapeutic protein production and are mentioned in Table 3.

Table 3 Immobilization of mammalian and insect cells (Shirai et al. 1988; Mano et al. 1992; Yamaji and Fukuda 1994; Lüdemann et al. 1995; Yamaji et al. 2006, 2000; Jen et al. 2000; Park et al. 2000; Balysheva et al. 2007; Shishido et al. 2007)

Cell Lines	Recombinant Proteins	Immobilised Approaches
BHK21, ψ2, and L929 cells	Erythropoietin	Entrapped in alginate gel particles
Chinese hamster ovary (CHO) cells	Tissue-type plasminogen activator	Adsorbed on glass beads (as microcarriers)
Hybridoma SS-3	IgG immunoglobulins	Immobilization in chitosan microcapsules and alginate polyethylene glycol microgranules
Mouse hybridoma cells	Monoclonal antibody	Entrapped in polyacrylate–alginate gels
Mouse myeloma MPC-11	Lactate dehydrogenase	Entrapped within porous polyvinyl formal resins (biomass support particles)
Murine hybridoma cell	Monoclonal antibody against penicillin-G-amidase	Immobilized in macroporous Siran(®)-carriers
Spodoptera frugiperda (Sf-9) cells	β-Galactosidase	Entrapped within reticulated polyvinyl formal (PVF) resin (biomass support particles)
Spodoptera frugiperda (Sf-9) cells	HSV-2 glycoprotein D (gD2)	Entrapped in silk fibroin hydrogel
Trichoplusia ni BTI-Tn-5B1-4 (High five) cells	L particles (L protein of the hepatitis B virus surface antigen)	Entrapped within porous polyvinyl formal resin (biomass support particles)

5.2 Regeneration of Bone-Like Tissue

Various approaches for bone regeneration involve either pharmacological approach such as administration of growth factors at the site of injury or through cell-based tissue engineering (Buza and Einhorn 2016). The aim of bone tissue engineering is to develop 3-D scaffolds, which can mimic the extracellular matrix and provide mechanical support for the regeneration of bone (De Witte et al. 2018). Such osteogenic scaffolds provide a surface for cell attachment and stimulate the functional bone tissue formation similar to in vivo through tailored biophysical cues to direct the organization and behavior of cells (Bose et al. 2013; Ferrand et al. 2014).

Biomaterials for bone regeneration are usually based on bone matrix components such as calcium phosphate, hydroxyapatite, silica and collagen/gelatin as these materials provide an in vivo mimicking environment for bone regeneration.

During the recent advances in the field of bone regeneration, the following points are considered (Ritz et al. 2018):

1. The choice of implant materials like steel or titanium versus new materials like polymers or composites.
2. The best possible molecular cues to induce vascularization and bone growth or pre-vascularization of the implant by cell-based tissue engineering.
3. Immobilization of cytokines or growth factors in the material

Various natural (alginate or agarose gel, gelatin/collagen) and synthetic materials (polyethylene glycol (PEG) and polylactic acid (PLA)/polyglycolic acid (PGA)) have been used for bone tissue engineering (Salinas and Anseth 2009).

In an example, Li et al. studied the effects of hASC-derived exosomes, which were immobilized on the polydopamine-coated poly(lactic-co-glycolic acid) (PLGA) scaffolds on the osteogenic proliferation and migratory capabilities of human bone marrow-derived mesenchymal stem cells in vitro. The results showed that the exosomes were slowly and consistently released from the scaffolds, and they significantly increased the bone regeneration capacity (Li et al. 2018). In another study by Ritz U et al., a co-culture setup of human osteoblasts with endothelial cells and stromal-derived growth factor (SDF-1) was performed, and the cells were immobilized by photo-cross-linking dextran-based hydrogels. These were then implanted into the calvarial defect of critical size in mice. The authors showed that bone regeneration was induced in the co-culture setup due to promotion in angiogenesis, which could be either due to SDF-1 or the monoculture of endothelial cells (Ritz et al. 2018). Another group of scientists immobilized the stem cells on titanium implants using human exfoliated deciduous teeth-conditioned medium. This resulted in the promotion of the bone formation and shortening of the osseointegration period (Omori et al. 2015).

Recently, studies have also been performed on injectable hydrogels based on synthetic biomaterials for bone tissue engineering. In a study, Jang et al. prepared an injectable hydrogel scaffold made of methoxy polyethylene glycol-b-polycaprolactone copolymer and encapsulated differentiated osteoblasts within the hydrogel. They demonstrated increased expression of osteonectin, osteopontin and osteocalcin (Jang et al. 2016). In another study, an *N*-isopropylacrylamide/gelatin microparticle-composite injectable hydrogel was designed, and the gelatin microparticles were incorporated in the hydrogel to enhance bony bridging and mineralization in the bony defect. Authors then encapsulated mesenchymal stem cells in the hydrogel and observed significant tissue infiltration and osteoid formation thereby suggesting that this hydrogel system facilitates bone ingrowth as well as integration (Vo et al. 2016).

5.3 Cell Immobilization and Cartilage Tissue Engineering

The need for cartilage replacement has arisen due to poor regenerative properties of cartilage tissue. To this, various traditional and alternative tissue engineering-based solutions for repair and regeneration of the cartilage can be adapted. Cartilage tissue engineering primarily focuses on repairing, regenerating and development of biomimetic substitutes for cartilage (Zhang et al. 2009). Cartilage tissue engineering is dependent on the appropriate cell selection, mechanically and biologically compatible scaffold for cell delivery and chondrogenic bioactive molecules toward the generation of a tissue construct (Kuo et al. 2006). In summary, scaffolds should aid adequate cell adhesion, proliferation, differentiation into cartilage phenotype, for example, the biomaterial used for supporting cartilage engineered tissue should have porosity around 90% for the diffusion of nutrients as well as removal of waste products (Stoop 2008). Most importantly, these materials should be biodegradable with sufficient mechanical strength and allow remodeling as the new tissue forms (Ahmed and Hincke 2010).

In a study, Hao et al. encapsulated sheep chondrocytes in injectable hydrogels consisting of chitosan, sodium glycerophosphate and hydroxyethyl cellulose. The construct was cultured for a day in vitro and then transplanted into the articular cartilage defects of sheep. It was observed that 24 weeks following implantation, the chitosan-based hydrogel accumulated cartilaginous matrix and repaired the defect completely (Hao et al. 2010). Likewise, Endrez M et al. fabricated the grafts of human autologous fibrin, PGA and human chondrocytes for regenerating the articular cartilage damage (Endres et al. 2007). Other types of hydrogels such as agarose, collagen, alginate, fibrin and hyaluronan have also been used to encapsulate mesenchymal stem cells (MSCs) and chondrocytes and have been shown to promote cell proliferation, differentiation as well as cartilage-based ECM production (Mauck et al. 2000; Yamaoka et al. 2006; Liao et al. 2007; Chung et al. 2009; Ho et al. 2010). In an interesting study by performed by Arya et al., the authors investigated the role of the stiffness of carboxylated agarose hydrogels and RGD integrin-binding peptides on the differentiation potential of human articular chondrocytes (HACs). The study demonstrated interplay between hydrogel stiffness and immobilized RGD peptides. As a result, the encapsulated HACs demonstrated enhanced chondrogenesis in RGD modified stiff hydrogels as compared to the softer ones. Moreover, this effect was not seen on unmodified hydrogels (Arya et al. 2019).

Although natural polymers are cost-effective, environment-friendly, non-toxic, renewable, they are highly biodegradable, which leads to rapid degradation (Dwivedi et al. 2020). The biodegradable synthetic polymers used in cartilage tissue engineering are poly-α-hydroxy esters, especially PLA and PGA (Yoon and Fisher 2006). However, it has been observed that synthetic scaffolds form poor cartilage construct due to poor cell immobilization and decreased production of cartilage-specific ECM (Chen et al. 2003). Surface modifications and immobilization of these biodegradable scaffolds lead to better cell adherence and biocompatibility of the polymer (Dhandayuthapani et al. 2011). In this regard, Chen et al. used electrospun PLA nanofibers with cationized gelatin for improving their compatibility to chondrocytes. In vitro studies indicated increase in the cell viability, proliferation and differentiation of rabbit articular chondrocytes on these modified scaffolds, and subcutaneous implantation of these scaffolds loaded with chondrocyte led to the formation of cartilage tissue in 28 days (Chen and Su 2011).

5.4 *Diabetes and Cell Immobilization*

Diabetic patients who suffer from various secondary complications have been shown to be benefitted by insulin administration. However, difficulties associated with obtaining insulin and its administration prompted the encapsulation of insulin-secreting cells (Al-Tabakha and Arida 2008). It was soon demonstrated that encapsulated insulin-secreting pancreatic islets cells (PICs) were superior in maintaining normoglycemia as compared to non-capsulated PICs. Unfortunately, availability of PICs is affected by donor availability. This has been taken care of by utilization of alternate cell sources such as pig islet cells, mesenchymal stem cells (MSCs), induced pluripotent stem cells (iPSCs) and human embryonic stem cells (ESCs) (Kim and Jun 2018). As an example, one study demonstrated the survival and insulin production by pig islet cells, which were transplanted about 10 years ago in human patients (Elliott et al. 2007).

Translation of preclinical immobilization/encapsulation techniques started as early as 1994, wherein islet transplantation was performed on a diabetic man (Soon-Shiong et al. 1994). Similarly, encapsulated porcine islets were transplanted into diabetic patients and showed no sign of porcine endogenous retrovirus. Further, human islets encapsulated into alginate beads have been used in various clinical trials in Italy and Australia (Calafiore et al. 2006; Tuch et al. 2009).

5.5 *Neural Differentiation*

Neural differentiation is the process when embryonic stem cells differentiate into the neural lineage, including functional neurons, glial cells as well as oligodendrocytes under appropriate culture conditions (Carpenter et al. 2001). Due to non-regenerative properties of brain cells, transplantation of stem cells is done which can later lead to neural differentiation (Sarnowska et al. 2013). However, most of the transplanted cells die because of immune rejection, hypoxia and oxidative stress (Sortwell et al. 2000; Emgård et al. 2003). Due to lack of contact with the ECM components, cells have been shown to undergo apoptosis (Frisch and Francis 1994). To overcome this, tissue engineering-based strategies are adapted for stem cell transplantation, such as entrapment of cells within natural or synthetic biomaterial or supportive 3-D biodegradable matrices, which can then be transplanted (Zhang et al. 2014). This section will discuss a few studies toward neural differentiation using tissue engineering and cell immobilization techniques.

A study by Franco et al. explored the potential of PEG hydrogels as an encapsulation system for immortalized neural stem cells. The authors produced emulsion using poly(ethylene glycol) diacrylate in mineral oil and photo-cross-linked it following the addition of a photoinitiator solution; later, the cells were combined with precursor solution, together resulting in the formation of a microsphere. To prove the adaptability of the system, authors selected three cell lines NIH-3T3 (fibroblasts), MHP36 (neural stem cells) and bEnd.3 (brain endothelial cells) relevant to neural stem cell therapies and investigated morphologies, adhesion preferences and MMP expression. This type of microencapsulation proved to be a good platform for microsphere production, and further modifications with cell peptides to hydrogels increased the cell spreading and migration of the encapsulated cells (Franco et al. 2011).

Research toward treating Parkinson's disease using encapsulated cells was started with rodents as a proof of concept (Tresco et al. 1992). In this regard, Terraf et al. performed a study in which the mouse embryonic stem cell lines were differentiated to dopaminergic (DAergic) neuron cells through their culture on 3-D Poly-ε-Caprolactone (PCL) nanofibrous scaffolds embedded in Matrigel along with specific signaling molecules. Mouse embryonic stem cell lines generated here co-expressed Nurr1 (transcription factor in DAergic neuron development) and GPX-1 (a neuroprotective enzyme against oxidative stress). Authors concluded that the 3-D scaffold promoted differentiation into functional DAergic-like cells and demonstrates potential in treatment of Parkinson's disease (Terraf et al. 2017). Similarly, there have been studies regarding the potential treatment of Alzheimer's disease by encapsulating genetically modified cells within polymeric hydrogels. In this regard, Winn et al. encapsulated modified BHK cells, which released human nerve growth factor, and transplanted them into both fimbria-fornix-lesioned rat brains and naive controls. Chronic release of neural growth factor from the encapsulated cells led to complete protection of axotomized medial septal cholinergic neurons which is classic in Alzheimer's disease. Authors also proved the long-term survival of these encapsulated cells (Winn et al. 1994).

In a study, collagen 3-D matrices supported the neural differentiation into cortical neurons, astrocytes and neural progenitor cells (Dhara and Stice 2008). In this regard, the comparative differentiation of neural precursor cells in 3-D collagen gels modified with poly(ornithine)/Laminin, growth factor-reduced matrigel and PuraMatrix® gels. The authors concluded that the two-fold significant change in enhancement of the neurogenic differentiation was observed in cells encapsulated in matrigel as compared to 3-D collagen gels and PuraMatrix® gels (Uemura et al. 2010). Li et al. concluded that the differentiation of neural stem/progenitor cells (NSPC) into various neural cell lineages such as neurons, oligodendrocytes and astrocytes was directly associated with the porosity of the 3-D scaffold. Toward this, the authors fabricated a porous methacrylamide chitosan scaffold, and NSPC were cultured on to them (Li et al. 2012).

5.6 Angiostatin Release for Cancer Therapy

Angiostatin is a naturally occurring circulating endogenous proteolytic fragment of a protein plasminogen, specific for angiogenesis inhibition generally derived from tumors. It inhibits blood supply to tumors at the primary and metastatic level and also suppresses endothelial cell proliferation. Due to potent antitumor activity, angiostatin is considered for clinical trials in cancer therapy (Cao and Xue 2004). In this regard, Visted T and his colleagues increased the production efficacy of encapsulated cells having antineoplastic properties. More specifically, they genetically modified HEK 293EBNA cells to produce angiostatin and optimized high yielding alginate microcapsules by varying bead size, cellular density, homogeneity and ion composition of gel for continuous and sustained release of recombinant angiostatin. Here, the authors concluded that the modification in the encapsulation procedure of alginate microcapsules demonstrates potential in cancer therapy (Visted et al. 2003). In another study, Cirone P et al. worked on comparative analysis of the effect of continuous release of IL-2 and angiostatin individually on tumor cells and found that IL-2 led to profuse inflammation, whereas angiostatin did not have long-term effect on tumor cells. They then co-encapsulated interleukin-2 fusion protein (sFvIL-2) and angiostatin secreting cells within alginate-poly-L-lysine-alginate microcapsule and implanted the same in B16-F0/neu tumour-bearing mice. They found a synergistic improvement in tumor suppression and improved survival rate in tumor-bearing mice as compared to single delivery of either IL-2 or angiostatin thereby demonstrating the potential of the established system in cancer therapy (Cirone et al. 2005).

Endostatin is another potent inhibitor of angiogenesis and tumor growth. Joki et al. encapsulated modified BHK endostatin releasing cells (0.2 million) in alginate-PLL microcapsules for delivery of endostatin in glioblastoma xenografts, and the authors demonstrated that local delivery of endostatin through microcapsules is an effective therapeutic approach for treating various tumors (Joki et al. 2001).

5.7 *Bioartificial Organs and Cell Immobilization*

Bioartificial organs are defined as designing, growing and maintaining the living tissue on a 3-D scaffold for performing complex biochemical functions like adaptive control and replacement of normal tissue (Prokop 2006). One of the most common and well-studied examples of bioartificial organs is the liver. Herein, the focus has been on culturing hepatocytes using PBRs (discussed earlier in Sect. 4.4) resulting in a bioartificial liver (BAL) device. Figure 4 shows the flow diagram for the operation of a typical BAL. BAL devices use metabolically active liver cells (hepatocytes), which are incorporated within the bioreactor; the metabolically active liver cells are then induced to perform the hepatic functions by processing the blood or plasma of patients with liver failure (Kumar et al. 2011). Since the proliferation of hepatocytes is very difficult in vitro, therefore, culture conditions, packing materials and entrapment procedures are important for immobilizing hepatocytes in PBRs (Ohshima et al. 2008). A BAL device is expected to perform the multiple functions of a liver including carbohydrate metabolism, synthesis of proteins, amino acid metabolism, urea synthesis, lipid metabolism, drug biotransformation as well as removal of waste.

There are various porous materials which have been used as immobilization matrices toward the construction of the PBRs for biomedical applications; these include glass beads, fibers, commercially available materials such as Fibra-Cel®, SIRAN®, ceramics, Cellsnow®, Cytodex 3, non-woven fibers (NWF), polyurethane membrane (PUM), polyurethane foam (PUF) and polyvinyl fluoride (PVF). The choice of immobilization matrices for BAL devices depends on various factors such as required pore size, immobilization efficiency and cell density. It has been reported that matrices such as PUM have close-ended pores which result in clogging; this can be solved by using open-pore matrices such as PVF & PUF (Meuwly et al. 2007). The first-ever BAL device was built using nonporous glass beads of 1.5 mm diameter and 30 ml of packed bed BAL. It had more than 80% immobilization efficiency and cultured hepatocytes for more than 4 days (Li et al. 1993). Further, Yanagi et al. developed PVF cubes-based BAL device which was a small-scale PBR with 2-4 ml packed bed volume and generated high cell densities of 1.2×10^7 within 26 h of the culture period (Yanagi et al. 1992). Further studies included hepatocyte culture for a period of nine days in serum-containing medium; the cells were able to retain the metabolic activities similar to in vivo (Yanagi et al. 1992, 1999). In later studies, Kurosawa et al. and Ijima et al. showed the use of PUM and PUF for achieving higher cell densities up to 10^7 hepatic cells. In case of PUM, the growth was performed in a serum-free environment for 1 week, and the expression levels of hepatocyte were maintained. On the other hand, PUF-based 300 ml BAL device was developed using 30 g hepatocytes and was compared with in vivo studies performed on dogs. It was found that such a module had to be freshly prepared every 24 h (Ijima et al. 2000a, 2000b; Kurosawa et al. 2000). In another study, the hepatocytes retained the metabolic activity for up to 40 days of culture. The authors used a matrix termed Cellsnow®, which is based on porous microcarriers. High cell density of 10^7 cells was obtained using 40 ml of packed bed volume (Pörtner et al. 2007). Recently, Seldom et al. designed and tested a clinical scale BAL using hepatoblastoma cell line. The cells were cultured as three-dimensional organoids in a fluidized bed bioreactor with single-use bioprocessing equipment. The nutrient supply was done with BioXpert recipe processes. This commercial device demonstrated good metabolic and phenotypic liver functions (Selden et al. 2017). Although there have been many studies working toward the generation of a successful BAL device, one of the major advantages of the system is the low availability of primary hepatocytes. However, the hepatocytes from discarded donor liver and porcine hepatocytes are good alternatives and have demonstrated potential as comparable counterparts toward the generation of BAL (Nicolas et al. 2017).

Apart from BAL, other bioartificial organs such as bioartificial kidney (Oo et al. 2011) and bioartificial lung (Oh et al. 2010) have been designed using hollow fiber membrane reactors toward the maintenance of appropriate biochemical function of the living tissues. However, replacing the primary cells in a bioreactor-based bioartificial organ with xenografts or stem cells would provide better future prospects for clinical applications.

5.8 Microfluidics for Cell Immobilization

Microfluidics may be defined as “the science and technology of systems that process or manipulate small (10^{-9} to 10^{-18} L) amounts of fluids, using channels with dimensions of tens to hundreds of micrometers” (Whitesides 2006). Current applications of microfluidics range from separation coupled to mass spectroscopy, high-throughput screening in drug development, bioanalyses, examination and manipulation of samples consisting of a single cell or a single molecule and synthesis of 18F-labeled organic compounds for positron emission tomography (PET) (Kang et al. 2008).

Microfluidics-based cell culture systems provide a better environment control over the traditional cell culture systems and also enable the same operations such as sub-culturing and changing the medium (Gómez-Sjöberg et al. 2007). 3-D cell culture integrated with microfluidics has been studied to demonstrate the improved biological relevance of microfluidic systems. Encapsulating the cells into hydrogels (natural or synthetic) is a common strategy to mimic the in vivo conditions. The gel embedded cells can easily be loaded onto the microfluidic devices; however, the system is associated with some limitations such as insufficient supply of oxygen and nutrients (Choi et al. 2007; Wong et al. 2008). In a study, Izuta et al. immobilized mammalian cells on the polyethylene glycol-coated surface of the substrate in a microfluidic device for enabling the rupture of the cells using shear forces and exposure of the cytoplasmic content as shown in Fig. 5. Authors then used the cytoplasmic cell membrane sheets for studying phosphorylation of tyrosine kinase receptors and enzymatic modification of G-protein coupling receptors. This platform can be used for studying the novel targets in the molecular mechanism of Juxta membrane, which might ultimately help in drug discovery (Izuta et al. 2017). In another study, a gel-free microfluidic 3-D cell culture system was developed using intracellular polymeric linker, polyethyleneimine-hydrazide (PEI-hy) and microfabricated pillar arrays. On this device, aldehyde groups of the sodium periodate treated cells (A549 & C3A) reacted with the hydrazides group to form cell aggregates which then accumulated by the micropillar arrays. The two different kinds of cells cultured on this gel-free system displayed 3-D cellular morphology and cellular differentiation capability, proving it to be a better versatile perfusion culture platform (Ong et al. 2008).

Immobilizing cells in a microfluidic system is important since it allows the alteration in the solution conditions within the microenvironment of the device without disturbing the immobilized cells. Rapid immobilization of cells in a microfluidic device has been performed using various acoustic forces; dielectrophoresis (DEP) trapping is one of the most promising methods (Çetin and Li 2011; Fernandez et al. 2017; Zhang H et al. 2019). In a study, Forey et al. immobilized the suspended mammalian cells in a microfluidic device using Di electrophoresis (DEP) forces with polyelectrolyte multilayers (PEMS) treated surface of the device for studying the mammalian neural cells differentiation process. This treatment enabled the cellular attachment of the mammalian cells even after the removal of DEP force which otherwise caused the detachment. Authors concluded that DEP trapping in combination with PEMS treatment is a useful strategy to manipulate the cell adhesion, hence making it a reliable platform to study the differentiation process of various cell types (Forry et al. 2006).

6 Challenges and Future Perspective

The crucial role of tissue engineering in medical and pharmaceutical fields is well documented. The role of cell immobilization in a bioreactor system to generate tissues for various biomedical and pharmaceutical applications is crucial in clinical biology for treatment of various endocrine diseases following cell replacement therapy by using either allograft or xenograft. Therefore, cell encapsulation inside a matrix holds significant promises to research and development of new technology.

Each application from biomedicine to bio-industry requires a peculiar specification of the polymeric material to be utilized for encapsulation of specific cells. The main challenge in cell immobilization is to choose the "right" material that is stable and can be stored appropriately during transportation. In general, the material should be non-toxic for the active encapsulation of cells while possessing the ability to withstand the mechanical stress toward the development of a functional tissue within a bioreactor (De Castro et al. 2005; Opara 2017). It is important to use biocompatible and semipermeable materials, which do not interfere with the biological activity of cells encapsulated within the polymer matrix, and also support the diffusion of nutrients/gases/waste in and out of the polymer matrix. Moreover, it is required to standardize the purity grade, composition ratio (in case of composite polymers), reaction conditions, physicochemical parameters and source reproducibility in order to avoid lab-to-lab variability (Orive et al. 2004). However, even though immobilized system consisting of the right kind of cells and scaffolds is generated, the industrialization or commercialization of the immobilized systems is generally challenging as compared to suspended free cultures. This is because the preparation and handling of immobilized cell systems are often complex, time-consuming and often non-economical. Another issue is the instability of the support matrices, which may lead to leakage of cells from matrices, resulting in loss of cells in high numbers. Finally, there is another limitation in the diffusion of oxygen and/or other nutrients to the core of complex 3-D polymeric matrices.

In recent times, capsule size used for cell encapsulation has been the focus of research, and smaller capsules demonstrate improved nutrient and oxygen supply resulting in enhanced cell viability as compared to larger capsules. Therefore, it is an important parameter to be considered during microencapsulation studies. Cell immobilization has a significant future in different areas of research, and the major future challenge is to make the technique more reproducible with regard to application in humans. Development of bioartificial organs is a recent application of immobilized cell-based tissue for efficient transplantation with safety and appropriate functionality (de Vos 2017).

The next step is to develop a special microcapsule that has multiple micro-compartments dynamically interconnected to each other and containing cells, which mimic primary cellular products. The final step is to continuously aim to improve the cell life expectancy within the polymeric scaffolds, accompanying the requisite functional activity and eventually a self-conserved cell regeneration capacity to address treatment for high social impact diseases (Orive et al. 2004; de Vos 2017). In this regard, the microfluidic device is a potential platform for cell immobilization, proliferation and differentiation toward the development of an organ-on-a-chip (Zhang et al. 2017). These devices are sensitive to the surrounding environment and highly precise in controlling the physical and soluble factors over space and time simultaneously (Zhang et al. 2017). Current microfluidics platforms, which are available, have usage to limited applications because of limited dimensions and requirement of controlled and standardized systems. Therefore, implementations of these systems with the integration of new materials would lead to development of commercial tools for cell immobilization-based tissue engineering applications.

7 Conclusion

Cell immobilization has demonstrated potential for manufacturing of various bio-based products such as recombinant proteins, monoclonal antibodies and immunoglobulins. This technology is advantageous over free cell systems and supports the operation of cell culture in continuous mode, especially in case of bioreactors (Ge et al. 2016). Mammalian cell immobilization within the 3-D environment provides long-term viability and cell function similar to in vivo and has demonstrated application in the field of tissue engineering. In this regard, cell immobilization by adhesion, matrix entrapment or microencapsulation helps us to understand the cellular development and expression of phenotypes for specific applications (Jen et al. 2000). The type of immobilization technique and the polymer in which the mammalian cells are immobilized are chosen based on the intended downstream application. In addition to the use of support materials such as 3-D scaffolds, bioreactors have demonstrated potential in the production of low as well as high-value bioproducts. Although cell immobilization on 3-D scaffolds finds application in the field of tissue engineering, one needs to consider various factors such as easily scale up, commercialization potential, FDA approval as well as acceptability by surgeons, patients and health-care providers. Nevertheless, the development of bioartificial organs opens avenues toward improved human life.

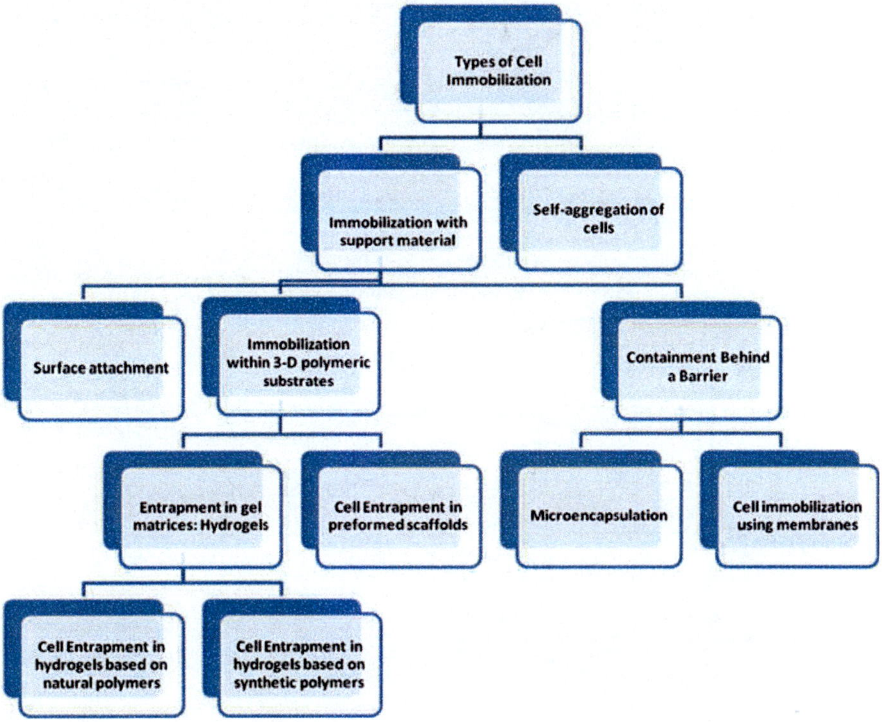

Fig. 1 Scheme depicting various types of immobilization in tissue engineering

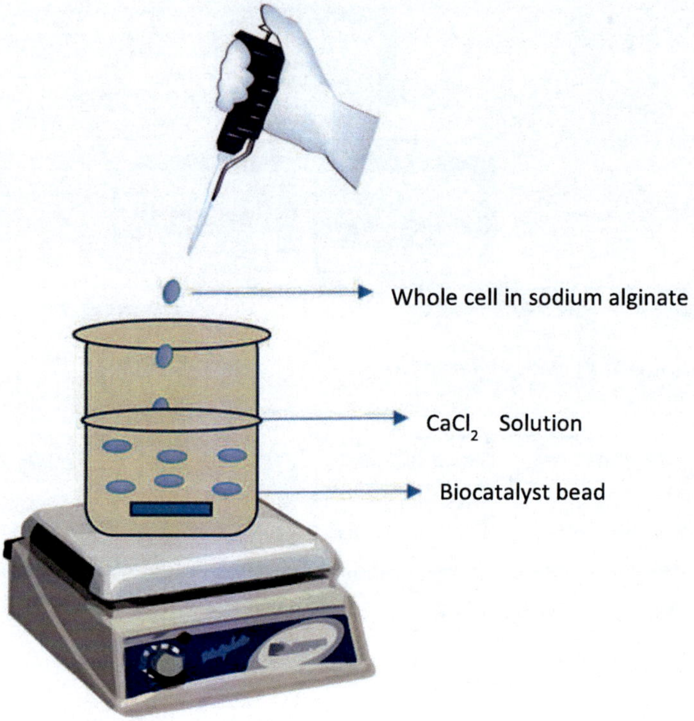

Fig. 2 Scheme representing cell entrapment into alginate polymer

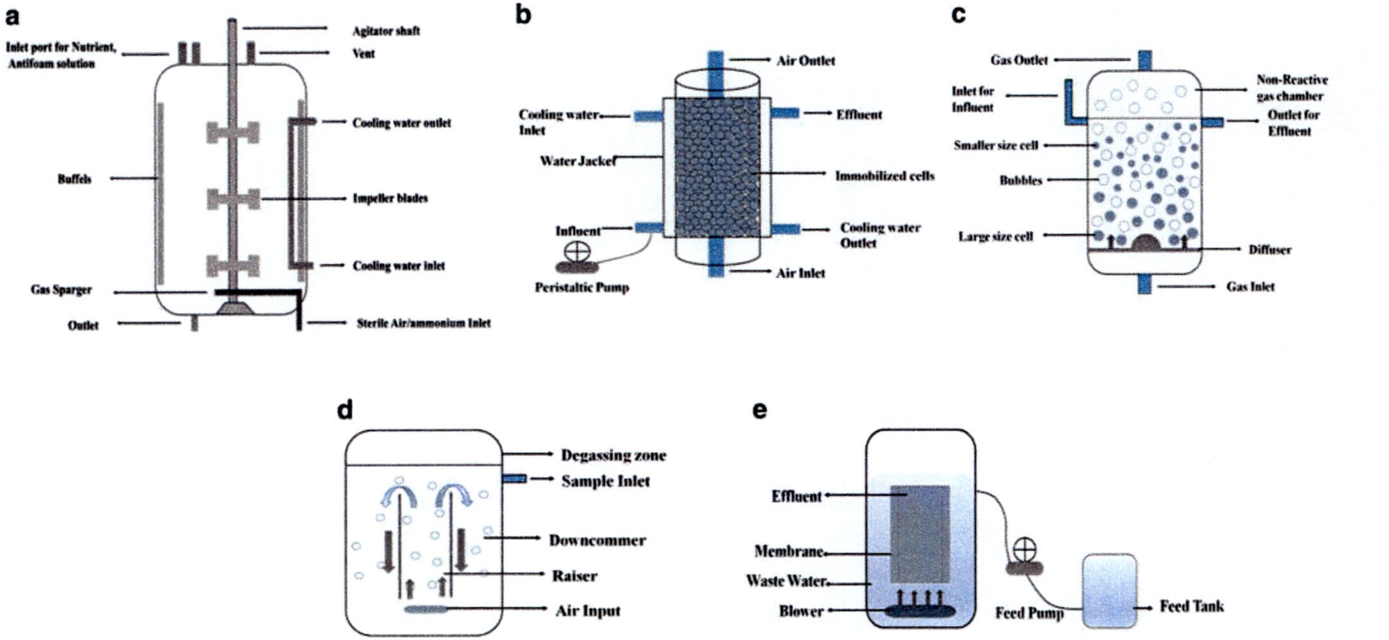

Fig. 3 Various types of bioreactors for cell immobilization. **a** Stirred tank bioreactor, **b** Packed bed reactor, **c** Fluidized bed reactor, **d** Airlift bioreactor, **e** Membrane bioreactor

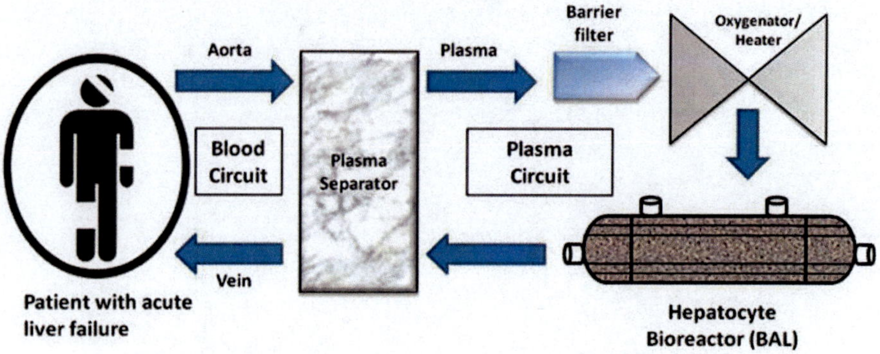

Fig. 4 Flow setup of bioartificial liver device connected to the patient

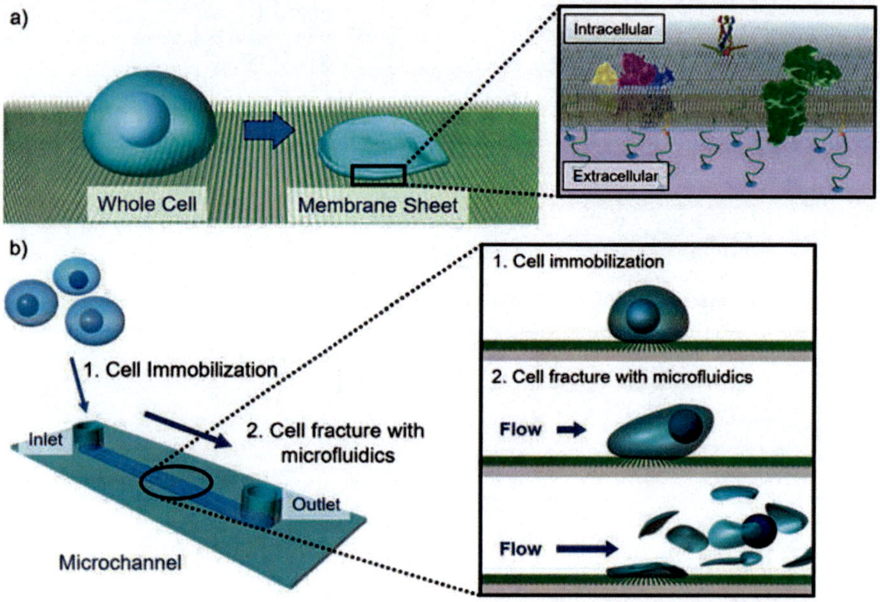

Fig. 5 Cell membrane sheets and a microfluidic device. **a** Fabrication of cell membrane sheets from live cells on a glass substrate and **b** Fabrication of cell membrane sheets from live cells using a microfluidic device. Briefly, bottom surface of microchannels was coated with lipids followed by immobilization of cells through interaction between cell membrane and immobilized lipids. These immobilized cells were then fractured via shear stress forces resulting in the formation of intact cell membrane sheets. Reproduced from Izuta et al. (2017) licensed under https://creativecommons.org/licenses/by/4.0/

References

Abercrombie M (1970) Contact inhibition in tissue culture, In Vitro. Soc In Vitro Biol 128–142. doi: https://doi.org/10.2307/4291530

Ahmed TAE and Hincke MT (2010) 'Strategies for articular cartilage lesion repair and functional restoration'. Tissue Eng Part B: Rev 305–329. doi: https://doi.org/10.1089/ten.teb.2009.0590

Ahrens L et al (2019) Biotin-avidin-mediated capture of microspheres on polymer fibers. Molecules. MDPI AG 24(11):2036. https://doi.org/10.3390/molecules24112036

Akhmanova M et al (2015) 'Physical, spatial, and molecular aspects of extracellular matrix of in vivo niches and artificial scaffolds relevant to stem cells research. Stem Cells Int doi: 10.1155/2015/167025

Al-Tabakha M, Arida A (2008) 'Recent challenges in insulin delivery systems: a review. Ind J Pharm Sci. 70:278–286. doi: https://doi.org/10.4103/0250-474X.42968.

Alinejad Y et al (2020) Chitosan microbeads produced by one-step scalable stirred emulsification: a promising process for cell therapy applications. ACS Biomater Sci Eng 6(1):288–297. https://doi.org/10.1021/acsbiomaterials.9b01638

Andersen T, Auk-Emblem P, Dornish M (2015) 3D cell culture in alginate hydrogels. Microarrays. MDPI AG 4(2):133–161. https://doi.org/10.3390/microarrays4020133

Ansari S et al (2016) Muscle tissue engineering using gingival mesenchymal stem cells encapsulated in alginate hydrogels containing multiple growth factors. Ann Biomed Eng 44(6):1908–1920. https://doi.org/10.1007/s10439-016-1594-6

Antolli PG, Liu Z (2012) Bioreactors : design, properties, and applications. Nova Science Publishers

Arima Y, Iwata H (2007) Effects of surface functional groups on protein adsorption and subsequent cell adhesion using self-assembled monolayers. J Mater Chem 17(38):4079–4087. https://doi.org/10.1039/b708099a

Arya N et al (2019) RGDSP functionalized carboxylated agarose as extrudable carriers for chondrocyte delivery. Mater Sci Eng 99:103–111. https://doi.org/10.1016/j.msec.2019.01.080

Attanasio C, Netti PA (2017) Bioreactors for cell culture systems and organ bioengineering. In: Kidney transplantation, bioengineering, and regeneration: kidney transplantation in the regenerative medicine era. Elsevier Inc., pp 889–899. doi: https://doi.org/10.1016/B978-0-12-801734-0.00064-3

Bacabac RG et al (2008) Round versus flat: bone cell morphology, elasticity, and mechanosensing. J Biomech 41(7):1590–1598. https://doi.org/10.1016/j.jbiomech.2008.01.031

Bai FW et al (2011) Immobilization Technology. In: Moo-Young M (ed) Comprehensive biotechnology, 2nd edn, vol 2, pp 477–489. Elsevier, Dublin. https://doi.org/10.1016/B978-0-08-088504-9.00115-X

Balysheva VI et al (2007) Immobilization of animal cells in microgranules and microcapsules. Russian Agric Sci 33(3):195–197. https://doi.org/10.3103/s1068367407030184

Bayati V et al (2011) The evaluation of cyclic uniaxial strain on myogenic differentiation of adipose-derived stem cells. Tissue Cell 43(6):359–366. Elsevier. https://doi.org/10.1016/j.tice.2011.07.004

Belfort G (1989) Membranes and bioreactors: a technical challenge in biotechnology. Biotechnol Bioeng 33(8):1047–1066. https://doi.org/10.1002/bit.260330814

Bey O, Eigenberger G (1997) Fluid flow through catalyst filled tubes. Chem Eng Sci 52(8):1365–1376. Elsevier Science Ltd. https://doi.org/10.1016/S0009-2509(96)00509-X

Bhatia S, Bera T (2015) Classical and nonclassical techniques for secondary metabolite production in plant cell culture. In: Modern applications of plant biotechnology in pharmaceutical sciences. Elsevier Inc., pp 231–291. doi: https://doi.org/10.1016/B978-0-12-802221-4.00007-8

Blocki A et al (2017) Engineering of cell-laden gelatin-based microgels for cell delivery and immobilization in regenerative therapies. Clin Hemorheol Microcirc 67(3–4):251–259. https://doi.org/10.3233/CH-179206

Blose KJ et al (2014) Bioreactors for tissue engineering purposes. Regenerative Med Appl Organ Transp. doi: https://doi.org/10.1016/B978-0-12-398523-1.00013-6

Bose S, Vahabzadeh S, Bandyopadhyay A (2013) Bone tissue engineering using 3D printing. Mater Today 16:496–504. doi: https://doi.org/10.1016/j.mattod.2013.11.017

Bratch K, Al-Rubeai M (2001) Culture of primary rat hepatocytes within a flat hollow fibre cassette for potential use as a component of a bioartificial liver support system. Biotechnol Lett 23(2):137–141. https://doi.org/10.1023/A:1010387612109

Bugarski B, Goosen MFA, Vunjak-Novakovic G (1999) Principles of bioreactor design for encapsulated cells. Cell Encapsul Technol Therapeut 395–416. doi: https://doi.org/10.1007/978-1-4612-1586-8_30

Buza JA, Einhorn T (2016) Bone healing in 2016. Clinical Cases in Mineral and Bone Metabolism. CIC Edizioni Internazionali s.r.l., pp 101–105. doi: https://doi.org/10.11138/ccmbm/2016.13.2.101

Calafiore R et al (2006) Standard technical procedures for microencapsulation of human islets for graft into nonimmunosuppressed patients with type 1 diabetes mellitus. Transpl Proc Transpl Proc 38(4):1156–1157. https://doi.org/10.1016/j.transproceed.2006.03.014

Cao Y, Xue L (2004) 'Angiostatin', seminars in thrombosis and hemostasis. Semin Thromb Hemost 83–93. doi: https://doi.org/10.1055/s-2004-822973

Carpenter MK et al (2001) Enrichment of neurons and neural precursors from human embryonic stem cells. Exp Neurol 172(2):383–397. doi: https://doi.org/10.1006/exnr.2001.7832

Çetin, B. and Li, D. (2011) Dielectrophoresis in microfluidics technology. Electrophoresis 32:2410–2427. doi: https://doi.org/10.1002/elps.201100167

Chang H-I, Wang Y (2011) Cell responses to surface and architecture of tissue engineering scaffolds.In: Regenerative medicine and tissue engineering—cells and biomaterials. InTech. doi: 10.5772/21983

Chen G et al (2003) The use of a novel PLGA fiber/collagen composite web as a scaffold for engineering of articular cartilage tissue with adjustable thickness. J Biomed Mater Res Part A. 67(4): 1170–1180. doi: https://doi.org/10.1002/jbm.a.10164

Chen JP, Su CH (2011) Surface modification of electrospun PLLA nanofibers by plasma treatment and cationized gelatin immobilization for cartilage tissue engineering. Acta Biomater 7(1):234–243. https://doi.org/10.1016/j.actbio.2010.08.015

Chibata I et al (1987) Immobilization of cells in carrageenan. Methods Enzymol 135(C):189–198. doi: https://doi.org/10.1016/0076-6879(87)35077-3

Choe G et al (2018) Hydrogel biomaterials for stem cell microencapsulation polymers. MDPI AG 10(9):997. https://doi.org/10.3390/polym10090997

Choi CH et al (2007) Generation of monodisperse alginate microbeads and in situ encapsulation of cell in microfluidic device. Biomed Microdevice 9(6):855–862. https://doi.org/10.1007/s10544-007-9098-7

Chung C et al (2009) The influence of degradation characteristics of hyaluronic acid hydrogels on in vitro neocartilage formation by mesenchymal stem cells. Biomaterials 30(26):4287–4296. https://doi.org/10.1016/j.biomaterials.2009.04.040

Cirone P, Shen F, Chang PL (2005) A multiprong approach to cancer gene therapy by coencapsulated cells. Cancer Gene Ther 12(4):369–380. https://doi.org/10.1038/sj.cgt.7700786

Collins MN, Birkinshaw C (2007) Comparison of the effectiveness of four different crosslinking agents with hyaluronic acid hydrogel films for tissue-culture applications. J Appl Polym Sci 104(5):3183–3191. Wiley. https://doi.org/10.1002/app.25993

Cui FZ et al (2006) Hyaluronic acid hydrogel immobilized with RGD peptides for brain tissue engineering. J Mater Sci Mater Med 17:1393–1401. doi: https://doi.org/10.1007/s10856-006-0615-7

Dadsetan M et al (2011) The effects of fixed electrical charge on chondrocyte behavior. Acta Biomater 7(5):2080–2090. https://doi.org/10.1016/j.actbio.2011.01.012

De Castro M et al (2005) Comparative study of microcapsules eleborated with three polycations (PLL, PDL, PLO) for cell immobilization. J Microencapsul 22(3):303–315. https://doi.org/10.1080/02652040500099893

De Isla N et al (2010) Introduction to tissue engineering and application for cartilage engineering. Bio-med Mater Eng 20:127–133. doi: https://doi.org/10.3233/BME-2010-0624

De Meester SL et al (1998) Stress-induced fractal rearrangement of the endothelial cell cytoskeleton causes apoptosis. Surgery 124(2):362–71. Available at: https://www.ncbi.nlm.nih.gov/pubmed/9706160. Accessed 3 Feb 2020

de Vos P (2017) Historical perspectives and current challenges in cell microencapsulation, pp 3–21. doi: https://doi.org/10.1007/978-1-4939-6364-5_1

De Witte T-M et al (2018) Bone tissue engineering via growth factor delivery: from scaffolds to complex matrices. Regenerative Biomater 5(4):197–211. https://doi.org/10.1093/rb/rby013

Denbigh KG, Turner JCR (1984) Chemical reactor theory: an introduction

Dervakos GA, Webb C (1991) On the merits of viable-cell immobilisation. Biotechnol Adv. https://doi.org/10.1016/0734-9750(91)90733-C

Detzel CJ, van Wie BJ, Ivory CF (2010) Fluid flow through a high cell density fluidized-bed during centrifugal bioreactor culture. Biotechnol Progr 26(4):1014–1023. https://doi.org/10.1002/btpr.395

Dhandayuthapani B et al (2011) Polymeric scaffolds in tissue engineering application: a review. Int J Polym Sci. doi: https://doi.org/10.1155/2011/290602.

Dhara SK, Stice SL (2008) Neural differentiation of human embryonic stem cells. J Cell Biochem 105:633–640. doi: https://doi.org/10.1002/jcb.21891

dos Belgrano F, S. et al (2018) Cell immobilization on 3D-printed matrices: a model study on propionic acid fermentation. Bioresour Technol 249:777–782. https://doi.org/10.1016/j.biortech.2017.10.087

Drugmand JC, Schneider YJ, Agathos SN (2012) Insect cells as factories for biomanufacturing. Biotechnol Adv 30:1140–1157. doi: https://doi.org/10.1016/j.biotechadv.2011.09.014

Duarte JC et al (2013) Effect of immobilized cells in calcium alginate beads in alcoholic fermentation. AMB Express. https://doi.org/10.1186/2191-0855-3-31

Dwivedi R et al (2020) Polycaprolactone as biomaterial for bone scaffolds: review of literature. J Oral Biol Craniofac Res 10:381–388. doi: https://doi.org/10.1016/j.jobcr.2019.10.003

Elliott RB et al (2007) Live encapsulated porcine islets from a type 1 diabetic patient 9.5 yr after xenotransplantation. Xenotransplantation 14(2):157–161. https://doi.org/10.1111/j.1399-3089.2007.00384.x

Ellis MJ, Chaudhuri JB (2007) Poly(lactic-co-glycolic acid) hollow fibre membranes for use as a tissue engineering scaffold. Biotechnol Bioeng Biotechnol Bioeng 96(1):177–187. https://doi.org/10.1002/bit.21093

Ellis M, Jarman-Smith M, Chaudhuri JB (2005) Bioreactor systems for tissue engineering: a four-dimensional challenge. In: Bioreactors for tissue engineering: principles, design and operation. 1–18. doi: https://doi.org/10.1007/1-4020-3741-4_1

Emgård M et al (2003) Both apoptosis and necrosis occur early after intracerebral grafting of ventral mesencephalic tissue: a role for protease activation. J Neurochem 86(5):1223–1232. https://doi.org/10.1046/j.1471-4159.2003.01931.x

Endres M et al (2007) Human polymer-based cartilage grafts for the regeneration of articular cartilage defects. Tissue Cell 39(5):293–301. https://doi.org/10.1016/j.tice.2007.05.002

Fernandez RE et al (2017) Review: microbial analysis in dielectrophoretic microfluidic systems. Anal Chim Acta. 966:11–33. doi: https://doi.org/10.1016/j.aca.2017.02.024

Ferrand A et al (2014) Osteogenetic properties of electrospun nanofibrous PCL scaffolds equipped with chitosan-based nanoreservoirs of growth factors. Macromol Biosci 14(1):45–55. https://doi.org/10.1002/mabi.201300283

Ferrari M, Cirisano F, Morán MC (2019) Mammalian cell behavior on hydrophobic substrates: influence of surface properties. Colloids Interfaces 3(2):48. https://doi.org/10.3390/colloids3020048

Forry SP et al (2006) Cellular immobilization within microfluidic microenvironments: dielectrophoresis with polyelectrolyte multilayers. J Am Chem Soc 128(42):13678–13679. https://doi.org/10.1021/ja0627951

Franco CL, Price J, West JL (2011) Development and optimization of a dual-photoinitiator, emulsion-based technique for rapid generation of cell-laden hydrogel microspheres. Acta Biomater 7(9):3267–3276. https://doi.org/10.1016/j.actbio.2011.06.011

Frisch SM, Francis H (1994) Disruption of epithelial cell-matrix interactions induces apoptosis. J Cell Biol 124(4):619–626. Rockefeller University Press. https://doi.org/10.1083/jcb.124.4.619

Funayama A et al (2008) Repair of full-thickness articular cartilage defects using injectable type II collagen gel embedded with cultured chondrocytes in a rabbit model. J Orthopaed Sc 13(3):225–232. https://doi.org/10.1007/s00776-008-1220-z

Gao L (2012) Modeling and dynamics analyses of immobilized CSTR bioreactor using transfer function model. In: Proceedings of 2012 international symposium on information technologies in medicine and education, ITME 2012. IEEE, pp 692–695. doi: https://doi.org/10.1109/ITiME.2012.6291398

Gao Y et al (2018) Biomineralized poly (L-lactic-co-glycolic acid)-tussah silk fibroin nanofiber fabric with hierarchical architecture as a scaffold for bone tissue engineering. Mater Sci Eng C 84:195–207. https://doi.org/10.1016/j.msec.2017.11.047

Gasperini L, Mano JF, Reis RL (2014) Natural polymers for the microencapsulation of cells. J R Soc Interf. doi: https://doi.org/10.1098/rsif.2014.0817

Gawkowska D, Cybulska J, Zdunek A (2018) Structure-related gelling of pectins and linking with other natural compounds: a review. Polymers 10(7):762. https://doi.org/10.3390/polym10070762

Ge X, Yang L, Xu J (2016) Cell immobilization: fundamentals, technologies, and applications. Indus Biotechnol. 205–235. doi: https://doi.org/10.1002/9783527807833.ch7.

Glinel K et al (2012) Antibacterial surfaces developed from bio-inspired approaches. Acta Biomater 8:1670–1684. doi: https://doi.org/10.1016/j.actbio.2012.01.011

Gloeckner H, Lemke HD (2001) New miniaturized hollow-fiber bioreactor for in vivo like cell culture, cell expansion, and production of cell-derived products. Biotechnol Progress 17(5):828–831. https://doi.org/10.1021/bp010069q

Gómez-Sjöberg R et al (2007) Versatile, fully automated, microfluidic cell culture system. Anal Chem 79(22):8557–8563. https://doi.org/10.1021/ac071311w

Gonzlez G, Prieto N, Fabio O (2012) Fluid dynamics of gas—solid fluidized beds. Adv Fluid Dyn InTech. https://doi.org/10.5772/25791

Górecka E, Jastrzębska M (2011) Review article: immobilization techniques and biopolymer carriers. Biotechnol Food Sci 75(1):65–86

Gramer MJ, Poeschl DM (2000) Comparison of cell growth in T-flasks, in micro hollow fiber bioreactors, and in an industrial scale hollow fiber bioreactor system. Cytotechnology 34(1–2):111–119. https://doi.org/10.1023/A:1008167713696

Gungormusler-Yilmaz M et al (2016) Cell immobilization for microbial production of 1,3-propanediol. Critic Rev Biotechnol 482–494. doi: https://doi.org/10.3109/07388551.2014.992386

Guo P, Yuan Y, Chi F (2014) Biomimetic alginate/polyacrylamide porous scaffold supports human mesenchymal stem cell proliferation and chondrogenesis. Mater Sci Eng C 42:622–628. https://doi.org/10.1016/j.msec.2014.06.013

Guo S et al (2016) Parallel control over surface charge and wettability using polyelectrolyte architecture: effect on protein adsorption and cell adhesion. ACS Appl Mater Interfaces 8(44):30552–30563. https://doi.org/10.1021/acsami.6b09481

Hao T et al (2010) The support of matrix accumulation and the promotion of sheep articular cartilage defects repair in vivo by chitosan hydrogels. Osteoarthritis Cartilage 18(2):257–265. https://doi.org/10.1016/j.joca.2009.08.007

Highfill JG, Haley SD, Kompala DS (2000) Large-scale production of murine bone marrow cells in an airlift packed bed bioreactor. Biotechnol Bioeng 50(5):514–520. doi: https://doi.org/10.1002/(SICI)1097-0290(19960605)50:5<514::AID-BIT5>3.0.CO;2-I

Ho STB et al (2010) The influence of fibrin based hydrogels on the chondrogenic differentiation of human bone marrow stromal cells. Biomaterials 31(1):38–47. https://doi.org/10.1016/j.biomaterials.2009.09.021

Hoesli CA et al (2011) Pancreatic cell immobilization in alginate beads produced by emulsion and internal gelation. Biotechnol Bioeng 108(2):424–434. https://doi.org/10.1002/bit.22959

Huang X et al (2012) Microenvironment of alginate-based microcapsules for cell culture and tissue engineering. J Biosci Bioeng 1–8. doi: https://doi.org/10.1016/j.jbiosc.2012.02.024

Huber D et al (2018) Hydrodynamics in cell studies. Chem Rev 118:2042–2079. doi: https://doi.org/10.1021/acs.chemrev.7b00317

Ijima H et al (2000a) Conditions required for a hybrid artificial liver support system using a PUF/hepatocyte-spheroid packed-bed module and it's use in dogs with liver failure. Int J Artif Org 23(7):446–53. Available at: https://www.ncbi.nlm.nih.gov/pubmed/10941638. Accessed: 29 Jan 2020

Ijima H et al (2000b) Development of a hybrid artificial liver using a polyurethane foam/hepatocyte-spheroid packed-bed module. Int J Artif Org 23(6):389–97. Available at: https://www.ncbi.nlm.nih.gov/pubmed/10919756. Accessed: 29 Jan 2020

Ito Y (2007) Covalently immobilized biosignal molecule materials for tissue engineering. Soft Matter 4(1):46–56. https://doi.org/10.1039/b708359a

Izuta S et al (2017) Microfluidic preparation of anchored cell membrane sheets for in vitro analyses and manipulation of the cytoplasmic face. Sci Rep 7(1):1–11. https://doi.org/10.1038/s41598-017-14737-7

Jagani H et al (2010) An overview of fermenter and the design considerations to enhance its productivity, Pharmacologyonline

Jang JY et al (2016) In vivo osteogenic differentiation of human dental pulp stem cells embedded in an injectable in vivo-forming hydrogel. Macromol Biosci 16(8):1158–1169. https://doi.org/10.1002/mabi.201600001

Janke D et al (2013) The "artificial artery" as in vitro perfusion model. PLoS ONE. In: Pesce M (ed) Public Library of Science, vol 8, no 3, p e57227. doi: https://doi.org/10.1371/journal.pone.0057227

Jasińska UT et al (2018) Immobilization of bifidobacterium infantis cells in selected hydrogels as a method of increasing their survival in fermented milkless beverages. J Food Qual. https://doi.org/10.1155/2018/9267038

Jen AC, Wake MC, Mikos AG (2000) Review: hydrogels for cell immobilization. Biotechnol Bioeng 50(4):357–364. https://doi.org/10.1002/(sici)1097-0290(19960520)50:4%3c357::aid-bit2%3e3.0.co;2-k

Joki T et al (2001) Continuous release of endostatin from microencapsulated engineered cells for tumor therapy. Nat Biotechnol 19(1):35–39. Nature Publishing Group. https://doi.org/10.1038/83481

Jutila AA et al (2015) Encapsulation of chondrocytes in high-stiffness agarose microenvironments for in vitro modeling of osteoarthritis mechanotransduction. Ann Biomed Eng 43(5):1132–1144. https://doi.org/10.1007/s10439-014-1183-5

Kadic E, Heindel TJ (2010) Mixing considerations in stirred tank bioreactors when using fluid property altering microorganisms. Am Soc Mech Eng 859–870. doi: https://doi.org/10.1115/FEDSM-ICNMM2010-30366

Kai D et al (2011) Guided orientation of cardiomyocytes on electrospun aligned nanofibers for cardiac tissue engineering. J Biomed Mater Res B Appl Biomater 98B(2):379–386. https://doi.org/10.1002/jbm.b.31862

Kandow CE et al (2007) Polyacrylamide hydrogels for cell mechanics: steps toward optimization and alternative uses. Methods Cell Biol 29–46. doi: https://doi.org/10.1016/S0091-679X(07)83002-0

Kang L et al (2008) Microfluidics for drug discovery and development: From target selection to product lifecycle management. Drug Discov Today 1–13. doi: https://doi.org/10.1016/j.drudis.2007.10.003

Kanojia V, et al (2017) An overview of microbial cell culture. J Pharm Phytochem 6(6):1923–1928

Karel SF, Libicki SB, Robertson CR (1985) The immobilization of whole cells: engineering principles. Chem Eng Sci 40(8):1321–1354. https://doi.org/10.1016/0009-2509(85)80074-9

Kargi F (1992) Membrane bioreactors for animal cell culture. Recent Adv Biotechnol 449–454. doi: https://doi.org/10.1007/978-94-011-2468-3_30

Keselowsky BG, Collard DM, García AJ (2003) Surface chemistry modulates fibronectin conformation and directs integrin binding and specificity to control cell adhesion. J Biomed Mater Res Part A 66(2):247–259. https://doi.org/10.1002/jbm.a.10537

Khan KH (2013) Gene expression in mammalian cells and its applications. Adv Pharm Bull 3(2):257–263. https://doi.org/10.5681/apb.2013.042

Khan SA, Siddiqui MH, Osama K (2019) Bioreactors for hairy roots culture: a review. Current Biotechnol 7(6):417–427. https://doi.org/10.2174/2211550108666190114143824

Kierstan M, Bucke C (1977) The immobilization of microbial cells, subcellular organelles, and enzymes in calcium alginate gels. Biotechnol Bioeng 19(3):387–97. doi: https://doi.org/10.1002/bit.260190309

Kim D, Jun HS (2018) In vivo imaging of transplanted pancreatic islets. Front Endocrinol 382. doi: https://doi.org/10.3389/fendo.2017.00382

Kim S, English AE, Kihm KD (2009) Surface elasticity and charge concentration-dependent endothelial cell attachment to copolymer polyelectrolyte hydrogel. Acta Biomater 5(1):144–151. https://doi.org/10.1016/j.actbio.2008.07.033

Klein J, Kressdorf B (1989) Polymers for the immobilization of whole cells and their application in biotechnology. Angewandte Makromolekulare Chemie 166(1):293–309. Wiley. https://doi.org/10.1002/apmc.1989.051660120

Koch M et al (2018) Tissue engineering of large full-size meniscus defects by a polyurethane scaffold: accelerated regeneration by mesenchymal stromal cells.In:Sánchez M (ed) Stem Cells International. Hindawi, p 8207071. doi: 10.1155/2018/8207071

Kontturi LS et al (2014) An injectable, in situ forming type II collagen/hyaluronic acid hydrogel vehicle for chondrocyte delivery in cartilage tissue engineering. Drug Deliv Trans Res 4(2):149–158. https://doi.org/10.1007/s13346-013-0188-1

Kootenaei F (2014) Membrane biological reactors (MBR) and their applications for water reuse. pdfs.semanticscholar.org. Available at: https://pdfs.semanticscholar.org/76f6/efabe09ef249b3c91eb067d2acdcb8d14730.pdf. Accessed: 18 Feb 2020

Kourkoutas Y et al (2004) Immobilization technologies and support materials suitable in alcohol beverages production: a review. Food Microbiol 21:377–397. doi: https://doi.org/10.1016/j.fm.2003.10.005

Kovárová-Kovar K, Egli T (1998) Growth kinetics of suspended microbial cells: from single-substrate-controlled growth to mixed-substrate kinetics. Microbiol Mol Biol Rev 62(3):646–666. https://doi.org/10.1128/mmbr.62.3.646-666.1998

Krishnamoorthy S, Noorani B, Xu C (2019) Effects of encapsulated cells on the physical-mechanical properties and microstructure of gelatin methacrylate hydrogels. Int J Mol Sci 20(20):5061. https://doi.org/10.3390/ijms20205061

Krzeminski P et al (2017) Membrane bioreactors—a review on recent developments in energy reduction, fouling control, novel configurations, LCA and market prospects. J Membr Sci 527:207–227. doi: https://doi.org/10.1016/j.memsci.2016.12.010

Kulanthaivel S et al (2017) Gum tragacanth-alginate beads as proangiogenic-osteogenic cell encapsulation systems for bone tissue engineering. J Mater Chem B 5(22):4177–4189. https://doi.org/10.1039/c7tb00390k

Kumar A, Tripathi A, Jain S (2011) Extracorporeal bioartificial liver for treating acute liver diseases. J Extra-Corporeal Technol 195–206

Kuo CK et al (2006) Cartilage tissue engineering: its potential and uses. Current Opin Rheumatol 18(1):64–73. https://doi.org/10.1097/01.bor.0000198005.88568.df

Kuo KC et al (2015) Bioengineering vascularized tissue constructs using an injectable cell-laden enzymatically crosslinked collagen hydrogel derived from dermal extracellular matrix. Acta Biomater 27:151–166. https://doi.org/10.1016/j.actbio.2015.09.002

Kurosawa H et al (2000) Polyurethane membrane as an efficient immobilization carrier for high-density culture of rat hepatocytes in the fixed-bed reactor. Biotechnol Bioeng 70(2):160–166. doi: https://doi.org/10.1002/1097-0290(20001020)70:2<160::AID-BIT5>3.0.CO;2-C

Lalonde ME, Durocher Y (2017) Therapeutic glycoprotein production in mammalian cells. J Biotechnol 251:128–140. doi: https://doi.org/10.1016/j.jbiotec.2017.04.028

Langer R, Vacanti JP (1993) Tissue engineering. Science 260(5110):920–926. https://doi.org/10.1126/science.8493529

Lee JH et al (1994) Cell behaviour on polymer surfaces with different functional groups. Biomaterials 15(9):705–711. https://doi.org/10.1016/0142-9612(94)90169-4

Lemon G et al (2006) Mathematical modelling of engineered tissue growth using a multiphase porous flow mixture theory. J Math Biol 52(5):571–594. https://doi.org/10.1007/s00285-005-0363-1

Li AP et al (1993) Culturing of primary hepatocytes as entrapped aggregates in a packed bed bioreactor: a potential bioartificial liver. Vitro Cel Dev Biol Animal: J Soc In Vitro Biol 29(3):249–254. https://doi.org/10.1007/BF02634192

Li H, Wijekoon A, Leipzig ND (2012) 3D differentiation of neural stem cells in macroporous photopolymerizable hydrogel scaffolds. PLoS ONE. In: Gelain F (ed). Public Library of Science 7(11):e48824. doi: https://doi.org/10.1371/journal.pone.0048824

Li W et al (2018) Tissue-engineered bone immobilized with human adipose stem cells-derived exosomes promotes bone regeneration. ACS Appl Mater Interf Am Chem Soc 10(6):5240–5254. https://doi.org/10.1021/acsami.7b17620

Liao E et al (2007) Tissue-engineered cartilage constructs using composite hyaluronic acid/collagen I hydrogels and designed poly(propylene fumarate) scaffolds. Tissue Eng Tissue Eng 13(3):537–550. https://doi.org/10.1089/ten.2006.0117

Liouni M, Drichoutis P, Nerantzis ET (2008) Studies of the mechanical properties and the fermentation behavior of double layer alginate-chitosan beads, using Saccharomyces cerevisiae entrapped cells. World J Microbiol Biotechnol 24(2):281–288. https://doi.org/10.1007/s11274-007-9467-7

Lock LT, Laychock SG, Tzanakakis ES (2011) Pseudoislets in stirred-suspension culture exhibit enhanced cell survival, propagation and insulin secretion. J Biotechnol 151(3):278–286. https://doi.org/10.1016/j.jbiotec.2010.12.015

Loh QL, Choong C (2013) Three-dimensional scaffolds for tissue engineering applications: role of porosity and pore size. Tissue Eng Part B: Rev 19:485–502. doi: https://doi.org/10.1089/ten.teb.2012.0437

Lu J et al (2016) A new fluidized bed bioreactor based on diversion-type microcapsule suspension for bioartificial liver systems. PLOS ONE. In: Laconi E (ed) Public Library of Science, vol 11, no 2, p e0147376. doi: https://doi.org/10.1371/journal.pone.0147376

Lüdemann I et al (1995) Improvement of the culture stability of non-anchorage-dependent animal cells grown in serum-free media through immobilization. Cytotechnol Int J Cell Culture Biotechnol 19(2):111–124. doi: https://doi.org/10.1007/BF00749766

Ma Z et al (2005) Cartilage tissue engineering PLLA scaffold with surface immobilized collagen and basic fibroblast growth factor. Biomaterials 26(11):1253–1259. https://doi.org/10.1016/j.biomaterials.2004.04.031

Mano T et al (1992) New immobilization method of mammalian cells using alginate and polyacrylate. J Ferment Bioeng 73(6):486–489. https://doi.org/10.1016/0922-338X(92)90143-I

Mao, Y. et al. (2019) 'Endogenous viable cells in lyopreserved amnion retain differentiation potential and anti-fibrotic activity in vitro', Acta Biomaterialia. Acta Materialia Inc, 94, pp. 330–339. doi: https://doi.org/10.1016/j.actbio.2019.06.002.

Margaritis A, Wallace JB (1984) Novel bioreactor systems and their applications. Bio/Technol 2:447–453. doi: https://doi.org/10.1038/nbt0584-447

Martinsen A, Skjåk-Bræk G, Smidsrød O (1989) Alginate as immobilization material: I. Correlation between chemical and physical properties of alginate gel beads. Biotechnol Bioeng 33(1):79–89. https://doi.org/10.1002/bit.260330111

Mauck RL et al (2000) Functional tissue engineering of articular cartilage through dynamic loading of chondrocyte-seeded agarose gels. J Biomech Eng ASME 122(3):252–260. https://doi.org/10.1115/1.429656

Mavituna F (2004) Pre-formed carriers for cell immobilisation, pp 121–139. doi: https://doi.org/10.1007/978-94-017-1638-3_7

Mazzei D et al (2010) A low shear stress modular bioreactor for connected cell culture under high flow rates. Biotechnol Bioeng. doi: https://doi.org/10.1002/bit.22671

McClelland KS, Ng ET, Bowles J (2016) Agarose/gelatin immobilisation of tissues or embryo segments for orientated paraffin embedding and sectioning. Differentiation 91(4–5):68–71. https://doi.org/10.1016/j.diff.2015.12.001

Mehrali M et al (2019) Pectin Methacrylate (PEMA) and gelatin-based hydrogels for cell delivery: converting waste materials into biomaterials. ACS Appl Mater Interf 11(13):12283–12297. https://doi.org/10.1021/acsami.9b00154

Meuwly F et al (2007) Packed-bed bioreactors for mammalian cell culture: Bioprocess and biomedical applications. Biotechnol Adv. https://doi.org/10.1016/j.biotechadv.2006.08.004

Mihaila SM et al (2014) Fabrication of endothelial cell-laden carrageenan microfibers for microvascularized bone tissue engineering applications. Biomacromolecules. Am Chem Soc 15(8):2849–2860. https://doi.org/10.1021/bm500036a

Mota M, Teixeira JA, Yelshin A (2001) Immobilized particles in gel matrix-type porous media. Homogeneous porous media model. Biotechnol Prog 17(5):860–865. https://doi.org/10.1021/bp010064t

Moussavou G et al (2015) Production of monoclonal antibodies in plants for cancer immunotherapy. BioMed Res Int doi: https://doi.org/10.1155/2015/306164

Nicolas CT et al (2017) Concise review: liver regenerative medicine: from hepatocyte transplantation to bioartificial livers and bioengineered grafts. Stem Cells 35:42–50. doi: https://doi.org/10.1002/stem.2500

Nielsen LK (1999) Bioreactors for hematopoietic cell culture. Ann Rev Biomed Eng Ann Rev 1(1):129–152. https://doi.org/10.1146/annurev.bioeng.1.1.129

Nikolova MP, Chavali MS (2019) Recent advances in biomaterials for 3D scaffolds: A review. Bioact Mater 4:271–292. https://doi.org/10.1016/j.bioactmat.2019.10.005

Oh HI et al (2010) Hemocompatibility assessment of carbonic anhydrase modified hollow fiber membranes for artificial lungs. Artif Org 34(5):439–442. doi: https://doi.org/10.1111/j.1525-1594.2009.00882.x

Ohshima N, Yanagi K, Miyoshi H (2008) Packed-bed type reactor to attain high density culture of hepatocytes for use as a bioartificial liver. Artif Org 21(11):1169–1176. https://doi.org/10.1111/j.1525-1594.1997.tb00470.x

Olmer R et al (2012) Suspension culture of human pluripotent stem cells in controlled, stirred bioreactors. Tissue Eng Part C: Methods 18(10):772–784. doi: https://doi.org/10.1089/ten.tec.2011.0717

Omori M et al (2015) A new application of cell-free bone regeneration: Immobilizing stem cells from human exfoliated deciduous teeth-conditioned medium onto titanium implants using atmospheric pressure plasma treatment. Stem Cell Res Ther 6(1). doi: https://doi.org/10.1186/s13287-015-0114-1.

Ong SM et al (2008) A gel-free 3D microfluidic cell culture system. Biomaterials 29(22):3237–3244. https://doi.org/10.1016/j.biomaterials.2008.04.022

Oo ZY et al (2011) The performance of primary human renal cells in hollow fiber bioreactors for bioartificial kidneys. Biomaterials 32(34):8806–8815. https://doi.org/10.1016/j.biomaterials.2011.08.030

Opara EC (2017) Applications of cell microencapsulation. Methods Mol Biol 23–39. doi: https://doi.org/10.1007/978-1-4939-6364-5_2

Orive G et al (2004) History, challenges and perspectives of cell microencapsulation. Trends Biotechnol 22:87–92. doi: https://doi.org/10.1016/j.tibtech.2003.11.004

Osmoković A et al (2006) Development of a packed bed bioreactor for cartilage tissue engineering. FME Trans 34:65–70

Pandey AR et al (2017) Chitosan: application in tissue engineering and skin grafting. J Polym Res doi: https://doi.org/10.1007/s10965-017-1286-4

Papadopoulos A et al (2011) Injectable and photopolymerizable tissue-engineered auricular cartilage using poly(ethylene glycol) dimethacrylate copolymer hydrogels. Tissue Eng Part A 17(1–2):161–169. https://doi.org/10.1089/ten.TEA.2010.0253

Papas KK et al (2007) A stirred microchamber for oxygen consumption rate measurements with pancreatic islets. Biotechnol Bioeng 98(5):1071–1082. https://doi.org/10.1002/bit.21486

Park BG et al (2000) Development of high density mammalian cell culture system for the production of tissue-type plasminogen activator. Biotechnol Bioproces Eng 5(2):123–129. https://doi.org/10.1007/BF02931883

Park JH et al (2014) Nanocoating of single cells: from maintenance of cell viability to manipulation of cellular activities. Adv Mater 26(13):2001–2010. https://doi.org/10.1002/adma.201304568

Park JK, Chang HN (2000) Microencapsulation of microbial cells. Biotechnol Adv 18(4):303–319. https://doi.org/10.1016/S0734-9750(00)00040-9

Petry F et al (2018) Three-dimensional bioreactor technologies for the cocultivation of human mesenchymal stem/stromal cells and beta cells. Stem Cells Int doi: https://doi.org/10.1155/2018/2547098

Plieva FM et al (2008) Macroporous gel particles as robust macroporous matrices for cell immobilization. Biotechnol J 3(3):410–417. https://doi.org/10.1002/biot.200700134

Polacheck WJ, Charest JL, Kamm RD (2011) Interstitial flow influences direction of tumor cell migration through competing mechanisms. Proc Natl Acad Sci USA 108(27):11115–11120. https://doi.org/10.1073/pnas.1103581108

Pörtner R et al (2007) Fixed bed reactors for the cultivation of mammalian cells: design, performance and scale-up. Open Biotechnol J

Prabhawathi V et al (2014) Antibiofilm properties of interfacially active lipase immobilized porous polycaprolactam prepared by LB Technique. PLoS ONE 9(5):e96152. https://doi.org/10.1371/journal.pone.0096152

Prokop A (2006) Bioartificial organs in the twenty-first century. Ann N Y Acad Sci 944(1):472–490. doi: 10.1111/j.1749-6632.2001.tb03856.x

Rahmany MB, Van Dyke M (2013) Biomimetic approaches to modulate cellular adhesion in biomaterials: a review. Acta Biomater 5431–5437. doi: https://doi.org/10.1016/j.actbio.2012.11.019

Ramasubramanyan K, Venkatasubramanian K (1990) Large-scale animal cell cultures: design and operational considerations. Adv Biochem Eng Biotechnol 42:13–26. https://doi.org/10.1007/bfb0000729

Rana D, Tabasum A, Ramalingam M (2016) Cell-laden alginate/polyacrylamide beads as carriers for stem cell delivery: preparation and characterization. RSC AdvR Soc Chem 6(25):20475–20484. https://doi.org/10.1039/c5ra26447b

Rao C et al (2013) The effect of microgrooved culture substrates on calcium cycling of cardiac myocytes derived from human induced pluripotent stem cells. Biomaterials 34(10):2399–2411. https://doi.org/10.1016/j.biomaterials.2012.11.055

Ray NG et al (1990) Continuous cell cultures in fluidized-bed bioreactors: cultivation of hybridomas and recombinant chinese hamster ovary cells immobilized in collagen microspheres. Ann N Y Acad Sci 589(1):443–457. https://doi.org/10.1111/j.1749-6632.1990.tb24263.x

Ritz U et al (2018) Photocrosslinked dextran-based hydrogels as carrier system for the cells and cytokines induce bone regeneration in critical size defects in mice. Gels 4(3):63. https://doi.org/10.3390/gels4030063

Roberts I et al (2012) Scale-up of human embryonic stem cell culture using a hollow fibre bioreactor. Biotechnol Lett 34(12):2307–2315. https://doi.org/10.1007/s10529-012-1033-1

Rosiak JM, Ulański P (1999) Synthesis of hydrogels by irradiation of polymers in aqueous solution. Radiat Phys Chem 55(2):139–151. https://doi.org/10.1016/S0969-806X(98)00319-3

Rouwkema J et al (2009) Supply of nutrients to cells in engineered tissues. Biotechnol Genet Eng Rev 26(1):163–178. https://doi.org/10.5661/bger-26-163
Salehi-Nik N et al (2013) Engineering parameters in bioreactor's design: a critical aspect in tissue engineering. Biomed Res Int. https://doi.org/10.1155/2013/762132
Salinas CN, Anseth KS (2009) Mesenchymal stem cells for craniofacial tissue regeneration: designing hydrogel delivery vehicles. J Dental Res 88:681–692. doi: https://doi.org/10.1177/0022034509341553
Salter GJ, Kell DB (1991) New materials and technology for cell immobilization. Current Opin Biotechnol 2(3):385–389. https://doi.org/10.1016/S0958-1669(05)80143-0
Sarem M et al (2018) Interplay between stiffness and degradation of architectured gelatin hydrogels leads to differential modulation of chondrogenesis in vitro and in vivo. Acta Biomaterialia. 69:83–94. doi: https://doi.org/10.1016/j.actbio.2018.01.025
Sarnowska A et al (2013) Encapsulation of mesenchymal stem cells by bioscaffolds protects cell survival and attenuates neuroinflammatory reaction in injured brain tissue after transplantation. Cell Transpl 22(1):67–82. doi: https://doi.org/10.3727/096368913X672172
Sawyer SW et al (2018) Conductive gelatin methacrylate-poly(aniline) hydrogel for cell encapsulation. Biomed Phys Eng Expr 4(1):015005. https://doi.org/10.1088/2057-1976/AA91F9
Schneider GB et al (2004) The effect of hydrogel charge density on cell attachment. Biomaterials 25(15):3023–3028. https://doi.org/10.1016/j.biomaterials.2003.09.084
Scott CD (1987) Immobilized cells: a review of recent literature. Enzyme Microb Technol 9:66–72. doi: https://doi.org/10.1016/0141-0229(87)90145-1
Selden C et al (2017) A clinical-scale BioArtificial Liver, developed for GMP, improved clinical parameters of liver function in porcine liver failure. Sci Rep 7(1). doi: https://doi.org/10.1038/s41598-017-15021-4
Sen P, Nath A, Bhattacharjee C (2017) Packed-bed bioreactor and its application in dairy, food, and beverage industry. Current Dev Biotechnol Bioeng 235–277. doi: https://doi.org/10.1016/B978-0-444-63663-8.00009-4
Shachar M et al (2011) The effect of immobilized RGD peptide in alginate scaffolds on cardiac tissue engineering. Acta Biomat 7(1):152–162. https://doi.org/10.1016/j.actbio.2010.07.034
Shirai Y et al (1988) Continuous production of erythropoietin with immobilized animal cells. Appl Microbiol Biotechnol 29(6):544–549. https://doi.org/10.1007/BF00260982
Shishido T et al (2007) Production of bionanocapsules in immobilized insect cell culture using porous biomass support particles. J Biosci Bioeng 103(6):572–574. https://doi.org/10.1263/jbb.103.572
Silini AR et al (2018) Immunological and differentiation properties of amniotic cells are retained after immobilization in pectin gel. Cell Transpl 27(1):70–76. https://doi.org/10.1177/0963689717738786
Singh D et al (2011) Proliferation of chondrocytes on a 3-d modelled macroporous poly(hydroxyethyl methacrylate)-gelatin cryogel. J Biomater Sci Polym Edn 22(13): 1733–1751. doi: https://doi.org/10.1163/092050610X522486
Singh D et al (2014) Synthesis of composite gelatin-hyaluronic acid-alginate porous scaffold and evaluation for in vitro stem cell growth and in vivo tissue integration. Colloids Surf B: Biointerf 116:502–509. https://doi.org/10.1016/j.colsurfb.2014.01.049
Soon-Shiong P et al (1994) Insulin independence in a type 1 diabetic patient after encapsulated islet transplantation. The Lancet 343(8903):950–951. https://doi.org/10.1016/S0140-6736(94)90067-1
Sortwell CE, Pitzer MR, Collier TJ (2000) Time course of apoptotic cell death within mesencephalic cell suspension grafts: Implications for improving grafted dopamine neuron survival. Exp Neurol 165(2):268–277. doi: https://doi.org/10.1006/exnr.2000.7476
Stoop R (2008) Smart biomaterials for tissue engineering of cartilage. Injury 39(1 SUPPL.):77–87. https://doi.org/10.1016/j.injury.2008.01.036
Storm MP et al (2016) Hollow fiber bioreactors for In Vivo-like mammalian tissue culture. J Visual Exp 2016(111). doi: https://doi.org/10.3791/53431

Suarez-Arnedo A et al (2018) 3D alginate hydrogels with controlled mechanical properties for mammalian cell encapsulation. In: 2018 9th international seminar of biomedical engineering, SIB 2018—conference proceedings. doi: https://doi.org/10.1109/SIB.2018.8467745

Szekalska M et al (2016) Alginate: current use and future perspectives in pharmaceutical and biomedical applications. Int J Polym Sci doi: https://doi.org/10.1155/2016/7697031

Tallawi M et al (2015) Strategies for the chemical and biological functionalization of scaffolds for cardiac tissue engineering: a review. J R Soc Interf. doi: https://doi.org/10.1098/rsif.2015.0254

Tavoulareas ES (1991) Fluidized-bed combustion technology. Ann Rev Energy Environ 16(1):25–57. https://doi.org/10.1146/annurev.eg.16.110191.000325

Terraf P, Babaloo H, Kouhsari SM (2017) Directed differentiation of dopamine-secreting cells from Nurr1/GPX1 expressing murine embryonic stem cells cultured on matrigel-coated PCL scaffolds. Mol Neurobiol. 54(2):1119–1128. doi: https://doi.org/10.1007/s12035-016-9726-4

Thakar H et al (2019) Biomolecule-conjugated macroporous hydrogels for biomedical applications. ACS Biomater Sci Eng Am Chem Soc 5:6320–6341. doi: https://doi.org/10.1021/acsbiomaterials.9b00778

Ting Y-P, Sun G (2000) Use of polyvinyl alcohol as a cell immobilization matrix for copper biosorption by yeast cells. J Chem Technol Biotechnol 75(7):541–546. https://doi.org/10.1002/1097-4660(200007)75:7%3c541::AID-JCTB247%3e3.0.CO;2-9

Trelles JA, Rivero CW (2013) Whole cell entrapment techniques. Methods Molecul Biol 1051:365–374. doi: https://doi.org/10.1007/978-1-62703-550-7_24

Tresco PA, Winn SR, Aebischer P (1992) Polymer encapsulated neurotransmitter secreting cells. Potential treatment for Parkinson's disease. ASAIO J 38(1):17–23. Available at: https://www.ncbi.nlm.nih.gov/pubmed/1348191. Accessed: 2 Mar 2020

Tripathi A, Kathuria N, Kumar A (2009) Elastic and macroporous agarose-gelatin cryogels with isotropic and anisotropic porosity for tissue engineering. J Biomed Mater Res Part A. 90(3):680–694. doi: https://doi.org/10.1002/jbm.a.32127

Tripathi A, Kumar A (2011) Multi-featured macroporous agarose-alginate cryogel: synthesis and characterization for bioengineering applications. Macromol Biosci 11(1):22–35

Tripathi A, Melo JS (2015) Preparation of a sponge-like biocomposite agarose-chitosan scaffold with primary hepatocytes for establishing an in vitro 3D liver tissue model. RSC Adv R Soc Chem 5(39):30701–30710. https://doi.org/10.1039/c5ra04153h

Tripathi A, Melo JS (2019) Cryostructurization of polymeric systems for developing macroporous cryogel as a foundational framework in bioengineering applications. J Chem Sci 131(92):1–11. https://doi.org/10.1007/s12039-019-1670-1

Tripathi A, Vishnoi T, Singh D, Kumar A (2013) Modulated crosslinking of macroporous polymeric cryogel affect in-vitro cell adhesion and growth. Macromol Biosci 13(7):838–850

Tuch BE et al (2009) Safety and viability of microencapsulated human islets transplanted into diabetic humans. Diabetes Care 32(10):1887–1889. https://doi.org/10.2337/dc09-0744

Two step culture for production of recombinant herpes simplex virus type 2 glycoprotein D in immobilized Spodoptera frugiperda cells (2012) African J Biotechnol 11(101):16655–16660. doi: https://doi.org/10.5897/AJB11.3356

Uemura M et al (2010) Matrigel supports survival and neuronal differentiation of grafted embryonic stem cell-derived neural precursor cells. J Neurosci Res 88(3):542–551. https://doi.org/10.1002/jnr.22223

US4032407A—Tapered bed bioreactor—Google Patents (no date). Available at: https://patents.google.com/patent/US4032407. Accessed: 6 Feb 2020

US4910228A—Methanol—Google Patents (no date). Available at: https://patents.google.com/patent/US4910228A/en. Accessed 6 Feb 2020

US5073491A—Immobilization of cells in alginate beads containing cavities for growth of cells in airlift bioreactors—Google Patents (no date) Available at: https://patents.google.com/patent/US5073491A/en. Accessed 29 Jan 2020

Vatsa A, Smit TH, Klein-Nulend J (2007) Extracellular NO signalling from a mechanically stimulated osteocyte. J Biomechan 40(1). doi: https://doi.org/10.1016/j.jbiomech.2007.02.015

Vilariño-Feltrer G et al (2016) Schwann-cell cylinders grown inside hyaluronic-acid tubular scaffolds with gradient porosity. Acta Biomater 30:199–211. https://doi.org/10.1016/j.actbio.2015.10.040

Visted T et al (2003) Prospects for delivery of recombinant angiostatin by cell-encapsulation therapy. Hum Gene Ther 14(15):1429–1440. https://doi.org/10.1089/104303403769211646

Vo TN et al (2016) Injectable dual-gelling cell-laden composite hydrogels for bone tissue engineering. Biomaterials 83:1–11. https://doi.org/10.1016/j.biomaterials.2015.12.026

Wan Y et al (2012) Nanotextured substrates with immobilized aptamers for cancer cell isolation and cytology. Cancer 118(4):1145–1154. https://doi.org/10.1002/cncr.26349

Wang S-J, Zhong J-J (2007) Bioreactor engineering. In: Bioprocessing for value-added products from renewable resources. pp 131–161. doi: https://doi.org/10.1016/B978-044452114-9/50007-4

Warnock JN, Bratch K, Al-Rubeai M (2005) Packed bed bioreactors. In: Bioreactors for tissue engineering: principles, design and operation. Springer, Berlin/Heidelberg, pp 87–113. doi: https://doi.org/10.1007/1-4020-3741-4_4

Whitesides GM (2006) The origins and the future of microfluidics. Nature 442:368–373. doi: https://doi.org/10.1038/nature05058

Willaert R (2011) Cell immobilization and its applications in biotechnology. In: Fermentation microbiology and biotechnology, 3rd edn. doi: 10.1201/b11490-13

Willaert RG, Baron GV (1996) Gel entrapment and micro-encapsulation gel entrapment and micro-encapsulation: methods, applications and engineering principles

Winn SR et al (1994) Polymer-encapsulated cells genetically modified to secrete human nerve growth factor promote the survival of axotomized septal cholinergic neurons. In: Proceedings of the national academy of sciences of the United States of America. National Academy of Sciences, vol 91, no 6, pp 2324–2328. doi: https://doi.org/10.1073/pnas.91.6.2324

Wong AP et al (2008) Partitioning microfluidic channels with hydrogel to construct tunable 3-D cellular microenvironments. Biomaterials 29(12):1853–1861. https://doi.org/10.1016/j.biomaterials.2007.12.044

Wu I, Elisseeff J (2014) Biomaterials and tissue engineering for soft tissue reconstruction. Nat Synth Biomed Polym. doi: https://doi.org/10.1016/B978-0-12-396983-5.00015-6

Wung N et al (2014) Hollow fibre membrane bioreactors for tissue engineering applications. Biotechnol Lett 36:2357–2366. doi: https://doi.org/10.1007/s10529-014-1619-x

Wurm FM (2004) Production of recombinant protein therapeutics in cultivated mammalian cells. Nat Biotechnol 22:1393–1398. doi: https://doi.org/10.1038/nbt1026

Wüst S et al (2014) Tunable hydrogel composite with two-step processing in combination with innovative hardware upgrade for cell-based three-dimensional bioprinting. Acta Biomater 10(2):630–640. https://doi.org/10.1016/j.actbio.2013.10.016

Yamaji H et al (2000) Production of recombinant protein by baculovirus-infected insect cells in immobilized culture using porous biomass support particles. J Biosci Bioeng 89(1):12–17. doi: https://doi.org/10.1016/S1389-1723(00)88044-5

Yamaji H et al (2006) Efficient production of recombinant protein in immobilized insect cell culture using serum-free basal media after baculovirus infection. Biochem Eng J 28(1):67–72. https://doi.org/10.1016/j.bej.2005.09.003

Yamaji H, Fukuda H (1994) Growth kinetics of animal cells immobilized within porous support particles in a perfusion culture. Appl Microbiol Biotechnol 42(4):531–535. https://doi.org/10.1007/BF00173916

Yamaoka H et al (2006) Cartilage tissue engineering using human auricular chondrocytes embedded in different hydrogel materials. J Biomed Mater Res a 78(1):1–11. https://doi.org/10.1002/jbm.a.30655

Yanagi K et al (1992) A packed-bed reactor utilizing porous resin enables high density culture of hepatocytes. Appl Microbiol Biotechnol 37(3):316–320. https://doi.org/10.1007/BF00210985

Yanagi K, Miyoshi H, Ohshima N (1999) A high-density culture of hepatocytes using porous substrate for use as a bioartificial liver. J Artif Org 2(2):124–128. https://doi.org/10.1007/BF02480053

Yates EW et al (2012) Ligament tissue engineering and its potential role in anterior cruciate ligament reconstruction. Stem Cells Int. https://doi.org/10.1155/2012/438125

Yoon DM, Fisher JP (2006) Chondrocyte signaling and artificial matrices for articular cartilage engineering. Adv Exp Med Biol 67–86. doi: https://doi.org/10.1007/978-0-387-34133-0_5

Yuan L et al (2016) Effects of composition and mechanical property of injectable collagen I/II composite hydrogels on chondrocyte behaviors. Tissue Eng Part A. Mary Ann Liebert Inc 22(11–12):899–906. doi: https://doi.org/10.1089/ten.tea.2015.0513

Zarrintaj P et al (2018) Agarose-based biomaterials for tissue engineering. Carbohydrate Polymers 66–84. doi: https://doi.org/10.1016/j.carbpol.2018.01.060

Zhang Z et al (2014) Biomaterials and Stem Cells for Tissue Engineering. Expert Opin Biol Ther 13(4):527–540. http://doi.org/10.1517/14712598.2013.756468

Zhang J et al (2017) Stem cell culture and differentiation in microfluidic devices toward organ-on-a-chip. Fut Sci OA. 3(2):FSO187. doi: https://doi.org/10.4155/fsoa-2016-0091

Zhang L et al (2010) Chondrogenic differentiation of human mesenchymal stem cells: a comparison between micromass and pellet culture systems. Biotechnol Lett 32(9):1339–1346. https://doi.org/10.1007/s10529-010-0293-x

Zhang R et al (2019) Hybridization of a phospholipid polymer hydrogel with a natural extracellular matrix using active cell immobilization. Biomater Sci R Soc Chem 7(7):2793–2802. https://doi.org/10.1039/c9bm00093c

Zhang T et al (2012) Channelled scaffolds for engineering myocardium with mechanical stimulation. J Tissue Eng Regener Med 6(9):748–756. https://doi.org/10.1002/term.481

Zhang H, Chang H, Neuzil P (2019) DEP-on-a-chip: dielectrophoresis applied to microfluidic platforms. Micromachines. https://doi.org/10.3390/mi10060423

Zhang L, Hu J, Athanasiou KA (2009) The role of tissue engineering in articular cartilage repair and regeneration. Crit Rev Biomed Eng 37:1–57. doi: https://doi.org/10.1615/CritRevBiomedEng.v37.i1-2.10

Zhao XS et al (2006) Immobilizing catalysts on porous materials. Mater Today 9(3):32–39. https://doi.org/10.1016/S1369-7021(06)71388-8

Zhao Y et al (2018) Ultra-tough injectable cytocompatible hydrogel for 3D cell culture and cartilage repair. J Mater Chem B R Soc Chem 6(9):1351–1358. https://doi.org/10.1039/c7tb03177g

Zhu J (2012) Mammalian cell protein expression for biopharmaceutical production. Biotechnol Adv 30(5):1158–1170. https://doi.org/10.1016/j.biotechadv.2011.08.022

Zhu Y (2007) Immobilized cell fermentation for production of chemicals and fuels. In: Bioprocessing for value-added products from renewable resources. Elsevier, pp 373–396. doi: https://doi.org/10.1016/B978-044452114-9/50015-3

Zoro BJH et al (2008) The impact of process stress on suspended anchorage-dependent mammalian cells as an indicator of likely challenges for regenerative medicines. Biotechnol Bioeng 99(2):468–474. Wiley. https://doi.org/10.1002/bit.21544

Strategies and Advancement in Growth Factor Immobilizable ECM for Tissue Engineering

Y. Ikegami and H. Ijima

Abstract Immobilization of functional molecules in biotechnology is extremely important. Growth factors are biomolecules with high bioactivity on cells and are very useful for culture engineering and regenerative medicine applications. However, their poor stability and high cost are challenges for practical application. Immobilization of growth factors has been conducted by various methods to localize growth factors and to keep high local concentrations. In this chapter, previous methods to immobilize growth factors are outlined. Thereafter, organ-specific extracellular matrix (ECM) and ECM-modelized material as growth factor immobilizable materials are mentioned with focus on their properties and their applications such as regenerative medicine.

Keywords Growth factor · Decellularization · Liver-specific extracellular matrix · ECM-modelized material · Heparin · Tissue engineering

Abbreviations

ECM	Extracellular matrix
NGF	Nerve growth factor
HGF	Hepatocyte growth factor
EGF	Epidermal growth factor
bFGF	Basic fibroblast growth factor
VEGF	Vascular endothelial growth factor
GAG	Glycosaminoglycan
L-ECM	Liver-specific extracellular matrix
TX100	Triton X-100
PBS	Phosphate buffered saline

Y. Ikegami · H. Ijima (✉)
Department of Chemical Engineering, Kyushu University, 744 Motooka, Nishi-ku, Fukuoka 819-0395, Japan
e-mail: ijima@chem-eng.kyushu-u.ac.jp

A. Tripathi and J. S. Melo (eds.), *Immobilization Strategies*,
Gels Horizons: From Science to Smart Materials,
https://doi.org/10.1007/978-981-15-7998-1_3

H&E staining	Hematoxylin & Eosin staining
DNA	Deoxyribonucleic acid
ELISA	Enzyme-linked immuno-sorbent assay
PUF	Polyurethane foam
Hep-col	Heparin-conjugated collagen
Hep-gela	Heparin-conjugated gelatin
NHS	N-hydroxysuccinimide
MES	2-morpholinoethanesulfonic acid, monohydrate
EDC	1-Ethyl-3-(3-dimethylaminopropyl) carbodiimide

1 Introduction

Functional molecules, such as growth factors, are used widely in the field of regenerative medicine and as stimulatory molecules within cell cultures. For instance, nerve growth factor (NGF) enhances the cells' function of peripheral nerve and promotes their regeneration (Sofroniew et al. 2001). Moreover, hepatocyte growth factor (HGF) up regulates the function of hepatocytes (Semler et al. 2000). Many useful functional molecules, namely growth factors, have high bioactivity and offer potential in therapeutics. However, their poor stability and high cost are limitations for practical application.

Nonetheless, it is essential to immobilize functional molecules in biotechnology as it leads to improving their stability, prolonging half-life, and protecting against heat stress. Previously, it was reported that immobilized growth factors had higher stability than free growth factors in solution (Mizumachi and Ijima 2013). As a result, studies have been conducted for the development of substrates immobilizing growth factors, and of recombinant growth factors, which have the ability to bind specifically to substrates (Ikada and Tabata 1998; Nishi et al. 1998).

In this section, we outline materials capable of immobilizing growth factors with a focus on culture engineering and tissue engineering for regenerative medicine application.

2 Immobilization of Functional Molecules

2.1 Effectiveness of Immobilizing Functional Molecules in Biotechnology

In the field of biotechnology, animal cells, bacteria, and biomolecules derived from these sources have been used for drug discovery and industrial scale production of materials. For example, enzymes, which are biocatalysts, are used for the material production through a homogeneous reaction by mixing with a reactive substrate. This reaction system ignores the effect of mass transfer and thus has a high reaction rate. However, the added enzyme, which can be used repeatedly and is expensive, is discarded in this system due to the challenge of separating it from the mixture (Fig. 1a).

Therefore, the enzyme can be immobilized to effectively re-use the expensive biocatalyst. The following methods have been adapted for immobilizing the enzyme: binding method, in which an enzyme is immobilized on a carrier surface; encapsulating method, in which an enzyme is fixed in the carrier; crosslinking method, in which the enzyme is polymerized to prevent leakage from the bioreactor (Fig. 1**b**). Properly immobilized enzymes can be easily separated and retrieved from the mixture with the product after the reaction, allowing repeated use (Fig. 1a). Moreover, when the enzyme is filled and fixed in the plug flow bioreactor, the bioactivity per unit volume can be enhanced by increasing the density.

Thus, it is imperative to select an appropriate immobilizing method for the efficient use of functional molecules.

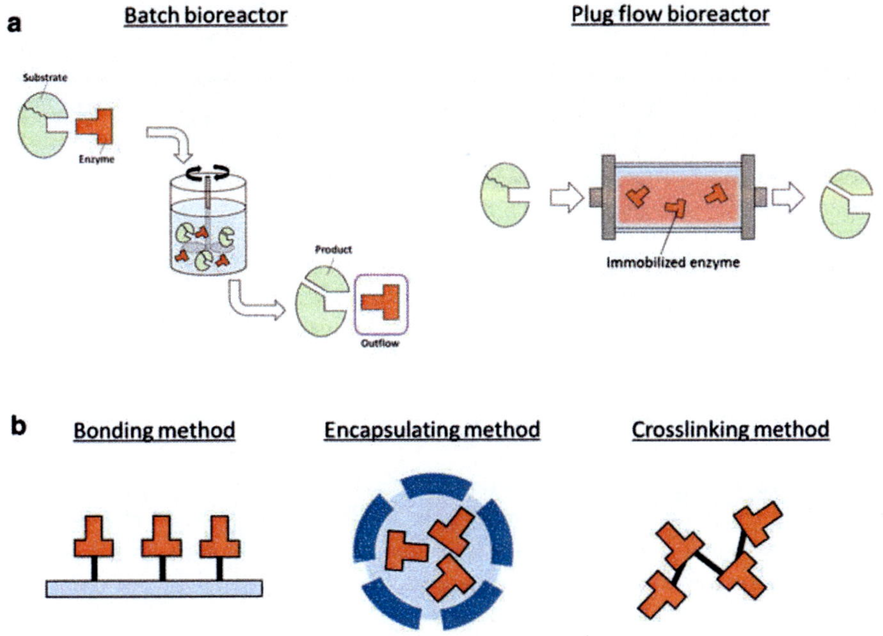

Fig. 1 Immobilization of functional molecules. **a** Advantage of immobilizing functional molecules; **b** immobilizing method

2.2 *Immobilization of Growth Factors*

For cell culture and regenerative medicine, immobilization of functional molecules, such as growth factors, is also effective. Growth factors have various bioactivities for cells and regulate the cell function by binding to the receptor on the cells surface, with a proper conformation. However, growth factors easily lose their bioactivity potential due to their poor stability, and its local concentration must be kept high so that the growth factors can elicit the desired effect on cells. Furthermore, in many cases for regenerating tissues, it is important to keep the growth factor localized at the injured site for day-ordered term.

The typical method for immobilizing growth factors is shown below (Fig. 2 and Table 1).

Direct Immobilization Using Chemical Crosslinking: Immobilization of growth factors on a carrier using chemical crosslinking is performed by forming a covalent bond between the growth factor and the carrier with a chemical crosslinker (Fig. 2a). In this method, when the carrier has a functional group necessary for the crosslinking reaction, the growth factor can be immobilized on the carrier. Lee et al. fabricated electrospun nanofibers composed of poly lactic-co-glycolic acid and immobilized NGF using a chemical crosslinker (Lee et al. 2012). The immobilized NGF induced neurite extension of neural model cells (i.e., PC12 cells). However, this method potentially causes denaturation of the immobilized growth factor as a result of the crosslinking reaction and requires a large amount of expensive growth factors. In addition, any remaining crosslinker is harmful for cells. Collectively, the issues highlighted here reduce the possibility of applying this method to applications such as cell culture and regenerative medicine.

Recombinant Growth Factor: Recently, recombinant growth factors that bind spontaneously to a carrier have been developed, considering the negative effect of chemical crosslinking on growth factor and cells (Fig. 2b). These recombinant growth factors can be crosslinked between the binding site on the carrier just by exposing them to the carrier; this is less harmful to the growth factor and cells. Nishi et al. developed epidermal growth factor (EGF) and basic fibroblast growth factor (bFGF) that have a collagen-binding domain (Nishi et al. 1998). This allows the growth factors to spontaneously bind to collagen, which is often used as a cell culture carrier. As a result, immobilization can occur without denaturation by crosslinker and can maintain biological activity on cells. Although these well-designed growth factors are effective for applications in cell culture and regenerative medicine, recombinant growth factor requires enormous time and is extremely expensive to develop. Thus, it is not practical at this time.

Immobilization Using Affinity: The immobilizing method using affinity realizes that the growth factor is spontaneously immobilized on the carrier by add a material that has high affinity for the growth factor (Fig. 2c). This immobilizing method does not require a crosslinking reaction nor does it have a negative effect on immobilized a growth factor. Furthermore, it is very versatile because any growth factor having affinity for the carrier can be immobilized. In the past, we conjugated

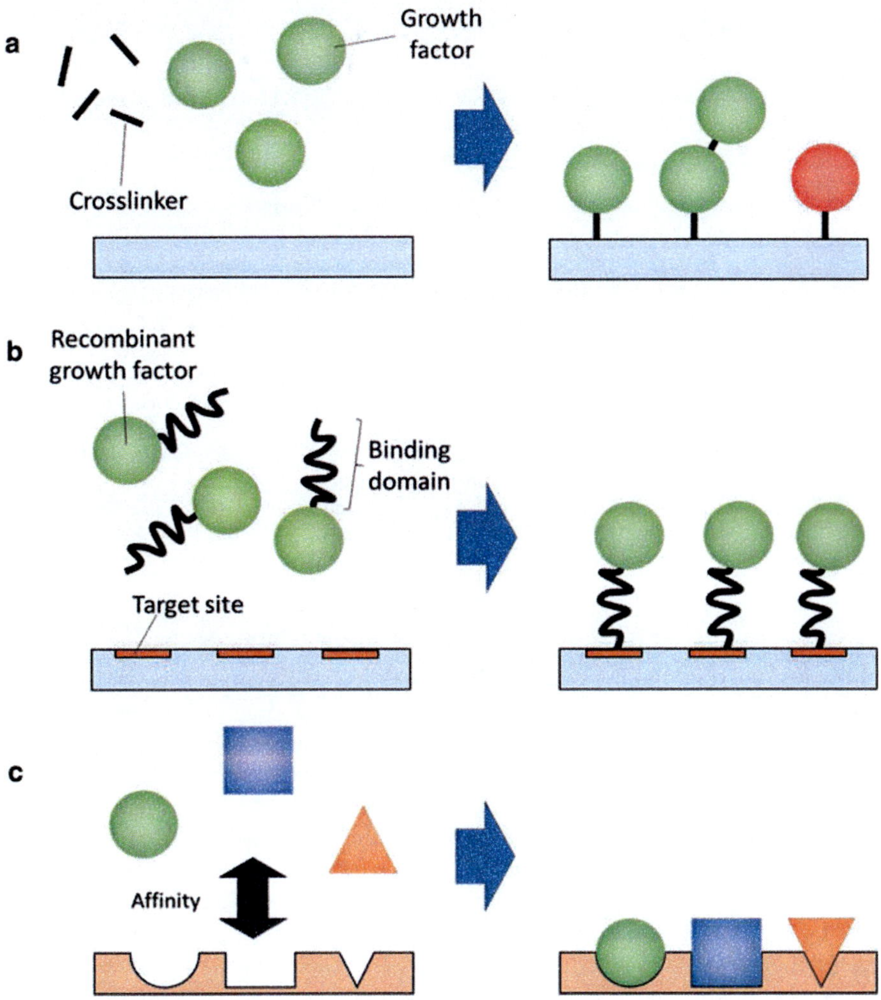

Fig. 2 Immobilizing methods of growth factors. **a** Direct immobilization using chemical crosslinking; **b** recombinant growth factor; **c** immobilization using affinity

heparin, which has high affinity for various growth factors and is used as a medium component to stabilize growth factors, into a culture carrier. This substrate could spontaneously immobilize growth factors such as HGF and vascular endothelial growth factor (VEGF). Moreover, the immobilized growth factor using this method has shown to have higher bioactivity on cells, compared to that of directly immobilized growth factor using the chemical crosslinker method (Ijima et al. 2009a, b; Hou et al. 2010; Nakamura et al. 2013a, b).

Which Method is the Most Suitable for Immobilization of Growth Factors? Earlier, the various methods to immobilize bioactive growth factors were outlined.

Table 1 Characteristics of immobilizing methods

	Advantages	Disadvantages
Direct immobilization using chemical crosslinking	• Flexible choice of immobilizable growth factors • Firmly immobilizing	• Denaturation of growth factors by chemical crosslinking • Possible cytotoxicity by remaining chemical crosslinkers
Recombinant growth factor	• Specific binding between recombinant growth factors and substrates • Without denaturing growth factors	• Numerous time and cost for development
Immobilization using affinity	• Spontaneous binding due to the affinity between growth factors and substrates • Relatively flexible choice of immobilizable growth factors • Use of natural growth factors without modification	• Environment-dependent binding strength between growth factors and substrates

Basically, when immobilizing a growth factor, an immobilizing method using an affinity without crosslinking reaction should be preferred as it does not denature the growth factor. However, it is important to note that depending on the condition, the growth factor may not be sufficiently immobilized by the method.

For example, in the immobilizing method using hydrophobic interactions, the hydrophobic region of the growth factor and the material constituting the carrier are bound to each other. Additionally, the growth factor can be easily immobilized on the carrier, and the binding becomes stronger in an environment where the salt concentration is high. Conversely, the hydrophobic interaction is weakened and the immobilized growth factor is easily detached from the carrier in an environment where a water-soluble organic solvent is present. Lastly, it is necessary to pay attention to the deactivation and the denaturation of growth factors under high salt concentrations.

In contrast, when using the immobilizing method that relies upon ionic bonds, ionic bonds are formed between ionizing functional groups in the growth factor and the material constituting the carrier. Similar to the method relying upon hydrophobic interactions, the growth factor can be easily immobilized on the carrier, but the weakly immobilized growth factor is easily detached under high salt concentrations, such as certain conditions that can arise in vivo.

In vivo, the growth factor immobilized by hydrophobic interactions will be eliminated from the surface of the carrier after a competitive adsorption/desorption equilibrium interaction with albumin, which is an abundant protein in the body. The ionic bond between the immobilized growth factor and the carrier is weakened due to the high salt concentrations present in the body. However, the use of a material having a strong binding force to the growth factor for the carrier will enable the growth factor to be retained on the carrier for a relatively long time even under high salt conditions in vivo. Furthermore, the immobilizing method using affinity makes it possible to immobilize a growth factor on the carrier without the need for modification of the native growth factor, making it useful for medical application. This is further backed up as there is elimination of the possibility of denaturation of the growth factor, such as those seen through the crosslinking reaction. Moreover, there is a significantly lower development cost than that of recombinant growth factor.

Immobilization of growth factors using affinity is preferable for the application in regenerative medicine. Additionally, the direct immobilizing method using crosslinking reaction can firmly immobilize the growth factor on the carrier, whereas the recombinant growth factor can be immobilized without crosslinking reaction. Hence, it is important to select and combine methods immobilizing growth factor according to the situation and desired result.

3 Organ-Specific Extracellular Matrix (ECM) and ECM-Modelized Material as a Substrate-Immobilizing Growth Factor

In the body, almost all cells are surrounded by an extracellular matrix (ECM). ECM not only serves as a structure that fills the space between cells but also acts as a scaffold for cell adhesion and to regulate cell functions by immobilizing and retaining in vivo functional molecules. As a result of its natural capabilities, ECM is considered as a material suitable for immobilizing growth factors.

The components of ECM include proteins (e.g., collagen, laminin, and fibronectin), proteoglycans composed of glycosaminoglycans (GAG) such as heparan sulfate and chondroitin sulfate, and glycoproteins. Among these, collagen is the main component of ECM and provides a scaffold for cell adhesion. In contrast, heparan sulfate immobilizes functional molecules such as in vivo growth factors and supplies biological signals to cells by increasing the local concentration of signaling molecules. Here, like the heparan sulfate, the above-mentioned heparin is a kind of heparan sulfate and is known to have an affinity for various growth factors. Therefore, we developed an ECM-modelized material using components to mimic the main function of ECM (i.e., collagen, gelatin, and heparin).

In this part, we explain the effectiveness and application of the organ-specific ECM and ECM-modelized material that has been developed as a growth factor immobilizing substrate.

3.1 Liver-Specific ECM (L-ECM)

As described above, almost all cells which constitute tissue are surrounded by the ECM. ECM composition is different depending on the organ/tissue, and there should be a suitable scaffold composition for each cell type. Therefore, if cell components that can become immune antigens can be removed from the tissue and ECM can be isolated, the material having not only a composition preferable for each cell type, but also a proteoglycan-derived growth factor immobilizability, can be effectively used.

This part explains about the preparation method for producing L-ECM as an organ-specific ECM, the basic characteristics, and the biological effectiveness.

Preparation Method of L-ECM: L-ECM is prepared through decellularization of a native liver tissue using surfactant solution, and enzymatically solubilizing it (Fig. 3). Specifically, cell components in the native liver tissue are removed by being exposed to Triton X-100 (TX100, non-ionic surfactant) in phosphate buffered saline (PBS). Then, the decellularized liver tissue is washed with PBS, lyophilized, and finally, solubilized enzymatically using pepsin in hydrochloric acid (Nakamura et al. 2013a, b; Ijima et al. 2018, 2019).

Basal Characteristics of L-ECM: The basic characteristics of L-ECM are shown below (Fig. 4). The decellularized liver tissue using TX100 was dyed by H&E staining, whereby the nucleus is stained with hematoxylin and the cytoplasm is stained with eosin. Nuclei remaining in decellularized tissues were observed because nuclei were relatively difficult to remove from tissues among cellular components that could elicit an immune response. As a result, the decellularized liver tissue was not stained as compared with the native liver, and it was confirmed that the nucleus and cytoplasm were almost removed from the liver tissue (Fig. 5a, b) (Ijima et al. 2019). Similarly, when the remaining DNA in L-ECM was quantified, a minute volume of nuclear components remained in the decellularized liver tissue compared to native liver, which implies that immune antigens in the liver tissue were efficiently

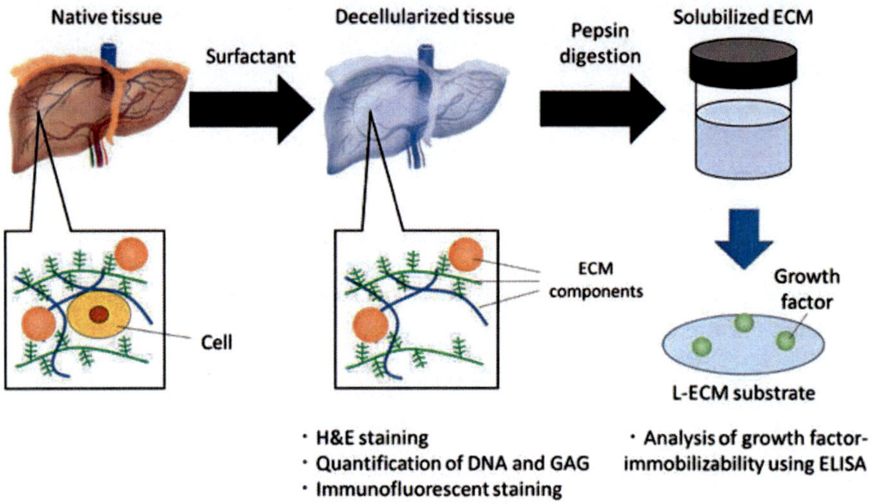

Fig. 3 Preparation method of L-ECM

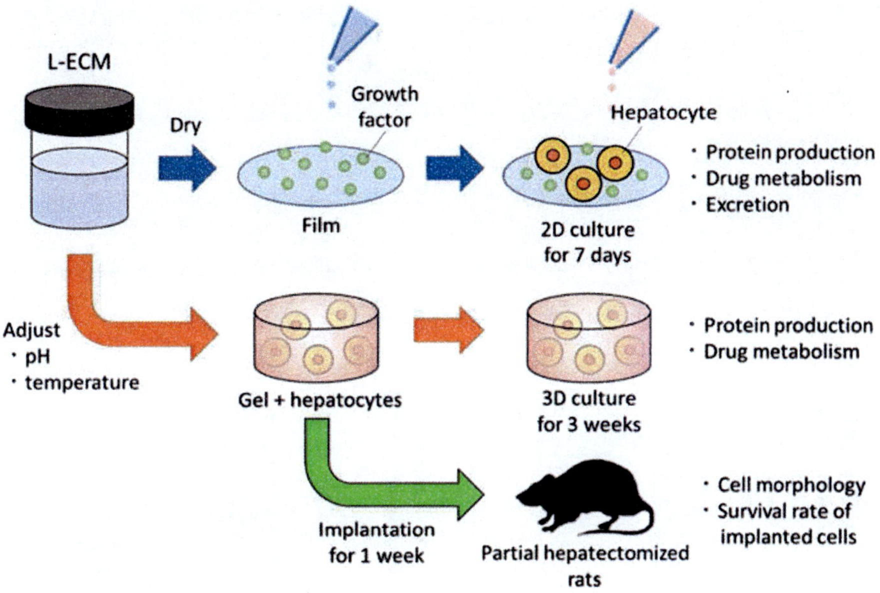

Fig. 4 Evaluation scheme of L-ECM

removed by decellularization using TX100 (Fig. 5c) (Ijima et al. 2019). Furthermore, component analysis showed that decellularized liver tissue preserved the main components of ECM, such as GAG, a series of collagens, and laminin (Fig. 5d–i) (Ijima et al. 2018, 2019).

Moreover, we added HGF as an example of growth factor on substrates using L-ECM and evaluated immobilizing density of HGF on the substrate by ELISA. Results indicated that L-ECM had the ability to immobilize HGF spontaneously, and with high efficiency (Fig. 5j) (Ijima et al. 2019). Subsequently, remaining growth factor on the substrate was detected after washing the substrate-immobilizing-HGF. As a result, it was found that L-ECM has the potential to retain approximately 90% of growth factors for several days, whereas almost all growth factors were released under the collagen condition, only one of the ECM' main component (Ijima et al. 2019). Further, L-ECM could form a gel by just changing to physiological pH and temperature, similar to collagen, which is generally used as a substrate for gel-embedded culture (Fig. 5k, l) (Ijima et al. 2019).

Biological Effectiveness of L-ECM: L-ECM had the following biological effects on primary rat hepatocytes (Fig. 4). Generally, in 2D culture, primary rat hepatocytes extend flat on the substrate with the culture time and lose their functions (Kitajima et al. 2007; Nahmias et al. 2007). However, primary rat hepatocytes on L-ECM scaffold did not flatten and instead formed tissue-like structure compared to those in collagen condition. Additionally, during the seven-day-cell culture, hepatocytes on the L-ECM scaffold showed higher hepatic functions compared to those of

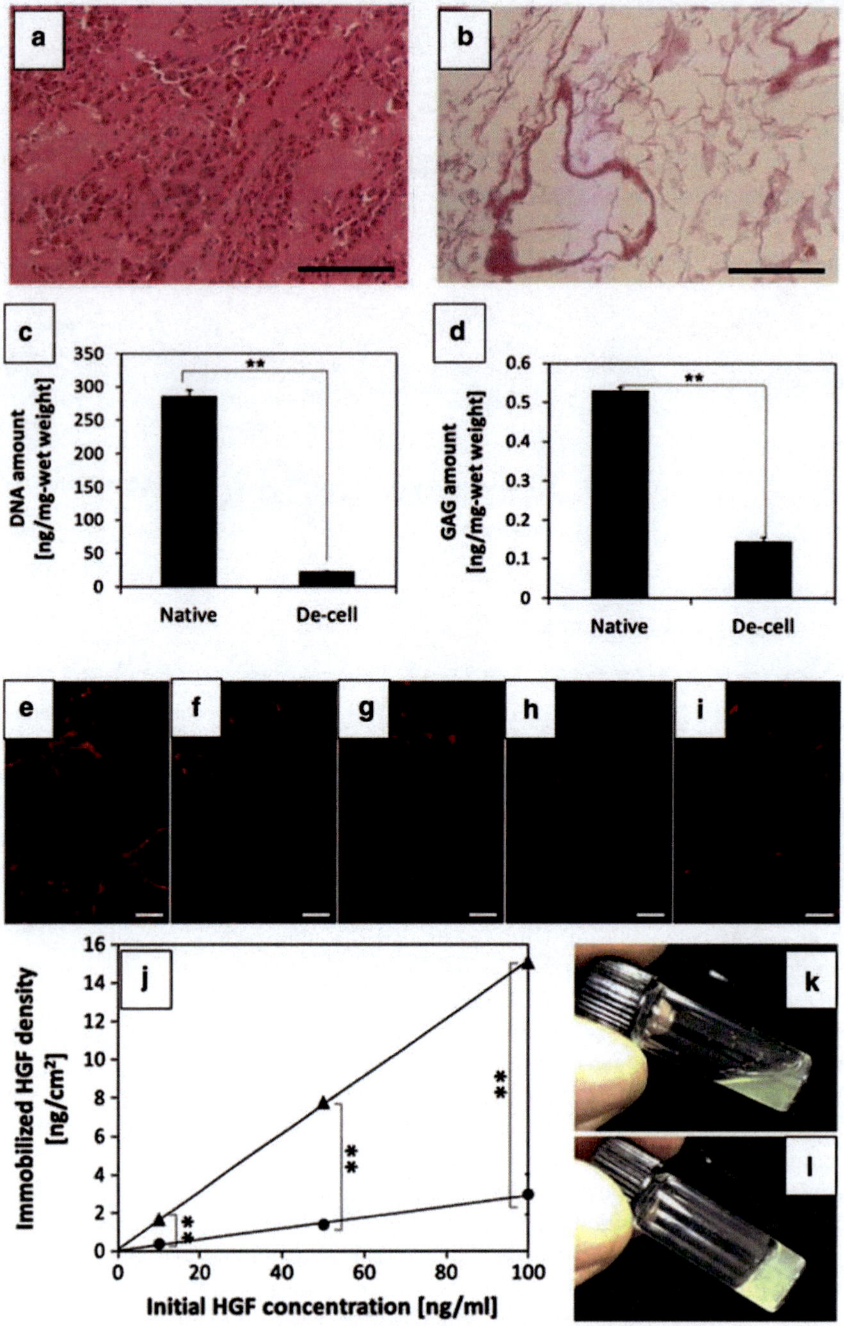
a
b
c
DNA amount [ng/mg-wet weight]
350
300
250
200
150
100
50
0
**
Native
De-cell
d
GAG amount [ng/mg-wet weight]
0.6
0.5
0.4
0.3
0.2
0.1
0
**
Native
De-cell
e
f
g
h
i
j
Immobilized HGF density [ng/cm²]
16
14
12
10
8
6
4
2
0
**
**
**
0
20
40
60
80
100
Initial HGF concentration [ng/ml]
k
l

◂**Fig. 5** Characteristics of L-ECM. Histological evaluation of natural liver (**a**) and decellularized liver (**b**) by H&E staining. **a, b** Scale bars represent 100 μm. DNA (**c**) and GAG (**d**) contents of decellularized liver ($n = 3$, bars represent S.D., **: $p < 0.01$). Immunofluorescent staining of decellularized liver for collagen types I (**e**), III (**f**), IV (**g**), V (**h**), and laminin (**i**). **e–i** Scale bars represent 50 μm. HGF immobilizability of Type I collagen and L-ECM films (**j**). Triangles, L-ECM film; circles, Type I collagen film ($n = 3$, bars represent S.D., **: $p < 0.01$). Gross morphology of L-ECM sol (**k**) and L-ECM gel (**l**). Figure **e–i** were reproduced from Ref. Ijima et al. (2018), and the others were reproduced from Ref. Ijima et al. (2019) with permission from Elsevier

collagen condition in typical functions, such as protein synthesis, drug metabolism, and excretion (Fig. 6a–c) (Nakamura et al. 2013a, b).

Similarly, in 3D culture, the ability of primary rat hepatocytes in L-ECM gel were maintained for three weeks to evaluate protein synthesis and drug metabolism. It was found that the drug metabolism activity under L-ECM condition was higher than that of the collagen condition (Fig. 6d, e) (Ijima et al. 2019). Furthermore, to verify the effect of L-ECM for liver regeneration, L-ECM gel, filled into polyurethane foam (PUF) with primary rat hepatocytes, was implanted subcutaneously to 70% partially hepatectomized rats. The implanted hepatocytes formed clusters in the specimen, and the number and size of clusters under the condition using L-ECM were higher than those under the collagen condition (Fig. 6f) (Ijima et al. 2019). Moreover, the survival rate of hepatocytes transplanted with L-ECM gel was sustained, compared to that of the condition with collagen gel (Fig. 6g) (Ijima et al. 2019).

To facilitate tissue regeneration, several studies have been conducted, which administered growth factors and cells that support regeneration in vivo. These additives are expected to not only act on the surrounding lesion site individually, but also promote the regeneration further by the combinatorial effect. The above results indicate that the L-ECM component and the ability to immobilize growth factors attributed to the L-ECM component improve the function and survival rate of implanted hepatocytes and promote tissue formation for liver regeneration. Hence, L-ECM is expected to be a very useful material for liver repair after trauma.

3.2 *ECM-Modelized Material*

We developed heparin-conjugated collagen/gelatin (Hep-col/Hep-gela) as the ECM-modelized material by chemically crosslinking heparin to collagen and gelatin (Fig. 7) (Ijima et al. 2009a, b; Hou et al. 2010; Mizumachi and Ijima 2013; Nakamura et al. 2013a, b; Nagai et al. 2017). This material is composed of components responsible for the main function of ECM and can immobilize growth factors by electrostatic affinity via the sulfate groups in heparin, similar to proteoglycans present in ECM. For this reason, the substrate using ECM-modelized material enables immobilization of growth factors and is less detrimental for cells and growth factors compared with the immobilizing method using chemical crosslinkers. In addition, this substrate is

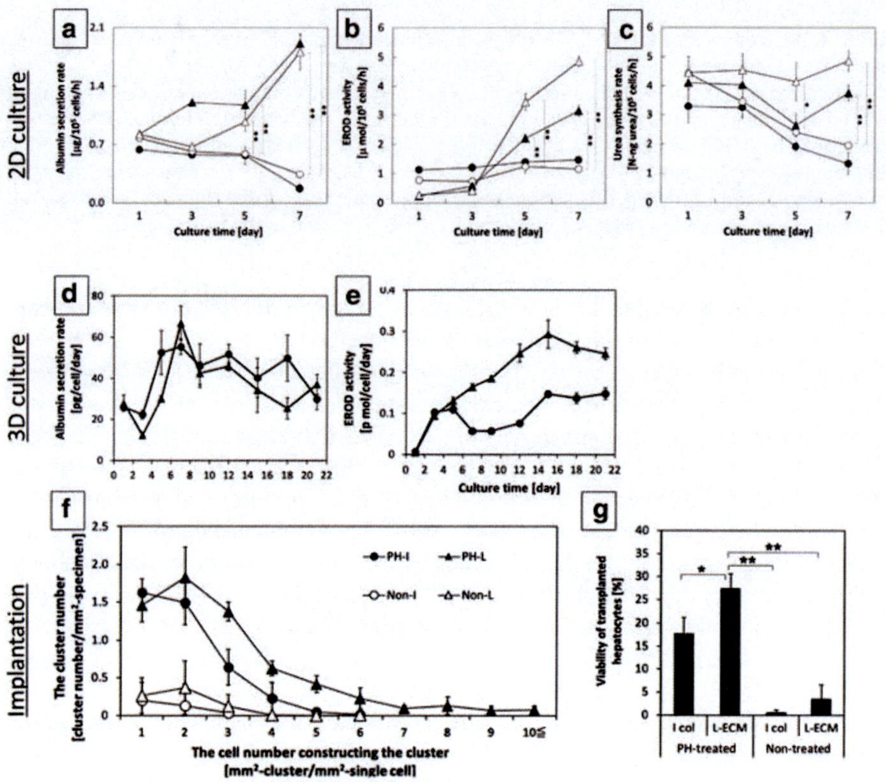

Fig. 6 Liver-specific functions of primary rat hepatocytes on various films. **a** Albumin secretion rates; **b** EROD activity, drug metabolic activity; **c** urea synthesis rates. Closed circles, HGF-immobilized collagen film; open circles, collagen film; closed triangles, HGF-immobilized L-ECM film; open triangles, L-ECM film ($n = 3$, bars represent S.D., **: $p < 0.01$). Liver-specific functions of primary rat hepatocytes in type I collagen and L-ECM gels. **d** albumin secretion rates; **e** EROD activity ($n = 3$, bars represent S.D.) Number of clusters classified by the constituent cell number (**f**) and viability (**g**) of transplanted hepatocytes into non-treated and PH-treated rats ($n = 3$, bars represent S.D., *: $p < 0.05$, **: $p < 0.01$). All figures were reproduced from ref. (Nakamura et al. 2013a, b; Ijima et al. 2019) with permission from Elsevier

expected to construct a microenvironment more suitable for cells by mimicking the mechanism when growth factors are immobilized in ECM.

Preparation Method of ECM-modelized Material: For the synthesis of ECM-modelized material, we used a reaction to form an amide bond between succinate ester and amine, which is a commonly used method for protein modification (Ijima et al. 2009a, b; Hou et al. 2010; Mizumachi and Ijima 2013; Nagai et al. 2017). Specifically, the carboxyl group of heparin was N-hydroxysuccinimide (NHS)-esterified by adding it to 2-morpholinoethanesulfonic acid, monohydrate (MES) buffer containing 1-Ethyl-3-(3-dimethylaminopropyl) carbodiimide (EDC) and NHS. Thereafter, the amino group of collagen/gelatin forms an amide bond directly with NHS-activated

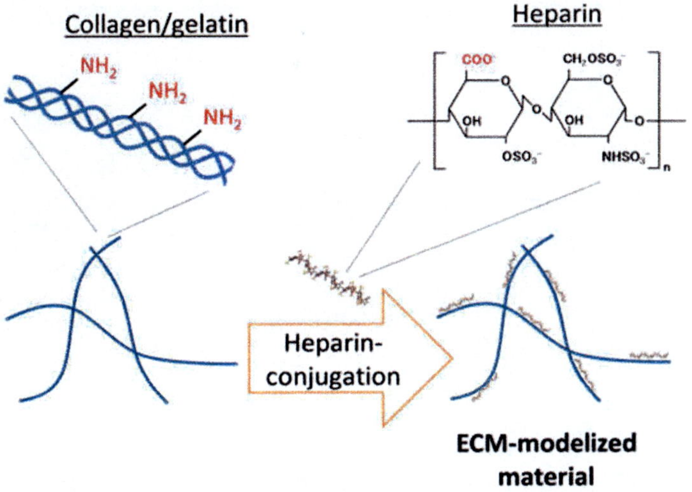

Fig. 7 Preparation method of ECM-modelized material

carboxyl group of a heparin when exposed to NHS-esterified heparin. The chemical crosslinker is harmful for cells and growth factors can be satisfactorily removed by washing after the crosslinking reaction has occurred.

Basal Characterization of ECM-modelized Material: To suppress over-crosslinking between amino and carboxyl groups in collagen, we removed the excess chemical crosslinkers from the heparin/EDC/NHS solution by dialysis and then mixed it with a collagen solution. Like collagen and L-ECM, Hep-col solution could form a gel by changing the environment to have physiologically representative temperature and pH levels. This also preserves the properties of collagen as a raw material.

In contrast, Hep-gela is prepared by chemically crosslinking heparin to gelatin, which is derived from the acid- or alkali-treated collagen (Nakamura et al. 2013a, b). Unlike Hep-col, Hep-gela has a different sol–gel transition temperature and crystal structure from raw gelatin (Nakamura et al. 2013a, b). In addition, the mechanical strength and biodegradation rate of Hep-gela gel crosslinked with glutaraldehyde was evaluated. It was seen that Hep-gela has lower mechanical strength and faster degradation rates than gelatin prepared at the same concentration, confirming the influence of the conjugated heparin on the stability of the gel. However, in comparison between collagen gel and Hep-gela gel, the Hep-gela condition showed higher mechanical strength and lower degradation rates (Nakamura et al. 2013a, b). Henceforth, in situations where high mechanical strength is required for substrates, the use of Hep-gela is preferable as a raw material.

Alternatively, it is confirmed that Hep-col and Hep-gela could immobilize various growth factors such as VEGF (Mizumachi and Ijima 2013; Nakamura et al. 2013a, b), HGF (Hou et al. 2010), and bFGF (Nagai et al. 2017) with higher efficiency and retained them for a longer term in comparison with heparin-free condition by

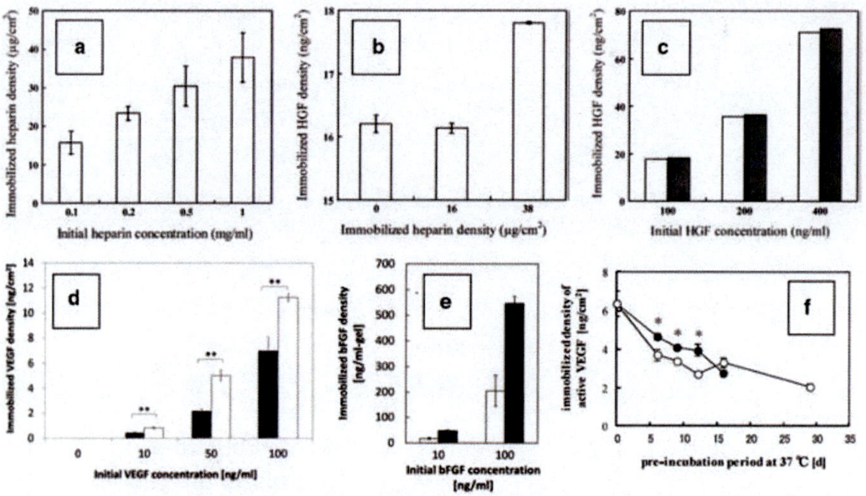

Fig. 8 **a** Immobilized heparin densities on collagen films. **b** Immobilized HGF densities on Hep-col films. **c** Immobilized HGF density on Hep-col films. Open and hatched bars indicated the results at day 1 and day 7, respectively. $n = 3$, bars represent S.D. **d** Immobilized VEGF density on gelatin and Hep-gela films. Closed bars, gelatin film; open bars, Hep-gela films. $n = 3$, bars represent S.D., **: $p < 0.01$. **e** Immobilized b FGF densities in type I collagen and Hep-col gels. Open and hatched bars indicated the results of collagen and Hep-col gels, respectively. $n = 3$, bars represent S.D. **f** Stability of VEGF at 37 °C. Closed circles, immobilized VEGF; Open circles, VEGF solution; $n = 3$. Bars represent S.D. *: $p < 0.05$. Figure **f** was reproduced from Ref. (Mizumachi and Ijima 2013) with permission from Advanced Biomedical Engineering Editorial Office, and the others were reproduced from Ref. (Hou et al. 2010; Nakamura et al. 2013a, b; Nagai et al. 2017) with permission from Elsevier

ELISA (Fig. 8c–e) (Hou et al. 2010). Moreover, the composition of ECM-modelized material could be manipulated by the amount of added heparin, which affected the immobilizing efficiency of growth factors (Fig. 8a, b) (Hou et al. 2010). Furthermore, in culture evaluation of human umbilical vein endothelial cells (HUVECs), it was shown that VEGF immobilized on the scaffold using ECM-modelized material had higher stability than VEGF in culture media (Fig. 8f) (Mizumachi and Ijima 2013).

Biological Effectiveness of ECM-modelized Material: ECM-modelized material had the following biological effectiveness, which is highlighted in Fig. 9. In 2D culture, like the case using L-ECM, primary rat hepatocytes on the scaffold using ECM-modelized material suppressed flatting and upheld the production rate of albumin, one the main functions of the liver (Fig. 9**a**). Moreover, hepatocytes on the scaffold immobilized HGF via heparin exhibited a more vigorous albumin secretion than those on the scaffold which directly immobilized HGF using chemical crosslinkers. The albumin production rate of hepatocytes under the condition directly immobilizing growth factors was comparable with that of growth factor-free condition (Ijima et al. 2009a, b; Hou et al. 2010). These results imply that the direct immobilization of growth factors on the scaffold using chemical crosslinking leads to denaturation of growth factors. And conversely, growth factor immobilized on a

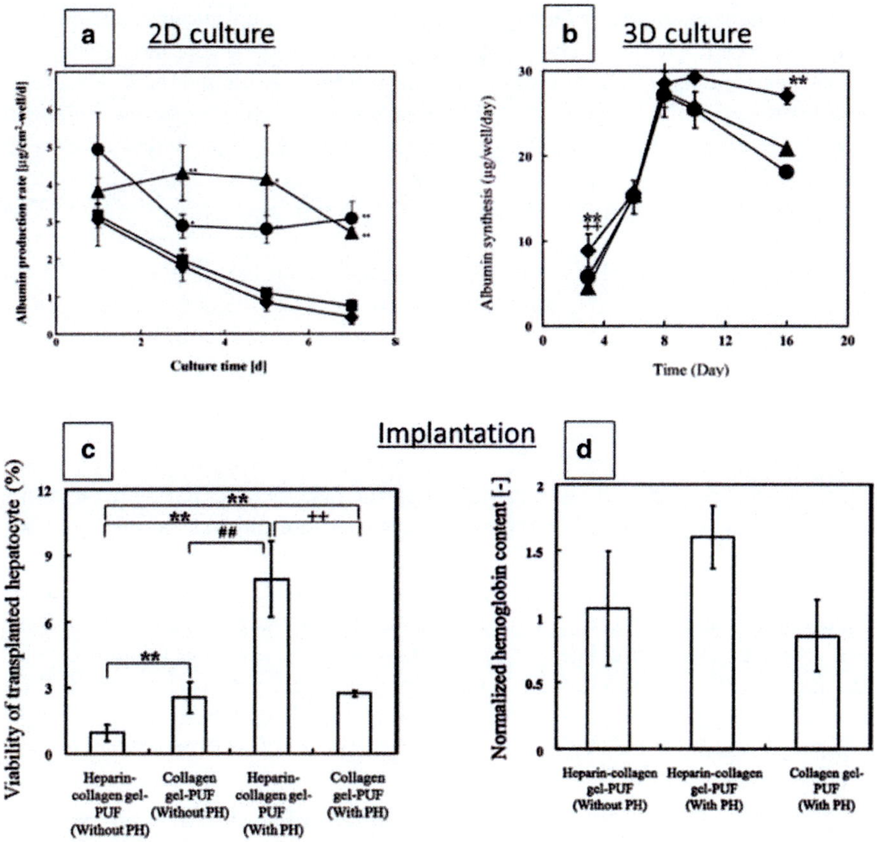

Fig. 9 Albumin production rate of primary rat hepatocytes on various films (**a**) in gels (**b**). **a** Results on films. Diamonds, cross-linked collagen film; triangles, collagen film in HGF-supplemented medium; squares, HGF directly immobilized collagen film; circles, HGF/heparin-immobilized collagen film. Bars represent S.D. Asterisk indicates statistically significant differences from crosslinked collagen. *: $p < 0.05$, **: $p < 0.01$. **b** Results in gels. Diamonds, HGF/heparin-immobilized collagen gel; triangles, HGF-immobilized collagen gel; circles, collagen gel. *: $p < 0.05$, +: $p < 0.05$, **: $p < 0.01$, ++: $p < 0.01$, significant difference versus collagen gel and HGF-immobilized collagen gel, respectively. $n = 3$, bars represent S.D. **c** Viability of transplanted hepatocytes in various scaffolds with or without partial hepatectomy (PH). **: $p < 0.01$, +: $p < 0.05$, ++: $p < 0.01$, #: $p < 0.05$, ##: $p < 0.01$, significant difference versus (1) heparin-immobilized, collagen-gel-filled PUF, (2) collagen-gel-filled PUF with PH, and (3) collagen-gel-filled PUF, separately. $n = 3$, bars represent S.D. **d** Hemoglobin content from various types of hepatocyte-embedded scaffolds after implantation. *: $p < 0.05$, +: $p < 0.05$. $n = 3$, bars indicate S.D. All figures were reproduced from Ref. (Ijima et al. 2009a, b; Hou et al. 2010, 2012) with permission from Elsevier

scaffold via heparin improves the function of hepatocytes. Furthermore, hepatocytes under the condition with heparin-mediated immobilization of growth factors had hepatic functions equivalent to those under the condition using the medium supplemented with growth factors (Fig. 9a). This result suggests that the ECM-modelized material can reduce the amount of growth factors required to achieve desired effects.

Even in 3D culture, the same tendency was observed in the culture evaluation of primary rat hepatocytes (Fig. 9b) (Hou et al. 2010). Neural stem cells were embedded into Hep-col gel premised with bFGF, and the number of viable cells in the gel was evaluated after one week. The result confirmed that the proliferation of neural stem cells in Hep-col gel was remarkably enhanced compared to that under the condition using collagen (Nagai et al. 2017). It is hypothesized that this could be due to the concentration of growth factors inside Hep-col gel being upheld for a longer time than that in the collagen gel.

Furthermore, when primary rat hepatocytes seeded in a collagen gel were subcutaneously implanted into rats, the survival rate of the implanted cells did not change, regardless if this was on rats with or without partial hepatectomy before the implantation (Fig. 9c). By contrast, in the case using Hep-col gel, partial hepatectomy improved the survival rate of the implanted cells and promoted angiogenesis in the recipient rat (Fig. 9c, d) (Hou et al. 2012). In addition, hepatocytes were subcutaneously implanted into partial hepatectomized rat following vascularization using VEGF-immobilized Hep-col gel. As a result, tissue regeneration was facilitated more under the condition in which pre-vascularization was performed (Hou et al. 2011). These results suggest that the secreted growth factor, along with liver regeneration and the vigorous mass transfer via blood vessels induced by the immobilized growth factor, promotes the liver regeneration.

Based on the above results, the ECM-modelized material has an immobilizability of growth factors like L-ECM and is effective for applications in cell culture and cell transplantation.

4 Applications of Substrates Capable of Immobilizing Growth Factors

4.1 Effectiveness of Growth Factor Immobilization for Tissue Regeneration

There is currently a plethora of evidence available that highlights the potential of various growth factors in regenerative medicine. Especially, bFGF has various functions such as promoting angiogenesis, activation of fibroblast proliferation, promotion of ECM production, and induction of neural differentiation. Therefore, in the early stages of the regenerative medicine field, bFGF was administered freely and systemically by injecting within an aqueous solution. However, as mentioned above, growth factor has low stability when free and its local concentration decreases at the

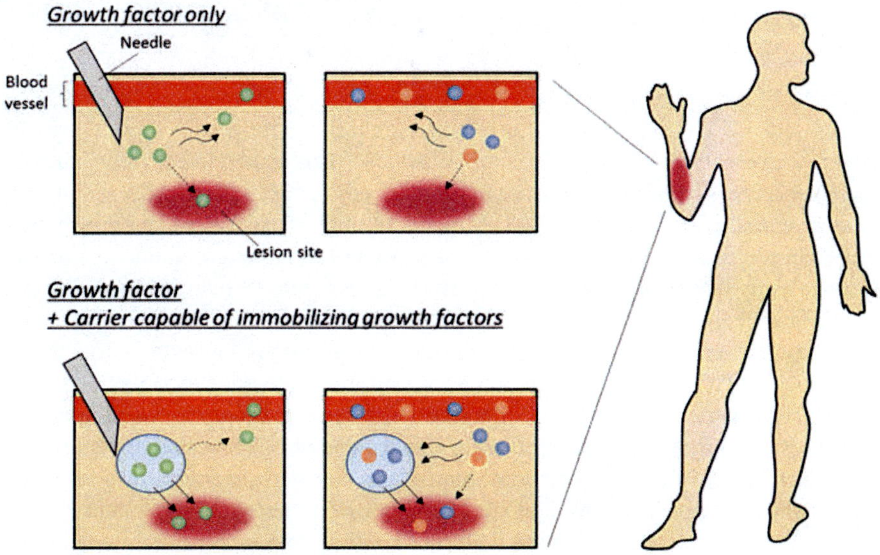

Fig. 10 Growth factors after administration in vivo with or without a carrier capable of immobilizing growth factors

lesion site due to dilution with blood or tissue fluid (Fig. 10). For these reasons, it is necessary to use a large excess of expensive growth factors to obtain their effects, which may increase the treatment cost and increase the possibility of side effects on tissues other than at the injured site.

Therefore, in order to make the best use of the effective growth factor, it is important to maintain a high local concentration of the growth factors at the lesion site for the desired period. Tabata et al. developed a method to release bFGF slowly using an acidic gelatin hydrogel along with biodegradation of the substrate (Fig. 10) (Ikada and Tabata 1998). In this method, bFGF is immobilized and retained in the hydrogel by electrostatic forces, which prevent the immobilized growth factor from diffusing easily throughout the body. Consequently, it also upholds the local concentration of growth factors and subsequently decreases the risk of side effects. Furthermore, bFGF immobilized on a substrate not only affects the injury directly, but it also induces angiogenesis at the injured site, which will improve the survival rate of implanted cells facilitating tissue regeneration. Besides, gelatin has been clinically used for a long time and has been proven to be safe and biodegradable in vivo.

4.2 Various Applications of L-ECM and ECM-Modelized Material

L-ECM and ECM-modelized material can immobilize and retain various growth factors, such as bFGF with high efficiency, as described above. In previous studies, many researchers reported the combinatorial effect of different growth factors on tissue regeneration (Lang et al. 2008; Ijima et al. 2009a, b; Shimosaka and Bhawal 2013). Thus, it is considered the co-immobilization of multiple growth factors on the substrate using our material is very effective for regenerative medicine applications.

During the regenerative process, growth factors that promote the regeneration of the tissue are often produced. For instance, the concentration of growth factors in the blood, such as HGF and VEGF, increases in partial hepatectomized rats (Lindroos et al. 1991; Taniguchi et al. 2001). In addition, in the regenerative stage of peripheral nerves, a Büngner band through Wallerian degeneration is formed at the injured site, where growth factors to promote peripheral nerve regeneration are secreted (Griffin and Thompson 2008). Because of the affinity between our material and growth factors, it is expected that our materials continuously trap the secreted growth factors. Consequently, it causes that the local concentration of growth factors naturally increases at the repairing site after implantation of the substrate, which further promotes tissue regeneration.

Our materials are composed of biocompatible and biodegradable material and can be transformed into various forms such as film (Ijima et al. 2009a, b; Hou et al. 2010; Nakamura et al. 2013a, b), gel (Hou et al. 2010, 2011, 2012; Nakamura et al. 2013a, b; Nagai et al. 2017; Ijima et al. 2018), fiber (Bual et al. 2018; Ikegami et al. article in press), and are applicable to various scenarios with regenerative medicine (Fig. 11).

In drug screening, it is required to culture cells with high functionality and high throughput. Hydrogel-embedding culture, a typical 3D culture system, can construct a microenvironment that is relatively similar to that in vivo, whereas it is difficult to perform high-throughput evaluation due to poor diffusivity of polymers such as antibodies for staining. In the past, we developed a culture system using a thinned Hep-col gel with the thickness reduced from 2 mm to several hundred μm (Nagai et al. 2017). Fluorescently labeled antibodies used for cell staining were added to the thin-layered gel, and the sustained release behavior was evaluated. As a result, the majority of the fluorescent substance remained after 12 h in a typical 2-mm-thick gel. On the contrary, almost all the fluorescent substance was immediately removed from the thin-layered gel within 1 h. Moreover, when immunofluorescent staining was performed, the cells in the thin-layered gel could be clearly observed compared to the cells in the 2-mm-thick gel. These results show that this system can overcome problems of hydrogel-embedded culture (e.g., unclear observation due to light scattering and too long evaluation time as a result of slow substance diffusion) (Fig. 12a–e). Furthermore, growth factor immobilization via heparin can enhance the function of embedded cells and also reduce costs by improving the stability of growth factors and enhancing the titer of growth factors to cells.

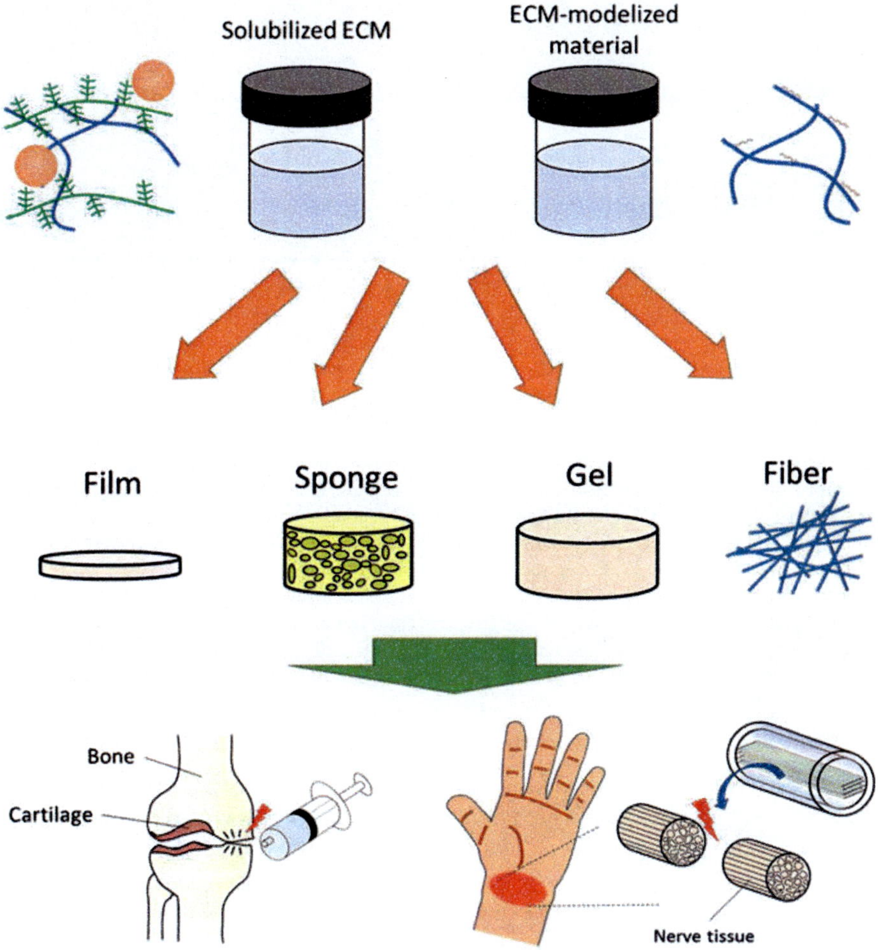

Fig. 11 Various applications of L-ECM and ECM-modelized material

Therefore, L-ECM and ECM-modelized material that can be used for both drug screening and in regenerative medicine applications are very promising materials.

4.3 Decellularized Liver as a Template for Organ-Scaled Tissue Regeneration

In recent years, we have been working on organ-scaled liver tissue regeneration using organ-specific ECM. There has been global interest for the regeneration of organs or tissues. The liver is one of the most commonly studied organs as the demand and

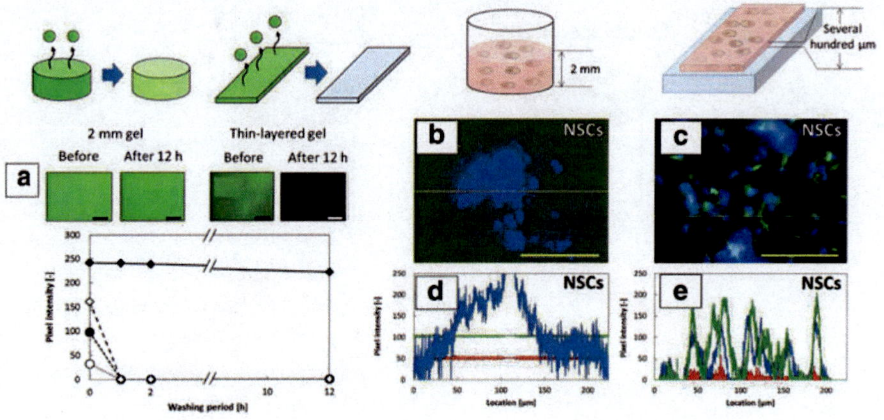

Fig. 12 Characteristics of the culture system using thin-layered gel and 2 mm gel. **a** Diffusion behavior of fluorescent substances from thin- layered gel and 2 mm gel. Changes in gel intensity over time at various gel thickness are shown. Closed diamonds, 2 mm gel; open diamonds, 300-μm-thin-layered gel; closed circles, 200-μm-thin-layered gel; open circle, 100-μm-thin-layered gel. $n = 3$, bars represent S.D. Fluorescent photographs of 2 mm gel and thin-layered gel before and after washing are shown above the line graph. **b–e** Fluorescence microscopic observation of neural cells embedded in a collagen gel. **b, c** Fluorescent microphotographs of neural cells in 2 mm gel (**b**) and thin-layered gel (**c**). Bars indicate 30 μm. **d, e** Distribution histogram of pixel intensity along the yellow dashed line of (**b**) and (**c**). All figures were reproduced from Ref. (Nagai et al. 2017) with permission from Elsevier

priority of liver tissue regeneration is extremely high, the liver is a central organ for metabolism in vivo. In liver tissue engineering for the purpose of repairing severe liver failure, it is important to construct a functional liver tissue substitute. However, since the hepatic parenchymal cells constituting the liver tissue have high oxygen demand, a scaffold having a dense and extensive vascular network to provide an adequate oxygen supply is necessary for constructing organ-scaled liver tissue in which cells are seeded at a high density.

Therefore, we aim to create an organ-scaled template having the dense vascular network of the liver using decellularization and to create an organ-scaled liver substitute by seeding hepatocytes into it (Fig. 13a) (Shirakigawa et al. 2013). Previously, we have shown that properly treated rat decellularized liver has a dense vascular network (Fig. 13b–e) (Shirakigawa et al. 2012, 2013). The decellularized liver is composed of liver-specific ECM components and is expected to be an ideal scaffold for hepatocytes to exhibit their functions. In addition, it is expected to promote regeneration by immobilizing growth factors produced during liver regeneration by ECM components contained in the decellularized liver. When hepatocytes were seeded inside the decellularized liver and cultured, hepatic functions such as albumin synthesis and ammonia metabolism were confirmed (Fig. 13f, g) (Shirakigawa et al. 2013; Sakamoto et al. 2019). We have succeeded in the location-specific arrangement of

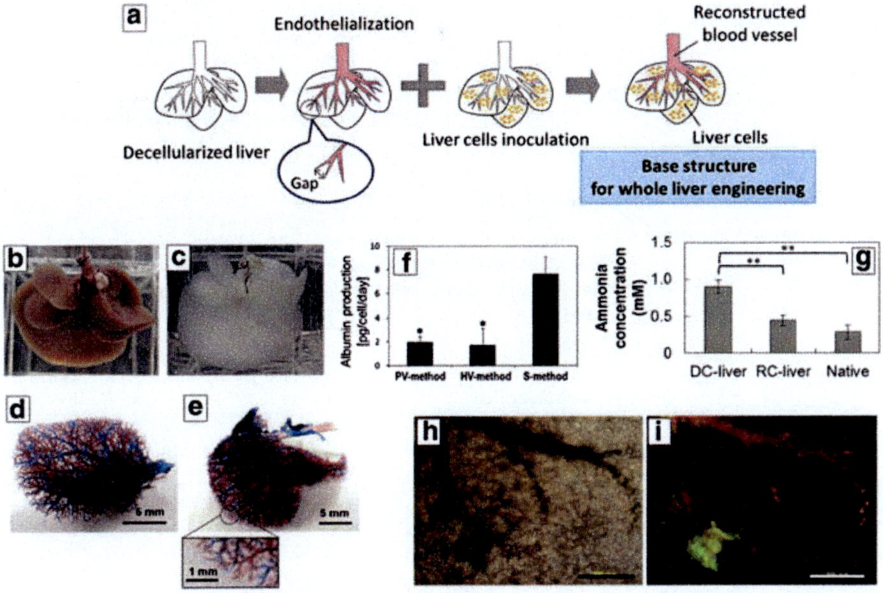

Fig. 13 Organ-scaled scaffold for liver tissue regeneration. **a** Concept of the study. **b, c** Appearances of the native rat liver (**b**) and decellularized rat liver (**c**). **d, e** The cast of the tubular structure in native liver (**d**) and decellularized liver (**e**). Red and blue resins were injected via portal vein and hepatic vein, respectively. **f, g** Liver-specific function of hepatocytes cultured in decellularized liver on albumin production, **f** ($n = 3$, bars indicate S.D.); ammonia metabolism, **g** ($n = 2$, **: $p < 0.01$). **h, i** Microphotographs of liver and endothelial cells cultured in decellularized liver for three days. **e** Phase-contrast micrograph. **i** Fluorescent micrograph. Endothelial cells are red, and liver cells are green. Scale bars indicate 500 µm. Figure **g** was reproduced from Ref. (Sakamoto et al. 2019) with permission from Springer Nature, and the others were reproduced from Ref. (Shirakigawa et al. 2012, 2013) with permission from Elsevier

hepatocytes and vascular endothelial cells within the decellularized liver (Fig. 13h, i) (Shirakigawa et al. 2013).

Hence, there are still many issues to be solved, but further research on the construction of organ-scale liver tissue using decellularized liver as a scaffold will lead to the creation of treatments for severe liver failure.

5 Pros and Cons of L-ECM and ECM-Modelized Material

So far, we have described L-ECM and ECM-modelized materials as materials capable of immobilizing growth factors. In this part, we discuss the advantages and disadvantages of L-ECM and ECM-modelized material (Table 2).

Both L-ECM and ECM-modelized materials are biocompatible and biodegradable and have the GAG-derived immobilizability of growth factors. Moreover, these are

Table 2 Characteristics of L-ECM and ECM-modelized material

	Advantages	Disadvantages
L-ECM	• Immobilizability of growth factors • Liver-specific ECM components • Moldability • Biocompatibility • Biodegradability	• Relatively expensive • Uncertain safety in vivo
ECM-modelized material	• Immobilizability of growth factors • Possible industrial production • A lot of knowledge about raw materials • Moldability • Biocompatibility • Biodegradability	• Possible lack of ingredients • Not optimized

excellent in moldability and can be applied to various fields by transforming into various forms such as films, gels, and fibers.

Among them, L-ECM contains informational proteins, such as laminin, in addition to collagen and GAG, which perform some of the main functions of ECM, and is expected to show various effects on cells. However, L-ECM is relatively expensive, and the long-term safety in vivo has not been proven although telopeptide chains that can serve as immune antigens are removed by pepsin digestion.

In contrast, ECM-modelized material is composed only of collagen, gelatin, and heparin, and is considered to be capable of industrial production. In addition, these components have been used clinically for a long time, and their effects are well investigated by many researchers. For example, heparin not only stabilizes growth factors by binding to growth factors but also increases the sensitivity of the receptors on the cell surface to growth factors. However, the composition of the ECM-modelized material is so simple that its ingredients may be insufficient depending on the goal, and further, improvements will be required.

Therefore, both L-ECM and ECM-modelized materials have sufficient functionality as materials capable of immobilizing growth factors. They show high potential for applications such in regenerative medicine and culture engineering. Further investigation of how liver-specific components affect cells may help to realize the clinical use of L-ECM and improve the ECM-modelized material.

6 Conclusions

For cell culture and regenerative medicine, growth factors are very important, and their local concentration should be kept high in order to exhibit bioactivity to the surrounding cells. As a result, research has been conducted on immobilizing growth factors by various methods. Among them, a method using organ-specific ECM and ECM-modelized material can imitate the immobilizing mechanism of growth factors in vivo and can spontaneously immobilize natural growth factors without modification.

This chapter described the effectiveness of L-ECM and ECM-modelized material as a material capable of immobilizing growth factors and their application to regenerative medicine and cell culture carriers. Substrates using L-ECM and ECM-modelized material were able to immobilize various growth factors spontaneously with high efficiency and improved the cell functions. In addition, these materials can be molded according to the application to various shapes, such as film, gel, and fiber. Furthermore, immobilization of growth factors via heparan sulfate/heparin not only reduces the amount of growth factor used but also increases the stability of the growth factor and the binding strength between the growth factor and the receptor present on the cells. Consequently, it will enable the use of growth factors for various applications. In addition, liver-specific ECM components contained in L-ECM will have a further effect on the functional expression of cells.

References

Bual R, Kimura H, Ikegami Y, Shirakigawa N, Ijima H (2018) Fabrication of liver- derived extracellular matrix nanofibers and functional evaluation in in vitro culture using primary hepatocytes. Materialia 4:518–528

Griffin JW, Thompson WJ (2008) Biology and pathology of nonmyelinating Schwann cells. Glia 56(14):1518–1531

Hou YT, Ijima H, Matsumoto S, Kubo T, Takei T, Sakai S, Kawakami K (2010) Effect of a hepatocyte growth factor/heparin-immobilized collagen system on albumin synthesis and spheroid formation by hepatocytes. J Biosci Bioeng 110(2):208–216

Hou YT, Ijima H, Takei T, Kawakami K (2011) Growth factor/heparin-immobilized collagen gel system enhances viability of transplanted hepatocytes and induces angiogenesis. J Biosci Bioeng 112(3):265–272

Hou YT, Ijima H, Shirakigawa N, Takei T, Kawakami K (2012) Development of growth factor-immobilizable material for hepatocyte transplantation. Biochem Eng J 69:172–181

Ijima H, Kubo T, Hou YT (2009a) Primary rat hepatocytes form spheroids on hepatocyte growth factor/heparin-immobilized collagen film and maintain high albumin production. Biochem Eng J 46:227–233

Ijima H, Mizumoto H, Nakazawa K, Kajiwara T, Maysushita T, Funatsu K (2009b) Hepatocyte growth factor and epidermal growth factor promote spheroid formation in polyurethane foam/hepatocyte culture and improve expression and maintenance of albumin production. Biochem Eng J 47:19–26

Ijima H, Nakamura S, Bual R, Shirakigawa N, Tanoue S (2018) Physical properties of the extracellular matrix of decellularized porcine liver. Gels 4(2)

Ijima H, Nakamura S, Bual RP, Yoshida K (2019) Liver-specific extracellular matrix hydrogel promotes liver-specific functions of hepatocytes in vitro and survival of transplanted hepatocytes in vivo. J Biosci Bioeng 128(3):365–372

Ikada Y, Tabata Y (1998) Protein release from gelatin matrices. Adv Drug Deliv Rev 31(3):287–301

Ikegami Y, Ijima H (article in press) Development of heparin-conjugated nanofibers and a novel biological signal by immobilized growth factors for peripheral nerve regeneration. J Biosci Bioeng. https://doi.org/10.1016/j.jbiosc.2019.09.004

Kitajima T, Terai H, Ito Y (2007) A fusion protein of hepatocyte growth factor for immobilization to collagen. Biomaterials 28(11):1989–1997

Lang EM, Schlegel N, Reiners K, Hofmann GO, Sendtner M, Asan E (2008) Single-dose application of CNTF and BDNF improves remyelination of regenerating nerve fibers after C7 ventral root avulsion and replantation. J Neurotrauma 25(4):384–400

Lee JY, Bashur CA, Milroy CA, Forciniti L, Goldstein AS, Schmidt CE (2012) Nerve growth factor-immobilized electrically conducting fibrous scaffolds for potential use in neural engineering applications. IEEE Trans Nanobiosci 11(1):15–21

Lindroos PM, Zarnegar R, Michalopoulos GK (1991) Hepatocyte growth factor hepatopoietin A rapidly increases in plasma before DNA synthesis and liver regeneration stimulated by partial hepatectomy and carbon tetrachloride administration. Hepatology 13(4):743–750

Mizumachi H, Ijima H (2013) Measuring stability of vascular endothelial growth factor using an immobilization technique. Adv Biomed Eng 2:130–136

Nagai T, Ikegami Y, Mizumachi H, Shirakigawa N, Ijima H (2017) Development of an in-situ evaluation system for neural cells using extracellular matrix-modeled gel culture. J Biosci Bioeng 124(4):430–438

Nahmias Y, Berthiaume F, Yarmush ML (2007) Integration of technologies for hepatic tissue engineering. Adv Biochem Eng Biotechnol 103:309–329

Nakamura S, Ijima H (2013a) Solubilized matrix derived from decellularized liver as a growth factor-immobilizable scaffold for hepatocyte culture. J Biosci Bioeng 116(6):746–753

Nakamura S, Kubo T, Ijima H (2013b) Heparin-conjugated gelatin as a growth factor immobilization scaffold. J Biosci Bioeng 115(5):562–567

Nishi N, Matsushita O, Yuube K, Miyanaka H, Okabe A, Wada F (1998) Collagen- binding growth factors: production and characterization of functional fusion proteins having a collagen-binding domain. Proc Natl Acad Sci USA 95(12):7018–7023

Sakamoto H, Shirakigawa N, Bual RP, Fukuda Y, Nakamura S, Miyata T, Yamao T, Yamashita YI, Baba H, Ijima H (2019) A novel evaluation system for whole-organ-engineered liver graft by ex vivo application to a highly reproducible hepatic failure rat model. J Artif Organs 22(3):222–229

Semler EJ, Ranucci CS, Moghe PV (2000) Mechanochemical manipulation of hepatocyte aggregation can selectively induce or repress liver-specific function. Biotechnol Bioeng 69(4):359–369

Shimosaka M, Bhawal UK (2013) bFGF Upregulates the expression of NGFR in PC12 cells. J Hard Tissue Biol 22(1):19–24

Shirakigawa N, Ijima H, Takei T (2012) Decellularized liver as a practical scaffold with a vascular network template for liver tissue engineering. J Biosci Bioeng 114(5):546–551

Shirakigawa N, Takei T, Ijima H (2013) Base structure consisting of an endothelialized vascular-tree network and hepatocytes for whole liver engineering. J Biosci Bioeng 116(6):740–745

Sofroniew MV, Howe CL, Mobley WC (2001) Nerve growth factor signaling, neuroprotection, and neural repair. Annu Rev Neurosci 24:1217–1281

Taniguchi E, Sakisaka S, Matsuo K, Tanikawa K, Sata M (2001) Expression and role of vascular endothelial growth factor in liver regeneration after partial hepatectomy in rats. J Histochem Cytochem 49(1):121–130

Prospects of Cell Immobilization in Cancer Research and Immunotherapy

Remya Komeri, H. P. Syama, G. U. Preethi, B. S. Unnikrishnan, R. Shiji, M. G. Archana, Deepa Mohan, Anuj Tripathi, and T. T. Sreelekha

Abstract Cell immobilization is upgraded from a status of fundamental technique to an advanced approach for therapeutic applications. Thinking beyond the conventional surgical, radiation and chemotherapy for cancer management, the advent of immunotherapy via transplanting tumor antigens or engineered cancer cells has a promising role. Highlighting the significance of transplanting cellular vaccines and genetically engineered cells, creating a dynamic barrier through immobilization technique appreciates the physical isolation of cells from the systemic factors, continuous delivery of therapeutic molecules and formation of a local immune stimulatory environment. Transplantation of potential stem cells encapsulated in suitable matrices is also explored for cancer therapy. Irrespective of the therapeutic applications, cell immobilization approaches have prospects in generating reconstructive tumor models for cancer drug screening. In this chapter, the methods of cell immobilization, potential matrices for cell encapsulation, relevance in advanced cancer immunotherapy, and applications in reconstructive tumor models are discussed.

Keywords Cell immobilization · Cancer · Polymer · Cancer vaccines · Immunotherapy

List of Abbreviations

Ca Calcium

R. Komeri · H. P. Syama · G. U. Preethi · B. S. Unnikrishnan · R. Shiji · M. G. Archana · D. Mohan · T. T. Sreelekha (✉)
Laboratory of Biopharmaceuticals and Nanomedicine, Division of Cancer Research, Regional Cancer Centre (Research Centre, University of Kerala), Thiruvananthapuram, Kerala, India
e-mail: sreelekhatt@rcctvm.gov.in; ttsreelekha@gmail.com

A. Tripathi
Nuclear Agriculture and Biotechnology Division, Bhabha Atomic Research Centre, Mumbai, India

Homi Bhabha National Institute, Mumbai, India

A. Tripathi and J. S. Melo (eds.), *Immobilization Strategies*, Gels Horizons: From Science to Smart Materials,
https://doi.org/10.1007/978-981-15-7998-1_4

PVA	Polyvinyl alcohol
IgG	Immunoglobulin G
PSS	Polystyrene sulfonic acid
PEI	Polyethylene imine
PEG	Polyethylene glycol
PVA	Polyvinyl alcohol
PLGA	Polylactic-co-glycolic acid
DNA	Deoxyribonucleic acid
mAbs	Monoclonal antibodies
FDA	Food and drug administration
PD-1	Programmed death 1
PDL-1	Programmed death ligand 1
CTLA	Cytotoxic t-lymphocyte antigen
HBV	Hepatitis B virus
HPV	Human papillomavirus
RNA	Ribonucleic acid
APC	Atigen-presenting cells
DC	Dendritic cell
GM-CSF	Granulocyte-macrophage colony-stimulating factor
IL-2	Interleukin 2
PAP	Prostatic acid phosphates
MART-1	Melanoma associated antigen recognized by T cells
EVA	Ethyl vinyl acetate
MIP	Macrophage inflammatory protein
CD	Cluster of differentiation
HER	Human epidermal growth factor receptor
MSR	Mesoporous silica rods
OVA	Ovalbumin
ECM	Extracellular matrix
TME	Tumor microenvironment
CNC	Cellulose nanocrystals

1 Basics of Cell Immobilization

Immobilization of cells can be defined as the physical captivity of intact cells in a suitable matrix by preserving its biological activity. Retaining the viability, proliferation and functional characteristics of cells are imperative for an immobilized system. Cell immobilization techniques have promising applications in biomedicine, pharmacology, cosmetology, food and agricultural sciences, beverage production, analytical applications, and biologics production. Biomedical applications are reviewed in the chapter giving significance to cancer research. Contribution of various immobilization techniques is well appreciated in cell therapy, tissue engineering, and high throughput drug screening. Cleverly engineered immobilization technique offers protection to the cell from unfavorable conditions, washout, shear as well as immunological rejection. Immobilization also enables cell–cell interaction and functional reorganization into tissue-like architecture. Surface area to volume ratio, porosity, hydrophilicity, and the adopted methodology influence the bioactivity of the immobilized cells.

1.1 Methods of Cell Immobilization

Adsorption, entrapment, encapsulation, interfacial polymerization, coacervation, and emulsification are the common methods used for cell immobilization (Elakkiya et al. 2016; Tripathi and Melo 2017). Interaction of cells with a supporting matrix through vander Waals, electrostatic, hydration or hydrophobic forces are exploited for adsorption. Physical entrapment of cells inside a suitable gel matrix is practiced for entrapment as one of the safest and simplest methods for immobilization. However, a more versatile method is cell encapsulation, in which cells are immobilized in liquid form within a protective outer membrane of polymeric, lipo-protein based or non-ionic nature. Encapsulation can be achieved via emulsification, coacervation, or interfacial polymerization.

1.2 Materials for Cell Immobilization

Both natural and synthetic materials with controllable sol–gel transition are employed as support materials for cell immobilization. Natural polymers are preferred for better bioavailability and biocompatibility suitable for cell survival, whereas, difficult to control the physico-chemical properties (Kumar and Tripathi 2012). Synthetic polymers are advantageous with tunable physico-chemical characteristics, ease of modifications, less batch-to-batch variation, and better availability with known composition. The following biomaterials have been explored for cell immobilization.

1.2.1 Natural Polymers for Cell Immobilization

Alginate: Alginates are widely being used as immobilizing materials for cells or tissue in the development of artificial organs and with the potential to be used in the treatment of a variety of diseases, such as Parkinson's disease, chronic pain, liver failure, and hypocalcemia (Melvik and Dornish 2004). Alginate is a natural anionic polysaccharide obtained from several organisms including Azotobactervinelandii, several Pseudomonas species, and a variety of brown seaweed. Alginate gels could be developed by an ionic network in the presence of cations such as Ca^{2+} or other multivalent counter ions. Alginate gel beads are used as a versatile platform for entrapment of cells for immobilization, as an implantation material and also for immobilizing drugs or nutrients for delivery systems. Owing to the remarkable and distinguished structural property of alginate, it can form hydrogels and capsules of different size and can also be clubbed with other natural polymer like agarose and gelatin for achieving amiable elasticity with modulated surface chemistry to control the cell adhesion and growth (Tripathi and Kumar 2011; Tripathi et al. 2013a). Recently, Suarez-Arnedo et al. (2019) and the team have developed a protocol to fabricate three-dimensional matrices based on alginate hydrogels for mammalian cell encapsulation. Alginate possesses a broad range of functional properties, and due to that reason, their success as immobilization matrices relies on an appropriate choice of materials and methodology for each application (Smidsrød and Skja 1990). The biomaterial is made up of varying amounts of two monomers (α-L-guluronic acid and β-D-mannuronic acid). Recent studies revealed the impact of monomeric composition, molecular structure as well as gel formation kinetics of the hydrogel on the stability, mechanical resistance, biodegradability, permeability, and biodegradability of the material. As per recent reports, hydrogels having higher concentration of alginate tend to be stiffer and exhibit higher elastic components (Suarez-Arnedo et al. 2018). The biocatalyst activity recovery and mechanical strength of polyvinyl alcohol (PVA) and sodium alginate were used to synthesize cell immobilized beads (Wei et al. 2018).

Agarose: Agarose is derived from agar which is mainly associated with minimal immune responses. Agar is a complex polysaccharide obtained from various red algae. The main property of agar is its capability to liquefy upon heating to 96° and hardening into jelly on cooling to 40–45°. Agarose has been used for the immobilization of cells in the form of spherical beads, blocks, and membranes. Agarose gels could be easily engineered in beaded forms such as microspheres whose size and porosity could be tuned by carefully controlling the preparation procedures; among these, spray or suspension gelation is the most popular (Zucca et al. 2016). However, cryogelation technology is one of the advanced greener approaches to create porous scaffold of various natural and synthetic polymers for the immobilization of cells and biomolecules (Tripathi and Melo 2019). Also, reports show agarose microencapsulated mouse islets induced normoglycemia for up to 56 days without inflammatory cell infiltration in diabetic Balb/c mice (Salter and Kell 1991). The foremost challenge with agarose was to create a gel with sufficient immunoprotection due to its inability to block diffusion of cytotoxic immunoglobulin G (IgG). The immunoprotective properties of agarose gels are determined by the concentration of agar solution to form perm-selective membranes. Incorporation of 5% polystyrene sulfonic acid (PSS) onto 5% agarose to form the core of microcapsules in combination with polybrene layer coating to prevent the leakage of PSS and a layer of carboxyl methyl cellulose as the outermost shell offered biocompatibility of microcapsules (Bahulekar et al. 1991).

Chitosan: Chitosan, a linear polysaccharide, is the second most abundant support compound that can be obtained from nature after cellulose. It is a polymer derived from deacetylated chitin chains and has been proposed as a substitute for alginate due to the only difference in few functional groups. The promising features that make it an important polymer are its natural origin, non-toxicity, biocompatibility, biodegradability, and low cost (Kumar and Tripathi 2012). Several studies have proven the successful reduction of pericapsular fibrosis by coating chitosan as a layer for alginate-based microcapsule (Yang et al. 2010). Baruch and Machluf (2006) reported that chitosan-alginate complexes could improve long-term mechanical stability. However, the low solubility of chitosan under physiological pH limits its application of in islet encapsulation (Ruel-Gariépy et al., 2002; Yang et al. 2010). Chitosan has low pH values as 4 are needed to dissolve the polymer (Tripathi and Melo 2015, 2016). Islets are very susceptible to low pH. Hence, many attempts have been made to modify the polymer as such that it is soluble under more physiological pH. As a result, novel water-soluble chitosan derivatives have been developed Sobol et al. (2013) which could be dissolved at pH 7.0. Yang et al. (2010) generated chitosan hydrogel that allows capsule formation at physiological pH values by adding glycerol 2-phosphate disodium salt hydrate into acetic chitosan solution. Rat islets microencapsulated in this hydrogel reversed hyperglycemia in diabetic mice with a progressive increase in body weight as a consequence (Yang et al. 2016).

Cellulose: Cellulose, usually composed of 1,4-linked b-D-glucopyranosyl chains, is an excellent polymer for cell immobilization that occurs very abundant on earth. The low cost and commercial availability in both fibrous, as well as granular forms, makes it an approachable candidate for immobilization. Cellulose needs to be modified chemically to form hydroxypropyl cellulose, carboxy methyl cellulose, and ethyl cellulose to enhance its solubility in water. Cellulose has been exploited as encapsulation material with rat, porcine, and mouse islets. There are many controversies prevailing about biocompatibility of cellulose derivatives. Some groups stated the absence of host reactions to cellulose-based capsules, whereas others reported visible tissue reactions involving immune infiltrates and fibrous capsular formation in vivo (Yang et al. 2016). Risbud and Bhonde (2001); Risbud et al. (2003) reported that cellulose membranes avoid contact between activated complement proteins and the encapsulated islets.

1.2.2 Synthetic Polymers for Cell Immobilization

Polyacrylamide: Polyacrylamide is a polymer obtained by polymerization of acrylamide monomers. The first used matrix for cell immobilization is polyacrylamide. The main advantage of polyacrylamide is that of being nonionic, and also, the added feature is that within the matrix gel, properties of immobilized enzymes are only minimally altered. Studies reported that the thermal stability of polyacrylamide might be due to the presence of strong interaction of the hydrogen bonds present within this matrix. But dimethyl aminopropionitrile which is the initiator of the polymerization process is highly toxic and requires great care (Borin et al. 2018).

Poly (ethylene imine) (PEI): Polyethyleneimine (PEI), with the highest concentration of amino groups, has gained importance as an excellent carrier in a number of industrially immobilized biosystems. The use of poly (ethylene imine) has attained approval as a secondary direct food additive by the US Food and Drug Administration under the Federal Food, Drug, and Cosmetic Act. Due to that reason, unlike chemical agents like glutaraldehyde, the use of poly (ethylene imine) has attained great importance within the immobilized cell industry (Bahulekar et al. 1991). There are a number of patents that have been filed on the use of PEI to bind a wide variety of enzymes and whole cells. PEI is used in a wide range variety of ways with and without other synthetic and natural polymers. It imparts a hydrophilic character and mechanical strength to immobilized cell preparation as a minor constituent. It is widely used in cross-linked or derivatized forms as the primary constituent of the carrier. The islet cells were encapsulated in the form of biocompatible semi-permeable membranes by using a combination of alginate/polylysine/polyethyleneimine. The cells remained viable in culture for four months and later transplanted into the animal body also led to the survival for several months. In a study by Joung et al. (1987), PEI was used to modify alginate, and the resulting polymer aggregate was studied as a cell carrier. The immobilization technique which they opted depended on reversible gelation of the PEI-modified alginate. The modification by PEI offered an increase in mechanical strength, improved cell retention, better catalyst life, ease of pelletization.

Polyethylene glycol (PEG): Polyethylene glycol (PEG) is considered as one among the most important versatile synthetic polymer used in cell immobilization and also used as a coating in cell encapsulated microcapsules. Water-soluble nature of PEG makes its application microencapsulation in the absence of too harsh solvents (Hu and de Vos 2019). Reátegui et al. (2014) Incorporated PEG to silica to encapsulate human foreskin fibroblasts and mouse embryonic fibroblasts. PEG added mechanical stability to the matrix thereby attained a complete immobilization which restricted the motility, growth, and proliferation of encapsulated cells. Koh et al. (2002) exploited the photolithography technique to develop poly(ethylene glycol) (PEG)-based hydrogel microstructures encapsulating viable mammalian cells on glass and silicon substrates. PEG can also act as an immunoprotective membrane to prolong functional islet survival (Weber et al. 2009; Knobeloch et al. 2017). Unlike most of the synthetic polymers, PEG forms hydrogels with a high water content that offers a mild microenvironment for encapsulated cells inside and a protein-resistant surface outside (Nuttelman et al. 2008; Lutolf and Hubbell 2005).

Polyvinyl alcohol (PVA): Polyvinyl alcohol (PVA) is a biodegradable, non-toxic, water-soluble, and biocompatible thermoplastic polymer. Polyvinyl alcohol is obtained from polyvinyl acetate by the hydrolysis of vinyl acetate to form vinyl alcohol groups. PVA is completely insoluble in almost all organic solvents except ethanol, in which it exhibits low solubility. Being a thermoplastic polymer, PVA can be converted to various structures by freeze-thawing processes, and as such, the gelation of PVA is performed through repeated freeze-thawing cycles. Polyacrylonitrile is a synthetic polymer that has gained significant attention due to some distinctive features like chemical resistance, abrasion resistance, and thermal stability. Biodegradability, as well as low hydrophilicity, is considered as a major obstacle to the long-term use of a polyacrylonitrile matrix (Borin et al. 2018).

Poly (lactic-co-glycolic acid) (PLGA): Cell encapsulation using aliphatic polyesters has already been proposed for, but its mechanical instability and difficulty in permeation tunability owing to its biodegradability has limited its application. Poly (lactic-co-glycolic acid) (PLGA) is a linear, polymerized aliphatic polyester, thought to overcome some issues due to its better biostability. However, PLGA still undergoes hydrolysis under physiological conditions and produces two monomers named lactic acid and glycolic acid. However, these two monomers are reported to be non-toxic at normal physiological dose. PLGA microencapsulated porcine islets have been xenotransplanted into diabetic rats and reduced hyperglycemia considerably (Hu and de Vos 2019). Xu et al. (2017) suggested that the degradation of PLGA could be inhibited by modifying its structure, or its degree of crystallinity or amount of residual monomer that could pave a way to make it as a promising material for cell encapsulation due to its biocompatibility.

2 Prospects of Cell Immobilization in Cancer Therapy

Cancer is the second major cause of death in the world and is responsible for an estimated 9.6 million deaths in 2018 (Siegel et al. 2018). The global cancer statistics of 2018 revealed the alarming situations while considering the 18.1 million new cancer cases in 185 countries (Bray et al. 2018). Lung cancer is the most prevalent and caused 18.4% of cancer deaths (Bray et al. 2018). The number of new cancer patients is significantly higher in Asian countries (Fig. 1). The rise in smoking population, change in life style, food habits, and an increase in life expectancy with an

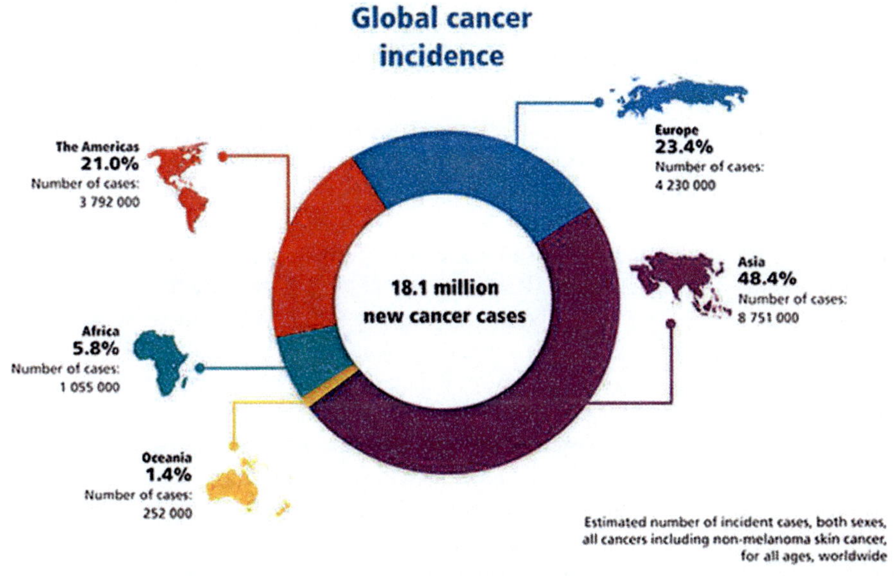

Fig. 1 Global statistics of cancer incidence (Bray et al. 2018)

advanced healthcare system sensitize people more to the incidence of cancer. Still, with the advanced medical care, a well acceptable, ultimate treatment solution for cancers is not yet identified. More to that, the genetic variability, the extent of susceptibility, and aggressive nature of several cancers including glioma further complicate the advent of a universal therapeutic strategy for cancer patients irrespective of the anatomical location of cancer. A mini review of current therapeutic approaches of cancer treatment and the possible limitations is explained as follows.

2.1 Current Therapeutic Approaches for Cancer Management

2.1.1 Chemotherapy

Chemotherapy is the use of any drug to kill cancer cells. Surgery and radiation therapy remove, destroy, or damage cancer cells in a particular area, but chemotherapy can work throughout the whole body. However, chemotherapeutic drugs cannot differentiate cancer cells from normal cells and affect fast-growing healthy cells of skin, hair, intestines, and bone marrow. Side effects of chemotherapeutic drugs vary with the type of cancer, location, dose, and the general health of the person. Nausea, vomiting, alopecia, fatigue, neutropenia, hearing loss, mucositis, loss of appetite, permanent damage to the heart, lung, liver, kidneys, or reproductive system, etc. are some other side effects of chemotherapy (Aggarwal et al. 2009). Some chemotherapeutic drugs can damage the tissues if they leak out of the blood vessel (Villalobos 2006). Aggressive chemotherapy can also cause massive destruction of the ovarian reserve in female and leads to premature ovarian failure and infertility (Maltaris et al. 2007).

2.1.2 Radiation Therapy

The use of high doses of radiation to kill cancer cells and shrink tumors is called radiation therapy or radiotherapy. Radiotherapy slows tumor growth by damaging the DNA (Tubiana 1971). Internal radiation therapy with a solid source of radiation is used in cervical, prostate, breast, esophageal, skin cancer. Internal radiation with a liquid source is called systemic therapy, in which radioactive substance travels in the blood throughout the body. Radioactive iodine or a radioactively labeled monoclonal antibody is the liquid sources of radiation which is swallowed or injected into the body. The type of radiation therapy, part of the body being treated, and a dose of radiation, treatment schedule, and overall health status of individuals determine the side effects of radiation therapy. Fatigue is the most common side effect that may remain for several weeks or months after radiation. Parts of bone marrow (such as the pelvis) when exposed to radiation therapy can lead to anemia (Albuquerque et al.

2011). Skin problems are another major side effect of external radiation therapy (Al-Mefty et al. 1990). The skin becomes dry, red, itchy, and even become tanned. Hair loss (alopecia) or thinning of hair, loss of appetite, dry mouth, difficulty in swallowing, changes in taste or smell, nausea, vomiting, low blood cell count, etc. are the major side effects of radiotherapy. Radiation therapy to the brain at very high doses leads to long-term problems like swelling of the brain, seizures, ear aches, headache, cognitive problems, extreme fatigue, and hormone problems (Al-Mefty et al. 1990). Radiation therapy to the upper abdomen can damage the esophagus and stomach and leads to radiation enteritis and kidney problems (Gupta et al. 2013). Radiotherapy on pelvis causes skin problems, hair loss, diarrhea, bladder infection, blood in the urine, urinary incontinence, fertility problems and sexual problems for women and men. Some of the late side effects are fertility problems, heart problems, changes in the skin, osteoporosis, mental, or emotional problems, and more seriously the secondary cancers.

2.1.3 Targeted Therapy

Targeting specific genes, proteins or the tumor microenvironment using small-molecule drugs or monoclonal antibodies is promising to improve the efficacy of treatment. Some monoclonal antibodies target and mark cancer cells so that they will be better seen and destroyed by the immune system (Mittal et al. 2014) or directly stop or destroy cancer cells. Skin problems are the most important side effects of targeted therapy. Still, the availability of targeted proteins and heterogeneity of cancer affects the efficacy of treatment. Even targeted therapy can cause dry skin, itching, skin rashes, red sore cuticles, hand–foot syndrome, hair becomes dry and brittle, yellow-colored hair, dryness of eye leads to prolonged damage of cornea (Remacle 2019).

2.1.4 Immunotherapy

Cancer immunotherapy relays on the mechanism by which patient's own immune system is activated and modulated to fight against the cancer cells (Zhang and Chen 2018). Cancer cells are detected by the immune system through abnormally expressed host proteins or neoantigens associated with abnormal proliferation of cancer cells. Immunotherapy for cancer treatment diverges into two categories such as passive and active. Passive immunotherapy depended on the direct attack on cancer cells by the administration of monoclonal antibodies (mAbs), cytokines and lymphocytes targeted to specific antigens. However, active immunotherapy aimed to elevate self-immune response against cancer cells achieved by vaccination and immunomodulation (Munhoz and Postow 2016). Rituximab, Trastuzumab, Cetuximab, Bevacizumab are FDA approved mAbs for different cancer treatments (Chiavenna et al. 2017). Another kind of immune therapy known as immune checkpoint blockade therapy is an attractive and promising strategy to enhance the immunity against cancer cells by blocking certain receptors in immune cells. The clinically available checkpoint blockade therapy is based on programmed death receptor-ligand (PD-PDL1) interaction and cytotoxic T-lymphocyte-associated protein 4 (anti-CTLA-4). Immune checkpoint blockade therapies are approved by the US Food and Drug Administration (FDA) approved and among which ipilimumab, monoclonal antibodies (mAbs) against cytotoxic T-lymphocyte antigen 4 (CTLA4), nivolumab, mAbs against programmed cell death protein 1 (PD1) and atezolizumab, mAbs against PDL1 are widely employed in clinics. Despite the rapid advance in immune checkpoint therapy, the meta-analysis of clinical trials showed that 27% of patients treated with PD-1/PD-L1 inhibitors and 72% of patients treated with CTLA-4 inhibitors, developed immune-related adverse effects ranging from mild skin rash, diarrheaand colitis to severe thyroiditis, hepatitis, primary adrenal insufficiency, colon perforation associated with enterocolitis, type 1 diabetes, and renal/cardiac/ ocular toxicity (Wang et al. 2017; Bertrand et al. 2015). Moreover, the expensive treatment regimen, organ-specific cross-reactivity, and influence of intrinsic and extrinsic factors at tumor and tumor microenvironment delimit the clinical success.

2.2 *Necessity to Develop Cancer Vaccines*

Cancer vaccines encourage the immune system to recognize the tumor-specific antigen as non-self and promote immune reaction to destroy the abnormal cells. Evolution of antibiotic resistance in oncogenic pathogens, identifying the significance of immune-surveillance and immune stimulation, the advent of monoclonal antibody and reduced clinical success of other immunotherapeutics enlighted the thought of cancer vaccine. Cancer vaccines designed to treat existing cancer are therapeutic vaccines and to prevent the onset of cancer is prophylactic (Speiser and Flatz 2014). Prophylactic cancer vaccines mainly take account of antigens carried by an infectious microorganism which is able to provoke the immune system of healthy human beings to inhibit cancer growth. Examples include the Hepatitis B virus (HBV) and human papillomavirus (HPV) vaccine to prevent liver and cervical cancer, respectively. Therapeutic cancer vaccines are made to activate the immune system using patient's own cells or cancer cell lines by incorporating specific tumor antigen in ex vivo and reunited to the same patient bloodstream (Zhang and Chen 2018).

2.3 *Significance of Cellular Cancer Vaccines and the Role of Cell Immobilization*

2.3.1 Current Status of Cellular Cancer Vaccine

Classifying the therapeutic cancer vaccines, whole-cell cancer vaccines (Autologous and allogeneic), dendritic cell vaccine, peptide vaccine, and genetic vaccine (DNA/RNA vaccine) shows promises in the clinical arena (Guo et al. 2013).

Cellular vaccines are developed using killed cancer cells derived from a patient's tumor, inactivated cancer cell lines or the tumor-specific antigen-loaded antigen-presenting cells (APCs). The mechanism of induction of tumor-specific immune response through dendritic cell (DC) vaccination is depicted in Fig. 2. Autologous tumor vaccines are cancer cell vaccines established from patient's tumor cells which are re-administered to the same patient to enhance the immune system reaction against cancer (Guo et al. 2013). Several clinical trials were gone with an autologous tumor cell vaccine powered with immunostimulatory properties (Asada et al. 2002; Fishman et al. 2008). These vaccines can present all the existing tumor-specific antigens to the immune system but the need for adequate amount of tumor sample is warranted. GVAX is such a vaccine fashioned by transfecting granulocyte–macrophage colony-stimulating factor (GM-CSF) gene in tumor cells. GM-CSF secreted by the GVAX vaccine upon administration leads to the activation of dendritic cells followed by antigen presentation to B and T cells that can lead to the production of cytokine IL-2 (Nemunaitis 2005). Even though this vaccine worked fine with mouse models, the clinical trials response was limited (Hollingsworth and Jasen

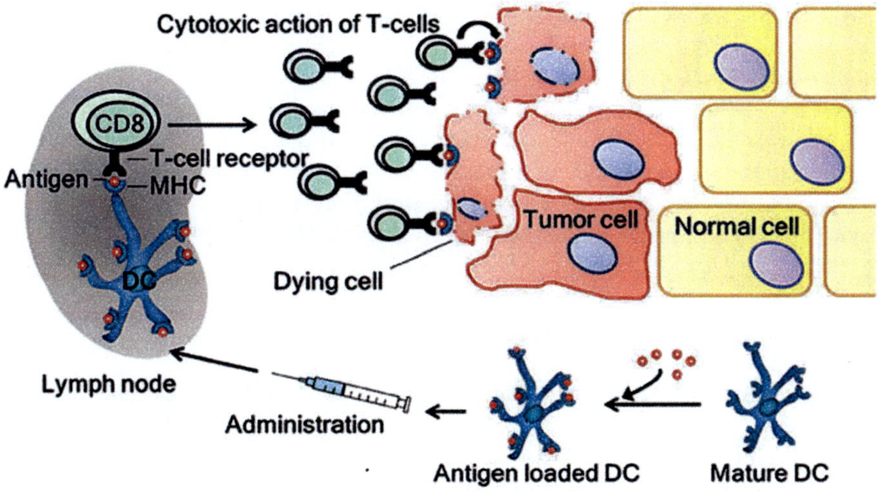

Fig. 2 Activation of anti-tumor immunity and destruction of tumor cells followed by cellular cancer vaccination (van Willigen et al. 2018)

2019). Allogeneic tumor cell vaccines using exogenous cancer cells or cell lines are advantageous over autologous vaccines in scale-up production, commercialization, management for specific antigen and availability for appropriate immune enrichment. Canavaxin™ vaccine is such a vaccine constructed using three melanoma cell lines and offered promising results in five-year overall survival percentage compared to control in phase II trials; however, phase III trials failed to establish its efficacy in melanoma patients (Guo et al. 2013).

Dendritic cells (DCs), the powerful antigen-presenting cells, capable to express tumor antigen to boost the immune system and attack cancer cells are well explored as a cellular cancer vaccine. The DCs isolated from patients were transduced with a fusion protein of GM-CSF and antigen of prostatic acid phosphates (PAP), which are administrated back to the patient was capable to generate PAP targeted T cells. In phase III clinical trial, the PAP specific DC vaccine showed better survival in patients with metastatic prostrate cancer. This vaccine was the first cancer vaccine approved by the FDA known as sipuleucel-T with a trading name of Provenge in 2010 (Longo 2010). Melanoma associated antigen recognized by T cells (MART-1) pulsed or whole gene transduced DC vaccine was evaluated in phase I and II trials. However, the number of DC vaccinations was limited in patients due to the difficulty of DC isolation and culture from the same patients (Guo et al. 2013).

Despite the tiresome efforts, only one vaccine, a dendritic cell vaccine termed Dendreon's Provenge (sipuleucel-T), was approved by the FDA for treatment of metastatic castration-resistant prostate cancer (DeFrancesco 2019). However, the complexity of ex vivo manipulation, difficulty in transportation of live cells and compromised production facility, not only raised the cost of treatment regimen ($93,000 in US market per treatment course of three infusions/month) but also reduced the commercial success of sipuleucel-T (News article on Dendreon Bankrupt 2014). In this scenario, more significance has to be given to engineer cancer vaccine with less ex vivo modification, simplified treatment regimen, and ease of commercialization.

Irradiated autologous/allogeneic tumor cell vaccine is considered cost-effective with the potential to present the entire spectrum of tumor antigens to the immune system, however, reported with discouraging clinical outcomes due to the inadequate T cell response, inappropriate temporal control, and limited ability to overcome tumor-associated immune suppression (Keenan and Jaffee 2012).

2.3.2 Role of Cell Immobilization in Cellular Cancer Vaccine

Immobilization strategy has been introduced to ensure the efficient and site-specific delivery of vaccine components to enhance the anti-tumor immunity. The faster clearance of vaccine components by the phagocytic system, dilution of vaccine components by the physiological fluids to an in-sensitive concentration, and obstacles in crossing biological barriers are the major reasons responsible forlower efficacy of cellular vaccines. To address these issues, several biomaterials were explored to design delivery platforms in various forms ranging from nanoscale vehicles to larger

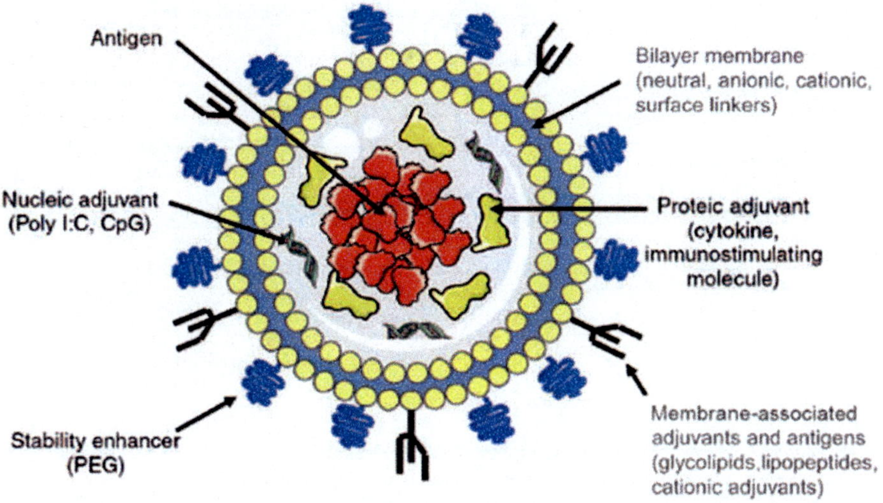

Fig. 3 Unilamellar liposome with antigenic vaccine components for vaccine delivery (Schwendener 2014)

implantable or injectable scaffolds (Tripathi et al. 2009,2013b; Tripathi and Melo 2017; Zhang et al. 2018; Tripathi and Melo 2019).

Immobilization of Vaccine Components in Nanoscale Level

Liposomes and polymeric poly (lactic-co-glycolic) acid (PLGA) nanoparticles are FDA approved delivery platforms for cancer vaccines. Liposomes are well suited to immobilize both hydrophilic and hydrophobic components in the core and bilayer respectively; however, PLGA nanovesicles are preferred for the immobilization of lipophilic components (Zhang et al. 2018). Liposome-based delivery system is suitable to immobilize water-soluble antigenic peptides or tumor lysate in the inner core and lipophilic adjuvants or linker molecules in bilayer along with a provision of surface modification with suitable antigenic components or DC attractants (Fig. 3). More to that, size-dependant tissue penetration, easier access to the lymph nodes, stability and integrity of structure providing a gradual release of antigenic components can enhance the performance of the vaccine (Schwendener 2014). The location of imobilized antigenic components in liposomes determines the immunogenicity of the vaccine.

Both encapsulated and surface-attached antigens are immunogenic, whereas entrapped antigens are released only after phagocytic intake by APC followed by vesicle disruption. However, antigens attached to the surface of liposomes will be readily available for B-cell recognition and antibody-mediated immune response. Still, incorporation of adjuvants such as CpGs in dioleoyl-3-trimethyl ammonium propane (Bal et al. 2011), chemical ligation of selected ligands for receptors on

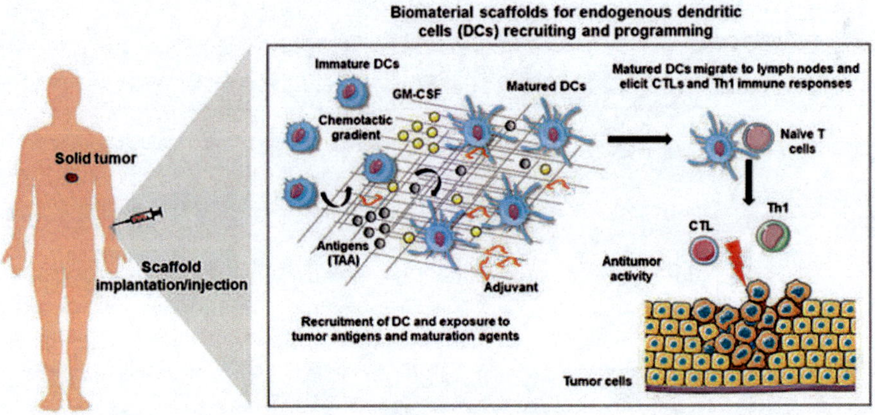

Fig. 4 Mechanism of subcutaneously implanted scaffold-based cancer vaccine immobilized with DC chemoattractants, adjuvants, and tumor antigens (Calmeiro et al. 2019)

immune cells (Cruz et al. 2014), or preparation of liposome with innate adjuvant characterestics (Ichihashi et al. 2013) can ensure strong immune activation and anti-tumor immunity. Cruz et al. (2014) have reported a targeted nano-liposomal formulation encapsulated with a protein fragment NY-ESO-1, which is expressed in a wide variety of tumors, including esophageal, prostate, head and neck, lung, ovarian, breast, stomach, and melanoma. The liposomes were modified to target the Fc receptor on DC via conjugating the thiolated Fc fragment on the maleimide group of poly(ethylene glycol) (PEG) tagged with phosphatidyl ethanolamine on the surface of the liposome. NY-ESO-1 protein fragment was combined with tetanus toxoid, which is the clinically relevant adjuvant to enhance the immune response. In addition, palmitoylated IL-1β peptide, an inflammatory cytokine to activate the DC population, was also incorporated on the liposomal surface to establish a strong immune response against tumor cells. Cruz et al. (2014) identified this strategy as a promising tool to enhance the anti-tumor immunity when the immune system is frequently compromised especially in advanced cancers.

Implantable Biomaterial Scaffold to Immobilize Cellular Vaccines

Scaffold-based vaccines were attempted in 2002 using ethyl vinyl acetate (EVA)-based tubes. Tumor-associated antigen and DC attractant macrophage inflammatory protein 3b (MIP-3b) were immobilized in EVA tubes and implanted subcutaneously in mice model of E.G7-OVA, fibrosarcoma, and Lewis lung sarcoma (Kumamoto et al. 2002). With the promising results of this in situ Langerhans cell vaccine, more researchers were enthusiastic to develop cancer vaccine scaffolds. The mechanism of action of the scaffold-based cancer vaccine is depicted in Fig. 4.

Porous poly(lactide-coglycolide) (PLG) scaffold prepared from the gas-forming process was explored to deliver immobilized cancer vaccine components. Surgically implantable porous poly(lactide-co-glycolide) (PLG) scaffold loaded with GM-CSF and CpG oligonucleotides (CpG-ODN) was employed to immobilize autologous tumor cell lysate and termed as WDVAX vaccine for the treatment of stage IV metastatic melanoma (Calmeiro et al. 2019). GM-CSF was immobilized into the PLG scaffold using a high-pressure CO_2 foaming process 50% encapsulation efficiency, which is gradually released upto 60% within five days of administration (Mooney et al. 1996). CpG-ODN condensed with cationic polyethyleneimine was also immobilized to the anionic PLG scaffold matrices, with favorable release kinetics and more than 80% retention over 25 days (Ali et al. 2009). The WDVAX vaccine containing melanoma tumor lysates was studied in the syngeneic B16-F10 murine melanoma model. WDVAX vaccine was reported with remarkable activation on CD11c+, pDCs, CD^{8+} DC subsets, and tumor-specific CTL responses. Complete regression of tumor was observed in 47% of mice with 50% survival, 33% long-term survival, and 90% prophylactic efficacy (Ali et al. 2014). Phase I clinical trial of WDVAX(NCT01753089) was attempted in 23 patients with 4 scaffolds by implantation in various time intervals. The vaccine immobilized in the PLGA scaffold is subcutaneously implanted on the patient's arm. CT scan and/or MRI exams are performed in every three months even after the end vaccination. The outcome of the clinical study is expected to be out in 2020.

Injectable Hydrogels to Immobilize and Deliver Cellular Cancer Vaccine

Surgical implantation of scaffold-based cancer vaccine is considered to be a risk regarding the higher chances of infection, especially in patients of advanced cancer with the compromised immune system. Therefore, cancer vaccines immobilized in injectable preformed scaffolds are encouraged over the implantable vaccine. Cryogels are the first developed injectable scaffolds, in which aqueous polymers are crosslinked at −20 °C to facilitate the formation of ice crystals inside, and allowed to thaw at room temperature to create macropores throughout the scaffold (Bencherif et al. 2015). Cryogels can retain shape memory even after injection through a 16 or 18 gauge needle. Still, the stability and retention of cryogels depend on the extent of crosslinking and elasticity of chemical bonds involved in polymerization (Bencherif et al. 2012).

The cryogels of alginate-gelatin hydrogels were prepared to deliver the cellular vaccine. The cryogel encapsulated with live attenuated HER-2/neu-overexpressing breast cancer cells, GM-CSF, and CpG-ODN were injected in a mouse model of breast cancer that resulted in the activation of DCs and enhanced antitumor response. 100% survival is observed in vaccinated mice with a 70-fold enhancement in antibody production (Bencherif et al. 2013). Alginate cryogels loaded with irradiated tumor cells, CpG-ODN, and GM-CSF were evaluated in a mouse melanoma model, reported with recruitment and activation of CD8+ DCs, CD11+ DCs, and pDCs that confirmed the prophylactic and therapeutic efficacy of the vaccine against cancer (Bencherif et al. 2015).

Other stimuli-responsive hydrogels were also investigated to design an injectable system to deliver the vaccine components. Monomethoxypoly (ethylene glycol)-co-poly(lactic-co-glycolic acid) copolymer (mPEG-PLGA) exhibits thermoresponsive property with the faster sol–gel transition from 4 to 37 °C that demonstrated great promise in cancer vaccine delivery. mPEG-PLGA polymer injected with GM-CSF ensured the stable release of cytokine sufficient to establish DC activation and anti-tumor immune response (Liu et al. 2014). Recently, these hydrogels were employed to inject the nanoparticle loaded with tumor antigens to improve the CTL response and reduce the need for repeated vaccination (Sun et al. 2018).

DC programming scaffold vaccines are constructed with mesoporous silica rods (MSR) with a hexagonal mesoporous structure via a silica sol–gel reaction (Schmidt-Winkel et al. 1999). The internal porous morphology of spontaneously assembled MSRs loaded with ovalbumin (OVA), CpG-ODN, and GM-CSF can facilitate cell infiltration. Recruitment of large numbers of CD11c+ DCs, B220+ B cells, and CD14 + monocytes was observed at the site of injection of MSR-based vaccine. Activation of Th1 and Th2 immune responses followed by vaccination resulted in significant anti-tumor property in mice challenged with EG7-OVA lymphoma cells (Kim et al. 2015).

2.3.3 Future Prospects of Cell Immobilization in Cancer Therapy

Introducing biomaterial strategies and immobilization techniques in vaccine engineering seems promising with superior future prospects. Addressing the following queries may improve the efficacy of immobilized cellular vaccines in the clinical arena.

a. Designing scaffolds using more selective chemotactic agents (CX3CL1, CCL2, and CCL7) to recruit specific DC subpopulations such as Langerhans cells and cDC1 cells (CD141 + CLEC9A + XCR1 +) can enhance the efficacy of the vaccine. Attempting various immobilization strategies will be warranted to immobilize the various chemotactic agents with retained function.
b. The immense scope of intravenously injectable immobilized vaccines targeted to the lymphoid organs or lymph nodes is appreciated in the field. Synthetic high-density lipoprotein (sHDL) nanodiscs immobilized with neoantigen was designed for the delivery to lymphoid organs to sustain antigen presentation on DCs (Kuai et al. 2017). This minimally invasive method was tested in a murine MC38 colon carcinoma model that enhanced neoantigen-specific CTLs and resulted in slowed tumor growth and complete tumor regression in 88% of tested mice.
c. With the promising advent of biomaterial-based immobilization strategy in cancer vaccine formulation, a universal therapeutic modality for personalized immunotherapy will be possible with ease of clinical application in near future irrespective of the type of cancer.

3 Prospects of Cell Immobilization in Cancer Research

3.1 Cell Immobilization in 3D Tumeroid Generation

3.1.1 Introduction to 3D Tumeroid in Cancer Research

Studies on two-dimensional (2D) cancer cell monolayer are routinely used for cancer drug discovery and screening, even though it cannot replicate the complex tumor micro-environment. The 3D tumor models can mimic the in vivo tumor microenvironment for proper analysis of drug or active compounds on cancer regression. Smaller tumeroids (400–600 μm) would contain viable cells at the rim and core region, while larger tumeroids would possess a characteristic necrotic core. 3D culture of large tumeroid can create a physiological gradient in multiple layers due to the slower diffusion of nutrients which are essential for signaling, angiogenesis, and growth (Bartlett et al. 2014). The differences between 2D and 3D tumor models are represented in Table 1.

The most validated 3D model in cancer research is the multicellular tumor spheroid. Cellular spheroids use the natural tendency of cells to form aggregates by re-establishing cellular connections initially through integrin-ECM connections and later with hemophilic cadherin–cadherin interactions. Spherical aggregates of cells can display mass transport and differentiation properties appropriate for cell–cell and cell–matrix interactions. Spheroids usually contain a necrotic center surrounded by and quiescent zone and outer proliferating cells based on the availability of nutrients (Cui et al. 2017). Spheroids models can mimic cellular heterogeneity and drug resistance property observed with most of the tumors in humans (Nunes et al. 2019). Spheroids are amenable to the co-culture of different cell types such as endothelial cells, immune cells, and fibroblasts to recreate the complexity of stromal elements in the tumor microenvironment in vitro (Lazzari et al. 2017). Organotypic multicellular explant cultures by gentle mechanical or enzymatic dissociation of cancer tissues are followed to reproduce the tumor microenvironment (TME) and retain

Table 1 Comparison of 2D and 3D tumor model in cancer research

Parameters	2D tumor model	3D tumor model
Morphology	Elongated; Forced apical-based polarity	Rounded and loss polarity
Viability	Lower	Higher
Drug metabolism	Artificial	Reflects in vivo mode
In vivo Relevance	Low	High
Transport	Gradients absent; unlimited access	Gradient diffusional transport; limited as in vivo
Focal adhesions	Basal surface only	Distributed throughout
Cost of production	Cheap and easy to maintain	Expensive and time-consuming

histological features in tumeroids. These spheroids contain non-tumor cells and other stromal components representing the heterogeneity of TME (Pampaloni et al. 2017). Multicellular tumor spheroids are derived from cell lines grown in suspension culture which will promote cell–cell adhesion, metabolic and proliferative pathways are mimicked and show clinically relevant multicellular chemo-resistance. Tumor-derived organoids developed by Sato et al. (2011) are a powerful ex vivo model of organogenesis. They have reconstituted a variety of tissues using non-tumor cells and later introduced oncogene or tumor suppressor genes to investigate cancer development under in vitro culture conditions (Sato et al. 2011). Tumor-derived spheroids can be fabricated with an enriched population of cancer stem cells that can serve as a surrogate system to evaluate CSC-related characteristics in vitro (Achilli et al. 2012).

3.2 *Advantages of In Vitro Tumor Models*

The various types of in vitro tumor models intended for different research applications are described in the table with listed advantages (Table 2).

3.3 *Significance of In Vitro Tumor Models*

In vitro and in vivo studies have a significant role in cancer research, especially to analyze the carcinogenicity of materials, formulate and standardize new cancer therapies, screening of chemotherapeutics, and investigating the molecular mechanism of tumor growth. A wide range of animal models of primary and metastatic tumors have been developed in small laboratory animals including mice and rats. Environmentally induced models, human tumor xenografts in immunocompromised mice, and genetically engineered mice are well appreciated for the evaluation of cancer therapeutics. However, higher cost of experiments, lesser chances for standardization of variables, the influence of uncontrolled variables, and an effort to reduce the cruelty toward laboratory animals encourages well established in vitro models to replace the in vivo tumor models in cancer research. Reconstructing complexity and heterogeneity in in vitro models of solid tumors confers better extrapolation to in vivo conditions (Katt et al. 2016). Complexity of tumor models relies on recapitulating the tumor growth/proliferation, cell migration and invasion, matrix remodeling on tumor progression, and rapid angiogenic property.

Table 2 In vitro tumor models and applications

Method of fabrication	Description	Advantages
Tumor spheroid models		
Cell suspension culture	Spheroids are cultured in specific suspension culture	Helps to avoid sedimentation and adherence, mass production is possible
Non-adherent surfaces	Cell aggregation is promoted on a planar non-adherent surfaces	High throughput, using microwell plates. uniform spheroid size
Hanging drop technique	Cells are aggregated to spheroids in hanging droplets under gravity	Allows coculture with defined cell types, uniform size
Microfluidic devices	Spheroids are generated within microfluidic channels	Control on spheroid size and growth parameters Faster spheroid formation
Hybrid tumor models		
Patient-specific tumor model	Ex vivo primary tumor section, or biopsies embedded in a suitable hydrogel	Patient-specific assay Mimics tumor invasion and maintains tumor heterogeneity
3D invasion models	Clusters of Primary tumor cells or cell line embedded in a hydrogel	Balance of complexity and experimental control
Avascular microfluidic model	Tumor cells were grown in a 2D microfluidic device with continuous media perfusion	Easy to isolate the effect of variables, allows real-time tracking of cells, ideal for invasion and migration assay
Tumor microvessel model		
Vascular tumor model	Construct of tumorospheres with endothelial cells embedded in predefined ECM scaffold	Suitable for studies on angiogenesis, the influence of shear stresses on vascularisation
Microvessel self-assembly	Tumeroids with self-assembled endothelial cells and microvessels (5–50 μm)	Suitable for studies on chemotherapeutic screening and tumor cell dormancy

3.4 Prospects of Cell Immobilization in 3D Tumeroid Generation

Growth and migratory potential of cancer cells is influenced by the composition of the extracellular matrix associated with the tumor microenvironment. Cancer cells showing aggressive migratory and proliferative properties may not exhibit a similar effect in vivo. The behavior and growth of cancer cells solely depend on the biophysical and biochemical properties of tumor-associated extracellular matrix (ECM) (Zent and Pozzi 2010). Therefore, spheroids of cancer cells aggregated by natural driven cues will not be sufficient to mimic all characteristics as that of the native tumor. A supporting matrix simulating tumor micro-environment is inevitable for the proper patterning and anchoring of cells in a multicellular tumeroids (Gu and Mooney 2016).

3.4.1 Hydrogels for Cell Immobilization in 3D Tumeroid Generation

Physically or chemically crosslinked, water inflated hydrogels are well appreciated to simulate the extracellular microenvironment of tumor tissue. Flexibility to control the biomechanical property and biophysical characteristics by chemical or physical strategies ensures the suitability of hydrogels as an ideal candidate for 3D tumeroid generation. Several biomaterials with both natural and synthetic origin are used to prepare hydrogels provided with specific chemical or physical cues.

Matrigel is the "golden standard" biomaterial for the generation of 3D tumeroids. Matrigel is extracted from Engelbreth-Holm-Swarm mouse sarcoma tumors composed of laminin, type IV collagen, entactin, proteoglycans, and essential multiple growth factors (Benton et al. 2014). Matrigel has been used for the generation of multicellular tumeroids of breast cancer (Lee et al. 2007), colorectal cancer (Yeung et al. 2010), and prostate cancer (Lang et al. 2001) using respective cell lines. The role of the β1integrin receptor and epidermal growth factor receptor in the transition of breast cancer cells to the normal mammary epithelial cells was identified using matrigel-based tumeroid (Weaver et al. 1997).

Collagen hydrogels are extensively used to grow tumeroids of osteosarcoma, breast cancer cells, prostate cancer cell lines, human colorectal cancer cells (Kondo et al. 2011), and primary cancer cells from colorectal cancer patients (Szot et al. 2011). Polysaccharide-based hydrogels including hyaluronic acid (Pedron et al. 2013; Gurski et al. 2009) agarose (Helmlinger et al. 1997), or alginate (Alessandri et al. 2013) modified with arginyl glycyl aspartic acid (RGD) peptides (Fischbach et al. 2009) have used to provide controlled biophysical conditions for tumeroid growth. As the degradation of polysaccharides is independent of cell-secreted proteolytic enzymes, these hydrogels provide better mechanical stability, which is explored to determine the effect of stress on tumor progression in tumeroids (Helmlinger et al. 1997; Alessandri et al. 2013). The interpenetrating network of polysaccharides and proteins is preferred for the controlled growth of tumeroids regarding the mechanical stability and cell adhesion property. In this aspect, an alginate-collagen-agarose hydrogel was designed to grow human ovarian cancer cells in the tumeroid model (Shin et al. 2016). The influence of hydrogel stiffness in the malignant progression of the mammary epithelium was studied in alginate-matrigel hydrogel (Chaudhuri et al. 2014).

Even with excellent biocompatibility and suitable biophysical property, the use of natural scaffolds in tumeroids is limited by their batch-to-batch variation, uncontrolled and fast degradation, and unidentified composition which may affect cancer cell fate and reproducibility of the experiments (Alemany-Ribes and Semino 2014). Therefore, biocompatible, biodegradable, and cell adhesive-modified synthetic hydrogel scaffolds are explored for cell spheroid generation (Tripathi et al.

2013a). Poly (ethylene glycol)-based hydrogels modified with RGD peptide and matrix metalloproteinase degradable sites have used to grow tumeroids of the lung, ovarian, and brain cancer cell lines (Gill et al. 2012; Loessner et al. 2010; Wang et al. 2014). Synthetic self-assembled, fibrous peptide hydrogel formed from RADA16-I (AcN-RADARADARADARADA-CNH2) (commercially name: BD PuraMatrix) have been used to grow ovarian cancer cell in tumeroid (Yang and Zhao 2011).

The hybrid hydrogels derived from natural and synthetic materials are advantageous in the fabrication of ideal tumeroids. Hydrogels formed from aldehyde-modified HA and bis(oxyamine)-PEG by oxime click are such a formulation with hybrid nature (Nunes et al. 2019). Nanofibrillar hydrogel formed of cellulose nanocrystals (CNCs) end-grafted with poly(N-isopropylacrylamide) was employed to grow breast cancer cells, which is reported with growth profiles similar to those in matrigel (Chau et al. 2015).

3.4.2 Influence of Physicochemical Behavior of Hydrogels in Tumeroid Generation

The composition, stiffness, and porosity of hydrogels strictly influence the growth and migratory potential of tumeroids. While evaluating the growth of B16-F10 Melanoma cells in fibrin hydrogels of various stiffness ranging from 90–1050 Pa revealed the inhibitory effect of increasing stiffness on the cell proliferation and tumeroid size (as given in Fig. 5) (Liu et al. 2012).

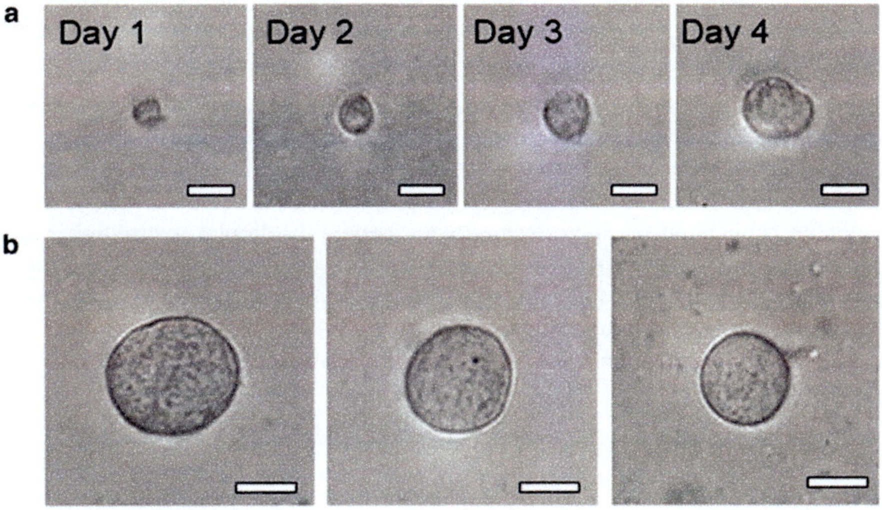

Fig. 5 Growth of B16-F10 cells in fibrin hydrogen of stiffness 90 Pa (**a**). Size of tumeroids grown in fibrin hydrogen with the stiffness of 90 Pa (Left), 420 Pa (middle), and 1050 Pa (Right)

A similar observation was evident in generation of tumeroids of hepatocellular carcinoma in PEG-Collagen-based hydrogels (Liang et al. 2011). Higher concentration of polymer, reduced porosity, and closer cell adhesion points in stiffer hydrogels can compromise the cell proliferation and migratory pattern of cancer cells. The composition of hydrogels has a role to determine the growth of tumeroids especially considering the synthetic hydrogels. Incorporation of RGD peptides in PEG hydrogels (Loessner et al. 2010) and alginate hydrogels (Fischbach et al. 2009) was reported with a significantly higher spheroid number and size that underlies the need for standardization of apt formulation of synthetic matrices before tumeroid generation. The nature of crosslinking also has to be considered while selecting hydrogel for tumeroid generation. Hydrogels with reversible gelation are greatly appreciated in the context of the post-growth release of tumeroids. Thermoreversible pNIPAAM hydrogels provided with suitable modifications are well appreciated for the growth of tumeroids at 37 °C and safe release of tumeroids by slightly lowering the temperature to below the lower critical solution temperature (Li et al. 2017).

3.5 *Future Prospects of Hydrogel-Based 3D Tumeroids*

Fulfilling the following challenges in the field of 3D tumor model can benefit the field of cancer research in various means.

a. Construction of vascular tumeroids is a challenging scenario, which is less attempted except the approaches of simple co-culture of cancer cells with microvascular endothelial cells. Bioprinting strategies can be explored to introduce pre-designed vasculature in the tumor model which can value over the current co-culture method.
b. The release of multicellular tumeroids from hydrogel structure after proper maturation is recommended to ease the genotypic and proteomic analysis. However, the complete release of tumeroid without any damage is difficult with the existing hydrogel systems.
c. Viable application of micro-technologies preferably droplet microfluidics or inkjet bioprinting strategies can enhance the efficacy of hydrogel-based tumeroids in high throughput drug screening (Fig. 6). Tumor model on chip can be designed which may also benefit to check the feasibility of a specific therapeutic approach in personalized mode.

4 Conclusion

Cell immobilization is a well-established technique with various applications in the medical and research field. The scope of immobilizing cells in a controlled and protective matrix is well appreciated in the field of cancer therapy and cancer research. The significance of immobilized cells or cellular components derived from

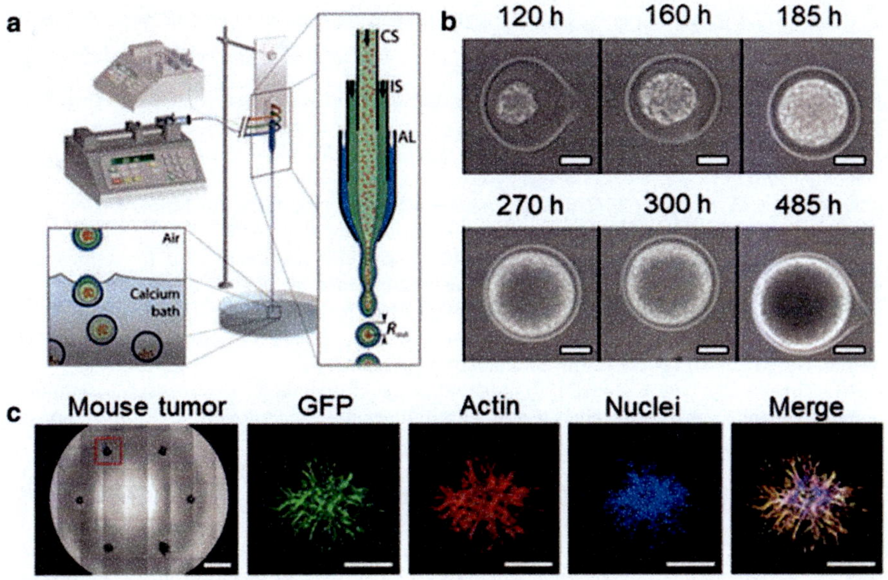

Fig. 6 Microfluidic generation of alginate microdroplets immobilized with tumor cells in the calcium bath. Growth of tumeroids inside the alginate beads and immunostained images of multicellular tumeroid are represented (Li and Kumacheva, 2018)

a patient's tumor included with selected cytokines is recently accepted for targeted, prolonged, and enhanced activation of a specific population of dendritic cells as a cellular cancer vaccine. More to that, the controlled immobilization of cancer cells and other supporting stromal cells in high density is well adapted for the generation of functional multicellular tumeroids that have immense application in the evaluation of cancer therapeutics and identification of mechanism regarding the tumor initiation, progression, and migration. Being a fundamental technique, advanced strategies in cell immobilization are still rising with exciting upshots on cancer research and cancer immune therapeutics.

Acknowledgements Authors would like to acknowledge Board of Research in Nuclear Sciences, Department of Atomic Energy, Government of India for the financial support.

References

Achilli T-M, Meyer J, Morgan JR (2012) Advances in the formation, use and understanding of multi-cellular spheroids. Expert Opin Biol Ther 12:1347–1360. https://doi.org/10.1517/14712598.2012.707181

Aggarwal BB, Danda D, Gupta S, Gehlot P (2009) Models for prevention and treatment of cancer: problems vs promises. Biochem Pharmacol 78:1083–1094. https://doi.org/10.1016/j.bcp.2009.05.027

Albuquerque K, Giangreco D, Morrison C et al (2011) Radiation-related predictors of hematologic toxicity after concurrent chemoradiation for cervical cancer and implications for bone marrow-sparing pelvic IMRT. Int J Radiat Oncol Biol Phys 79:1043–1047. https://doi.org/10.1016/j.ijrobp.2009.12.025

Alemany-Ribes M, Semino CE (2014) Bioengineering 3D environments for cancer models. Adv Drug Deliv Rev 79–80:40–49. https://doi.org/10.1016/j.addr.2014.06.004

Alessandri K, Sarangi BR, Gurchenkov VV et al (2013) Cellular capsules as a tool for multicellular spheroid production and for investigating the mechanics of tumor progression in vitro. Proc Natl Acad Sci USA 110:14843–14848. https://doi.org/10.1073/pnas.1309482110

Ali OA, Huebsch N, Cao L et al (2009) Infection-mimicking materials to program dendritic cells in situ. Nat Mater 8:151–158. https://doi.org/10.1038/nmat2357

Ali OA, Verbeke C, Johnson C et al (2014) Identification of immune factors regulating antitumor immunity using polymeric vaccines with multiple adjuvants. Cancer Res 74:1670–1681. https://doi.org/10.1158/0008-5472.CAN-13-0777

Al-Mefty O, Kersh JE, Routh A, Smith RR (1990) The long-term side effects of radiation therapy for benign brain tumors in adults. J Neurosurg 73:502–512. https://doi.org/10.3171/jns.1990.73.4.0502

Asada H, Kishida T, Hirai H et al (2002) Significant antitumor effects obtained by autologous tumor cell vaccine engineered to secrete interleukin (IL)-12 and IL-18 by means of the EBV/lipoplex. Mol Ther 5:609–616. https://doi.org/10.1006/mthe.2002.0587

Bahulekar R, Ayyangar NR, Ponrathnam S (1991) Polyethyleneimine in immobilization of biocatalysts. Enzyme and Microbial Technology 13:858–868. https://doi.org/10.1016/0141-0229(91)90101-F

Bal SM, Hortensius S, Ding Z et al (2011) Co-encapsulation of antigen and Toll-like receptor ligand in cationic liposomes affects the quality of the immune response in mice after intradermal vaccination. Vaccine 29:1045–1052. https://doi.org/10.1016/j.vaccine.2010.11.061

Bartlett R, Everett W, Lim S et al (2014) Personalized In Vitro Cancer Modeling—Fantasy or Reality? Transl Oncol 7:657–664. https://doi.org/10.1016/j.tranon.2014.10.006

Baruch L, Machluf M (2006) Alginate-chitosan complex coacervation for cell encapsulation: effect on mechanical properties and on long-term viability. Biopolymers 82:570–579. https://doi.org/10.1002/bip.20509

Bencherif S, Draganov D, Lewin S et al (2013) Immunologically active cryogels for breast cancer therapy (P4329). J Immunol 190: 126.1

Bencherif SA, Sands RW, Bhatta D et al (2012) Injectable preformed scaffolds with shape-memory properties. Proc Natl Acad Sci USA 109:19590–19595. https://doi.org/10.1073/pnas.1211516109

Bencherif SA, Warren Sands R, Ali OA et al (2015) Injectable cryogel-based whole-cell cancer vaccines. Nat Commun 6:7556. https://doi.org/10.1038/ncomms8556

Benton G, Arnaoutova I, George J et al (2014) Matrigel: from discovery and ECM mimicry to assays and models for cancer research. Adv Drug Deliv Rev 79–80:3–18. https://doi.org/10.1016/j.addr.2014.06.005

Bertrand A, Kostine M, Barnetche T et al (2015) Immune related adverse events associated with anti-CTLA-4 antibodies: systematic review and meta-analysis. BMC Med 13:211. https://doi.org/10.1186/s12916-015-0455-8

Borin GP, de Melo RR, Crespim E et al (2018) An overview on polymer gels applied to enzyme and cell immobilization. In: Thakur VK, Thakur MK (eds) Polymer gels: science and fundamentals. Springer, Singapore, pp 63–86

Bray F, Ferlay J, Soerjomataram I et al (2018) Global cancer statistics 2018: GLOBOCAN estimates of incidence and mortality worldwide for 36 cancers in 185 countries. CA Cancer J Clin 68:394–424. https://doi.org/10.3322/caac.21492

Calmeiro J, Carrascal M, Gomes C et al (2019) Biomaterial-based platforms for in situ dendritic cell programming and their use in antitumor immunotherapy. J Immunother Cancer 7. https://doi.org/10.1186/s40425-019-0716-8

Chau M, Sriskandha SE, Thérien-Aubin H, Kumacheva E (2015) Supramolecular nanofibrillar polymer hydrogels. In: Seiffert S (ed) Supramolecular polymer networks and gels. Springer International Publishing, Cham, pp 167–208

Chaudhuri O, Koshy ST, Branco da Cunha C et al (2014) Extracellular matrix stiffness and composition jointly regulate the induction of malignant phenotypes in mammary epithelium. Nat Mater 13:970–978. https://doi.org/10.1038/nmat4009

Chiavenna SM, Jaworski JP, Vendrell A (2017) State of the art in anti-cancer mAbs. J Biomed Sci 24:15. https://doi.org/10.1186/s12929-016-0311-y

Cruz LJ, Rueda F, Simón L et al (2014) Liposomes containing NY-ESO-1/tetanus toxoid and adjuvant peptides targeted to human dendritic cells via the Fc receptor for cancer vaccines. Nanomedicine (Lond) 9:435–449. https://doi.org/10.2217/NNM.13.66

Cui X, Hartanto Y, Zhang H (2017) Advances in multicellular spheroids formation. J R Soc Interface. https://doi.org/10.1098/rsif.2016.0877

DeFrancesco L (2019) Provenge twists again. In: Nature Biotechnology. https://www.nature.com/articles/nbt0910-882a. Accessed 14 Sep

Dendreon Bankrupt (2014) Nature Biotechnology 32: 1176. https://doi.org/10.1038/nbt1214-1176a

Elakkiya M, Prabhakaran D, Thirumarimurugan M (2016) Methods of cell immobilization and its applications. Methods 5(4):211–216

Fischbach C, Kong HJ, Hsiong SX et al (2009) Cancer cell angiogenic capability is regulated by 3D culture and integrin engagement. Proc Natl Acad Sci USA 106:399–404. https://doi.org/10.1073/pnas.0808932106

Fishman M, Hunter TB, Soliman H et al (2008) Phase II trial of B7–1 (CD-86) transduced, cultured autologous tumor cell vaccine plus subcutaneous interleukin-2 for treatment of stage IV renal cell carcinoma. J Immunother 31:72–80. https://doi.org/10.1097/CJI.0b013e31815ba792

Gill BJ, Gibbons DL, Roudsari LC et al (2012) A synthetic matrix with independently tunable biochemistry and mechanical properties to study epithelial morphogenesis and EMT in a lung adenocarcinoma model. Cancer Res 72:6013–6023. https://doi.org/10.1158/0008-5472.CAN-12-0895

Gu L, Mooney DJ (2016) Biomaterials and emerging anticancer therapeutics: engineering the microenvironment. Nat Rev Cancer 16:56–66. https://doi.org/10.1038/nrc.2015.3

Guo C, Manjili MH, Subjeck JR et al (2013) Therapeutic cancer vaccines: past, present and future. Adv Cancer Res 119:421–475. https://doi.org/10.1016/B978-0-12-407190-2.00007-1

Gupta M, Davis M, LeGrand S et al (2013) Nausea and vomiting in advanced cancer: the Cleveland clinic protocol. J Support Oncol 11:8–13. https://doi.org/10.1016/j.suponc.2012.10.002

Gurski LA, Jha AK, Zhang C et al (2009) Hyaluronic acid-based hydrogels as 3D matrices for in vitro evaluation of chemotherapeutic drugs using poorly adherent prostate cancer cells. Biomaterials 30:6076–6085. https://doi.org/10.1016/j.biomaterials.2009.07.054

Helmlinger G, Netti PA, Lichtenbeld HC et al (1997) Solid stress inhibits the growth of multicellular tumor spheroids. Nat Biotechnol 15:778–783. https://doi.org/10.1038/nbt0897-778

Hollingsworth RE, Jansen K (2019) Turning the corner on therapeutic cancer vaccines. NPJ Vaccines 4: 1–10. https://doi.org/10.1038/s41541-019-0103-y

Hu S, de Vos P (2019) Polymeric approaches to reduce tissue responses against devices applied for islet-cell encapsulation. Front Bioeng Biotechnol 7:134. https://doi.org/10.3389/fbioe.2019.00134

Ichihashi T, Satoh T, Sugimoto C, Kajino K (2013) Emulsified phosphatidylserine, simple and effective peptide carrier for induction of potent epitope-specific T cell responses. PLoS ONE 8:e60068. https://doi.org/10.1371/journal.pone.0060068

Immune Checkpoint Blockade Therapy: Merits and Demerits. In: ResearchGate. https://www.researchgate.net/publication/309267727_Immune_Checkpoint_Blockade_Therapy_Merits_and_Demerits. Accessed 14 Mar 2019

Joung JJ, Akin C, Royer GP (1987) Immobilization of growing cells by polyethyleneimine-modified alginate. Appl Biochem Biotechnol 14:259–275. https://doi.org/10.1007/BF02800313

Katt ME, Placone AL, Wong AD et al (2016) In vitro tumor models: advantages, disadvantages, variables, and selecting the right platform. Front Bioeng Biotechnol 4. https://doi.org/10.3389/fbioe.2016.00012

Keenan BP, Jaffee EM (2012) Whole cell vaccines–past progress and future strategies. Semin Oncol 39:276–286. https://doi.org/10.1053/j.seminoncol.2012.02.007

Kim J, Li WA, Choi Y et al (2015) Injectable, spontaneously assembling, inorganic scaffolds modulate immune cells in vivo and increase vaccine efficacy. Nat Biotechnol 33:64–72. https://doi.org/10.1038/nbt.3071

Knobeloch T, Abadi SEM, Bruns J et al (2017) Injectable polyethylene glycol hydrogel for islet encapsulation: an in vitro and in vivo characterization. Biomed Phys Eng Express 3. https://doi.org/10.1088/2057-1976/aa742b

Koh W-G, Revzin A, Pishko MV (2002) Poly(ethylene glycol) hydrogel microstructures encapsulating living cells. Langmuir 18:2459–2462

Kondo J, Endo H, Okuyama H et al (2011) Retaining cell-cell contact enables preparation and culture of spheroids composed of pure primary cancer cells from colorectal cancer. Proc Natl Acad Sci USA 108:6235–6240. https://doi.org/10.1073/pnas.1015938108

Kuai R, Ochyl LJ, Bahjat KS et al (2017) Designer vaccine nanodiscs for personalized cancer immunotherapy. Nat Mater 16:489–496. https://doi.org/10.1038/nmat4822

Kumamoto T, Huang EK, Paek HJ et al (2002) Induction of tumor-specific protective immunity by in situ Langerhans cell vaccine. Nat Biotechnol 20:64–69. https://doi.org/10.1038/nbt0102-64

Kumar A, Tripathi A (2012) Biopolymeric scaffolds for Tissue Engineering. In: Tiwari A, Srivastava RB (Ed) Biotechnology in Biopolymers Developments, Applications and Challenging Areas, i-Smithers Repra Publication Ltd. United Kingdom, p 233–285

Lang SH, Sharrard RM, Stark M et al (2001) Prostate epithelial cell lines form spheroids with evidence of glandular differentiation in three-dimensional Matrigel cultures. Br J Cancer 85:590–599. https://doi.org/10.1054/bjoc.2001.1967

Lazzari G, Couvreur P, Mura S (2017) Multicellular tumor spheroids: a relevant 3D model for the in vitro preclinical investigation of polymer nanomedicines. Polym Chem 8:4947–4969. https://doi.org/10.1039/C7PY00559H

Lee GY, Kenny PA, Lee EH, Bissell MJ (2007) Three-dimensional culture models of normal and malignant breast epithelial cells. Nat Methods 4:359–365. https://doi.org/10.1038/nmeth1015

Li Y, Khuu N, Gevorkian A et al (2017) Supramolecular nanofibrillar thermoreversible hydrogel for growth and release of cancer spheroids. Angew Chem Int Ed 56:6083–6087. https://doi.org/10.1002/anie.201610353

Li Y, Kumacheva E (2018) Hydrogel microenvironments for cancer spheroid growth and drug screening. Sci Adv 4: eaas8998. https://doi.org/10.1126/sciadv.aas8998

Liang Y, Jeong J, DeVolder RJ et al (2011) A cell-instructive hydrogel to regulate malignancy of 3D tumor spheroids with matrix rigidity. Biomaterials 32:9308–9315. https://doi.org/10.1016/j.biomaterials.2011.08.045

Liu J, Tan Y, Zhang H et al (2012) Soft fibrin gels promote selection and growth of tumorigenic cells. Nature Mater 11:734–741. https://doi.org/10.1038/nmat3361

Liu Y, Xiao L, Joo K-I et al (2014) In situ modulation of dendritic cells by injectable thermosensitive hydrogels for cancer vaccines in mice. Biomacromol 15:3836–3845. https://doi.org/10.1021/bm501166j

Loessner D, Stok KS, Lutolf MP et al (2010) Bioengineered 3D platform to explore cell-ECM interactions and drug resistance of epithelial ovarian cancer cells. Biomaterials 31:8494–8506. https://doi.org/10.1016/j.biomaterials.2010.07.064

Longo DL (2010) New therapies for castration-resistant prostate cancer. N Engl J Med 363:479–481. https://doi.org/10.1056/NEJMe1006300

Lutolf MP, Hubbell JA (2005) Synthetic biomaterials as instructive extracellular microenvironments for morphogenesis in tissue engineering. Nat Biotechnol 23:47–55. https://doi.org/10.1038/nbt1055

Maltaris T, Seufert R, Fischl F et al (2007) The effect of cancer treatment on female fertility and strategies for preserving fertility. Eur J Obstet Gynecol Reprod Biol 130:148–155. https://doi.org/10.1016/j.ejogrb.2006.08.006

Melvik JE, Dornish M (2004) Alginate as a carrier for cell immobilisation. In: Nedović V, Willaert R (eds) Fundamentals of cell immobilisation biotechnology. Springer, Netherlands, Dordrecht, pp 33–51

Mittal D, Gubin MM, Schreiber RD, Smyth MJ (2014) New insights into cancer immunoediting and its three component phases–elimination, equilibrium and escape. Curr Opin Immunol 27:16–25. https://doi.org/10.1016/j.coi.2014.01.004

Mooney DJ, Baldwin DF, Suh NP et al (1996) Novel approach to fabricate porous sponges of poly(d, l-lactic-co-glycolic acid) without the use of organic solvents. Biomaterials 17:1417–1422. https://doi.org/10.1016/0142-9612(96)87284-X

Munhoz RR, Postow MA (2016) Recent advances in understanding antitumor immunity. F1000Res 5:2545. https://doi.org/10.12688/f1000research.9356.1

Nemunaitis J (2005) Vaccines in cancer: GVAX, a GM-CSF gene vaccine. Expert Rev Vaccines 4:259–274. https://doi.org/10.1586/14760584.4.3.259

Nunes AS, Barros AS, Costa EC et al (2019) 3D tumor spheroids as in vitro models to mimic in vivo human solid tumors resistance to therapeutic drugs. Biotechnol Bioeng 116:206–226. https://doi.org/10.1002/bit.26845

Nuttelman CR, Rice MA, Rydholm AE et al (2008) Macromolecular monomers for the synthesis of hydrogel niches and their application in cell encapsulation and tissue engineering. Prog Polym Sci 33:167–179. https://doi.org/10.1016/j.progpolymsci.2007.09.006

Pampaloni F, Reynaud EG, Stelzer EHK (2007) The third dimension bridges the gap between cell culture and live tissue. Nat Rev Mol Cell Biol 8:839–845. https://doi.org/10.1038/nrm2236

Pedron S, Becka E, Harley BAC (2013) Regulation of glioma cell phenotype in 3D matrices by hyaluronic acid. Biomaterials 34:7408–7417. https://doi.org/10.1016/j.biomaterials.2013.06.024

Reátegui E, Kasinkas L, Kniesz K et al (2014) Silica–PEG gel immobilization of mammalian cells. J Mater Chem B 2:7440–7448. https://doi.org/10.1039/C4TB00812J

Remacle C (2019) The patients' journey with targeted therapies. In: Charnay-Sonnek F, Murphy AE (eds) Principle of nursing in oncology: new challenges. Springer International Publishing, Cham, pp 65–82

Risbud MV, Bhargava S, Bhonde RR (2003) In vivo biocompatibility evaluation of cellulose macrocapsules for islet immunoisolation: Implications of low molecular weight cut-off. J Biomed Mater Res a 66:86–92. https://doi.org/10.1002/jbm.a.10522

Risbud MV, Bhonde RR (2001) Suitability of cellulose molecular dialysis membrane for bioartificial pancreas: in vitro biocompatibility studies. J Biomed Mater Res 54:436–444. https://doi.org/10.1002/1097-4636(20010305)54:3%3c436::aid-jbm180%3e3.0.co;2-8

Ruel-Gariépy E, Leclair G, Hildgen P et al (2002) Thermosensitive chitosan-based hydrogel containing liposomes for the delivery of hydrophilic molecules. J Control Release 82:373–383. https://doi.org/10.1016/S0168-3659(02)00146-3

Salter GJ, Kell DB (1991) New materials and technology for cell immobilization. Curr Opin Biotechnol 2:385–389. https://doi.org/10.1016/S0958-1669(05)80143-0

Sato T, Stange DE, Ferrante M et al (2011) Long-term expansion of epithelial organoids from human colon, adenoma, adenocarcinoma, and Barrett's epithelium. Gastroenterology 141:1762–1772. https://doi.org/10.1053/j.gastro.2011.07.050

Schmidt-Winkel P, Yang P, Margolese DI et al (1999) Fluoride-induced hierarchical ordering of mesoporous silica in aqueous acid-syntheses. Adv Mater 11:303–307. https://doi.org/10.1002/(SICI)1521-4095(199903)11:4%3c303::AID-ADMA303%3e3.0.CO;2-M

Schwendener RA (2014) Liposomes as vaccine delivery systems: a review of the recent advances. Therapeutic Adv Vaccines 2:159–182. https://doi.org/10.1177/2051013614541440

Shin S, Ikram M, Subhan F et al (2016) Alginate–marine collagen–agarose composite hydrogels as matrices for biomimetic 3D cell spheroid formation. RSC Adv 6:46952–46965. https://doi.org/10.1039/C6RA01937D

Siegel RL, Miller KD, Jemal A (2018) Cancer statistics, 2018. CA Cancer J Clin 68:7–30. https://doi.org/10.3322/caac.21442

Smidsrød O, Skja G, (1990) Alginate as immobilization matrix for cells. Trends Biotechnol 8:71–78. https://doi.org/10.1016/0167-7799(90)90139-O

Sobol M, Bartkowiak A, de Haan B, de Vos P (2013) Cytotoxicity study of novel water-soluble chitosan derivatives applied as membrane material of alginate microcapsules. J Biomed Mater Res A 101:1907–1914. https://doi.org/10.1002/jbm.a.34500

Speiser DE, Flatz L (2014) Cancer immunotherapy drives implementation science in oncology. Hum Vaccin Immunother 10:3107–3110. https://doi.org/10.4161/21645515.2014.983000

Suarez-Arnedo A, Sarmiento P, Cruz JC et al (2018) 3D alginate hydrogels with controlled mechanical properties for mammalian cell encapsulation. IX Int Seminar Biomed Eng (SIB) 2018:1–5. https://doi.org/10.1109/sib.2018.8467745

Suarez-Arnedo A, Narváez DM, Sarmiento P et al (2019) Tridimensional alginate disks of tunable topologies for mammalian cell encapsulation. Anal Biochem 574:31–33. https://doi.org/10.1016/j.ab.2019.03.008

Sun Z, Liang J, Dong X et al (2018) Injectable hydrogels coencapsulating granulocyte-macrophage colony-stimulating factor and ovalbumin nanoparticles to enhance antigen uptake efficiency. ACS Appl Mater Interfaces 10:20315–20325. https://doi.org/10.1021/acsami.8b04312

Szot CS, Buchanan CF, Freeman JW, Rylander MN (2011) 3D in vitro bioengineered tumors based on collagen I hydrogels. Biomaterials 32:7905–7912. https://doi.org/10.1016/j.biomaterials.2011.07.001

Tripathi A, Kathuria N, Kumar A (2009) Elastic and macroporous agarose-gelatin cryogels with isotropic and anisotropic porosity for tissue engineering. J Biomed Mater Res A 90(3):680–694. https://doi.org/10.1002/jbm.a.32127

Tripathi A, Kumar A (2011) Multi-featured macroporous agarose-alginate cryogel: synthesis and characterization for bioengineering applications. Macromol Biosci 11(1):22–35

Tripathi A, Melo JS (2015) Preparation of sponge-like biocomposite agarose-chitosan scaffold with primary hepatocytes for establishing an in-vitro 3D liver tissue model. RSC Adv 5:30701–30710

Tripathi A, Melo JS (2016) Synthesis of low-density biopolymeric chitosan-agarose cryomatrix and its surface functionalization with bio-transformed melanin for the enhanced recovery of uranium(VI) from aqueous subsurfaces. RSC Adv 6:37067–37078

Tripathi A (2019) Melo JS (2019) Cryostructurization of polymeric systems for developing macroporous cryogel as a foundational framework in bioengineering applications. J Chem Sci 131:92. https://doi.org/10.1007/s12039-019-1670-1

Tripathi A, Melo JS (eds) (2017) Advances in biomaterials for biomedical applications. Springer-Nature, Singapore

Tripathi A, Melo JS D'Souza SF (2013a) Magnetic nanoparticles in tissue regeneration. In: Tiwari A Tiwari A (eds) Nanomaterials in drug delivery, imaging, and tissue engineering. WILEY-Scrivener Publisher, USA, pp 443–492. ISBN 9781118290323

Tripathi A, Vishnoi T, Singh D, Kumar A (2013) Modulated crosslinking of macroporous polymeric cryogel affects in vitro cell adhesion and growth. Macromol Biosci 13(7):838–850. https://doi.org/10.1002/mabi.201200398

Tubiana M (1971) The kinetics of tumour cell proliferation and radiotherapy. Br J Radiol 44:325–347. https://doi.org/10.1259/0007-1285-44-521-325

van Willigen WW, Bloemendal M, Gerritsen WR et al (2018) Dendritic cell cancer therapy: vaccinating the right patient at the right time. Front Immunol 9. https://doi.org/10.3389/fimmu.2018.02265

Villalobos A (2006) Dealing with chemotherapy extravasations: a new technique. J Am Anim Hosp Assoc 42:321–325. https://doi.org/10.5326/0420321

Wang C, Tong X, Yang F (2014) Bioengineered 3D brain tumor model to elucidate the effects of matrix stiffness on glioblastoma cell behavior using PEG-based hydrogels. Mol Pharm 11:2115–2125. https://doi.org/10.1021/mp5000828

Wang P-F, Chen Y, Song S-Y et al (2017) Immune-related adverse events associated with anti-PD-1/PD-L1 treatment for malignancies: a meta-analysis. Front Pharmacol 8:730. https://doi.org/10.3389/fphar.2017.00730

Weaver VM, Petersen OW, Wang F et al (1997) Reversion of the malignant phenotype of human breast cells in three-dimensional culture and in vivo by integrin blocking antibodies. J Cell Biol 137:231–245. https://doi.org/10.1083/jcb.137.1.231

Weber LM, Lopez CG, Anseth KS (2009) Effects of PEG hydrogel crosslinking density on protein diffusion and encapsulated islet survival and function. J Biomed Mater Res A 90:720–729. https://doi.org/10.1002/jbm.a.32134

Wei G, Ma W, Zhang A et al (2018) Enhancing catalytic stability and cadaverine tolerance by whole-cell immobilization and the addition of cell protectant during cadaverine production. Appl Microbiol Biotechnol 102:7837–7847. https://doi.org/10.1007/s00253-018-9190-3

Xu Y, Kim C-S, Saylor DM, Koo D (2017) Polymer degradation and drug delivery in PLGA-based drug-polymer applications: A review of experiments and theories. J Biomed Mater Res Part B Appl Biomater 105:1692–1716. https://doi.org/10.1002/jbm.b.33648

Yang HK, Ham D-S, Park H-S et al (2016) Long-term efficacy and biocompatibility of encapsulated islet transplantation with chitosan-coated alginate capsules in mice and canine models of diabetes. Transplantation 100:334–343. https://doi.org/10.1097/TP.0000000000000927

Yang K-C, Qi Z, Wu C-C et al (2010) The cytoprotection of chitosan based hydrogels in xenogeneic islet transplantation: an in vivo study in streptozotocin-induced diabetic mouse. Biochem Biophys Res Commun 393:818–823. https://doi.org/10.1016/j.bbrc.2010.02.089

Yang Z, Zhao X (2011) A 3D model of ovarian cancer cell lines on peptide nanofiber scaffold to explore the cell–scaffold interaction and chemotherapeutic resistance of anticancer drugs. Int J Nanomedicine 6:303–310. https://doi.org/10.2147/IJN.S15279

Yeung TM, Gandhi SC, Wilding JL et al (2010) Cancer stem cells from colorectal cancer-derived cell lines. Proc Natl Acad Sci USA 107:3722–3727. https://doi.org/10.1073/pnas.0915135107

Zent R, Pozzi A (2010) Cell-extracellular matrix interactions in cancer. Springer-Verlag, New York

Zhang H, Chen J (2018) Current status and future directions of cancer immunotherapy. J Cancer 9:1773–1781. https://doi.org/10.7150/jca.24577

Zhang R, Billingsley MM, Mitchell MJ (2018) Biomaterials for vaccine-based cancer immunotherapy. J Control Release 292:256–276. https://doi.org/10.1016/j.jconrel.2018.10.008

Zucca P, Fernandez-Lafuente R, Sanjust E (2016) Agarose and its derivatives as supports for enzyme immobilization. Molecules 21. 10.3390/molecules21111577

Nanosystems for Repairing Retinal Degeneration

Deepti Singh, Pierre C. Dromel, Shao-bin Wang, and Anuj Tripathi

Abstract Retinal degenerations are the leading causes of visual impairment in aging populations. Advances in retinal stem cell biology provide an avenue for studying and treating these diseases. Remarkably, retinal tissues are self-organizing, which simplifies the challenge of creating three-dimensional culture models. The problem that confronts the investigator is that current models are small, multilayered spheres, which limits experimental access to both sides of the multilayer. Furthermore, for grafting into a diseased retina, these neurospheres must be dissociated with the hope that they will reorganize into a multilayered sheet in the environment of the diseased retina. This chapter discusses the scaffolds under investigation that can promote formation of a planar, multilayered model of the retina. Such a tissue would allow investigators to study the mechanisms of retinal diseases, identify therapeutic targets, develop therapeutic agents, and serve as an engraftable tissue.

Keywords Retinal regeneration · Stem cells · Scaffold · Engraftment

D. Singh (✉) · P. C. Dromel
Schepens Eye Research Center, MASS. Eye and Ear Infirmary, Harvard Medical School, 20 Staniford street, Boston 02114, USA
e-mail: deepti_singh@meei.harvard.edu

P. C. Dromel
e-mail: colombep@mit.edu

P. C. Dromel
Department of Material Science and Engineering, Massachusetts Institute of Technology, Cambridge, MA, USA

S. Wang
University of Virginia, 21 Hospital, Charlottesville, VA 22903, USA
e-mail: sw3ta@virginia.edu

A. Tripathi
Nuclear Agriculture and Biotechnology Division, Bhabha Atomic Research Centre, Mumbai, India
e-mail: anujtri@barc.gov.in

A. Tripathi and J. S. Melo (eds.), *Immobilization Strategies*,
Gels Horizons: From Science to Smart Materials,
https://doi.org/10.1007/978-981-15-7998-1_5

List of Abbreviations

AMD	Age-related macular degeneration
APOE	Apolipoprotein E
ARPE-19	A human retinal pigment epithelial cell line
ECM	Extracellular matrix
HEK	Human embryonic kidney
hESC	Human embryonic stem cells
HuCNS-SC	Human central nervous system stem cell
HUVEC	Human umbilical vein endothelial cells
PCL	Polycaprolactone
PDMS	Polydimethylsiloxane
PLGA	Poly lactic-co-glycolic acid
PVA	Polyvinyl alcohol
RCS	Royal college of surgeons
RP	Retinitis pigmentosa
RPCs	Retinal progenitor cells
RPE	Retinal pigmented epithelium
RPE	Retinal pigment epithelium

1 Introduction

1.1 Retina Neurogenesis and RPE

Organogenesis is directly related to local cellular population and its functionality.. The organ that illustrates this is the nervous system in which proliferating neural cells give rise to various glia and neurons as it exit the cell cycle, they further differentiate into its matured form to set up connection between themselves (Abe 2002). Retina follows some of these steps in which dividing neural cells in neural plates is designated to form eye field which further splits into two lateral domains (Centanin and Wittbrodt 2014). Each of these evaginates to form an optic vesicle that further invaginates to form the cup architecture of retina, with outer epithelial layer of cells forming retinal pigmented epithelium (RPE) and the inner neuro-epithelial becomes the retina (Yip 2014; Miyake and Araki 2014). The neural retina exit progenitors cells after undergoing several cell cycle to convert retina into a three layers structure consisting of 6 major types of differentiated neuronal cells (each of these cells can further subdivide to form many subtypes) and one form of glial cells (Chow and Lang 2001) (Fig. 1). Early progenitor cells (RPCs) are responsible for tissue genesis. Proliferating property of these cells comes partly due to the inheritance of dividing temperament from ancestral RPCs and partly owing to microenvironments of retina that favors cell division. At a particular point, during embryogenesis/organogenesis of each species,

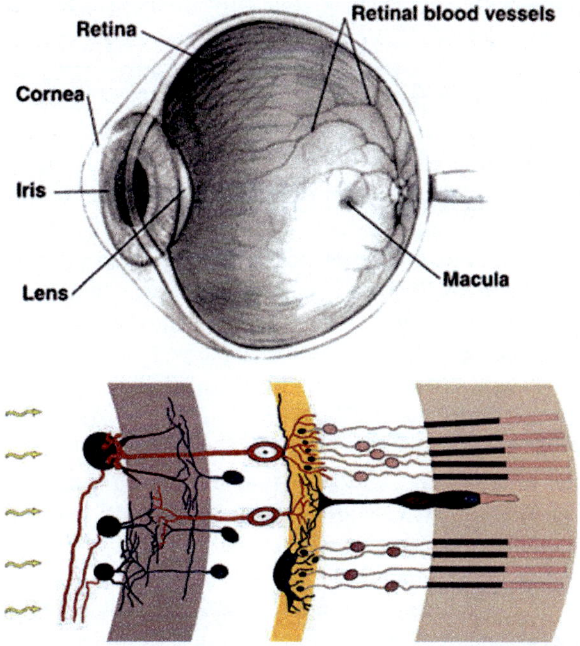

Fig. 1 Structure of eyes showing the microarchitecture of retina and different layers within in this tissue. Reproduced from WIKIPEDIA

RPCs starts to drop out of cell cycle and differentiate permanently with few exceptions like Muller glial cells that retain considerable features of multiplication, have the ability to reenter cells cycle at any time point if provided in appropriate conditions, and can give rise to neurons. RPCs division is heterogeneous in a way that it can undergo multiple cell division cycle to produce clones of different sizes and cellular structures. Each specie possesses a different time points for the retinal development, like in the case of frog: retinal generation is faster within 2 days as it goes from the onset of its first post-mitotic division to complete retina generation (central retina). Clonal labeling of RPCs in central retina shows 1–16 cells/clone within few hours before onset of differentiation cascade (Holt et al. 1988). However, in mice, clone size ranges from 1 to 200 cells occurring for over 2 weeks and an average clone size is approximately 40–50 cells depending upon RPC location (Turner et al. 1990).There is apparent randomness in clone size and cellular fate suggesting strong section of stochasticity in progenitor's life choice (Holt et al. 1988; Turner et al. 1990; Wetts and Fraser 1988). However, it is known that there are conserved patterns in cell birth and fate as it follows histogenetic order in retinal production in which ganglion cells emerge first and Muller glial cells are formed last (Wetts and Fraser 1988). As retina develops, progenitors take much longer to divide and produce more post-mitotic cells regularly. There are certain cells that retain considerable features of dividing cells which under appropriate conditions can give rise to neurons and, depending upon the orientation of these cells, each differentiate and take up a unique role (Turner et al. 1990; Ackman and Crair 2014; Martinez-De Luna et al. 2011). The microenvironment for RPC is provided by the retinal pigmented epithelium (RPE) which is a unique tissue in eyes that is embryologically derived from the same neural lineage as sensory retina but differentiate in secretory epithelium without any photoreceptive or neural functions. However, they are critical for providing structural support and viability of photoreceptors (Martinez et al. 2011; McGill et al. 2012; Zhao et al. 2007; Wen et al. 2015; Uygun et al. 2009). These RPE controls growth factor synthesis, neural environment, responding to humoral agents with regenerative capacity that can be harmful or beneficial depending upon the nature of injury (Engelsberg and Ghosh 2011). Cells within these RPE appear cuboidal in cross-section and hexagonal in architecture when viewed from top and these are all joined near apical side by tight junctions (zonula occludens) that block the free movement of fluids and ions (Tu and Palczewski 2014; Liu et al. 2011; Charlton-Perkins and Cook 2010; Kuo et al. 2009; Afshari and Fawcett 2009; Lin-Jones et al. 2009; Mousa et al. 1999; Samie et al. 1995). These junctions also form one part of the blood-retinal barrier while other being formed by capillary endothelium of intrinsic retinal vessels (Lin et al. 2009; Liu et al. 2008; Cavallotti et al. 2005). RPE cells are roughly around 10–14 μm in diameter in macular regions, but becomes broader and narrower (60 μm) toward periphery (Lehmann et al. 2015; Bonilha 2014; Hamilton and Leach 2011; Wimmers et al. 2007; Rosenthal and Strauss 2002; Mikhailov et al. 2002). As the density of photoreceptors differs across the retina, the quantity of photoreceptors overlying each RPE roughly remains constant which is in range of 45. There is great relevance of this constancy as each RPE cell is metabolically accountable for a variety of functional support to the overlying retina. With the invention of induced pluripotent stem cells, a model for early stage of human ontogenesis could be easily studied. These cells undergo normal development timeline in a controlled targeted stepwise differentiation process (Sun et al. 2015; Eiraku and Sasai 2012; Yoshida et al. 2014; Zhong et al. 2014; Brandl et al. 2014; Westenskow et al. 2014; Wahlin et al. 2014; Liu et al. 2014a, b, c, d; Kanemura et al. 2014; Park et al. 2014; Tucker et al. 2013a, b; Minegishi et al. 2013; Melville et al. 2013; Tucker et al. 2013a, b; Jin et al. 2012; Hu et al. 2012; West et al. 2012; Torrez et al. 2012; Krohne et al. 2012; Rowland et al. 2013; Phillips et al. 2012; Qiu et al. 2012; Das and Pal 2010; Buchholz et al. 2009). Work done by Gamm and groups systemically explored retinal development and factors that play critical role in human retinogenesis (Phillips et al. 2012). During the first phase of stepwise retinal phenotypic differentiation from undifferentiated stem cells is the appearance of eye field cells. It was previously thought that hESCs-derived neuroectodermal cells adopt anterior neuroepithelial morphology in absence of any external signaling molecules, however, in this study, free-floating hESCs aggregate and adhere to laminin-coated culture dishes that allow neural rosette formation which were mechanically removed to grow as neurospheres. During this process, hESCs quickly lose pluripotency genes such as Nanog and Oct4 and acquire transcription factors associated with eye field phenotype (like Rx, Lhx2, Six 3, six6, Tll), general neural induction such as Sox1, Sox2, and Pax6 along with anterior neural specifications Otx2. This work showed that hESCs could serve as a comprehensive model for in vitro retinogenesis and open field for understanding retinal generation with aim to achieve its repair. RPE is involved in many retinal disorders mainly due to its ability to scar, i.e., hyper or depigmentation which is often manifestation of disease.

Retinal degeneration is a major cause of untreatable blindness in USA and UK and a very small percentage of it can be attributed to heredity. It could be due to age-related macular degeneration (AMD) which is known to affect 1 in 10 especially in age group of 60 or retinitis pigmentosa (RP), a genetic disorder. Nevertheless, whether its age or genetic or both, these conditions culminate to the same final pathway which is loss of the light-sensing photoreceptors and, if left untreated, could lead to sever visual impairment (Garcia et al. 2015). There are different signs that disease exhibits such as distorted visions (grid lines appears wavy to the affected person), pigmentary changes, and exudative changes such as hemorrhages in eyes. There are various approaches by which the retinal repair and macular regeneration have been targeted like transplanting cells to the degenerated area. The technique of replacing lost RPE cells with completely differentiated human ESCs derived equivalents is one of the way that precise area of the retina could be targeted (Reichman and Goureau 2016; Pfeffer and Philp 2014) (Fig. 2). Stem cell's neural stem cell therapy is believed to act by providing functional support to the retinal photoreceptors by releasing growth factors at different time points. This study showed HuCNS-SC was able to satisfy number of criteria that proved its potential for clinical application (McGill et al. 2004). Not only did these cells elicit any immune reaction in host RCS rat but prevented retinal disease progression which could be noticed in both functional and anatomical measures. The HuCNS-SC transplanted into host prevented typical retinal pathology which is normally noticed in RCS rat over time. It was observed that single injection of HuCNS-SC preserved cone and rod photoreceptors and provided long-term functional advantages that strongly support human testing in patient who suffer vision loss due to degeneration of photoreceptors. However, in this study, the subretinal space in which these cells were transplanted does not provide cues or microenvironment for directing neural stem cells into glial cells or neurons due to the degenerated area, being unable to form the bridge for complete regeneration. Besides these, there are number of problems associated with this approach, such as attaining precise delivery of cells, survival rate of injected cells and its integration with surrounding tissue along with achieving controlled cellular differentiation. These are critical before the success of this method could be realized. Treating AMD is more complex since it requires smaller concentration of progenitor cells, differentiated cells and neuronal cells and achieving this by simple cell therapy seems to be a difficult task. Combining the principles of tissue engineering in which materials are designed to support progenitor's cells, provide cues for differentiation and would be thin enough to fit in sub-retinal space.

1.2 AMD and Embryonic Stem Cells

AMD can be characterized by the presence of extracellular deposition known as "drusen" which gathers between the Bruch's membrane and basal surface of "retinal

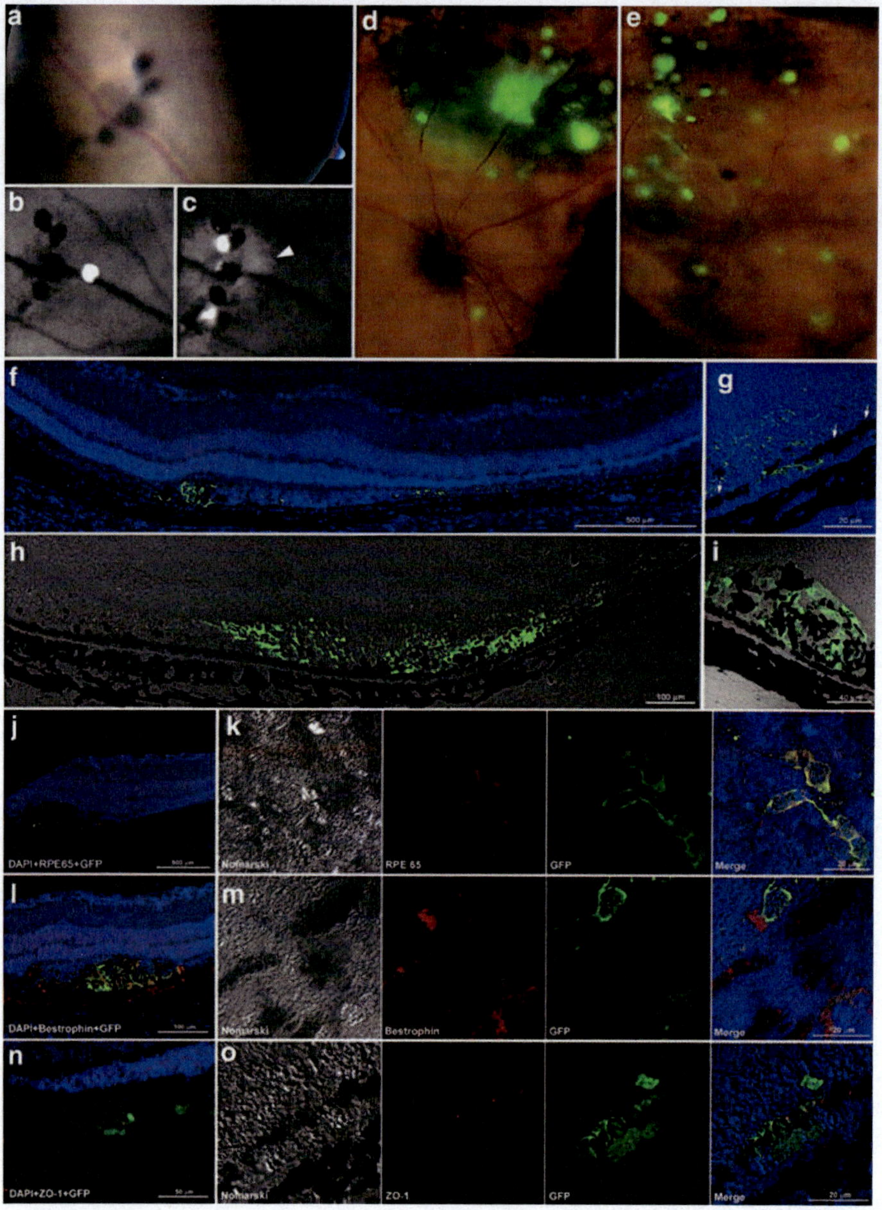
a
b
c
d
e
f
500 μm
g
20 μm
h
100 μm
i
j
k
DAPI+RPE65+GFP
Nomarski
RPE 65
GFP
Merge
l
m
DAPI+Bestrophin+GFP
Nomarski
Bestrophin
GFP
Merge
20 μm
n
o
DAPI+ZO-1+GFP
Nomarski
ZO-1
GFP
Merge
20 μm

◂**Fig. 2** hESC-Derived RPE Cells Survive, Integrate, and Express RPE Markers after Subretinal Transplantation in RCS Rat Eyes (**a**–**c**). Fundus imaging of subretinal grafts of pigmented cells in eyes of live RCS rats; fundus photograph (**a**) and red-free photograph (**b**) (note retinal vessels crossing over the pigmented areas). **c** The grafted eGFP + cells emit fluorescence, as evident by the hyperfluorescent areas (arrowhead) surrounding the pigmented grafts. **d, e** Fluorescence microscope images of eyecup preparations, showing large clusters (**d**) as well as multiple dispersed smaller clusters (**e**) of subretinal eGFP + cells. **f**–**i** Immunostaining with anti-GFP showing the subretinal location of grafts dispersed in large areas (**f, h**) and pigmented cells coexpressinge GFP (**g, i**). Note eGFP+ cells integrating within the host RPE layer (**g**); arrows showing host RPE. (**j**–**o**) Immunostaining showing eGFP+ transplanted cells co-expressing RPE65 (**j, k**), Bestrophin (**l, m**), and ZO-1 (**n, o**). **j, l, n** Left-most panel in each row shows low-magnification fluorescent image of grafts co-expressing eGFP and the relevant marker. **k, m, o** High-magnification confocal images in each row show from left to right pigment (by Nomarski optics) and co-expression of the different RPE markers and eGFP. Note the polygonal shape of eGFP+ cells in **m, o**. Reproduced with permission from Idealson et al. (2009)

pigmented epithelia" (RPE) and over a period of time causes retinal detachment. Retina is a complex of nerves and visual receptors that lie within choroid along with large networks of blood vessels and Bruch's membrane which is the inner ECM that separates choroid capillary network from neural retina by maintaining blood–brain barrier and controls the movement of the fluids within the tissue. AMD due to damaged retina results in the loss of vision in the macula (center region of visual field). Early studies performed suggested it was the formation of drusen that could possibly lead to degeneration of RPE initiated by shedding membranous debris from its basal surface. These results have been further confirmed by recent finding of various groups. More current investigations are being performed at molecular level to understand structure, composition, and biogenesis of drusen (Johnson et al. 2011). Proteomic and immuno-histological analysis show that drusen is made up on lipid and protein components, furthermore, it was found that these circulating molecules are in smaller number synthesized locally by RPE cells. Beside these, complement system and its component have also been found to play some role in drusen genesis. This could be the result of local chronic inflammation at RPE/Bruch's membrane interface and its formation is directly proportional to AMD pathogenesis. Despite progresses and understanding of AMD, the trigger that activates complement cascade and downstream molecular interactions leading to AMD pathogenesis is still elusive which is due to non-availability of suitable model either in vivo or in vitro that can reproduce the silent features of AMD.

Various experiments to overcome this bottleneck and provide a platform that could facilitate in-depth analysis and understanding of event that leads to drusen formation and AMD pathogenesis have been engineered. Cell therapy for replenishing degenerating RPE cells would potentially halt the progression of disease. Few researchers working on animal models have shown that sub-retinal transplantation of RPE would rescue photoreceptors and preserve visual functions. Partial visual restoration using autologous RPE has also been reported in humans, however, before finding treatment, it is of utmost important to understand the conditions that results in AMD disease. Eye formation has attracted lot of attention in both classic and modern-day engineers and development biologist. The retinal anlage is demarcated by Rax (R_x) expression which first appears as optical vesicle, laterally evaginating epithelial vesicle from diencephalon subsequently, the distal portion folds within to form two-walled cups like configuration, the optic cup which goes on to become outer (pigmented) and inner (neurosensory) stratum of retina. To mimic this system in laboratory is complex, however, Eiraku and Sasai (2012) showed that use of embryonic stem cells (ESCs) in three-dimensional system could successfully reduce the complexity of this organogenesis process (Fig. 3). ES cells cultured in serum free-floating culture formed embryoid-body like aggregates and it was possible to induce retinal differentiation in this modified SFEBq culture using different transient activin treatment. However, there was no clear retinal epithelial formation with this procedure hence, further modification of culture medium was done along with culturing ECs on Matrigel (basement-membrane matrix components) that aided in formation of rigid epithelial structure (Fig. 3b–d). On the 6th day of experiment, ECs aggregated to form a hollow sphere that contained polarized apical neuro-epithelium N-cadherin$^+$ that instinctively divided into Rx-GFP positive and negative cells (Fig. 1a) and on 7th day, the positive cells formed hemispherical epithelial vesicles emerging from main body within one of the 4 vesicles (Fig. 3b, c) and on 8th to 10th day, it undergoes dynamic shape change and formed two cup-like structure (Fig. 3f, g). The distal portion of this expressed neural retinal markers (Fig. 3h, i) and reminiscent of marker prolife was that of retinal pigment epithelium (RPE) progenitors' cells which ultimately becomes pigmented (Fig. 3i–m). This was unexpected behaviors of ESCs and provided newer prospects for studying and understanding RPE as this study revealed complex morphogenesis of retinal anlage in the in vitro condition and 3D neural retina tissue is heralds as next generation of regenerative medicine for retinal and macular degeneration therapeutics. Researchers around the world have tried to fabricate 3D culture system that can mimic the Bruch's membrane as closely as possible to provide an ideal environment for macular regeneration (Wang et al. 2003).

1.3 Nanomatrix and Controlled Cell Expressions

Regenerative medicine is ideal for replacing or repairing tissues or organ affected by some disease or trauma. One could use principles of materials science for fabricating advanced materials that could act as template for cellular growth and also one that would provide physical and chemical cue for cellular migration and differentiation (Singh et al. 2014, 2013; Tripathi and Melo 2015). Transplantation of cultured progenitor or autologous limbal stem cells has shown promises by aiding in recovery of corneal integrity and improved visual functions after chemical injury to cornea. In particular, regenerative nanomedicine which involves usage of nanoparticles, nanosheets, or disc loaded with genetic transcriptional factors and other transforming molecules for reprogramming of cells in the in vivo system has thrown the

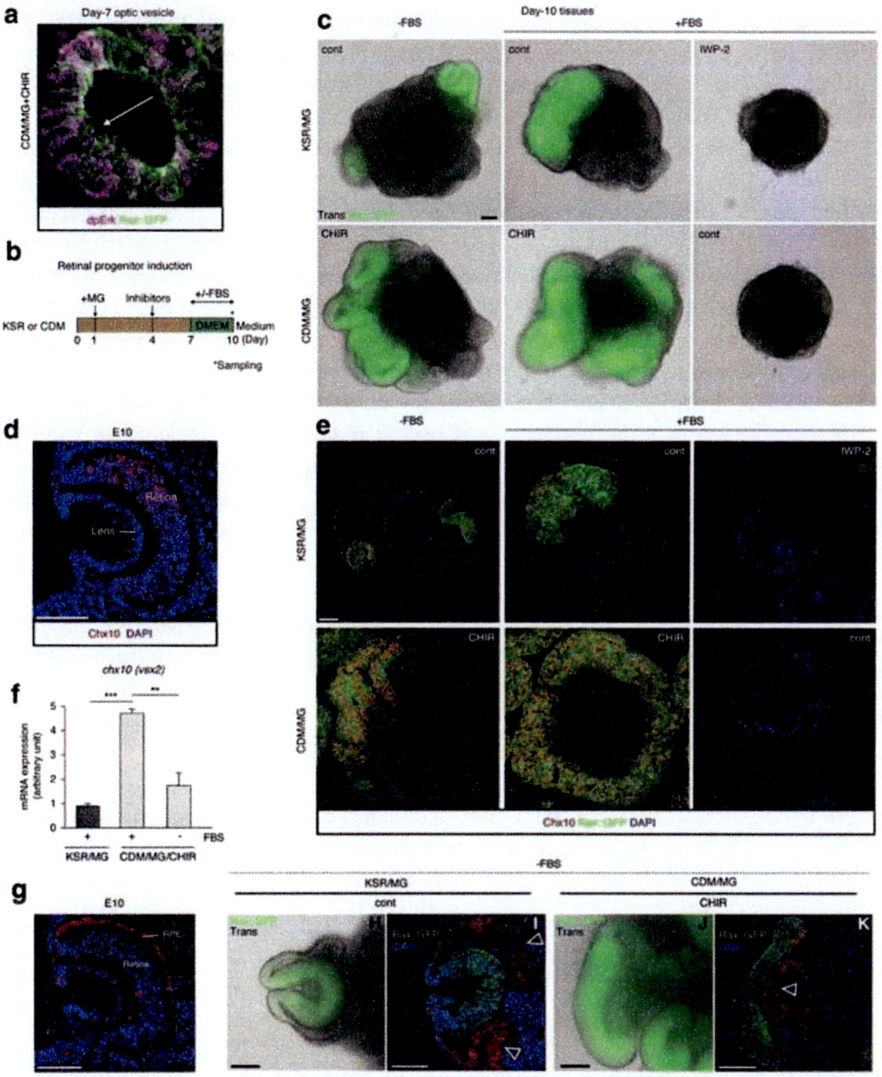

Fig. 3 Retinal progenitor marker *chx10* was highly induced in the CDM/MG+ CHIR condition. The ES culture showing different stages of eyes cup formation. Reproduced with permission from Sakakuraa (2016)

field wide open as engineers, physicians, and biologist all play an equal role during the fabrication of biologically functional tissue substitute (Cassidy et al. 2014; Kuppan et al. 2013). One of the early example of this is the use of polyplexes for treating retinal pigmented epithelial and for restoring light-sensitive neurons by transfecting cone cells with rAAV-HaloR and induction of light-sensitive neurons by acrylamide azobenzene quaternary ammonium (without transfection) (Lung et al. 2012). As the nanomedicine field advances, the improved treatment for RP, macular, AMD, and allied diseases could be engineered. Regeneration of damaged cells in case of fish by Muller cells and for amphibians from RPE is spontaneous process achieved by endogenous mechanisms (Martinez et al. 2011; Goldman 2014; Jayaram et al. 2014; Lenkowski and Raymond 2014; Powell et al. 2013; Mizuno et al. 2012; Vergara and Rio-Tsonis 2009; Yoshii et al. 2007). Microscale topographical cues can influence the behavior of retinal cells for attachment, migration, and differentiation indecently of biochemical cue could be provided by culturing cells of matrix. The shape, size, lateral shaping, and surface chemistry and geometry of the matrix ligands are physical features that mimic natural ECM thereby influencing cellular behavior. Scaffold with apt nanoscale features can improve efficacy of transplanted graft by preventing anoikis and apoptotic reaction that occurs due to the absences of cellular adhesions sites on ECM substitute, promoting and maintenance of differentiated cells, offering a 3D organization of cell–matrix assemble and induces a positive host-response. Nanofiber scaffolds are generally used for creating niches for stem cells self-renewal characteristics or as substrate used for cell sheet delivery due to its high surface area to volume ratio and also can present several epitopes for cell attachment thus promoting neural stem cells differentiation which is one of the most important aspects for RPE culture and macular regeneration (Alamein et al. 2013; Chen et al. 2011; Pritchard et al. 2010a, b). In case of mammals, there is central marginal zone which is source for RPE cells however, this region does not support retinal regeneration (Rizzolo 2014), hence, using different cells such as Muller cells (endogenous source for RPE cells) or embryonic stem cells can used as alternative cells (Peng et al. 2013) source along with artificial ECM that plays crucial role in retinal regeneration.In case of keratocytes, the shaped, alignment, and migration of cells could be altered by nanotopographical cues of nanomaterials (Liu et al. 2014a, b, c, d; Kang et al. 2014; Elsayed and Merkel 2014). As it is known that stem cells are prevented from exiting the mitotic cycle by residing in the favorable environment called niches (cellular and non-cellular part of the system) (Leri et al. 2014; Szekely et al. 2014; Quesenberry et al. 2015; Oh et al. 2014) and when it was mimicked artificially, the same natural phenomena for attaining desired stem cell differentiation could be repeated.

2 Nanosystems for Macular Regeneration

Polymeric nanomaterial has been used as delivery vehicle for different therapeutical applications and imaging processes. A set of experiment performed by Roy et al. (2013) showed preferential internalization of hydrogel nanodiscs by cells over nanorods and explored shaped specific uptake mechanisms. Materials composition, size, surface charges, and topography all are known to play critical role during cell–matrix interaction or nanoparticle uptake by cells. Understanding the phenomena behind how particle properties could affect the cellular uptake process is not only crucial for charting the treatment plan but also for improving the efficacy of therapy. In this work, it was shown the size and shape of the particle are critical during cellular uptake since the particles are large or medium size were internalized by cells more effectively in comparison to nanorods and discs. Use of different types of cells such as HEK, HeLa, HUVEC and RPE showed that each cell processed the nanorods, discs, and sheets. In case of HVVEC's nanorods and discs were internalized via macropinocytosis and clathrin-mediated pathways but RPE cells showed no involvement of clathrin pathways like other epithelial cells. Moreover, in case of hydrophilic polymeric nanoparticle, since it influenced the interplay with cells in 3 ways (1) deciding the contact area or strength of adhesion force between the nanoparticles and cellular membrane, (2) strain energy that is required to effectively deform membrane around the particle and finally (3) particle concentration (locally) or sedimentation effect at the surface of cells. This work showed that depending upon the geometry of the matrix, cells could smartly "sense" and thereby trigger very specific uptake pathway and these results provided a fundamental insight toward exploiting nanoparticle geometry as design criterion for controlling cellular response which can be used for engineering new treatments and attaining high cell targeting and delivery.

2.1 RPE on Nanoporous Material

To find a treatment for AMD, it is utmost important to explore and extensively research on the disease prognosis. RPE cell-culture working model that mimics critical facets of drusen formation and associated complement activation was studied by Johnson et al. (2011). In this study, they showed that it was possible to fabricate 3D culture of human retinal pigmented epithelial (RPE) that could mimic different key aspects of early AMD including accumulation of sub-RPE deposit which constitutes of human drusen and also could activate complement cascade that leads to formation of associated terminal complement complexes. RPE cultured on porous support materials showed prominent drusen constituents apolipoprotein E (APOE) formation and upon exposure to human serum, there was accumulation of additional drusen-associate components formation such as vitronectin, serum amyloid P and clusterin (Fig. 4) along with activation of complement cascade indicating that specific protein interaction could possible accelerate drusen formation. Highly differentiated RPE cultured on permeable ECM produce sub-RPE deposits that builds up within the pores of materials over the course of time and among the different ECM proteins one of these was APOE (ubiquitous protein associated with RPE as these cells both synthesize and secrete them) and other proteins (clusterin, APSCs and vitronectin) were found to selectively accumulate in APOE only upon exposure to human serum. This work provided basic platform and model system in three-dimensional set up in which cellular, molecular, and genetically analysis could be performed for deeper understanding of AMD pathogenesis and provided a chance for testing new therapeutically agents that could be used for its treatment. Since work done before this used in vitro tissue culture plates for ARPE-19 cells, however, drusen-like properties were neither mimicked nor identified. The porous scaffold used was HA-mixed cellulose esters with pore size of 400 nm and modified for high cell attachments,

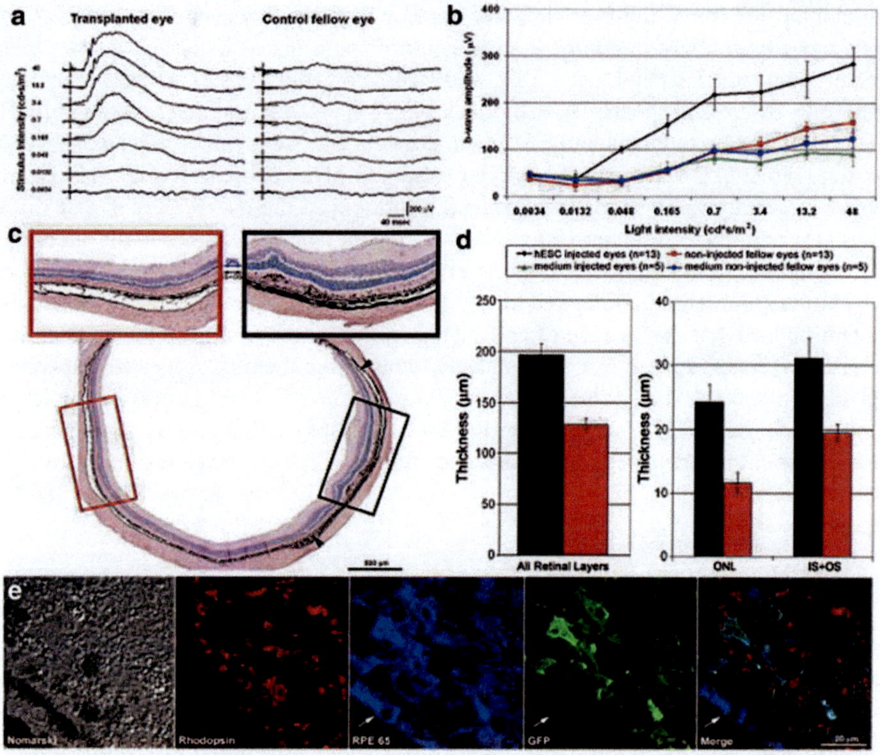

Fig. 4 Transplanted hESCs derived RPE cells provide functional and structural rescue in RCS rat model of retinal degeneration Reproduced with permission from Idealson et al. (2009)

proliferation, and cell surface receptor for culturing different progenitor cells. The porous scaffold with nanopores was found to be ideal to facilitate nutrient transport and cell migration and more importantly it could recapitulate some key pathological process that play critical role in AMD such as disruption of RPE is primarily reason for AMD and if the subretinal transplantation of RPE at site of degeneration could be achieved it offers an ideal treatment plan (Xiang et al. 2014). Transplanted RPE could generate neo monolayer, inhibit development of choroidal vessel and aid in reconstruction of RPE-photoreceptor interface. However, using conventional tissue engineering matrix for doing this present its own complication such as large incision of sclera and retinal could result in leakage of vitreous fluid which can cause post-operative infections. Hence innovative therapeutics that combines the principles of biology and materials science is required for treating AMD.

2.2 Nanosheet for Targeted Delivery of RPE

Nanobiotechnology plays a critical role in advances of diagnostics and clinical therapeutics, for example, drug delivery, tissue engineering, and biosensor. Tissue engineering is described as branch of regenerative medicine which includes principles of biology, material science, engineering converging together to generate tailor-made transplantable tissue substitute usually achieved by engineering cellular polymeric matrix. These matrices are synthesized using different natural and synthetic polymers and can be modulated to attain specific property depending upon its end use, however, so drawbacks have been noticed in this innovative approach such as contaminations or toxic byproduct buildup during long-term cell culture. Therefore, bioengineers have now started to invest on fabrication of smart biomaterials with enhanced healing ability for wound tissue or biomaterials with unique ability to recruit cells from host tissue thereby assist in integration of implanted material with host tissue along with directing the cellular migration and organization without eliciting any immune response. One of the most recent advances made in this area is fabrication of ultrathin polymeric sheet often referred as nanomembranes, nanofilms or nanosheets which are new class of nanomaterials. Recent advance in nanotechnology has enabled researchers to produce free-standing nanopolymeric sheet as macromolecular organization. In comparison to its bulk film counterparts, these nanosheets have large aspect ratio (~10^6) providing unique physical property like transparency, high flexibility, and non-covalent adhesion. These nanosheets are measures around 1 cm laterally and few tens of nanometer in thickness which resembles the dimensions of extra cellular matrices or outer layer of cellular membrane (Giannaccini et al. 2014; Liu et al. 2014a, b, c, d; Peng et al. 2013; Fujie et al. 2014; Warnke et al. 2013; Ulbrich et al. 2011).

Free-standing polymeric nanosheet fabricated recently using natural biodegradable polysaccharides by spin-coating-assisted LBL (layer by layer) technique by alternatively depositing oppositely charged polyelectrolyte for macromolecular organization via electrostatic interactions thereby avoiding the use of chemical cross-linkers (Villares et al. 2014). Using this technique, Fujie et al. (2011) have shown that this nanosheet could be used for repairing different tissue defects and it was further explored by different groups for fabrication of peptide nanosheet, rods, and discs. In case of AMD, the main complication is generating subretinal choroidal neovascularization along with RPE cells hence engineering cell delivery vehicle was ideal choice for treatment of AMD. Nanosheet provides a much-needed platform that could be manipulated to obtain the results that is required for AMD and in step toward this micro-patterned nanosheet fabricated by using PLGA and collagen nano was explored (Fig. 5). Micro-patterned sheet was prepared by combining spin coating and micro-CP technique in which PDMS master stamp was prepared with columnar convex portions using conventional photolithography Su-8 molds. PLGA solution was separately prepared and mixed with magnetic nanoparticles for visualizing nanosheet and this mixture was further spin coated onto PDMS stamp resulting in PLGA/MNPs which was transferred on a glass coated with PVA

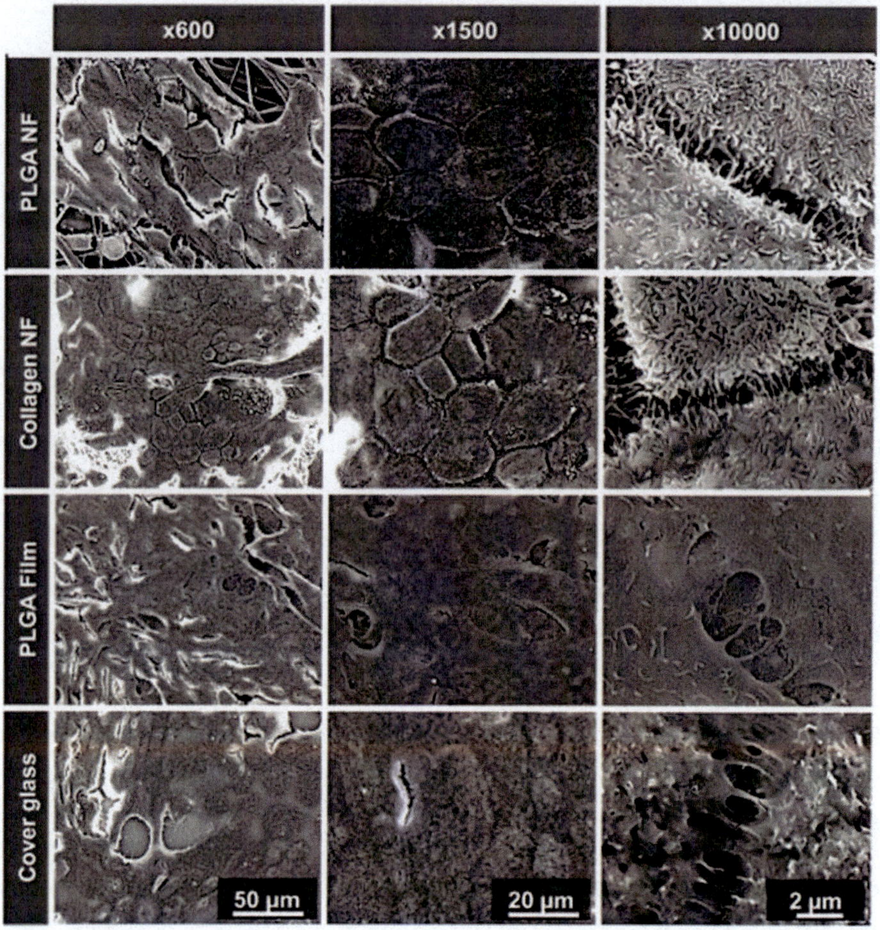

Fig. 5 RPE culture in different nanofibrous scaffold shows RPE maintain or form in vivo like cell phenotype with well-formed tight junctions strikingly similar hexa/polygonal structure in both PLGA and collagen nanofibers. Reproduced with permission from Warnke et al. (2013)

and for enhancing cellular adhesion collagen was spin coated to this substrate. The obtained PLGA/MNPs were circular shaped and dissolution of PVA allowed the release of brown-colored nanosheets of 170 nm thickness and could be easily aspirated into intravenous catheter. RPE monolayer were selectively seeded onto micro-patterned nanosheet to aid local deliver of cellular organization and characterized using confocal laser scanning microscope after Rhodamine B staining. Results confirmed the formation RPE monolayer which is important structural characteristics of epithelial cells. This nanosheet was further injected into sub-retinal space via intravenous catheter of swine ocular globe. Results obtained shows the sheet has successfully released and spread into sub-retinal space without causing any structural distortion to macular. This result opened new area for engineering smart biomaterials especially combining the techniques of nanobiotechnology.

More recent modification of the above technique researchers has been able to fabricate 2D nanomaterials with atomic precision that has led to engineering an ultrathin material with exceptional functionality. Specially, using organic nanomaterials holds promise as biocompatible matrix which can be chemically modified and built from bottom-up by self-assembling of small molecules, polymers/monomers, or proteins. This polymeric nanomaterial is exciting new prospect for bioengineers since it has shown to be excellent template for bottom-up assembly of semi-conductors, organic–inorganic composite or circuits as well as provide high surface area for membrane-based filtration, sensing, and catalysis. Materials with such precisely defined porous structure, protein or metal recognition sites along with different reactive groups such that it can be patterned directly into the polymeric backbone sequence. One such example is peptoids that highly designable class of polymers assembled into 2D peptoid nanosheets (Sanii et al. 2014; Robertson et al. 2014).

2.3 *Nanofiber Topographical Cues for RPE Culture*

Replacing RPE monolayer can be one of option for treatment of AMD patient; however, other than neovascular form, there are not many remedies available in current scenario. One attempt in past has been made to transplant suspension of RPE but this process faced several technical and biological challenges. Yet, these attempts have resulted in some advances as the researchers have entered phase 1 clinical trials which used ESCs suspended subretinally for AMD. However, tissue engineering provides an excellent opportunity for overcoming the limitation of cell therapy. The fundamental concept is to use polymeric material as base for culturing RPE to aid in maturation phase of cells into functional monolayer and this cell–matrix construct can be transplanted into the complexes underneath the retina. Furthermore, these matrixes would facilitate facile precise delivery and integration of graft allowing it to function more aptly than cell therapy. In this case, free cell suspension grafts, the cells are more likely to immigrate to different site and difficult to attain its full functionality. An ideal cell scaffold fabricated using principles of tissue engineering RPE monolayers transplantation should be able to stimulate conditions and micro-environment of healthy macular tissue along with being a biocompatible and safe in regulatory aspects. Rizzolo et al. (2008) have worked extensively on retinal cells, induced pluripotent stem cells and embryonic stem cells for retinal differentiation. The PCL was cast as a flat, solid sheet that was etched or decorated fabrication of nanofibers using polycaprolactone (PCL). If transplanted into the sub-retinal space, the sheet would initially block contact with the RPE. In an untested attempt to overcome this problem, holes were engineered into the sheet to permit limited RPC-RPE contact (Sodha et al. 2011). Among different models created, the crisscrossing nanofibers of PCL was synthesized by layer-by-layer assembly showed when co-cultured, the RPE contacts the RPC to recreate a sub-retinal space via cell–cell contact and secreted trophic factors, RPE should induce the differentiation of photoreceptors and could be potential model for treating AMD and retinitis pigmentosa patient.

3 Future Directions and Conclusion

There are two fundamental problems with transplantation as a human therapy:

1. Ideally, there would be a medical option for a patient with a mild burn. Even if grafts were ultimately required, a medical therapy could serve to augment transplantation. And recall, grafts in rodents fail when retinal degeneration is moderate or advanced, but this is when a therapy is most likely needed in humans. Would the human eye respond as rodents do?
2. Transplantation is technically difficult and would be hard to disseminate in practice. Alternatively, if a medical therapy was available, intravitreal or even subretinal injections would be much easier to employ in practice, for both the provider and the patient. As a parallel, witness the rapid, wide-spread dissemination of intravitreally injected anti-angiogenic agents to treat the wet form of AMD.

To address these problems, consider that monocultures of RPE or retinal progenitor cells (RPC) require retina and RPE to differentiate and mature independently of one another. However, animal models demonstrate that the neural layers profoundly affect the maturation of the RPE and vice-versa (Rizzolo 2007; Sun et al. 2008; Kolomeyer et al. 2011; Zhu et al. 2011; Nasonkin et al. 2013). The two tissues should be studied as a unit because there are significant differences between human and non-primate RPE, (Peng et al. 2011; Rizzolo et al. 2011).We believe when novel, three-dimensional (3D) nanomodel is engineered to study how human RPE and retinal cells interact. This model allows us to explore the complex interactions of human RPE with the neural retina, test the putative therapeutic agents, and determine their mechanism of action. Stem cells-based therapy is most efficient and promising approach when used along with bio-engineered scaffold as this scaffold prevent the stress cell experience during transplantation allows cells to migrate and integrate in the right layer by maintaining the orientation of the graft during surgery. Biodegradable scaffold with precise stiffness and degradation rate could significantly improves the efficacy of the transplantation thereby increasing the chance successful regeneration of disease retina by replenishing with healthy cells.

Acknowledgements We would like to thank all the researchers and authors whose work we have cited here.

References

Abe T (2002) Regeneration of the retina using pigment epithelial cell transplantation. Nippon Ganka Gakkai zasshi 106(12):778–803. Discussion 804
Ackman JB, Crair MC (2014) Role of emergent neural activity in visual map development. Curr Opin Neurobiol 24(1):166–175
Afshari FT, Fawcett JW (2009) Improving RPE adhesion to Bruch's membrane. Eye 23(10):1890–1893
Agarwal R, Singh V, Jurney P, Shi L, Sreenivasan SV, Roy K (2013) Mammalian cells preferentially internalize hydrogel nanodiscs over nanorods and use shape-specific uptake mechanisms. Proc Natl Acad Sci USA 110(43):17247–17252
Alamein MA, Stephens S, Liu Q, Skabo S, Warnke PH (2013) Mass production of nanofibrous extracellular matrix with controlled 3D morphology for large-scale soft tissue regeneration. Tissue Eng Part C, Methods 19(6):458–472
Anderson DH, Radeke MJ, Gallo NB, Chapin EA, Johnson PT, Curletti CR, Hancox LS, Hu J, Ebright JN, Malek G, Hauser MA, Rickman CB, Bok D (2010) The pivotal role of the complement system in aging and age-related macular degeneration: hypothesis re-visited. Progr Retinal Eye Res 29(2):95–112
Bonilha VL (2014) Retinal pigment epithelium (RPE) cytoskeleton in vivo and in vitro. Exp Eye Res 126:38–45
Brandl C, Zimmermann SJ, Milenkovic VM, Rosendahl SM, Grassmann F, Milenkovic A, Hehr U, Federlin M, Wetzel CH, Helbig H, Weber BH (2014) In-depth characterisation of Retinal Pigment Epithelium (RPE) cells derived from human induced pluripotent stem cells (hiPSC). NeuroMol Med 16(3):551–564
Buchholz DE, Hikita ST, Rowland TJ, Friedrich AM, Hinman CR, Johnson LV, Clegg DO (2009) Derivation of functional retinal pigmented epithelium from induced pluripotent stem cells. Stem Cells 27(10):2427–2434
Cassidy JW, Roberts JN, Smith CA, Robertson M, White K, Biggs MJ, Oreffo RO, Dalby MJ (2014) Osteogenic lineage restriction by osteoprogenitors cultured on nanometric grooved surfaces: the role of focal adhesion maturation. Acta Biomater 10(2):651–660
Cavallotti C, Corrado BG, Feher J (2005) The human choriocapillaris: evidence for an intrinsic regulation of the endothelium. J Anat 206(3):243–247
Centanin L, Wittbrodt J (2014) Retinal Neurogenesis. Development 141(2):241–244
Charlton-Perkins M, Cook TA (2010) Building a fly eye: terminal differentiation events of the retina, corneal lens, and pigmented epithelia. Curr Top Dev Biol 93:129–173
Chen H, Fan X, Xia J, Chen P, Zhou X, Huang J, Yu J, Gu P (2011) Electrospun chitosan-graft-poly (varepsilon-caprolactone)/poly (varepsilon-caprolactone) nanofibrous scaffolds for retinal tissue engineering. Int J Nanomed 6:453–461
Chow RL, Lang RA (2001) Early eye development in vertebrates. Annu Rev Cell Dev Biol 17:255–296
Das AK, Pal R (2010) Induced pluripotent stem cells (iPSCs): the emergence of a new champion in stem cell technology-driven biomedical applications. J Tissue Eng Regener Med 4(6):413–421
Eiraku M, Sasai Y (2012) Self-formation of layered neural structures in three-dimensional culture of ES cells. Curr Opin Neurobiol 22(5):768–777
Elsayed M, Merkel OM (2014) Nanoimprinting of topographical and 3D cell culture scaffolds. Nanomedicine 9(2):349–366
Engelsberg K, Ghosh F (2011) Human retinal development in an in situ whole eye culture system. Dev Neurosci 33(2):110–117
Fujie T, Mori Y, Ito S, Nishizawa M, Bae H, Nagai N, Onami H, Abe T, Khademhosseini A, Kaji H (2014) Micropatterned polymeric nanosheets for local delivery of an engineered epithelial monolayer. Adv Mater 26(11):1699–1705

Fujie T, Ricotti L, Desii A, Menciassi A, Dario P, Mattoli V (2011) Evaluation of substrata effect on cell adhesion properties using freestanding poly(L-lactic acid) nanosheets. Langmuir: ACS J Surf Colloids 27(21):13173–13182

Garcia JM, Mendonca L, Brant R, Abud M, Regatieri C, Diniz B (2015) Stem cell therapy for retinal diseases. World J Stem Cells 7(1):160–164

Giannaccini M, Giannini M, Calatayud MP, Goya GF, Cuschieri A, Dente L, Raffa V (2014) Magnetic nanoparticles as intraocular drug delivery system to target retinal pigmented epithelium (RPE). Int J Mol Sci 15(1):1590–1605

Goldman D (2014) Muller glial cell reprogramming and retina regeneration. Nat Rev Neurosci 15(7):431–442

Gupta A, Vara DS, Punshon G, Sales KM, Winslet MC, Seifalian AM (2009) In vitro small intestinal epithelial cell growth on a nanocomposite polycaprolactone scaffold. Biotechnol Appl Biochem 54(4):221–229

Hamilton RD, Leach L (2011) Isolation and properties of an in vitro human outer blood-retinal barrier model. Methods Mol Biol 686:401–416

Holt CE, Bertsch TW, Ellis HM, Harris WA (1988) Cellular determination in the Xenopus retina is independent of lineage and birth date. Neuron 1(1):15–26

Hu G, Huang K, Yu J, Gopalakrishna-Pillai S, Kong J, Xu H, Liu Z, Zhang K, Xu J, Luo Y, Li S, Sun YE, Iverson LE (2012) Identification of miRNA signatures during the differentiation of hESCs into retinal pigment epithelial cells. PLoS ONE 7(7):e37224

Idealson M et al (2009) Directed differentiation of human embryonic stem cells into functional retinal pigment epithelium cells. Cell Stem cells 5(4):396–408

Jayaram H, Jones MF, Eastlake K, Cottrill PB, Becker S, Wiseman J, Khaw PT, Limb GA (2014) Transplantation of photoreceptors derived from human Muller glia restore rod function in the P23H rat. Stem Cells Transl Med 3(3):323–333

Jin ZB, Okamoto S, Xiang P, Takahashi M (2012) Integration-free induced pluripotent stem cells derived from retinitis pigmentosa patient for disease modeling. Stem Cells Transl Med 1(6):503–509

Johnson LV, Forest DL, Banna CD, Radeke CM, Maloney MA, Hu J, Spencer CN, Walker AM, Tsie MS, Bok D, Radeke MJ, Anderson DH (2011) Cell culture model that mimics drusen formation and triggers complement activation associated with age-related macular degeneration. Proc Natl Acad Sci USA 108(45):18277–18282

Kanemura H, Go MJ, Shikamura M, Nishishita N, Sakai N, Kamao H, Mandai M, Morinaga C, Takahashi M, Kawamata S (2014) Tumorigenicity studies of induced pluripotent stem cell (iPSC)-derived retinal pigment epithelium (RPE) for the treatment of age-related macular degeneration. PLoS ONE 9(1):e85336

Kang K, Kim MH, Park M and Choi IS (2014) Neurons on nanotopographies: behavioral responses and biological implications. J Nanosci Nanotechnol 14(1):513–521

Kolomeyer AM, Sugino IK, Zarbin MA (2011) Characterization of conditioned media collected from aged versus young human eye cups. Invest Ophthalmol Vis Sci 52(8):5963–5972

Koo S, Muhammad R, Peh GS, Mehta JS, Yim EK (2014) Micro- and nanotopography with extracellular matrix coating modulate human corneal endothelial cell behavior. Acta Biomater 10(5):1975–1984

Krohne TU, Westenskow PD, Kurihara T, Friedlander DF, Lehmann M, Dorsey AL, Li W, Zhu S, Schultz A, Wang J, Siuzdak G, Ding S, Friedlander M (2012) Generation of retinal pigment epithelial cells from small molecules and OCT4 reprogrammed human induced pluripotent stem cells. Stem Cells Transl Med 1(2):96–109

Kuo MW, Wang SH, Chang JC, Chang CH, Huang LJ, Lin HH, Yu AL, Li WH, Yu J (2009) A novel puf-A gene predicted from evolutionary analysis is involved in the development of eyes and primordial germ-cells. PLoS ONE 4(3):e4980

Kuppan P, Sethuraman S, Krishnan UM (2013) PCL and PCL-gelatin nanofibers as esophageal tissue scaffolds: optimization, characterization and cell-matrix interactions. J Biomed Nanotechnol 9(9):1540–1555

Lehmann B, Heimes B, Gutfleisch M, Spital G, Pauleikhoff D and Lommatzsch A (2015) Serous vascularized pigment epithelial detachment in exudative AMD. Morphological typing and risk of tears in the RPE]. Der Ophthalmologe: Zeitschrift der Deutschen Ophthalmologischen Gesellschaft 112(1):49–56

Lenkowski JR, Raymond PA (2014) Muller glia: Stem cells for generation and regeneration of retinal neurons in teleost fish. Progr Retinal Eye Res 40:94–123

Leri A, Rota M, Hosoda T, Goichberg P and Anversa P (2014) Cardiac stem cell niches. Stem Cell Res 13(3 Pt B):631–646

Li WJ, Jiang YJ, Tuan RS (2006) Chondrocyte phenotype in engineered fibrous matrix is regulated by fiber size. Tissue Eng 12(7):1775–1785

Lin H, Zhang Z, Zhang H, Yan P, Wang Q, Bai L (2009) Primary culture of human blood-retinal barrier cells and preliminary study of APOBEC3 expression: an in vitro study. Invest Ophthalmol Vis Sci 50(9):4436–4443

Lin-Jones J, Sohlberg L, Dose A, Breckler J, Hillman DW, Burnside B (2009) Identification and localization of myosin superfamily members in fish retina and retinal pigmented epithelium. J Comp Neurol 513(2):209–223

Liu F, Wong MM, Chiu SK, Lin H, Ho JC, Pang SW (2014) Effects of nanoparticle size and cell type on high sensitivity cell detection using a localized surface plasmon resonance biosensor. Biosens Bioelectron 55:141–148

Liu H, Yang R, Tinner B, Choudhry A, Schutze N, Chaqour B (2008) Cysteine-rich protein 61 and connective tissue growth factor induce deadhesion and anoikis of retinal pericytes. Endocrinology 149(4):1666–1677

Liu Q, Wang W, Zhang L, Zhao L, Song W, Duan X, Zhang Y (2014) Involvement of N-cadherin/beta-catenin interaction in the micro/nanotopography induced indirect mechanotransduction. Biomaterials 35(24):6206–6218

Liu Y, Song X, Han Y, Zhou F, Zhang D, Ji B, Hu J, Lv Y, Cai S, Wei Y, Gao F, Jia X (2011) Identification of anthocyanin components of wild Chinese blueberries and amelioration of light-induced retinal damage in pigmented rabbit using whole berries. J Agr Food Chem 59(1):356–363

Liu Z, Jiang R, Yuan S, Wang N, Feng Y, Hu G, Zhu X, Huang K, Ma J, Xu G, Liu Q, Xue Z, Fan G (2014) Integrated analysis of DNA methylation and RNA transcriptome during in vitro differentiation of human pluripotent stem cells into retinal pigment epithelial cells. PLoS ONE 9(3):e91416

Liu Z, Yu N, Holz FG, Yang F, Stanzel BV (2014) Enhancement of retinal pigment epithelial culture characteristics and subretinal space tolerance of scaffolds with 200 nm fiber topography. Biomaterials 35(9):2837–2850

Lung MS, Pilowsky P, Goldys EM (2012) Activation of the mammalian cells by using light-sensitive ion channels. Methods Mol Biol 875:241–251

Martinez-De Luna RI, Kelly LE, El-Hodiri HM (2011) The Retinal Homeobox (Rx) gene is necessary for retinal regeneration. Developmental Biology 353(1):10–18

McGill TJ, Cottam B, Lu B, Wang S, Girman S, Tian C, Huhn SL, Lund RD, Capela A (2012) Transplantation of human central nervous system stem cells—neuroprotection in retinal degeneration. Eur J Neurosci 35(3):468–477

McGill TJ, Lund RD, Douglas RM, Wang S, Lu B, Prusky GT (2004) Preservation of vision following cell-based therapies in a model of retinal degenerative disease. Vision Res 44(22):2559–2566

Melville H, Carpiniello M, Hollis K, Staffaroni A, Golestaneh N (2013) Stem cells: a new paradigm for disease modeling and developing therapies for age-related macular degeneration. J Transl Med 11:53

Mikhailov A, Cole RW, Rieder CL (2002) DNA damage during mitosis in human cells delays the metaphase/anaphase transition via the spindle-assembly checkpoint. Current Biol: CB 12(21):1797–1806

Minegishi Y, Iejima D, Kobayashi H, Chi ZL, Kawase K, Yamamoto T, Seki T, Yuasa S, Fukuda K, Iwata T (2013) Enhanced optineurin E50K-TBK1 interaction evokes protein insolubility and initiates familial primary open-angle glaucoma. Hum Mol Genet 22(17):3559–3567

Misra RD, Depan D, Shah JS (2012) Structure-process-functional property relationship of nanostructured carbon mediated cellular response for soft-tissue reconstruction and replacement. Acta Biomater 8(5):1908–1917

Miyake A, Araki M (2014) Retinal stem/progenitor cells in the ciliary marginal zone complete retinal regeneration: a study of retinal regeneration in a novel animal model. Dev Neurobiol 74(7):739–756

Mizuno A, Yasumuro H, Yoshikawa T, Inami W, Chiba C (2012) MEK-ERK signaling in adult newt retinal pigment epithelium cells is strengthened immediately after surgical induction of retinal regeneration. Neurosci Lett 523(1):39–44

Mousa SA, Lorelli W, Campochiaro PA (1999) Role of hypoxia and extracellular matrix-integrin binding in the modulation of angiogenic growth factors secretion by retinal pigmented epithelial cells. J Cell Biochem 74(1):135–143

Nasonkin IO, Merbs SL, Lazo K, Oliver VF, Brooks M, Patel K, Enke RA, Nellissery J, Jamrich M, Le YZ, Bharti K, Fariss RN, Rachel RA (2013) Conditional knockdown of DNA methyltransferase 1 reveals a key role of retinal pigment epithelium integrity in photoreceptor outer segment morphogenesis. Development 140(6):1330–1341

Nocini PF, Zanotti G, Castellani R, Grasso S, Cristofaro MG, De Santis D (2013) Bi-layered collagen nano-structured membrane prototype (collagen matrix 10826((R))) for oral soft tissue regeneration: an "in vitro" study. Clin Oral Implant Res 24(6):612–617

Oh J, Lee YD, Wagers AJ (2014) Stem cell aging: mechanisms, regulators and therapeutic opportunities. Nat Med 20(8):870–880

Park TS, Bhutto I, Zimmerlin L, Huo JS, Nagaria P, Miller D, Rufaihah AJ, Talbot C, Aguilar J, Grebe R, Merges C, Reijo-Pera R, Feldman RA et al (2014) Vascular progenitors from cord blood-derived induced pluripotent stem cells possess augmented capacity for regenerating ischemic retinal vasculature. Circulation 129(3):359–372

Peng S, Rao VS, Adelman RA, Rizzolo LJ (2011) Claudin-19 and the barrier properties of the human retinal pigment epithelium. Invest Ophthalmol Vis Sci 52:1392–1403

Peng S, Gan G, Qiu C, Zhong M, An H, Adelman RA, Rizzolo LJ (2013) Engineering a blood-retinal barrier with human embryonic stem cell-derived retinal pigment epithelium: transcriptome and functional analysis. Stem Cells Transl Med 2(7):534–544

Pfeffer BA and Philp NJ (2014) Cell culture of retinal pigment epithelium: Special Issue. Exp Eye Res 126:1–4

Phillips MJ, Wallace KA, Dickerson SJ, Miller MJ, Verhoeven AD, Martin JM, Wright LS, Shen W, Capowski EE, Percin EF, Perez ET, Zhong X, Canto-Soler MV (2012) Blood-derived human iPS cells generate optic vesicle-like structures with the capacity to form retinal laminae and develop synapses. Invest Ophthalmol Vis Sci 53(4):2007–2019

Powell C, Grant AR, Cornblath E, Goldman D (2013) Analysis of DNA methylation reveals a partial reprogramming of the Muller glia genome during retina regeneration. Proc Natl Acad Sci USA 110(49):19814–19819

Pritchard CD, Arner KM, Langer RS, Ghosh FK (2010) Retinal transplantation using surface modified poly(glycerol-co-sebacic acid) membranes. Biomaterials 31(31):7978–7984

Pritchard CD, Arner KM, Neal RA, Neeley WL, Bojo P, Bachelder E, Holz J, Watson N, Botchwey EA, Langer RS, Ghosh FK (2010) The use of surface modified poly(glycerol-co-sebacic acid) in retinal transplantation. Biomaterials 31(8):2153–2162

Qiu X, Yang J, Liu T, Jiang Y, Le Q, Lu Y (2012) Efficient generation of lens progenitor cells from cataract patient-specific induced pluripotent stem cells. PLoS ONE 7(3):e32612

Quesenberry PJ, Goldberg LR, Dooner MS (2015) Concise reviews: A stem cell apostasy: a tale of four H words. Stem Cells 33(1):15–20

Reichman S, Goureau O (2016) Production of retinal cells from confluent human iPS cells. Methods Mol Biol 1357:339–351. https://doi.org/10.1007/7651_2014_143

Rizzolo LJ (2007) Development and role of tight junctions in the retinal pigment epithelium. Int Rev Cytol 258:195–234

Rizzolo LJ (2014) Barrier properties of cultured retinal pigment epithelium. Exp Eye Res 126:16–26

Rizzolo LJ, Peng S, Luo Y, Xiao W (2011) Integration of tight junctions and claudins with the barrier functions of the retinal pigment epithelium. Prog Retin Eye Res 30(5):296–323

Robertson EJ, Oliver GK, Qian M, Proulx C, Zuckermann RN, Richmond GL (2014) Assembly and molecular order of two-dimensional peptoid nanosheets through the oil-water interface. Proc Natl Acad Sci USA 111(37):13284–13289

Rosenthal R, Strauss O (2002) Ca2+-channels in the RPE. Adv Exp Med Biol 514:225–235

Rowland TJ, Blaschke AJ, Buchholz DE, Hikita ST, Johnson LV, Clegg DO (2013) Differentiation of human pluripotent stem cells to retinal pigmented epithelium in defined conditions using purified extracellular matrix proteins. J Tissue Eng Regener Med 7(8):642–653

Sakakuraa E, Eiraku M, Takataab N (2016) Specification of embryonic stem cell-derived tissues into eye fields by Wnt signaling using rostral diencephalic tissue-inducing culture. Mech Dev 141:90–99

Samie FH, Jinks RN, Weiner WW, Chamberlain SC (1995) The morphology and physiology of a "mini-ommatidium" in the median optic nerve of Limulus polyphemus. Vis Neurosci 12(1):69–76

Sanii B, Haxton TK, Olivier GK, Cho A, Barton B, Proulx C, Whitelam S, Zuckermann RN (2014) Structure-determining step in the hierarchical assembly of peptoid nanosheets. ACS Nano 8(11):11674–11684

Singh D, Tripathi A, Zo S, Singh D, Han SS (2014) Synthesis of composite gelatin-hyaluronic acid-alginate porous scaffold and evaluation for in vitro stem cell growth and in vivo tissue integration. Colloids Surf, B 116:502–509

Singh D, Zo SM, Kumar A, Han SS (2013) Engineering three-dimensional macroporous hydroxyethyl methacrylate-alginate-gelatin cryogel for growth and proliferation of lung epithelial cells. J Biomater Sci Polym Ed 24(11):1343–1359

Sodha S, Wall K, Redenti S, Klassen H, Young MJ, Tao SL (2011) Microfabrication of a Three-Dimensional Polycaprolactone Thin-Film Scaffold for Retinal Progenitor Cell Encapsulation. J Biomater Sci Polym Ed 22(4–6):443–456

Su H, Liu Y, Wang D, Wu C, Xia C, Gong Q, Song B, Ai H (2013) Amphiphilic starlike dextran wrapped superparamagnetic iron oxide nanoparticle clsuters as effective magnetic resonance imaging probes. Biomaterials 34(4):1193–1203

Sun J, Mandai M, Kamao H, Hashiguchi T, Shikamura M, Kawamata S, Sugita S and Takahashi M (2015) Protective effects of human iPS-derived retinal pigmented epithelial cells in comparison with human mesenchymal stromal cells and human neural stem cells on the degenerating retina in rd1 mice. Stem cells

Sun R, Peng S, Chen X, Zhang H, Rizzolo LJ (2008) Diffusible retinal secretions regulate the expression of tight junctions and other diverse functions of the retinal pigment epithelium. Mol Vis 14:2237–2262

Szekely T Jr, Burrage K, Mangel M, Bonsall MB (2014) Stochastic dynamics of interacting haematopoietic stem cell niche lineages. PLoS Comput Biol 10(9):e1003794

Torrez LB, Perez Y, Yang J, Zur Nieden NI, Klassen H, Liew CG (2012) Derivation of neural progenitors and retinal pigment epithelium from common marmoset and human pluripotent stem cells. Stem Cells Int 2012:417865

Tripathi A, Kathuria N, Kumar A (2009) Elastic and macroporous agarose-gelatin cryogels with isotropic and anisotropic porosity for tissue engineering. J Biomed Mater Res 90(3):680–694. https://doi.org/10.1002/jbm.a.32127

Tripathi A, Kumar A (2011) Multi-featured macroporous agarose-alginate cryogel: synthesis and characterization for bioengineering applications. Macromol Biosci 11(1):22–35

Tripathi A, Melo JS (2015) Preparation of sponge-like biocomposite agarose-chitosan scaffold with primary hepatocytes for establishing an in-vitro 3D liver tissue model. RSC Adv 5:30701–30710

Tripathi A, Melo JS, D'Souza SF (2013) Magnetic nanoparticles in tissue regeneration. In: Tiwari A, Tiwari A (eds) Nanomaterials in drug delivery, imaging, and tissue engineering, WILEY-Scrivener Publisher, USA, pp 443–492. ISBN: 9781118290323
Tu X, Palczewski K (2014) The macular degeneration-linked C1QTNF5 (S163) mutation causes higher-order structural rearrangements. J Struct Biol 186(1):86–94
Tucker BA, Anfinson KR, Mullins RF, Stone EM, Young MJ (2013) Use of a synthetic xeno-free culture substrate for induced pluripotent stem cell induction and retinal differentiation. Stem Cells Trans Med 2(1):16–24
Tucker BA, Mullins RF, Streb LM, Anfinson K, Eyestone ME, Kaalberg E, Riker MJ, Drack AV, Braun TA and Stone EM (2013) Patient-specific iPSC-derived photoreceptor precursor cells as a means to investigate retinitis pigmentosa. eLife 2:e00824.
Turner DL, Snyder EY, Cepko CL (1990) Lineage-independent determination of cell typein the embryonic mouse retina. Neuron 4(6):833–845
Ulbrich S, Friedrichs J, Valtink M, Murovski S, Franz CM, Muller DJ, Funk RH, Engelmann K (2011) Retinal pigment epithelium cell alignment on nanostructured collagen matrices. Cells, Tissues, Organs 194(6):443–456
Uygun BE, Sharma N, Yarmush M (2009) Retinal pigment epithelium differentiation of stem cells: current status and challenges. Crit Rev Biomed Eng 37(4–5):355–375
Vergara MN, Del Rio-Tsonis K (2009) Retinal regeneration in the Xenopus laevis tadpole: a new model system. Molecular Vision 15:1000–1013
Villares A, Moreau C, Capron I, Cathala B (2014) Impact of ionic strength on chitin nanocrystal-xyloglucan multilayer film growth. Biopolymers 101(9):924–930
Wahlin KJ, Maruotti J, Zack DJ (2014) Modeling retinal dystrophies using patient-derived induced pluripotent stem cells. Adv Exp Med Biol 801:157–164
Wang H, Ninomiya Y, Sugino IK, Zarbin MA (2003) Retinal pigment epithelium wound healing in human Bruch's membrane explants. Invest Ophthalmol Vis Sci 44(5):2199–2210
Wang W, Itoh S, Konno K, Kikkawa T, Ichinose S, Sakai K, Ohkuma T, Watabe K (2009) Effects of Schwann cell alignment along the oriented electrospun chitosan nanofibers on nerve regeneration. J Biomed Mater Res Part A 91(4):994–1005
Wang YF, Guo HF, Ying DJ (2013) Multilayer scaffold of electrospun PLA-PCL-collagen nanofibers as a dural substitute. J Biomed Mater Res B Appl Biomater 101(8):1359–1366
Warnke PH, Alamein M, Skabo S, Stephens S, Bourke R, Heiner P, Liu Q (2013) Primordium of an artificial Bruch's membrane made of nanofibers for engineering of retinal pigment epithelium cell monolayers. Acta Biomater 9(12):9414–9422
Wen W, Pillai-Kastoori L, Wilson SG, Morris AC (2015) Sox4 regulates choroid fissure closure by limiting Hedgehog signaling during ocular morphogenesis. Dev Biol 399(1):139–153
West EL, Gonzalez-Cordero A, Hippert C, Osakada F, Martinez-Barbera JP, Pearson RA, Sowden JC, Takahashi M, Ali RR (2012) Defining the integration capacity of embryonic stem cell-derived photoreceptor precursors. Stem Cells 30(7):1424–1435
Westenskow PD, Kurihara T, Friedlander M (2014) Utilizing stem cell-derived RPE cells as a therapeutic intervention for age-related macular degeneration. Adv Exp Med Biol 801:323–329
Wetts R, Fraser SE (1988) Multipotent precursors can give rise to all major cell types of the frog retina. Science 239(4844):1142–1145
Wimmers S, Karl MO, Strauss O (2007) Ion channels in the RPE. Progr Retinal Eye Res 26(3):263–301
Xiang P, Wu KC, Zhu Y, Xiang L, Li C, Chen DL, Chen F, Xu GT, Wang AJ, Li M, Jin ZB (2014) A novel Bruch's membrane-mimetic electrospun substrate scaffold for human retinal pigment epithelium cells. Biomaterials 35(37):9777–9788
Yip HK (2014) Retinal stem cells and regeneration of vision system. Anat Rec 297(1):137–160
Yoshida T, Ozawa Y, Suzuki K, Yuki K, Ohyama M, Akamatsu W, Matsuzaki Y, Shimmura S, Mitani K, Tsubota K, Okano H (2014) The use of induced pluripotent stem cells to reveal pathogenic gene mutations and explore treatments for retinitis pigmentosa. Mol Brain 7:45

Yoshii C, Ueda Y, Okamoto M, Araki M (2007) Neural retinal regeneration in the anuran amphibian Xenopus laevis post-metamorphosis: transdifferentiation of retinal pigmented epithelium regenerates the neural retina. Dev Biol 303(1):45–56

Zhao L, Wang Z, Liu Y, Song Y, Li Y, Laties AM, Wen R (2007) Translocation of the retinal pigment epithelium and formation of sub-retinal pigment epithelium deposit induced by subretinal. Deposit Mol Vis 13:873–880

Zhong X, Gutierrez C, Xue T, Hampton C, Vergara MN, Cao LH, Peters A, Park TS, Zambidis ET, Meyer JS, Gamm DM, Yau KW, Canto-Soler MV (2014) Generation of three-dimensional retinal tissue with functional photoreceptors from human iPSCs. Nat Commun 5:4047

Zhu D, Deng X, Spee C, Sonoda S, Hsieh CL, Barron E, Pera M, Hinton DR (2011) Polarized secretion of PEDF from human embryonic stem cell-derived RPE promotes retinal progenitor cell survival. Invest Ophthalmol Vis Sci 52(3):1573–1585

Systemic Drug Delivery to the Posterior Segment of the Eye: Overcoming Blood–Retinal Barrier Through Smart Drug Design and Nanotechnology

Sudhir H. Ranganath, M. Y. Thanuja, C. Anupama, and T. D. Manjunatha

Abstract The human eye being a highly privileged organ with complex barriers, reaching the posterior part of the eye via systemic routes, is still a major clinical challenge. The most critical of these cellular barriers is the blood–retinal barrier (BRB) formed by the retinal pigment epithelium (RPE) and the retinal endothelium. In this chapter, we comprehensively review the interplay of tight junctions, carrier-mediated transport proteins, influx/efflux transporters and the molecular interactions involved in controlling biomolecular transport across BRB. An effort is made to elucidate how these cellular machineries synergistically form very tight protective barricade against the entry of drugs administered through the systemic route. Thus, we have emphasized the need to identify and understand the complexity of BRB which is key to achieving effective drug delivery to the posterior part of the eye via systemic route. Next, we have comprehensively reviewed the role of drug-immobilized nanostructures and smart prodrugs in overcoming BRB and transient modulation of BRB. So far, very few formulations have reached bedside, and the reasons for such poor rate of clinical translation have been identified and critically analyzed and potential solutions discussed. Deeper investigations of the pathophysiology and understanding of BRB leading to the development of efficient drug delivery systems are warranted. Finally, we provide futuristic perspectives to achieving clinical translation of systemic therapies to the posterior eye via development of state-of-the-art and precise diagnosis tools fortified by AI and nanocontrast agents, innovative drug delivery systems based on biomolecule immobilization to overcome BRB, realistic preclinical models and rationally designed clinical trials.

M. Y. Thanuja and C. Anupama have equally contributed to this work.

S. H. Ranganath (✉) · M. Y. Thanuja · C. Anupama · T. D. Manjunatha
Bio-INvENT Laboratory, Department of Chemical Engineering, Siddaganga Institute of Technology, Tumakuru, Tumkur 572103, Karnataka, India
e-mail: sudhirh@sit.ac.in

C. Anupama
Department of Biotechnology, Siddaganga Institute of Technology, Tumakuru, Tumkur 572103, Karnataka, India

A. Tripathi and J. S. Melo (eds.), *Immobilization Strategies*,
Gels Horizons: From Science to Smart Materials,
https://doi.org/10.1007/978-981-15-7998-1_6

Keywords Posterior eye · Blood–retinal barrier · Tight junctions · Membrane transporters · Systemic therapy · Drug immobilization · Nanomedicine · Permeability · Prodrugs · Controlled drug delivery

List of abbreviations

AAT	Amino acid transporter
ABCAs	ATP-binding cassette transporters
ABCC4	ATP-binding cassette subfamily number 4
ABCG2	ATP-binding cassette subfamily G member 2
AI	Artificial intelligence
AJs	Adherens junctions
AMD	Age-related macular degeneration
BAB	Blood–aqueous barrier
BBB	Blood–brain barrier
BM	Bruch's membrane
BRB	Blood–retinal barrier
CAT1	Cationic amino acid transporter 1
CB	Cannabinoid
CMV	Cytomegalovirus
CNN	Convolutional neural networks
CNV	Choroid neovascularization
CNT	Concentrative nucleoside transporter
CPP	Cell-penetrating peptides
CRISPR	Clustered regularly interspaced short palindromic repeats
CSMF	Confocal scanning microfluorometer
DDS	Drug delivery system
DL	Deep learning
DME	Diabetic macular edema
DMAP-GFX	Dimethylamino-propyl derivative of gatifloxacin
DR	Diabetic retinopathy
EAAT	Excitatory amino acid receptors
ENT	Equilibrative nucleoside transporter
ER	Endoplasmic reticulum
FD	Frequency domain
FNP	Flash nanoprecipitation
FR	Folate receptor
GA	Geographic atrophy
GCV	Ganciclovir
GLUT	Glucose transporter
IOP	Intraocular pressure
iBRB	Inner BRB
JAM	Junctional adhesion molecule

LAT	Large neutral amino acid transporter
LCNV	Laser-induced choroidal neovascularization
LMN	Lumefantrine
MAGUK	Membrane-associated guanylate kinase
MCTs	Monocarboxylate transporter
ML	Machine learning
MSCs	Mesenchymal stem cells
MRI	Magnetic resonance imaging
MRP	Multidrug resistant protein
NPC	Nanoparticle conjugate
OATs	Organic anion transporters
oBRB	Outer BRB
OCTN2	Organic cation/Carnitine transporter 2
OCT	Optical coherence tomography
PAM	Photoacoustic microscopy
PDR	Proliferative diabetic retinopathy
PEPT	Peptide transporter
PET	Positron emission tomography
P-gp	P-glycoprotein
POT	Proton-coupled oligopeptide transporter
PTOCT	Photothermal OCT
RAC	Retinal astroglial cells
RB	Retinoblastoma
RD	Retinal degeneration
RFC	Reduced folate carrier
RFVT	Riboflavin transporter
RPE65	Retinoid isomerase
RP	Retinitis pigmentosa
RUI	Retinal uptake index
RPE	Retinal pigment epithelium
SLC	Solute carrier organic anion transporter
SLC6A14	Sodium- and chloride-dependent neutral and basic amino acid transporter
SMCTs	Sodium-coupled electrogenic monocarboxylate transporters
SMPCV	Symptomatic macular polypoidal choroidal vasculopathy
SMVT	Sodium-dependent multivitamin transporter
SOPT	Sodium-coupled oligopeptide transporter
SR-B1	Scavenger receptor class B type 1
SVCT	Sodium–ascorbate cotransporter
TAUT	Taurine transporter
TJs	Tight junctions
TER	Transepithelial resistance
THC	Tetrahydrocannabinoid
TALENs	Transcription activator-like effector nucleases
UTMD	Ultrasound-targeted microbubble destruction

ZFNs	Zinc finger nucleases
ZO	Zonula occludens

1 Introduction

The retina is a light-sensitive layer of the eye located in the posterior globe of the eye comprising of ten layers and plays a key role in vision. Multiple disorders of the retina are known to impair vision and if not treated early, leads to blindness. Aging exacerbates retinal health. For instance, it is predicted that by 2040, the number of cases of age-related macular degeneration (AMD) in humans could reach 288 million worldwide (Wong et al. 2014; Cunha-Vaz 2017). In addition, glaucoma and diabetic retinal disorders are also prevalent in the working-age population and are the leading causes of blindness. It is estimated that 28 million people worldwide currently suffer from vision threatening diabetic retinal disorders (Yau et al. 2012). Multiple, rare genetic disorders leading to the degeneration of retina and blindness have also been reported in younger population (del Amo et al. 2017). Rigorous investigations of the pathophysiology of these retinal disorders have led to the identification of newer targets and development of novel drugs. Nevertheless, drug delivery to the retina and the posterior part of the eye is still challenging given the presence of the blood–ocular barriers including the blood–aqueous barrier (BAB) and the blood–retinal barrier (BRB).

Of all the different routes of drug administration, systemic route is the most non-invasive and could be achieved orally or intravenously. However, in the context of retinal drug delivery, systemic route is the hardest because the drugs have to overcome BRB to reach the target site. In this context, five mechanisms of biomolecular transport across the BRB have been identified (Liu and Liu 2019), which are: (i) paracellular transport regulated by the tight junctions between the cells of the retina; (ii) facilitated transport of select solutes regulated by transmembrane transport proteins of the retinal cell membranes via concentration gradient; (iii) active transport of solutes against concentration gradient regulated by transmembrane transport proteins via ATP consumption (or) directed transport of solutes via electrochemical gradient regulated by antiporters and cotransporters; (iv) transcytosis via invagination of vesicles followed by cell crossing and fusion with membrane of the opposite cell to release the contents; (v) solute transformation or degradation into different forms as it traverses the cell.

Out of these mechanisms, paracellular and facilitated/active transport mechanisms dominate and are controlled by static and dynamic barriers, respectively. Hence, it is imperative that we understand the anatomy of the BRB and the mechanisms in which it controls transport of small and large molecules across the retina.

2 Anatomy and the Barrier Functions of the Blood–Retinal Barrier

Neuronal cells of the retina are the most sensitive and critical cells in the eye analogous to the neurons of the brain. There is limited but controlled contact of the neuronal cells of the retina and the blood, because the retinal neuronal cells require specialized microenvironment including a stable ionic environment for their efficient functioning. The restriction of non-specific transport of ions, drugs and nutrients between the neural retina and blood is primarily controlled by the posterior blood–retinal barrier (BRB).

The existence and site of the BRB in the retina were first demonstrated by J. G. Cunha-vaz et al., by highlighting morphological differences in ocular blood vessels and retinal blood vessels using electron microscope (Cunha-Vaz et al. 1966). Retina has direct contact with blood at two positions: (i) at the level of the retinal blood vessels and (ii) at the chorioretinal interface. The BRB consists of the endothelia of the retinal capillaries and the retinal pigment epithelium (RPE). Based on the location of the barriers, BRB can be classified into outer BRB (oBRB) and inner BRB (iBRB). The oBRB is made up of the retinal pigment epithelium, and the iBRB is composed of endothelial cells of the retinal vessels.

2.1 The Outer BRB

The chorioretinal interface consists of the choriocapillaris, Bruch's membrane (BM) and the retinal pigment epithelium (RPE). The choriocapillaris significantly varies from the retinal vessels, for they contain fenestrated endothelial cells, and the junctional structures are made up of macula occludens. Due to this kind of endothelial structure, blood can easily pass through the choriocapillaris and hence they are not associated with any blood–retinal barrier function.

Multilayered BM serves as the basement membrane for both RPE and the choriocapillaris, and is present below the choriocapillaris. It has a relatively minor contribution toward BRB's barrier function, since it allows small molecules (<1.0–6.15 nm) to diffuse across (Zayas-Santiago et al. 2011), but hinders the transport of macromolecules (proteins, particles) (Tratta et al. 2014; Hazlett and Meyer 1974). However, the properties that define the BM's barrier function (elasticity (Booij et al. 2010), permeability (Cankova et al. 2011) and hydraulic conductivity (Lee et al. 2015), are found to decrease with age. This reduction in the BM's barrier function would lead to accumulation of waste material in the subretinal space and nearby structures.

2.1.1 The Static Outer BRB and its Permeability Characteristics

The primary barrier present in the choriocapillary interface is the retinal pigment epithelium (RPE) present below the Bruch's membrane and is a tight cellular monolayer. It is continuous, uniform and extends throughout the entire retina. RPE is the main cellular structure of the oBRB between the sensitive neuronal cells such as rods and cones and blood vessels of the choroid (Cunha-Vaz 1976). It opposes the free movement of molecules across the blood and neuronal cells via paracellular pathway, which represents the static oBRB. The main components of the static oBRB are gap junctions, desmosomes, adherens junctions and tight junctional complexes present between RPE cells as depicted by Fig. 1a. Under physiological conditions, this paracellular route is highly regulated without any leakage. Its important functions constitute retinal homeostasis and barrier function.

Tight junctions (TJs) are complex, dynamic structures which interact with other proteins to regulate cell proliferation, polarity of the membrane proteins, and transepithelial diffusion through the paracellular spaces (Rizzolo et al. 2011; Díaz-Coránguez et al. 2017). They act as gatekeepers for the paracellular route and comprise a combination of intrinsic and extrinsic proteins including zonula occludens (ZO), junctional adhesion molecule (JAM), occludins, cadherins and claudins which form highly selective barriers (Campbell and Humphries 2013). ZO are scaffolding peripheral phosphoproteins of tight junctions which aid anchoring of the junctional macromolecular complexes to perijunctional actin cytoskeleton. Specifically, ZO-1/2/3 are intrinsic proteins of TJs belonging to the large family of membrane-associated guanylate kinase (MAGUK)-like proteins comprising a number of subfamilies based on its active domain sequence. ZO proteins contain three PDZ active domains, which help to interact with cell adhesion molecules, cytoskeletal proteins and receptors, hence, involved in signaling pathways. These ZO proteins also help in maintaining polarity of the membrane (Campbell and Humphries 2013).

Transmission electron microscopic view of TJs of RPE reveals pentalaminar structures formed by a network of strands containing gap junctions (Rizzolo et al. 2011). Caveolin-1, omega-shaped plasmalemma vesicles are highly expressed on the surface of RPE to promote both paracellular and transcellular transport (Mora et al. 2006). Occludins are intrinsic, transmembrane phosphoproteins which are ubiquitously expressed by epithelial cells. The extracellular loops of occludins interact with the adjacent cells and form junctional strands to regulate permeability and selectivity toward ions. (Rizzolo et al. 2011; Xia and Rizzolo 2017). While the first extracellular loop is responsible for cell–cell pairing, the second loop interacts with ZO-1/2 and the last loop interacts with F-actin. The interaction between occludin, ZO-1 and JAM regulates the paracellular permeability (Campbell and Humphries 2013). Discontinuous distribution of occludin leads to increased paracellular permeability and diabetic retinopathy. Tight junctional integrity is also maintained by another class of transmembrane proteins called claudins. They interact with ZO-1 at the C-terminal side containing tyrosine and valine. The supramolecular assembly of claudins and occludins together contributes to the overall functions of TJs (Campbell and Humphries 2013) as depicted in Fig. 1a.

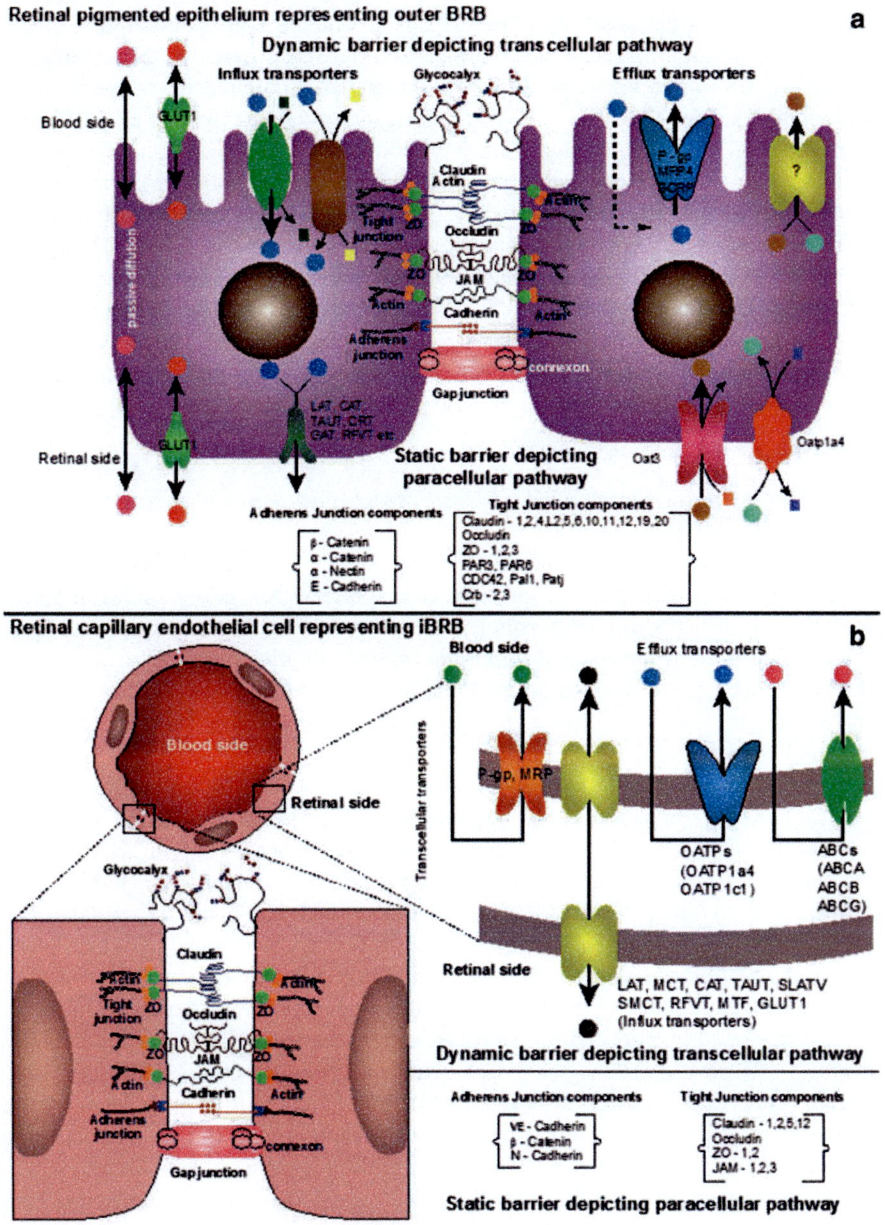

Fig. 1 Schematic representation of the molecular components of static and dynamic barriers of **(a)** the outer BRB and **(b)** the inner BRB and the depiction of paracellular and transcellular pathways

The paracellular permeability of the RPE is significantly greater than that of the BM and the choroidal tissue (Pitkänen et al. 2005). Measurement of transepithelial resistance (TER) of in vitro and ex vivo cultures of confluent RPE suggests wide range of values from 30 to 1050 Ω cm^2 (Ablonczy et al. 2011; Lynn et al. 2017) due to difference in culture conditions and RPE source. A corroborating in vivo study in monkeys has confirmed that 2 nm sized microperoxidase could not permeate through the RPE (Smith and Rudt 1975). Pitkanen and coworkers have shown that the RPE is highly permeable to lipophilic molecules (betaxolol: 16×10^{-6} cm/s), but almost impermeable to hydrophilic molecules including carboxyfluorescein and FITC-dextran (Pitkänen et al. 2005). They also confirmed that the permeability of RPE was dependent on the molecular size (permeability to FITC of MW 77 kDa was more than three and two orders of magnitude higher than that of betaxolol and carboxyfluorescein, respectively).

2.1.2 The Dynamic Outer BRB and its Permeability Characteristics

Drug delivery to the retina is not only influenced by the static oBRB, but also by dynamic factors including carrier proteins (transporters) across both sides of RPE. Depending on the direction of transport, the transporter proteins are classified into efflux and influx transporters, and important ones are depicted in Fig. 1a.

Efflux transporters belong to the ATP binding cassette superfamily, and they are involved in drug resistance, protection of tissues from toxic compounds, metabolism and signal transduction. RPE cells expel drugs through these efflux transporters and avoid build-up of toxic materials inside the retina. The first identified active efflux protein in human retina is P-glycoprotein (P-gp) encoded by MDR1 gene and functions via ATP hydrolysis. However, the expression of P-gp in oBRB is not consistent in all organisms. For example, P-gp expression was not detected by LC-MS/MS in human fetal RPE and ARPE-19 (Pelkonen et al. 2017), but found to be 2.01 fmol/μg protein in porcine oBRB which was significantly less than that of the iBRB (8.70 fmol/μg protein) (Zhang et al. 2017). Similarly, the only human RPE cell line that expresses functional P-gp is D407 (Constable et al. 2006). P-gp performs efflux of various drugs including antibiotics, steroids, anticancer drugs, immunosuppressants across the cell membrane (Sanchez-Covarrubias et al. 2014), and it shows broad substrate (drug) specificity (Lee and Pelis 2016) and thus multidrug resistance. For instance, P-gp is over expressed with the use of Daunomycin in the treatment for proliferative vitreoretinopathy and results in multidrug resistance (Esser et al. 1998). Other multidrug-resistant proteins MRP1, MRP4 and MRP5 are also present in the RPE, and these are involved in the transport of anions and conjugated compounds (Mannermaa et al. 2009). Another class of efflux transporter is the organic anion transporters (OATs) which are expressed on RPE. Studies using pravastatin in rats show that OATP1a4 expressed on the apical membrane of RPE facilitated pravastatin efflux (Fujii et al. 2015). The retinal uptake index (RUI) of [^{3}H]pravastatin was significantly reduced when it competed with digoxin (0.01 mM) and probenecid (1 mM), which are two OATP1a4 substrates with higher affinity than pravastatin.

2.1.3 Mechanisms of Efflux of Drugs and their Inhibition

Several models have been proposed which suggest the presence of affinity between the substrates (drugs) toward the active core of efflux transporters or pumps (enzymes). Upon binding, the efflux pumps send the drugs out of the cell against concentration gradient (Varma et al. 2003). Table 1 lists other important efflux transporters and their corresponding drug substrates delivered across the oBRB (Barza et al. 1983; Tachikawa et al. 2008; Hosoya and Tachikawa 2009; Aukunuru et al. 2001).

In the context of systemic drug delivery, these transporters play critical roles in the development of multidrug resistance (Tachikawa et al. 2012), and hence regulating their expression is an important therapeutic approach. In addition, the inhibition of drug efflux is one of the major approaches to enhance the success of systemic drug delivery to oBRB. In this direction, various drugs have been employed as efflux inhibitors by: (i) modulating substrate binding sites by allosteric inhibition (Varma et al. 2003); (ii) inhibiting hydrolysis of ATP (Shapiro and Ling 1997); and (iii) destabilizing the structural assembly of membrane lipids (Drori et al. 1995). For instance, 5 mM of Verapamil was used to attenuate P-gp function to transport [^{3}H]digoxin out of oBRB (Hosoya et al. 2010). Few important inhibitors of OATs are *p*-aminohippuric acid (Hosoya et al. 2010), benzyl penicillin, digoxin and probenecid (Fujii et al. 2015).

On the other hand, influx transporters of the oBRB belong to the solute carrier (SLC) superfamily, involved in the transport of molecules into RPE from the blood. The major influx proteins expressed in RPE are the amino acid transporter (AAT), folate receptor (FR)–alpha, peptide transporter and sodium-dependent multivitamin transporter (SMVT) (Vadlapatla et al. 2013). FR-alpha (a membrane anchored glycosylphosphatidylinositol protein) has greater affinity for folic acid as well as methyltetrahydrofolate. Vitamin C transporter and SMVT proteins are responsible for the translocation of vitamin C and cofactors including biotin, pantothenic acid and lipoic

Table 1 Important efflux transporters expressed on the membrane of RPE with their corresponding substrates

Efflux transporters	Substrates including drugs	References
OAT3, SLC22A8, SLC21A, SLCO	Organic anions	Barza et al. (1983), Tachikawa et al. (2008)
MRP4, ABCC4, BHEAS	B-lactum antibiotics, 6-mercaptopurine, anionic drugs	Tachikawa et al. (2008), Hosoya and Tachikawa (2009)
MRPs	Fluorescein, N[4-(benzoylamino) phenyl sulfonyl] glycine	Tachikawa et al. (2008), Aukunuru et al. (2001)
ABCG2	Mitoxantrone, Doxorubicin, Pheophorbide	Tachikawa et al. (2008)
ABCAs: ABCA3, ABCA9	Lipids (Sterols, phospholipids, retinoids)	Tachikawa et al. (2008)

acid (Vadlapatla et al. 2013). AATs are involved in transporting drugs in addition to their physiological role of transporting various amino acids and nutrients to the retinal tissue. L-carnitine, gabapentin are substrates analogues for AATs, and they are used in the determination of membrane permeability of drugs. Other AATs present in oBRB include glutamate transporters, neutral amino acid transporters (ASCT1 and ASCT2), Na^+-independent large neutral amino acid transporter (LAT2), and carnitine transporter. Peptide transporters belong to the proton-coupled oligopeptide transporter (POT) family and includes peptide transporter 1 and 2 (PEPT1, PEPT2) involved in the transport of peptide-based drugs across the oBRB. Peptide transporter substrate analogues d-amphetamine, cytarabine, decitabine, DMAE, ganciclovir, glucosamine, and L-DOPA are used to study drug permeability. Unique distribution of transport proteins at the apical and basolateral side shows the significance of RPE compared to other epithelial cells (Tachikawa et al. 2012). For example, AAT isoforms such as LAT1/LAT2 helps in the exchange of amino acids with different affinity. LAT1 exhibits more affinity to neutral bulky functional group amino acids, in contrast to LAT2. Hence, efflux transporters are coupled to influx of certain amino acids and ions. Isoforms of H^+-coupled electroneutral monocarboylate transporter (MCTs) and Na^+-coupled electrogenic monocarboxylate transporters (SMCTs) (Tachikawa et al. 2012) are expressed at the apical membrane and basolateral membrane of RPE, which helps in the transport of several monocarboxylic drugs including foscarnet, salicylate and benzoate. Thus, influx transporters are potential routes for the systemic delivery of drugs across oBRB. Table 2 lists other important influx transporters and their corresponding drug substrates delivered across the oBRB.

2.2 *The Inner BRB*

iBRB plays a primary role in the transport of nutrients and efflux of drugs and neurotransmitters to the neural retina. In the systemic route of drug delivery, drugs have to cross oBRB (RPE) first followed by iBRB to reach the posterior part of the eye. Similar to oBRB, the iBRB is comprised of the static iBRB and the dynamic iBRB. The static iBRB allows paracellular transport and is characterized by the presence of junctional proteins including TJs, adherens junctions (AJs), gap junctions and JAM proteins. The dynamic iBRB comprises membrane proteins such as channels and pumps and allows transcellular transport of molecules.

2.2.1 The Static Inner BRB and its Permeability Characteristics

The retinal vessels are made up of a continuous layer of endothelium surrounded by a thick basement membrane which represents the iBRB. The non-fenestrated endothelial cells are held together by tight junctions as represented by Fig. 1b. Cunha-Vaz suggested that the characteristic features of iBRB are due to the tight junctional

Table 2 Important influx transporters expressed on the membrane of RPE with their corresponding substrates

Influx transporters	Substrates including drugs	References
SMCT1	Propionate, B-complex vitamin nicotinate, Gabapentin, γ-Hydroxybutyrate, 5-aminosalicylate, 3-bromopyruvate, L-2-oxothiazolidine carboxylate	Miyauchi et al. (2004), Gopal et al. (2005), Ganapathy et al. (2008), Gopal et al. (2007), Thangaraju et al. (2009), Babu et al. (2011)
LAT1/LAT2	L-Dopa and amino acid mustards	Nakauchi et al. (2003)
RFC	Folate and its analogues	Zhao et al. (2009), Huang et al. (1997), Smith et al. (1999), Lee and Pelis (2016)
SVCT1/SVCT2	Reduced form of vitamin C	Tombran-Tink and Barnstable (2008)
Facilitative GLUT1	Glucose, DHA	
OCTN2	Carnitine, B-lactum antibiotics, tetraethylammonium, pyrilamine, quinidine, verapamil, valprote	Tamai et al. (1998), Wu et al. (1999), Ohashi et al. (1999)
SLC6A14	Nitric oxide synthase inhibitors	Hatanaka et al. (2001)
	Valacyclovir and valganciclovir (antiviral agent)	Hatanaka et al. (2004a)
SMVT	Biotin, pantothenate, lipoic acid	Voloboueva et al. (2005), Jia et al. (2007)
OCT3	Prazocin, clonidine, cimetidine, imipramine, desipramine, quinine, nicotine and methylene-dioxymethamphetamine	Koepsell et al. (2007)
	Diphenhydramine, pyrilamine, quinidine, quinacrine	Han et al. (2001)
SR-B1	Xanthophylls, carotenoids	During et al. (2008)
SOPT1/SOPT2	Opioid peptides	Huankai et al. (2003), Chothe et al. (2010)

complexes present between the endothelial cells of the retinal blood vessels (Cunha-Vaz 1966). The endothelial cells are connected with each other with ZO, zonula adherens and desmosomes which completely seal the intercellular space (Shakib and Cunha-Vaz 1966a). The presence of ZO in the retinal vessels between the endothelial cells is responsible for the epithelium-like structure of the retinal endothelial layer and corroborates the permeable nature of the retinal vessels (Shakib and Cunha-Vaz 1966b). The highly organized multiple transmembrane signaling proteins, namely occludins, claudins and junctional adhesion molecule (JAM), constitute the TJs. In iBRB, the PDZ sequence of JAM binds with PDZ domain of ZO-1 and the extracellular domains of JAM and claudins interact with adjacent cells (trans-interaction) or with the same cells (cis-interaction) (Bazzoni et al. 2000). The cytoplasmic tails of claudin, occludin and JAM proteins are linked to actin cytoskeleton via ZO-1/2/3, respectively. Since ZO connects the intracellular actin with TJs, its alteration leads to impairment in junctional complexes leading to increase in paracellular permeability. Further corroboration of this fact is demonstrated in many tracer experiments in various species (pig, monkey, rat). For example, the iBRB exhibits resistance to intravenously administered 20 nm carbon nanoparticles, 8 nm ^{125}I-albumin, 4 nm horseradish peroxidase and 2 nm microperoxidase (Ashton and Cunha-Vaz 1965; Smith and Rudt 1975; Törnquist et al. 1990). On the other hand, molecules including fluorescein and mannitol which are much smaller than the paracellular spaces of the iBRB (Hämäläinen et al. 1997) were clearly permeable across the iBRB (Cunha-Vaz and Maurice 1967; Tachikawa et al. 2010; Thornit et al. 2010).

The iBRB is also made up of AJs comprising of adherens junctional proteins including VE-cadherin and catenin. These junctions separate luminal membrane (blood side) and abluminal membrane (retinal side) from each other. VE-cadherin is a Ca^{2+}-dependent cell adhesion protein which binds with β-catenin. Another type of junction of iBRB is the gap junction which consists of hemichannels (connexon) and connects two adjacent cells. They facilitate cross-talk between the cells and also allows small molecules having molecular weight below 1000 Da to traverse (Klaassen et al. 2013).

2.2.2 The Dynamic Inner BRB and its Permeability Characteristics

The dynamic iBRB regulates transcellular transport (Klaassen et al. 2013). Few small lipophilic molecules passively diffuse across the retinal endothelial cells (Toda et al. 2011). However, larger lipophilic and hydrophilic molecules require ATP-dependent processes to traverse the dynamic iBRB including receptor-mediated vesicular transport, non-receptor-mediated pinocytosis, transporters and pumps as represented by Fig. 1b. The regulation of transcellular movement of molecules from the blood to the neural retinal tissue is via controlled expression of molecules at both, luminal and abluminal sides. Retinal endothelial cells express high density of efflux/influx pumps in comparison to receptors, transporters and vesicle formation mediators (Sagaties et al. 1987), which collectively contribute to the dynamic iBRB.

The iBRB expresses transporters of D-glucose, monocarboxylates, amino acids and vitamins (Kubo et al. 2013). These transporters also contribute to the iBRB permeability of drugs signifying their potential to deliver drugs to the posterior eye (Tachikawa et al. 2012). For instance, leucine-referring amino acid transporter1 (LAT1) (Tomi et al. 2005), Na^+-dependent multivitamin transporter (SMVT) (Ohkura et al. 2010), organic cation transporter N2 (OCTN2) (Tachikawa et al. 2010), and monocarboxylate transporter 1 (MCT1) (Hosoya et al. 2001) facilitate the influx transport of systemically administered drugs into the retina. Amino acid analogues, nucleoside analogs, antioxidants, and substrate-mimetic drugs are carried across iBRB (Tachikawa et al. 2012). A recent study reveals that Na^+-independent L-type amino acid transporter 1 (LAT1), such as 2-aminobicyclo[2.2.1]heptane-2-carboxylic acid (BCH), is responsible for the transport of antiseizure drug gabapentin across both iBRB and oBRB in vitro (Akanuma et al. 2018). Table 3 lists important

Table 3 Important influx transporters expressed on the membrane of retinal capillary endothelium (RCE) with their corresponding substrates

Influx transporters	Substrates including drugs	References
Na^+-dependent L-ascorbic acid transporter	Vitamin C, ascorbic acid, L-phenylalanine, L-arginine and taurine	Hosoya and Tomi (2005)
LAT1, CAT1, β/TAUT	L-DOPA, neutral/cationic/β-amino acids	
Creatine transporter	Creatine	
TAUT, GAT-2, GAT-3, GAT-1, BGT-1	Taurine/GABA, B-alanine	
RFVT (Riboflavin transporter)	Riboflavin	Kubo et al. (2017)
Choline transporter	Choline	Tachikawa et al. (2012)
SR-BI	α-tocopherol	
OCTN2	Acetyl-l-carnitine	
RFC1	Formyltetrahydrofolate/methyl-tetrahydrofolate	Hosoya and Tachikawa (2012)
SR-BI	HDL	
ENT1/ENT2	Equilibrative nucleoside substrates	
CNT1/CNT2	Concentrative nucleoside substrates	Liu and Liu (2019)

influx transporters of iBRB and their corresponding drug substrates delivered across the iBRB.

The iBRB expresses efflux transporters, such as P-glycoprotein (P-gp/MDR1/ABCB1), organic anion transporting polypeptide 1a4 (OATP1a4/2), organic anion transporter 3 (OAT3) and multidrug resistance protein 4 (MRP4) which confirm its effluxing capability of drugs (Shen et al. 2003; Kennedy and Mangini 2002; Hosoya et al. 2009; Tomi and Hosoya 2004; Tagami et al. 2009). ABC transporters, P-gp and MRP4 (multidrug-resistant protein) are promising efflux transporters against anionic drugs as well. OATP transporters such as OATP1a4 and OATP1c1 are anionic efflux transporters, and they possess substrate specificity toward digoxin and estradiol 17β-glucuronide, respectively. Similarly, ABCA, ABCB and ABCG are ATP dependent group of efflux transporters that have broad range of substrate specificity.

Under physiological state, the efflux transporters remove drugs by means of ATP-mediated active transport mechanism. These efflux transporters prevent the entry of systemically administered drugs, and therefore, it is imperative to identify their complete 3D complexity and substrate selectivity. In this direction, further research is needed to understand the diversity of the efflux transporters. A clear systematic understanding of the coordination of influx transporters and efflux transporters of iBRB is warranted. Table 4 lists out important efflux transporters and their corre-

Table 4 Important efflux transporters expressed on RCE membrane and corresponding substrates

Efflux transporters	Substrates including drugs	References
P-gp/MDR1	Cyclosporine A	Hosoya et al. (2009)
MRP4	Estradiol 17β glucuronide and organic ions	
OATPla4	Estradiol 17β glucuronide and organic ions	
OATPlc1	Organic anions	
OAT3	PAH, PCG & 6-MP	
ABCG2	Mitoxantrone, Doxorubicin, Pheophorbide a	Tachikawa et al. (2012)
ABCA	Sterols, phospholipids & retinoids	
ABCB1	(R)-^{11}c- verapamil	Bauer et al. (2017)

sponding drug substrates delivered across the iBRB (Hosoya et al. 2009; Tachikawa et al. 2012; Bauer et al. 2017).

2.3 *Changes in Permeability of oBRB and iBRB in Retinal Diseases*

Major posterior eye diseases including macular edema, posterior uveitis, diabetic retinopathies, hereditary dystrophies, age-related macular degeneration and injuries due to trauma and intraocular surgery are associated with modifications in BRB permeability. Breakdown of molecular assembly of the barrier proteins discussed earlier leads to the alteration in paracellular and transcellular permeability of oBRB and iBRB. The resistance of the paracellular route is ten times superior compared to the transcellular route (Edelhauser et al. 2010). The primary mediators in the alteration of BRB permeability are pro-inflammatory factors (TNF-α, IL-1β)(Joussen et al. 2002; Gonçalves et al. 2013), adhesion molecules (Joussen et al. 2001), growth factors (Hosoya and Tachikawa 2013) and exogeneous stresses including NO/superoxide radicals, hypoxia and ischemia (Leal et al. 2007; Gonçalves et al. 2012; Zhang et al. 2014).

For instance, increase in VEGF-induced vascular permeability via NO signaling by reduced expression of occludin in retinal endothelial cells (Hosoya and Tachikawa 2013) has been recorded. High glucose levels in diabetic retinopathy affect the dynamic structure of GLUT1 transporter and results in the reduction of cotransportation of glucose and DHA (Kubo and Hosoya 2012). VEGF and pro-inflammatory cytokines modulate the expression of claudin-5 in RPE and increases occludin phosphorylation resulting in a leaky retinal vasculature and barrier dysfunction (Antonetti et al. 1998; Murakami et al. 2009). Retinal endothelial cells are also affected by endoplasmic reticulum (ER) stress in proliferative diabetic retinopathy, and the end result is downregulation of claudin-5 expression and increased iBRB permeability (Adachi et al. 2011). In case of oBRB, high glucose concentration influences the expression of occludin, claudin-1 and ZO-1 proteins in ARPE-19 cell lines (Villarroel et al. 2009). In diabetic retinopathy, oxidative stress induces upregulation of Taurine transporter in human RPE cells (Nakashima et al. 2005).

In the context of systemic drug delivery to the posterior eye, there are evidences of exploitation of the increased permeability of the BRB to deliver drugs. For instance, the enhanced permeability of the choroidal neovasculature and RPE was found to be partly responsible for targeting of the neovasculature by mitomycin immunoconjugate (Kamizuru et al. 2001), interferon-β-DTPA-dextran (Yasukawa et al. 2002), TNP-470–PVA conjugate (Yasukawa et al. 1999) and anti-VEGF intraceptor plasmid conjugated to transferrin/RGD peptide-functionalized PLGA nanoparticles (Singh et al. 2009).

3 Recent and Advanced Approaches of Overcoming BRB and Targeted Drug Delivery to the Posterior Segment of the Eye

Despite the fact that the BRB is compromised in most of the retinal diseases, systemic route of drug administration to the retina is still a huge unmet clinical need. Hence, many groups have investigated and developed strategies to overcome this permeability barrier across the retina. In this direction, we will now discuss various novel strategies including (i) smart drug design which includes drugs mimicking physiological substrates; prodrugs and cotransporters; (ii) drug-immobilized nanocarriers; and (iii) transient modulation of permeability of the BRB using gene delivery (Campbell and Humphries 2013) and induction by hyperthermia (Tabatabaei et al. 2016).

3.1 Smart Drugs for Overcoming BRB

Modification of drugs to impart BRB permeable properties is crucial for delivering drugs to targeted sites across the retina. In this regard, prodrug approach is an emerging area where the active moiety of the drug is modified to increase membrane permeability, transporter targeting, solubility, and site specificity resulting in enhanced retinal bioavailability (Patel et al. 2018). Desirable properties of prodrugs are: (i) solubility in water; (ii) lipophilicity ($\log P = 2 - 3$) to ensure efficient crossing of BRB; (iii) bioreversibility resulting in the release of the active parent drug from the pro-moiety at the site of action; (iv) chemical stability for optimal pharmacological action; and (v) minimal toxicity.

Most prodrugs are chemically modified using substrates for influx/efflux transporters and receptor-targeting molecules. The membrane influx/efflux transporter includes transporters of peptide (PEPT1 and PEPT2), amino acids (LAT1 and LAT2), nucleoside and nucleobase, GLUT1, monocarboxylic acid (MCT), organic anions, organic cations, vitamin (SVCT2), aquaporins and P-gps (MDR1) (Boddu and Nesamony 2013). Membrane receptors for molecules including insulin like growth factor, bradykinin/tachykinin, histamine, estrogen/progesterone/serotonin, glucocorticoid/mineralocorticoid, TNF, VEGF and hyaluronan (Mehta et al. 2018) are often expressed in the retinal endothelial cells and RPE. Another prodrug approach is to improve the lipophilicity of hydrophilic drug molecules and enhance BRB permeation. In this approach, the drug molecule is covalently bound to a lipid moiety such as fatty acid, diglyceride or phosphoglyceride that facilitates diffusion of the drug across the cell membrane (Patel and Patel 2018). Ophthalmic prodrugs are synthesized by modifying functional groups including carboxyl, hydroxyl, amine and carbonyl groups (Barot et al. 2012). Permeability of the prodrug can be enhanced by modifying carboxylic acids or alcohols to esters because of increased lipophilicity. For drugs that lack functional groups, derivates of ketones such as oxime are employed to create prodrugs, which hydrolyzes in the body and releases the drug (Burgess et al. 2018). Different routes of transport of prodrugs across oBRB and iBRB upon systemic administration are depicted in Fig. 2.

3.1.1 Transporter-Targeting Prodrug Approach

The transporter-targeting prodrugs are designed to bind to specific influx transporters in cells such as GLUT1 and ASCT1 and not bind to efflux transporters including P-gp/MDR proteins to facilitate crossing of ocular barriers upon oral or intravenous administration (Barot et al. 2012). The retinal peptide transport system was effectively exploited by Majumdar et al. to allow the release of the active drug in the retinal tissue (Majumdar et al. 2006). They reported that dipeptide monoester prodrugs of ganciclovir (GCV), val-GCV and gly-val-GCV serve as substrates for retinal peptide transport system and were stable in vitreous humor. The prodrug rapidly released the active parent drug GCV in retinal homogenates as well as in the retinal tissue in rabbits. In another study, prodrug of gatifloxacin (Dimethylamino-propyl derivative of gatifloxacin or DMAP-GFX)-targeting organic cation transporter (OCT) transporter was synthesized by Vooturi et al., and they reported enhanced accumulation in the retina 1 h after topical dosing (Vooturi et al. 2012). Folate-conjugated nanoparticles delivering doxorubicin was exploited as a prodrug to target the folate transporter system and deliver the drug into retinoblastoma cells in vitro (Boddu et al. 2010). Katragadda et al. demonstrated the in vivo systemic absorption of amino acid prodrug of acyclovir after oral administration in rats owing to their affinity toward amino acid transporters (ASC and $B^{0,+}$) as well as peptide transporters (PEPT1) (Katragadda et al. 2008). Hatanaka et al. demonstrated transport of amino acid prodrug (valacyclovir) via amino acid transporter $ATB^{0,+}$ coupled with Na^+ and Cl^- (Hatanaka et al. 2004). Vadlapudi and coworkers reported cellular translocation of biotinylated prodrug (Biotin-Acyclovir (B-ACV) in RPE cells in vitro due to the presence of functional SMVTs (Vadlapudi et al. 2012). Another study also proved the role of SMVTs on the retinal uptake of biotin-ganciclovir (biotin-GCV) in rabbit retina in vivo (Janoria et al. 2009). Some of the transporter prodrugs in clinical use are

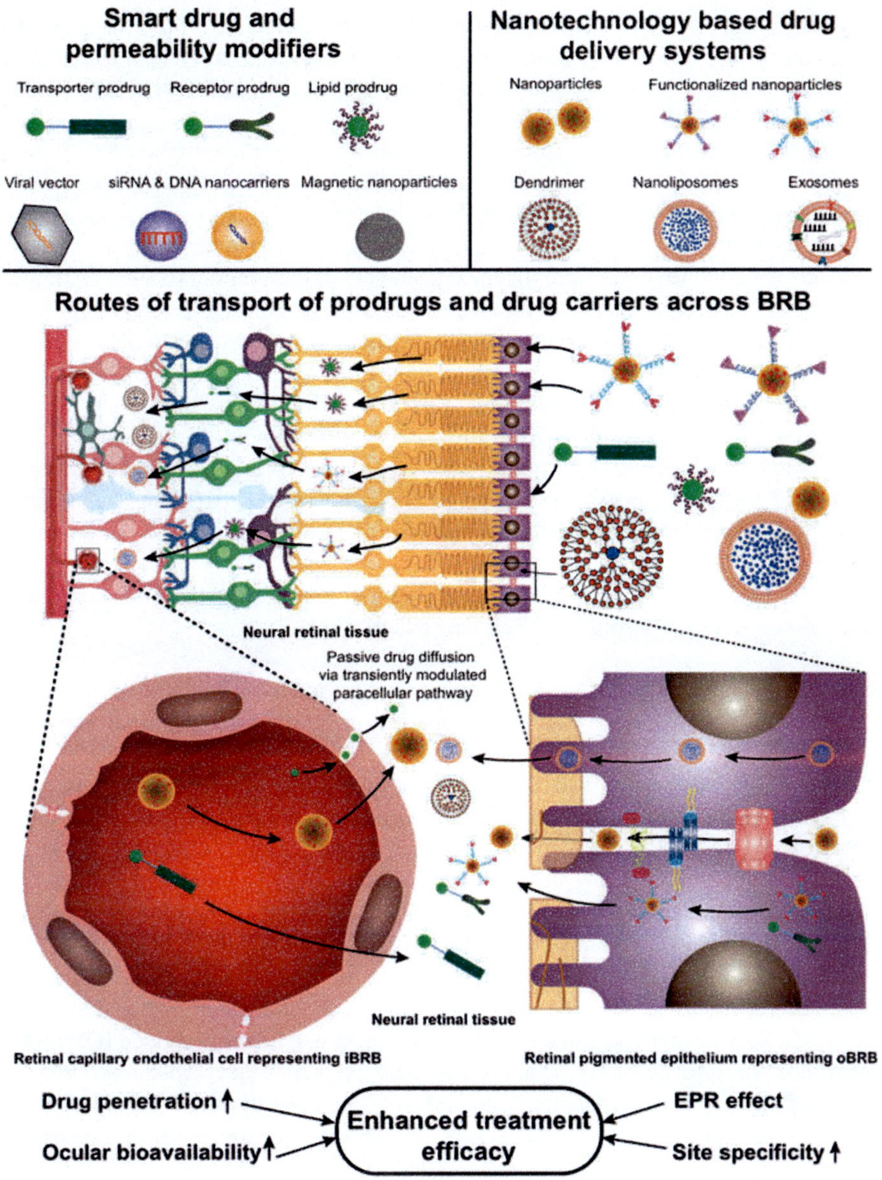

Fig. 2 Smart drug design, drug immobilization-, nanotechnology-enabled and transient modulation-based recent advances in the systemic therapy for posterior eye diseases

dipivefrin (oral) for open-angle glaucoma (Egorov et al. 2000), prednisone (oral) for uveitis (Babu and Mahendradas 2013), pegaptanib for neovascular age-related macular degeneration (Vinores 2006) and valganciclovir (oral) for cytomegalovirus (CMV) retinitis (Patil et al. 2010; Kabata et al. 2012).

Another type of transporter prodrug is the stereoisomeric dipeptide prodrug which consists of two amino acid isoforms (L-isomer or D-isomer) conjugated to the parent drug via an ester bond (Patil et al. 2018). These prodrugs can be designed to not only degrade into the parental drug upon cell entry, but also to target specific transporters on the cell membrane to enhance the chances of cellular uptake. For example, antiherpes virus drug acyclovir has been modified into a prodrug to target oligopeptide transporter (PEPT) in corneal epithelial cells. This prodrug was delivered via nanoparticles to increase corneal penetration and bind specifically to the PEPT transporter compared to the standard treatment (Anand et al. 2003). There are no dipeptide prodrugs specifically reported in the context of treatment of retinal disorders. However, there is evidence that PEPTs such as sodium-coupled oligopeptide transporter (SOPT2) are expressed in human RPE (Chothe et al. 2010), and hence, this prodrug approach has potential to treat retinal diseases.

3.1.2 Receptor-Targeting Prodrug Approach

The receptor-targeting prodrugs are specifically designed to bind to specific receptors on either RPE cells or retinal endothelial cells such as receptors for IGF, bradykinin/tachykinin, histamine, estrogen/progesterone/serotonin, glucocorticoid, TNF and VEGF (Mehta et al. 2018) to facilitate transcellular transport of drugs upon oral or intravenous administration. These receptors are polarized to allow transport of solutes. PF-04217329 (Taprenepag isopropyl), a prodrug of CP-544326 (active acid metabolite) which is a selective prostaglandin EP(2) agonist, was reported to activate EP(2) receptor and as a result, overexpress cAMP of the iris–ciliary body upon topical delivery (Prasanna et al. 2011). The prodrug formulation enhanced the corneal penetration and ocular bioavailability of the drug in an in vivo glaucoma model and resulted in a sustained and significant reduction in the intraocular pressure (IOP). Another receptor that has been identified as a target to reduce IOP is the cannabinoid receptor (CB1/2). Mainolfi and coworkers have developed a prodrug of the CB1/2 agonist and demonstrated significant enhancement in its distribution in iris/ciliary body and aqueous humor of rats 3 h after topical administration compared to the parent drug itself (Mainolfi et al. 2013). Recently, ONO-9054, the prodrug of ONO-AG-367 (dual agonist of EP3 and FP Receptors) was tested for its pharmacokinetics and pharmacodynamics in healthy, normotensive adults in healthy, normotensive adults in a randomized, double-masked, placebo-controlled, single-dose escalating study (Suto et al. 2015). The clinical trial reported that a single dose of the prodrug was safe and well tolerated, and IOP was significantly reduced after 9 h. Another example is the recently identified selective, non-Prostanoid EP2 receptor agonist called omidenepag and its prodrug OMDI (Iwamura et al. 2018). In normal osmotensive monkeys, Iwamura and coworkers demonstrated that low doses of OMDI lowered IOP and hence, selected as a clinical candidate for the treatment of glaucoma. Receptor-targeting prodrugs for treating posterior eye diseases such as age-related macular degeneration (AMD) have been recently reported. For instance, vision loss in AMD is due to choroid neovascularization (CNV). Its pathogenesis has been attributed to the HIF-1α/VEGF/VEGF receptor 2 pathway, and hence, a prodrug targeting this receptor has been investigated (Xu et al. 2020). They reported that a prodrug of epigallocatechin-3-gallate (pro-EGCG) mitigated the laser-induced CNV leakage by downregulating the HIF-1α/VEGF/VEGFR2 pathway.

3.1.3 Lipid Prodrug Approach

The lipid prodrugs are designed to render drugs lipophilic by covalently binding them with a lipid moiety such as fatty acid, diglyceride and phosphoglyceride (Patel and Patel 2018) and allow facilitated diffusion of drugs via the transcellular route. Lipid conjugation would also enhance the PK/PD profile of the drug. By manipulating various factors including chain length of the lipids, double bond configuration and lipid–drug linkage, one can achieve varieties of lipid prodrugs (Jespersen et al. 2011). Optimization of lipophilicity of prodrugs is critical to achieving maximum permeability. This is because, very high lipophilicity will result in low permeability since the prodrugs tends to bind to the lipid bilayer of the cell membrane rather than permeating through it (Chien and Schoenwald 1986). In this context, maximum permeability could be achieved when the lipophilicity (log*P*) value is between 2 and 4, where *P* is the partition coefficient represented by the ratio of the concentration of the drug in octanol to the concentration in an aqueous buffer of pH 7.4 (Chien and Schoenwald 1986). In vitro studies in human RPE cell line (ARPE-19) revealed that a long-chain acyl ester lipid-conjugated prodrug of ganciclovir was non-toxic and possessed improved permeability in RPE cell line (Cholkar et al. 2014). Vadlapudi et al. demonstrated enhanced cell absorption of biotinylated lipid prodrug of acryclovir (ACV) via SMVT in corneal cells in vitro and increased transepithelial permeability in rabbit corneas when topically administered for the treatment of herpetic keratitis (Vadlapudi et al. 2013). In another study, cidofovir and its cyclic analogue were conjugated into a lipophilic molecule (via long chain of carbon) and modified to target SMVT. The modified lipid prodrug had improved physicochemical properties such as lipophilicity, partition coefficient, cell permeability, well-controlled bioreversion kinetics, and interaction with SMVT transporter. The modification also resulted in a significant increase in its cellular accumulation in vitro and in ex vivo choroid and retinal tissue of rabbit eyes for the treatment of human cytomegalovirus retinitis (Gokulgandhi et al. 2012). Recently, a biotinylated lipid prodrug of cyclic cidofovir (B–C12-cCDF) was formulated within polymeric nanomicelles and utilized as a carrier for targeted retinal delivery in various retinal cell lines including HRPE (human retinal pigment epithelial, D407), HCE-T (human corneal epithelial) and CCL 20.2 (human conjunctival epithelial) cells (Mandal et al. 2017). Significant transport of lipid prodrug-loaded nanomicelles across HCE-T cells further corroborated that the prodrug enhanced the permeability and their subsequent uptake into D407 cells.

3.1.4 Other Prodrug Approaches

Most of the drugs have been formulated using mineral oil or surfactant-based solutions or emulsions for topical delivery applications (Adelli et al. 2017). These formulations have improved drug solubility; however, the cell membrane permeation characteristics remain relatively unaltered. By virtue of their lipophilicity, most drugs would rapidly partition into the lipophilic cell membrane and get trapped due to interaction with the lipids and fail to efficiently partition into the more hydrophilic aqueous layers and thus, hardly reach the posterior tissues of the eye (Majumdar et al. 2010). Therefore, various prodrugs have been developed to demonstrate enhanced ocular tissue permeation characteristics compared to the native drugs. For instance, Δ^9-tetrahydrocannabinol (THC) which is an antiglaucoma drug but with low aqueous solubility has been modified into relatively hydrophilic prodrugs, using mono- and divaline esters (THC-Val and THC-Val-Val) and the amino acid (valine)-dicarboxylic acid (hemisuccinate) ester (THC-Val-HS) by Adelli and coworkers (Adelli et al. 2017). However, they reported that the amino acid prodrugs of THC showed poor aqueous solubility and the dual amino acid-dicarboxylic acid ester-based prodrug was significantly superior in comparison. They also reported that the prodrug significantly improved the penetration of THC into the anterior segment of the eye following topical application that resulted in significantly improved IOP-lowering activity. In another study, the synthetic analogue to the visual chromophore 11-cis-retinal namely 9-cis-retinal was conjugated with chitosan to create a prodrug (Gao et al. 2018). Its intravitreal administration in RPE-specific 65 retinoid isomerase (RPE65)-deficient dogs improved photoreceptor function for several weeks. Recently, a nanoparticle–conjugate (NPC) prodrug of a smaller fragment of the anti-angiogenic 34-mer peptide of PEDF, namely 9-mer displayed longer intraocular residence upon intravitreal administration in laser-induced mouse and rabbit models of choroidal neovascularization (Sheibani et al. 2019). This NPC showed extended efficacy, even when injected 14 days before laser treatment.

In summary, most of the prodrugs developed thus far have been employed for crossing the corneal barrier and reaching the anterior chamber. In vitro RPE models and in vivo models of posterior diseases including glaucoma, AMD, CMV retinitis, retinoblastoma have been utilized to show efficacy of the prodrugs. However, very few clinical trials exist that utilize orally or intravenously administered prodrugs for treating posterior eye diseases. Hence, it will be of great interest to assess distribution of these prodrugs to the posterior of the eye as well, especially via systemic administration.

3.2 *Drug Immobilization and Nanotechnology-Enabled Approaches of Overcoming BRB*

Nanotechnology-enabled systemic drug delivery to the posterior segment of the eye via overcoming the BRB involves immobilization of the drug inside or on the surface and targeting moiety nanocarriers on the surface of nanocarriers. These nanocarriers are of several kinds including nanoparticles, liposomes, dendrimers, metal nanoparticles, hydrogels and micelles as depicted in Fig. 2. Immobilization of the drug inside the nanocarriers facilitates higher bioavailability at the target site as well as protection from degradation and tissue binding. On the other hand, the targeting moieties are surface immobilized on the nanocarriers to enable their direct interaction with the receptors or transporters on the target cells responsible for drug influx. Immobilization strategies involve physical entrapment or chemical conjugation via covalent, ionic, hydrogen bond and hydrophobic–hydrophilic interactions. The immobilized entities can then be made functional or released using a variety of mechanisms including enzymatic degradation, photoactivated cleavage, ligand–receptor binding, and pH/temperature-mediated cleavage. In addition to the above stated advantages, the nanotechnology-based drug delivery systems can exploit the enhanced permeation and retention effect and bypass the reticuloendothelial system to allow rapid internalization of the therapeutic entities. Few recent examples of immobilization of therapeutic entities using nanocarriers for overcoming BRB and drug delivery to the posterior segment of the eye are discussed next.

Different nanotechnology approaches that enable overcoming BRB and drug delivery across oBRB and iBRB upon systemic administration are depicted in Fig. 2. Singh et al. have demonstrated enhanced, targeted retinal delivery of antivascular endothelial growth factor (anti-VEGF) intraceptor plasmid into CNV lesions (Singh et al. 2009). They employed intravenously injected intraceptor-loaded PLGA nanoparticles functionalized with dual complexes, i.e., transferrin and arginine-glycine-aspartic acid (RGD) peptide which are targets of the integrins $\alpha v\beta 3$ and $\alpha v\beta 5$. These integrins are specifically overexpressed in ocular tissues in AMD patients and vascular tissues in proliferative diabetic retinopathy patients (Friedlander et al. 1996). In a brown Norway CNV-induced rat model, the dual-functionalized PLGA nanoparticles targeted the neovascular eye and inhibited the progression of laser-induced CNV. Another similar study reported that an integrin $\alpha\nu\beta 3$-targeting ligand bound to dominant negative Raf mutant gene-loaded cationic nanoparticles could effectively target choroidal neovascular membrane (CNV) (Salehi-Had et al. 2011). Intravenous injection of these nanoparticles into CNV-induced Norway male rats exhibited reduction in size of the lesions due to efficient crossing of CNVs across BRB and targeted gene delivery. Pathological changes in BRB allow the accumulation of inflammatory microglia/macrophages in retinal tissues in diseases including AMD and diabetic retinopathy, and the current treatment includes frequent intravitreal injections (Moisseiev et al. 2014), which due to chronic treatment may further damage the eye. An alternative, safer and simpler strategy of intravenous administration of microglia-targeting PAMAM dendrimers was reported, where in the dendrimers efficiently partitioned from the vitreous humor into the retina and targeted the activated microglial cells at the injury site by escaping the damaged BRB (Kambhampati et al. 2015). In BALB/c albino mice models of ischemia/reperfusion injury, they demonstrated that the dendrimers accumulated in the Iba-1^+ microglial cells of the diseased retina and remained for at least 21 days, but cleared off in healthy retina within 24 h. Thus, this intravenous approach is expected to be more effective than the current standard of care (monthly intravitreal injections) for treating neurodegenerative and diabetic retinal diseases given its targeting ability and reduced off target effects and sustained drug release capabilities.

Retinal degeneration (RD) is a genetically heterogenous disease lacking pharmacological targets, and the presence of BRB renders the clinical translation of novel therapies very difficult (Rip et al. 2014). The presence of Na^+-dependent glutathione (GSH) transporter in the apical membrane of retinal cells (Kannan et al. 2001; Garcia et al. 2011) was exploited to allow effective crossing of the BRB using a GSH-targeted PEGylated liposomal formulation in murine RD models (Vighi et al. 2018). Here, Vighi and coworkers report targeted delivery of an inhibitory cGMP analogue (CN03) using GSH-PEG liposomes to neuroretina in addition to prolonged circulation time, increased CN03 bioavailability at the target site and increased photoreceptor numbers in RD murine models in vivo. In another study, avastin-loaded magnetite (Fe_3O_4) nanocomposites were developed for the treatment of AMD via systemic administration (Zargarzadeh et al. 2018). Here, the authors report that Fe_3O_4 nanoparticles were coated with dextran to facilitate adhesion to retinal endothelium. On similar lines, Tabatabaei and coworkers have reported that hyperthermia induced transient leakiness in BRB when human transferrin glycoprotein-coated Fe_3O_4 nanoparticles were intravenously administered in a rat model and subjected to AC magnetic field (Tabatabaei et al. 2016). Transferrin glycoprotein aided the adhesion of nanoparticles to the retinal endothelium and hyperthermia created by the nanoparticles transiently disrupted the BRB (for about 2 h) and allowed the diffusion of intravenously administered dyes into the retinal tissue.

Another nanotechnology-based approach of overcoming BRB and penetrating retinal cells involves surface modification of nanoparticles with cell-penetrating peptides (CPP). Multiple types of CPPs (cationic, amphipathic and hydrophobic) have been designed to target cells via direct penetration, endocytosis or other mechanisms (Guo et al. 2016). These CPPs have also proven to be efficient in mediating the uptake of plasmid DNA and DNA containing lipoplexes and polyplexes by surpassing the cell membrane barrier (Trabulo et al. 2010). For instance, recently, Wang et al. have developed Tat-C (48–57) CPP-functionalized and doxorubicin-loaded PEG-PLA nanoparticles (Wang et al. 2019). In addition, the CPP is bound to a photo-cleavable group DEACM ((7-diethylamino) coumarin-4-yl methyl carboxyl), which inhibits the cellular uptake of the nanoparticles because of inactive CPP. However, upon light irradiation, DEACM is cleaved and the CPP is activated allowing targeted uptake of the nanoparticles by the neovascular lesions in mouse models of CNV upon intravenous injection of the nanoparticles. Consequently, the targeted delivery of doxorubicin to the lesions significantly reduced their size.

In addition to synthetic nanocarriers, cell-derived vesicles or exosomes can also be employed to penetrate BRB and target cells of the retina. Exosomes are bilayered membrane nanovesicles enriched with bioactive components including proteins, lipids, miRNA, mRNA and DNA (Colombo et al. 2014). Exosomes facilitate intercellular transfer of these bioactive components and act as potent mediators of intercellular communication in various physiological and pathological conditions (Gurunathan et al. 2019). Knickelbein et al. investigated the effect of RPE-derived extracellular vesicles (EVs) on T cell and human monocytes and reported that they did not induce T cell death and increased expression of CD14 and CD16 in human monocytes. The immunosuppressive cytokine TGF-β1 was upregulated due to the presence of EV and apparently responsible for immunomodulatory effects. Hence, EVs secreted by RPE cells are potential therapeutic entities for posterior uveitis (Knickelbein et al. 2016). Periocular injection of human umbilical cord-derived mesenchymal stem cell (MSC) exosomes in Lewis rats induced with autoimmune uveoretinitis for 7 days resulted in the downregulation of CD4, IFN-γ, CD4 and IL-17 in the retina. The exosomes also inhibited the migration of leukocytes and inhibited chemo-attractive effect of CCL2 and CCL21 on inflamed cells and hence have a great potential to treat autoimmune uveoretinitis (Bai et al. 2017). Exosomes from retinal astroglial cells (RAC) have been reported to inhibit the migration of macrophages and formation of tubules in mouse retinal microvascular endothelia cells. The exosomes also suppressed retinal vessel leakage and inhibited choroidal neovascularization in laser-induced CNV model due to the presence of anti-angiogenic components such as ET-1, PIGF-2 and PTX3 (Hajrasouliha et al. 2013). Even though exosomes are potential therapeutic agents for many posterior eye diseases, the specific mechanism for anti-inflammatory and immunomodulatory effects has to be investigated in detail before they can realize clinical translation. The understanding of the mechanisms of exosomal penetration of the BRB is necessary to realize their potential as drug delivery carriers.

3.3 Controlled and Transient Modulation of BRB using Gene Delivery and Physical Forces

An indirect way to facilitate targeted drug delivery to the retinal components is via transient and controlled modulation of BRB permeability. Non-invasive methods including electroporation, iontophoresis, sonophoresis and microneedles are used for modulating BRB permeability to transfer siRNA, drugs or other therapeutic agents to the posterior part of the eye (Huang et al. 2018). Iontophoresis utilizes electromotive force to push ionic drugs to move across ocular barriers, whereas electrophoresis relies on creation of transmembrane potential resulting in pores on the cell membrane of the barrier to facilitate diffusion across (Chizmadzhev et al. 1998; Apollonio et al. 2012). Ultrasound-targeted microbubble destruction (UTMD)-mediated gene transfer is a non-invasive approach to transfect retina (Wan et al. 2015). It is a promising approach compared to liposome-based gene transfer to treat posterior eye diseases, and it increases transfection efficiency in human RPE cell line (Zheng et al. 2012). In another study, Park et al. studied the influence of retinal vascular permeability by UTMD in rat models and proved that UTMD induces temporary disruption of BRB (applied at 10 ms burst applied at 1 Hz for 60 s), without affecting cellular integrity and recovered after 3 h (Park et al. 2012). siRNA-loaded mPEG-PLGA-PLL NPs in rat retina were successfully and safely delivered by UTMD method (Du et al. 2017). The complete mechanism of UTMD facilitated disruption of BRB is still not clear, but is apparently due to the disassembly of the molecular structure of TJs resulting in the loss of barrier function owing to the ultrasound bursts combined with a gas contrast agent in BRB microvessels (Sheikov et al. 2008). The junctional proteins of TJs and AJs of BRB were altered, i.e., the claudins were phosphorylated by RNAi facilitating the modulation of paracellular permeability to chloride (Yamauchi et al. 2004). Other studies have reported that cell–cell interactions and paracrine effects increase the permeability of BRB in vitro (Tretiach et al. 2005; Behzadian et al. 2001). In addition, hypertonicity of the medium also enhances GABA uptake by cultured rat retinal capillary endothelial cells (Yahara et al. 2010). These above approaches may pave the way for transport of drugs by transient modulation of BRB for effective therapies to the posterior eye diseases.

4 Clinical Translation of Systemic Therapies to the Retina: What Lies Ahead?

The success of systemic therapies for the treatment of various posterior eye diseases relies on four critical factors as depicted in Fig. 3: (i) understanding the pathophysiology leading to efficient identification of targets and corresponding drugs through realistic preclinical models; (ii) development of innovative, efficient, natural and scalable drug delivery vehicles and/or prodrugs and/or safer methods of transient BRB manipulation, for overcoming resistance from BRB and for targeting; (iii) non-invasive diagnostic tools for characterizing BRB and pathophysiology in humans through the use of advanced tools including artificial intelligence, deep learning and single-molecule sensing in addition to state-of-the art instrumentation; (iv) highly regulated scale-up process for large-scale manufacturing of prodrugs and/or nanocarriers for systemic delivery and well-defined clinical trials. It is noteworthy that these challenges can be overcome or success achieved only by cross-disciplinary contributions from medical doctors, engineers and basic scientists. Each of the four points is elaborated next.

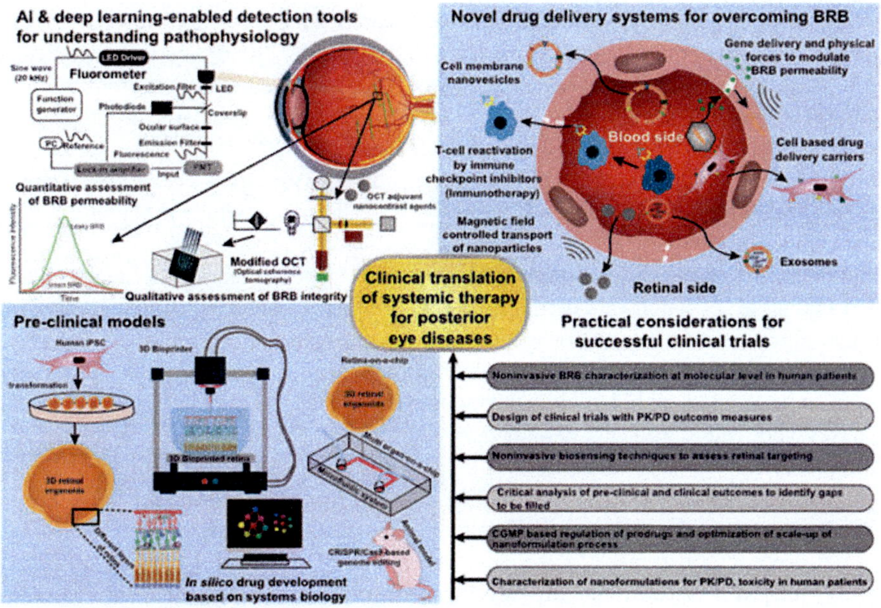

Fig. 3 Future of clinical translation of systemic therapy for posterior eye diseases depicting detection tools to understand pathophysiology, novel drug delivery systems to overcome BRB, realistic preclinical models and practical considerations for successful clinical trials

4.1 Mimicking Human Retina for Better Understanding of Pathophysiology and Drug Design

Preclinical models that represent the pathologies in reality, i.e., with all the secondary conditions that prevail in human patients along with the primary pathology need to be developed. The future of these models depends on technological advances as shown in Fig. 3: (i) generation of advanced animal models for various retinal disorders using genome-editing technology (CRISPR); (ii) establishment of alternative models of retinal disorders including ex vivo organoids using iPSCs, 3D bioprinting technology, retina-on-a-chip technology and complemented by systems biology or in silico drug design and development.

The generation of animal models that mimic the pathologies of the posterior segment of the eye has been advancing rapidly since the advent of the clustered regularly interspaced short palindromic repeats (CRISPR)/CRISPR-associated protein 9 (Cas9)-based genome-editing technology (Peng et al. 2017). In comparison to other gene-editing tools including transcription activator-like effector nucleases (TALENs) and zinc-finger nucleases (ZFNs), CRISPR/Cas9 system offers multiplexed gene-editing features and significantly higher efficiency (Pennisi 2013). For instance, retinitis pigmentosa (RP) is a genetic disorder of the retina caused due to multiple mutations and current animal models only mimic photoreceptor degeneration. Thus, using CRISPR technology, scientists have created animal models for RP which mimics the clinical phenotypes (Arno et al. 2016; Wu et al. 2016; Feehan et al. 2017) demonstrating not only photoreceptor degeneration, but also dysfunction of the rod photoreceptors. Other such examples include advanced animal models for retinoblastoma (Naert et al. 2016) and Leber congenital amaurosis (LCA) (Zhong et al. 2015). Continuous development of CRISPR genome-editing systems is needed to circumvent any off-target events and immune responses and mimic clinical phenotypes in animal models.

The number of retinal disorders multiply rapidly, but the rate of therapies developed is unable to match this demand. In this context, three-dimensional structures mimicking human retina, also termed as retinal organoids represent an alternative to animal models. It is also a promising platform for understanding developmental aspects and capable of high-throughput drug screening and mimicking human diseases when integrated with patient-derived human pluripotent stem cells and gene-editing technology (Llonch et al. 2018). They can be derived from human-induced pluripotent stem cells and have been demonstrated to mimic human retina in terms of cell heterogeneity, cell–cell, cell–matrix interactions, retinal disease pathology and physiological responses (Aasen and Vergara 2019). These features have significantly improved the current drug development process and expedited the rate and efficiency of bringing therapeutic candidates into clinics. However, there are a number of barriers that need to be circumvented to achieve full realization of retinal organoids for drug development. They include: (i) culture time: Time required for developing 3D retinal organoids is much larger compared to their 2D counterparts; however, the advent of rotating-wall bioreactors has slightly reduced culture time (DiStefano et al. 2018), and until these culture protocols are fully optimized and scaled up, one can maintain organoid banks facilitating rapid and on-demand studies; (ii) lack of extra-retinal structures: Vascularization and polarized RPE representing the BRB are currently lacking in retinal organoids and must be developed for effective drug screening for relevant diseases such as diabetic retinopathy and AMD. Very recently, Achberger and coworkers have generated complex multilayer tissue models in a human retina-on-a-chip platform by merging organoid and organ-on-a-chip technology (Achberger et al. 2019). This is a novel microphysiological model of the human retina which integrates more than seven different retinal cell types derived from hiPSCs and mimics physiological phenomena including vasculature-like perfusion among others. Overactivation of excitatory amino acid receptors (EAAT) located on the cell membranes has been implicated in many retinal diseases (Kanai and Hediger 1992). The presence of such transport system on the retina and the retinal epithelium opens up opportunities for designing prodrugs targeting posterior segment of the eye (Pow 2001). Hence, we need retinal organoids which can mimic these changes in BRB to investigate how differently prodrugs or nanoformulations overcome the resistance from BRB. (iii) Maturation: The current retinal organoids do not mimic mature retinal components and are hence not suitable for drug screening in diseases like AMD or retinal degeneration. Hence, protocols for retinal organoid maturation need to be established; (iv) genetic and transcriptional variability of retinal organoids derived from various iPSC clones (Capowski et al. 2019) are found to originate due to interdonor variability (Burrows et al. 2016). Hence, more studies focused on developing systems with lesser inconsistencies in retinal organoid cultures and increased rigor during disease modeling and drug design/screening are warranted; (v) mimicking complexity: The evaluation of target tissue in response to drugs must not be done in isolation, rather on an interactive level with other tissues and organs. Such a possibility has been achieved to a certain extent by the advent of organ-on-a-chip microfluidic technologies (Maschmeyer et al. 2015; Miranda et al. 2018). It is hence, critically required to develop multi organ-on-a-chip models mimicking retinal disorders and interactions with other tissues/organs; (vi) scalability and automation: The current methods of manual generation of retinal organoids must be completely automated for efficient liquid handling and tissue construction using advanced 3D bioprinting technologies. This would enormously aid in the scalability and also enhance reproducibility. So far, attempts have been made to 3D bioprint retinal ganglion cells and glial cells (Lorber et al. 2016) and human RPE cells (Shi et al. 2017). The future lies in 3D printing of a viable and fully functional retina with all its cellular and extracellular components. (v) Cost: While addressing issues of culture time, scalability and automation would bring down the cost, the improvements in tissue complexity will indirectly reduce cost because it would result in fewer unsuccessful drugs going into clinical trials. Finally, mathematical models that can predict disease onset, progression, in addition to the identification of biomarkers, delineating pathological mechanisms and response to therapy can complement the above approaches. These systems biology or in silico tools are robust because they integrate bioimaging, omics, pharmacological and clinical data for prediction (Handa et al. 2019).

4.2 *The Proof is in the Pudding: Putting Innovative Nanotechnology-based Drug Delivery to Test*

Nanotechnology-based drug delivery systems have proved their mettle in the treatment and detection of diseases of the posterior segment of the eye. However, they have been limited to only the preclinical stage. Several challenges still remain to be addressed for successful clinical translation of these systems including: (i) robust understanding and control over the physicochemical properties of nanoparticles; (ii) nanotoxicity and biodistribution characteristics; and (iii) ability to mimic and target the host environment; (iv) biocompatibility; and (v) scale-up.

Aggregation of nanoparticles inside the tissue after administration is a common phenomenon and could be avoided by manipulating the physicochemical properties of the nanomaterials and could help drug distribution at the target site (Weng et al. 2017). This is particularly critical because the aggregated nanoparticles are cytotoxic and cause inactivation of cell metabolism, disruption/alteration of cell membranes and inhibit tissue functions. The toxic effects of nanoparticles have been reported to emanate from oxidative stress and probable initiation of the activation of different transcription factors (Medina et al. 2007). In general, nanoparticles can be taken up by lymphatic nodes and distributed through the lymphatic system in parallel with the blood vascular system. An additional consideration is the potential inflammatory, immunostimulatory and immunosuppressive properties described for different kinds of nanoparticles (Medina et al. 2007) and draws scrupulous evaluation of immunotoxic effects. Many studies have been conducted in an effort to standardize the protocol for evaluation of nanoparticles for immunotoxicity (Putman et al. 2003; Stone et al. 2009). Almeida et al. have suggested modification of nanoparticles to prolong blood circulation and enhance treatment efficacy, but non-specific distribution of nanoparticles are unavoidable. Longer in vivo studies are required to assess the distribution of nanoparticles as well as innate and adaptive immune responses, excretion over time and toxicity levels (Almeida et al. 2011).

More pressing issues in the context of systemic drug delivery to the posterior part of the eye are efficient crossing of the BRB and targeting the tissue/cell of interest for the drug to take effect. Conventional synthetic nanomaterials need very complex chemistry of functionalization to achieve BRB overcoming and/or targeting ability. Active targeting using functionalized NPs are limited by complexity of the NP architecture, ligand-conjugation chemistry, ligand–target interaction, immune responses, choice of the target and drug resistance. In addition, the accumulation of the NPs at the target site is controlled and limited by NP architecture, size, ligand chemistry and conjugation, ligand size, and target cell properties. On the other hand, a new class of nanostructures has been explored which are derived from cell membranes (Thanuja et al. 2018) as shown in Fig. 3. They are particularly becoming more popular due to their innate-tissue targeting characteristics, adaptive sensing and host integration. Over the last decade, a number of cells and cell-derived membranes have been employed for actively targeted drug delivery ranging from erythrocytes, metastatic cancer cells, macrophages, mesenchymal stem cells (MSCs) and platelets and their potential have been comprehensively reviewed elsewhere (Thanuja et al. 2018). Hence, we will assess if they can be employed as drug carriers which can cross the BRB and target retinal tissues. In fact, there are evidences that MSCs penetrate the blood–brain barrier (BBB) via multiple mechanisms including inhibition of TIMP-3 (Menge et al. 2012), upregulation of aquaporin-4 (Tang et al. 2014), temporary opening of BBB using mannitol (Gonzales-Portillo et al. 2014) and probably under diseased conditions where the BBB is disrupted (Prakash and Carmichael 2015). So far, there are no studies to show that MSCs or other stem cells or cell membrane-derived nanovesicles can penetrate the BRB. Fortunately, the BBB and BRB are functionally similar to the extent that they are affected by similar stressors, and are apparently compromised years prior to the onset of their respective diseases (Ikram et al. 2012). For instance, BBB and BRB showed similar drug permeability characteristics due to the presence of common efflux transporters such as P-gp (Toda et al. 2011) and tight junctions (Greene and Campbell 2016). Hence, it will be exciting to investigate if MSCs or MSC membrane-derived nanovesicles or other relevant targeting cellular vehicles will be able to penetrate BRB and reach target cells. These cells and their membranes can also be engineered to facilitate BRB penetration by specifically upregulating receptors that can bind to influx transporters on the BRB components. Furthermore, novel tools must be developed for safer and controllable disruption of BRB to facilitate drug and drug-loaded nanocarrier penetration across the BRB. For instance, magnetic nanoparticles with RPE or RCEC targeting functionality could be employed to photothermally disrupt the BRB transiently. In summary, ingenuity in selection and manipulation of natural drug carriers to penetrate BRB and target retinal components will be the future, and it will be interesting to observe how they perform in preclinical stage to start with.

In the context of transient modulation of BRB permeability and gene therapy, gene delivery (RNA interference) should be employed extensively. Small interfering RNA (siRNA) can interfere with defective genes leading to post translational gene silencing strategies. Specifically, microRNA (miRNA), oligodeoxynucleotides, catalytically active small RNA molecules and siRNAs are potent sequence specific gene silencers (Thakur et al. 2012). However, the negative phosphate backbone of the genes is electrostatically repelled with negatively charged lipid bilayers of cells, ending up with limited penetration efficiency of siRNA. This is worsened by the physical properties of drugs including high molecular weight, shorter half-life and poor uptake in target tissue. Therefore, delivery of genes (RNAi) to the retina is quite a challenging task. Systemically administered genetic vectors have many hurdles to overcome including BRB and degradation by DNAses in the serum before reaching their target site. On the other hand, viral vectors carrying genes exert immune responses, get entrapped by Kupffer cells leading to liver toxicity. Furthermore, the delivered genes are encountered by lysosomes and degraded and the genes have to penetrate the nucleus to exert its effects. In order to address these issues, non-viral vectors (synthetic carriers), plasmids encoding for siRNA, and in the future cell membranes or exosomes can be used as vehicles to deliver genes for both transient modulation of BRB and for gene therapy.

Even though immune-related disorders are relatively rare in the posterior eye, some of the diseases including autoimmune uveitis and AMD stem from immune-compromise. Presence of inflammatory cells and persistent cytokines and other immune-regulatory proteins in the eye results in autoimmune uveitis (Egwuagu et al. 2015). Its management consists of downregulating the immune response to preserve ocular architecture and removal of the stimuli via topical or systemic corticosteroids. These drugs have serious adverse effects which necessitate the development of less toxic and targeted therapies. T cells are central to the development of uveitis, and hence, targeting lymphocyte signaling pathways represents promising interventions. Recent efforts are focused on blocking effector pathways or their costimulatory molecules at immune response checkpoints. For instance, targeting IL-2 cytokine and CD28/B7 molecular pathway, bypassing the antagonistic CTLA-4 signal transduction have been promising (Papotto et al. 2014). Another recent example is using DNA methylation inhibitor zebularine which was shown to target $CD4^+$ T cells and control intraocular inflammation (Zou et al. 2019). In case of AMD, targeting divergent microglial functions, modulation of cellular bioenergetics, inflammasome activation, or autophagy pathways may facilitate immunity and tissue health, and thus these approaches are considered as novel therapeutic paradigms (Copland et al.

2018). Thus, targeting immune checkpoints using their inhibitors is the future of immunotherapy for posterior segment of the eye, and the previously discussed delivery strategies would enhance the penetration of these inhibitors across the BRB, ensuring treatment efficacy.

4.3 Harnessing Innovative Tools for Non-invasive Diagnosis of the Retina and the BRB

Evidence shows that BRB breakdown is naturally induced due to trauma and various pathologies (Occhiutto et al. 2012). Currently, there are several approaches employed for the diagnosis of retinal diseases including optical coherence tomography (OCT), fundus photography, fluorescein angiography, positron emission tomography (PET), magnetic resonance imaging (MRI), ultrasonography and confocal microscopy. For instance, MRI is employed to monitor the progress of posterior eye diseases including diabetic retinopathy and AMD by in vivo imaging of neovasucularization (Yang et al. 2002; Townsend et al. 2008). Despite being used clinically, these tools have some inherent limitations such as poor image quality, sensitivity and resolution. Particularly, MRI has good spatial resolution but low sensitivity, whereas PET offers high sensitivity but limited spatial resolution (Finger et al. 2005; Kiyosawa et al. 1996). OCT in particular offers high-resolution images and is very effective at imaging retinal diseases including AMD, DR and glaucoma (Schmidt-Erfurth et al. 2017; de Barros Garcia et al. 2017; Hood 2017). Despite these promises, OCT has limited applications in detecting molecular changes because they are not measurable by employing tissue reflectivity only. Hence, various OCT adjuvants possessing very high contrast such as gold nanorods have been developed (Lapierre-Landry et al. 2017), but without the ability to target molecular changes. Gordon and coworkers have recently developed gold nanorods as OCT adjuvants that can target ICAM2 in the retina of laser-induced choroidal neovascularization (LCNV) mouse retina model in an effort to procure additional image contrast within the lesion (Gordon et al. 2019). Clinically relevant, anti-VEGF-induced changes were detected and imaged with high resolution by this photothermal OCT (PTOCT) imaging that are not visible using standard OCT systems. Gold nanoparticles (AuNPs) possess robust optical scattering and strong absorption owing to their surface plasmon resonance. This was exploited in another study, where choroidal and retinal microvasculature in a rabbit model were visualized with high resolution using photoacoustic microscopy (PAM) and OCT using PEG-conjugated AuNPs (Nguyen et al. 2019) as multimodal contrast agents. A unique class of naturally evolved gas-filled protein nanostructures which are also the first genetically encodable OCT contrast agents have been developed by Lu and coworkers (Lu et al. 2019). These nanovesicles possess differential refractive index relative to surrounding aqueous tissue due to the presence of gas. In addition, they can be detected at picomolar levels given their nanoscale motion properties via static and dynamic OCT. Upon reaching the target, the gas vesicles can be selectively burst using ultrasound allowing erasing of OCT contrast. Using these unique traits, the authors have demonstrated the use of gas nanovesicles as purified contrast agents in vivo in mouse retina. We predict that in the future, more such targeted and high-contrast adjuvants to OCT will be developed which would facilitate faster and accurate clinical diagnosis of retinal diseases.

In addition to these qualitative diagnostic tools, there is a critical need for tools which give quantitative information, particularly about the permeability of the BRB as depicted in Fig. 3. In this direction, ocular fluorometers employing fluorescence spectroscopy are highly sensitive for the measurement of permeability across the cornea. However, they can only perform steady state fluorescence measurement at an axial (depth) resolution of ~500 μm (Masters 1990; de Kruijf et al. 1987). An improvement in the depth resolution to about 7 μm was demonstrated Srinivas and Maurice using a confocal scanning microfluorometer (CSMF) to assess the transcorneal fluorescence (Srinivas and Maurice 1992). This technique was implemented for in vivo measurement of physiological parameters such as pO_2 and electrolyte levels (Na^+ and Cl^-) using fluorescence lifetime of fluorophores (Wang et al. 2009). Recently, the conventional ocular fluorometer has been modified to perform fluorescence lifetime spectroscopy and termed as ocular spot fluorometer in frequency domain (FD) approach for depth resolution measurement inside the eye (Sridhar et al. 2019). The ocular spot fluorometer has also been customized for the measurement of intensity of light scattering resulting in aqueous flare from the anterior chamber which is indicative of the breakdown of blood–aqueous barrier and release of serum proteins in uveitis (Sudhir et al. 2018). It provided improved precision, sensitivity and dynamics of measurement of aqueous flare. On similar lines, fluorometers can be employed with suitable modifications to measure permeability of the BRB in conjunction with fluorescein angiography and OCT angiography (Taha et al. 2018; Spaide et al. 2015) and probably other targeted fluorophores to quantify changes in BRB permeability before and after therapeutic interventions.

Finally, artificial intelligence (AI) has the potential to rapidly and comprehensively analyze digital images comprising millions of morphological datasets. In the context of retina, the datasets include fundus photographs, optical coherence tomography and visual fields. Machine learning (ML) and deep learning (DL) are able to identify, localize and quantify pathological features pertaining to most retinal diseases including DR, glaucoma, DME and AMD (Ting et al. 2019). For example, fully automated AI-based systems have recently been approved for screening of DR. These methods provide a whole gamut of features ranging from target identification for novel therapies, drug screening, disease grading, therapy guidance, recurrences and quantification of therapeutic effects (Schmidt-Erfurth et al. 2017). Some examples of automated screening based on retinal images are from a large multiethnic cohorts of patients with diabetes (Ting et al. 2017), DR screening (Tufail et al. 2017) and a convolutional neural network (CNN) applied to OCT for screening advanced AMD or DME (Kermany et al. 2018). In addition, the screening, diagnosis and monitoring of retinal diseases in primary care and community settings may be performed using DL in conjunction with telemedicine. Despite these revolutionary features, AI and DL have to overcome technical challenges, interpretation of algorithmic data and clinical legal issues. Nonetheless, the future of AI and DL for retinal diseases holds great promise especially because it can provide personalized healthcare and large-scale management of very complex and new generation of retinal diseases.

4.4 Bench to Bedside: Practical Considerations for Harnessing Systemic Drug Delivery to Posterior Eye in Clinical Settings

It is evident from historical and ongoing clinical trials that systemic administration of drugs for the treatment of posterior eye diseases constitute to only about 5–10% of the total interventions. This is despite the fact that systemic route is simpler and safer in comparison to intravitreal and other local routes. The obvious reason for this lack of clinical translation is the presence of BRB which restricts the entry of the therapeutic entities to cross the BRB and reach retinal targets. To overcome this, if the systemic dose is increased, there are increased risks of toxicity and side effects.

The most common posterior eye diseases including posterior uveitis, DR, PDR and DME need frequent intravitreal injections or implants of corticosteroids and/or anti-VEGF antibody and are the standard of care. However, chronic intravitreal injections pose the risks of infections and retinal detachment (Falavarjani and Nguyen 2013). Hence, the development of systemic therapies (oral or intravenous) is the critical need of the hour. The fact that the BRB is compromised in most of these diseases must be exploited in delivering drugs to the retinal tissue across BRB. Table 5 lists out a few clinical trials employing systemic drug administration (oral or intravenous) for the treatment of important diseases of the posterior segment of the eye. It is however noteworthy that most of these interventions are either adjuvant or for treating symptoms secondary to the primary pathology. A very obvious lack of PK/PD studies of the drugs as outcome measures upon administration can also be observed.

The development of prodrugs and nanocarriers with improved BRB-crossing capabilities has only realized limited clinical success. For example, some of the transporter prodrugs in clinical use include dipivefrin (oral) for open-angle glaucoma (Egorov et al. 2000), prednisone (oral) for uveitis (Babu and Mahendradas 2013), pegaptanib for neovascular age-related macular degeneration and valganciclovir (oral) for cytomegalovirus (CMV) retinitis (Patil et al. 2010; Kabata et al. 2012). The pegaptanib clinical trial (phase III) revealed that the prodrug was effective against AMD in early stages or with other drugs, and not so much against late state AMD with mature blood vessels (Vinores 2006). Despite these sporadic cases of positive clinical outcomes, there is an urgent need for more smart prodrugs and nanocarriers with BRB permeable properties and sustained and targeted drug delivery capabilities. Moreover, these therapeutic entities must be comprehensively evaluated for their PK/PD characteristics as primary outcome measures in clinical trials. So far, the characterization of BRB permeability has been limited to only

Table 5 Clinical trials involving systemic administration of therapeutics for the treatment of various posterior eye diseases

Clinical Trial ID	Phase	Condition	No. of patients	Intervention	Outcome measures	Basis of the clinical trial design
NCT03702374	III	DR	180	Combined antioxidant therapy (i.v.)	Intraretinal microvascular abnormalities	Reduction of oxidative stress, mitochondrial dysfunction and/or grade of retinopathy via antioxidants
NCT04140201	IV	DR	80	Simvastatin, Fenofibrate, Omega 3 fatty acid	Reduction of macular edema	Lipid-lowering agents inhibit progression of DR and restore visual acuity
NCT00917553	II	DR	33	Doxycycline	Change in macular volume, progression to PDR, and visual acuity	Reduced systemic inflammation, inhibition of retinal microglial activation leading to DR regression
NCT03962296	IV	DR	153	Entelon® (Vitis vinifera extract), Doxium® (calcium dobesilate)	Change in total macular volume, visual acuity	Binding of the drug to blood vessel wall leading to reduced capillary permeability and edema formation
NCT03238963	II	DR	100	BI 1467335	Ocular adverse events, proof of mechanism, safety	Inhibition of amine oxidase, copper containing 3 (AOC3)
NCT00604383	III	DR	685	Ruboxistaurin	Sustained moderate visual loss, % of participants experiencing DME and PDR	Inhibition of activation of PKC, increased neovascularization and retinal vascular permeability
NCT03161652	II	DR	120	Levosulpiride	Visual acuity, retinal thickness, prolactin serum and vitreous levels	Elevation of prolactin in the serum (which has a protective role) by blocking dopamine D2 receptor

(continued)

Table 5 (continued)

Clinical Trial ID	Phase	Condition	No. of patients	Intervention	Outcome measures	Basis of the clinical trial design
NCT02753400	II	PDR	24	Emixustat hydrochloride	Change in aqueous humor concentration of IL-6, IL-8, IP-10, PDGF-AA, TGFβ-1, etc.	Modulation of visual cycle activity and reduction in retinal O_2 demand and slow down of PDR progression
NCT01921192	IV	PDR	160	Folic Acid, vitamins B6, B12	Homocysteine levels in serum	Lowering of homocysteine levels inhibits progression of PDR
NCT00090519	III	DME	731	Ruboxistaurin	Macula-involved diabetic macular edema, sustained moderate visual loss	Selective PKC β inhibition slows down DME and may also indicate a neuroprotective effect of the drug
NCT01506895	II	DME	54	Darapladib	Visual acuity, changes in retinal anatomy, PK/PD parameters of the drug	Lipoprotein-associated phospholipase A2 inhibitory mechanism for the treatment of DME
NCT00505947	III	DME	12	Infliximab	Visual acuity, anatomical improvement of DME	Inhibition of TNF-α
NCT01821677	III	DME	355	Danazol	Visual acuity	Actin and capillary permeability modulation
NCT01782989	II/III	GA-AMD	286	ORACEA®	Rate of enlargement in area of geographic atrophy & visual acuity	Anti-inflammatory effects
NCT00138632	I/II	CNV/AMD	50	PTK787	Visual acuity, changes in macular edema	Anti-angiogeneis via VEGFR2 tyrosine kinase inhibition
NCT00304954	II	AMD	13	Daclizumab (i.v.) Infliximab (i.v.)	Visual acuity, changes in retinal thickness	Immunomodulation leading to anti-angiogenic effect

(continued)

Table 5 (continued)

Clinical Trial ID	Phase	Condition	No. of patients	Intervention	Outcome measures	Basis of the clinical trial design
NCT03744767	II	AMD	20	Spironolactone	Visual acuity, changes in retinal thickness, RPE detachment	Inhibition of the MR pathway or endothelial-specific deletion of MR
NCT03558061	II	AMD	30	ALK4290	Visual acuity, retinal thickness, RPE detachment, drug plasma concentration	C-C chemokine receptor type 3 (CCR3) inhibition
NCT00674323	IV	SMPCV-AMD	61	Verteporfin photodynamic therapy	Complete regression of polyps, visual acuity, retinal thickness	Photosensitization and inhibition of abnormal blood vessels
NCT02870907	II	RB	185	Etoposide, vincristine, carboplatin, cyclophosphamide	Rate of extra ocular relapses, acute toxicities	Combined chemotherapy and radiotherapy
NCT00335738	III	RB	331	Liposomal vincristine sulfate (i.v.)	Event-free survival, toxicity	Adjuvant chemotherapy

preclinical models. Hence, there is an urgent need for the development of non-invasive techniques to characterize BRB permeability and also assess drug targeting at the molecular level in human patients. To this end, single-molecule sensing could be effectively harnessed, along with the development of biosensors with very high selectivity, sensitivity and resolution and non-invasive instrumentation. For example, plasmon-enhanced detection which allows direct optical detection of weakly emitting and completely non-fluorescent species (Taylor and Zijlstra 2017) is the future of non-invasive biomedical diagnostics. Generally, promising preclinical studies end up being unsuccessful in clinical studies due to many heterogeneities and complex reasons. This necessitates a very thorough investigative comparison of preclinical vs clinical studies on a case-by-case basis to identify the gaps as indicated in Fig. 3.

Beyond the conventional regulatory approvals (based on CGMP manufacturing practices) needed for the newer class of therapeutic entities, the scale-up issues with respect to the nanoformulations need to be addressed. Significant changes in mass and momentum transfer rates during scale-up lead to variations in the nanoparticle assembly process (McNeil 2009). Colombo et al. have reported that increasing impeller speeds and time of agitation reduced the nanoparticle size (Colombo et al. 2001). In another study, the drug loading of the nanoparticles was reduced due to scale-up (Galindo-Rodríguez et al. 2005). Scaling-up the synthesis of nanocarrier's constitutive materials poses another critical challenge despite the existence of CGMP guidelines because of the heterogenous and complex nature of the nanomaterials. Characterization of individual pharmaceutical entities along with a variety of carriers including polymers, dendrimers, micelles, liposomes and cell membranes is another cumbersome and complicated process. Hence, well-defined and robust quality control systems to facilitate batch-to-batch consistency and biological equivalence must be established. In this direction, a good practice represents identification of critical parameters affecting the reproducibility of nanomaterial properties and monitoring them early on to minimize reproducibility issues later on (Crist et al. 2012).

A few successful examples of scale-up of nanoformulations are discussed next. One of the first successful scale-ups of a theranostic nanoemulsion has been reported by Liu and coworkers (Liu et al. 2015), where they demonstrate that a 1 L capacity microfluidizer could produce Celecoxib-loaded near-IR-labeled perfluorocarbon nanoemulsions. They also report that as the scale increased, the number of passes decreased to prevent heat build-up in the product. To accommodate for the reduction in number of passes, they increased the temperature by about 15 °C, but did not observe any discernable changes in the properties (size, zeta potential, pH, stability, functionality) of the nanoemulsions. In a recent study, Dormont and coworkers reported that scale-up of bromohydrine synthesis to squalene adenosine changed the impurity profile of the nanoparticles and the impurity was highly cytotoxic (Dormont et al. 2019). In addition, these impure nanoparticles could not be separated using conventional chromatographic techniques and caused significant variations to the colloidal properties, internal structure and biological properties. Nevertheless, these issues were successfully sorted out by their industrial partner and later obtained highly reproducible nanoparticles on a large scale. A novel flash nanoprecipitation (FNP) process was developed by Feng and coworkers as a highly scalable and continuous nanoparticle production process (Feng et al. 2019). FNP offers a continuous, stabilizer-directed rapid precipitation process, where in an antimalaria drug, Lumefantrine (LMN) was formulated into 200 nm nanoparticles with enhanced bioavailability and dissolution kinetics. In addition, a large-scale multi-inlet vortex mixer was employed for the formulation process with a production rate of around 1 kg/day and highly reproducible size and polydispersity. Finally, a scalable continuous spray drying process was also employed to obtain dried nanoparticle powder to ensure long-term storage and stability. In summary, the poor translational capacity of nanomedicines is implicated to (i) scale-up issues, (ii) poorly defined regulatory guidelines for comprehensive characterization, and (iii) inaccurate preclinical to clinical correlation due to lack of realistic animal models and alternative approaches. Hence, these issues need urgent attention and solutions.

5 Conclusions

Relatively unknown but, underlying pathophysiology of ocular diseases of the posterior segment of the eye is a major drawback in the development of systemic therapies. A thorough understanding of anatomy and physiology of posterior eye is prerequisite to this cause. Targeting posterior segment diseases in retina/choroid is difficult due to the presence of BRB, and the level of the drug reaching this target is significantly low. Evidence shows the presence of dynamic and static barriers comprising influx/efflux transporters and junctions, respectively, in the cells of the retina. There is a complex interplay between different components of these barriers in controlling transport of various biomolecules across the BRB. So far, studies on the role of these different barrier components have helped researchers develop BRB overcoming drugs and drug delivery carriers including prodrugs and target functionalized nanoparticles. Here, the principles of immobilization of drugs into various nanostructures play an important role. Transient modulation of BRB using gene delivery using vector immobilization strategies and physical forces also is in vogue.

Based on these, we think that the success of systemic therapies for the treatment of various posterior eye diseases relies firstly on the development and use of realistic preclinic models to simulate BRB and the underlying pathology. They include generation of retinal organoids using human iPSCs, CRISPR/Cas9-based gene editing-based animal models, 3D bioprinted and multi organ-on-a-chip-based preclinical models and fortified by systems biology and in silico drug screening tools. Secondly, we need to develop and test highly efficient BRB overcoming and targeted drug delivery tools. This includes cell- and cell membrane-based nanovesicles, and exosomes with natural BRB-crossing and targeting capabilities, innovative guided transport of drugs/drug-loaded nanocarriers and targeted immunotherapies in addition to safe and efficient BRB modulating modalities. Here, the role of immobilization of drugs into nanostructures plays a key role, and we need to come up with highly scalable nanoformulations and immobilization processes. Thirdly, the understanding of BRB characteristics in health and disease in addition to the pathophysiology of various posterior eye disorders is possible with highly precise and state-of-the-art non-invasive diagnostic tools. To this end, AI and deep learning tools can be integrated with modified OCT and nanocontrast agents for qualitative assessment and fluorescence spectroscopy-based quantitative assessment of the retinal tissue is the future. Finally, few practical considerations for clinical trials involving systemic therapies for posterior eye diseases include the need for non-invasive characterization of BRB and retinal targeting of drugs and nanomedicine in patients during the trial, well-defined PK/PD of drugs as outcome measures, comparative analysis of preclinical vs clinical data and efficient scale-up of nanoformulations and prodrug manufacturing processes to meet the clinical need.

Acknowledgements This work is supported by the Science & Engineering Research Board (SERB) research grant to Dr. Sudhir Ranganath from the Department of Science & Technology (DST), Government of India, (EMR/2017/002449 dated 04-09-2018) and by the intramural research grant from Sree Siddaganga Education Society, Tumkur (3150/1516). Ms. Thanuja M Y acknowledges Ph.D. Research Assistantship from SSES, Tumakuru. The authors acknowledge Mr. Harish Kumar N., for his artwork contribution.

Publication Statement This submission has not been published anywhere previously and is not simultaneously being considered for any other publication.

Author Disclosure Statement Dr. Sudhir H. Ranganath (SHR) is a director of Tvastra InnoTech Solutions Private Limited, a company that has an option to license IP generated by SHR. SHR may benefit financially if the IP is licensed and further validated. SHR's interests were reviewed and are subject to a management plan overseen by Siddaganga Institute of Technology, Tumakuru, in accordance with their conflict of interest policies. None of the other authors have any proprietary interests or conflicts of interest related to this submission.

References

Aasen DM, Vergara MN (2019) New drug discovery paradigms for retinal diseases: a focus on retinal organoids. J Ocul Pharmacol Ther 36(1):18–24

Ablonczy Z, Dahrouj M, Tang PH, Liu Y, Sambamurti K, Marmorstein AD, Crosson CE (2011) Human retinal pigment epithelium cells as functional models for the RPE in vivo. Invest Ophthalmol Vis Sci 52(12):8614–8620

Achberger K, Probst C, Haderspeck J, Bolz S, Rogal J, Chuchuy J, Nikolova M, Cora V, Antkowiak L, Haq W, Shen N, Schenke-Layland K, Ueffing M, Liebau S et al (2019) Merging organoid and organ-on-a-chip technology to generate complex multi-layer tissue models in a human retina-on-a-chip platform. Elife 8:e46188

Adachi T, Yasuda H, Nakamura S, Kamiya T, Hara H, Hara H et al (2011) Endoplasmic reticulum stress induces retinal endothelial permeability of extracellular-superoxide dismutase. Free Radical Res 45(9):1083–1092

Adelli GR, Bhagav P, Taskar P, Hingorani T, Pettaway S, Gul W, ElSohly MA, Repka MA et al (2017) Development of a delta9-tetrahydrocannabinol amino acid-dicarboxylate prodrug with improved ocular bioavailability. Invest Ophthalmol Vis Sci 58(4):2167–2179

Akanuma SI, Yamakoshi A, Sugouchi T, Kubo Y, Hartz AMS, Bauer B, Ki H (2018) Role of L-type amino acid transporter 1 at the inner blood-retinal barrier in the blood-to-retina transport of gabapentin. Mol Pharm 15(6):2327–2337

Almeida JP, Chen A, Foster A, Drezek R (2011) vivo biodistribution of nanoparticles. Nanomedicine 6(5):815–835

Anand BS, Hill JM, Dey S, Maruyama K, Bhattacharjee PS, Myles ME, Nashed YE et al (2003) In vivo antiviral efficacy of a dipeptide acyclovir prodrug, val-val-acyclovir, against HSV-1 epithelial and stromal keratitis in the rabbit eye model. Invest Ophthalmol Vis Sci 44(6):2529–2534

Antonetti DA, Barber AJ, Khin S, Lieth E, Tarbell JM, Gardner TW (1998) Vascular permeability in experimental diabetes is associated with reduced endothelial occludin content: vascular endothelial growth factor decreases occludin in retinal endothelial cells. Penn State Retina Research Group. Diabetes 47(12):1953–1959

Apollonio F, Liberti M, Marracino P, Mir L (2012) Electroporation mechanism: Review of molecular models based on computer simulation. In: Proceedings of 6th European conference on antennas and propagation, EUCAP 2012, pp 356–358

Arno G, Agrawal SA, Eblimit A, Bellingham J, Xu M, Wang F, Chakarova C, Parfitt DA, Lane A, Burgoyne T, Hull S, Carss K, Fiorentino A, Hayes MJ, Munro PM, Nicols R, Pontikos N, Holder GE, Ukirdc Asomugha C, Raymond FL, Moore AT, Plagnol V, Michaelides M, Hardcastle AJ, Li Y, Cukras C, Webster AR et al (2016) Mutations in REEP6 cause autosomal-recessive retinitis pigmentosa. Am J Hum Genet 99(6):1305–1315

Ashton N, Cunha-Vaz JG (1965) Effect of histamine on the permeability of the ocular vessels. Arch Ophthalmol 73(2):211–223

Aukunuru JV, Sunkara G, Bandi N, Thoreson WB, Kompella UB (2001) Expression of multidrug resistance-associated protein (MRP) in human retinal pigment epithelial cells and its interaction with BAPSG, a novel aldose reductase inhibitor. Pharm Res 18(5):565–572

Babu K, Mahendradas P (2013) Medical management of uveitis–current trends. Indian J Ophthalmol 61(6):277

Babu E, Ananth S, Veeranan-Karmegam R, Coothankandaswamy V, Smith SB, Boettger T, Ganapathy V, Martin PM (2011) Transport via SLC5A8 (SMCT1) is obligatory for 2-oxothiazolidine-4-carboxylate to enhance glutathione production in retinal pigment epithelial cells. Invest Ophthalmol Vis Sci 52(8):5749–5757

Bai L, Shao H, Wang H, Zhang Z, Su C, Dong L, Yu B, Chen X, Li X, Zhang X (2017) Effects of mesenchymal stem cell-derived exosomes on experimental autoimmune uveitis. Sci Rep 7(1):4323–4323

Barot M, Bagui M, Gokulgandhi R, Mitra A (2012) Prodrug strategies in ocular drug delivery. Med Chem 8(4):753–768

Barza M, Kane A, Baum J (1983) Pharmacokinetics of intravitreal carbenicillin, cefazolin, and gentamicin in rhesus monkeys. Invest Ophthalmol Vis Sci 24(12):1602–1606

Bauer M, Karch R, Tournier N, Cisternino S, Wadsak W, Hacker M, Marhofer P, Zeitlinger M, Langer O (2017) Assessment of P-glycoprotein transport activity at the human blood-retina barrier with (R)-11C-verapamil PET. J Nucl Med 58(4):678–681

Bazzoni G, Martínez-Estrada OM, Orsenigo F, Cordenonsi M, Citi S, Dejana E (2000) Interaction of junctional adhesion molecule with the tight junction components ZO-1, Cingulin, and Occludin. J Biol Chem 275(27):20520–20526

Behzadian MA, Wang X-L, Windsor LJ, Ghaly N, Caldwell RB (2001) TGF-β increases retinal endothelial cell permeability by increasing MMP-9: Possible role of glial cells in endothelial barrier function. Invest Ophthalmol Vis Sci 42(3):853–859

Boddu SH, Nesamony J (2013) Utility of transporter/receptor (s) in drug delivery to the eye. World J Pharmacol 2(1):1–17

Boddu SH, Jwala J, Chowdhury MR, Mitra AK (2010) In vitro evaluation of a targeted and sustained release system for retinoblastoma cells using Doxorubicin as a model drug. J Ocul Pharmacol Ther 26(5):459–468

Booij JC, Baas DC, Beisekeeva J, Gorgels TG, Bergen AA (2010) The dynamic nature of Bruch's membrane. Progr Retinal Eye Res 29(1):1–18

Burgess A, Azad T, Pathak N, Amin V, Gupta SV (2018) Transporter-targeted prodrug approach for retina and posterior segment disease. In: Drug delivery for the retina and posterior segment disease. Springer, pp 309–316

Burrows CK, Banovich NE, Pavlovic BJ, Patterson K, Gallego Romero I, Pritchard JK, Gilad Y (2016) Genetic variation, not cell type of origin, underlies the majority of identifiable regulatory differences in iPSCs. PLoS Genet 12(1):e1005793–e1005793

Campbell M, Humphries P (2013) The blood-retina barrier. In: Biology and regulation of blood-tissue barriers. Springer, Berlin, pp 70–84

Cankova Z, Huang J-D, Kruth HS, Johnson M (2011) Passage of low-density lipoproteins through Bruch's membrane and choroid. Exp Eye Res 93(6):947–955

Capowski EE, Samimi K, Mayerl SJ, Phillips MJ, Pinilla I, Howden SE, Saha J, Jansen AD, Edwards KL, Jager LD, Barlow K, Valiauga R, Erlichman Z, Hagstrom A, Sinha D, Sluch VM, Chamling X, Zack DJ, Skala MC, Gamm DM (2019) Reproducibility and staging of 3D human retinal organoids across multiple pluripotent stem cell lines. Development 146 (1):dev171686

Chien DS, Schoenwald RD (1986) Improving the ocular absorption of phenylephrine. Biopharm Drug Dispos 7(5):453–462

Chizmadzhev YA, Indenbom AV, Kuzmin PI, Galichenko SV, Weaver JC, Potts RO (1998) Electrical properties of skin at moderate voltages: contribution of appendageal macropores. Biophys J 74(2 Pt 1):843–856

Cholkar K, Trinh HM, Vadlapudi AD, Mitra AK (2014) Synthesis and characterization of ganciclovir long chain lipid prodrugs. Adv Ophthalmol Vis Syst 1(2):00007

Chothe PP, Thakkar SV, Gnana-Prakasam JP, Ananth S, Hinton DR, Kannan R, Smith SB, Martin PM, Ganapathy V (2010) Identification of a novel sodium-coupled oligopeptide transporter (SOPT2) in mouse and human retinal pigment epithelial cells. Invest Ophthalmol Vis Sci 51(1):413–420

Colombo A, Briançon S, Lieto J, Fessi H (2001) Project, design, and use of a pilot plant for nanocapsule production. Drug Dev Ind Pharm 27(10):1063–1072

Colombo M, Raposo G, Théry C (2014) Biogenesis, secretion, and intercellular interactions of exosomes and other extracellular vesicles. Annu Rev Cell Dev Biol 30:255–289

Constable PA, Lawrenson JG, Dolman DEM, Arden GB, Abbott NJ (2006) P-Glycoprotein expression in human retinal pigment epithelium cell lines. Exp Eye Res 83(1):24–30

Copland DA, Theodoropoulou S, Liu J, Dick AD (2018) A perspective of AMD through the eyes of immunology. Investig Ophthalmol Vis Sci 59(4):AMD83-AMD92

Crist RM, Grossman JH, Patri AK, Stern ST, Dobrovolskaia MA, Adiseshaiah PP, Clogston JD, McNeil SE (2012) Common pitfalls in nanotechnology: lessons learned from NCI's nanotechnology characterization laboratory. Integrative Biology 5(1):66–73

Cunha-Vaz J (1966) Studies on the permeability of the blood-retinal barrier. 3. Breakdown of the blood-retinal barrier by circulatory disturbances. British J Ophthalmol 50(9):505

Cunha-Vaz JG (1976) The blood-retinal barriers. Doc Ophthalmol 41(2):287–327

Cunha-Vaz J (2017) The blood-retinal barrier in the management of retinal disease: EURETINA award lecture. Ophthalmologica 237(1):1–10

Cunha-Vaz JG, Maurice DM (1967) The active transport of fluorescein by the retinal vessels and the retina. J Physiol 191(3):467–486

Cunha-Vaz JG, Shakib M, Ashton N (1966) Studies on the permeability of the blood-retinal barrier. I. On the existence, development, and site of a blood-retinal barrier. Br J Ophthalmol 50(8):441

De Barros Garcia JMB, Isaac DLC, Avila M (2017) Diabetic retinopathy and OCT angiography: clinical findings and future perspectives. Int J Retina Vitreous 3(1):14

De Kruijf EJFM, Boot JP, Laterveer L, van Best JA, Ramselaar JAM, Oosterhuis JA (1987) A simple method for determination of corneal epithelial permeability in humans. Curr Eye Res 6(11):1327–1334

Del Amo EM, Rimpelä A-K, Heikkinen E, Kari OK, Ramsay E, Lajunen T, Schmitt M, Pelkonen L, Bhattacharya M, Richardson D, Subrizi A, Turunen T, Reinisalo M, Itkonen J,

Toropainen E, Casteleijn M, Kidron H, Antopolsky M, Vellonen K-S, Ruponen M, Urtti A (2017) Pharmacokinetic aspects of retinal drug delivery. Progr Retinal Eye Res 57:134–185
Díaz-Coránguez M, Ramos C, Antonetti DA (2017) The inner blood-retinal barrier: cellular basis and development. Vision Res 139:123–137
DiStefano T, Chen HY, Panebianco C, Kaya KD, Brooks MJ, Gieser L, Morgan NY, Pohida T, Swaroop A (2018) Accelerated and improved differentiation of retinal organoids from pluripotent stem cells in rotating-wall vessel bioreactors. Stem Cell Reports 10(1):300–313
Dormont F, Rouquette M, Mahatsekake C, Gobeaux F, Peramo A, Brusini R, Calet S, Testard F, Lepetre-Mouelhi S, Desmaële D (2019) Translation of nanomedicines from lab to industrial scale synthesis: the case of squalene-adenosine nanoparticles. J Controlled Release 307:302–314
Drori S, Eytan GD, Assaraf YG (1995) Potentiation of anticancer-drug cytotoxicity by multidrug-resistance chemosensitizers involves alterationsin membrane fluidity leading to increased membrane permeability. Eur J Biochem 228(3):1020–1029
Du J, Sun Y, Li F-H, Du L-F, Duan Y-R (2017) Enhanced delivery of biodegradable mPEG-PLGA-PLL nanoparticles loading Cy3-labelled PDGF-BB siRNA by UTMD to rat retina. J Biosci 42(2):299–309
During A, Doraiswamy S, Harrison EH (2008) Xanthophylls are preferentially taken up compared with β-carotene by retinal cells via a SRBI-dependent mechanism. J Lipid Res 49(8):1715–1724
Edelhauser HF, Rowe-Rendleman CL, Robinson MR, Dawson DG, Chader GJ, Grossniklaus HE, Rittenhouse KD, Wilson CG, Weber DA, Kuppermann BD (2010) Ophthalmic drug delivery systems for the treatment of retinal diseases: basic research to clinical applications. Invest Ophthalmol Vis Sci 51(11):5403–5420
Egorov E, Alekseev V, Erichev V, Usachev V, Simonova S, Iakubova L (2000) Therapy of primary open-angle glaucoma by oftan-dipivefrin. Vestn oftalmol 116(1):32–33
Egwuagu CE, Sun L, Kim SH, Dambuza IM (2015) Ocular inflammatory diseases: molecular pathogenesis and immunotherapy. Curr Mol Med 15(6):517–528
Esser P, Tervooren D, Heimann K, Kociok N, Bartz-Schmidt KU, Walter P, Weller M (1998) Intravitreal daunomycin induces multidrug resistance in proliferative vitreoretinopathy. Invest Ophthalmol Vis Sci 39(1):164–170
Falavarjani KG, Nguyen Q (2013) Adverse events and complications associated with intravitreal injection of anti-VEGF agents: a review of literature. Eye 27(7):787
Feehan JM, Chiu CN, Stanar P, Tam BM, Ahmed SN, Moritz OL (2017) Modeling dominant and recessive forms of retinitis pigmentosa by editing three rhodopsin-encoding genes in xenopus Laevis using Crispr/Cas9. Sci Rep 7(1):6920
Feng J, Markwalter CE, Tian C, Armstrong M, Prud'homme RK (2019) Translational formulation of nanoparticle therapeutics from laboratory discovery to clinical scale. J Trans Med 17(1):200
Finger P, Kurli M, Reddy S, Tena L, Pavlick A (2005) Whole body PET/CT for initial staging of choroidal melanoma. Br J Ophthalmol 89(10):1270–1274
Friedlander M, Theesfeld CL, Sugita M, Fruttiger M, Thomas MA, Chang S, Cheresh DA (1996) Involvement of integrins alpha v beta 3 and alpha v beta 5 in ocular neovascular diseases. Proc Natl Acad Sci United USA 93(18):9764–9769
Fujii S, Setoguchi C, Kawazu K, Hosoya K-i (2015) Functional characterization of carrier-mediated transport of pravastatin across the blood-retinal barrier in rats. Drug Metab Dispos 43(12):1956–1959
Galindo-Rodríguez SA, Puel F, Briançon S, Allémann E, Doelker E, Fessi H (2005) Comparative scale-up of three methods for producing ibuprofen-loaded nanoparticles. Eur J Pharm Sci 25(4–5):357–367
Ganapathy V, Thangaraju M, Gopal E, Martin PM, Itagaki S, Miyauchi S, Prasad PD (2008) Sodium-coupled monocarboxylate transporters in normal tissues and in cancer. J Am Assoc Pharm Sci 10(1):193
Gao S, Kahremany S, Zhang J, Jastrzebska B, Querubin J, Petersen-Jones SM, Palczewski K (2018) Retinal-chitosan conjugates effectively deliver active chromophores to retinal photoreceptor cells in blind mice and dogs. Mol Pharmacol 93(5):438–452

Garcia TB, Oliveira KRM, do Nascimento JLM, Crespo-López ME, Picanço-Diniz DLW, Mota TC, Herculano AM (2011) Glutamate induces glutathione efflux mediated by glutamate/aspartate transporter in retinal cell cultures. Neurochem Res 36 (3):412–418

Gokulgandhi MR, Barot M, Bagui M, Pal D, Mitra AK (2012) Transporter-targeted lipid prodrugs of cyclic cidofovir: a potential approach for the treatment of cytomegalovirus retinitis. J Pharm Sci 101(9):3249–3263

Gonçalves A, Leal E, Paiva A, Teixeira Lemos E, Teixeira F, Ribeiro C, Reis F, Ambrósio A, Fernandes R (2012) Protective effects of the dipeptidyl peptidase IV inhibitor sitagliptin in the blood–retinal barrier in a type 2 diabetes animal model. Diabetes Obes Metab 14(5):454–463

Gonçalves A, Ambrósio AF, Fernandes R (2013) Regulation of claudins in blood-tissue barriers under physiological and pathological states. Tissue Barriers 1(3):e24782

Gonzales-Portillo GS, Sanberg PR, Franzblau M, Gonzales-Portillo C, Diamandis T, Staples M, Sanberg CD, Borlongan CV (2014) Mannitol-enhanced delivery of stem cells and their growth factors across the blood-brain barrier. Cell Transplant 23(4–5):531–539

Gopal E, Fei Y-J, Miyauchi S, Zhuang L, Prasad P, Ganapathy V (2005) Sodium-coupled and electrogenic transport of B-complex vitamin nicotinic acid by slc5a8, a member of the Na/glucose co-transporter gene family. Biochem J 388:309–316

Gopal E, Miyauchi S, Martin PM, Ananth S, Roon P, Smith SB, Ganapathy V (2007) Transport of nicotinate and structurally related compounds by human SMCT1 (SLC5A8) and its relevance to drug transport in the mammalian intestinal tract. Pharm Res 24(3):575–584

Gordon AY, Lapierre-Landry M, Skala MC, Penn JS (2019) Photothermal optical coherence tomography of anti-angiogenic treatment in the mouse retina using gold nanorods as contrast agents. Trans Vis Sci Technol 8(3):18–18

Greene C, Campbell M (2016) Tight junction modulation of the blood brain barrier: CNS delivery of small molecules. Tissue Barriers 4(1):e1138017–e1138017

Guo Z, Peng H, Kang J, Sun D (2016) Cell-penetrating peptides: possible transduction mechanisms and therapeutic applications. Biomed Rep 4(5):528–534

Gurunathan S, Kang M-H, Jeyaraj M, Qasim M, Kim J-H (2019) Review of the isolation, characterization, biological function, and multifarious therapeutic approaches of exosomes. Cells 8(4):307

Hajrasouliha AR, Jiang G, Lu Q, Lu H, Kaplan HJ, Zhang H-G, Shao H (2013) Exosomes from retinal astrocytes contain antiangiogenic components that inhibit laser-induced choroidal neovascularization. J Biol Chem 288(39):28058–28067

Hämäläinen KM, Kananen K, Auriola S, Kontturi K, Urtti A (1997) Characterization of paracellular and aqueous penetration routes in cornea, conjunctiva, and sclera. Invest Ophthalmol Vis Sci 38(3):627–634

Han Y-H, Sweet DH, Hu D-N, Pritchard JB (2001) Characterization of a novel cationic drug transporter in human retinal pigment epithelial cells. J Pharmacol Exp Ther 296(2):450–457

Handa JT, Bowes Rickman C, Dick AD, Gorin MB, Miller JW, Toth CA, Ueffing M, Zarbin M, Farrer LA (2019) A systems biology approach towards understanding and treating non-neovascular age-related macular degeneration. Nat Commun 10(1):3347

Hatanaka T, Nakanishi T, Huang W, Leibach FH, Prasad PD, Ganapathy V, Ganapathy ME (2001) Na^+ -and Cl^--coupled active transport of nitric oxide synthase inhibitors via amino acid transport system B0,+. J Clin Investig 107(8):1035–1043

Hatanaka T, Haramura M, Fei Y-J, Miyauchi S, Bridges CC, Ganapathy PS, Smith SB, Ganapathy V, Ganapathy ME (2004) Transport of amino acid-based prodrugs by the Na+ -and Cl–coupled amino acid transporter ATB0, + and expression of the transporter in tissues amenable for drug delivery. J Pharmacol Exp Ther 308(3):1138–1147

Hazlett LD, Meyer DB (1974) Ferritin uptake in the Japanese quail retina. Exp Eye Res 19(4):303–309

Hood DC (2017) Improving our understanding, and detection, of glaucomatous damage: an approach based upon optical coherence tomography (OCT). Progr Retinal Eye Res 57:46–75

Hosoya K-i, Tachikawa M (2009) Inner blood-retinal barrier transporters: role of retinal drug delivery. Pharm Res 26(9):2055–2065
Hosoya K-i, Tachikawa M (2012) The inner blood-retinal barrier: molecular structure and transport biology. Adv Exp Med Biol 763:85–104
Hosoya K-I, Tachikawa M (2013) The inner blood-retinal barrier. In: Biology and regulation of blood-tissue barriers. Springer, Berlin, pp 85–104
Hosoya K-i, Tomi M (2005) Advances in the cell biology of transport via the inner blood-retinal barrier: establishment of cell lines and transport functions. Biol Pharm Bull 28(1):1–8
Hosoya K-i, Kondo T, Tomi M, Takanaga H, Ohtsuki S, Terasaki T (2001) MCT1-mediated transport of L-lactic acid at the inner blood–retinal barrier: a possible route for delivery of monocarboxylic acid drugs to the retina. Pharm Res 18(12):1669–1676
Hosoya K-i, Makihara A, Tsujikawa Y, Yoneyama D, Mori S, Terasaki T, Akanuma S-i, Tomi M, Tachikawa M (2009) Roles of inner blood-retinal barrier organic anion transporter 3 in the vitreous/retina-to-blood efflux transport of aminohippuric acid, benzylpenicillin, and 6-mercaptopurine. J Pharmacol Exp Ther 329(1):87–93
Hosoya K-i, Yamamoto A, Akanuma S-i, Tachikawa M (2010) Lipophilicity and transporter influence on blood-retinal barrier permeability: a comparison with blood-brain barrier permeability. Pharm Res 27(12):2715–2724
Huang W, Prasad PD, Kekuda R, Leibach FH, Ganapathy V (1997) Characterization of N5-methyltetrahydrofolate uptake in cultured human retinal pigment epithelial cells. Invest Ophthalmol Vis Sci 38(8):1578–1587
Huang D, Chen Y-S, Rupenthal ID (2018) Overcoming ocular drug delivery barriers through the use of physical forces. Adv Drug Deliv Rev 126:96–112
Huankai H, Miyauchi S, Bridges CC, Smith SB, Ganapathy V (2003) Identification of a novel Na+- and Cl^--coupled transport system for endogenous opioid peptides in retinal pigment epithelium and induction of the transport system by HIV-1 Tat. Biochem J 375(1):17–22
Ikram MK, Cheung CY, Wong TY, Chen CPLH (2012) Retinal pathology as biomarker for cognitive impairment and Alzheimer's disease. J Neurol Neurosurg Psychiatry 83(9):917
Iwamura R, Tanaka M, Okanari E, Kirihara T, Odani-Kawabata N, Shams N, Yoneda K (2018) Identification of a selective, non-prostanoid EP2 receptor agonist for the treatment of glaucoma: omidenepag and its prodrug omidenepag isopropyl. J Med Chem 61(15):6869–6891
Janoria KG, Boddu SH, Wang Z, Paturi DK, Samanta S, Pal D, Mitra AK (2009) Vitreal pharmacokinetics of biotinylated ganciclovir: role of sodium-dependent multivitamin transporter expressed on retina. J Ocul Pharmacol Ther 25(1):39–49
Jespersen H, Andersen J, Ditzel H, Mouritsen O (2011) Lipids, curvature stress, and the action of lipid prodrugs: free fatty acids and lysolipid enhancement of drug transport across liposomal membranes. Biochimie 94:2–10
Jia L, Liu Z, Sun L, Miller SS, Ames BN, Cotman CW, Liu J (2007) Acrolein, a toxicant in cigarette smoke, causes oxidative damage and mitochondrial dysfunction in RPE cells: protection by (R)-α-lipoic acid. Invest Ophthalmol Vis Sci 48(1):339–348
Joussen AM, Murata T, Tsujikawa A, Kirchhof B, Bursell S-E, Adamis AP (2001) Leukocyte-mediated endothelial cell injury and death in the diabetic retina. Am J Pathol 158(1):147–152
Joussen AM, Poulaki V, Mitsiades N, Kirchhof B, Koizumi K, Dohmen S, Adamis AP (2002) Nonsteroidal anti-inflammatory drugs prevent early diabetic retinopathy via TNF-α suppression. Fed Am Soc Exp Biol 16(3):438–440
Kabata Y, Takahashi G, Tsuneoka H (2012) Cytomegalovirus retinitis treated with valganciclovir in Wegener's granulomatosis. Clin Ophthalmol (Auckland, NZ) 6:521
Kambhampati SP, Clunies-Ross AJM, Bhutto I, Mishra MK, Edwards M, McLeod DS, Kannan RM, Lutty G (2015) Systemic and intravitreal delivery of dendrimers to activated microglia/macrophage in ischemia/reperfusion mouse retina. Invest Ophthalmol Vis Sci 56(8):4413–4424

Kamizuru H, Kimura H, Yasukawa T, Tabata Y, Honda Y, Ogura Y (2001) Monoclonal antibody-mediated drug targeting to choroidal neovascularization in the rat. Invest Ophthalmol Vis Sci 42(11):2664–2672

Kanai Y, Hediger MA (1992) Primary structure and functional characterization of a high-affinity glutamate transporter. Nature 360(6403):467–471

Kannan R, Tang D, Hu J, Bok D (2001) Glutathione transport in human retinal pigment epithelial (HRPE) cells: apical localization of sodium-dependent GSH transport. Exp Eye Res 72(6):661–666

Katragadda S, Jain R, Kwatra D, Hariharan S, Mitra AK (2008) Pharmacokinetics of amino acid ester prodrugs of acyclovir after oral administration: interaction with the transporters on Caco-2 cells. Int J Pharm 362(1–2):93–101

Kennedy BG, Mangini NJ (2002) P-glycoprotein expression in human retinal pigment epithelium. Mol Vision 8:422–430

Kermany DS, Goldbaum M, Cai W, Valentim CC, Liang H, Baxter SL, McKeown A, Yang G, Wu X, Yan F (2018) Identifying medical diagnoses and treatable diseases by image-based deep learning. Cell 172(5):1122–1131. e1129

Kiyosawa M, Inoue C, Kawasaki T, Tokoro T, Ishii K, Ohyama M, Senda M, Soma Y (1996) Functional neuroanatomy of visual object naming: a PET study. Graefe's Archive Clin Exp Ophthalmol 234(2):110–115

Klaassen I, Van Noorden CJF, Schlingemann RO (2013) Molecular basis of the inner blood-retinal barrier and its breakdown in diabetic macular edema and other pathological conditions. Progr Retinal Eye Res 34:19–48

Knickelbein JE, Liu B, Arakelyan A, Zicari S, Hannes S, Chen P, Li Z, Grivel J-C, Chaigne-Delalande B, Sen HN, Margolis L, Nussenblatt RB (2016) Modulation of immune responses by extracellular vesicles from retinal pigment epithelium. Invest Ophthalmol Vis Sci 57(10):4101–4107

Koepsell H, Lips K, Volk C (2007) Polyspecific organic cation transporters: structure, function, physiological roles, and biopharmaceutical implications. Pharm Res 24(7):1227–1251

Kubo Y, Hosoya K-i (2012) Inner blood-retinal barrier transporters: relevance to diabetic retinopathy. Diabetic Retinopathy Intech Open: 91–108

Kubo Y, Shimizu Y, Kusagawa Y, Akanuma S-I, Hosoya K-I (2013) Propranolol transport across the inner blood–retinal barrier: potential involvement of a novel organic cation transporter. J Pharm Sci 102(9):3332–3342

Kubo Y, Yahata S, Miki S, Akanuma S-i, Hosoya K-i (2017) Blood-to-retina transport of riboflavin via RFVTs at the inner blood-retinal barrier. Drug Metab Pharmacokinet 32(1):92–99

Lapierre-Landry M, Gordon AY, Penn JS, Skala MC (2017) In vivo photothermal optical coherence tomography of endogenous and exogenous contrast agents in the eye. Sci Rep 7(1):9228–9228

Leal EC, Manivannan A, Hosoya K-I, Terasaki T, Cunha-Vaz J, Ambrósio AF, Forrester JV (2007) Inducible nitric oxide synthase isoform is a key mediator of leukostasis and blood-retinal barrier breakdown in diabetic retinopathy. Invest Ophthalmol Vis Sci 48(11):5257–5265

Lee J, Pelis RM (2016) Drug transport by the blood-aqueous humor barrier of the eye. Drug Metab Dispos 116:069369

Lee Y, Hussain AA, Seok J-H, Kim S-H, Marshall J (2015) Modulating the transport characteristics of bruch's membrane with steroidal glycosides and its relevance to age-related macular degeneration (AMD). Invest Ophthalmol Vis Sci 56(13):8403–8418

Liu L, Liu X (2019) Roles of drug transporters in blood-retinal barrier. In: Drug transporters in drug disposition, effects and toxicity. Springer, Berlin, pp 467–504

Liu L, Bagia C, Janjic JM (2015) The first scale-up production of theranostic nanoemulsions. BioResearch 4(1):218–228

Llonch S, Carido M, Ader M (2018) Organoid technology for retinal repair. Dev Biol 433(2):132–143

Lorber B, Hsiao W-K, Martin KR (2016) Three-dimensional printing of the retina. Curr Opin Ophthalmol 27(3):262–267

Lu GJ, Chou L-d, Malounda D, Patel AK, Welsbie DS, Chao DL, Ramalingam T, Shapiro MG (2019) Biomolecular contrast agents for optical coherence tomography. bioRxiv:595157

Lynn SA, Ward G, Keeling E, Scott JA, Cree AJ, Johnston DA, Page A, Cuan-Urquizo E, Bhaskar A, Grossel MC, Tumbarello DA, Newman TA, Lotery AJ, Ratnayaka JA (2017) Ex-vivo models of the Retinal Pigment Epithelium (RPE) in long-term culture faithfully recapitulate key structural and physiological features of native RPE. Tissue Cell 49(4):447–460

Mainolfi N, Powers J, Amin J, Long D, Lee W, McLaughlin ME, Jaffee B, Brain C, Elliott J, Sivak JM (2013) An effective prodrug strategy to selectively enhance ocular exposure of a cannabinoid receptor (CB1/2) agonist. J Med Chem 56(13):5464–5472

Majumdar S, Kansara V, Mitra AK (2006) Vitreal pharmacokinetics of dipeptide monoester prodrugs of ganciclovir. J Ocul Pharmacol Ther 22(4):231–241

Majumdar S, Hingorani T, Srirangam R (2010) Evaluation of active and passive transport processes in corneas extracted from preserved rabbit eyes. J Pharm Sci 99(4):1921–1930

Mandal A, Cholkar K, Khurana V, Shah A, Agrahari V, Bisht R, Pal D, Mitra AK (2017) Topical formulation of self-assembled antiviral prodrug nanomicelles for targeted retinal delivery. Mol Pharm 14(6):2056–2069

Mannermaa E, Vellonen K-S, Ryhänen T, Kokkonen K, Ranta V-P, Kaarniranta K, Urtti A (2009) Efflux protein expression in human retinal pigment epithelium cell lines. Pharm Res 26(7):1785–1791

Maschmeyer I, Lorenz AK, Schimek K, Hasenberg T, Ramme AP, Hübner J, Lindner M, Drewell C, Bauer S, Thomas A, Sambo NS, Sonntag F, Lauster R, Marx U (2015) A four-organ-chip for interconnected long-term co-culture of human intestine, liver, skin and kidney equivalents. Lab Chip 15(12):2688–2699

Masters BR (1990) Noninvasive diagnostic techniques in ophthalmology. Springer 1(6):254–265

McNeil SE (2009) Nanoparticle therapeutics: a personal perspective. Wiley Interdiscip Rev Nanomed Nanobiotechnol 1(3):264–271

Medina C, Santos-Martinez MJ, Radomski A, Corrigan OI, Radomski MW (2007) Nanoparticles: pharmacological and toxicological significance. Br J Pharmacol 150(5):552–558

Mehta T, Patel V, Sharma OP (2018) Receptor-targeted prodrug approach for retina and posterior segment disease. In: Drug delivery for the retina and posterior segment disease. Springer, Berlin, pp 363–382

Menge T, Zhao Y, Zhao J, Wataha K, Gerber M, Zhang J, Letourneau P, Redell J, Shen L, Wang J, Peng Z, Xue H, Kozar R, Cox CS, Jr., Khakoo AY, Holcomb JB, Dash PK, Pati S (2012) Mesenchymal stem cells regulate blood-brain barrier integrity through TIMP3 release after traumatic brain injury. Sci Trans Med 4(161):161ra150

Miranda CC, Fernandes TG, Diogo MM, Cabral JMS (2018) Towards multi-organoid systems for drug screening applications. Bioengineering (Basel) 5(3):49

Miyauchi S, Gopal E, Fei Y-J, Ganapathy V (2004) Functional identification of SLC5A8, a tumor suppressor down-regulated in colon cancer, as a Na^{+}-coupled transporter for short-chain fatty acids. J Biol Chem 279(14):13293–13296

Moisseiev E, Waisbourd M, Ben-Artsi E, Levinger E, Barak A, Daniels T, Csaky K, Loewenstein A, Barequet IS (2014) Pharmacokinetics of bevacizumab after topical and intravitreal administration in human eyes. Graefes Arch Clin Exp Ophthalmol 252(2):331–337

Mora RC, Bonilha VL, Shin B-C, Hu J, Cohen-Gould L, Bok D, Rodriguez-Boulan E (2006) Bipolar assembly of caveolae in retinal pigment epithelium. Am J Physiol-Cell Physiol 290(3):C832–C843

Murakami T, Felinski EA, Antonetti DA (2009) Occludin phosphorylation and ubiquitination regulate tight junction trafficking and vascular endothelial growth factor-induced permeability. J Biol Chem 284(31):21036–21046

Naert T, Colpaert R, Van Nieuwenhuysen T, Dimitrakopoulou D, Leoen J, Haustraete J, Boel A, Steyaert W, Lepez T, Deforce D, Willaert A, Creytens D, Vleminckx K (2016) CRISPR/Cas9 mediated knockout of rb1 and rbl1 leads to rapid and penetrant retinoblastoma development in Xenopus tropicalis. Sci Rep 6:35264–35264

Nakashima E, Pop-Busui R, Towns R, Thomas TP, Hosaka Y, Nakamura J, Greene DA, Killen PD, Schroeder J, Larkin DD (2005) Regulation of the human taurine transporter by oxidative stress in retinal pigment epithelial cells stably transformed to overexpress aldose reductase. Antioxid Redox Signal 7(11–12):1530–1542
Nakauchi T, Ando A, Ueda-Yamada M, Yamazaki Y, Uyama M, Matsumura M, Ito S (2003) Prevention of ornithine cytotoxicity by nonpolar side chain amino acids in retinal pigment epithelial cells. Invest Ophthalmol Vis Sci 44(11):5023–5028
Nguyen VP, Li Y, Qian W, Liu B, Tian C, Zhang W, Huang Z, Ponduri A, Tarnowski M, Wang X, Paulus YM (2019) Contrast agent enhanced multimodal photoacoustic microscopy and optical coherence tomography for imaging of rabbit choroidal and retinal vessels in vivo. Sci Rep 9(1):5945
Occhiutto ML, Freitas FR, Maranhao RC, Costa VP (2012) Breakdown of the blood-ocular barrier as a strategy for the systemic use of nanosystems. Pharmaceutics 4(2):252–275
Ohashi R, Tamai I, Yabuuchi H, Nezu J-I, Oku A, Sai Y, Shimane M, Tsuji A (1999) Na^{+}-dependent carnitine transport by organic cation transporter (OCTN2): its pharmacological and toxicological relevance. J Pharmacol Exp Ther 291(2):778–784
Ohkura Y, Akanuma S-i, Tachikawa M, Hosoya K-i (2010) Blood-to-retina transport of biotin via Na + -dependent multivitamin transporter (SMVT) at the inner blood-retinal barrier. Exp Eye Res 91(3):387–392
Papotto PH, Marengo EB, Sardinha LR, Goldberg AC, Rizzo LV (2014) Immunotherapeutic strategies in autoimmune uveitis. Autoimmun Rev 13(9):909–916
Park J, Zhang Y, Vykhodtseva N, Akula JD, McDannold NJ (2012) Targeted and reversible blood-retinal barrier disruption via focused ultrasound and microbubbles. PLoS ONE 7(8):e42754–e42754
Patel DM, Patel JK (2018) Lipid prodrug approach for retina and posterior segment disease. In: Patel JK, Sutariya V, Kanwar JR, Pathak YV (eds) Drug delivery for the retina and posterior segment disease. Springer International Publishing, Cham, pp 317–325
Patel A, Patel JK, Pathak YV (2018) Injectable pro-drugs approach for retina and posterior segment disease. In: Drug delivery for the retina and posterior segment disease. Springer, Berlin, pp 327–349
Patil AJ, Sharma A, Kenney MC, Kuppermann BD (2010) Valganciclovir in the treatment of cytomegalovirus retinitis in HIV-infected patients. Clin Ophthalmol (Auckland, NZ) 4:111
Patil S, Pathak N, Pathak YV (2018) Stereoisomeric dipeptide prodrug approach for retina and posterior segment disease. In: Drug delivery for the retina and posterior segment disease. Springer, pp 351–360
Pelkonen L, Sato K, Reinisalo M, Kidron H, Tachikawa M, Watanabe M, Uchida Y, Urtti A, Terasaki T (2017) LC–MS/MS based quantitation of ABC and SLC transporter proteins in plasma membranes of cultured primary human retinal pigment epithelium cells and immortalized ARPE19 cell line. Mol Pharm 14(3):605–613
Peng Y-Q, Tang L-S, Yoshida S, Zhou Y-D (2017) Applications of CRISPR/Cas9 in retinal degenerative diseases. Int J Ophthalmol 10(4):646–651
Pennisi E (2013) The CRISPR Craze. Science 341(6148):833
Pitkänen L, Ranta V-P, Moilanen H, Urtti A (2005) Permeability of retinal pigment epithelium: effects of permeant molecular weight and lipophilicity. Invest Ophthalmol Vis Sci 46(2):641–646
Pow DV (2001) Amino acids and their transporters in the retina. Neurochem Int 38(6):463–484
Prakash R, Carmichael ST (2015) Blood-brain barrier breakdown and neovascularization processes after stroke and traumatic brain injury. Curr Opin Neurol 28(6):556–564
Prasanna G, Carreiro S, Anderson S, Gukasyan H, Sartnurak S, Younis H, Gale D, Xiang C, Wells P, Dinh D (2011) Effect of PF-04217329 a prodrug of a selective prostaglandin EP2 agonist on intraocular pressure in preclinical models of glaucoma. Exp Eye Res 93(3):256–264
Putman E, van der Laan JW, van Loveren H (2003) Assessing immunotoxicity: guidelines. Fundam Clin Pharmacol 17(5):615–626

Rip J, Chen L, Hartman R, van den Heuvel A, Reijerkerk A, van Kregten J, van der Boom B, Appeldoorn C, de Boer M, Maussang D, de Lange ECM, Gaillard PJ (2014) Glutathione PEGylated liposomes: pharmacokinetics and delivery of cargo across the blood-brain barrier in rats. J Drug Target 22(5):460–467
Rizzolo LJ, Peng S, Luo Y, Xiao W (2011) Integration of tight junctions and claudins with the barrier functions of the retinal pigment epithelium. Progr Retinal Eye Res 30(5):296–323
Sagaties MJ, Raviola G, Schaeffer S, Miller C (1987) The structural basis of the inner blood-retina barrier in the eye of *Macaca mulatta*. Invest Ophthalmol Vis Sci 28(12):2000–2014
Salehi-Had H, Roh MI, Giani A, Hisatomi T, Nakao S, Kim IK, Gragoudas ES, Vavvas D, Guccione S, Miller JW (2011) Utilizing targeted gene therapy with nanoparticles binding alpha v beta 3 for imaging and treating choroidal neovascularization. PLoS ONE 6(4):e18864
Sanchez-Covarrubias L, Slosky L, Thompson B, Davis T, Ronaldson P (2014) Transporters at CNS barrier sites: obstacles or opportunities for drug delivery? Current Pharm Des 20(10):1422–1449
Schmidt-Erfurth U, Klimscha S, Waldstein SM, Bogunović H (2017) A view of the current and future role of optical coherence tomography in the management of age-related macular degeneration. Eye (Lond) 31(1):26–44
Shakib M, Cunha-Vaz J (1966a) Studies on the permeability of the blood-retinal barrier: IV. Junctional complexes of the retinal vessels and their role in the permeability of the blood-retinal barrier. Exp Eye Res 5(3):229-IN216
Shakib M, Cunha-Vaz JG (1966b) Studies on the permeability of the blood-retinal barrier: IV. Junctional complexes of the retinal vessels and their role in the permeability of the blood-retinal barrier. Exp Eye Res 5(3):229-IN216
Shapiro AB, Ling V (1997) Effect of quercetin on Hoechst 33342 transport by purified and reconstituted P-glycoprotein. Biochem Pharmacol 53(4):587–596
Sheibani N, Wang S, Darjatmoko SR, Fisk DL, Shahi PK, Pattnaik BR, Sorenson CM, Bhowmick R, Volpert OV, Albert DM, Melgar-Asensio I, Henkin J (2019) Novel anti-angiogenic PEDF-derived small peptides mitigate choroidal neovascularization. Exp Eye Res 188:107798
Sheikov N, McDannold N, Sharma S, Hynynen K (2008) Effect of focused ultrasound applied with an ultrasound contrast agent on the tight junctional integrity of the brain microvascular endothelium. Ultrasound Med Biol 34(7):1093–1104
Shen J, Cross ST, Tang-Liu DDS, Welty DF (2003) Evaluation of an immortalized retinal endothelial cell line as an in vitro model for drug transport studies across the blood-retinal barrier. Pharm Res 20(9):1357–1363
Shi P, Tan EYS, Yeong WY, Laude A (2017) Hybrid three-dimensional (3D) bioprinting of retina equivalent for ocular research. Int J Bioprint 3(2):138–146
Singh S, Grossniklaus H, Kang S, Edelhauser H, Ambati B, Kompella U (2009) Intravenous transferrin, RGD peptide and dual-targeted nanoparticles enhance anti-VEGF intraceptor gene delivery to laser-induced CNV. Gene Ther Nat 16(5):645
Smith RS, Rudt LA (1975) Ocular vascular and epithelial barriers to microperoxidase. Invest Ophthalmol Vis Sci 14(7):556–560
Smith SB, Kekuda R, Gu X, Chancy C, Conway SJ, Ganapathy V (1999) Expression of folate receptor alpha in the mammalian retinol pigmented epithelium and retina. Invest Ophthalmol Vis Sci 40(5):840–848
Spaide RF, Klancnik JM Jr, Cooney MJ (2015) Retinal vascular layers imaged by fluorescein angiography and optical coherence tomography angiography. JAMA Ophthalmol 133(1):45–50
Sridhar A, Ramesh Babu DR, Pandya K, Kompella UB, Srinivas SP (2019) Phase-resolved fluorometer for fluorescence lifetime measurements in the human eye. Mater Today: Proc 10:16–19
Srinivas S, Maurice D (1992) A microfluorometer for measuring diffusion of fluorophores across the cornea. IEEE Trans Biomed Eng 39(12):1283–1291
Stone V, Johnston H, Schins RPF (2009) Development of in vitro systems for nanotoxicology: methodological considerations. Crit Rev Toxicol 39(7):613–626

Sudhir RR, Murthy PP, Tadepalli S, Murugan S, Padmanabhan P, Krishnamurthy A, Dickinson SL, Karthikeyan R, Kompella UB, Srinivas SP (2018) Ocular spot fluorometer equipped with a lock-in amplifier for measurement of aqueous flare. Trans Vis Sci Technol 7(6):32–32

Suto F, Rowe-Rendleman CL, Ouchi T, Jamil A, Wood A, Ward CL (2015) A novel dual agonist of EP3 and FP receptors for OAG and OHT: safety, pharmacokinetics, and pharmacodynamics of ONO-9054 in healthy volunteers. Invest Ophthalmol Vis Sci 56(13):7963–7970

Tabatabaei SN, Tabatabaei MS, Girouard H, Martel S (2016) Hyperthermia of magnetic nanoparticles allows passage of sodium fluorescein and Evans blue dye across the blood–retinal barrier. Int J Hyperth 32(6):657–665

Tachikawa M, Toki H, Tomi M, Hosoya K-i (2008) Gene expression profiles of ATP-binding cassette transporter A and C subfamilies in mouse retinal vascular endothelial cells. Microvasc Res 75(1):68–72

Tachikawa M, Takeda Y, Tomi M, Hosoya K-i (2010) Involvement of OCTN2 in the transport of acetyl-l-carnitine across the inner blood-retinal barrier. Invest Ophthalmol Vis Sci 51(1):430–436

Tachikawa M, Hosoya K, Smith SB, Martin PM, Ganapathy V (2012) Transport of drugs across the inner and outer bloodretinal barriers: Relevance of transporters in the retinal blood vessel endothelium and the retinal pigment epithelium. In: Mitra AK (ed) Advances in ocular drug delivery, Research Signpost, Kerala, pp 1–31

Tagami M, Kusuhara S, Honda S, Tsukahara Y, Negi A (2009) Expression of ATP-binding cassette transporters at the inner blood–retinal barrier in a neonatal mouse model of oxygen-induced retinopathy. Brain Res 1283:186–193

Taha NM, Asklany HT, Mahmoud AH, Hammoda L, Attallah HR, Kamel AM, AbdelWahab MA (2018) Retinal fluorescein angiography: a sensitive and specific tool to predict coronary slow flow. Egyptian Heart J 70(3):167–171

Tamai I, Ohashi R, Nezu J-i, Yabuuchi H, Oku A, Shimane M, Sai Y, Tsuji A (1998) Molecular and functional identification of sodium ion-dependent, high affinity human carnitine transporter OCTN2. J Biol Chem 273(32):20378–20382

Tang G, Liu Y, Zhang Z, Lu Y, Wang Y, Huang J, Li Y, Chen X, Gu X, Wang Y, Yang G-Y (2014) Mesenchymal stem cells maintain blood-brain barrier integrity by inhibiting aquaporin-4 upregulation after cerebral ischemia. Stem Cells 32(12):3150–3162

Taylor AB, Zijlstra P (2017) Single-molecule plasmon sensing: current status and future prospects. Am Chem Soc Sens 2(8):1103–1122

Thakur A, Fitzpatrick S, Zaman A, Kugathasan K, Muirhead B, Hortelano G, Sheardown H (2012) Strategies for ocular siRNA delivery: Potential and limitations of non-viral nanocarriers. J Biol Eng 6(1):7

Thangaraju M, Karunakaran SK, Itagaki S, Gopal E, Elangovan S, Prasad PD, Ganapathy V (2009) Transport by SLC5A8 with subsequent inhibition of histone deacetylase 1 (HDAC1) and HDAC3 underlies the antitumor activity of 3-bromopyruvate. Cancer 115(20):4655–4666

Thanuja MY, Anupama C, Ranganath SH (2018) Bioengineered cellular and cell membrane-derived vehicles for actively targeted drug delivery: So near and yet so far. Adv Drug Deliv Rev 132:57–80

Thornit DN, Vinten CM, Sander B, Lund-Andersen H, la Cour M (2010) Blood-retinal barrier glycerol permeability in diabetic macular edema and healthy eyes: estimations from macular volume changes after peroral glycerol. Invest Ophthalmol Vis Sci 51(6):2827–2834

Ting DSW, Cheung CY-L, Lim G, Tan GSW, Quang ND, Gan A, Hamzah H, Garcia-Franco R, San Yeo IY, Lee SY, Wong EYM, Sabanayagam C, Baskaran M, Ibrahim F, Tan NC, Finkelstein EA, Lamoureux EL, Wong IY, Bressler NM, Sivaprasad S, Varma R, Jonas JB, He MG, Cheng C-Y, Cheung GCM, Aung T, Hsu W, Lee ML, Wong TY (2017) Development and validation of a deep learning system for diabetic retinopathy and related eye diseases using retinal images from multiethnic populations with diabetes. J Am Med Assoc 318(22):2211–2223

Ting DSW, Pasquale LR, Peng L, Campbell JP, Lee AY, Raman R, Tan GSW, Schmetterer L, Keane PA, Wong TY (2019) Artificial intelligence and deep learning in ophthalmology. Br J Ophthalmol 103(2):167–175

Toda R, Kawazu K, Oyabu M, Miyazaki T, Kiuchi Y (2011) Comparison of drug permeabilities across the blood&2013; Retinal Barrier, Blood&2013; Aqueous Humor Barrier, and Blood 2013; Brain Barrier. J Pharm Sci 100(9):3904–3911

Tombran-Tink J, Barnstable CJ (2008) Ocular transporters in ophthalmic diseases and drug delivery. Springer Science & Business Media, vol 30, no 1, pp 1–467

Tomi M, Hosoya K-i (2004) Application of magnetically isolated rat retinal vascular endothelial cells for the determination of transporter gene expression levels at the inner blood–retinal barrier. J Neurochem 91(5):1244–1248

Tomi M, Mori M, Tachikawa M, Katayama K, Terasaki T, Hosoya K-i (2005) L-Type Amino Acid Transporter 1–Mediated l-Leucine Transport at the Inner Blood-Retinal Barrier. Invest Ophthalmol Vis Sci 46(7):2522–2530

Törnquist P, Alm A, Bill A (1990) Permeability of ocular vessels and transport across the blood-retinal-barrier. Eye 4(2):303–309

Townsend KA, Wollstein G, Schuman JS (2008) Clinical application of MRI in ophthalmology. NMR in Biomed Int J Devoted Dev Appl Magn Resonan Vivo 21(9):997–1002

Trabulo S, Cardoso AL, Mano M, De Lima MCP (2010) Cell-penetrating peptides-mechanisms of cellular uptake and generation of delivery systems. Pharmaceuticals (Basel) 3(4):961–993

Tratta E, Pescina S, Padula C, Santi P, Nicoli S (2014) In vitro permeability of a model protein across ocular tissues and effect of iontophoresis on the transscleral delivery. Eur J Pharm Biopharm 88(1):116–122

Tretiach M, Madigan MC, Wen L, Gillies MC (2005) Effect of Müller cell co-culture on in vitro permeability of bovine retinal vascular endothelium in normoxic and hypoxic conditions. Neurosci Lett 378(3):160–165

Tufail A, Rudisill C, Egan C, Kapetanakis VV, Salas-Vega S, Owen CG, Lee A, Louw V, Anderson J, Liew G, Bolter L, Srinivas S, Nittala M, Sadda S, Taylor P, Rudnicka AR (2017) Automated diabetic retinopathy image assessment software: diagnostic accuracy and cost-effectiveness compared with human graders. Ophthalmology 124(3):343–351

Vadlapatla RK, Vadlapudi AD, Ponnaluri VC, Pal D, Mukherji M, Mitra AK (2013) Molecular expression and functional activity of efflux and influx transporters in hypoxia induced retinal pigment epithelial cells. Int J Pharm 454(1):444–452

Vadlapudi AD, Vadlapatla RK, Pal D, Mitra AK (2012) Functional and molecular aspects of biotin uptake via SMVT in human corneal epithelial (HCEC) and retinal pigment epithelial (D407) cells. AAPS J 14(4):832–842

Vadlapudi AD, Vadlapatla RK, Earla R, Sirimulla S, Bailey JB, Pal D, Mitra AK (2013) Novel biotinylated lipid prodrugs of acyclovir for the treatment of herpetic keratitis (HK): transporter recognition, tissue stability and antiviral activity. Pharm Res 30(8):2063–2076

Varma MV, Ashokraj Y, Dey CS, Panchagnula R (2003) P-glycoprotein inhibitors and their screening: a perspective from bioavailability enhancement. Pharmacol Res 48(4):347–359

Vighi E, Trifunović D, Veiga-Crespo P, Rentsch A, Hoffmann D, Sahaboglu A, Strasser T, Kulkarni M, Bertolotti E, van den Heuvel A, Peters T, Reijerkerk A, Euler T, Ueffing M, Schwede F, Genieser H-G, Gaillard P, Marigo V, Ekström P, Paquet-Durand F (2018) Combination of cGMP analogue and drug delivery system provides functional protection in hereditary retinal degeneration. Proc Natl Acad Sci U S A 115(13):E2997–E3006

Villarroel M, García-Ramírez M, Corraliza L, Hernández C, Simó R (2009) High glucose concentration leads to differential expression of tight junction proteins in human retinal pigment epithelial cells. Endocrinol Nutr 56(2):53–58

Vinores SA (2006) Pegaptanib in the treatment of wet, age-related macular degeneration. Int J Nanomed 1(3):263

Voloboueva LA, Liu J, Suh JH, Ames BN, Miller SS (2005) (R)-α-lipoic acid protects retinal pigment epithelial cells from oxidative damage. Invest Ophthalmol Vis Sci 46(11):4302–4310

Vooturi SK, Kadam RS, Kompella UB (2012) Transporter targeted gatifloxacin prodrugs: synthesis, permeability, and topical ocular delivery. Mol Pharm 9(11):3136–3146

Wan C, Li F, Li H (2015) Gene therapy for ocular diseases meditated by ultrasound and microbubbles (Review). Mole Med Rep 12(4):4803–4814

Wang AL, Lukas TJ, Yuan M, Du N, Tso MO, Neufeld AH (2009) Autophagy and exosomes in the aged retinal pigment epithelium: possible relevance to drusen formation and age-related macular degeneration. PLoS ONE 4(1):e4160–e4160

Wang Y, Liu C-H, Ji T, Mehta M, Wang W, Marino E, Chen J, Kohane DS (2019) Intravenous treatment of choroidal neovascularization by photo-targeted nanoparticles. Nat Commun 10(1):804–804

Weng Y, Liu J, Jin S, Guo W, Liang X, Hu Z (2017) Nanotechnology-based strategies for treatment of ocular disease. Acta Pharmaceutica Sinica B 7(3):281–291

Wong WL, Su X, Li X, Cheung CMG, Klein R, Cheng C-Y, Wong TY (2014) Global prevalence of age-related macular degeneration and disease burden projection for 2020 and 2040: a systematic review and meta-analysis. Lancet Glob Health 2(2):e106–e116

Wu X, Huang W, Prasad PD, Seth P, Rajan DP, Leibach FH, Chen J, Conway SJ, Ganapathy V (1999) Functional characteristics and tissue distribution pattern of organic cation transporter 2 (OCTN2), an organic cation/carnitine transporter. J Pharmacol Exp Ther 290(3):1482–1492

Wu W-H, Tsai Y-T, Justus S, Lee T-T, Zhang L, Lin C-S, Bassuk AG, Mahajan VB, Tsang SH (2016) CRISPR repair reveals causative mutation in a preclinical model of retinitis pigmentosa. Mol Ther 24(8):1388–1394

Xia T, Rizzolo LJ (2017) Effects of diabetic retinopathy on the barrier functions of the retinal pigment epithelium. Vision Res 139:72–81

Xu J, Tu Y, Wang Y, Xu X, Sun X, Xie L, Zhao Q, Guo Y, Gu Y, Du J, Du S, Zhu M, Song E (2020) Prodrug of epigallocatechin-3-gallate alleviates choroidal neovascularization via down-regulating HIF-1α/VEGF/VEGFR2 pathway and M1 type macrophage/microglia polarization. Biomed Pharmacother 121:109606

Yahara T, Tachikawa M, Akanuma S-i, Hosoya K-i (2010) Hypertonicity enhances GABA uptake by cultured rat retinal capillary endothelial cells. Drug Metab Pharmacokinet 25(6):611–615

Yamauchi K, Rai T, Kobayashi K, Sohara E, Suzuki T, Itoh T, Suda S, Hayama A, Sasaki S, Uchida S (2004) Disease-causing mutant WNK4 increases paracellular chloride permeability and phosphorylates claudins. Proc Natl Acad Sci U S A 101(13):4690

Yang M-S, Hu Y-J, Lin KC-R, Lin CC-L (2002) Segmentation techniques for tissue differentiation in MRI of ophthalmology using fuzzy clustering algorithms. Magn Reson Imaging 20(2):173–179

Yasukawa T, Kimura H, Tabata Y, Miyamoto H, Honda Y, Ikada Y, Ogura Y (1999) Targeted delivery of anti-angiogenic agent TNP-470 using water-soluble polymer in the treatment of choroidal neovascularization. Invest Ophthalmol Vis Sci 40(11):2690–2696

Yasukawa T, Kimura H, Tabata Y, Kamizuru H, Miyamoto H, Honda Y, Ogura Y (2002) Targeting of interferon to choroidal neovascularization by use of dextran and metal coordination. Invest Ophthalmol Vis Sci 43(3):842–848

Yau JW, Rogers SL, Kawasaki R, Lamoureux EL, Kowalski JW, Bek T, Chen S-J, Dekker JM, Fletcher A, Grauslund J (2012) Global prevalence and major risk factors of diabetic retinopathy. Diabetes Care 35(3):556–564

Zargarzadeh M, MadaahHosseini HR, Delavari H, Irajirad R, Aghaie E (2018) Synthesis of magnetite (Fe_3O_4)–Avastin nanocomposite as a potential drug for AMD treatment. Micro & Nano Letters 13(8):1141–1145

Zayas-Santiago A, Marmorstein AD, Marmorstein LY (2011) Relationship of stokes radius to the rate of diffusion across Bruch's membrane. Invest Ophthalmol Vis Sci 52(7):4907–4913

Zhang X, Zeng H, Bao S, Wang N, Gillies MC (2014) Diabetic macular edema: new concepts in patho-physiology and treatment. Cell Biosci 4(1):27

Zhang Z, Uchida Y, Hirano S, Ando D, Kubo Y, Auriola S, Akanuma S-i, Hosoya K-i, Urtti A, Terasaki T, Tachikawa M (2017) Inner blood-retinal barrier dominantly expresses breast cancer resistance protein: comparative quantitative targeted absolute proteomics study of CNS barriers in pig. Mol Pharm 14(11):3729–3738

Zhao R, Matherly LH, Goldman ID (2009) Membrane transporters and folate homeostasis: intestinal absorption and transport into systemic compartments and tissues. Expert Rev Mol Med 11

Zheng M-M, Zhou X-Y, Wang L-P, Wang Z-G (2012) Experimental research of RB94 gene transfection into retinoblastoma cells using ultrasound-targeted microbubble & destruction. Ultrasound Med Biol 38(6):1058–1066

Zhong H, Chen Y, Li Y, Chen R, Mardon G (2015) CRISPR-engineered mosaicism rapidly reveals that loss of Kcnj13 function in mice mimics human disease phenotypes. Sci Rep 5:8366–8366

Zou Y, Hu X, Schewitz-Bowers LP, Stimpson M, Miao L, Ge X, Yang L, Li Y, Bible PW, Wen X, Li JJ, Liu Y, Lee RWJ, Wei L (2019) The DNA methylation inhibitor zebularine controls CD4+ T cell mediated intraocular inflammation. Front Immunol 10(1950)

The Effects of Irradiation with Cold Atmospheric-Pressure Plasma on Cellular Function

Katsuya Iuchi

Abstract Plasma is the fourth state of matter. Cold atmospheric-pressure plasma (CAP) is one of the weakly ionized electrical discharges, which is artificially produced by certain devices at atmospheric pressure. CAP has been used in various biological and medical applications. CAP generates a large number of reactive oxygen and nitrogen species (RONS), including hydroxyl radical, superoxide anion, singlet oxygen, hydrogen peroxide, peroxynitrite, nitric oxide, nitrogen dioxide, nitrogen trioxide, nitrous oxide, and dinitrogen tetroxide. RONS, produced by CAP, are the key molecules responsible for its biological activities and beneficial effects. However, the details of CAP-induced RONS, and their effects on intracellular processes (organelle response to CAP treatment), are not fully understood. Several studies have shown that CAP up- or down-regulate cellular functions through the modification of biomolecules such as nucleic acids (DNA and RNA), proteins, and lipids. Thereby modulating cell proliferation, differentiation, and death. In this chapter, we discuss the biological activities of CAP in terms of the cellular response to it. Understanding of the CAP-mediated regulation of cellular function would be useful in the developing novel treatment modalities for diseases associated with the redox signaling.

Keywords Cold atmospheric-pressure plasma · Cellular function · Biomolecules · Reactive oxygen and nitrogen species

Abbreviations

BAPTA-AM	1,2-Bis(2-aminophenoxy)ethane-*N*,*N*,*N*′,*N*′-tetraacetic acid tetrakis(acetoxymethyl ester)
HEPES	4-(2-Hydroxyethyl)-1-piperazineethanesulfonic acid

K. Iuchi (✉)
Department of Materials and Life Science, Faculty of Science and Technology, Seikei University, 3-3-1 Kichijojikitamachi, Musashino-shi, Tokyo 180-8633, Japan
e-mail: iuchi@st.seikei.ac.jp

A. Tripathi and J. S. Melo (eds.), *Immobilization Strategies*,
Gels Horizons: From Science to Smart Materials,
https://doi.org/10.1007/978-981-15-7998-1_7

RAS	A class of protein called small GTPase
ATF3	Activating transcription factor 3
AP-1	Activator protein 1
ADP	Adenosine diphosphate
ATP	Adenosine triphosphate
ASCs	Adipose tissue-derived stem cells
AD	Alzheimer's disease
Aβ	Amyloid-β
Ar	Argon
O	Atomic oxygen
Bcl-2	B-cell lymphoma 2
BAX	BCL2-associated X protein
BPH	Benign prostatic hyperplasia
BM-MSCs	Bone marrow-derived stem cells
BrdU	Bromodeoxyuridine
CHOP	C/EBP-homologous protein
Ca^{2+}	Calcium
CRT	Calreticulin
CSCs	Cancer stem cells
CCL4	Chemokine (C–C motif) ligands 4
CXCL1	Chemokine (C-X-C motif) ligand 1
YKL-40	Chitinase-3-like protein 1
JNK	C-jun NH2-terminal kinase
CD115	Cluster of differentiation 115
CD133, ABCB5	Cluster of differentiation 133
CD28	Cluster of differentiation 28
CAP	Cold atmospheric-pressure plasma
PAM	Cold plasma-activated medium
Cx26	Connexin 26
cNOS	Constitutive NOS
CDK	Cyclin-dependent kinase
xCT	Cystine/glutamate antiporter solute carrier family 7 member 11
(SCL7a11	System Xc-, SXC)
DAMPs	Damage-associated molecular patterns
DNA	Deoxyribonucleic acid
N_2O_3	Dinitrogen trioxide
DMEM	Dulbecco's Modified Eagle Medium
ER	Endoplasmic reticulum
eNOS	Endothelial NOS
E-cadherin	Epithelial cadherin
ERK	Extracellular signal-regulated kinase
GBM	Glioblastoma multiforme
GRP78	Glucose-regulated protein 78
GSH	Glutathione
GM-CSF	Granulocyte–macrophage colony stimulating factor

GADD45α	Growth arrest and DNA-damage-inducible protein, Alpha
GADD45β	Growth arrest and DNA-damage-inducible protein, Beta
H2A.X	H2A histone family member X
HSP	Heat shock protein
HSPs	Heat shock proteins
He	Helium
HSCs	Hematopoietic stem cells
HO-1	Heme oxygenase 1
HBS	HEPES-buffered saline
HMGB1	High-mobility group box 1
A549	Human adeno carcinoma cell line
MCF7	Human breast adenocarcinoma
Ca9-22	Human carcinoma of the gingiva
HeLa	Human cell line derived from cervical cancer
Ho-1-u-1	Human cell line derived from squamous cell carcinoma at mouth floor
HCT-116	Human colon cancer cell line
HEK293T	Human embryonic kidney cell line
U251SP	Human glioblastoma cell line
U373MG	Human glioblastoma multiforme cell line
HSCs	Human hepatic stellate cells
HepG2	Human hepatoma cancer cell line
HaCaT	Human keratinocyte cell line
U937	Human lymphoma
Molt-4	Human lymphoma cell line
A375	Human malignant melanoma
HBL	Human mamma carcinoma cell line
U87MG	Human neuronal glioblastoma
HSC	Human oral squamous carcinoma
PANC-1	Human pancreatic cancer cell line
PC-3	Human prostate cancer cell line
SK-MEL-28	Human skin melanoma cell line
Hmel1	Human sporadic melanoma biopsy specimens with BRAF mutations
HSC-4	Human squamous carcinoma cell line from tongue
Sa3	Human squamous carcinoma cells derived from oral cancer.
FaDu	Human squamous cell carcinoma of the hypopharynx
H_2O_2	Hydrogen peroxide
H·	Hydrogen radicals
HOO·	Hydroperoxy radicals
·OH	Hydroxyl radicals
HIF	Hypoxia inducible factor
ICD	Immunogenic cell death
nNOS	Including neuronal NOS
iNOS	Inducible NOS

INFγ	Interferon-γ
IL-1β	Interleukin-1β
IL-8	Interleukin-8
ISEMFs	Intestinal subepithelial myofibroblasts
ISEMFs	Intestinal subepithelial myofibroblasts
$[Ca^{2+}]$	Intracellular Ca^{2+}
MNPs	Iron oxide-based magnetic nanoparticles
KEAP1	Kelch-like ECH associated protein 1
LDH	Lactate dehydrogenase
Mel Im	Melanoma cell lines derived from a metastasized specimen
Mel Juso	Melanoma cell lines derived from a primary tumor
MEL	Melittin
MAMs	Mitochondria-associated ER membranes
MPT	Mitochondrial membrane permeability transition
3T3L1	Mouse embryonic fibroblast cell line
B16	Mouse melanoma cell line
B16F10	Mouse melanoma cell line
N2a	Mouse neuroblastoma cell line
MCTSs	Multicellular tumor spheroids
C17.2-NSCs	Murine neural stem cells
NPs	Nanoparticles
N-cadherin	Neural cadherin
NSCs	Neural stem cells
NRF2	NF-E2-related actor 2
NAMPT	Nicotinamide phosphoribosyltransferase
NO·	Nitric oxide
NO_2^-	Nitrite, Nitrous ions
NOx	Nitrogen oxides
3-NT	Nitrotyrosine
IMR-90-SV	Normal human fibroblast
HS-K	Normal human fibroblast
NF-κB	Nuclear factor kappa-light-chain-enhancer of activated B cells
OSCC	Oral cavity squamous cell carcinoma
SAS	Oral squamous carcinoma cell lines
SLC22A16	Organic cationic transporter
O_3	Ozone
PTP	Permeability transition pores
PPAR-γ	Peroxisome proliferator-activated receptor-γ
$ONOO^-$	Peroxynitrite
PBS	Phosphate-buffered salin
PAM	Plasma-activated medium
PAL	Plasma-activated Ringer's lactate solutions
PT-PBS	Plasma-treated phosphate-buffered saline solution
Pt-NPs	Platinum nanoparticles
AKT	Protein kinase B

RHS	Reactive hydrogen species
RNS	Reactive nitrogen species
RONS	Reactive oxygen and nitrogen species
ROS	Reactive oxygen species
RNA	Ribonucleic acid
SN	Silymarin nano-emulsion
$^1O_2$	Singlet oxygen
SNAI1	Snail family transcriptional repressor 1
SSZ	Sulfasalazine
O_2^-	Superoxide anion
3-D	Three-dimensional
TGFβ	Transforming growth factor-β
TRP	Transient receptor potential
TNF-α	Tumor necrosis factor α
TRAIL	Tumor necrosis factor-related apoptosis-inducing ligand
2D	Two-dimensional
UV-A	Ultraviolet light
VEGF-A	Vascular endothelial growth factor-A
BRAF	V-raf murine sarcoma viral oncogene homolog B

1 Introduction

Cold atmospheric-pressure plasma (CAP) is referred to as a weakly ionized electrical discharge at atmospheric pressure. CAP is produced and maintained by increasing the energy level of a substance. The solid state is transferred to the liquid and gaseous states, and finally to plasma (Fig. 1). To date, CAP has been used in various fields including food hygiene, dentistry, and medical research (Hoffmann et al. 2013), for the sterilization of medical instruments and food containers (Shintani et al. 2010), and in blood coagulation, wound healing, and cancer treatment

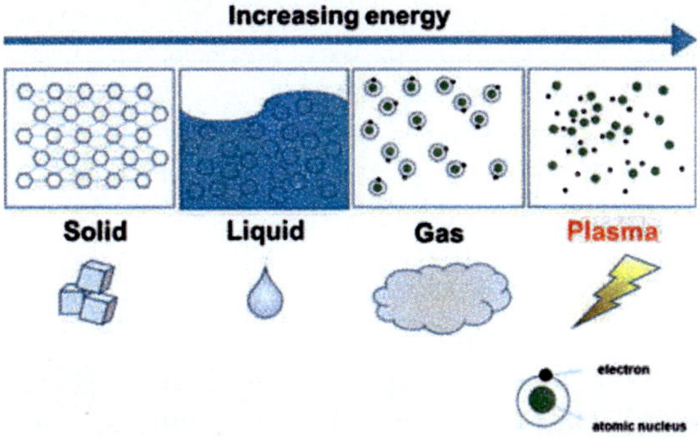

Fig. 1 Generation of plasma via the increase in energy level of a substance. The solid state transformed to the liquid and gaseous states, and finally to a plasma

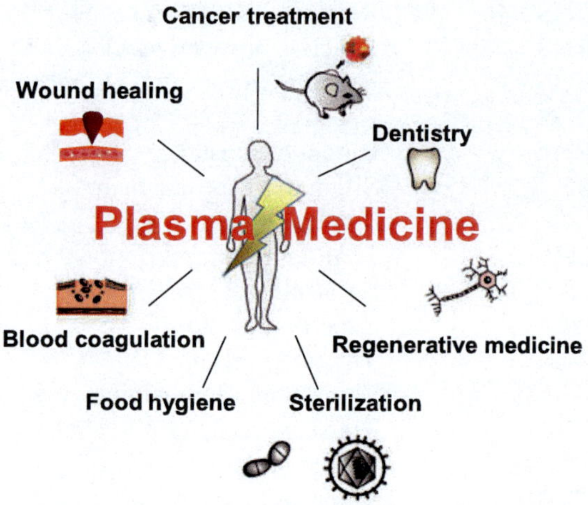

Fig. 2 In vitro, in vivo, and clinical research using plasma. Plasma medicine is used in various fields, including plasma physics, chemical tools, biological techniques, and clinical medicine

(Fridman et al. 2008; Graves 2017; Yan et al. 2017) (Fig. 2). The reactive oxygen and nitrogen species (RONS) produced by CAP are important factors responsible for its numerous beneficial biological activities. The effects of CAP are distinct from those of the corresponding reactive species, such as hydrogen peroxide (H_2O_2) and peroxynitrite ($ONOO^-$). The detailed mechanisms of CAP-induced chemical phenomenon between gas- and liquid-phase reactions and biological processes are not yet fully understood. CAP induces the production of numerous RONS such as hydrogen peroxide (H_2O_2) and nitrous ions (NO_2^-). These RONS have been detected in plasma-irradiated Dulbecco's Modified Eagle Medium (DMEM) and induce apoptosis of tumor cells not but normal cells (Kurake et al. 2016). The underlying mechanisms of tumor-specific anti-proliferative activity of CAP remain unknown. Excluding its use in skin treatment, the complex biological effects of CAP make its clinical application difficult to manage.

The development of CAP devices for various applications has accelerated. CAP is artificially generated by a range of different plasma devices, such as the plasma jet, dielectric barrier discharge, and gliding arc discharges (Schutze et al. 1998; Stryczewska et al. 2013; Yan et al. 2017). As CAPs are produced in ambient humid air, a broad range of ROS and RNS is formed in the afterglow (Murakami et al. 2012, 2013, 2014). The carrier gasses (i.e., helium (He), argon (Ar), or nitrogen (N_2)) are used to produce the CAP. A small amount of oxygen (O_2) enhances the ROS production in the core of the CAPs. In this chapter, Ar, He, N_2, air, Ar-O_2, and Ar-N_2 plasma are denoted as ArCAP, He-CAP, N_2CAP, AirCAP, ArO_2CAP, and Ar-N_2CAP, respectively.

Since 2005, numerous studies have been published on the biological and medical applications of CAPs. Most of these studies used air, He, or Ar as the carrier gas in the CAP-producing device (AirCAP, He-CAP, and ArCAP, respectively), while N_2 has been used less frequently (Dubuc et al. 2018). Since each carrier gas produces different levels of RONS, the biological activity of CAP depends on the type of gas used as the carrier (Bekeschus et al. 2017). Bekeschus et al. showed that He-O_2 CAP, but not He, He–H_2O, or He–O_2–H_2O CAP, inactivated THP-1 cells, which were derived from acute monocytic leukemia patient. Moreover, the experimental parameters, including the applied voltage, discharge frequency, treatment time, and flow rate of the carrier gas, influence the generation of the CAP and its anti-microbial activity (Mohd Nasir et al. 2016). Although the anti-microbial and anti-cancer activities of CAP have been studied over several decades, the mechanisms underlying the cellular regulation-induced by CAP are not yet fully understood.

2 Types of CAP Treatment in Biomedical Research

Direct and indirect treatment with CAP have been reported in plasma research. Direct plasma treatments are currently integrated in clinical practices (Woedtke et al. 2019), in food and clinical applications, such as the removal of contaminants from food, the packaging materials, and the sterilization of medical instruments and devices (Lopez et al. 2019). Indirect plasma treatments are used in cancer immunotherapy and plasma-activated medium therapy (Tanaka et al. 2018).

3 RONS Produced by CAP

CAP produces a large number of reactive components, including radicals and RONS, and reactive hydrogen species, such as atomic oxygen (O), hydroxyl radicals (·OH), singlet oxygen ($^1O_2$), superoxide anion (O_2^-), ozone (O_3), and/or H_2O_2, nitrogen oxides (NO· and NOx), peroxynitrite ($ONOO^-$), and hydrogen radicals (H·) (Fig. 3) (Hirst et al. 2016). These reactive species are the key mediators of CAP treatment in vitro and in vivo, resulting in changes to cellular functions. CAP induces liquid-phase RONS formation, causing an increase in the RONS penetration into cells; however, these reactive species cannot enter cells and thus act in extracellular matrix.

3.1 ROS Levels by CAP

Intracellular ROS play an important role as a functional mediators of various signaling pathways. ·OH, H_2O_2, NO_2^-, and NO_3^- are formed in deionized water irradiated with Ar, Ar–O_2, and Ar–N_2 plasma (Park et al. 2016). The highly reactive ·OH immediately reacts with molecules at the site of its generation. In addition to their production in CAP, ·OH radicals are also formed from the reaction between ferrous iron (Fe(II)) and H_2O_2 (i.e., the Fenton reaction). Adachi et al. showed the generation of ·OH and increased levels of intracellular ferrous ions in PAM-treated cells. Further, ·OH generation and increased levels of ferrous iron were inhibited by a Fe(II) chelator.

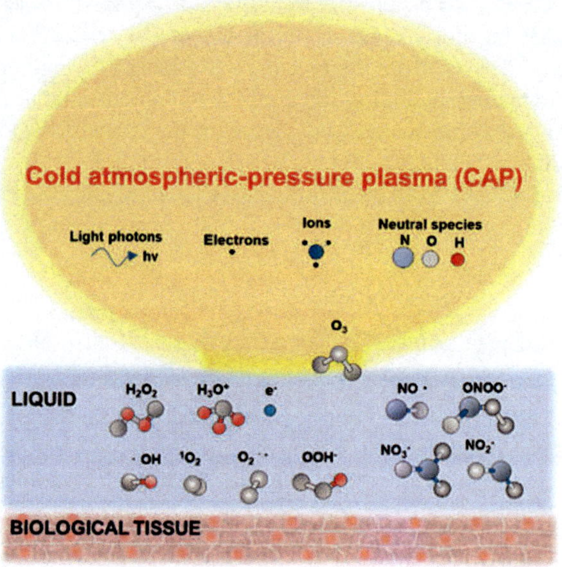

Fig. 3 Cold atmospheric-pressure plasmas have a weakly ionized electrical discharge and occur at atmospheric pressure. The main components include ions, photons, and neutral species in the plasma phase. Various reactive oxygen and nitrogen species (RONS) permeate the plasma–liquid interface and move toward biological tissues. The schematic of plasma species are adapted from Hirst et al. (2016)

Fe(II) is released from ferrtin, an intracellular iron store (Adachi et al. 2016). The ·OH produced in CAP-treated water can be detected indirectly using a fluorescent probe (hydroxyterephthalic acid), which is formed by the reaction of terephthalic acid with ·OH (Kanazawa et al. 2011).

Takai et al. (2013) reported unique molecular mechanism underlying the CAP action during sterilization of various solutions (Takai et al. 2013). They showed that the permeation of hydroperoxyl radicals (HOO·) into the cell membrane was important for its effect and the sterilization via HOO· action depended on the pH.

3.2 *RNS Levels Produced Following CAP*

RNS, as well as ROS, alters in cellular functions via protein modifications leading to pathophysiological consequences. The NO_2^- (nitrite) concentration in DMEM and He-CAP-treated deionized water increases with treatment duration. NO· s exerts its toxic to cancer cells, predominantly by generating pro-apoptotic intermediates, such as peroxynitrite $ONOO^-$ and N_2O_3 and inhibits DNA repair enzymes, including poly (ADP-ribose) polymerase (Hirst and Robson 2007). Nitric oxide (NO) has a very short half-life and can react with superoxides, leading to the formation of $ONOO^-$, which modifies the structure and function of proteins via post-translational modification.

S-nitrosylation, one of the post-translational modifications, plays a role in major molecular signaling pathways in mammalian cells. Controlling mitochondrial dynamics and mitophagy via *S*-nitrosylation regulates cell senescence and aging in mammals (Rizza et al. 2018). Sha and Marshall reviewed the transcription factors, including NF-κB, HIF-1, and AP-1, and their activating pathways as the primary targets of *S*-nitrosylation, (Sha and Marshall 2012). In addition, *S*-nitrosylation also regulates gene expressions indirectly by manipulating other cell signaling pathways, such as the JNK and RAS pathways (Sha and Marshall 2012).

4 Effects of CAP on Intracellular Calcium Levels

The endoplasmic reticulum (ER) stores calcium (Ca^{2+}) intracellularly. It regulates the intracellular levels of Ca^{2+} $[Ca^{2+}]_i$, as well as protein and lipid synthesis. Mitochondria also plays a role in regulating the $[Ca^{2+}]_i$ levels (Fig. 4) (Syntichaki and Tavernarakis 2003). CAP can modulate the intracellular levels of Ca^{2+}.

4.1 Changes in Intracellular Ca^{2+} Levels After CAP Treatment

Ca^{2+} is an essential intracellular second messenger. The level of $[Ca^{2+}]_i$ is regulated by intracellular organelles, such as the ER and mitochondria. CAP induced liquid-phase RONS formation, resulting in apoptotic cell death through caspase activation and increase in the level of $[Ca^{2+}]_i$ (Iuchi et al. 2018; Jawaid et al. 2016; Zhao et al. 2013). CAP treatment disturbs intracellular Ca^{2+} homeostasis in HepG2 cells and by inducing accumulation of the intracellular free Ca^{2+} in a time-dependent manner (Zhao et al. 2013). Furthermore, He-CAP treatment increases $[Ca^{2+}]_i$ by approximately 20-fold, leading to apoptotic cell death (Jawaid et al. 2016). Platinum nanoparticles (Pt-NPs) are potent anti-oxidants that inhibit the He-CAP-induced increase in $[Ca^{2+}]_i$ in human lymphoma U937 cells (Jawaid et al. 2016). We recently reported that a cell-permeable Ca^{2+} chelator, BAPTA-AM, inhibited morphological altercations and cell death induced by N_2CAP in human embryonic kidney cell line, HEK293T (Iuchi et al. 2018). Thus, an increase in $[Ca^{2+}]_i$ following CAP treatment is one of the underlying mechanisms of CAP-induced cell death.

In addition to CAP, cold plasma-activated medium (PAM) too exhibits anti-tumor activities, thereby emerging as a promising novel tool for cancer chemotherapy. PAM modulates the membrane potential and calcium dynamics in TRAIL-resistant human malignant cells and promotes caspase-independent cell death (Tokunaga et al. 2018). Schneider et al. showed that direct and indirect CAP treatment increased the levels of cytoplasmic Ca^{2+} in melanoma cell lines Mel Im (derived from a metastasized specimen) and Mel Juso (derived from a primary tumor). They also showed that the increase was the absence of extracellular Ca^{2+}, indicating that CAP treatment led to calcium efflux is from intracellular stores (Schneider et al. 2018).

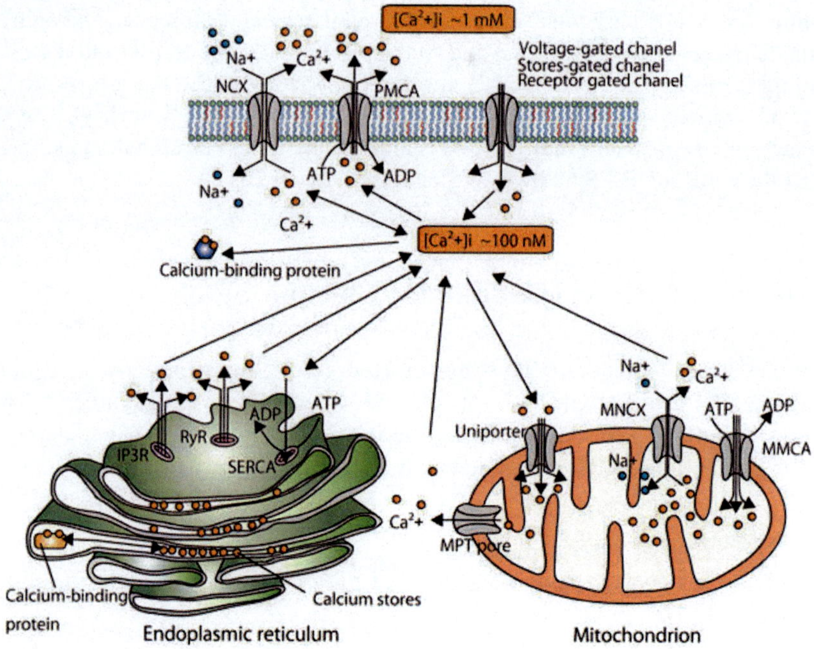

Fig. 4 Mechanisms of calcium homeostasis. The intracellular calcium concentration ($[Ca^{2+}]_i$) is closely regulated within narrow limits. Under various stress conditions, $[Ca^{2+}]_i$ increases through Ca^{2+} influx from extracellular pools through various channels (voltage-, ligand-, or concentration-gated channels), including the sodium/calcium exchanger (NCX). Under normal conditions, NCX is the main channel for calcium efflux, but it can also contribute to Ca^{2+} influx (reverse mode exchange), especially during strong depolarization, and with increased levels of intracellular sodium, Na^+. $[Ca^{2+}]_i$ can also increase via its release from ER stores, counterbalancing the mechanisms against the increase in Ca^{2+} concentration in the cytoplasm. The plasma membrane Ca^{2+} pump (PMCA), NCX, and sarco-endoplasmic reticulum Ca^{2+} ATPase (SERCA) are essential to restore normal Ca^{2+} levels. Increased $[Ca^{2+}]_i$ drives Ca^{2+} overload in the mitochondria via mitochondrial membrane Ca^{2+} ATPase (MMCA) and Uniporter. In turn, Ca^{2+} overload triggers the secondary release of Ca^{2+} from mitochondrial stores via mitochondrial NCX (MNCX) and mitochondrial pores, which are opened during mitochondrial permeability transition (MPT). Ca^{2+}-binding proteins in the cytoplasm and in the ER play a role of buffering Ca^{2+}. The schematic of the intracellular calcium homeostasis was adapted from Syntichaki (2003)

Sasaki et al. showed that plasma-irradiated HEPES-buffered saline (HBS) induced an increase in $[Ca^{2+}]_i$ in mouse fibroblast 3T3L1 cells (Sasaki et al. 2016). This was caused by Ca^{2+} influx through transient receptor potential (TRP) channels triggered by the CAP-induced reactive species. Further, they reported that the ROS produced by the ·OH-initiated reaction was responsible for the increase in $[Ca^{2+}]_i$ in response to plasma-irradiated HBS. Furthermore, another study showed that CAP-induced Ca^{2+} influx in melanoma cells was higher in acidic conditions compared with physiological conditions (Schneider et al. 2019). While CAP induced NO formation in a dose- and pH-dependent manner, the protein nitration occurred only under acidic conditions (Schneider et al. 2019).

5 The Modification of Biomolecules by CAP

Biomolecules are divided into several types based on their size: small molecules, monomers, oligomers, and polymers. Deoxyribonucleic acid (DNA) and ribonucleic acid (RNA) are long polymers of nucleotides. In addition, peptides and proteins are formed from twenty standard and two non-standard amino acids. Lipid peroxidation is an important mechanism of cell membrane destruction (Fig. 5). CAP not only disrupts biomolecules but also activates them, leading to formation of nucleotides, proteins, and lipid radicals, which can be detected by chemical probes (Alcock et al. 2018; Sato and Nakamura 2019; Yamanaka et al. 2012) and immobilizing CAP-treated cells.

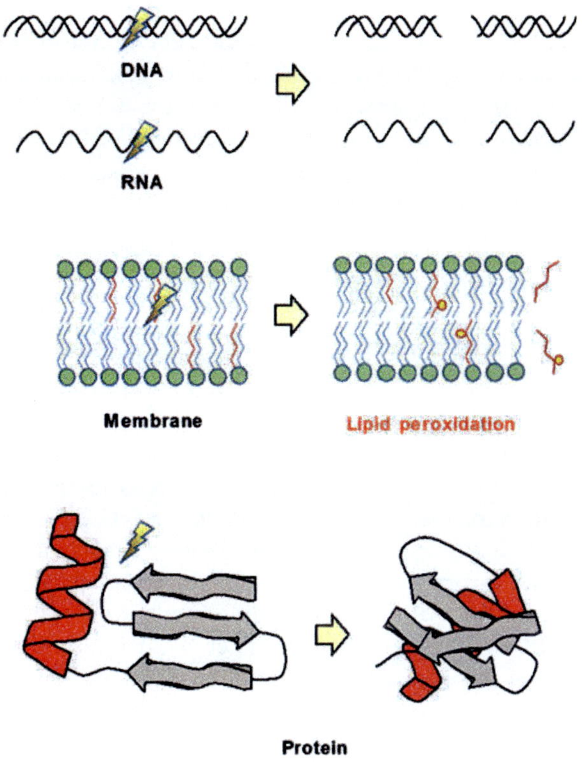

Fig. 5 Biomolecules modified by CAP. DNA, RNA, lipids, and proteins are modified by CAP

5.1 *Nucleic Acids (RNA and DNA)*

The plasma medicine has demonstrated that CAP induces DNA damage in a dose-dependent manner in both prokaryotic and eukaryotic systems (Arjunan et al. 2015). However, CAP-induced DNA damage of cancer cells could be a potential benefit of this mechanism. Inactivation of influenza viruses by modifying the viral genomic RNA and degrading viral proteins has been previously shown (Sakudo et al. 2013). Furthermore, Sakudo et al. showed that liquid-phase reactive chemical products (e.g., H_2O_2, NO_3, and NO_2) under ultraviolet light (UV-A) and slightly increased temperature can inactivate the adenovirus (Sakudo et al. 2016). Further, isolated and cellular DNA can be damaged by CAP treatment in vitro (Arjunan et al. 2015). Finally, CAP also induces DNA damage in glioblastoma U87MG and colorectal carcinoma HCT-116 cells (Vandamme et al. 2012).

5.2 *Plasma Membrane (Lipid)*

CAP does not induce lipid peroxidation in human keratinocytes (HaCaT), human colorectal carcinoma cells (HCT-116), or skin melanoma cells (SK-MEL-28) (Dezest et al. 2017). In contrast, culture medium activated by CAP using pure Ar gas in ambient air induced the apoptotic death of HeLa cells via lipid peroxidation. PAM induced intracellular RONS generation in HeLa cells, leading to caspase activation and lipid peroxidation of intracellular membranes followed by induction of apoptosis (Furuta et al. 2017).

5.3 *Amino Acid Modification and Protein Degradation*

Proteins are large biological molecules consisting of amino acid residues and perform a wide range of functions. The functions of each protein are the result of its structure and conformation, which depend on the covalent, ionic, and hydrogen bonds, in addition to other chemical interactions. As described below, CAP directly modifies and destroys the protein structure in vitro, leading to loss of their function.

The sensitivity of Cysteine (Cys) to RONS is well documented. Lackmann et al. studied the CAP-mediated modification of Cys by using Fourier-transform infrared (FTIR) spectroscopy (Lackmann et al. 2018). They detected the various modifications of Cys (e.g., cysteine sulfenic acid, *S*-nitrosocysteine and hydrogen-abstracted cysteine) following CAP treatment. Tyrosine (Tyr) is another amino acid with documented sensitivity to ROS and RNS. Nitrotyrosine (3-NT) is a biomarker of RNS-dependent stress. In proteins, 3-NT is formed under conditions of oxidative and nitrosative stress. Tyrosine nitration is detected in many organs under diseased conditions (Souza et al. 2008). Takai et al. investigated the CAP-mediated amino acids modification in vitro (Takai et al. 2014a). They reported that CAP treatment led to the following chemical modifications in the several amino acids (i) hydroxylation and nitration of tyrosine, phenylalanine, and tryptophan; (ii) sulfonation and disulfide linkage formation of cysteine; (iii) sulfoxidation of methionine; (iv) amidation and ring-opening of histidine and proline.

RONS produced by CAP regulate the protein activities via direct amino acid modifications. Accordingly, CAP modifies the protein structure and/or degrade proteins (e.g., lipase (Li et al. 2011), lysozyme (Takai et al. 2012), hemoglobin, polyphenoloxidase, and peroxidase (Surowsky et al. 2013), α-chymotrypsin (Attri and Choi 2013), and MTH1880, a thermophilic protein from *Methanobacterium thermoautotrophicum* (Attri et al. 2018). Furthermore, CAP induced heme degradation in horseradish peroxidase (Ke and Huang 2013), lactate dehydrogenase (LDH) (Zhang et al. 2015), and myoglobin (Park et al. 2016). The α-helix and β-sheet of MTH1880 decreased and increased slightly, respectively, after plasma treatment in a time-dependent manner (Attri et al. 2018).

Gap junctions are intercellular channels formed by connexin proteins, such as connexin 26, 30.3, 31, 32, 37, 43, and 45. The mRNA expression of connexins is associated with cancer prognosis (Aasen et al. 2019). Reactive species (·OH and HO_2·) produced by CAP interacts with the N-terminal of human connexin 26 (Cx26), thereby disrupting the gap junction and channels containing Cx26 (Xu et al. 2018b).

The amyloid-β protein (Aβ), a mediator of Alzheimer's disease (AD), forms fibrils in the plaques found in the brains of patients with AD. They are critical components of the AD pathogenesis (Panza et al. 2019). Treatment with CAP affects the protease (trypsin) resistance of the Aβ fibrils, making them prone to degradation (Takai et al. 2014b).

6 Changes in Protein Expression After CAP Treatment

CAP directly modifies the amino acids found in proteins, disturbs protein folding, and indirectly modulates gene expression levels in cells. The expression of chaperone-related proteins, inflammatory cytokines, redox-related transcriptional factors, and DNA damage-response proteins are both up- or downregulated by CAP.

6.1 Chaperone Proteins

Chaperone-related proteins, including heat shock proteins (HSPs), are modulated by CAP treatment. The HSPs are expressed in response to the intracellular and extracellular stress conditions and are involved in the regulation of ROS-driven apoptosis. The HSP90 chaperone was cleaved after CAP treatment in SW480, MDA-MB-231, MCF7, and NCF3 cells (Bekeschus et al. 2019). Further, CAP increased the amount of HSP27 in the human ovarian cancer cell line OVCAR-3, but not SKOV-3, the ovarian cancer cells (Bekeschus et al. 2018b). The expression of HSP genes and the formation of HSP104 aggregates in the budding yeast *Saccharomyces cerevisiae* were induced by CAP treatment(Itooka et al. 2016).

Calreticulin (CRT) acts as a molecular chaperone for the correct folding of glycosylated proteins in the ER. Cell membrane-bound CRT is released as a damage-associated molecular pattern (DAMP) molecules and induces immune responses to cancer cells. Further, it is increased by CAP treatment in both viable and dead CT26 colon cancer cells (Bekeschus et al. 2018a). CAP induces oxidative stress in human lung carcinoma cells A549 leading to surface exposure of CRT (Lin et al. 2017b). Treatment with CAP induced ER stress and modulated intracellular localization of CRT.

6.2 *Inflammation-Related Proteins*

CAP increased the copy numbers of interleukin-8 (IL-8) mRNA and the secretion of IL-8 in THP-1 cells, while other inflammatory molecules (IL-1β, IL-6, IL-10, IL-17A, IL-22, granulocyte–macrophage colony stimulating factor (GM-CSF), interferon-γ (INFγ), tumor necrosis factor α (TNF-α), and transforming growth factor-β (TGFβ)) were not modulated (Bekeschus et al. 2016). In addition, CAP significantly induced the expression of VEGF-A in HaCaT human keratinocytes, resulting in an anti-inflammatory effect via a reduction in the levels of IL-1β and IL-6 expression by TNF-α (Lee et al. 2019). Melanoma B16F10 cells treated with CAP secrete TNFα, IL-10, chemokine (C–C motif) ligands 4 (CCL4), and IL-1β into the culture medium (Rödder et al. 2019). CAP treatment promotes pro-inflammatory cytokine levels in the tissue during wound healing. Peroxisome proliferator-activated receptor (PPAR)-γ is a transcription factor mainly involved in lipid metabolism and inflammation (Varga et al. 2011). He-CAP increases the mRNA and protein expressions of PPAR-γ, a regulator of inflammatory response in primary cultured human intestinal subepithelial myofibroblasts (ISEMFs) and hepatic stellate cells (HSCs) (Brun et al. 2014). Therefore, CAP can regulate inflammation via modulating inflammation-related proteins.

6.3 *Redox Response Proteins*

Kelch-like ECH associated protein 1 (KEAP1) is one of the CUL3-based E3 ubiquitin ligases. The activation of the NF-E2-related actor 2 (NRF2) was controlled by KEAP1. NRF2/KEAP1 signaling pathway is indicated in the adaptive response to extra- and intracellular stresses and protects against chemicals and cellular stress (Kensler and Wakabayashi 2010). NRF2-related signaling proteins are affected by CAP treatment. CAP-induced cell death is inhibited by a NRF2-related antioxidant system (Ishaq et al. 2014). The KEAP1-NRF2 pathway regulates the expression of numerous cytoprotective genes, including *HO-1* and *NADPH-quinone oxidoreductase 1*. *HO-1* is upregulated in response to a number of stimuli, including heme, NO, heavy metals, growth factor, cytokines, and modified lipids (Loboda et al. 2016). Schmidt et al. showed that NRF2 and the components of the phase II enzyme pathway acted as the key regulators of the cellular response to CAP in keratinocytes (Schmidt et al. 2015).

NRF2 not only mediates cellular defense against increased ROS levels, but also regulates wound healing. Recently, Schmidt and Bekeschus showed that CAP treatment improves the wound-healing processes via redox-regulated pathways (Woedtke et al. 2019). They reviewed that the improved wound healing by CAP involved multiple cellular events, a transient and reversible modification of proteins and the lipid bilayer, regulation of inflammation signaling, enhanced expression of angiogenetic factors, and modulation of cell proliferation and death.

CAP treatment facilitated the accumulation of RONS via intracellular glutathione depletion, inactivation of superoxide dismutase and catalase, and nitric oxide synthase (NOS) families [iNOS (inducible NOS), cNOS (constitutive NOS), neuronal NOS (nNOS) and endothelial NOS (eNOS)] in HepG2 cells (Zhao et al. 2013). Moreover, CAP significantly increased the expression of the anti-inflammatory gene *HO-1* in THP-1 cells (Bekeschus et al. 2016).

CAP significantly induced the expression of *VEGF-A* in HaCaT cells (Lee et al. 2019) by activating extracellular signal-regulated kinase (ERK) protein. Lee et al. showed that the activation of ERK was essential for the hypoxia inducible factor (HIF) 1 alpha. The increase in NO in the media treated with CAP might be the main reason for the increases in HIF-1 protein activity.

6.4 DNA Damage-Related Proteins

CAP leads to DNA damage (phosphorylation of H2A.X), increases the levels of intracellular ROS, and modulates the oxidative stress signaling in the human T-lymphoblastoid cell line Jurkat (Turrini et al. 2017). Further, CAP induces the transcriptional regulations of copper-zinc superoxide dismutase 1, catalase, and glutathione-disulfide reductase in these cells.

Tanaka et al. showed that the plasma-activated Ringer's lactate solutions (PAL) and PAM (Tanaka et al. 2019) downregulated phosphor-AKT in the human glioblastoma cell line U251SP. However, PAM induced higher levels of intracellular ROS in U251SP cells than PAL. PAM upregulates the expression of GADD45α and other genes involved in stress-induction pathway, in contrast PAL cannot upregulate the expression of GADD45α significantly. PAM induces oxidative stress to a greater extent than PAL. Notably, PAL does not upregulate the expression of ATF3 and c-JUN, which are downstream of GADD45α.

6.5 CAP Induces Dysfunction of Cellular Organelles

The oxidative and nitrosative stress causes dysfunction of cell, thereby resulting in the modifications to biomolecules. Organelles can sense perturbations in the cellular homeostasis caused by oxidative and nitrosative stress (Galluzzi et al. 2014). These organelles, specially the mitochondria, ER, lysosomes, and the Golgi apparatus, are primary components to integrate the cell damage sensing and cell death signaling (Ferri and Kroemer 2001).

The underlying mechanisms behind CAP-induced cell death include the structural and functional alteration of organelles (Yan et al. 2017, 2015). CAP increases $[Ca^{2+}]_i$ level and causes loss of mitochondrial membrane potential ($\Delta\Psi m$) in human cultured cells. The intracellular level of calcium is regulated by organelles under normal conditions. The treatment of cells with CAP induces the dysfunction of the organelles, thereby resulting in perturbations of $[Ca^{2+}]_i$ levels (Fig. 6). The CAP-induced morphological and functional changes of organelles are considered as cell death-initiating events. The mechanisms of CAP-induced organelle dysfunction are discussed in the following subsections.

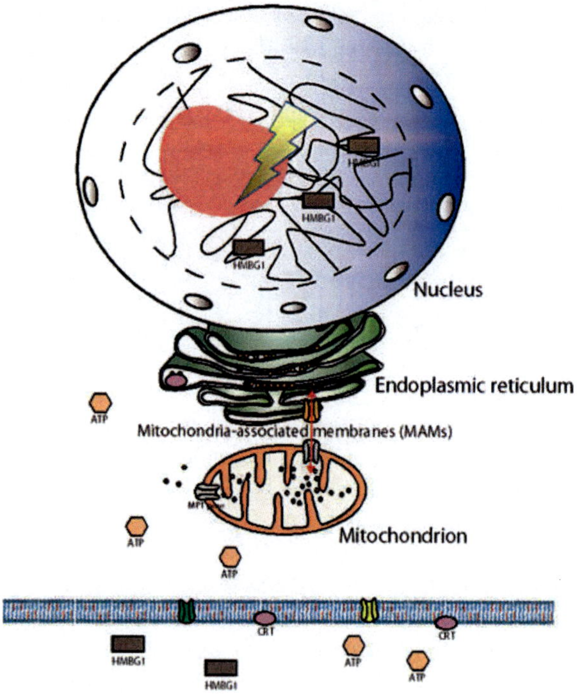

Fig. 6 Organelle modification by CAP. The ER and mitochondria are the main organelles modulated by CAP

6.6 *Effect of CAP Treatment on Mitochondrial Function*

The most important function of mitochondria is to produce energy, in the form of ATP. In addition, mitochondria have capacity to buffer Ca^{2+} between 50 and 500 nM in cells. The underlying mechanisms of Ca^{2+} transport by mitochondria are very systemic and are therefore easily perturbed by a redox imbalance. The mitochondrial membrane permeability transition (MPT) occurs under oxidative and nitrative stress, and high $[Ca^{2+}]_i$ conditions. The Ca^{2+} overload and RONS open or close membrane permeability transition pores (PTP), thereby altering the levels of substrates necessary for energy production in mitochondria (Rasola and Bernardi 2011). The mitochondrial dysregulation induced by CAP treatment has been previously evaluated.

- He-O_2 CAP treatment induced apoptosis and the loss of $\Delta\Psi$m in LP-1 multiple myeloma cells (Xu et al. 2018a).
- The depolarization of the mitochondria membrane potential and the generation of ROS were induced by N_2- and AirCAP in the human cervical cancer cell line HeLa (Ahn et al. 2011).
- Both extracellular and intracellular RONS were produced by AirCAP in HeLa cells (Ahn et al. 2014). These RONS induced the depolymerization of the $\Delta\Psi$m and mitochondrial ROS accumulation, along with activations of c-jun NH_2-terminal kinase (JNK) and p38 kinase, followed by apoptotic cell death in HeLa cells.

However, the underlying mechanism of mitochondrial dysfunction by CAP treatment remains unclear. It is speculated that the RONS produced by CAP treatment may inhibit mitochondrial functions (e.g., ATP production and Ca^{2+} storage ability).

6.7 *Effects of CAP Treatment on the Function of the ER*

As described above, the expression and localization of stress response proteins in the ER are regulated by intracellular and extracellular RONS, whereby ER-related stress induces Ca^{2+}-dependent MPT and apoptotic cell death (Deniaud et al. 2008). He-CAP triggers ER stress in HepG2 cells. Homeostasis of $[Ca^{2+}]_i$ is disturbed along with alterations in the expression of glucose-regulated protein 78 (GRP78), C/EBP-homologous protein (CHOP), and pro-caspase 12. These effects can accumulate over time and eventually cause cellular dysfunction and ER stress-mediated cell death (Zhao et al. 2013).

A relationship between the Ca^{2+} homeostasis in mitochondrial and ER and cell death has been previously demonstrated (Marchi et al. 2018). ER-mitochondrial Ca^{2+} transfer is possible via mitochondria-associated ER membranes (MAMs). Studies have suggested that CAP may regulate the function of these membranes, thereby modulating the Ca^{2+} transfer.

7 CAP Treatment Causes Cell Cycle Arrest

The cell cycle is controlled by three major checkpoints and is tasked with maintaining the genomic stability. Cell cycle arrest is a stopping point in the cell cycle (Fig. 7). In a previous study using HCT-116 cells, the CAP treatment at two doses (10 and 20 J/cm^2) led to a significant reduction in the number of cells in G_0/G_1-phase, while an increase in the number of cells in S phase was observed in response treatment with 20 J/cm^2 CAP. In U87MG cells, CAP induced a 92% and 97% decrease in BrdU incorporation at the doses of 10 and 20 J/cm^2, respectively (Vandamme et al. 2012). In addition, the cell cycle distribution of THP-1 cells treated with CAP showed a greater number of cells in the G_2 phase (Bekeschus et al. 2016).

PAM is produced by exposing a liquid medium to CAP. The indirect irradiation of cells with PAM has a cytotoxic effect on tumor cells, similar to direct irradiation with CAP (Kaushik et al. 2018). In a previous study using human lung adenocarcinoma

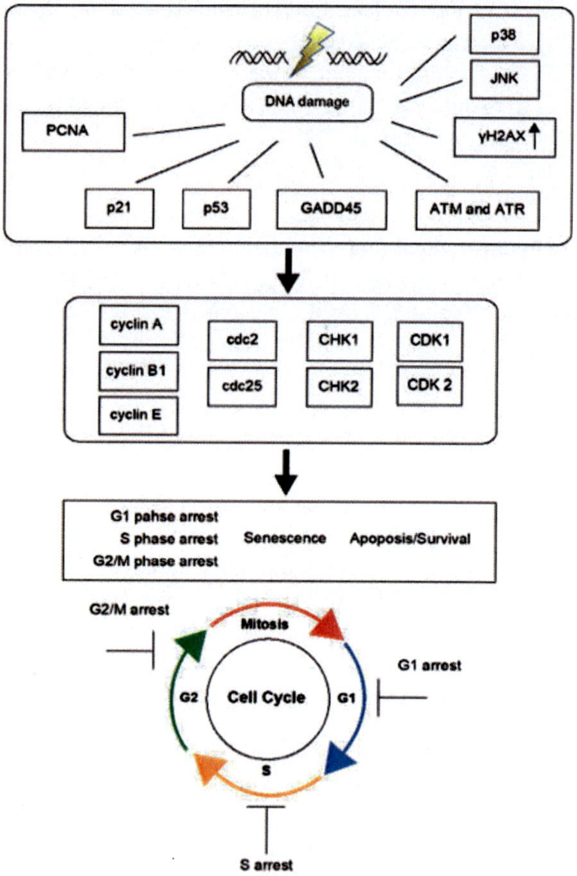

Fig. 7 Schematic diagram depicting points of cell cycle arrest by CAP along with cell cycle-regulating proteins

epithelial A549 cells, PAM reduced the percentage of cells in the G_0/G_1 phase, but increased the number of cells in G_2/M phase in (Hara et al. 2019).

Both CAP and ionizing radiation individually induced G_2/M phase arrest alone, while their combination significantly increased cell cycle arrest in HeLa and HepG2 cells (Lin et al. 2018). Further, CAP reduced the cell viability of murine melanoma B16 cells, following the induction of G_1/S and G_2/M cell cycle arrest and apoptosis (Shi et al. 2014). CAP-induced alteration of the cell cycle was due to the down-regulation of the cell cycle-related protein, such as cyclin B1 and cyclin D1 (Shi et al. 2014). Yan et al. found that treatment with CAP increased the apoptotic cell death associated with cell cycle arrest at the G_2/M phase in human hepatocellular carcinoma cells (HepG2) (Yan et al. 2010). Moreover, CAP increased the levels of the cyclin-dependent kinase (CDK) inhibitor, p21, and the tumor suppressor protein, p53.

8 Cell Death-Inducing Activity of CAP

Although the underlying mechanisms of CAP-induced cell death are not yet fully understood, the apoptotic and non-apoptotic cell death pathways are reported to play a role in CAP-induced cell death (Fig. 8).

CAP induces both apoptotic and non-apoptotic cell death in a time- and dose-dependent manner (Hoffmann et al. 2013). Although He- and He–N_2–CAPs induced apoptotic cell death in human primary keratinocytes, human keratinocytes HaCaT, and human colorectal cancerous cells HCT-116, low levels of apoptosis were observed in human primary fibroblasts and skin melanoma SK-MEL-28 cells (Dezest et al. 2017). This indicated that He- and He-N_2CAP induce apoptotic cell death in HaCaT and HCT-116 cells, while SK-MEL-28 cells are resistant to treatment with He- and He-N_2CAP.

Ahn et al. reported that Air- and N_2CAP treatment of HeLa cells induced apoptosis via caspase activation and mitochondrial dysfunction (Ahn et al. 2011). Similarly, in HEK293T cells, N_2CAP activated caspase −3, −8, and −9 related pathways involved in apoptotic cell death (Iuchi et al. 2018).

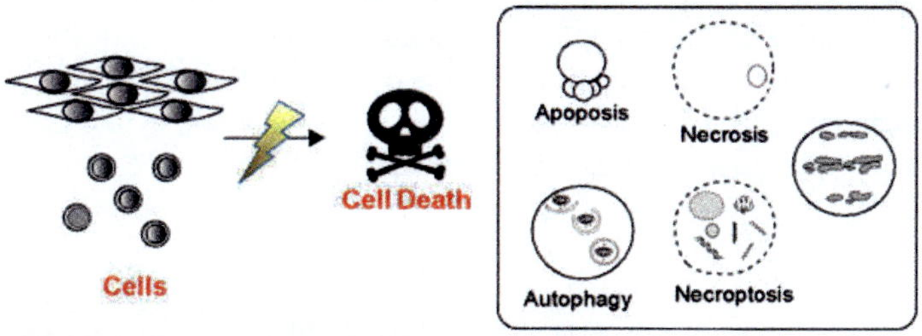

Fig. 8 Apoptotic and non-apoptotic cell death induced by CAP

Chang et al. showed that CAP induced apoptotic cell death in oral cavity squamous cell carcinoma (OSCC). Moreover, apoptosis of p53 wild-type OSCCs was mediated through a novel mechanism involving DNA damage induced by CAP. They also showed that the cell cycle arrest happened at sub-G_1 via the ATM/p53 pathway in p53 wild-type OSCCs, but not in p53-mutated OSCCs (Chang et al. 2014).

In Jurkat cells, CAP induces apoptotic cell death via the activation of caspase-8 following upregulation of BAX and Bcl-2 (Turrini et al. 2017). In addition to apoptotic and necrotic cell death, autophagy is another cellular response to stress in benign prostatic hyperplasia (BPH) epithelial BPH-1 cells and prostate cancer PC-3 cells (Hirst et al. 2015).

CD95 (APO-1/Fas) is a member of TNF receptor superfamily and a key cell death receptor (Peter and Krammer 2003). Xu et al. showed that He–O_2 CAP treatment significantly induced the expression of CD95 and phospho-p53 in LP-1 multiple myeloma cells (Xu et al. 2018a). He–O_2 CAP treatment resulted in apoptosis and the loss of mitochondrial membrane potential in LP-1 multiple myeloma cells (Xu et al. 2018a).

Conway et al. showed that CAP treatment induced accumulation of lysosomes and caspase-independent cell death in glioblastoma multiforme U373MG cells (Conway et al. 2019). Furthermore, rapid accumulation of acidic vesicles and cell death following CAP treatment in p53-mutated glioblastoma multiforme (GBM) cells was described. The authors showed that autophagy was not involved in CAP-induced cell death. Instead, the rapid accumulation of late stage endosomes/lysosomes followed by membrane permeabilization, loss of $\Delta\Psi m$, and caspase-independent cell death was reported. Lunov et al. showed that He-CAP induced necroptosis, which was inhibited by necrostatin-1 (Nec-1, a well-known inhibitor of necroptosis) (Lunov et al. 2017).

9 The Regulation of Cellular Function by CAP

CAP is known to enhance osteoblast differentiation from a pre-osteoblastic cell line, MC3T3-E1, along with activation of alkaline phosphatase (Tominami et al. 2017). Moreover, treatment with air-based CAP and N_2CAP leads to apoptotic cell death through mitochondrial dysfunction and caspase activation in HeLa cells (Ahn et al. 2011). N_2CAP activated several pathways related to apoptotic cell death, while He-CAP triggered ER stress in HepG2 cells (Zhao et al. 2013). CAP, in combination with gold nanoparticles and hyperthermia, showed synergistic cytotoxic effects on cultured cancer cell lines such as U937, MOLT-4, and HCT-116 (Moniruzzaman et al. 2017).

9.1 CAP Regulates Cell Proliferation

CAP has been used for wound healing and tissue regeneration. The enhancement of cell proliferation mediated by CAP treatment enhances cell cycle progression; the ROS produced by CAP facilitate the cell cycle progression.

He-CAP treatment induces the proliferation and migration of primary cultured human cells, such as hepatic stellate cells (HSCs) and intestinal subepithelial myofibroblasts (ISEMFs), mainly through the formation of intracellular ROS (Brun et al. 2014).

Plasma causes hypertrophy and hyperplasia of THP-1 cells in a time-dependent manner. A short-term plasma treatment stimulates cell proliferation, while a long-term exposure compromises cell viability modestly, with a concomitantly enhanced metabolic activity (Bekeschus et al. 2016).

CAP activates cell proliferation of various mesodermal-derived human adult stem cells via promoting the G1-S transition, as they retain their pluripotent and stem cell properties (Park et al. 2019). Park et al. showed that CAP enhanced the cell proliferation of adipose tissue-derived stem cells (ASCs), bone marrow-derived mesenchymal stem cells (BM-MSCs), and hematopoietic stem cells. Further, CAP upregulated the protein expression of CD44 and CD105, the stem cell-specific surface markers. Exposure to CAP also increased the expressions of *Oct4*, *Sox2*, and *Nanog*, the well-known pluripotent genes, in ASCs and BM-MSCs compared to untreated cells. Park et al. indicated that these effects were mediated by the acceleration of G1-S phase and the increase in the expression of PCNA in CAP-exposed ASCs.

9.2 CAP Induces Cell Differentiation

Cell proliferation is an opposing phenomenon to cell differentiation. As described above, CAP treatment enhances cell proliferation. However, CAP has also been found to enhance osteoblast differentiation in MC3T3-E1 cells (Tominami et al. 2017) and effectively directs the differentiation of neural stem cells (NSCs) into neuronal lineage in vitro (Xiong et al. 2014). Xiong et al. showed that CAP induced rapid proliferation and differentiation with longer neurites and cell bodies in murine neural stem cells, eventually forming neuronal networks (Xiong et al. 2014). CAP induced neural differentiation in mouse neuroblastoma Neuro 2A (N2a) cells and promoted neural maturation in living zebrafish (Jang et al. 2018). Jang et al. showed that extracellular NO, mitochondrial O_2^-, and cytosolic H_2O_2 played important roles in CAP-induced neural differentiation. The tyrosine kinase receptor Ras and the ERK signaling pathway are also triggered via intracellular H_2O_2, thereby inducing neural differentiation. Collectively, these findings suggest that CAP can determine cell fate by regulating cell proliferation and differentiation.

Neuronal differentiation has been studied in a variety of systems, such as embryonic carcinoma, neuroblastoma, and pluripotent stem cells, and sophisticated animal systems. The regulation of cell differentiation is tissue-specific, which leads to the complexity of the cell fate between proliferation and differentiation (Ruijtenberg and van den Heuvel 2016). The induction of neural differentiation by CAP might be dependent on the amount and types of ROS and/or RNS to stimulate gene expression.

9.3 The Cellular Response

The cell migration rate of primary mouse fibroblast cells was reportedly altered by a mild plasma treatment (Volotskova et al. 2011). CAP enhanced migration of macrophage-like RAW264.7 cells to attempt healing of an artificial wound (Miller et al. 2014). The cellular migration may be stimulated by CAP-produced RONS. Indeed, ROS was involved in the cell migration and adhesion (Hurd et al. 2012).

10 The Application of CAP in the Treatment of Cancer

CAP exerts a strong anti-proliferative effects on cultured cancerous cells in vitro. In the past decade, CAP have shown promising applications in cancer therapy (Yan et al. 2017). The reactive molecules produced by CAP have been suggested for cancer treatment. The clinical application of direct CAP treatment for head and neck cancer treatment has been reported (Metelmann et al. 2015).

10.1 Role of CAP in Cancer Therapy

CAP has the potential for use as a novel cancer therapeutic modality (Hirst et al. 2016). Several reports have shown that cancer cells are more sensitive to treatment of CAP than normal cells (Volotskova et al. 2012), due to differences in their cell cycle distribution. The expression of γH2A.X (pSer139), a DNA damage marker during S-phase, is specifically increased in CAP-treated cells. In vitro and in vivo studies have shown the selective effect of CAP against cancer cells without damaging healthy cells (Keidar et al. 2011). Exposure to CAP induces cell death in oral squamous carcinoma cell lines such as SAS, Ca9-22, HSC-2, HSC-3, HSC-4, Sa3 and, Ho-1-u-1, more readily than in fibroblasts such as HS-K and IMR-90-SV, which is dependent on the levels of catalytic Fe(II) (Sato et al. 2019). CAP treatment also inhibited the growth of established tumors in a murine melanoma model. However, the underlying mechanism of the anti-proliferative effects of CAP has not yet been elucidated.

10.2 The Regulation of the Immune System by CAP

Inducing killing of cancer cells using immunogenic cell death (ICD) is latest in line of cancer treatment approaches (Zhou et al. 2019). ICD is a form of regulated cell death that improves the effectiveness of anti-cancer treatment and relieves the suffering of patients. ICD is defined by the exposure of damage-associated molecular patterns (DAMPs) in the tumor microenvironment and stimulating the dysfunctional anti-tumor immune system. DAMPs [high-mobility group box 1 (HMGB1), ATP, surface-exposed CRT, and HSP] are associated with ICD (Fig. 9) (Kroemer et al. 2013; Terenzi et al. 2016).

CAP induces ICD for a specific and systemic response (Miller et al. 2015). PAM treatment induces the CRT expression in cells and increases the quantity of ATP released from the human mamma carcinoma cell line (HBL), human sporadic melanoma biopsy specimens with BRAF mutations (Hmel1), and the human pancreatic cancer cell line PANC-1 (Azzariti et al. 2019).

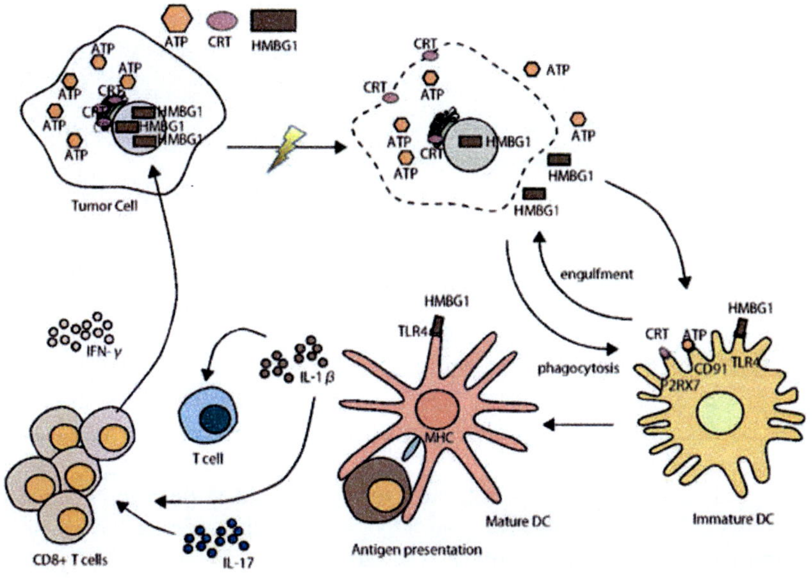

Fig. 9 Biological mechanism of immunogenic cell death (ICD). Intracellular and extracellular stress and/or cell death lead to ER stress and autophagy. Calreticulin (CRT), chaperone protein in the ER, is exposed on the outer membrane of cells during ICD. Adenosine triphosphate (ATP) is not only an energy currency but also a secreted molecule during apoptosis. As the cellular membrane becomes permeabilized during secondary necrosis, high-mobility group box 1 (HMGB1) is released from cells during ICD. CRT, ATP, and HMGB1 bind to receptors on immature dendritic cells (DCs). This initiates the maturation and recruitment of DCs into the tumor bed the engulfment of tumor antigens by DCs and optimal antigen presentation to T cells (Kroemer et al. 2013). Taken together, these processes cause a potent immune response which can potentially lead to the eradication of tumor cells. IFN, interferon; IL, interleukin; TLR, Toll-like receptor. The illustration was adapted from Terenzi et al. (2016)

Lin et al. reported that ROS, generated by nanosecond-pulsed dielectric barrier discharge plasma, triggered ICD in lung carcinoma cell line, A549 (Lin et al. 2017b). Further, CAP induced the release of surface CRT and ATP secretion, with the concomitant production of ROS. They also showed that CAP induced both cell death and CRT upregulation in mouse and human melanoma cell lines, B16F10 and A375, respectively (Lin et al. 2019). Plasma regulates the functions of macrophages, including cytokine secretion, migration, antigen presentation to T and B cells, and cancer killing. The treatment of macrophages with a nanosecond-pulsed dielectric barrier discharge directly enhanced their cytotoxic activity against tumor cells, but not against normal cells (Lin et al. 2017a).

B16F10 melanoma cells treated with CAP are recognized immunologically and exert stimulating effects on immune cells (Rödder et al. 2019). B16F10 melanoma cells exposed to CAP show reduced metabolic and migratory activity, as well as an increased release of danger signals [ATP and chemokine (C-X-C motif) ligand 1 (CXCL1)]. The levels of interleukin-1β (IL-1β) and CCL4 are increased in CAP-treated mono- and co-cultures with immune cells. Plasma-treated melanoma cells induce ERK phosphorylation and increase CD28 expression in T cells, suggesting their activation. In monocytes, the expression of CD115, a marker for activation, is elevated.

10.3 CAP Suppresses Cancer Stem Cells

Cancer stem cells (CSCs) are a distinct subpopulation of tumor cells and possess stem-like properties. The homeostasis of iron is dysregulated in CSCs, which show a high concentration of intracellular iron (Recalcati et al. 2019). In biological studies, two-dimensional (2D) cell cultures have been used as in vitro models. However, the properties of 2D cells are different from the 3D cellular structure in the tissues and organs (Antoni et al. 2015). Lately, the 3D culture models have been used in the search of novel cancer chemotherapeutic targets (Bielecka et al. 2017; Maliszewska-Olejniczak et al. 2019).

Judee et al. reported the anti-proliferative effects of PAM on colon adenocarcinoma multicellular tumor spheroids (MCTSs), a model that mimics the 3D organization and microtumor regions (Judee et al. 2016). They found that H_2O_2 plays a major role in MCTS DNA damage. Sagwal et al. elucidated the effects of CAP on the cytotoxicity of traditional agents (doxorubicin, epirubicin, and oxaliplatin) on the tumor spheroids prepared from human (SK-MEL-28) melanoma cells (Sagwal et al. 2018); CAP treatment increased cytotoxicity in SK-MEL 28 tumor spheroids pretreated with cytotoxic agents.

PAM inhibits the viability of head and neck FaDu MCTSs in a time-dependent manner (Chauvin et al. 2018). The growth of spheroids generated from HCT-116 was delayed depending on the time of exposure to PAM (Chauvin et al. 2019). PAM impaired the viability of HCT-116 spheroids via intracellular ATP depletion, leading to apoptotic and necrotic cell death with mitochondrial oxidative stress.

10.4 The Synergistic Effects of CAP on Cytotoxicity in Combination with Other Therapeutic Tools

Nanoparticles (NPs) are increasingly being used in various biomedical applications (Zottel et al. 2019). Metal nanoparticles induce hyperthermic cytotoxicity when exposed to near-infrared radiation or radiofrequency fields (Cherukuri et al. 2010). The synergistic effects cytotoxic effects of CAP in combination with gold NPs and hyperthermia, have been previously demonstrated (Moniruzzaman et al. 2017). Li et al. described the synergistic effect of CAP and iron oxide-based magnetic nanoparticles (MNPs) as anti-proliferative, ROS-producing, and cell cycle arrest inducing in A549 cells (Li et al. 2019). They also showed that iron oxide-based MNPs and CAP inhibited the migration and invasiveness of A549 cells. Moreover, iron oxide-based MNPs and CAP attenuated tumor growth in xenograft nude mice models.

Sulfasalazine (SSZ) is a well-known anti-inflammatory drug that is commonly used as a first-line treatment for many rheumatic diseases (Narayan et al. 2017). SSZ is also an inhibitor of the glutamate transporter xCT (SCL7a11, system Xc-, SXC), which is known to reduce the levels of intracellular glutathione (GSH), leading to increased cellular oxidative stress. The effects of SSZ at a non-toxic concentration enhanced the CAP-induced death of human lymphoma Molt-4 (p53 WT) cells (Moniruzzaman et al. 2018).

In human and murine tumorigenic melanoma cell lines, SK-MEL-28 and B16, respectively, combined with doxorubicin, epirubicin, or oxaliplatin, and plasma treatment significantly increased anti-cancer activity in both 2D and 3D cultures. The synergistic effects in tumor cells was due to an enhanced intracellular accumulation of anti-cancer drugs via the upregulation of the organic cationic transporter SLC22A16 by CAP (Sagwal et al. 2018).

Melittin (MEL) is an amphipathic 26 amino acid α-helical peptide obtained from bee venom, which exhibits anti-cancer effects in vitro and in vivo. Shaw et al. showed that plasma-treated phosphate-buffered saline solution (PT-PBS) can enhance MEL-induced cell death with a concomitant increase in lipid peroxidation in MCF7 and A375 cells (Shaw et al. 2019). The combination of PT-PBS and MEL treatments reduced the size of tumors, as well as the expression of Ki-67, in melanoma cancer tumors *in ovo.*

Silymarin is a flavonolignan from the "milk thistle" (*Silybum marianum*) plant that can be incorporated into a nano-emulsion formulation for therapeutic applications. Adhikari et al. showed that this silymarin nano-emulsion (SN) has synergistic activity similar to that of CAP in G-361 human melanoma cells (Adhikari et al. 2019). The combination of SN and AirCAP increased the cytotoxicity in a time-dependent manner. Intracellular RONS significantly increased the levels of the DNA damage marker γ-H2AX, the potent melanoma cell surface marker PD-1, and the DNA methyltransferase DNMT in G-361 cells. Co-treatment with SN and AirCAP resulted in the increase in the levels of caspase −8, −9, and −3/−7, and PARP, and the expression of apoptotic genes, thereby blocking the HGF/c-MET pathway. Decreases in the levels of EMT markers (E-cadherin, YKL-40, N-cadherin, SNAI1) were found to occur simultaneously with a decline in the number of melanoma cells (BRAF, NAMPT) and stem cells markers (CD133, ABCB5). In vivo results have shown a significant reduction in SN with PAM in combination with a reduction in tumor weight and size.

11 Conclusion

This review discusses the latest information available regarding the regulatory activities of CAP and PAM in cellular functions. The unique characteristics of CAP are useful for in clinical applications and as a biological research tool for investigating their novel cellular mechanisms.

Acknowledgements I would like to thank my colleagues, Tomoyuki Murakami, Hisashi Hisatomi, Yoji Saito, and Takahiro Himuro, for their stimulating discussions. Tomoyuki Murakami was a great help in the preparation of this manuscript. I would also like to thank Katsuhisa Kitano and Kentaro Shiraki for their help in creating this manuscript. Finally, I thank Yuko Sakurada, Hazuki Kashiwa, and Takashi Yasuhiro for their assistance with figure preparation.

References

Aasen T et al (2019) Connexins in cancer: bridging the gap to the clinic. Oncogene 38:4429–4451. https://doi.org/10.1038/s41388-019-0741-6

Adachi T, Nonomura S, Horiba M, Hirayama T, Kamiya T, Nagasawa H, Hara H (2016) Iron stimulates plasma-activated medium-induced A549 cell injury Sci Rep 6: 20928. https://doi.org/10.1038/srep20928

Adhikari M et al (2019) Cold atmospheric plasma and silymarin nanoemulsion synergistically inhibits human melanoma tumorigenesis via targeting HGF/c-MET downstream pathway. Cell Commun Signal 17:52. https://doi.org/10.1186/s12964-019-0360-4

Ahn HJ et al (2014) Targeting cancer cells with reactive oxygen and nitrogen species generated by atmospheric-pressure air plasma. PLoS ONE 9:e86173. https://doi.org/10.1371/journal.pone.0086173

Ahn HJ, Kim KI, Kim G, Moon E, Yang SS, Lee JS (2011) Atmospheric-pressure plasma jet induces apoptosis involving mitochondria via generation of free radicals. PLoS ONE 6:e28154. https://doi.org/10.1371/journal.pone.0028154

Alcock LJ, Perkins MV, Chalker JM (2018) Chemical methods for mapping cysteine oxidation. Chem Soc Rev 47:231–268. https://doi.org/10.1039/c7cs00607a

Antoni D, Burckel H, Josset E, Noel G (2015) Three-dimensional cell culture: a breakthrough in vivo. Int J Mol Sci 16:5517–5527. https://doi.org/10.3390/ijms16035517

Arjunan KP, Sharma VK, Ptasinska S (2015) Effects of atmospheric pressure plasmas on isolated and cellular DNA-a review. Int J Mol Sci 16:2971–3016. https://doi.org/10.3390/ijms16022971

Attri P, Choi EH (2013) Influence of reactive oxygen species on the enzyme stability and activity in the presence of ionic liquids. PLoS ONE 8:e75096. https://doi.org/10.1371/journal.pone.0075096

Attri P, Han J, Choi S, Choi EH, Bogaerts A, Lee W (2018) CAP modifies the structure of a model protein from thermophilic bacteria: mechanisms of cap-mediated inactivation. Sci Rep 8:10218. https://doi.org/10.1038/s41598-018-28600-w

Azzariti A et al (2019) Plasma-activated medium triggers cell death and the presentation of immune activating danger signals in melanoma and pancreatic cancer cells. Sci Rep 9:4099. https://doi.org/10.1038/s41598-019-40637-z

Bekeschus S, Lippert M, Diepold K, Chiosis G, Seufferlein T, Azoitei N (2019) Physical plasma-triggered ROS induces tumor cell death upon cleavage of HSP90 chaperone. Sci Rep 9: 4112. https://doi.org/10.1038/s41598-019-38580-0

Bekeschus S, Mueller A, Miller V, Gaipl U, Weltmann K-D (2018) Physical plasma elicits immunogenic cancer cell death and mitochondrial singlet oxygen. IEEE Radiation Plasma Medical Sci 2:138–146. https://doi.org/10.1109/trpms.2017.2766027

Bekeschus S, Schmidt A, Bethge L, Masur K, von Woedtke T, Hasse S, Wende K (2016) Redox stimulation of human THP-1 monocytes in response to cold physical plasma. Oxid Med Cell Longev 2016: 5910695. https://doi.org/10.1155/2016/5910695

Bekeschus S et al. (2017) Oxygen atoms are critical in rendering THP-1 leukaemia cells susceptible to cold physical plasma-induced apoptosis. Sci Rep 7: 2791. https://doi.org/10.1038/s41598-017-03131-y

Bekeschus S, Wulf C, Freund E, Koensgen D, Mustea A, Weltmann K-D, Stope M (2018b) Plasma treatment of ovarian cancer cells mitigates their immuno-modulatory products active on THP-1. Monocytes Plasma 1: 201–217. https://doi.org/10.3390/plasma1010018

Bielecka ZF, Maliszewska-Olejniczak K, Safir IJ, Szczylik C, Czarnecka AM (2017) Three-dimensional cell culture model utilization in cancer stem cell research. Biol Rev Camb Philos Soc 92:1505–1520. https://doi.org/10.1111/brv.12293

Brun P et al (2014) Helium generated cold plasma finely regulates activation of human fibroblast-like primary cells. PLoS ONE 9:e104397. https://doi.org/10.1371/journal.pone.0104397

Chang JW et al. (2014) Non-thermal atmospheric pressure plasma induces apoptosis in oral cavity squamous cell carcinoma: Involvement of DNA-damage-triggering sub-G(1) arrest via the ATM/p53 pathway. Arch Biochem Biophys 545: 133–140. https://doi.org/10.1016/j.abb.2014.01.022

Chauvin J, Gibot L, Griseti E, Golzio M, Rols MP, Merbahi N, Vicendo P (2019) Elucidation of in vitro cellular steps induced by antitumor treatment with plasma-activated medium. Sci Rep 9:4866. https://doi.org/10.1038/s41598-019-41408-6

Chauvin J, Judee F, Merbahi N, Vicendo P (2018) Effects of plasma activated medium on head and neck FaDu cancerous cells: comparison of 3D and 2D response anticancer agents. Med Chem 18: 776–783. https://doi.org/10.2174/1871520617666170801111055

Cherukuri P, Glazer ES, Curley SA (2010) Targeted hyperthermia using metal nanoparticles. Adv Drug Deliv Rev 62:339–345. https://doi.org/10.1016/j.addr.2009.11.006

Conway GE et al. (2019) Cold atmospheric plasma induces accumulation of lysosomes and caspase-independent cell death in U373MG glioblastoma multiforme cells. Sci Rep 9: 12891. https://doi.org/10.1038/s41598-019-49013-3

Deniaud A, Sharaf el dein O, Maillier E, Poncet D, Kroemer G, Lemaire C, Brenner C (2008) Endoplasmic reticulum stress induces calcium-dependent permeability transition, mitochondrial outer membrane permeabilization and apoptosis. Oncogene 27: 285–299. https://doi.org/10.1038/sj.onc.1210638

Dezest M et al (2017) Mechanistic insights into the impact of cold atmospheric pressure plasma on human epithelial cell lines. Sci Rep 7:41163. https://doi.org/10.1038/srep41163

Dubuc A, Monsarrat P, Virard F, Merbahi N, Sarrette JP, Laurencin-Dalicieux S, Cousty S (2018) Use of cold-atmospheric plasma in oncology: a concise systematic review . Ther Adv Med Oncol 10:1758835918786475. https://doi.org/10.1177/1758835918786475

Ferri KF, Kroemer G (2001) Organelle-specific initiation of cell death pathways. Nat Cell Biol 3:E255-263. https://doi.org/10.1038/ncb1101-e255

Fridman G, Friedman G, Gutsol A, Shekhter AB, Vasilets VN, Fridman A (2008) Applied Plasma Medicine. Plasma Processes Polym 5:503–533. https://doi.org/10.1002/ppap.200700154

Furuta R et al. (2017) Intracellular responses to reactive oxygen and nitrogen species, and lipid peroxidation in apoptotic cells cultivated in plasma-activated medium. Plasma Processes Polym 14: 1700123. https://doi.org/10.1002/ppap.201700123

Galluzzi L, Bravo-San Pedro JM, Kroemer G (2014) Organelle-specific initiation of cell death. Nat Cell Biol 16:728–736. https://doi.org/10.1038/ncb3005

Graves DB (2017) Mechanisms of plasma medicine: coupling plasma physics Biochemistry, and Biology. IEEE Trans Radiation Plasma Med Sci 1:281–292. https://doi.org/10.1109/trpms.2017.2710880

Hara H, Kobayashi M, Shiiba M, Kamiya T, Adachi T (2019) Sublethal treatment with plasma-activated medium induces senescence-like growth arrest of A549 cells: involvement of intracellular mobile zinc. J Clin Biochem Nutr 65: 16–22. https://doi.org/10.3164/jcbn.19-17

Hirst AM, Frame FM, Arya M, Maitland NJ, O'Connell D (2016) Low temperature plasmas as emerging cancer therapeutics: the state of play and thoughts for the future. Tumour Biol 37:7021–7031. https://doi.org/10.1007/s13277-016-4911-7

Hirst AM, Simms MS, Mann VM, Maitland NJ, O'Connell D, Frame FM (2015) Low-temperature plasma treatment induces DNA damage leading to necrotic cell death in primary prostate epithelial cells. Br J Cancer 112:1536–1545. https://doi.org/10.1038/bjc.2015.113

Hirst D, Robson T (2007) Targeting nitric oxide for cancer therapy. J Pharm Pharmacol 59:3–13. https://doi.org/10.1211/jpp.59.1.0002

Hoffmann C, Berganza C, Zhang J (2013) Cold Atmospheric plasma: methods of production and application in dentistry and oncology. Med Gas Res 3:21. https://doi.org/10.1186/2045-9912-3-21

Hurd TR, DeGennaro M, Lehmann R (2012) Redox regulation of cell migration and adhesion. Trends Cell Biol 22:107–115. https://doi.org/10.1016/j.tcb.2011.11.002

Ishaq M, Evans MD, Ostrikov KK (2014) Atmospheric pressure gas plasma-induced colorectal cancer cell death is mediated by Nox2-ASK1 apoptosis pathways and oxidative stress is mitigated by Srx-Nrf2 anti-oxidant system. Biochim Biophys Acta 1843: 2827–2837. https://doi.org/10.1016/j.bbamcr.2014.08.011

Itooka K, Takahashi K, Izawa S (2016) Fluorescence microscopic analysis of antifungal effects of cold atmospheric pressure plasma in Saccharomyces cerevisiae. Appl Microbiol Biotechnol 100:9295–9304. https://doi.org/10.1007/s00253-016-7783-2

Iuchi K, Morisada Y, Yoshino Y, Himuro T, Saito Y, Murakami T, Hisatomi H (2018) Cold atmospheric-pressure nitrogen plasma induces the production of reactive nitrogen species and cell death by increasing intracellular calcium in HEK293T cells. Arch Biochem Biophys 654: 136–145. https://doi.org/10.1016/j.abb.2018.07.015

Jang JY, Hong YJ, Lim J, Choi JS, Choi EH, Kang S, Rhim H (2018) Cold atmospheric plasma (CAP), a novel physicochemical source, induces neural differentiation through cross-talk between the specific RONS cascade and Trk/Ras/ERK signaling pathway. Biomaterials 156: 258–273. https://doi.org/10.1016/j.biomaterials.2017.11.045

Jawaid P et al (2016) Helium-based cold atmospheric plasma-induced reactive oxygen species-mediated apoptotic pathway attenuated by platinum nanoparticles. J Cell Mol Med 20:1737–1748. https://doi.org/10.1111/jcmm.12880

Judee F, Fongia C, Ducommun B, Yousfi M, Lobjois V, Merbahi N (2016) Short and long time effects of low temperature plasma activated media on 3D multicellular tumor spheroids. Sci Rep 6: 21421. https://doi.org/10.1038/srep21421

Kanazawa S et al (2011) Observation of OH radicals produced by pulsed discharges on the surface of a liquid. Plasma Sources Sci Technol 20:034010. https://doi.org/10.1088/0963-0252/20/3/034010

Kaushik NK et al. (2018) Biological and medical applications of plasma-activated media, water and solutions. Biol Chem 400: 39–62. https://doi.org/10.1515/hsz-2018-0226

Ke Z, Huang Q (2013) Inactivation and heme degradation of horseradish peroxidase induced by discharge plasma. Plasma Processes Polym 10:731–739. https://doi.org/10.1002/ppap.201300035

Keidar M et al (2011) Cold plasma selectivity and the possibility of a paradigm shift in cancer therapy. Br J Cancer 105:1295–1301. https://doi.org/10.1038/bjc.2011.386

Kensler TW, Wakabayashi N (2010) Nrf2: friend or foe for chemoprevention? Carcinogenesis 31:90–99. https://doi.org/10.1093/carcin/bgp231

Kroemer G, Galluzzi L, Kepp O, Zitvogel L (2013) Immunogenic cell death in cancer therapy. Annu Rev Immunol 31:51–72. https://doi.org/10.1146/annurev-immunol-032712-100008

Kurake N et al. (2016) Cell survival of glioblastoma grown in medium containing hydrogen peroxide and/or nitrite, or in plasma-activated medium. Arch Biochem Biophys 605: 102–108. https://doi.org/10.1016/j.abb.2016.01.011

Lackmann JW et al (2018) chemical fingerprints of cold physical plasmas—an experimental and computational study using cysteine as tracer compound. Sci Rep 8:7736. https://doi.org/10.1038/s41598-018-25937-0

Lee HY, Lee HJ, Kim GC, Choi JH, Hong JW (2019) Plasma cupping induces VEGF expression in skin cells through nitric oxide-mediated activation of hypoxia inducible factor 1. Sci Rep 9: 3821. https://doi.org/10.1038/s41598-019-40086-8

Li H-P et al (2011) Manipulation of lipase activity by the helium radio-frequency, atmospheric-pressure glow discharge plasma jet. Plasma Processes Polym 8:224–229. https://doi.org/10.1002/ppap.201000035

Li W et al (2019) Cold atmospheric plasma and iron oxide-based magnetic nanoparticles for synergetic lung cancer therapy. Free Radic Biol Med 130:71–81. https://doi.org/10.1016/j.freeradbiomed.2018.10.429

Lin A et al (2019) Non-thermal plasma as a unique delivery system of short-lived reactive oxygen and nitrogen species for immunogenic cell death in melanoma cells. Adv Sci (Weinh) 6:1802062. https://doi.org/10.1002/advs.201802062

Lin A, Truong B, Fridman G, Fridman AA, Miller V (2017a) Immune cells enhance selectivity of nanosecond-pulsed dbd plasma against tumor cells. Plasma Med 7:85–96. https://doi.org/10.1615/PlasmaMed.2017019666

Lin A et al. (2017b) Nanosecond-pulsed DBD plasma-generated reactive oxygen species trigger immunogenic cell death in A549 Lung carcinoma cells through intracellular oxidative stress. Int J Mol Sci 18. https://doi.org/10.3390/ijms18050966

Lin L, Wang L, Liu Y, Xu C, Tu Y, Zhou J (2018) Nonthermal plasma inhibits tumor growth and proliferation and enhances the sensitivity to radiation in vitro and in vivo. Oncol Rep 40:3405–3415. https://doi.org/10.3892/or.2018.6749

Loboda A, Damulewicz M, Pyza E, Jozkowicz A, Dulak J (2016) Role of Nrf2/HO-1 system in development, oxidative stress response and diseases: an evolutionarily conserved mechanism. Cell Mol Life Sci 73: 3221–3247. https://doi.org/10.1007/s00018-016-2223-0

Lopez M, Calvo T, Prieto M, Mugica-Vidal R, Muro-Fraguas I, Alba-Elias F, Alvarez-Ordonez A (2019) A review on non-thermal atmospheric plasma for food preservation: mode of action. Determinants Effect Appl Front Microbiol 10:622. https://doi.org/10.3389/fmicb.2019.00622

Lunov O, Zablotskii V, Churpita O, Lunova M, Jirsa M, Dejneka A, Kubinova S (2017) Chemically different non-thermal plasmas target distinct cell death pathways. Sci Rep 7:600. https://doi.org/10.1038/s41598-017-00689-5

Maliszewska-Olejniczak K et al. (2019) Development of extracellular matrix supported 3D culture of renal cancer cells and renal cancer stem cells. Cytotechnology 71: 149–163. https://doi.org/10.1007/s10616-018-0273-x

Marchi S et al (2018) Mitochondrial and endoplasmic reticulum calcium homeostasis and cell death. Cell Calcium 69:62–72. https://doi.org/10.1016/j.ceca.2017.05.003

Metelmann H-R et al (2015) head and neck cancer treatment and physical plasma. Clin Plasma Med 3:17–23. https://doi.org/10.1016/j.cpme.2015.02.001

Miller V, Lin A, Fridman A (2015) Why target immune cells for plasma treatment of cancer. Plasma Chem Plasma Process 36:259–268. https://doi.org/10.1007/s11090-015-9676-z

Miller V, Lin A, Fridman G, Dobrynin D, Fridman A (2014) Plasma stimulation of migration of macrophages. Plasma Processes Polym 11:1193–1197. https://doi.org/10.1002/ppap.201400168

Mohd Nasir N, Lee BK, Yap SS, Thong KL, Yap SL (2016) Cold plasma inactivation of chronic wound bacteria. Arch Biochem Biophys 605:76–85. https://doi.org/10.1016/j.abb.2016.03.033

Moniruzzaman R et al (2018) Roles of intracellular and extracellular ROS formation in apoptosis induced by cold atmospheric helium plasma and X-irradiation in the presence of sulfasalazine. Free Radic Biol Med 129:537–547. https://doi.org/10.1016/j.freeradbiomed.2018.10.434

Moniruzzaman R et al (2017) Cold atmospheric helium plasma causes synergistic enhancement in cell death with hyperthermia and an additive enhancement with radiation. Sci Rep 7:11659. https://doi.org/10.1038/s41598-017-11877-8

Murakami T, Niemi K, Gans T, O'Connell D, Graham WG (2012) Chemical kinetics and reactive species in atmospheric pressure helium–oxygen plasmas with humid-air impurities. Plasma Sour Sci Technol 22: 015003. https://doi.org/10.1088/0963-0252/22/1/015003

Murakami T, Niemi K, Gans T, O'Connell D, Graham WG (2013) Interacting kinetics of neutral and ionic species in an atmospheric-pressure helium–oxygen plasma with humid air impurities. Plasma Sour Sci Technol 22: 045010. https://doi.org/10.1088/0963-0252/22/4/045010

Murakami T, Niemi K, Gans T, O'Connell D, Graham WG (2014) Afterglow chemistry of atmospheric-pressure helium–oxygen plasmas with humid air impurity. Plasma Sour Sci Technol 23: 025005. https://doi.org/10.1088/0963-0252/23/2/025005

Narayan N, Rigby S, Carlucci F (2017) Sulfasalazine induced immune thrombocytopenia in a patient with rheumatoid arthritis. Clin Rheumatol 36:477–479. https://doi.org/10.1007/s10067-016-3420-9

Panza F, Lozupone M, Logroscino G, Imbimbo BP (2019) A critical appraisal of amyloid-beta-targeting therapies for Alzheimer disease. Nat Rev Neurol 15:73–88. https://doi.org/10.1038/s41582-018-0116-6

Park J, Lee H, Lee HJ, Kim GC, Kim SS, Han S, Song K (2019) Non-thermal atmospheric pressure plasma is an excellent tool to activate proliferation in various mesoderm-derived human adult stem cells. Free Radic Biol Med 134:374–384. https://doi.org/10.1016/j.freeradbiomed.2019.01.032

Park JH, Kim M, Shiratani M, Cho AE, Choi EH, Attri P (2016) Variation in structure of proteins by adjusting reactive oxygen and nitrogen species generated from dielectric barrier discharge jet. Sci Rep 6:35883. https://doi.org/10.1038/srep35883

Peter ME, Krammer PH (2003) The CD95(APO-1/Fas) DISC and beyond. Cell Death Differ 10: 26–35. https://doi.org/10.1038/sj.cdd.4401186

Rödder K, Moritz J, Miller V, Weltmann K-D, Metelmann H-R, Gandhirajan R, Bekeschus S (2019) Activation of murine immune cells upon co-culture with plasma-treated B16F10 melanoma cells. Appl Sci 9: 660. https://doi.org/10.3390/app9040660

Rasola A, Bernardi P (2011) Mitochondrial permeability transition in Ca(2+)-dependent apoptosis and necrosis. Cell Calc 50: 222–233. https://doi.org/10.1016/j.ceca.2011.04.007

Recalcati S, Gammella E, Cairo G (2019) Dysregulation of iron metabolism in cancer stem cells. Free Radic Biol Med 133:216–220. https://doi.org/10.1016/j.freeradbiomed.2018.07.015

Rizza S et al (2018) S-nitrosylation drives cell senescence and aging in mammals by controlling mitochondrial dynamics and mitophagy. Proc Natl Acad Sci USA 115:E3388–E3397. https://doi.org/10.1073/pnas.1722452115

Ruijtenberg S, van den Heuvel S (2016) Coordinating cell proliferation and differentiation: antagonism between cell cycle regulators and cell type-specific gene expression. Cell Cycle 15:196–212. https://doi.org/10.1080/15384101.2015.1120925

Sagwal SK, Pasqual-Melo G, Bodnar Y, Gandhirajan RK, Bekeschus S (2018) Combination of chemotherapy and physical plasma elicits melanoma cell death via upregulation of SLC22A16. Cell Death Dis 9: 1179. https://doi.org/10.1038/s41419-018-1221-6

Sakudo A, Shimizu N, Imanishi Y, Ikuta K (2013) N 2 gas plasma inactivates influenza virus by inducing changes in viral surface morphology, protein, and genomic RNA. Biomed Res Int 2013: 694269. https://doi.org/10.1155/2013/694269

Sakudo A, Toyokawa Y, Imanishi Y (2016) Nitrogen gas plasma generated by a static induction thyristor as a pulsed power supply inactivates adenovirus. PLoS ONE 11:e0157922. https://doi.org/10.1371/journal.pone.0157922

Sasaki S, Kanzaki M, Kaneko T (2016) Calcium influx through TRP channels induced by short-lived reactive species in plasma-irradiated solution. Sci Rep 6:25728. https://doi.org/10.1038/srep25728

Sato K et al. (2019) Non-thermal plasma specifically kills oral squamous cell carcinoma cells in a catalytic Fe(II)-dependent manner. J Clin Biochem Nutr 65: 8–15. https://doi.org/10.3164/jcbn.18-91

Sato S, Nakamura H (2019) Protein chemical labeling using biomimetic radical chemistry. Molecules 24. https://doi.org/10.3390/molecules24213980

Schmidt A, Dietrich S, Steuer A, Weltmann KD, von Woedtke T, Masur K, Wende K (2015) Non-thermal plasma activates human keratinocytes by stimulation of antioxidant and phase II pathways. J Biol Chem 290:6731–6750. https://doi.org/10.1074/jbc.M114.603555

Schneider C, Gebhardt L, Arndt S, Karrer S, Zimmermann JL, Fischer MJM, Bosserhoff AK (2018) Cold atmospheric plasma causes a calcium influx in melanoma cells triggering cap-induced senescence. Sci Rep 8:10048. https://doi.org/10.1038/s41598-018-28443-5

Schneider C, Gebhardt L, Arndt S, Karrer S, Zimmermann JL, Fischer MJM, Bosserhoff AK (2019) Acidification is an essential process of cold atmospheric plasma and promotes the anti-cancer effect on malignant melanoma cells. Cancers (Basel) 11. https://doi.org/10.3390/cancers11050671

Schutze A, Jeong JY, Babayan SE, Jaeyoung P, Selwyn GS, Hicks RF (1998) The atmospheric-pressure plasma jet: a review and comparison to other plasma sources. IEEE Trans Plasma Sci 26:1685–1694. https://doi.org/10.1109/27.747887

Sha Y, Marshall HE (2012) S-nitrosylation in the regulation of gene transcription. Biochim Biophys Acta 1820:701–711. https://doi.org/10.1016/j.bbagen.2011.05.008

Shaw P, Kumar N, Hammerschmid D, Privat-Maldonado A, Dewilde S, Bogaerts A (2019) Synergistic effects of melittin and plasma treatment: a promising approach for cancer therapy. Cancers (Basel) 11. https://doi.org/10.3390/cancers11081109

Shi X-M, Zhang G-J, Chang Z-S, Wu X-L, Liao W-L, Li N (2014) Viability reduction of melanoma cells by plasma jet via inducing G1/S and G2/M cell cycle arrest and cell apoptosis. IEEE Trans Plasma Sci 42: 1640–1647. https://doi.org/10.1109/tps.2014.2320765

Shintani H, Sakudo A, Burke P, McDonnell G (2010) Gas plasma sterilization of microorganisms and mechanisms of action. Exp Ther Med 1:731–738. https://doi.org/10.3892/etm.2010.136

Souza JM, Peluffo G, Radi R (2008) Protein tyrosine nitration–functional alteration or just a biomarker? Free Radic Biol Med 45:357–366. https://doi.org/10.1016/j.freeradbiomed.2008.04.010

Stryczewska HD, Jakubowski T, Kalisiak S, Giżewski T, Pawłat J (2013) Power systems of plasma reactors for non-thermal plasma generation. J Adv Oxid Technol 16. https://doi.org/10.1515/jaots-2013-0105

Surowsky B, Fischer A, Schlueter O, Knorr D (2013) Cold plasma effects on enzyme activity in a model food system. Innov Food Sci Emerg Technol 19:146–152. https://doi.org/10.1016/j.ifset.2013.04.002

Syntichaki P, Tavernarakis N (2003) The biochemistry of neuronal necrosis: rogue biology? Nat Rev Neurosci 4:672–684. https://doi.org/10.1038/nrn1174

Takai E, Ikawa S, Kitano K, Kuwabara J, Shiraki K (2013) Molecular mechanism of plasma sterilization in solution with the reduced pH method: importance of permeation of HOO radicals into the cell membrane. J Phys D Appl Phys 46:295402. https://doi.org/10.1088/0022-3727/46/29/295402

Takai E et al (2014a) Chemical modification of amino acids by atmospheric-pressure cold plasma in aqueous solution. J Phys D Appl Phys 47:285403. https://doi.org/10.1088/0022-3727/47/28/285403

Takai E, Kitano K, Kuwabara J, Shiraki K (2012) Protein inactivation by low-temperature atmospheric pressure plasma in aqueous solution. Plasma Processes Polym 9:77–82. https://doi.org/10.1002/ppap.201100063

Takai E et al (2014b) Degeneration of amyloid-ß fibrils caused by exposure to low-temperature atmospheric-pressure plasma in aqueous solution. Appl Phys Lett 104:023701. https://doi.org/10.1063/1.4861842

Tanaka H, Mizuno M, Ishikawa K, Toyokuni S, Kajiyama H, Kikkawa F, Hori M (2018) New hopes for plasma-based cancer treatment. Plasma 1:150–155. https://doi.org/10.3390/plasma1010014

Tanaka H et al (2019) Oxidative stress-dependent and -independent death of glioblastoma cells induced by non-thermal plasma-exposed solutions. Sci Rep 9:13657. https://doi.org/10.1038/s41598-019-50136-w

Terenzi A, Pirker C, Keppler BK, Berger W (2016) Anticancer metal drugs and immunogenic cell death. J Inorg Biochem 165:71–79. https://doi.org/10.1016/j.jinorgbio.2016.06.021

Tokunaga T et al (2018) Plasma-stimulated medium kills TRAIL-resistant human malignant cells by promoting caspase-independent cell death via membrane potential and calcium dynamics modulation. Int J Oncol 52:697–708. https://doi.org/10.3892/ijo.2018.4251

Tominami K, Kanetaka H, Sasaki S, Mokudai T, Kaneko T, Niwano Y (2017) Cold atmospheric plasma enhances osteoblast differentiation. PLoS ONE 12:e0180507. https://doi.org/10.1371/journal.pone.0180507

Turrini E et al (2017) Cold atmospheric plasma induces apoptosis and oxidative stress pathway regulation in T-Lymphoblastoid Leukemia cells. Oxid Med Cell Longev 2017:4271065. https://doi.org/10.1155/2017/4271065

Vandamme M et al (2012) ROS implication in a new antitumor strategy based on non-thermal plasma. Int J Cancer 130:2185–2194. https://doi.org/10.1002/ijc.26252

Varga T, Czimmerer Z, Nagy L (2011) PPARs are a unique set of fatty acid regulated transcription factors controlling both lipid metabolism and inflammation. Biochim Biophys Acta 1812:1007–1022. https://doi.org/10.1016/j.bbadis.2011.02.014

Volotskova O, Hawley TS, Stepp MA, Keidar M (2012) Targeting the cancer cell cycle by cold atmospheric plasma. Sci Rep 2:636. https://doi.org/10.1038/srep00636

Volotskova O, Shashurin A, Stepp MA, Pal-Ghosh S, Keidar M (2011) Plasma-controlled cell migration: localization of cold plasma-cell interaction region. Plasma Med 1:85–92. https://doi.org/10.1615/PlasmaMed.v1.i1.70

Woedtke VON, Schmidt A, Bekeschus S, Wende K, Weltmann KD (2019) Plasma medicine: a field of applied redox biology. in Vivo 33:1011–1026. https://doi.org/10.21873/invivo.11570

Xiong Z et al (2014) Selective neuronal differentiation of neural stem cells induced by nanosecond microplasma agitation. Stem Cell Res 12:387–399. https://doi.org/10.1016/j.scr.2013.11.003

Xu D et al (2018) Cold atmospheric plasma as a potential tool for multiple myeloma treatment. Oncotarget 9:18002–18017. https://doi.org/10.18632/oncotarget.24649

Xu R-G, Chen Z, Keidar M, Leng Y (2018) The impact of radicals in cold atmospheric plasma on the structural modification of gap junction: a reactive molecular dynamics study. Int J Smart Nano Mater 10:144–155. https://doi.org/10.1080/19475411.2018.1541936

Yamanaka K, Saito Y, Sakiyama J, Ohuchi Y, Oseto F, Noguchi N (2012) A novel fluorescent probe with high sensitivity and selective detection of lipid hydroperoxides in cells . RSC Advances 2:7894. https://doi.org/10.1039/c2ra20816d

Yan D, Sherman JH, Keidar M (2017) Cold atmospheric plasma, a novel promising anti-cancer treatment modality. Oncotarget 8: 15977–15995. https://doi.org/10.18632/oncotarget.13304

Yan D, Talbot A, Nourmohammadi N, Cheng X, Canady J, Sherman J, Keidar M (2015) Principles of using cold atmospheric plasma stimulated media for cancer treatment. Sci Rep 5:18339. https://doi.org/10.1038/srep18339

Yan X et al (2010) On the mechanism of plasma inducing cell apoptosis. IEEE Trans Plasma Sci 38:2451–2457. https://doi.org/10.1109/tps.2010.2056393

Zhang H et al. (2015) Effects and mechanism of atmospheric-pressure dielectric barrier discharge cold plasma on lactate dehydrogenase (LDH) enzyme. Sci Rep 5: 10031. https://doi.org/10.1038/srep10031

Zhao S et al. (2013) Atmospheric pressure room temperature plasma jets facilitate oxidative and nitrative stress and lead to endoplasmic reticulum stress dependent apoptosis in HepG2 cells. PLoS One 8: e73665. https://doi.org/10.1371/journal.pone.0073665

Zhou J, Wang G, Chen Y, Wang H, Hua Y, Cai Z (2019) Immunogenic cell death in cancer therapy: present and emerging inducers. J Cell Mol Med 23:4854–4865. https://doi.org/10.1111/jcmm.14356

Zottel A, Videtic A, Jovcevska PI (2019) Nanotechnology meets oncology: nanomaterials in brain cancer research diagnosis and therapy. Materials (Basel) 12. https://doi.org/10.3390/ma12101588

Immobilization of Biomolecules on Plasma-Functionalized Surfaces for Biomedical Applications

M. C. Ramkumar, A. M. Trimukhe, R. R. Deshmukh, Anuj Tripathi, Jose Savio Melo, and K. Navaneetha Pandiyaraj

Abstract Immobilization of biomolecules using plasma-tailored surfaces of material has caught attention amongst many interdisciplinary scientists. The non-thermal plasma surface modification technique has evolved itself as a very promising candidate for biomedical applications. The daunting challenge in biocompatibility with respect to surface chemistry alterations is its intervention with host response in expected biological functions. Distinct features of non-thermal plasma such as dry state process, non-thermolability, presence of energetic reactive species, low cost, and rapid and controlled functionalization have been chosen over conventional surface modification techniques. Advances in immobilization of biomolecules using non-thermal plasma surface modification techniques have paved us the way to achieve the design of biologically inspired materials which can successfully mimic complex biological processes. Various plasma types in different modes such as deposition through chemical attachments (like hydrophilic interaction, ionic or covalent bonding) and graft polymerization have been utilized to rationalize the immobilization applications. Physical entrapment in immobilization applications on various surfaces is highlighted with respect to enzymes and various polymeric or particle surfaces. In this chapter, we have mainly discussed the latest advancements related to immobilization of enzymes, protein and other biomolecules on to the non-thermal

M. C. Ramkumar · K. Navaneetha Pandiyaraj (✉)
Research Division of Plasma Processing (RDPP), Department of Physics, Sri Shakthi Institute of Engineering and Technology, Coimbatore 641062, India
e-mail: dr.knpr@hotmail.com; dr.knpr@gmail.com

M. C. Ramkumar
Department of Physics, School of Basic Sciences, Vels Institute of Science, Technology and Advanced Studies, Chennai 600117, India

A. M. Trimukhe · R. R. Deshmukh
Department of Physics, Institute of Chemical Technology, Matunga, Mumbai 400019, India

A. Tripathi · J. S. Melo
Nuclear Agriculture and Biotechnology, Division Bhabha Atomic Research Centre, Mumbai 400085, India

Homi Bhabha National Institute, Mumbai 400094, India

A. Tripathi and J. S. Melo (eds.), *Immobilization Strategies*,
Gels Horizons: From Science to Smart Materials,
https://doi.org/10.1007/978-981-15-7998-1_8

plasma-modified surfaces. The important aspect of plasma surface modification is to understand the interaction between plasma-functionalized surface and binding molecules for immobilization applications. Different plasma processes in various modes such as graft polymerization and deposition through chemical attachments (like covalent bonding, ionic or hydrophilic interaction) are discussed in this chapter within the scope of immobilization. This chapter also enlightens about the fundamental understanding on the development of biocompatible surfaces on various polymeric substrates including 3D scaffolds by the use of non-thermal low-pressure and atmospheric pressure plasma-assisted polymerization, which have been subsequently used for immobilization of various biomolecules for advanced biomedical applications.

Keywords Non-thermal plasma · Surface functionalization · Immobilization · Biocompatibility · Biomedicine

Abbreviations

AAC	Acrylic acid
ADSC	Adipose-derived stem cell
AEMA	2-Amino-ethyl methacrylate
AFM	Atomic force microscopy
Ar	Argon
ATR-FTIR	Attenuated total reflectance -Fourier transform infrared
BOPP	Biaxially oriented polypropylene
CA	Contact angle
CE	Cornea epithelial
CO_2	Carbon dioxide
DBD	Dielectric barrier discharge
DC	Direct current
FEP	Poly(tetrafluoroethylene-co-hexafluoropropylene)
HCAEC	Human coronary artery endothelial cells
HDF	Human dermal fibroblast
HEP	Heparin
hTM	Human thrombomodulin
LDPE	Low-density polyethylene
MW	Microwave
N_2	Nitrogen
NH_3	Ammonia
O_2	Oxygen
OSC	O-stearoyl-chitosan
PCL	Poly(ε-caprolactone)
PDMS	Polydimethylsiloxane
PEG	Poly(ethylene glycol)

PEGMA	Poly(ethylene glycol) methacrylate
PET	Polyethylene terephthalate
PLGA	Poly(lactide-co-glycolide)
PLLA	Poly-L-lactic acid
PNIPAM	Poly(N-isopropylacrylamide)
PSF	Polysulfone
PTFE	Poly(tetrafluoroethylene)
PU	Polyurethane
ROB	Rat embryo osteoblastic
ROS	Rat osteosarcoma
SEM	Scanning electron microscopy
SR	Silicon rubber
UV	Ultraviolet
PEO	Polyethylene oxide
XPS	X-ray photoelectron spectroscopy

1 Introduction

Biomaterials are identified as the natural or synthetic material which comprises a section or an entire part of living body, that behaves, amplifies or substitutes inherent functions of body tissue (Tathe et al. 2010; Alexander et al. 1996). Biomaterial covers multiple domains like material science, chemistry, medicine and biology, and due to the incredible application potential, it has been popularized gradually and substantively in past few decades (Tripathi and Melo 2017). Porous scaffold is a class of biomaterials that exhibit enormous potential for multidisciplinary applications including biomedical, bioengineering and environmental biotechnology (Tripathi and Melo 2019a, b; Tripathi et al. 2010, 2013c, d). Currently, biomaterials are employed as bone plates, dental implants, blood vessels, artificial ligaments, heart valves, joint replacements, cardio vascular tubes and contact lenses (Unnikrishnan et al. 2020; Nair and Laurencin 2005). Numerous types of materials are employed as biomaterials, for instance metals, ceramics, glasses, alloys, composites and polymers. In precise, polymer holds various advantages compared to other biomaterials and can be an ideal substitute for metals as they play an essential role in biomedical industry (Lloyd et al. 2007; Lopez-Garcia et al. 2013).

Polymers (PP, LDPE, PET, etc.) are employed in various industries (automotive, food packing, biomedical and so on) for its widespread applications due to their remarkable material properties, for instance high chemical resistance, thermal stability, impact strength, low cost, etc (Pantoj et al. 2013; Morent et al. 2010a, b). Especially in medical field, polymers play a substantial role in fabrication of artificial heart valves, blood bags, stents, blood vessels, etc. (Xiang et al. 2009; Saxena et al. 2010), thanks to their unique material properties (high chemical and corrosion resistance, flexibility, low density and low cost) (Friedman and Walsh 2002; Sanchis et al. 2006). Polymeric materials are extensively replacing traditional bioengineering materials like stainless steel (SS) and titanium (Ti). In spite of the above desirable properties in an ideal biomaterial, the major hindrance of a polymeric material in biomedical field is the surface-induced thrombus formation when it comes in contact with blood components, and also its inadequate cell viability, adhesion and proliferation (Zhang et al. 2013; Yoshida et al. 2013), which are mainly due to their inherent surface hydrophobicity, lower surface energy and the deficiency of polar functional groups (Shenton et al. 2001; Morent et al. 2010a, b). Thus, it is vital to alter the surface properties (hydrophilicity, surface chemistry and topography) of biomedical materials before subjecting it to biological environment, because surface of the material primarily interacts with biological environment rather than bulk. Materials surface chemistry and topography play a substantial part in describing the biocompatible properties (Li et al. 2014; Mao et al. 2004; Kerri et al. 1989; Jacobs et al. 2012), which decides the biological response and governs their performance in vivo (Pratt et al. 1989). Moreover, the material should also possess substantial bulk properties so as to function appropriately in biological environment (Kumar et al. 2011). Hence, an appropriate surface modification technique is essential to tailor the surface of materials without varying the material bulk (Tripathi et al. 2013a; Chu et al. 2002; Pu et al. 2002).

For past few decades, numerous surface tailoring methods such as e-beam, wet chemical, corona discharge, UV irradiation and non-thermal plasma polymerization have been employed to improve the surface properties of polymers (Sodhi 1996; Gogolewski et al. 1996; Martin et al. 2007; Kumar and Tripathi 2012; Svorcik et al. 1998). Amongst, plasma surface modification has acquired extensive attention amongst researchers, as the traditional methods have certain limitations, for instance emission of VOC, time consuming, consumption of solvents and development of toxic residues, and also stability problems (Rossini et al. 2003; Heyse et al. 2007; Alexandrov and Hitchman 2005). It also provides a versatile route for specific yet controlled changes on the surface of material without altering the material bulk properties.

Plasma-based surface modification is a solvent-free single-step method which is proficient in providing homogeneous surface treatment and incorporating extensive range of functionalities in a short duration. By suitable alteration in the feed gas, precursor monomers, duty cycles and plasma parameters, one can create surfaces with rich wide range of functional groups, for instance hydroxyl, amines, epoxy, imine, carboxyl, etc (Biederman and Osada 1990; Goossens et al. 2001; Arefi-Khonsari and Tatoulian 2008; D'Agostino et al. 2005; Trimukhe et al. 2017; Yang et al. 2013). Functional groups obtained by plasma processing can induce biological properties onto the materials surface (Bhattacharyya et al. 2010; Xu et al. 2011) and can be employed to anchor specific groups for immobilization of biomolecules, i.e. whole cell, proteins, enzymes or any other biocatalysts that offer a more precise biological response. There are various kinds of plasma surface tailoring methods that can be employed to fabricate appropriate surfaces for immobilization of biomolecules (Fig. 1), such as (i) plasma treatment: plasma surface treatment by employing reactive and non-reactive gases (Ar, NH_3, O_2, N_2, air, etc.) may result in the development of

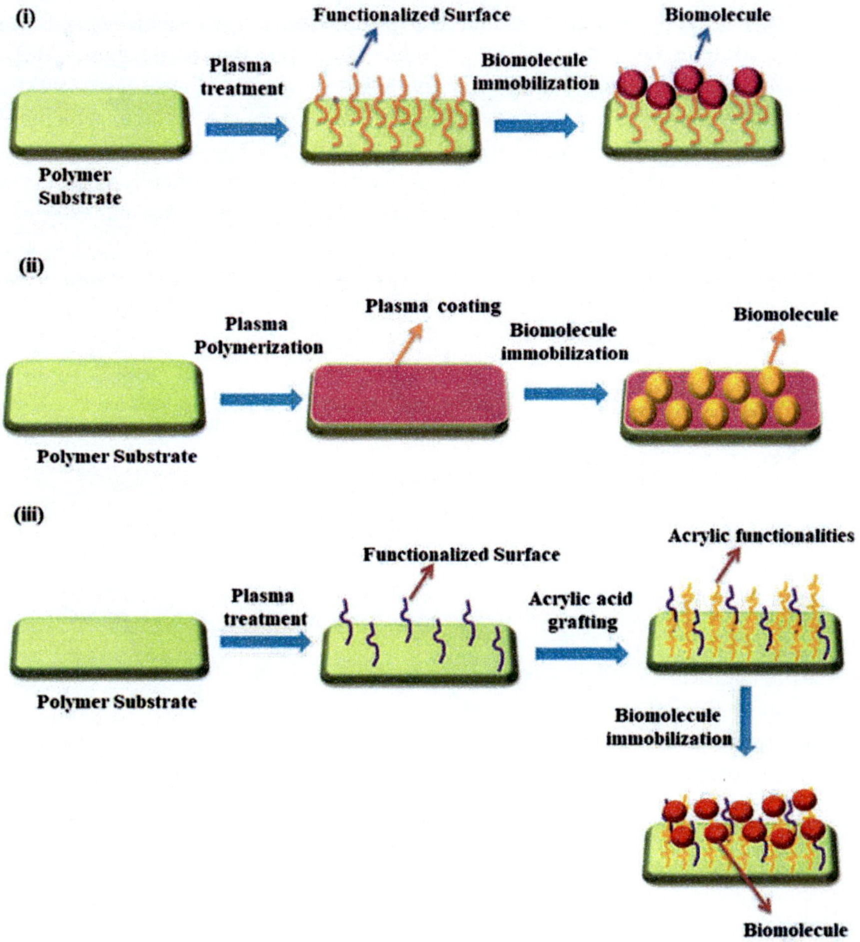

Fig. 1 Plasma-based surface modification techniques: (i) plasma treatment or activation, (ii) plasma polymerization and (iii) plasma-induced graft polymerization. The above-modified surfaces consist of various functionalities which can be employed as linkers for biomolecule immobilization

surfaces with large extent of O^- and N^- containing functionalities. Gaseous plasma treatment can also make rough surfaces. (ii) Plasma deposition or polymerization of precursors (acrylic acid, allyl alcohol, allylamine, etc.) can lead to the introduction of COOH, OH or NH_3 functional groups. (iii) Plasma-induced graft polymerization: The formation of active species on the surface of polymers and subsequent interaction with the precursor (Baquey et al. 1999); here, the free radicals produced due to the plasma treatment can initiate grafting or generate peroxides or hydroperoxides by the addition of oxidative gas.

Furthermore, these activated peroxides also cause grafting in the existence of monomer species. Subsequent to functionalization process, the surface-modified polymers can be subjected to wet chemical process in order to achieve immobilization of biomolecules, even by use of linker molecules that bound them to the polymer surface to sustain extended bioactivity and reusability (Favia et al. 1998).

Plasma processes used for the development of functionalized surfaces which are appropriate for immobilization of different biomolecules will be discussed further in this chapter. The outcome attained via various plasma surface modification processes and various biomolecules such as heparin, chitosan, collagen, peptides and enzymes on the surface of polymers and 3D polymer scaffolds will provide a comprehensive overview on cold plasma technology with a specific attention to the functionalized polymeric materials subjected to biological applications.

2 Advantages and Disadvantages of Plasma Technology for Surface Modification

2.1 Advantages

I. Illimitable chemical modifications with wide range of chemicals/precursors and gasses.
II. Applicable to broad variety of materials due to dry process of functionalization, which ultimately also conserve the water, energy and time of post-processing.
III. From surface erosion to variable thickness of tethering molecules can be produced.
IV. Material functionalization without affecting bulk property.
V. Highly controlled, precise and site specific.
VI. Minimize the generation of chemical waste.
VII. Lower processing costs due to the sophisticated instrumentation facility.
VIII. Eco-friendly process which does not allow emission of by-products of chemical or gaseous plasma due to the closed-system processing.

2.2 Disadvantages

I. Optimization of process parameters (like pressure, flow rate and power) is critical concern for process development and reproducibility which varies system to system.
II. Setting up a plasma processing facility is expensive.
III. Conversion from bench to field is challenging and delimits its potential application.
IV. Limiting to thin-surface modification (lower penetration limit) could be disadvantage for materials those needed bulk augmentation.

3 Immobilization on Polymers

3.1 Immobilization of Heparin

Heparin and heparin sulphate (HS) are anionic linear polysaccharide possessing utmost negative charge density compared to other biomolecules. For over nine decades, heparin is employed as an excellent anticoagulation agent and plays a substantial role in controlling immune system, cell attachment and spreading (Keselowsky et al. 2004). It can be covalently immobilized on various plasma-functionalized polymer surfaces by chemical grafting or via strong ionic bonding. The schematic representation of heparin biomolecules exhibiting anticoagulation property is depicted in Fig. 2. Currently, heparin is commonly employed on commercially existing various blood-contacting devices. Kim et al. (2000) activated the surface of poly(ethylene terephthalate) films by exposing it to O_2 glow discharge plasma followed by polymerization of acrylic acid (AAC) to incorporate COOH groups on the PET surfaces. Further, PEO was grafted onto the surface of AAC-polymerized PET films for co-immobilization of insulin and heparin. Their results unveiled the substantial elongation in activated partial thromboplastin time and plasma recalcification time for both heparin- and insulin-immobilized PET films. However, the adhesion of platelets was increased after polymerization of acrylic acid and the same was decreased with insertion of PEO and insulin and decreased further after heparin immobilization.

Another study showed the development of tailored PP non-woven fabrics (PPNWF) by plasma-assisted graft polymerization of acrylic acid and PEO and used for the immobilization of heparin. The authors reported that the surface-modified PP non-woven fabrics exhibited distinct surface hydrophilicity. In vitro analysis unveiled

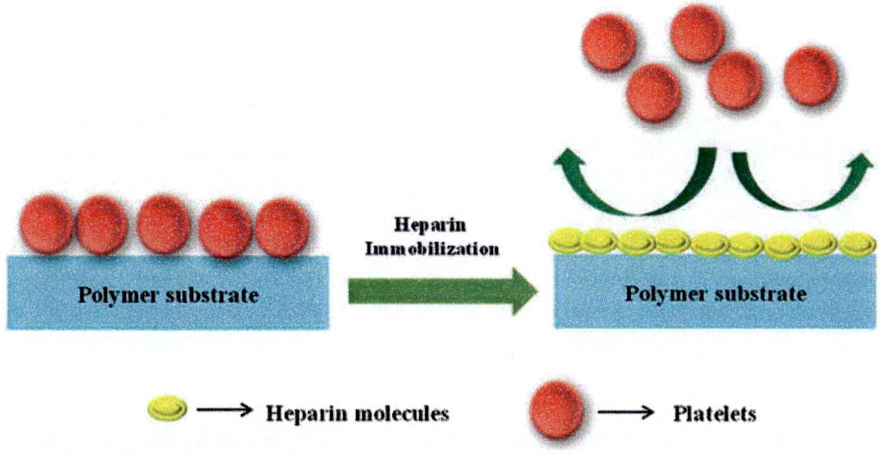

Fig. 2 Schematic of heparin molecule exhibiting anticoagulation property

that the heparinized PPNWFs exhibited lower haemolysis rates and excellent anti-platelet adhesion compared to untreated one (Li et al. 2013). RF O_2 plasma-assisted graft polymerization of acrylic acid on PU film surfaces has also been evaluated for the immobilization of heparin and chitosan. The ATR-FTIR outcome clearly confirmed the presence of heparin and chitosan functionalities on plasma-modified PU films. Contact angle analysis revealed that the immobilized PU films exhibit substantial hydrophilicity. Additionally, in vitro analysis revealed that the modified films possess outstanding biocompatibility which was confirmed by cyto-compatibility analysis by growing L929 fibroblast cells (Ayati-Najafabadiet al. 2012). A study showed the effect of various plasma operating parameters to influence the PEG chains grafted on plasma pre-treated PSF membranes for immobilization of heparin molecules. Their results unveiled that the heparinized membranes showed excellent hydrophilic property compared to other surface-modified films. Furthermore, XPS and ATR-FTIR outcome validated the presence of heparin functionalities on PSF membranes. The surface-modified membranes exhibited substantial enhancement in hemocompatibility compared to unmodified PSF membranes, confirmed by platelet adhesion, protein adsorption and coagulation analysis (Wang et al. 2017). Degoutin et al. (2012) aimed to enhance the antibacterial and anticoagulation properties of polypropylene fabrics by appropriate modification of the PP surface by plasma pre-treatment followed by grafting of acrylic acid. Finally, gentamicin and heparin were immobilized on the surface of acrylic grafted PP fabrics. The authors reported that the immobilization of gentamicin prevented *E. coli* adhesion as it possesses marked antibiotic property. In addition, the heparin-immobilized PP fabrics showed significant enhancement in anticoagulation effect. In another study, a series of biomolecules (heparin, chitosan and insulin) were immobilized on the surface of AAC- and PEG-coated PP films by Pandiyaraj and his colleagues. The XPS analysis revealed the presence of insulin and heparin functionalities on the surface of PP films. Furthermore, the biomolecule-immobilized films of AAC and PEG exhibited no traces of adhesion and activation of platelets on their surfaces (Pandiyaraj et al. 2016). Gao et al. (2013) tailored the surface properties of poly(tetrafluoroethylene-co-hexafluoropropylene) (FEP) by Ar plasma treatment and UV-induced graft polymerization of AAC. Conclusively, heparin was immobilized via esterification reaction on FEP film surfaces. Their outcome showed that the heparin-immobilized FEP films unveiled substantial antithrombogenic property. Water–O_2 radio frequency plasma has also been used to alter the surface properties of various polymeric materials for the incorporation of specific functional groups to assist covalent immobilization of heparin. XPS and SSIMS results established the effective incorporation of heparin on the surface of polymers with antithrombogenic property (Narayanan 1994). Another study aimed to enrich the biocompatible properties of PET and PTFE films. Initially, the films were treated by Ar plasma followed by the development of collagen and laminin coatings. Finally, poly(ethylene glycol) (PEG) and HEP were immobilized. In vitro analysis outcome unveiled the ample decrease in adsorption of fibrinogen and platelet adhesion (Chandy et al. 2000). O_2 plasma-activated polypropylene (PP) film was used to immobilize the heparin by the help of polyethylene glycol methacrylate (PEGMA), and its hemocompatibility test was performed, which confirmed the surface-modified

PP film has excellent anticoagulation property. In addition, the film inhibited the adhesion of platelets. The AT III assay outcome unveiled that the heparin-immobilized PP film has excellent bioactivity (Jin et al. 2012). Pandiyaraj et al. (2009) aimed to immobilize heparin or insulin molecules on the surface of polyethylene terephthalate (PET) films using DC glow discharge plasma activation followed by grafting of ACC and PEG. The authors validated the presence of amino and carboxyl functionalities on the surface of modified PET films by XPS analysis. Furthermore, blood compatibility studies unveiled that the biomolecule (heparin or insulin)-immobilized PET films showed reduction in platelet adhesion and protein adsorption, compared to other modified and untreated PET films.

3.2 Immobilization of Chitosan

Incorporation of antimicrobial property on the surface of polymers can be achieved by employing numerous antibiotics that has the ability to combat with bacteria populations (Trimukhe et al. 2017). Conversely, the major drawback in employing them is that they possess cytotoxic nature (Tripathi et al. 2013b). So it is a foremost challenge to select an antibiotic that has the ability to kill bacteria with no cytotoxic nature in order to perform immobilization on the surface of polymers. One such biomolecule that has no cytotoxic nature is chitosan (Tripathi and Melo 2015). It is a linear polysaccharide polymer, comprised of 1,4-linked d-glucosamine and *N*-acetyl-D-glucosamine and soluble in aqueous solution ($pH < 6.5$). It has widespread applications in biomaterials, medicine, drug-controlled release system, environment and agriculture (Tripathi and Melo 2016; Kumar and Tripathi 2012), owing to its fascinating biological properties, for instance, low immunogenicity, biocompatibility, non-toxicity and wound healing. Additionally, it can be employed in development of antibacterial coatings as it comprises excellent antimicrobial property (An and Friedman 1997; Raafat and Sahl 2009; Hori and Matsumoto 2010). Ding et al. (2004) used Ar plasma pre-treatment prior to chitosan immobilization on the surface of poly-L-lactic acid (PLLA) films. The outcome validated that the cell growth on the surface of chitosan-immobilized films was comparable to the cell culture plate. A two-step surface modification of poly(tetrafluoroethylene) (PTFE), i.e. low-pressure plasma activation followed by grafting of acrylic acid, was performed for the immobilization of chitosan on the surface of film under optimized condition. Contact angle (CA) analysis unveiled that the hydrophilic property of the PTFE film was improved after chitosan immobilization. XPS analysis clearly indicated the existence of N1s peak. In addition, increase in O/C ratio and decrease in F/C ratio clearly indicated the successful immobilization of chitosan on PTFE film surfaces. These results specified that the surface-modified films can be employed in biomedical applications (Sun et al. 2006). A study demonstrated that the O_2 plasma pre-treatment and subsequent immobilization of chitosan can significantly enhance the antibacterial property of PU films. The antibacterial property was tested against Gram-positive (*S.*

aureus) and Gram-negative (*Pseudomonas aeruginosa*) bacteria. The outcome validated that the chitosan-immobilized films exhibited substantial antibacterial activity (Kara et al. 2015). In another study, surface properties of polydimethylsiloxane (PDMS) films were tailored by employing a two-step process, i.e. (i) the PDMS films were pre-treated using O_2 plasma, after grafting of AAC, and (ii) plasma polymerization of AAC was performed on the surface of AAC-grafted PDMS films. Conclusively, chitosan and gelatin were immobilized. Results of cell culture analysis unveiled that the immobilized films showed substantial cell proliferation (Salati et al. 2011). Chen et al. (2013) used plasma-assisted graft polymerization of acrylic acid to immobilize chitosan molecules on the surface of poly(tetrafluoroethylene-co-hexafluoropropylene) (FEP) films to advance their surface properties. The authors reported that the surface-modified films exhibited excellent hydrophilicity. Moreover, L929 cell attachment was found to be significantly higher on chitosan-immobilized FEP films (Chen et al. 2013).

In another study, the surface of PP non-woven fabrics was activated by MW plasma followed by AAC grafting in order to immobilize molecules of chitosan. Thrombin time (TT), activated partial thromboplastin time (APTT) and fibrinogen concentration were performed to study the biocompatibility. The outcome validated that the surface-modified fabrics showed ample improvement in antithrombogenic property (Tyan et al. 2003). In another study, Vartiainen et al. tailored the surface of biaxially oriented polypropylene (BOPP) films by immobilizing chitosan. The immobilized BOPP films showed excellent antimicrobial properties, and the authors suggested that it can be used as food packing material (Vartiainen et al. 2005). A study demonstrated the alteration of the surface of polyethylene films by DBD plasma pre-treatment followed by immobilization of chitosan. The authors reported that surface modification substantially enhanced the hydrophilicity of PE with exceptional antibacterial activity against *S. aureus* and *E. Coli* (Theapsak et al. 2012). DC-pulsed O_2 plasma-induced grafting of AAC has also showed promising approach to tailor the surface properties of polypropylene (PP) non-woven fabric (NWF) for immobilization of poly(N-isopropylacrylamide) (PNIPAM) and chitosan. The relation between the AAC and chitosan was confirmed by the existence of amide bonds. The wound-dressing property of immobilized fabrics was analysed using Sprague Dawley (SD) rat as animal model, and their outcome was found to be outstanding, compared to unmodified fabrics (Chen et al., 2012). A novel way of incorporating polysaccharides on LDPE films was achieved by Popelka et al. (2002). Firstly, LDPE films were pre-treated by plasma and subsequently polyacrylic acid brushes were grafted. Further, immobilization of chitosan and chitosan/pectin was performed. The chitosan-immobilized films possessed higher surface roughness and surface free energy. Furthermore, the biomolecule-immobilized films exhibited exceptional antibacterial activity against *E. coli* and *S. aureus*.

Huang and his colleagues performed plasma-assisted grafting of acrylic acid on PLLA films in order to perform chitosan immobilization. In vitro analysis confirmed that the surface-modified PLLA films showed extensive antimicrobial property (Huang et al. 2006). O_2 plasma activation followed by UV-induced graft polymerization is another two-step approach for covalent immobilization of O-stearoyl-chitosan

(OSC) on the surface of polyethylene films. The modified films exhibited substantial hydrophilicity as confirmed by contact angle analysis. Moreover, the immobilized PE films revealed exceptional anti-fouling properties, compared to other PE films (Xin et al. 2013). Application of DC glow discharge plasma reactor (Fig. 3a) for activation of LDPE films for the tethering of polymer moieties like grafting of AAC, immobilization of poly(ethylene glycol) and chitosan, and for enhancement of their blood compatible properties with increase in surface free energy and hydrophilicity of the modified LDPE films has been successfully established. Pandiyaraj and his colleagues have demonstrated the presence of O—C—O, O—C = O, C—N and C—O functionalities on the surface of modified LDPE films by XPS analysis (Fig. 3b). The improvement in blood compatibility of LDPE films was validated by in vitro analysis, which showed reduction in adhesion of platelets and adsorption of proteins with prohibited thrombus formation (Pandiyaraj et al. 2015).

3.3 Immobilization of Collagen

Collagen is one of the most abundant proteins in human body and has a substantial role in different fields with excellent biocompatibility, non-toxicity and biodegradability. Owing to its distinctive qualities and its inherent biodegradable properties, collagen and its hydrolysed form, i.e. gelatin is widely employed in biomaterial developments

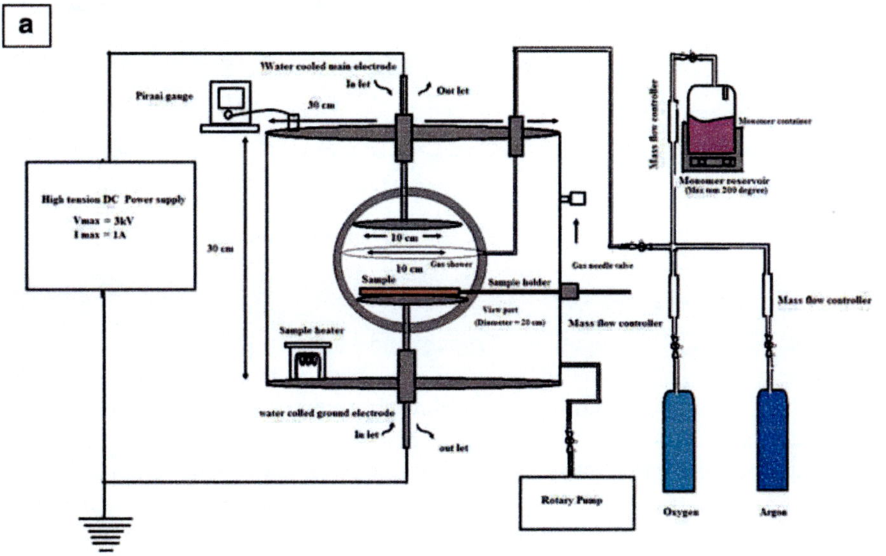

Fig. 3 **a** Schematic diagram of DC glow discharge plasma reactor and **b** C1s high-resolution XPS spectra of various surface-modified LDPE films (Reproduced with permission from Pandiyaraj et al. 2015)

Fig. 3 (continued)

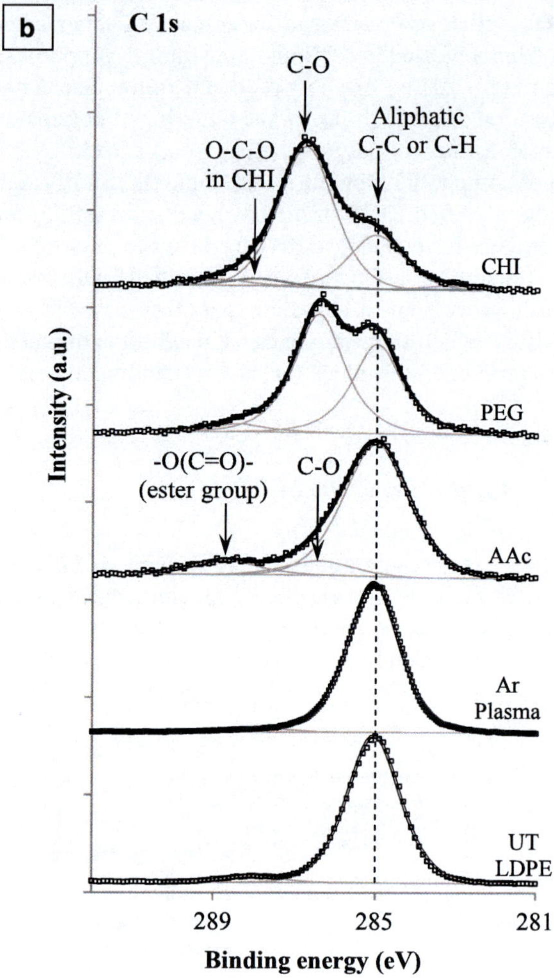

for various biomedical applications (Tripathi and Melo 2017; Tripathi et al. 2013a, 2009). The major applications are drug delivery systems, wound dressing, artificial blood valves, tissue engineering and antithrombogenicity surfaces (Sadeghi et al. 2016; Pence et al. 2014; Kumar and Tripathi 2012). It is a rope-like protein, contains nearly one-third of all proteins in biological tissues and plays a vital role in bringing the toughness and strength to the tissue. In addition, collagen and its derivatives can influence the cell growth in 3D micro-environment (Kozlowska et al. 2015; Inzana et al. 2014). The schematic representation of collagen-immobilized plasma surface-modified polymer-assisting cell adhesion is depicted in Fig. 4.

Juarez-Moreno et al. (2015) tailored the PDMS surfaces by O_2 plasma treatment followed by immobilization of collagen type I. Subsequently, after plasma treatment,

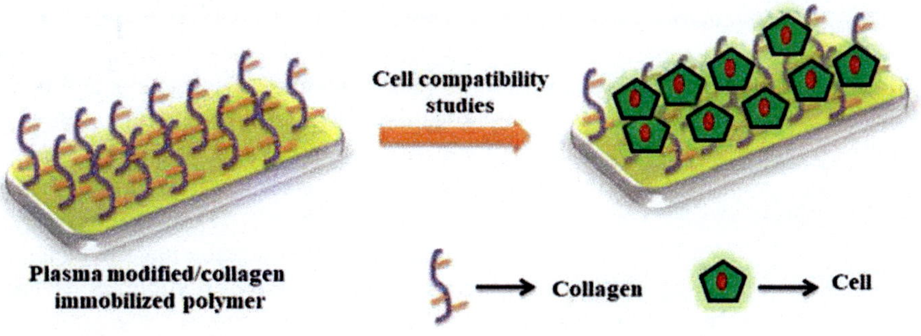

Fig. 4 Schematic of collagen-immobilized polymer-assisting cell adhesion

the PDMS films exhibited lower contact angle values and enhancement in surface roughness, validated by CA and AFM analysis. The existence of oxygen functionalities on PDMS surfaces was established by FTIR analysis. The modified PDMS films exhibited enhancement in adhesion strength, mainly due to the formation of hydrophilic functionalities and enhanced surface roughness. This study confirmed that the fabricated PDMS–collagen composite can be employed as wound dressing material or as skin substitute.

Gonzalez-Paz and his colleagues have developed advanced biomaterials using plasma technology. In this work, two different PUs were employed those are obtained from two fatty acids. Initially, the PU was modified by Ar plasma treatment for grafting of acrylic acid. Finally, the acrylic acid-grafted PU was functionalized with collagen type I. The amendments in surface chemistry and morphology were confirmed by XPS and SEM analysis. Further, in vitro analysis unveiled that the collagen-immobilized PU films showed increase in osteoblastic cell (MG63) adhesion, and the PU films having higher collagen content had excellent cytocompatibility compared to the control (Gonzalez-Paz et al. 2013). Bahrami et al. (2019) tailored the surface of polyurethane films using nitrogen plasma for the incorporation of peroxides and hydrogen peroxide groups on their surfaces. The plasma-treated PU films were grafted with acrylic acid for immobilization of chitosan and collagen on PU film surfaces. The presence of carboxyl, chitosan and collagen functionalities on surface of PU was validated by ATR-FTIR analysis. The surface-modified PU films showed increase in hydrophilic properties. The chitosan- and collagen-immobilized PU films enhanced the adhesion and growth of fibroblast cells, compared to pristine PU. In addition, the modified PU films showed considerable antibacterial activity against two pathogenic bacteria. In another study, plasma-immersion ion implantation (PIII) was employed to tailor the surface properties of PTFE films, for immobilization of collagen. This study demonstrated the possibility of covalent coupling of collagen onto the surface of PTFE without the help of any chemically linked molecules. Cell compatibility results unveiled that only the collagen-immobilized PTFE films assisted human dermal fibroblast attachment and spreading. The authors proposed that the PIII approach as a potential method to modify biomaterial can be employed in the designing of advanced prosthetics (Bax et al. 2010).

Plasma has shown potential application for the activation of PET films and graft polymerization of acrylic acid for immobilization of collagen (types I and II). The outcome of this study unveiled that the PET films can be suitable for improving adhesion and growth of human smooth muscle cells (Bisson et al. 2002). Cheng and Teoh (2004) aimed to modify PCL films by Ar plasma treatment, UV polymerization of AAC and immobilization of collagen. The outcome unveiled subsequent modification of the PCL films which showed increase in hydrophilicity and surface roughness. In vitro analysis was performed to study the cell compatibility of PCL films, which showed substantially enhanced adhesion and proliferation of the human dermal fibroblasts and myoblasts. A study demonstrated the immobilization of collagen type I on plasma treated non-woven poly(ε-caprolactone) (PCL) nanofibres. The existence of collagen on nanofibres was verified by XPS, SEM and CA measurements, as the nanofibres exhibited increase in surface wettability and chemical composition. MTT analysis unveiled that the surface-modified nanofibres can evidently increase human dermal fibroblast (HDF) attachment and proliferation, compared to untreated PCL (Duan et al. 2007). Shen et al. (2011) employed Ar plasma to incorporate peroxides on the surface of polyamide 6(PA6) membranes so as to perform graft polymerization of methacrylic acid. Conclusively, collagen type I was immobilized on their surfaces. Contact angle measurements and AFM analysis unveiled the increase in hydrophilicity and surface roughness, respectively. In addition, the PA6 membranes immobilized with collagen have shown improved proliferation of rat embryo osteoblastic (ROBs) cells. Suntornnond et al. (2016) studied the influence of plasma operating parameters (gas type, flow rate and treatment time) for enhancing the hydrophilicity of polycaprolactone (PCL) membranes. Further to plasma treatment, the PCL membranes were immobilized with collagen. Outcome unveiled that the surface modification process effectively minimized the hydrophobicity. The collagen-immobilized PCL membranes exhibited enhancement in L929 cell growth compared with cell culture plate, confirmed by MTT assay.

Modification in the surface properties of poly(D, L-lactide) films by NH_3 plasma has been performed for the grafting of collagen to enhance their cell affinity. Subsequently, enhancement in hydrophilicity was witnessed after modification. The authors reported that the amount of collagen immobilized was found to be higher on PDLLA films treated by plasma, compared to other traditional treatment. The cell compatibility of the modified PDLLA films was examined against L929 fibroblast cells. Their outcome unveiled that the collagen immobilization effectively increased the cell attachment and growth compared to other modified PDLLA films (Zhao et al. 2006). Keranov et al. (2008) used various procedures to tailor the surface properties of PDMS films. The multi-step process includes Ar plasma treatment, grafting of acrylic acid, coupling with a spacer molecule (PEG) and immobilization of collagen. The authors found that the spacer molecule with long polymer chain can better assist to incorporate more extent of collagen on PDMS surfaces. Bioactivity of modified PDMS films was examined against human fibroblast cells. The collagen-immobilized PDMS films showed excellent cell-compatible properties. Similarly,

He et al. (2005) aimed to develop nanofibre mesh (NFM) that assists human coronary artery endothelial cell (HCAEC) adhesion and growth. They modified poly(L-lactic acid)-co-poly(e-caprolactone) NFM via plasma treatment and immobilization of collagen. XPS analysis confirmed the existence of collagen of the NFM surfaces, and the same was quantified by colorimetric study. The cell compatibility analysis revealed that the collagen-immobilized NFM improved the HCAE cell adhesion, viability and spreading.

Introduction of amine functional moieties into polytetrafluoroethylene (PTFE) films via NH_3 plasma for the immobilization of collagen to assist cell adhesion and proliferation has been reported by Mahmoodi and his colleagues. ATR-FTIR analysis confirmed the incorporation of N_2 and O_2 containing functionalities on the modified PTFE films. Hydrophilic property was found to be increased after immobilization of collagen. In vitro analysis validated that the PTFE films immobilized with collagen exhibited higher cell viability and proliferation, compared to pristine PTFE films (Mahmoodi et al. 2017). Silicon rubber (SR) membranes have also been modified using Ar plasma treatment and further used for plasma-induced graft polymerization of acrylic acid. Later, collagen type III was linked onto the polyacrylic acid containing SR surfaces. Contact angle was found to be decreased subsequently to grafting of AAC and immobilization of collagen. The authors studied the effect of amendments in surface properties of SR against attachment and proliferation of cornea epithelial (CE) cells. Experimental outcome suggested that the collagen-linked SR surfaces were outstanding in cell adhesion and proliferation (Lee and Kao 1996). In similar study, Gupta et al. activated poly(ethylene terephthalate) via plasma treatment and grafted acrylic acid on plasma-treated PET films at different graft densities for collagen immobilization. The presence of carboxylic functional groups on PET films was confirmed by XPS analysis. The surface-modified films showed increase in surface roughness values. Their outcome validated that the presence of collagen was found to be increased with increase in acrylic graft density. Finally, the collagen-immobilized PET films exhibited excellent cell-compatible properties towards human smooth muscle cells (Gupta et al. 2002). Aflori et al. (2013) employed DC helium plasma for surface treatment of PET films to achieve immobilization of collagen on their surfaces. Increase in N_2 content on the surface of modified PET films was validated by XPS analysis. Grains and dendrites of collagen were developed on PET surfaces and were confirmed by AFM and SEM analysis (Fig. 5).

To improve the cell-compatible properties of medical polyurethane (PU) film, it was initially treated with O_2 plasma followed by immobilization of collagen type I using a single-step process. The results supported enhancement of hydrophilic properties of PU films after O_2 plasma treatment and collagen immobilization. Further, HeLa cells were grown on these functionalized surfaces which showed excellent cell attachment and proliferation compared to pristine PU (Fig. 6) (Li and Huang 2007). Yang et al. (2002) have improved the surface of poly(D, L-lactide) films by plasma treatment followed by immobilization of collagen. The authors reported that subsequent to modification the hydrophilicity was increased and the same was found to be

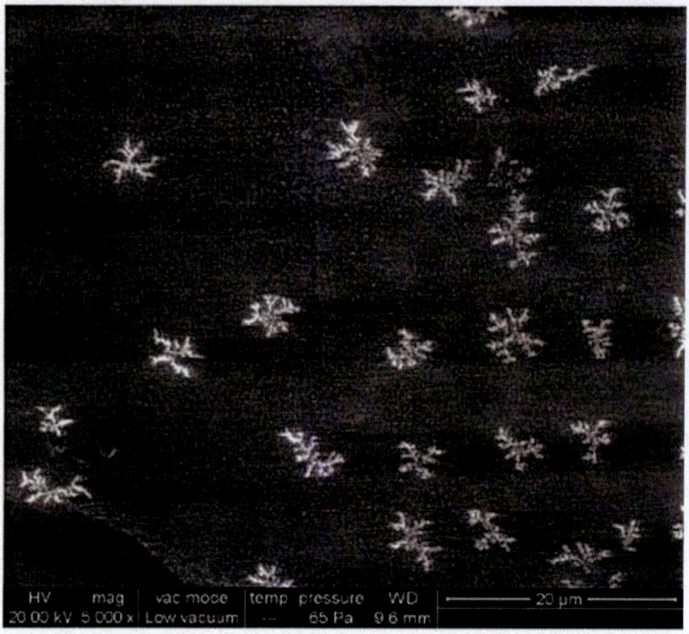

Fig. 5 SEM images of PET film subsequent to helium plasma treatment of 5 min and immobilization of collagen (Reproduced with the permission from Aflori et al. 2013)

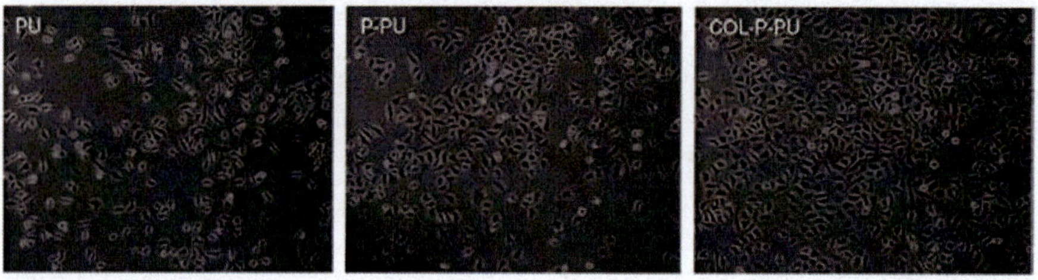

Fig. 6 Morphology of HeLa cells on untreated PU (PU), plasma-treated PU (P-PU) and collagen-immobilized plasma-treated PU (COL-P-PU) (Reproduced with the permission from Li and Huang 2007)

decreased after ageing. Plasma treatment significantly enhanced the surface roughness which helped the collagen molecules to stick firmly to the PDLLA films. Experimental evidence exhibited improvement in cell affinity after collagen immobilization on to the PDLLA film.

3.4 Immobilization of Peptides

In past few decades, peptides have been immobilized on the surface of various materials to enhance their biological performance like cell adhesion and tissue formation (Wojtowicz et al. 2010; Dupont et al. 2012). Peptides can be adsorbed or covalently grafted onto the surface of polymers or comprised in the bulk composition (Pierschbacher and Ruoslathi 1984; Hersel et al. 2003). In grafting technique, the adhesion of peptide was linked to the polymeric side chains or by an extra cross-linking intermediate. Amongst various peptides, RGD peptide (Arg-Gly-Asp) is widely used for surface immobilization to enhance the cell attachment and proliferation of various cell lines, which is characterized as one of the smallest adhesion domains comprises extra cellular matrix (ECM) like vitronectin and fibronectin. Jung and his colleagues have activated the surface of PLLA film by Ar plasma and performed in situ grafting of acrylic acid for improving the hydrophilicity. The COOH-containing PLLA films were immobilized with amine groups of Gly-Arg-Asp-Gly (GRDG) and a ligand peptide GRGD. The surface modification technique improved the hydrophilicity and smooth morphology of PLLA films. Cell culture studies unveiled that the PLLA films increased fibroblast adhesion. The authors concluded that the PLLA films immobilized with GRGD peptides presented outstanding adhesion and spreading of fibroblasts (Jung et al. 2005). Gentile et al. (2015) aimed to functionalize a novel composite POSS-PCU polymer with peptides via plasma polymerization. Initially, plasma polymerization of acrylic acid was performed on the surface of POSS-PCU polymers followed by grafting of KRSR and FHRRIKA peptides. XPS and IR spectroscopy confirmed the existence of carboxyl and peptide functionalities onto the surface of polymers. In vitro analysis was performed with bone marrow mesenchymal stem cells for demonstrating the efficiency of peptide-grafted POSS-PCU polymers which showed enhanced cell attachment on peptide-grafted composite POSS-PCU polymer. Another study showed the alteration of the surface of polycaprolactone (PCL) and poly-L-lactic acid (PLLA) films via pulsed plasma deposition of allylamine, followed by grafting of PEG adhesion peptides (RGD and YIGSR). Peptide-immobilized PCL and PLLA films showed higher mouse hepatocyte cell adhesion compared to control. The authors concluded that this surface modification technique can be employed to enhance the adhesion of peptides on polymers (Carlisle et al. 2000). Aucoin et al. (2002) employed MW plasma to tailor the surface of polydimethylsiloxane (PDMS) substrate. At first, plasma polymerization of allyl alcohol was performed on the surface of PDMS films. Conclusively, cell adhesion peptides such as YIGSR, PDSGR, RGDS and PHSRN were covalently attached to the modified PDMS substrate for improving hydrophilicity and amended chemical composition confirmed by CA measurement and XPS analysis, respectively. Human corneal epithelial cells were employed to assess the cell-compatible properties, which suggested that all the peptides exhibited substantial enhancement in cell adhesion. In another study, chitosan and PLGA membranes were modified using plasma and chemical methods, wherein membranes were first treated using O_2 plasma followed by attachment of laminin peptides by chemical route. The O_2 plasma treatment was found to be effective in introducing laminin on the surface of membranes. The cell assay showed that the laminin-immobilized membrane exhibited good cell affinity to Schwann cells (Huang et al. 2007). Lopez et al. (2005) executed plasma polymerization of AAC onto the surface of PET films using RF glow discharge plasma. The polymerized PET films were characterized with low density of COOH groups and exhibited uniform morphology. In addition, the films were found to be very stable. The PET films were further tailored by immobilization of RGD peptide via COOH as anchoring group. The surface-modified PET films exhibited substantial cyto-compatibility for 3T3 fibroblasts.

3.5 *Immobilization of Enzymes*

Enzymes are gaining interest amongst researchers owing to its substantial activity and selectivity, suitable to catalyse various complicated chemical reactions in a critical experimental environment (Kirk et al. 2002; Tripathi and Kumar 2012; Saini et al. 2015). Another important character is that enzymes exhibit their action in a very mild circumstance, for instance neutral pH, low temperature and pressure conditions, and finally they can be easily biodegraded (please refer introductory chapter). All these fascinating properties make enzyme an important candidate in numerous applications such as antibacterial surfaces in food packing industry, biomedical or pharmaceutical field (Arefi-Khonsari and Tatoulian 2008). Morshed et al. (2019) employed atmospheric pressure plasma for treatment of poly(ethylene terephthalate) non-woven fabric (PN) followed by chemical grafting of poly(amidoamine) or poly(ethylene glycol)-OH. The fabrics were finally immobilized with glucose oxidase (GOx) enzymes. The authors stated that the fabrics having higher concentration of amine functionalities exhibited active loading of enzymes. The enzyme-immobilized polyester presented extraordinary antibacterial activity against *Escherichia coli* and *Staphylococcus epidermidis*. By coupling of two-component polyurethane with horseradish peroxidase enzyme, a stable polymer conjugate was fabricated, which was further modified by combining with poly(dimethylsiloxane)(PDMS). The alkyl groups existed on the PDMS were converted into polar silanol functional groups by using an atmospheric pressure plasma jet system. The modified conjugate combined with PDMS showed higher stability compared to unmodified system. Conclusively, the enzyme-immobilized polymer conjugate exhibited substantial biological activity, compared to other modified samples (Kreider et al. 2013).

Vasilets et al. (1997) have activated polytetrafluoroethylene (PTFE) films by CO_2 plasma and acrylic acid graft polymerization. The modified PTFE films were finally functionalized with human thrombomodulin (hTM). The formation of various functional groups like COF, COOH, –C=C–, on the surface of PTFE was confirmed by XPS and ATR-FTIR analysis. Protein C activation test validated the immobilization of thrombomodulin onto the PTFE films further converted thrombin into an anticoagulant form. Abbas et al. reported the immobilization of trypsin onto the surface of silicon substrate using multiple step surface modification process. At first, plasma polymerization of allylamine was performed on the substrates. Later, covalent immobilization of enzyme was performed using a three-step procedure, which is as follows; (1) polymer surface was activated using a linker glutaraldehyde, (2) trypsin immobilization and (3) reduction of imino groups. The plasma-polymerized allylamine films showed higher adsorption of trypsin. The enzyme-immobilized substrates exhibited high catalytic activity (Abbas et al. 2009).

4 Immobilization on Polymer Scaffolds

Natural and synthetic polymeric scaffolds are playing vital role in many bioengineering applications majorly in tissue engineering and regenerative medicine (Tripathi and Melo 2017, 2019a; Tripathi and Kumar 2011). In this section, a comprehensive outlook is given about how the plasma surface modification technique and biomolecule immobilization can enhance the biocompatibility of polymer scaffolds for employing them in various biomedical applications. A schematic representation of the same is depicted in Fig. 7.

Surface modification of PCL scaffolds has been performed either by single-step modification process, i.e. O_2 plasma treatment, or by multi-step modification process that includes post-argon 2-amino-ethyl methacrylate grafting (AEMA). The modified PCL scaffold was further used for the immobilization of gelatin type B and physisorption of fibronectin. In addition, scaffolds of various designs are exposed to double protein coating. The enhancement in colonization by cells was higher for the PCL scaffolds modified by the multi-step surface modification method containing the protein coating, compared to other modified PCL scaffolds. This study concluded

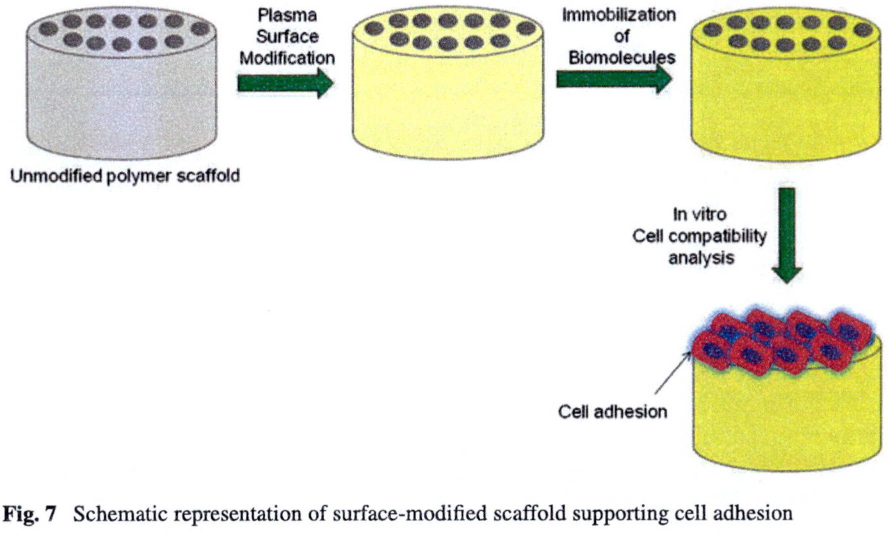

Fig. 7 Schematic representation of surface-modified scaffold supporting cell adhesion

that the employed surface modification technique influenced the cell colonization, compared to the design of the scaffold (Declercq et al. 2013). In another study, RGD peptides, i.e. RGES (Arg-Gly-Glu-Ser) and RGDS (Arg-Gly-Asp-Ser), were immobilized on the surface of PLLA scaffolds via plasma treatment. Cell culture outcome of rat osteosarcoma (OS), osteoblastic-like cells, validated that the RGDS immobilization could effectually improve the OS cell attachment and cell density on PLLA scaffolds. Additionally, the in vitro mineralization unveiled that the cell attachment was found exceptional and signifying that it has the tendency to form bone-like tissue, compared to pristine PLLA scaffolds. However, the cell attachment was found to be insignificant on RGES-immobilized PLLA scaffolds. The outcome of the work signifies that the plasma-assisted RGDS-immobilized PLLA scaffolds can be a potential candidate for production of bone-like tissues (Ho et al. 2006).

In another study, modified PLLA scaffolds were developed via plasma treatment followed by grafting of acrylic acid. Finally, the grafted PLLA scaffolds were immobilized with peptide ligands, either Gly-Arg-Asp-Gly(GRDG) or Gly-Arg-Gly-Asp (GRGD). The scaffolds possess the average porosity of about 90%. Cell culture analysis unveiled that the chondrocyte adhesion and proliferation were mainly influenced by the type of substrate. The maximum cell function and improved accumulation of total glycosaminoglycans (GAGs) was observed on RGD-immobilized scaffolds. Their findings suggested potential application of RGD-functionalized scaffold in cartilaginous tissue genesis (Jung et al. 2008). 3D porous PCL scaffold has also been surface functionalized with collagen biomolecules for improving its biocompatibility. Initially, O_2 plasma-assisted surface etching of PCL scaffold was performed followed by surface tethering of acrylic groups using acrylic acid as a precursor solution. Acrylic graft surface of PCL scaffold showed collagen binding and increased surface hydrophilicity confirmed by contact angle measurements. After functionalization, surface roughness was found to be increased which showed high binding and proliferation of M3CT3-E1 cells (Park et al. 2016). Similarly, another study demonstrated the surface functionalization of PCL scaffold by immobilization of recombinant human bone morphogenetic protein-2 (rhBMP-2), which plays an important role in the development of bone by inducing osteoblast differentiation in variety of cell types. Therefore, plasma polymerization of acrylic acid was performed on PCL scaffolds to form a thin film consisting of functional carboxylic groups for protein cross-linking. ATR-FTIR and XPS outcome validated the existence of carboxylic functional groups and rhBMP-2 on the surface of PCL scaffolds. The protein-2-immobilized PCL scaffolds enhanced the level of MG-63 cell differentiation, which was confirmed by the alkaline phosphate activity assay (Kim et al. 2013). In another study, allylamine was coupled to the surface of PCL by plasma polymerization, which provides high density of amine functional groups to the scaffold surface. Further, fibroblast growth factor-2 (FGF-2) and heparin were tagged to this aminated surface for improving the cell adhesion and functions. The stability and uniformity of the amine coatings on scaffolds were unveiled by XPS analysis and time-of–flight secondary ion mass spectrometry (ToF SIMS) images, respectively. The surface-modified scaffolds supported cell attachment and formation of different keratinocyte and fibroblast layers. The authors concluded that the surface modification technique exhibited outstanding cell attachment and organization in the development of skin composite in short duration (Robinson et al. 2016). For improving the bioactivity and biological performance of PCL scaffold, Sousa and his colleagues employed different strategies to tailor the surface of PCL scaffolds by activating it with plasma followed by grafting of acrylic acid via UV polymerization. Finally, collagen was immobilized on acrylic graft PCL scaffold. The collagen-immobilized PCL scaffolds showed enhancement in hydrophilicity. SEM analysis confirmed the amendments in surface morphology (see Fig. 8). Furthermore, the biological studies unveiled the increase in the adhesion of fibroblast cells. The authors highlighted that the surface modification of PCL scaffolds using collagen is an effective way to improve the hydrophilicity and also circumvent the inadequate cell contact of PCL scaffolds (Sousa et al. 2014).

Shen et al. (2009) examined the influence of different plasma atmosphere (ammonia, O_2 and CO_2) and compared their ability for incorporation of polar functionalities on the surface of poly(lactide-co-glycolide)(PLGA) scaffolds. They used rhBMP-2 protein as a model biomolecule. The study demonstrated that O_2 and CO_2 plasma improved the binding ability of PLGA towards rhBMP-2 and the same was found to be higher for O_2 plasma-treated PLGA scaffolds. The binding ability was mainly influenced by surface chemistry and topography that was induced by O_2 plasma. Furthermore, the rhmBP-2-immobilized O_2 plasma-treated PLGA scaffolds were found to be bioactive which was confirmed by stimulated differentiation of OCT-1 cell. In another study, Woo et al. (2010) have grafted β-glucan on the surface of PLGA scaffolds. The composite was further modified via MW-induced Ar plasma system at atmospheric pressure. It is evident from the result that the MW plasma increased the surface roughness of the β-glucan-immobilized PLGA scaffold. In addition, Ar plasma treatment offered cyto-compatible environment for adipose-derived stem cell (ADSC) attachment and growth. The authors proposed that plasma treatment could improve the cellular response of ADSC towards biodegradable polymer and can play a significant role in restoring soft tissues.

5 Conclusion

In this chapter, a wide overview has been provided on the role of plasma surface modification process followed by immobilization of biomolecules intended to attain a particular response in biological environment. Over the past few decades, research has been advanced from single-step plasma treatment or activation to well-designed surface modification methods that comprise plasma treatment or plasma polymerization or plasma-induced grafting and subsequent immobilization of proteins, enzymes or biomolecules. Polymeric materials were modified using these techniques that were able to assist the cell adhesion and growth substantially, and distinguished well compared to the pristine materials. Plasma is exclusively fascinating as it is environment-friendly and consistent and offers high control in tuning the surface properties without hampering the bulk properties. Heparin-immobilized plasma-modified polymer was found to have excellent anticoagulation property. Plasma

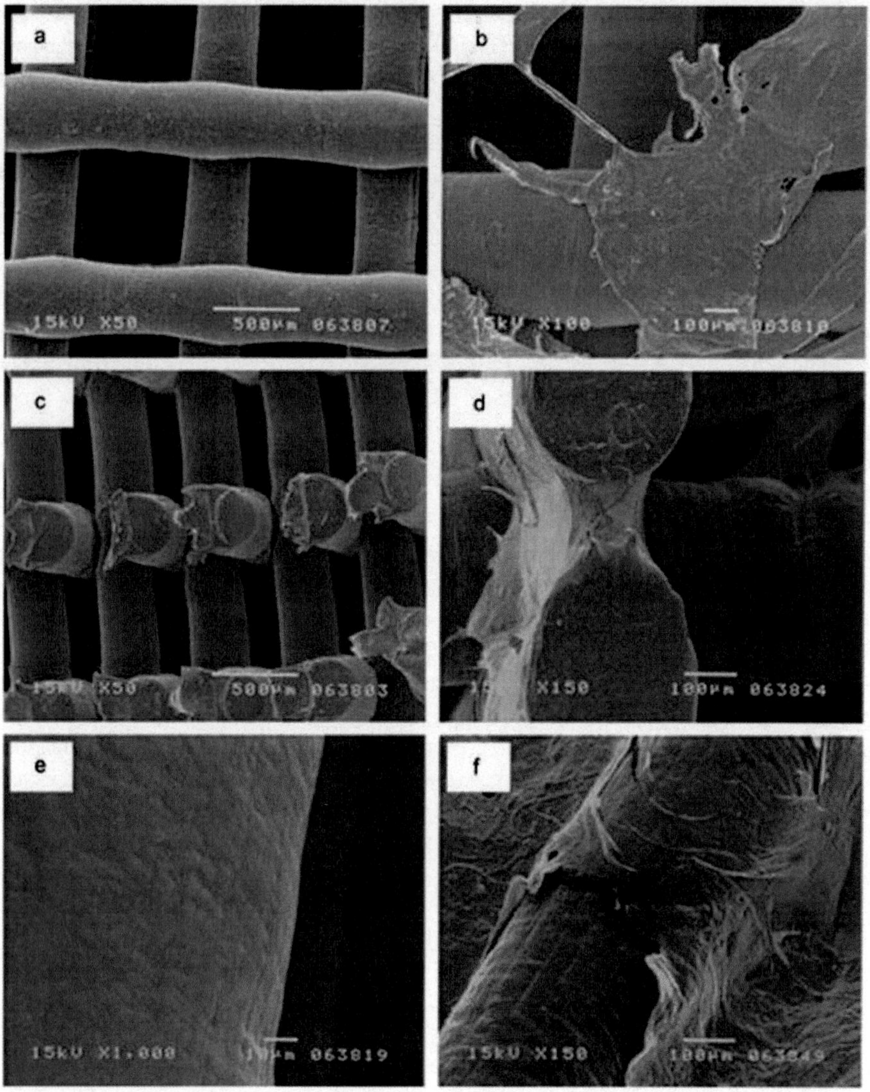

Fig. 8 SEM micrographs of PCL scaffolds before (left) and after (right) the collagen surface modification: top view (**a**) and (**b**), cross-sectional view (**c**) and (**d**) and filament (**e**) and (**f**) (Reproduced with the permission from Sousa et al. 2014)

polymers immobilized with chitosan exhibited substantial blood compatibility and antimicrobial properties against large number of microbes. Collagen- and peptide-immobilized films showed outstanding cell-compatible (cell adhesion, proliferation and spreading) properties. Enzyme-immobilized polymers showed substantial antimicrobial activity and antithrombogenic property by inhibiting the formation of thrombosis. In addition, various biomolecule-immobilized 3D polymer scaffolds unveiled excellent blood and cell compatibility. Conclusively, various advantages of plasma surface modification are followed by immobilization of biomolecules offering profound understanding and more widespread about their major role in enhancing the performance of biomaterials for the foreseeable future in biomedicine.

References

Abbas A, Vercaigne-Marko D, Supiot P, Bocquet B, Vivien C, Guillochon D (2009) Covalent attachment of trypsin on plasma polymerized allylamine. Colloids Surf B 73:315–324

Aflori M, Drobota M, Dimitriu DG, Stoica I, Simionescu B, Harabagiu V (2013) Collagen immobilization on polyethylene terephthalate surface after helium plasma treatment. Mater Sci Eng, B 178:1303–1310

Alexander H, Brunski HB, Cooper SL, Hench LL, Hergenrother RW, Hoffman AS, Kohn J, Langer R, Peppas NA, Ratner BD, Shalaby SW, Visser SA, Yannas IV (1996) Classes of materials used in medicine. In: Ratner BD, Hoffman AS, Schoen FJ, Lemons JE (eds) Biomaterials science: an introduction to materials in medicine, vol 2. Academic Press, Publication date: 20 May 1997. ISBN: 9780125824613, https://doi.org/10.1016/B978-0-08-050014-0.50007-9.

Alexandrov SE, Hitchman ML (2005) Chemical vapor deposition enhanced by atmospheric pressure non-thermal non-equilibrium plasmas. Chem Vap Deposition 11:457–468

An YH, Friedman RJ (1997) Laboratory methods for studies of bacterial adhesion. J Microbiol Methods 30(2):141–152

Arefi-Khonsari F, Tatoulian M (2008) Plasma processing of polymers by a low-frequency discharge with asymmetrical configuration of electrode. In : D'Agostino R, Favia P, Kawai Y, Ikegami H, Sato N, Arefi-Khonsari F (eds) Advanced plasma technology. Wiley VCH Publication, Germany, publication date: 12 Dec 2007. ISBN: 9783527405916, https://doi.org/10.1002/9783527622184.ch8

Aucoin L, Griffith CM, Pleizier G, Deslandes Y, Sheardown H (2002) Interactions of corneal epithelial cells and surfaces modified with cell adhesion peptide combinations. J Biomater Sci Polym Edn 13(4):447–462

Ayati-Najafabadi SA, Keshvari H, Ganji Y, Tahriri M, Ashuri M (2012) Chitosan/heparin surface modified polyacrylic acid grafted polyurethane film by two step plasma treatment. Surf Eng 28(9):710–714

Bahrami N, Khorasani SN, Mahdavi H, Ghiaci M, Mokhtari R (2019) Low-pressure plasma surface modification of polyurethane films with chitosan and collagen biomolecules. J Appl Polym Sci 136:47567

BaqueyC PalumboF, Porte-DurrieuMC LegeayG, TressaudA D'AgostinoR (1999) Plasma treatment of expanded PTFE offers a way to a biofunctionalization of its surface. Nucl Instrum Methods Phys Res Sect B 151:255–262

Bax DV, McKenzie DR, Weiss AS, Bilek MMM (2010) The linker-free covalent attachment of collagen to plasma immersion ion implantation treated polytetrafluoroethylene and subsequent cell-binding activity. Biomaterials 31:2526–2534

Bhattacharyya D, Xu H, Deshmukh RR, Timmons RB, Nguyen KT (2010) Surface chemistry and polymer film thickness effects on endothelial cell adhesion and proliferation. J Biomed Mater Res A 94(2):640–648

Biederman H, Osada Y (1990) Plasma chemistry of polymers. In: Biederman H (ed) Advances in polymer science. Springer-Verlag, Berlin. ISBN: 978-3-540-52159-4. https://doi.org/10.1007/3-540-52159-3_6

Bisson I, Kosinski M, Ruault S, Gupta B, Hilborn J, Wurm F, Frey P (2002) Acrylic acid grafting and collagen immobilization on poly(ethylene terephthalate) surfaces for adherence and growth of human bladder smooth muscle cells. Biomaterials 23:3149–3158

Carlisle ES, Mariappan MR, Nelson KD, Thomes BE, Timmons RB, Constantinescu A, Eberhart RC, Bankey PE (2000) Enhancing hepatocyte adhesion by pulsed plasma deposition and polyethylene glycol coupling. Tissue Eng 6(1):45–52

Chandy T, Das GS, Wilson RF, Rao GHR (2000) Use of plasma glow for surface engineering biomolecules to enhance blood compatibility of Dacron and PTFE vascular Prosthesis. Biomaterials 21:699–712

Cheng Z, Teoh SH (2004) Surface modification of ultra-thin poly (ε-caprolactone) films using acrylic acid and collagen. Biomaterials 25:1991–2001

Chen JP, Kuo CY, Lee WL (2012) Thermo-responsive wound dressings by grafting chitosan and poly(N-isopropylacrylamide) to plasma-induced graft polymerization modified non-woven fabrics. Appl Surf Sci 262:95–101

Chen Y, Yi J, Gao Q, Zhou X, Luo Y, Liu P (2013) Surface performance and cytocompatibility evaluation of acrylic acid-mediated carboxymethyl chitosan coating on poly(tetrafluoroethylene-co- hexafluoropropylene). Plasma Chem Plasma Process 33:1153–1165

Chu PK, Chen JY, Wang LP, Huang N (2002) Plasma-surface modification of biomaterials. Mater Sci Eng R Rep 36(5–6):143–206

D'Agostino R, Favia P, Oehr C, Wertheimer MR (2005) Low-temperature plasma processing of materials: past, present, and future. Plasma Processes Polym 2(1):7–15

Declercq HA, Desmet T, Berneel EEM, Dubruel P, Maria J (2013) Synergistic effect of surface modification and scaffold design of bioplotted 3-D poly-e-caprolactone scaffolds in osteogenic tissue engineering. Acta Biomater 9:7699–7708

Degoutin S, Jimenez M, Casetta M, Bellayer S, Chai F, Blanchemain N, Neut C, Kacem I, Traisnel M, Martel B (2012) Anticoagulant and antimicrobial finishing of non-woven polypropylene textiles. Biomed Mater 7(3):035001

Ding Z, Chen J, Gao S, Chang J, Zhang J, Kang ET (2004) Immobilization of chitosan onto poly-l-lactic acid film surface by plasma graft polymerization to control the morphology of fibroblast and liver cells. Biomaterials 25(6):1059–1067

Duan Y, Wang Z, Yan W, Wang S, Zhang S, Jia J (2007) Preparation of collagen-coated electrospunnanofibers by remote plasma treatment and their biological properties. J Biomater Sci Polym Edn 18(9):1153–1164

Dupont KM, Boerckel JD, Stevens HY, Diab T, Kolambkar YM, Takahata M, Schwarz EM, Guldberg RE (2012) Synthetic scaffold coating with adeno-associated virus encoding BMP2 to promote endogenous bone repair. Cell Tissue Res 347:575–588

Favia P, Palumbo F, D' Agostino R, Lamponi S, Magnini A, Barbucci R (1998) Immobilization of heparin and highly-sulphated hyaluronic acid on to the plasma-treated polyethylene. Plasmas Polym 3:77–96

Friedman M, Walsh G (2002) High performance films: review of new materials and trends. Polym Eng Sci 42:1756–1788

Gao Q, Chen Y, Wei Y, Wang X, Luo Y (2013) Heparin-grafted poly(tetrafluoroethylene-cohexafluoropropylene) film with highly effective blood compatibility via an esterification reaction. Surf Coat Technol 228:S126–S130

Gentile P, Ghione C, Tonda-Turo C, Kalaskarc DM (2015) Peptide functionalisation of nanocomposite polymer for bone tissue engineering using plasma surface polymerisation. RSC Adv 5:80039–80047

Gogolewski S, Varlet PM, Dillon JG (1996) Sterility, mechanical properties and molecular stability of polylactic internal-fixation devices treated with low-temperature plasmas. J Biomed Mater Res 32:227–235

Gonzalez-Paz RJ, Ferreira AM, Mattu C, Boccafoschi F, Lligadas G, Ronda JC, Galia M, Cadiz V, Ciardelli G (2013) Cytocompatible polyurethanes from fatty acids through covalent immobilization of collagen. React Funct Polym 73:690–697

Goossens O, Dekempeneer E, Vangeneugden D, Van de Leest R, Leys C (2001) Application of atmospheric pressure dielectric barrier discharges in deposition, cleaning and activation. Surf Coat Technol 142–144:474–481

Gupta B, Plummer C, Bisson I, Frey P, Hilborn J (2002) Plasma-induced graft polymerization of acrylic acid onto poly(ethylene terephthalate) films: characterization and human smooth muscle cell growth on grafted films. Biomaterials 23:863–871

Hersel U, Dahman C, Kessler H (2003) RGD modified polymers: biomaterials for stimulated cell adhesion and beyond. Biomaterials 24(24):4385–4415

Heyse P, Dams R, Paulussen S, Houthoofd K, Janssen K, Jacobs PA, Sels BF (2007) Dielectric barrier discharge at atmospheric pressure as a tool to deposit versatile organic coatings at moderate power input. Plasma Processes Polym 4(2):145–157

He W, Ma ZW, Yong T, Teo WE, Ramakrishna S (2005) Fabrication of collagen-coated biodegradable polymer nanofiber mesh and its potential for endothelial cells growth. Biomaterials 26:7606–7615

Hori K, Matsumoto S (2010) Bacterial adhesion: from mechanism to control. Biochem Eng J 48(3):424–434

Ho MH, Hou LT, Tu CY, Hsieh HJ, Lai JY, Chen WJ, Wang DM (2006) Promotion of Cell Affinity of Porous PLLA Scaffolds by Immobilization of RGD Peptides via plasma treatment. Macromol Biosci 6:90–98

Huang CL, Lin YY, Liao JD (2006) Immobilization of chitosan on the plasma-activated poly-L-lactic acid film surface using evaporated acrylic acid as the intermediate. Adv Sci Technol 49:197–202

Huang YC, Huang CC, Huang YY, Chen KS (2007) Surface modification and characterization of chitosan or plga membrane with laminin by chemical and oxygen plasma treatment for neural regeneration. J Biomed Mater Res Part 82A: 842–851

Inzana JA, Olvera D, Fuller SM, Kelly JP, Graeve OA, Schwarz EM, Kates SL, Awad HA (2014) 3D printing of composite calcium phosphate and collagen scaffolds for bone regeneration. Biomaterials 35(13):4026–4034

Jacobs T, Morent R, De Geyter N, Dubruel P, Leys C (2012) Plasma surface modification of biomedical polymers: Influence on cell-material interaction. Plasma Chem Plasma Process 32:1039–1073

Jin J, Jiang W, Shi Q, Zhao J, Yin J, Stagnaro P (2012) Fabrication of PP-g-PEGMA-gheparin and its hemocompatibility: from protein adsorption to anticoagulant tendency. Appl Surf Sci 258:5841–5849

Juarez-Moreno JA, Avila-Ortega A, Oliva AI, Aviles F, Cauich-Rodriguez JV (2015) Effect of wettability and surface roughness on the adhesion properties of collagen on PDMS films treated by capacitively coupled oxygen plasma. Appl Surf Sci 349:763–773

Jung HJ, Ahn KD, Han DK (2005) Surface characteristics and fibroblast adhesion behavior of RGD-immobilized biodegradable PLLA films. Macromol Res 13(5):446–452

Jung HJ, Park K, Kim JJ, Lee JH, Han KO, Han DK (2008) Effect of RGD-immobilized dual-pore poly(L-Lactic Acid) scaffolds on chondrocyte proliferation and extracellular matrix production. Artif Organs 32(12):981–989

Kara F, Aksoy EA, Yuksekdag Z, Aksoy S, Hasirci N (2015) Enhancement of antibacterial properties of polyurethanes by chitosan and heparin immobilization. Appl Surf Sci 357:1692–1702

Keranov I, Vladkova T, Minchev M, Kostadinova A, Altankov G (2008) Preparation, characterization, and cellular interactions of collagen-immobilized PDMS surfaces. J Appl Polym Sci 110:321–330

Kerri JP, Stuart KW, Jarrell BE (1989) Enhanced adherence of human adult endothelial cells to plasma discharge modified polyethylene terephthalate. J Biomed Mater Res 23:1131–1147

Keselowsky BG, Collard DM, Garcia AJ (2004) Surface chemistry modulates focal adhesion composition and signals through changes in integrin binding. Biomaterials 25:5947–5954

Kim YJ, Kang IK, Huh MW, Yoon SC (2000) Surface characterization and in vitro blood compatibility of poly(ethylene terephthalate) immobilized with insulin and/or heparin using plasma glow discharge. Biomaterials 21(2):121–130

Kim BH, Myung SW, Jung SC, Ko YM (2013) Plasma Surface Modification for Immobilization of Bone Morphogenic Protein-2 on Polycaprolactone Scaffolds. Jpn J Appl Phys 52(11S): 11NF01

Kirk O, Borchert TV, Fuglsang CC (2002) Industrial enzyme applications. Curr Opin Biotechnol 13(4):345–351

Kozlowska J, Sionkowska A, Skopinska-Wisniewska J, Piechowicz K (2015) Northern pike (Esoxlucius) collagen: Extraction, characterization and potential application. Int J Biol Macromol 81:220–227

Kreider A, Richter K, Sell S, Fenske M, Tornow C, Stenzel V, Grunwald I (2013) Functionalization of PDMS modified and plasma activated two-component polyurethane coatings by surface attachment of enzymes. Appl Surf Sci 273:562–569

Kumar A, Tripathi A (2012) Biopolymeric scaffolds for tissue engineering. In: Tiwari A, Srivastava RB (eds) Biotechnology in biopolymers developments, applications and challenging areas. i-Smithers Repra Publication Ltd. United Kingdom, pp 233–285

Lee SD, Kao CY (1996) Plasma-induced grafted polymerization of acrylic acid and subsequent grafting of collagen onto polymer film as biomaterial. Biomaterials 17:1599–1608

Li YH, Huang YD (2007) The study of collagen immobilization on polyurethane by oxygen plasma treatment to enhance cell adhesion and growth. Surf Coat Technol 201:5124–5127

Li R, Wang H, Wang W, Ye Y (2013) Immobilization of heparin on the surface of polypropylene non-woven fabric for improvement of the hydrophilicity and blood compatibility. J Biomater Sci Polym Edn 24(1):15–30

Li P, Li L, Wang W, Jin W, Liu X, Yeung KWK, Chu PK (2014) Enhanced corrosion resistance and hemocompatibility of biomedical NiTi alloy by atmospheric-pressure plasma polymerized fluorine-rich coating. Appl Surf Sci 297:109–115

Lloyd G, Friedman G, Jafri S, Schultz G, Fridman A, Harding K (2007) Gas plasma: medical uses and developments in wound care. Plasma Processes Polym 7(3–4):194–211

Lopez-Garcia J, Bilek F, Lehocky M, Junkar I, Mozetic M, Sowe M (2013) Enhanced printability of polyethylene through air plasma treatment. Vacuum 95:43–49

Lopez LC, Gristina TR, Ceccone G, Rossi F, Favia P, D'agostino R, (2005) Immobilization of RGD peptides on stable plasma-deposited acrylic acid coatings for biomedical devices. Surf Coat Technol 200:1000–1004

Mahmoodi M, Zamanifard M, Safarzadeh M, Bonakdar S (2017) *In vitro* evaluation of collagen immobilization on polytetrafluoroethylene through NH_3 plasma treatment to enhance endothelial cell adhesion and growth. Bio Med Mater Eng 28:489–501

Mao C, Yuan J, Mei H, Zhu A, Shen J, Lin S (2004) Introduction of photocrosslinkable chitosan to polyethylene film by radiation grafting and its blood compatibility. Mater Sci Eng C 24:479–485

Martin Y, Boutin D, Vermette P (2007) Study of the effect of process parameters for ***n***-heptylamine plasma polymerization on final layer properties. Thin Solid Films 515:6844–6852

Morent R, De Geyter N, Trentesaux M, Gengembre L, Dubruel P, Leys C, Payen E (2010) Stability study of polyacrylic acid films plasma-polymerized on polypropylene substrates at medium pressure. Appl Surf Sci 257: 372–380

Morent R, De Geyter N, Trentesaux M, Gengembre L, Dubruel P, Leys C, Payen E (2010) Influence of discharge atmosphere on the ageing behaviour of plasma-treated polylactic acid. Plasma Chem Plasma Process 30:525–536

Morshed MN, Behary N, Bouazizi N, Guan J, Chen G, Nierstrasz V (2019) Surface modification of polyester fabric using plasma-dendrimer for robust immobilization of glucose oxidase enzyme. Sci Rep 9:15730

Nair LS, Laurencin CT (2005) Polymers as biomaterials for tissue engineering and controlled drug delivery. In: Lee K, Kaplan D (eds) Tissue engineering I. Advances in biochemical engineering/biotechnology. Springer, Print publication date: 25 Oct 2005. ISBN 978-3-540-31944-3, https://doi.org/10.1007/b137240

Narayanan PV (1994) Surface functionalization by RF plasma treatment of polymers for immobilization of bioactive-molecules. J Biomater Sci Polym Edn 6(3):181–193

Pandiyaraj KN, Selvarajan V, Rhee YH, Kim HW, Shah SI (2009) Glow discharge plasma-induced immobilization of heparin and insulin on polyethylene terephthalate film surfaces enhances anti-thrombogenic properties. Mater Sci Eng C 29(3):796–805
Pandiyaraj KN, Ferrari AM, BotelhodoRego AM, Deshmukh RR, Su PG, Halleluyah M, Halim AS (2015) Low-pressure plasma enhanced immobilization of chitosan on low-density polyethylene for bio-medical applications. Appl Surf Sci 328: 1–12
Pandiyaraj KN, Ram Kumar MC, Arun Kumar A, Padmanabhan PVA, Deshmukh RR, Bah M, Shah SI, Su PG, Halim AS (2016) Tailoring the surface properties of polypropylene films through cold atmospheric pressure (CAPP) plasma assisted polymerization and immobilization of biomolecules for enhancement of anti-coagulation activity. Appl Surf Sci 370:545–556
Pantoja M, Encinas N, Abenojar J, Martínez MA (2013) Effect of tetraethoxysilane coating on the improvement of plasma treated polypropylene adhesion. Appl Surf Sci 280:850–857
Park YO, Myung SW, Kook MS, Jung SC, KimBH, (2016) Cell Proliferation on Macro/Nano Surface Structure and Collagen Immobilization of 3D Polycaprolactone Scaffolds. J Nanosci Nanotechnol 16:1415–1419
Pence JC, Gonnerman EA, Bailey RC, Harley BA (2014) Strategies to balance covalent and non-covalent biomolecule attachment within collagen-GAG biomaterials. Biomater Sci 2:1296–1304
Pierschbacher MD, Ruoslahti E (1984) Cell attachment activity of fibronectin can be duplicated by small synthetic fragments of the molecule. Nature 309:30–33
Popelka A, Novak I, Lehocky M, Junkar I, Mozetic M, Kleinova A, Janigova I, Slouf M, Bilek F, Chodak I (2002) A new route for chitosan immobilization onto polyethylene surface. Carbohyd Polym 90(4):1501–1508
Pratt KJ, Williams SK, Jarrell BE (1989) Enhanced adherence of human adult endothelial cells to plasma discharge modified polyethylene terephthalate. J Biomed Mater Res 23(10):1131–1147
Pu FR, Williams RL, Markkula TK, Hunt J (2002) Expression of leukocyte–endothelial cell adhesion molecules on monocyte adhesion to human endothelial cells on plasma treated PET and PTFE in vitro. Biomaterials 23:4705–4718
Raafat D, Sahl HG (2009) Chitosan and its antimicrobial potential—a critical literature survey. Microb Biotechnol 2(2):186–201
Robinson DE, Al-Bataineh SA, Farrugia BL, Michelmore A, Cowin AJ, Dargaville TR, Short RD, Smith LE, Whittle JD (2016) Plasma polymer and biomolecule modification of 3D scaffolds for tissue engineering. Plasma Processes Polym 13:678–689
Rossini P, Colpo P, Ceccone G, Jandt KD, Rossi F (2003) Surfaces engineering of polymeric films for biomedical applications. Mater Sci Eng C 23(3):353–358
Sadeghi AR, Nokhasteh S, Molavi A, Khorsand-Ghayeni M, Naderi-Meshkin H, Mahdizadeh A (2016) Surface modification of electrospun PLGA scaffold with collagen for bioengineered skin substitutes. Mater Sci Eng C 66:130–137
Saini AS, Tripathi A, Melo JS (2015) On-column enzymatic synthesis of melanin nanoparticles using cryogenic poly(AAM-co-AGE) monolith and its free radical scavenging and electro-catalytic properties. RSC Adv 5:87206–87215
Salati A, Keshvari H, Karkhaneh A, Taranejoo S (2011) Design and fabrication of artificial skin: Chitosan and gelatin immobilization on silicone by poly acrylic acid graft using a plasma surface modification method. J Macromol Sci Part B Phys 50:1972–1982
Sanchis MR, Blanes V, Blanes M, Garcia D, Balart R (2006) Surface modification of low density polyethylene (LDPE) film by low pressure O2 plasma treatment. Eur Polymer J 42(7):1558–1568
Saxena S, Ray AR, Kapil A, David GP, Letourneur D, Gupta B, Pelle AM (2010) Development of a new polypropylene-based suture: plasma grafting, surface treatment, characterization, and biocompatibility studies. Macromol Biosci 11(3):373–382
Shenton MJ, Lovell-Hoare MC, Stevens GC (2001) Adhesion enhancement of polymer surfaces by atmospheric pressure plasma treatment. J Phys D Appl Phys 34(18):342754–342760
Shen H, Hu X, Yang F, Bei J, Wang S (2009) The bioactivity of rhBMP-2 immobilized poly(lactide-co-glycolide) scaffolds. Biomaterials 30:3150–3157

Shen J, Li Y, Zuo Y, Zou Q, Zhang L, Liu H (2011) Surface modification of polyamide 6 immobilized with collagen: characterization and cytocompatibility. Int J Polym Mater 60:907–921
Sodhi RNS (1996) Application of surface analytical and modification techniques to biomaterial research. J Electron Spectroscopy Related Phenomena 81: 269–284
Sousa I, Mendes A, Pereira RF, Bartolo PJ (2014) Collagen surface modified poly(ε-caprolactone)scaffolds with improved hydrophilicity and cell adhesion properties. Mater Lett 134:263–267
Suntornnond R, An J, Chu CK (2016) Effect of gas plasma on polycaprolactone (PCL) membrane wettability and collagen type-I immobilized for enhancing cell proliferation. Mater Lett 171293–171296
Sun HX, Zhang L, Chai H, Chen HL (2006) Surface modification of poly (tetrafluoroethylene) films via plasma treatment and graft copolymerization of acrylic acid. Desalination 192:271–279
Svorcik V, Arenholz E, Rybka V, Ochsner R, Ryssel H (1998) AFM surface investigation of polyethylene modified by ion bombardment. Nucl Instrum Methods Phys Res Sect B 142:349–354
Tathe A, Ghodke M, Nikalje A (2010) A brief review: biomaterials and their application. Int J Pharm Pharmaceutical Sci 2(4):19–23
Theapsak S, Watthanaphanit A, Rujiravanit R (2012) Preparation of chitosan-coated polyethylene packaging films by DBD plasma treatment. ACS Appl Mater Interfaces 4(5):2474–2482
Trimukhe AM, Pandiyaraj KN, Tripathi A, Melo JS, Deshmukh RR (2017) Plasma surface modification of biomaterials for biomedical applications. In: Tripathi A, Melo J (eds) Advances in biomaterials for biomedical applications. Advanced structured materials. Springer-Nature, Singapore. ISBN: 978-981-10-3327-8, https://doi.org/10.1007/978-981-10-3328-5_3
Tripathi A, Melo JS (eds) (2017) Advances in biomaterials for biomedical applications. Springer-Nature, Singapore
Tripathi A, Kumar A (2011) Multi-featured macroporous agarose-alginate cryogel: synthesis and characterization for bioengineering applications. Macromol Biosci 11(1):22–35
Tripathi A, Melo JS (2015) Preparation of sponge-like biocomposite agarose-chitosan scaffold with primary hepatocytes for establishing an in-vitro 3D liver tissue model. RSC Adv 5:30701–30710
Tripathi A, Melo JS (2016) Synthesis of low-density biopolymeric chitosan-agarose cryomatrix and its surface functionalization with bio-transformed melanin for the enhanced recovery of uranium(VI) from aqueous subsurfaces. RSC Adv 6:37067–37078
Tripathi A, Melo JS (2019a) Cryostructurization of polymeric systems for developing macroporous cryogel as a foundational framework in bioengineering applications. J Chem Sci 131:92. https://doi.org/10.1007/s12039-019-1670-1
Tripathi A, Kathuria N, Kumar A (2009) Elastic and macroporous agarose-gelatin cryogels with isotropic and anisotropic porosity for tissue engineering. J Biomed Mater Res A 90(3):680–694. https://doi.org/10.1002/jbm.a.32127
Tripathi A, Sami H, Jain SR, Viloria-cols M, Zhuravleva N, Nilsson G, Jungvid H, Kumar A (2010) Improved bio-catalytic conversion by novel immobilization process using cryogel beads to increase solvent production. Enzyme Microb Technol 47:44–51. https://doi.org/10.1016/j.enzmictec.2010.03.009
Tripathi A, Kumar A (2012) Integrated approach for β-glucosidase purification from non-clarified crude homogenate using macroporous cryogel matrix. Sep Sci Technol 48(16):2410–2417
Tripathi A, Hadapad AB, Hire RS, Melo JS, D'Souza SF (2013a) Polymeric macroporous formulations for the control release of mosquitocidal Bacillus sphaericus ISPC-8. Enzyme Microb Technol 53(6–7):398–405. https://doi.org/10.1016/j.enzmictec.2013.08.006
Tripathi A, Melo JS, D'Souza SF (2013b) Uranium(VI) recovery from aqueous medium using novel floating macroporous alginate-agarose-magnetite cryobeads. J Hazard Mater 246–247:87–95
Tripathi A, Melo JS, D'Souza SF (2013c) Magnetic nanoparticles in tissue regeneration. In: Tiwari A, Tiwari A (eds) Nanomaterials in drug delivery, imaging, and tissue engineering. WILEY-Scrivener Publisher, USA, pp 443–492. ISBN: 9781118290323

Tripathi A, Vishnoi T, Singh D, Kumar A (2013d) Modulated crosslinking of macroporous polymeric cryogel affects in vitro cell adhesion and growth. Macromol Biosci 13(7):838–850. https://doi.org/10.1002/mabi.201200398

Tripathi A, Melo JS (2019b) Self-assembled biogenic melanin modulated surface-chemistry of biopolymers-colloidal silica composite porous matrix for the recovery of uranium. J Appl Polym Sci 136(5):46937. https://doi.org/10.1002/app.46937

Tyan YC, Liao JD, Lin SP (2003) Surface properties and *in vitro* analyses of immobilized chitosan onto polypropylene non-Woven fabric surface using antenna-coupling microwave plasma. J Mater Sci Mater Med 14:775–781

Unnikrishnan BS, Preethi GU, Joseph MM, Melo JS, Sreelekha TT, Tripathi A (2020) 3D printing in dental implants. In: Thomas DJ, Singh D (eds) 3D printing in medicine and surgery. Elsevier-Woodhead Publishing, USA. ISBN: 9780081025420

Vartiainen J, Rättö M, Tapper U, Paulussen S, Hurme E (2005) Surface modification of atmospheric plasma activated BOPP by immobilizing chitosan. Polym Bull 54:343–352

Vasilets VN, Hermel G, Knigt U, Werner C, Miiller M, Simon F, Grundke K, Kadas Y, Jacobasch HJ (1997) Microwave CO2 plasma-initiated vapour phase graft polymerization of acrylic acid onto polytetrafluoroethylene for immobilization of human Thrombomodulin. Biomaterials 18:1139–1145

Wang W, Zheng Z, Huang X, Fan W, Yu W, Zhang Z, Li L, Mao C (2017) Hemocompatibility and oxygenation performance of polysulfone membranes grafted with polyethylene glycol and heparin by plasma-induced surface modification. J Biomed Mater Res Part B Appl Biomater 105(7):1737–1746

Wojtowicz AM, Shekaran A, Oest ME, Dupont KM, Templeman KL, Hutmacher DW, Guldberg RE, Garcia AJ (2010) Coating of biomaterial scaffolds with the collagen-mimetic peptide GFOGER for bone defect repair. Biomaterials 31:2574–2582

Woo YI, Lee MH, Kim HL, Park JC (2010) Cellular responses and behaviors of adipose-derived stem cells onto β-Glucan and PLGA composites surface-modified by microwave-induced argon plasma. Macromol Res 18(1):90–93

Xiang B, Lam KS, Sun G (2009) Functional fibrous polypropylene solid support and its application in solid phase peptide synthesis and cell specific binding. React Funct Polym 69:905–914

Xin Z, Hou J, Ding J, Yang Z, Yan S, Liu C (2013) Surface functionalization of polyethylene via covalent immobilization of O-stearoyl-chitosan. Appl Surf Sci 279:424–431

Xu H, Deshmukh RR, Timmons RB, Nguyen KT (2011) Enhanced endothelialisation on surface modified poly(L-Lactic acid) substrates. Tissue Eng Part A 17:865–876

Yang J, Bei J, Wang S (2002) Enhanced cell affinity of poly (d, l-lactide) by combining plasma, treatment with collagen anchorage. Biomaterials 23:2607–2614

Yang SH, Wang EH, Gurak JA, Bhawal S, Deshmukh RR, Wijeratne AB, Edwards BL, Foss FW, Timmons RB, Schug KA (2013) Affinity mesh screen materials for selective extraction and analysis of antibiotics using transmission mode desorption electrospray ionization mass spectrometry. Langmuir 29:8046–8053

Yoshida S, Hagiwara K, Hasebe T, Hotta A (2013) Surface modification of polymers by plasma treatments for the enhancement of biocompatibility and controlled drug release. Surf Coat Technol 233:99–107

Zhang C, Jin J, Zhao J, Jiang W, Yin J (2013) Functionalized polypropylene non-woven fabric membrane with bovine serum albumin and its hemocompatibility enhancement. Colloids Surf B Bioinerfaces 102:45–52

Zhao JH, Wang J, Tu M, Luo BH, Zhou CR (2006) Improving the cell affinity of a poly(D, L-lactide) film modified by grafting collagen via a plasma technique. Biomed Mater 1(4):247–252

A Wide Portray of Upconversion Nanoparticles: Surface Modification for Bio-applications

Monami Das Modak and Pradip Paik

Abstract Present days have spotted many innovations in research on upconverting nanoparticles (UCNPs) co-doped with lanthanoids (mostly Er^{3+}, $Yb^{3+}/Tm^{3+}, Yb^{3+}$) through several synthetic approaches. A set of total 17 chemical elements (15 Lanthanides + Scandium + Yttrium) shows upconversion phenomena by exciting with NIR radiation (lower energy, longer wavelength) and finally results in visible light (higher energy, shorter wavelength) which makes it highly delightful for various applications. It is well known that for biological studies, fluorescence-imaging is an important phenomenon. Traditionally, as excitation radiation high energy lights are used which result in low energy emission by downconversion process. But, this phenomenon causes some difficulties like short tissue penetration depth, lower photochemical stability, lower sensitivity, cell death, DNA damage due to excitation with higher energy, whereas upconversion luminescence imaging gives better tissue penetration depth, good photochemical stability, improved sensitivity and very less probability of DNA damage and cell death as herein excitation is done with low energy photons and can be used in cancer therapy for destroying cancer cells with the emerging higher energy radiation. Herein, this reviews several current progresses with upconversion nanoparticles (UCNPs) in recent applications along with their different synthesis procedures, optical properties and growth of erection methods of different UCNPs applications have been highlighted based on their immobilization strategy.

Keywords Upconverting nanoparticles · Energy transfer · Emission · Immobilization · Biological applications

---

P. Paik (✉)
School of Biomedical Engineering, Indian Institute of Technology (BHU), Varanasi 221005, UP, India
e-mail: paik.bme@iitbhu.ac.in; ppse@uohyd.ac.in

M. Das Modak
School of Engineering Sciences and Technology, University of Hyderabad (UoH), Hyderabad 500046, TS, India

A. Tripathi and J. S. Melo (eds.), *Immobilization Strategies*,
Gels Horizons: From Science to Smart Materials,
https://doi.org/10.1007/978-981-15-7998-1_9

List of Abbreviations

UCNPs	Upconverting/upconversion nanoparticles
ESA	Excited state absorption
PA	Photon avalanche
ETU	Energy transfer upconversion
RE	Rare earth
TEM	Transmission electron microscopy
HRTEM	High resolution transmission electron microscopy
XRD	X-ray diffractometer
MRI	Magnetic resonance imaging
FRET	Fluorescent resonant energy transfer
PDT	Photodynamic therapy

1 Introduction

Upconverting nanoparticles (UCNPs) have already been established for prospective use as bio-labels and in biological analysis and bio-imaging which are the most promising technologies today. These promising and advanced technologies suggest that UCNPs could reconstruct the enduring technologies due to its ability to suspend as clear colloidal solutions (Boyer et al. 2007). Until now, the highest upconversion efficiencies have been remarked in $NaYF_4$ hexagonal phase co-doped with Er^{3+}/Yb^{3+} or Tm^{3+}/Yb^{3+} ion-pairs (Kramer et al. 2004; Suyver et al. 2005b).

Selection of host material is an important factor for achieving upconversion phenomenon properly. To achieve highest luminescence quantum yield and also less probability of occurring non-radiative relaxation, nowadays, fluoride is used in host materials such as REF_3 and $AREF_4$ (A = Alkali), and as a result, they help to increase refractive index and transparency, though Chlorides and Bromides can also enhance luminescence intensity but they are mostly sensitive to moisture, so difficult for imaging bio-molecules. Therefore, adding of fluoride in host materials is an increasing phenomenon today. Photon upconversion process is based on nonlinear optics, where the optical properties of material changes with the intensity of the incident exciting light. But are they noteworthy and adaptable enough to explain the difficulties of redirecting to a new technology, conventionally a lengthy and precious procedure? Could UCNPs become the upcoming wild materials reconstructing some of the contemporaneously used technologies and leading to the advanced research? Is it flexible as many as required to transmute many features in our life? By reviewing different proposed synthetic procedures allowing their several applications in different fields, those questions will be answered. Though there are several synthetic routes for preparing UCNP nanocrystals, among which hydrothermal synthesis has a lot of favourable advantages compared to the low reaction temp process (<250 °C); and it resulted in uniform distribution of size, shape and morphology. The procedure and the experimental set-up used in these methods are easy to handle and very simple. In such case, basically three conditions are responsible to convert alpha (α-$NaYF_4$) phase to beta (β-$NaYF_4$) phase; they are high hydrothermal temperature, long hydrothermal time and high fluoride to lanthanide molar ratio (Suyver et al. 2005b).

So, initially, the observed properties in UCNPs will be discussed.

By Raman spectroscopy, it has been confirmed that the dominant phonon modes in undopped $NaYF_4$ lie in the range 300–400 cm^{-1}. These low energy phonon modes describe the remarkable upconversion efficiency (Zhao et al. 2008). By the implementation of the plasmonic effects, it is possible to enhance upconversion fluorescence which has been already done by directed nano-assembly of $NaYF_4$: Yb^{3+}/Er^{3+} nanocrystals with gold nanospheres (Suyver et al. 2006). From literature, it has been observed that, crystallographic size, crystallographic phase (Schietinger et al. 2010) and optical emission properties (Feng et al. 2006; Heer et al. 2004; Sivakumar et al. 2005; Ehlert et al. 2008) of such resultant nanocrytals can be controlled simultaneously by influencing them with dopant ions. By controlling concentration of trivalent lanthanide dopant ions, it is quite possible to tune size (down to 10 nm), shape (cubic to hexagonal) and upconversion (Wang and Liu 2008; Auzel 2004; Suyver et al. 2005a) emission colour (green–blue) of $NaYF_4$ nanocrystals. There are two factors of dopant ions, their size and dipole polarizability, that can change the crystal size and shape of resultant nanocrystals. UCNPs can offer high photochemical stability, sharp emission bandwidths and large anti-stokes shift (Wang and Liu 2009). There are four experimental variables which are solvent in nature, reaction time and temperature and metal precursor concentration that inflict stiff sway over crystallization of resultant particles with completely explained crystal phase and size. Hexagonal UCNPs phase structure is always preferable compared to cubic structure in various applications. There is a huge difference (almost by a factor 7.5) (Wang et al. 2010a, b) in fluorescence intensity between its cubic and hexagonal phase—nanoparticles. Inspite of these drawbacks, in cubic-phase, sometimes, we prefer cubic phases in as-synthesized UCNPs depending on their particular sizes, shapes, crystallinity, fluorescence properties and easy dispersion in non-polar solvent.

Generally, it has been noticed that the structure of $NaREF_4$ system for both cubic and hexagonal phase differs on the basis of F^- cubes and ions present in them. In cubic structure, cations and vacancies sustain with uniform numbers of F^- cubes whereas for hexagonal structure, F^- ions are seated in an ordered array, to fit the structural change, electron cloud deformation of cations is crucial (Fig. 1). Basically, one ordered way is maintained for lanthanides. As lanthanide series starts from lanthanum (La), finishes with Lutetium (Lu) (Lanthanide series-La, Ce, Pr, Nd, Pm, Sm, Eu, Gd, Tb, Dy, Ho, Er, Tm, Yb, Lu), now if we move towards higher atomic no. and lower atomic mass, i.e. from La to Lu, weight will decrease and ionic radii will increase, dipole polarizibility increases, tendency towards e-distortion cloud increases which is the appreciative condition for hexagonal phase. So formation of hexagonal phases in lanthanides having higher atomic number will be more than that of having lower atomic number. Size of Y^{3+} ion is ($r = 1.159$ A°) in $NaYF_4$ host lattice. More hexagonal phases will be produced for doped lanthanide ions having

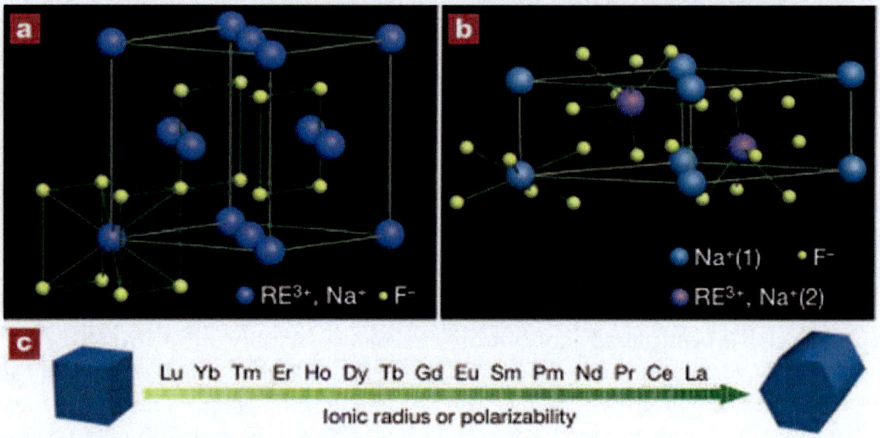

Fig. 1 Schematic of phase transformation in Na [Rare Earth (RE)] F_4 while doped with lanthanides, **a** cubic phase, cations and vacancies sustain with equal numbers of F^- cubes, **b** hexagonal phase, two types of cation sites sustain with an ordered F^- ion-array (Na^+, RE^{3+}/Na^+), **c** shows the fashion of cubic-to-hexagonal phase transition as a function of ionic radius/polarizability of elements present in lanthanoid-series. Reproduced with permission (Wang et al. 2010a, b). *Copyright* Nature publication

larger size (i.e. for Gd^{3+} ion $r = 1.193$ A°)than the size in Y^{3+} in host lattice and finally the growth of unit cell volume occurs due to the presence of larger-sized lanthanide dopant ions (Wang and Liu 2009).

2 Challenges in Production

There are several applications of UCNPs having different shapes, sizes and structures which can be significantly controlled by their development with a number of designed synthesis procedures and their dispersions in various media. Now, depending on the notable properties and the quality of UCNP (either in dried or in colloidal form) in as-synthesized materials, they can be used in different potential applications such as in silicon-based solar cell as spectral converters (Yi and Chow 2006; Xie and Liu 2012), in optical imaging and MRI as a multimodal contrast agent (Zou et al. 2012), in FRET bio-sensors (Park et al. 2009); as colour display (Wang et al. 2005a, b), versatile bio-probes in biomimetic surface—engineering (Kim et al., 2009), nanoprobes for sensing and imaging of pH, cells and small animals(Li et al., 2012), nano-transducers for killing diseased cells in deep tissues (Chatterjee et al., 2008), the most viable luminescent bio-labels in bio-conjugation and bio-imaging (Chatterjee and Yong, 2008). Therefore, the production of those UCNPs with required sizes, shapes and structures is very crucial for different applications. During their preparation, all the reaction parameters are being controlled in a precise manner due to the high sensitivity of the products through their formation periods. And thus the production of UCNPs is challenging.

If we go through the literature reviews, we will find that mostly several optical properties became prominent for their applications. This is the main property for which they are being broadly used in nanotransducers (Chatterjee et al. 2008; Mader et al. 2010), broad brands (Yi and Chow 2006; Xie and Liu 2012), Imaging contrast agent (Zou et al. 2012), multimodal imaging (Idris et al. 2012; Cheng et al. 2011a, b), superresolution-nanoscopy (Xu et al. 2011; Liu et al. 2017), imaging probes (Zhan et al. 2017) and so on. Therefore, review has been focused briefly on optical properties with possible photonic processes for UCNPs.

3 Optical Properties

Lanthanoids exist in lanthanide series and there are a total of 15 lanthanoids. Though transition metals (Scandium, Yttrium), where atoms have partially filling '*d*' subshell and elements in actinide series can exhibit upconversion phenomena, but specially RE elements having combination of 15 lanthanoids + Scandium + Yttrium, a set of total 17 chemical elements show upconversion phenomena very strongly, i.e. by exciting that lanthanide-doped nanocrystals with NIR radiation. In the upconversion process, we can get visible light with the emission of higher energy photons. Thus, upconversion process converts low energy, long wavelength NIR radiation to high energy, short wavelength visible light. The optical properties of upconversion nanoparticles arise due to conversion of near infrared radiation (NIR) into visible spectral range which is a very efficient phenomenon in rare earth-based materials. The nanocrystals based on rare earth ions showed strong upconversion emission with a continuous excitation wavelength at 973 nm (Suyver et al. 2006). Therefore, they have a great potential in applications of solid-state laser materials as well as several lighting, panel and colour display technologies (Yang et al. 2014; Scheps 1996). It is well known to us that $NaYF_4$ nanocrystals co-doped with Yb^{3+} and Er^{3+}/Tm^{3+} are the most promising upconverting nanomaterial today as these nanocrystals are sensitized by Yb^{3+} (sensitizer) and multicolour wavelength range can be distinguished from Er^{3+} or Tm^{3+} (activator) dopant ions.

Experimentally and theoretically it has been already proven that most of the lanthanide ions exhibit visible light under excitation with NIR radiation but sharp luminescence imaging can be observed by upconversion process under 980 nm excitation if some of them such as Er^{+3}, Tm^{+3} are co-doped with the host lattice like $NaYF_4$, hence the productive forms are written as $NaYF_4$: Yb^{3+}; $Er^{3+}/NaYF_4$: Yb^{3+}; Tm^{3+}.

Some important luminescence properties can be observed with using such UCNPs such as narrow band width with shorter wavelength compared to the excitation wavelength, i.e. anti-stokes-type emission (Wang and Liu 2008) and long-time emission. To get upconversion phenomenon properly, choosing of host material is an important factor. To achieve highest luminescence's quantum yield, and also less probability of occurring non-radiative relaxation nowadays, fluoride is used in host materials such as REF_3 and $AREF_4$ (A = Alkali), as a result they help to increase refractive index and transparency though chlorides/bromides can also enhance luminescence intensity. However, they are very sensitive to moisture and possess difficulty for imaging of bio-molecules and cells. To overcome this limitation, fluoride molecules are added in the system. The main difference between fluorescent and luminescent materials is their different characteristics. Fluorescent materials absorb high energy photons with the emission of visible light and low energy photons, which results in auto fluorescence, wide emission bands and limited sensitivity. To overcome these difficulties, upconverting luminescent nanoparticles with high quality developed promptly. Trivalent rare earth (RE) ion (Ho^{3+}, Er^{3+}, Tm^{3+}) follows upconversion process fairly (Downing et al. 2006). By incorporating a few energy transfer related mechanisms, we will try to understand the photon-generation mechanism in UCNPs.

There are some basic mechanisms which results in highest population in excited state and after that when they come in the ground state, emit high energy photons. Multiphoton absorption occurs by the excited state absorption mechanism (ESA) (Downing et al. 2006) (Fig. 2) which further involves multistep excitation by the same ion and finally highest population occurs at the excited state. This process is also known as phonon-assisted electronic transitions (Yang et al. 2014; Downing et al. 2006).

Suppose, energy of incident photons has flux ϕ (Say) is resonant with the energy difference $(E_1 - E_0)$, then some of the photons are absorbed from the ground state by the particular ion. Similarly, while the energy of the incident photons having certain flux resonant with the energy difference $(E_2 - E_1)$, then only incoming photons can be absorbed in the intermediate state by the same ion. Now, the population in the excited state becomes high and consequently the ion reaches to its high excited state, but this phase is not stable from this excited state and finally, upconversion luminescence occurs by radiating photons. We have considered this process for a single ion so we can assume that it is independent on ion-concentration. According

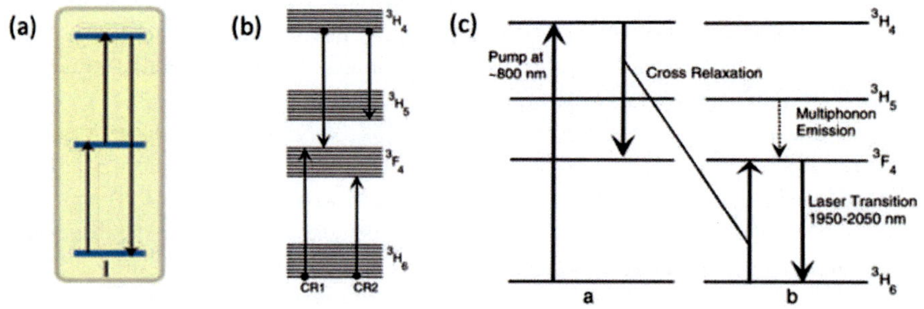

Fig. 2 **a** Excited state absorption mechanism (ESA). Reproduced with permission (Wang et al. 2011). Copyright: Nature Materials publication, **b** detailed energy-level diagram of the four multiplets of the Tm^{3+} ion. Reproduced with permission (Jackson 2004). Copyright: Optics Communications publication, **c** Two energy transfer upconversion processes (ETUs). Reproduced with permission (Jackson 2004). *Copyright* Optics Communications publication

to literature, anti-stokes fluorescence is proportional to the incident photon flux such that, for above-mentioned case, fluorescence is directly proportional to the square of the incident photon flux. ESA process is quite relative with the laser pumping process. As it avoids transfer losses, so it is assumed to be a suitable pumping process for upconversion single-doped mechanism. In ESA case, finally, we can conclude that by absorbing atleast two photons having sufficient energy, a single ion can reach its emitting level from which it undergoes luminescence.

Unlike ESA, (Fig. 2) where energy transfer upconversion process ETU is a pump power independent process and it involves two ions system—one is sensitizer ion and another one is activator ion. Sensitizer ion (S) can transfer its energy to activator ion (A) as they are placed much closer to each other. On excitation with suitable light, 'A' goes from its ground state to excited state and it is only possible if excited energies of two ions are equal or nearly equal to each other. After reaching the excited state, S ion can transfer its energy to the nearly placed A-ion whose excited state energy is equal. Here, it is to be notified that before emitting photons by S-ion, A goes from ground state to the excited state. Cross relaxation (CR) and energy transfer profess are very important in upconversion phenomenon. Cross Relaxation occurs due to overlapping of the spectra (Joubert 1999). For example, in Silica between $3H_4 > 3F_4$ Fluorescence spectra, there is a strong overlapping between them. Previously, it has already examined for Tm^{3+}-doped silica fibre lasers. The cross-relaxation mechanism with its energy-level diagram (Joubert 1999) of the four multiplets $3H_6$, $3F_4$, $3H_5$, $3H_4$ is shown in Fig. 2b.

Several spectral properties of RE-doped UCNPs have huge scope in biomedical imaging and therapeutics (Jackson 2004; van de Rijke et al. 2001; Hampl et al. 2001). Other exciting optical properties in the same material can be aroused by tuning their spectral properties in a proper way. Surface plasmon resonance caused by the interaction of metal with the incident light assists for different interesting optical events like radiative and non-radiative properties of nanocrystals (Lim et al. 2006; Zhang et al. 2010; Eustis and Sayed Eustis and El-Sayed 2006). In literature, the Raman spectroscopy with dominant phonon modes (300–400 cm^{-1}) of such RE-based materials has been already demonstrated (Zhao et al. 2008).

4 Different Synthetic Strategies for the Production of Upconversion Nanoparticles

Earlier, we have already discussed, choosing of host materials is a very important factor to UCNPs as they determine their quality. Now, to get better quality (such as better luminescence efficiency, chemical stability, low phonon energies giving low probability of non-radiative decay), synthesizing UCNPs in a proper way is very crucial key. Synthetic routes for upconversion nanoparticles are based on mainly three strategies: thermal decomposition, Hydrothermal or solvo-thermal synthesis, ionic—liquid based or ionothermal synthesis. Hence, we will discuss a variety of synthetic routes based on those three basic strategies.

4.1 Thermal Decomposition

To control the shapes and sizes of nanocrystals, thermal decomposition is one of the most familiar and popular method. As reagents,organic precursors (e.g. trifluoroacetate precursor), surfactants(oleic acid, oleylamine) and organic solvents are used. By product what we get is nuclei form of our desired nanocrystals which then goes under growth mechanism. There after dissolution and aggregation take place. Thermal decomposition follows overall four stages—Nucleation, Growth, Dissolution and Aggregation.

Most of the RE fluorides are synthesized by this route. Some examples of such prepared UCNPs are $NaYF_4$: Yb, Er; $NaYF_4$: Yb, Tm; $NaGdF_4$: Yb, Er; $NaGdF_4$: Yb, Tm; $NaLuF_4$: Yb, Er; NaLuF4: Yb, Tm, CaF_2, etc.

Drawbacks of thermal decomposition: Though thermal decomposition gives well shaped and sized monodispersed particles (Barnes et al. 2003), still it holds some disadvantages like

(a) Requires temperature in the range of 250–330 °C, which is quite high. During maintenance, such high temperature sometimes may cause burning, particle-aggregation and particle-enlarging.
(b) Surfactants having long hydrocarbon chains and polar capping groups associated with organic precursors during synthesis can yield difficulties for biological applications especially while we are using them for stabilizing nanomaterials.
(c) Requires high-boiling organic solvents to dissolve organic precursors.
(d) To overcome the difficulty of using surfactants, sometimes, we need surface modification/engineered modified surface which can be cost-effective.
(e) Ideal condition of this experimental set-up is always being used in an complete oxygen-free inert gas condition which is difficult to handle during synthesis, even a very small amount of oxygen during synthesis may damage and form an unsuccessful product.

Inspite of having the above-mentioned drawbacks in thermal decomposition procedure, this process has already been investigated with the production of Alpha-$NaYF_4$: 20%Yb, 2%Er (cubic-phase) and Beta-$NaYF_4$: 20%Yb, 2%Er (hexagonal phase) nanocrystals in a large-scale area where successful synthesis of co-doped $NaYF_4$ nanocrystals has been built-up with thermal decomposition procedures. The final shapes, sizes and structures of UCNPs can be changed by modifying or varying the reaction time, reaction temperature, reagent concentration and the resulted modified nanocrystals can be characterized using transmission electron microscopy(TEM), high-resolution transmission electron microscopy (HRTEM), field emission scanning electron microscopy (FESEM) and X-ray diffraction (XRD) patterns.

By introducing two synthetic procedures (Barnes et al. 2003; Boyer et al. 2006), collecting from different literature, hereby, I will describe the thermal decomposition with their required precursors, reaction temperature and reaction time. At a recent time, Gudel et al. (Boyer et al. 2007; Kramer et al. 2004; Zhao et al. 2008; Feng et al. 2006) spotted micrometre-sized hexagonal phases of $NaYF_4$: Yb^{3+}, Er^{3+}/Tm^{3+} crystals which are enable to show highest upconversion efficiencies. It is familiar to us that metal trifluoroacetates thermally decompose providing their corresponding metal fluorides, fluorinated and oxyfluorinated carbon species (Russel 1993; Rillings and Roberts 1974), whereas lanthanide trifluoroacetate precursors can be formulated from their corresponding lanthanide oxides and trifluoroacetic acid (Russel 1993).

TEM images of the colloidal UCNPs ($NaYF_4$ co-doped with Er^{3+}/Yb^{3+} and Tm^{3+}/Yb^{3+}) formed via thermal decomposition procedure are shown in Fig. 3 with two histograms of particle size distribution (histogram result between Number of Particles and Particle Diameter (nm) for Upconverting Nanocrystals).

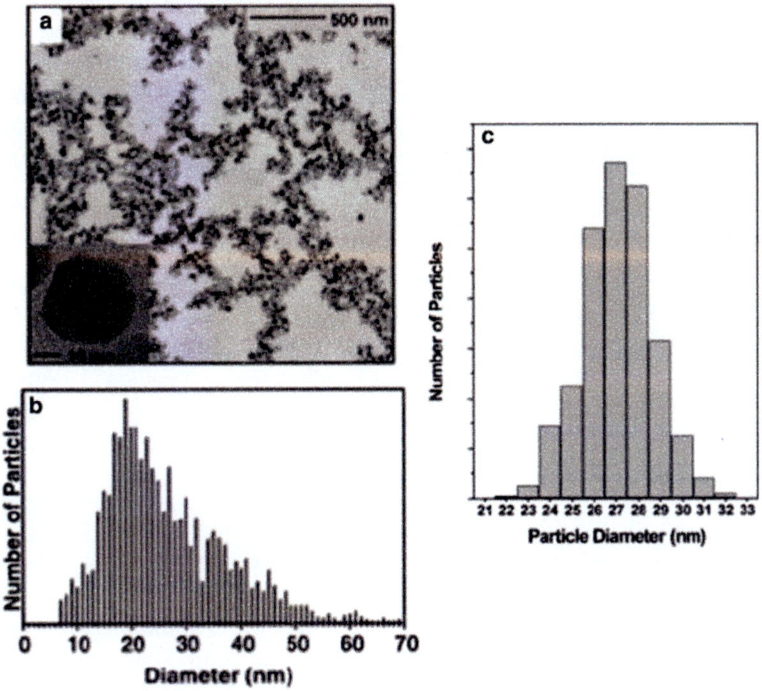

Fig. 3 **a** TEM image, **b** histogram for obtained particle sizes with ~1400 nanocrystals. Reproduced with permission (Boyer et al. 2006). Copy right: Journal of American Chemical Society publication, **c** histogram of obtained particle sizes with ~500 nanocrystals. Reprinted with permission (Boyer et al. 2007). *Copyright* Nano Letters publication

4.2 *Hydrothermal or Solvothermal Synthesis*

A convenient synthesis process for producing UCNPs with overcoming some of the difficulties associated with thermal decomposition process is termed as hydrothermal or solvothermal synthesis procedure. The advantages of this process over thermal decomposition are given by following.

(a) Temperature required is relatively low (160–220°) compared to that in thermal decomposition.
(b) Oxygen-free inert gas condition is not required and that's why it is easy to handle.
(c) Organic compounds are not required as this process involves water solution phenomena.
(d) Various nanocrystals with hexagonal and octodecahedral shapes can be formed or synthesized applying this method. By adding several fluoride sources and organic additives (e.g. trisodium citrate), we can get desired shapes and sizes with different morphologies and architectures.
(e) Synthesis through hydrothermal/solvothermal synthesis has been given in various literatures (Wang and Li 2007; Sun et al. 2006; Li et al. 2007; Mi et al. 2011).

Advantages of hydrothermal synthesis over others and morphological effect of resultant particles on hyderothermal time, tempereatures and on other parameters: Compared with the other synthetic routes, hydrothermal synthetic procedure is superior following some advantages like,

(a) At relatively lower temperature (<250 °C), synthesis can easily happen and during synthesis by changing concentration of precursors, nature of solvent reaction time and other experimental parameters size, structure and morphology of the resulting products can easily be modified.
(b) Experimentally, it has been observed that the purity of the resulting particles become significant in case of hydrothermal synthesis.
(c) The synthetic process and the required equipment used in such experiment are simpler and easier to handle.
(d) Sometimes, organic additives (EDTA, citrate) are used during synthesis to get small and uniform particles. The resultant size of particles differs with different organic additives due to their different chelating agent and molecular structure providing morphological product. The main reason behind that difference is the influence of chelator on the growth of particles (Yi et al. 2004). As for example, size of the particles synthesized with citrate is much less than that synthesized with EDTA.

There are generally four factors which substantially affect the size, structure and morphology of the resulting nanocrystals. They are described as follows:

(a) *Effect of fluoride–lanthanide molar ratio*: The amount of NaF content is counted in NaF: Ln molar ratio to investigate its effect on size, structure and morphology

Table 1 Different crystallization, shape and surface pattern with different amount of NaF:Ln ratio

NaF:Ln molar ratio	Crystallization	Shape	Surface pattern
4	Low crystallization of the sample providing more likely fcc (Suyver et al. 2005b; Wei et al. 2006)	Spherical	Coarse surface
8	Improved crystallization	More regular shape	More smoother surface
12	Further improved	Pure hexagonal-shaped sub-microplates	Further improved

of the resulting UCNPs. Larger the fluoride–lanthanide ratio provides the better crystallization, more regular and smoother surface, more hexagonal phases (i.e. influence the crystal structure as fcc or hexagonal) of the resulting nanocrystals. Now, the effect of NaF: Ln ratio on crystallization, shape and surface pattern of the synthesized Yb^{3+} and Er^{3+} co-doped $NaYF_4$ nanocrystals is shown in Table 1.

From Table 1, it is clear that the effect of NaF content on the crystal morphology plays an important role as the crystal phase is completely transformed into hexagonal phases at the NaF:Ln molar ratio of 12 or above.

(a) *Effect of citrate–lanthanide ratio*: In hydrothermal method, chelating agent has appreciable effect on size and aggregation of the particles. Therefore, by choosing proper chelating agent, it is quite possible to produce small-sized and dispersible nanoparticles (Suyver et al. 2005b; Sun et al. 2007). Therefore, citrate: Ln ratio plays an important role in forming different morphologies (Table 2).

In conclusion, the formation of cubic crystal structure is possible with citrate–lanthanide and ratio is greater than unity (citrate: lanthanide > 1) and with citrate–lanthanide ratio is less than unity there exist mixed phases (cubic and hexagonal) in resulting crystal structure (citrate: lanthanide < 1).

Table 2 Molar ratio of chelator: Lanthanide affecting the morphologies of $NaYF_4$: Yb^{3+}, Er^{3+} nanocrystals

Molar ratio of citrate–lanthanide	Morphology of the resulting particles
0:1 (i.e. without citrate)	Formulation of resulting nanoparticles as mixtures of nanorods as well as spherical nanoparticles
0.5	A mixture of cubic nanoparticles and hexagonal submicroplates
1.0 and 1.5	Development of pure nanoparticles having size distribution in the range between 30–50 nm (with the restriction of the crystal growth)

(a) *Effect of Hydrothermal temperature*: Generally, the hydrothermal required temperature in the range between 160 and 200 °C to eventuate phase transformation. At higher hydrothermal temperature, cubic to hexagonal phase transformation is possible owing to generation of energy. Lower hydrothermal temperature approves smaller-sized nanoparticles having cubic phases, whereas higher hydrothermal temperature approves the formation of nanocrystals and sub-microplate mixture having hexagonal phase.
(b) *Effect of Hydrothermal time*: Hydrothermal time affects the crystallization and growth of the resulting nanocrystals. Long hydrothermal duration is favourable for the transformation of cubic phases into hexagonal phases. The hydrothermal time increases the rate of several processes during synthesis such as dissolution, re-nucleation and crystallization processes. In literature, it has been described that pure hexagonal phase—microplates (our required phase) of $NaYF_4$ nanocrystals have been found with hydrothermal time around 2 or 2.5 h (keeping citrate: lanthanide ratio = 1:1; and temperature at 180 °C) (Suyver et al. 2005b).

TEM images of the upconversion nanoparticles (YVO_4: Er^{3+}) synthesized with hydrothermal/solvo-thermal treatment are shown in Fig. 4.

4.3 Ionic-Liquid Based or Ionothermal Synthesis

Although pure hexagonal phase (which can yield better upconversion than the cubic one) can be obtained by this method, however, it is comparatively less popular than the previous two methods because it holds certain limitations while synthesizing nanocrystals, e.g.

(a) Nanocrystals used must be non-flammable
(b) Does not require any organic solvent.
(c) Produced particles become broader in size with lower quality (less uniformity, less monodispersity, less chemical stability) than the other two processes.
(d) Various shapes and sizes cannot be found by using ionothermal synthesis, only water-soluble hexagonal phase formation is possible.

Due to these above-mentioned limitations, it can be used only for preparing a few selective nanocrystals. Inspite of having such limitations, few advantages are also present such as.

(a) Reaction occurs within a very short period of time.
(b) Requires low temperature range and vapour pressure which are easy to control.

Earlier, synthesis procedures for the production of β-phase $NaYF_4$ nanomaterials were developed (Wang et al. 2010a, b, 2011; Wang and Li 2007; Wei et al. 2006; Yi and Chow 2007; Zeng et al. 2005; Mai et al. 2006) with the formation of particles of size ~15 nm where most of the synthesis procedures required in high-boiling solvents or in their mixtures (e.g. oleic acid, oleylamine, 1-octadecene or their

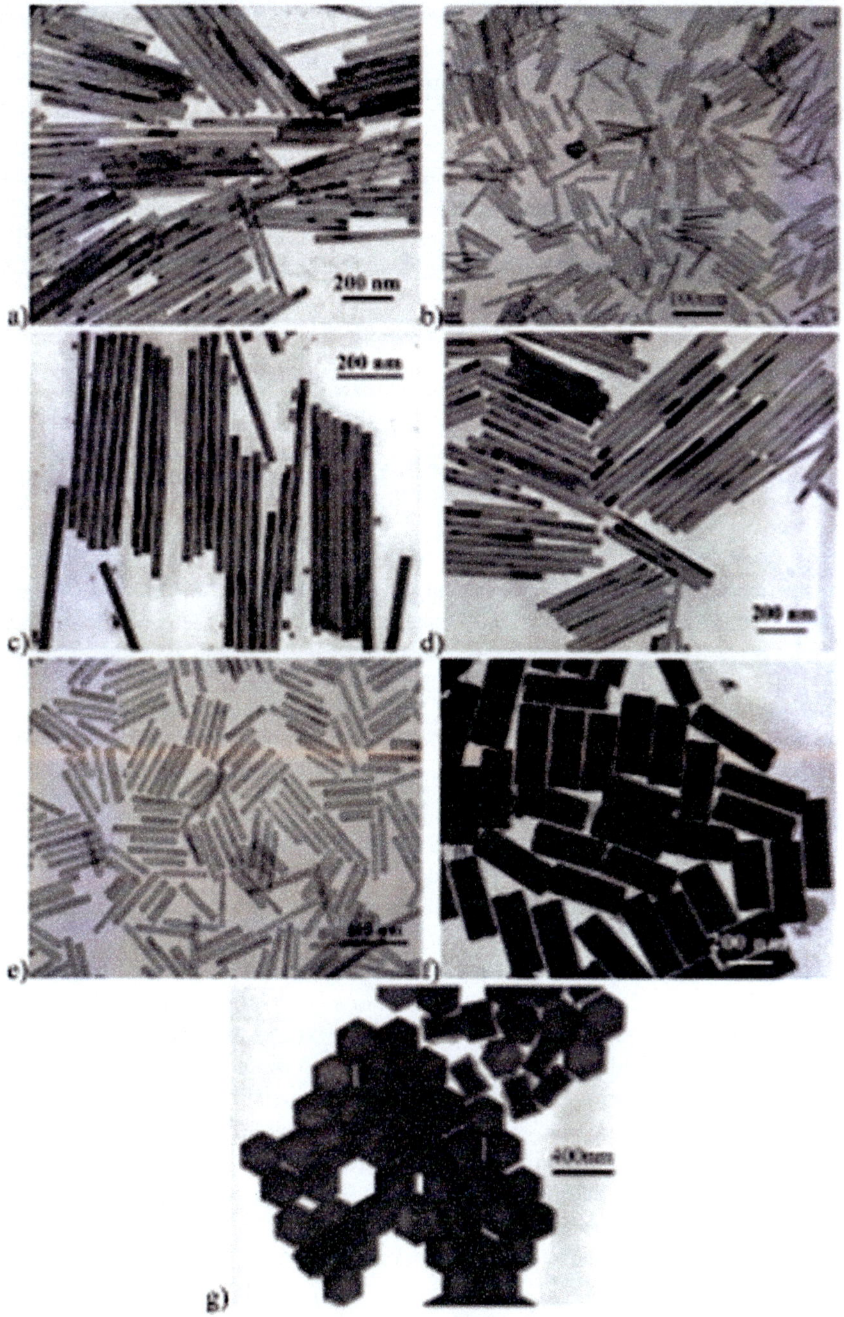

Fig. 4 TEM images of the resulted nanocrystals after solvo-thermal treatment. Reproduced with permission (Wang and Li 2007). *Copyright* Chemistry of Materials publication

solvent mixtures) owing to co-thermolysis of rare earth trifluoroacetates and sodium trifluoroacetate (Boyer et al. 2006). But these resulting luminescent nanomaterials seem to be difficult while using in bio-probes due to their sizes, not only that but also from literatures, it is revealed that they require a very high reaction temperature and yield rather very low compared to the other methods. Therefore, production of smaller size of such nanoparticles with high yield is a demanding matter in Bio-probe imaging. Chow et al. group synthesized 10-nm-sized hexagonal phase UCN nanocrystals in oleylamine at temperature above 300 °C having very narrow size distribution and introduced multiple precursors, e.g.—CF_3COONa, $(CF_3COO)_3Y$, $(CF_3COO)_3Yb$, $(CF_3COO)Er/(CF_3COO)_3Tm$, etc. (Wang et al. 2010a, b; Schafer et al. 2009). An average size of 7 nm of Yb^{3+}, Er^{3+} co-doped $NaYF_4$ nanocrystals were being synthesized at elevated temperature by Schafer group (Schafer et al. 2009).

For synthesizing, UCNPs precursor's properties also play a significant role in controlling size, shape and morphology. $NaYF_4$ nanocrystals co-doped with Er^{3+} and Yb^{3+} can be synthesized in the presence of precursors like-$Y_2(CO_3)_3$, $Yb_2(CO_3)_3$, $Er_2(CO_3)_3$, Na_2CO_3 and NH_4F as they can produce nanocrystals even at room temperature or at very low temperature having hexagonal phase (β-phase) giving better luminescence and high yield. This is one of the noteworthy advantages of these precursors providing a novel method of synthesis procedures at nanoscale production. Such novel synthesis method is based on the reaction of metal carbonates with Ammonium fluoride and synthesized in organic solvents in the range of temperature 20–280 °C. After decomposing of metal precursors pure, nanosized $NaYF_4$ nanocrystals co-doped with Er^{3+} and Yb^{3+} were formed having high yielded particles which are well separated and they contribute to broad size distribution (particle size ~4–10 nm). Thus, the transparent solution of these resulted nanoparticles exhibited visible upconversion emission on 978 nm of wave excitation. After that novel synthesis all the diffraction peaks appeared by XRD assign to hexagonal phases and no cubic phases are assigned, that means here, as all the phases belong to hexagonal phase, so, luminescence efficiency which is our desirable property becomes high. Cubic phase of Er^{3+} and Yb^{3+} co-doped $NaYF_4$ nanocrystals can be produced at elevator temperature (250 °C). All these reactions were done in pure oleylamine which can be replaced with oleic acid–oleylamine mixtures to discontinue the cubic phases formed in reaction. With the above-mentioned precursors under optimized conditions, 7-nm-sized particles (estimated as an average particle size) were formed having 84% yields (Schafer et al. 2009). Er^{3+}, Yb^{3+} ions coupled in hexagonal phase $NaYF_4$ matrix provides highest up conversion efficiency (Kramer et al. 2004; Yi and Chow 2007). Proper heat treatment is an important factor to increase the luminescence efficiency.

Thus, solid-state-reaction procedure is another way to form nanoscale hexagonal $NaYF_4$ powder with releasing ammonia at room temperature grinding dry powders (such as Na_2CO_3, NH_4F, and RE carbonates powders). But, owing to the presence of higher percentage of α-phases and other phases/impurities rather that β-phases in the as-synthesized sample, this procedure carries less importance compared to the others mentioned above. Not only that but also it requires high reaction temperature (~300 °C or above) during synthesizing upconversion nanocrystals.

Ionothermal synthesis is the procedure where ionic liquids are used as reaction media in thermal reactions and water-soluble hexagonal-phased $NaYF_4$: Yb^{3+}; Er^{3+}/Tm^{3+} nanoparticles are formed with. Mostly, this procedure occurs in a molecular solvent (Zhang and Chen 2010; Cooper et al. 2004).

A conventional strategy for synthesis of nanocrystals is based on a simple and agreeable methodology, known as liquid–solid solution process (LSS) which provides a variation of building blocks for assembling materials in nanotechnology. To obtain high quality nanocrystals, basically, noble metals have been chosen for achieving good uniformity, smooth surface and self-assembly. Three phases are developed during this process—solid phase, liquid phase and solution phase. It is possible to obtain uniform noble metal nanocrystals by the moderation (reduction) of noble metal ions from interfaces of solid, liquid and solutions at various classified temperature. Here, metal linoleate acts as solid phase, ethanol–linoleic acid acts as liquid phase and water–ethanol solution with noble metal ions serves as solution phase. To generate liquid and solution phases, ethanol is a common quantity. The greatest advantage of LSS process is to produce nanocrystals having various properties such as magnetic, semiconducting, fluorescence, dielectric and applications in solid-state lasers, luminescent probes and sensors (Wang et al. 2005a).

We can divide the whole LSS process into two sections, (i) phase transfer process and (ii) phase separation process. In phase transfer process, the aqueous solution with noble metal salt conjugate with ethanol which is also in a liquid form and produce water–ethanol solution containing noble metal ions. Again phase transformation based on ion exchange occurs between sodium linoleate and the water–ethanol solution accomodating noble metal ions resulting in noble metal linoleate and finally sodium ions enter into the aqueous phase. At certain selected temperature, ethanol in both phases (liquid phase and solution phase) alleviates noble metal ions at the interfaces of solid–liquid and solid–solution phases(Wang et al. 2005b). Here, the only remaining quantity which not reacting with other chemicals is the linoleic acid, which is absorbed on the noble metal—nanocrystal surface further with the reduction process (reduction of the noble metal ions). As a result, alkyl chains are developed outside the noble–metal surface enabling them to acquire hydrophobic surfaces. In LSS process, phase separation occurs spontaneously owing to the weight of nanocrystals as well as incompatibility between their hydrophobic surfaces and hydrophilic surroundings.

In Fig. 5, TEM micrograpgs of β-$NaYF_4$: Yb, Er nanoplates (Wei et al. 2006) prepared by liquid–solid dual phase approach are shown.

5 Few Recently Developed Synthetic Strategies

Now, in the subsequent section, some of the recently developed synthetic strategies for UCNPs and their special investigations have been summarized.

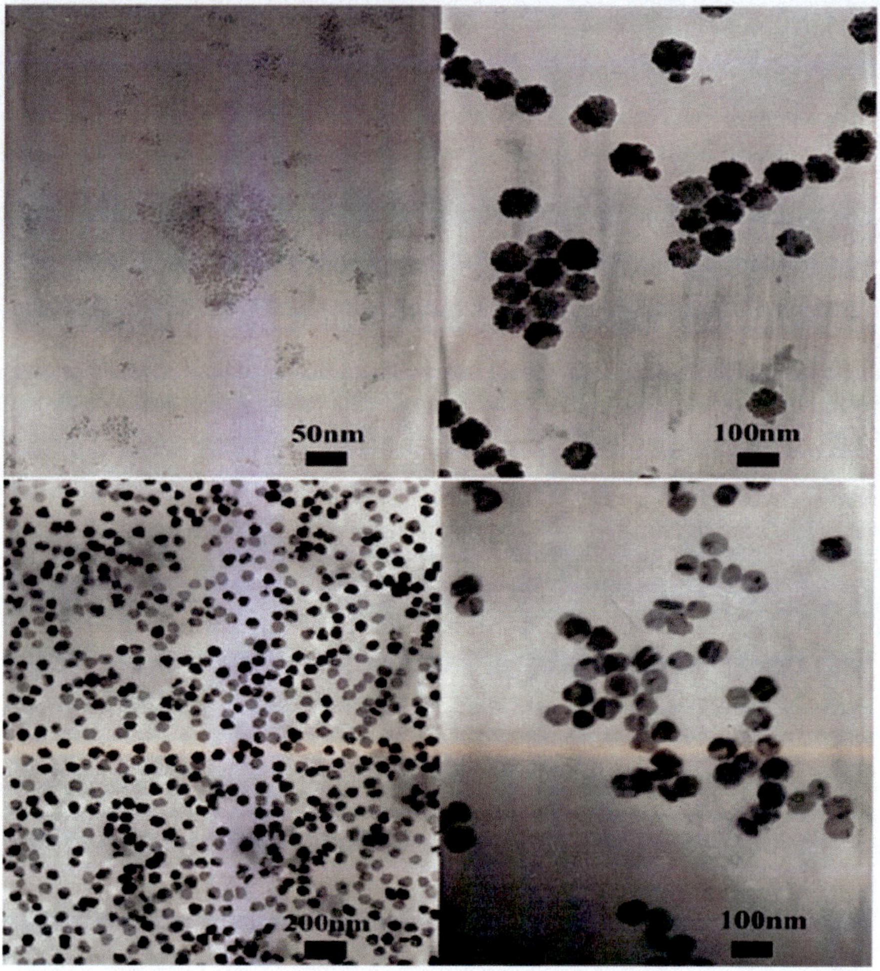

Fig. 5 TEM images of $NaYF_4$: Yb, Er UCNPs obtained at different reaction temperatures. Reported with permission (Wei et al. 2006). *Copyright* Chemistry of Materials publication

5.1 Microwave Assisted Synthesis

There are two leading mechanisms, dipole rotation and ionic conduction which determine the potentiality of a material for absorbing microwave radiation and transmuting it into heat. Though mostly, microwave-assisted synthesis was used in organic synthesis but nowadays to prepare monodispersed upconverting nanocrystals (Kubrakova and Toropchenova 2008; Zhu et al. 2004; Wang and Nann 2009; Panda et al. 2007), this method is using. Some advantages are there behind using such method, they are

(a) Convenient and reproducible method which allows preparing highly luminescent, small monodispersed nanocrystals (Wang and Nann 2009).
(b) Rapid reaction takes very less time to complete the reaction and to form luminescent nanocrystals.
(c) Provides particles having different required sizes and shapes.
(d) Hence, the polar reactants (accompanied by high microwave extinction coefficient) are excited by direct absorption of microwaves and as a result, the activation energy reduces, consequently reaction rate enhances. As outcome, lower temperature is required to perform the reactions, therefore, additional temperature in inhomogeneities is being eliminated and the leading experimental parameters (time, temperature, pressure) become easier.
(e) Growth and nucleation of resulting nanocrystals in such synthesis procedure can be distinctly described with clear explanation. Hence, growth and nucleation of nanocrystals occur in three phases, in first phase, slowly the reactant concentration increases and finally exceeds solubility, in second phase, reactant concentration extends the critical limit of super saturation, therefore, rapid nucleation is obtained and eventually this nucleation burst, reactant concentration reduces to the depletion of the solute for the growth of generated nuclei and here nucleation stage ends. In the third phase, gradually nuclei grow. Such processes follow the LaMer Mechanism. In such way, the whole reaction can be managed favourably and the resulting nanocrystals exhibit excellent monodispersity and crystallinity due to polar reactants, high microwave extinction coefficient.

5.2 Photopolymerization

Using this strategy, it is possible to induce different monomers with different desired functional properties (such as—Hydrophobic or hydrophilic, charged or neutral, chemically active or inactive) within up converting nanocrystals by coating them with thin polymer shells known as photopolymerization which utilizes internal UV or emitting visible light from them on NIR excitation. This is a simple, generic and straightforward and versatile approach to functionalize, conjugate, protect and make upconverting nanoparticles bio-compatible (Beyazit et al. 2014).

Utilization of NIR excitation in photopolymerization provides special advantages like, less probability of occurring photodamage to living cells owing to its high tissue penetration depth, ability (due to its higher wavelength) to avoid photobleaching, weak chemical stability as such difficulties arise with excitation having lower wavelength (Wang et al. 2010a, b; Chatterjee et al. 2010). With photopolymerization, upconverting nanocrystals draw some disadvantage while using in biological media as they are hydrophobic in nature and non-dispersible in nature. Therefore, after photopolymerization also their surfaces are needed to be modified and functionalized in a precisely controlled manner (Jiang et al. 2012).

5.3 Advanced and Recent Synthetic Strategies in UCNPs with Various Advanced Applications

Recently, unifying core and shell material became most attractive synthetic strategy yielding high quality and high sensitivity involving energy transfer process between Eu^{3+} and Gd^{3+} ions (Duhnen and Haase 2015). There exists a twisted relationship between upconversion efficiency and lattice geometry providing a strategy for designing the quantum efficiency of any lanthanide upconverter. This strategy has been published in literature in present day (Wisser et al. 2015). The chitosan—conjugation is the another method to increase the cell viability of upconverting nanocrystals in human breast cancer cells, in such approach upconverting nanocrystals exhibit bright upconversion fluorescence with controlled size and shape on 974 nm excitation (Ghosh 2014). Upconversion nanoparticles can also be served as luminescent nanotherometers over a wide temperature range and such nanothermometers can be processed by affiliation of liquid–solid solution hydrothermal strategy and thermal decomposition strategy and become excellent temperature sensor (Xu et al. 2015). Lanthanide-doped upconversion nanocomposites (such as—Ln^{3+}-doped $BiPO_4/BiVO_4$ nanocomposites) can also be used for photocatalysis applications. Lanthanides (e.g. Tm^{3+}) emit strong blue spectra which are transferred $BiVO_4$ resulting formation of excitations which produces reactive oxygen species. In literature survey, it has been suggested that upon NIR and solar irradiation photocatalysis activity of such resulting upconverting nanocomposites are more advantageous than the reported general nanocomposite photocatalysts. In recent past years, few advanced applications of UCNPs have been established, such as, a multifunctional nanocluster having dual compositions of gold nanorod (AuNR) and UCNPs showed applications in imaging of cancer cells and treatment. In neuroscience and engineering, stimulating deep brain neurons is an important desire. Molecularly tailored UCNPs (dendritic) have been served as less invasive optogenetic actuators to stimulate deep brain neurons (Chen et al. 2018a, b). UCNPs based theranostic system is very useful for multi-drug resistance reversing ability for chemotherapy (Chen et al. 2018a, b). A UCNP based system: $LiYF_4$: $Yb^{3+}/Tm^{3+}@SiO_2$ coated with chitosan (CH) hydrogel has become an excellent theranostic platform by controlling drug-delivery and deep-tissue imaging for several important applications, tissue engineering, bio-mapping and cellular imaging (Jalani et al. 2016). Highly specific tumour-imaging can be exhibited by the conjugate forms of Upconversion-nanoprobes with cancer cell (CC) membrane, such combined form (CC-UCNPs) represents novel materials presenting attractive class of advanced materials (Rao et al. 2016).

Table 3 Tunable morphology of UCNPs with controlling K^+ ions

Concentration of k^+ ions (in mol %)	Morphology of resulting nanocrystals	Diameter of the resulting particles
0	Spherical ($NaYF_4$)	26 nm
20	Regular hexagonal cross section (plate-like; uniform in size; $NaYF_4$)	-
40	Rectangular-like cross section ($NaYF_4$)	–
80–100	Spherical hexagonal (KYF_4)	

5.4 *Investigation of Phase Transformation and Morphology Tuning in UCNPs*

Phase and morphology of UCNPs can be tuned by using proper dopant ions. Zhang's group and Kal's group have been already depicted the crystal structure of such upconverting nanocrystals through K+ ions co doping (Liang et al. 2015) and there are various literature reported on the different morphologies and phases of the resulting nanocrystals (Wei et al. 2006; Yang et al. 2007; Ye et al. 2010; Dou and Zhang 2011). Phase transformation and morphology both are dependent on several experimental parameters, such as reactant's concentrations, reaction time and reaction temperature. Concentration of dopants can also play a significant role on phase transformation. For example, different concentrations of K + ion dopants in $NaYF_4$: Yb^{3+}, Er^{3+} nanocrystals display different XRD peaks. It has been reported that upto a certain limit (60 mol %) of such dopant concentration hexagonal $NaYF_4$ nanocrystals become dominant phase but beyond that limitation (when exceed 60 mol %: around 80–100 mol %) K^+ ions replace Na^+ ions and form hexagonal KYF_4 nanoparticles that mean phase transformation takes place (Liang et al. 2015). The reason behind that phase transformation is the imbalance of the intrinsic crystal structure. K^+ ions only settle in the substitution position of Na^+ ions and when number of occupying K^+ ions becomes large compared to Na^+ ions (at that moment Na^+ ions are too few), hexagonal KYF_4 appears having poor crystallinity.

Now, how the morphology of UCNPs ($NaYF_4$: Yb^{3+}, Er^{3+}) can be tuned with tuning concentration of dopant ions (K^+ ions), is shown in Table 3.

5.5 *Mechanistic Investigation of Photon Upconversion*

It is known, in most of the cases that the lanthanide-doped upconverting materials (ex: $NaYF_4$: Yb^{3+}, Er^{3+}) need 980 nm excitation. But this event creates problem which became a daunting challenge nowadays as absorption band of Yb^{3+} (980 nm), which acts as dopant in host lattice ($NaYF_4$), dangles with absorption band of water molecules used in the biological sample. Therefore, under 980 nm excitation overexposure of biological species results in cell death and tissue damage due to overheating issues. To overcome such drawback, 980 nm laser excitation can be replaced by 800 nm laser excitation occurring low absorption coefficients and it is possible through use of Nd^{3+} dopant ions as sensitizer due to their favourable characteristics (As their sharp absorption bands are centred around 800 nm) (Xie et al. 2013).

6 Biological Applications of UCNPs Based on Their Immobization to Assemblies

For several biological applications, surface modification of UCNPs plays an important part for upgrading their photostability and to attach with bio-molecules. Normally, UCNPs bear a good portion of surface dopant ions. Now presence of weakly bound impurities on surface and ligands can diminish the luminescence of dopant ions due to the creation of high energy oscillations. On the other side, the excitation energy of interior ions can be dissipated non-radiatively due to their transferred energy towards the crystal-surface. As a result, the crystal field strength decreases and finally the overall UC luminescence intensity reduces abruptly. To overcome this drawback, surface modification (surface passivation and surface functionalization) is required. By surface passivation, all the dopant ions can be confined in an interior core and therefore can dominate the energy transfer towards crystal-surface. An upconversion luminescence enhancement of about 30 times was successfully carried out by one research group with 1.5 nm thick $NaYF_4$ shell on 8 nm sized NaYF4: Yb/Tm nanocrystals (Mai et al. 2007). By varying the thickness of the shell, UC luminescence efficiency can be well tuned and by several research groups (Park et al. 2009; Yi and Chow 2007; Lu et al. 2008; Chen et al. 2008), this process has been successfully explained by immobizing the nanocrystals (UCNPs) inside a coated shell. Besides, surface passivation, surface functionalization is also an important part to use those UCNPs in biomedical and bio-detection. Hence, by ligand-exchange technique $NaYF_4$: Yb/Er nanoparticles can be made water soluble by utilizing bi-functional organic molecules which replaced amine ligands and modified the crystal surfaces as water-soluble carboxyl functionalised surface(Wang et al. 2010a, b).

Moreover, the immobilization concept of UCNPs has been confirmed by their several biological applications via bio-conjugate chemistry as these techniques are, while applied to UCNPs, capable of immobilization to bio-assemblies. Based on the immobilization concept, some specific bio-applications of UCNPs are discussed below.

In cells and small animals, the innovative UCNPs have been considered to be promising molecular probes for optical imaging. In 2013, Grebenik et al. group reported UCNP-labeled cancer lesion where the synthesized nanoparticles were capped with amphiphilic polymer. Further, mini-antibodies (scFv4D5) were attached onto UCNPs for allowing their specific binding to the human cells. As a result, the

UCNP based bio-complexes showed high specific immobilization on human breast adenocarcinoma cells SK-BR-3 (Grebenik et al. 2013).

In 2017, Shikha et al. group developed a UCNPs-based multiplexed detection system to encode PEGDA microbeads and to label antibodies. Hence, the multicolour codes were produced by mixing green and red emissions from UCNPs whereas its blue emissions were used to label antibody. By immobilizing probe antibodies on red-UCNPs and anti-human C reactive protein (hCRP) on green UCNPs, specific capturing of human serum albumin (HAS) protein and multiplexed detection of HCRP and HSA proteins were done, respectively (Shikha et al. 2017).

RGDS and TAT conjugated $NaYF_4$: Yb^{3+}/Er^{3+}, SiO_2 nanoparticles were targeted in HeLa cells by in vitro study where RGDS-conjugated probes were confined on cell-plasma membrane for the specific binding between the conjugated peptides and integrins. This application also clearly confirms the immobilization behaviour of UCNPs (Kostiv et al. 2016).

In 2016, T. sang et al. group proposed a conjugated system of $BaGdF_5$: Yb/Er UCNPs and AuNPs to increase the effective detection of limit for target Ebola virus from picomolar level to femtomolar level. The enhancement of this ultrasensitive detection exhibited a great potential for practical application due to the specificity between nanoprobes and Ebola virus oligonucleotides (Tsang et al. 2016).

The bare monodisperse UCNPs were immobilized for developing bio-sensing surface. Further, for developing bio-assays and bio-sensors, a high immobilization density for UCNPs was reported by Doughan et al. group in 2014 and the reported immobilization density was calculated of about ~1.3×10^{11} UCNP cm^{-2} where PEA—UCNPs were immobilized on functionalized coverslips (Doughan et al. 2014).

PEG-b-PAAc was immobilized on erbium-ion-doped Y_2O_3 [Y_2O_3: Er] particle—surface to enhance its dispersion and prevent adsorption. Further, co-immobilization of PEG-b-PAAc and BSA occurred due to the protein installation on particle—surface (Kamimura et al. 2008).

According to previous report (Freitag et al. 2019), the immobilized photoredox catalyst based on $NaYF_4$ nanoparticles was capable for in vivo applications. The immobilized system successfully performed under 980 nm NIR source.

In particular, based on the immobilization strategies, upconverting nanoparticles have given a huge response in bio-applications. In the applications of cellular and molecular biology, several bio-probes (Green fluorescent protein, (GFP), organic dyes) are used for identification of different bio-molecules, which is an indispensable step. To develop the sensitivity of some technical and analytical devices used in bio-field utilization of such available probes is a crucial step as they improve the efficiency of detection. Still some drawbacks (such as weak photostability of probes, measuring cell response having instrumental problem with resolution) are there with such bio-probes which limit their biological applications. To overcome such drawbacks nano particles with protein-imaging, nucleic acid-detection with nanodiamonds (Faklaris et al. 2008; Chang et al. 2008), metallic nanoparticles, dye-doped silica particles have been already discovered. As nanoparticles consist of a large number of ions, single particle and single molecule identification is favourable owing to large surface area implanting of different targeting groups at the surface and it can be done comfortably.

For more advanced applications, recently, RE-based nanoparticles have been proved to be most optimistic materials (Vetrone and Capobianco 2008) due to their especial properties such as long lifetimes, narrow emission lines, high photostability, low cytotoxicity and simplistic functionalization procedures which make them highly bio-compatible and recognizable compared with other Nano particles. In magnetic resonance imaging (MRI) due to their high magnetic moment and inoxidant detection applying reverse oxidoreduction process, the requirement of UCNPs is noticeable. In MRI, RE ions can be regarded as a powerful contrast agent. To fabricate nanoparticles (e.g. silica particles) (Shao et al. 2011) as high bio-compatible material various chemical properties could be induced by inducing different rare earth compounds as dopant-agent. Mesoporous silica shell nanocomposites used for bimodal imaging is the best example of such nanoparticles doped with RE compounds.

Surface modification or surface functionalization and bio-conjugation are crucial steps after synthesis known as post synthesis method before using in the biological field. Every nanoparticle has a certain surface chemistry and specific physical and chemical properties to govern nucleation or growth of resulting particles as well as to control their dispersion in the solvent used. Generally, after synthesis of upconversion nanoparticles, due to the presence of some organic ligands capped with them, they become non-dispersible which are not suitable for bio-applications. For using in different biomedical applications, upconversion nanoparticles based bio-probes are fabricated to functionalize their surfaces to ensure good dispersion in biological media and to uncover specific organic or bio-organic groups from the particles and to make them appropriate for accurate targeting receptor sites (Fig. 6).

Ensuring the stability of hydrophobic nanoparticles in aqueous media is a necessary condition in biological applications and that can be processed by applying

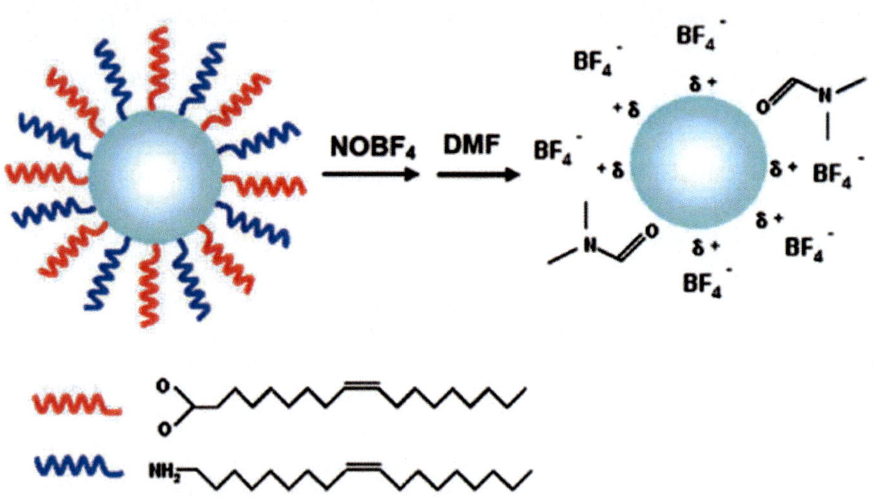

Fig. 6 Illustration of ligand-exchange with $NOBF_4$. Reproduced with permission (Dong et al. 2011). *Copyright* Journal of the American Chemical Society publication

diverse functional strategies which are described by following: In Ligand exchange, the original ligands are replaced by other molecules Citrate, PVP, PAA, TGA, PEG and finally become stable dispersible in biological media (Figs. 7, 8) (Boyer et al. 2010; Dong et al. 2011).

By direct surface grafting of molecules and polymer encapsulation method, several reactive groups (e.g. amino, thiols, etc.) can be attached on the resulting particle-surface (Nichkova et al. 2005). Streptavidin, antibodies are some biomolecules with which nanoparticles can be directly coupled. Because of surface hydroxyl groups acting as a coupling agent with silanes present in the oxide nanoparticles, by silica and organosilane coating functionalization can be done in a quite simple way.

Layer by layer assembly is another propitious method for surface modification, obtaining by sequential adsorption of oppositely charged polyelectrolytes on particle surface. Such sequential adsorption (i.e. sequential deposition of PAA and PAH) has

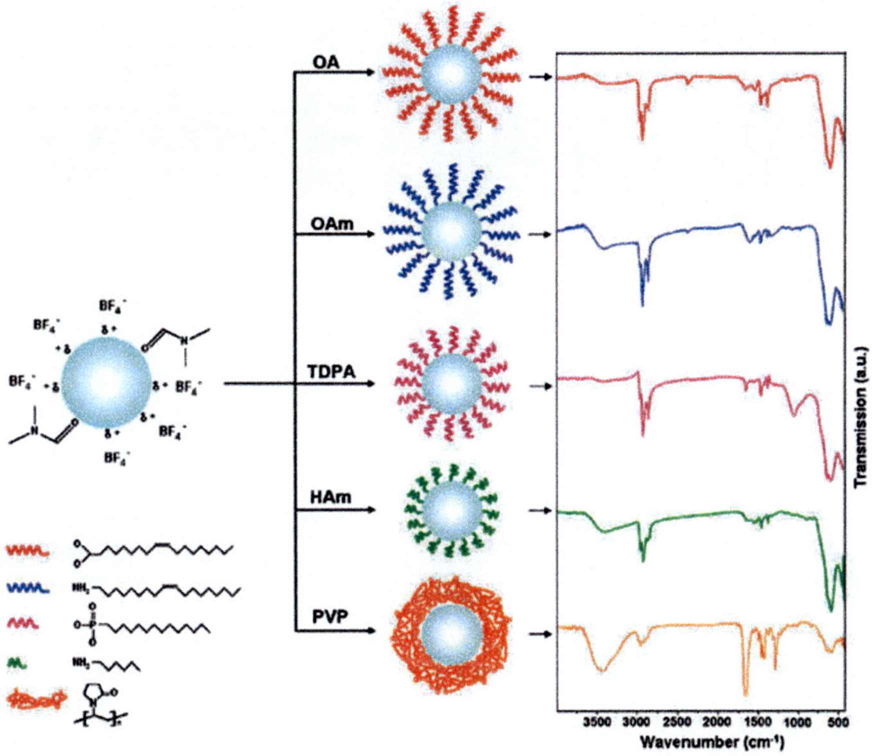

Fig. 7 Illustration of secondary ligand-exchange exhibiting surface functionalization of BF_4—modified Fe_3O_4 nanocrystals with several capping molecules and their corresponding FTIR spectra. Reproduced with permission (Dong et al. 2011). *Copyright* Journal of the American Chemical Society publication

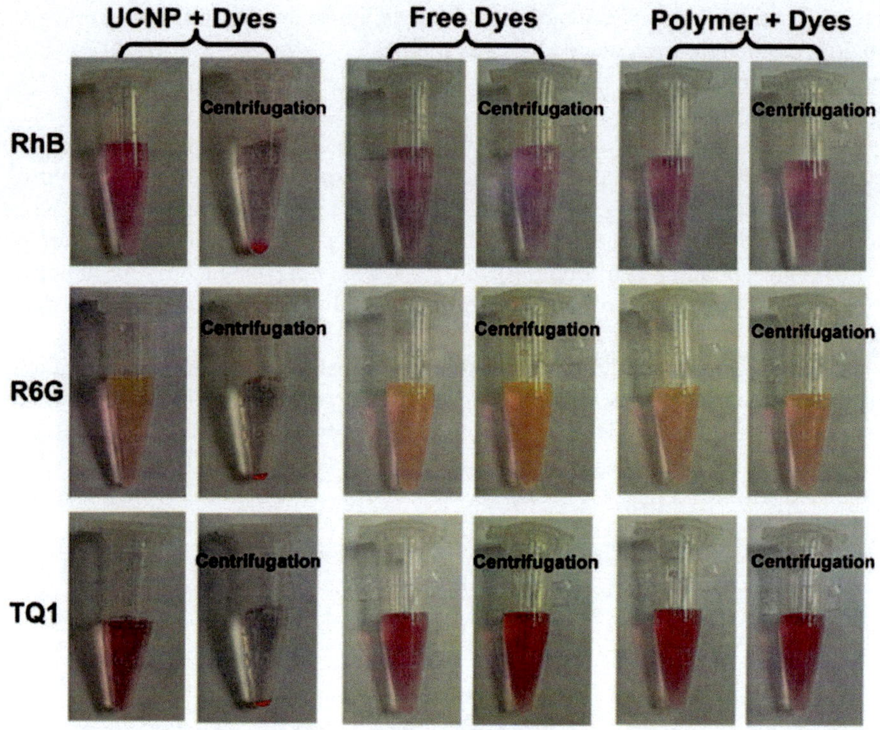

Fig. 8 UCNP/dye—images captured while centrifuged (before/after). UCNPs and dye molecules appeared as precipitate by centrifugation force while supernatant part appeared as clear colourless solutions in containers. Precipitation was not observed while centrifugation occurred with free dye molecules or PEG-PMHC$_{18}$+dye mixtures. These images indicated that UCNP-surface adsorbed three types of dye molecules rather than their encapsulation in UCNP-dye complexes by amphiphilic PEG-PMHC$_{18}$ polymer. Reproduced with permission (Cheng et al. 2011a, b). *Copyright* The Journal of Physical Chemistry C publication

been already observed in literature providing superior mechanical stability and a well-ordered NIR to visible upconversion luminescence. But here some drawbacks which limits its requirement as this method is only applicable for hydrophilic nanocomposites and time consuming, many washing steps are needed during that assembly. Surface silanization is the most frequently used surface functionalization method requiring the growth of silica shell on upconversion nanoparticles. The importance of this process is the use of silica as silica is considered to be bio-compatible and porosity is easily controllable. The encapsulation of silica layer on UCNPs is still under investigation to modify the difficulties(such as, most of the reagents used are toxic in nature, therefore before starting that encapsulating process they should be detached carefully) introduced by silica layer and other reagents used in such encapsulating process. If we desire to use silica encapsulation in FRET-bio-sensors, the increasing shape and size of the resulting particles will create difficulties during experiment. To suppress those difficulties, two approaches are useful, one is Stober method and other one is reverse microemulsion route.

Although controlling that Stober method is quite difficult and takes long time, thus it is straightforward and fruitful path in the presence of ethanol and ammonia, whereas reverse micro emulsion route can be controlled in the presence of homogeneous mixture of water, oil, surfactant and TEOS to dominate the increasing thickness, size and shape of the additional silica layer. One user-friendly surface modification for providing hydrophilic nanoparticles as well as hydrophobic layer for loading hydrophobic molecules(Drug, Dyes) is the Host–Guest Self-assembly method which has a lot of advantages over the other methods due to its simplicity and efficiency. As this procedure is very fast, over 95% of nanoparticles can be transferred to water using reagents (Liu et al. 2010).

We have already discussed that by proper surface modification, it is possible to affix various functional groups with UCNPs, like-carboxyl, amine and thiol groups. After such functionalization, these UCNPs with several attached groups can be decorated with various biomolecules which can contribute bio-identification of other targeted cells. There are diverse approaches of bio-conjugation. Among these, the greatest advantageous method is Direct Physisorption. This method occurs via non-covalent force of biomolecules. The activity of proteins and quantum yield of resulting nanoparticles remain unchanged after that bio-conjugation. Assisted Physisorption is of other crucial importance in bio-conjugation and uses pre-bound molecules and occurs by non-covalent coupling of two molecules. This is an essential method in bio-specificity and bio-sensitivity due to the proper orientation of molecules. By direct chemical coupling of bio-molecules also bio-conjugation can be induced in UCNPs. But this method cannot be controlled easily as other complex processes such as gel electrophoresis is required for separation after completion of this process. In UCNPs' surface modification with biotin-protein bond (streptavidin-coupling) can be useful for various biomedical application, which is the strongest and most stable non-covalent interaction.

A general overview on UCNPs applications is highlighted below.

6.1 DNA and Protein Detection

It has become a great challenge for application of RE-based nanoparticles for DNA and protein detection. Implementation of such nanoparticles in DNA detection is based on two strategies, DNA sensitive nanoparticles and DNA fragments nanoparticles. Here, we are describing one by one.

In case of the first approach, DNA concentration can be computed using fluorescent upconversion nanoparticles (Wang et al. 2009) which can be comparable to conventional measurement executed by spectrophotometer. As per optical properties, we already know that fluorescence property UCNPs make them valuable for their applications. Now, due to some specific dopant ions (such as: Ce^{3+}, Tb^{3+} dopants in LaF_3, fluorescence occur due to Tb^{3+} dopant ions) and their transitions in energy levels, the fluorescence of upconversion nanoparticles become very prominent and in the presence of DNA, the amount of fluorescence can be quenched by the generating generation of hydrogen bonds between DNA and nanoparticle. As a result, the energy can be transferred from excited dopant ions (Tb^{3+} in the given example above) to DNA. Finally, it is possible to quantify DNA concentration by determining the corresponding fluorescence intensity. In the second approach, to study mRNA pattern, DNA microarrays are used extensively for which there are two requirements, one for hybridization of probe DNA strands and other for labelling target DNA fluorescently. In such technology, also fluorescent intensity quantifies the DNA fragments in each probe. Later, other advantageous technology comparable to the depicted one, reported in literature according to which for labelling, target DNA is combined with UCNPs (e.g. Y_2O_2S: Yb, Er) and finally, it is possible to measure the concentration of the target DNA. The later one is superior technology than the former for determining DNA concentration, as the former technology weakly express mRNA.

UCNPs can also be used for diagnosis diseases such as Fe_3o_4/Gd_2o_3: Eu nanoparticles where Fe_3o_4 construct the magnetite core and Gdo_3 constructs the shell of the nanoparticles and the dopant ion Eu^{3+} make the particle fluorescent by transferring energy. Now, it is known, detection of single nucleotide polymorphisms (SNPs) is very significant phenomenon for recognition of various diseases like polycystic kidney disease (PKD) by using reverse transcription polymerase chain reaction (RT-PCR) but due to some shortcoming vis a vistime consuming and expensive, it cannot be applied in the biomedical field. In the present days, to solve these problems successfully, UCNPs have been introduced to detect SNPs (Son et al. 2008). In that approach, rare earth-based nanoparticles are used in such a systematic manner, that they can combine both magnetism and fluorescence properties to provide a strong and effective analysis.

Europium-doped gadolinium oxide nanoparticles can also be used for detection of protein micropatterns for coupling with antibodies rather than organic dyes. Such rare earth-based nanoparticles with organic cores have also been successfully utilized in immunoassays. As a result, sensitivity of the whole complex can be improved in the presence of rare earth-based UCPN nanoparticles.

6.2 Fluorescent Resonant Energy Transfer (FRET)

FRET has wide applications (among which protein–protein interaction is the most significant phenomenon with the usage of FRET) in biomedical field using other materials. For protein–protein interaction, one protein (labelled with fluorophore) acts as donor and other one acts as acceptor (labelled with fluorophore) using that fluorescence energy transfer method according to which an efficient energy is being transferred from donor to acceptor and as outcome detection of fluorescence become possible providing the basic principle of FRET process. If FRET can be done in the presence of lanthanide systems, then easily such nanoparticles can be served as donors, which gives some effective and favoured conditions like prevention from direct excitation, limitations in overlapping spectra and their easy separation, long lifetimes of excited state and less probability of photobleaching and there are certain reasons why lanthanide nanoparticles were taken in FRET experiment. Hence, the lanthanide systems with the large stokes shift permit excitation having shorter wavelength than the absorption by acceptor to prevent direct excitation and UCNPs with large anti-stokes shift permits excitation at much higher wavelength than the emission by acceptor to get a clean detection from FRET. Secondly, due to narrow emission spectrum occurred by lanthanide nanoparticles, there is a very less chance of overlapping between them and acceptor emission-spectrum. Thirdly, as lanthanide systems provide excited states having long lifetime, so conducts acceptor emission with longer time and as a result separation in time-resolved experiments occurs easily. Utilizing that longer-scale phenomenon, it is possible to apply quantum dots as acceptors in FRET applications.

Advantages of upconversion nanoparticles as donors in FRET application make them unique in biomedical applications. Most of the upconverting nanoparticles are co-doped with Er^{3+} and Yb^{3+} and gives remarkable fluorescence which make them an excellent donors having interesting potential in FRET applications. Protein detection with the help of nanoparticles is possible by coating (example of such coating, agent-streptavidin), mixing (example of such mixing, agent-biotinylated protein) and labelling (example of such labelling, compound-fluorescent acceptor) in a precisely controlled manner so that after IR excitation of donors, a clear detection of fluorescence becomes possible.

6.3 *In Vivo and in Vitro Biomedical Application*

Recently, lanthanide-doped UCNPs have gathered much attention in biomedicine. In 2012, Liang Cheng et al. showed in vivo imaging of UCNPs in his experiment (Cheng et al. 2010).Auto fluorescence of the UCL imaging with long exposure time allows in vivo detection of UCNPs. UCL emission spectra of different nanoparticles can be controlled with our requirement in biological systems by altering the concentration of the lanthanide dopant ions used during synthesis and this process is essential in multicolour vivo UCL imaging with organic dyes through hydrophobic force (Fig. 8) (Cheng et al. 2011a, b). In tumour diagnosis, UCL imaging plays significant role by tumour targeted molecular imaging using UCNP-based nanoprobes. Basically for synthesis of such nanoprobes, conjugating polymer-coated UCNPs are used, the aim of which is to bind different cancer cells having high specificity explore a new challenge in future studies and experiments. UCNPs have also been successfully synthesized to utilize in vitro applications such as labelling and tracking rabbit bone marrow mesenchymal stem cells (rBMSCs) (Ma et al. 2016). In 2008, both in vivo and in vitro studies were performed using aqueous dispersible rare earth-ion (Tm^{3+} and Yb^{3+}) co-doped fluoride ($NaYF_4$) nanocrystals, where upconversion process provided deeper penetration into biological specimen resulting high contrast optical imaging (Nyk et al. 2008). In 2011, in vitro and in vivo imaging was implemented by utilizing 915 nm LASER excited $NaYbF_4$ UCNPs co-doped with $Tm^{3+}/Er^{3+}/Ho^{3+}$– rare earth elements to avoid overheating irradiation caused by 980 nm laser excitation source.

Hence, the experimental results were successfully investigated and showed very high contrast upconversion bio-imaging and a successful performance for in vivo imaging where highly stable UCNPs encapsulated with DSPE-mPEG-5000 were injected into mice (Zhan et al. 2011). For delivering in vitro and in vivo operations, caged UCNPs are being served as excellent platforms to improve targets along with reduction of side effects from chemotherapy by using them as NIR-triggered targets (Chien et al. 2013).

To overcome the drawbacks of single imaging approach, nowadays, the multimodal imaging has gained enormous recognition in biomedical applications. There are several approaches (Zhu et al. 2012; He et al. 2011; Xing et al. 2012) to progress UCNPs based nanoprobes for multimodal bio-imaging. UCNPs as well as their nanocomposites can be utilized to fabricate such multimodal imaging bio-probes. Also in the field of cell labelling and in vivo tracking, UCNPs are sensible enough. Recently, investigation of mesenchymal stem cells (MSCs) has been started owing to their ability of recognition various types of cells (e.g. bone) under certain circumstances in various potential applications in bio-field such as in immunotherapy and gene therapy.

6.4 *Medicinal and Remedial Applications*

In present days, UCNPs and their composites can be considered as therapeutic agent in cancer treatment, drug gene delivery and for photodynamic therapy due to their various unique properties and functions as well as imaging capability. For imaging and therapy, chemotherapy drug molecules could be delivered with loading of UCNPs. For drug releasing system, polymer-coated UCNPs can also be used. For treatment of various gene-related diseases, gene therapy with gene encoding DNA or RNA has become a challenging matter. UV light emitted from UCNPs activates definite gene expression by operating DNA and RNA, though depth of penetration. The therapeutic efficiency is very high in case of UCNPs with NIR excitation compared with UCNPs with UV light and such PDT based on UCNPs exhibited greater potential in treatment of cancers as well as suppression of the size of large internal tumours.

To treat cancer diseases, Photodynamic Therapy (PDT) is a non-invasive and effective medical method which uses photosensitizers (PDT drugs) and light irradiation and they interact with molecular oxygen. There are several approaches to originate PDT reagents like silica encapsulation, polymer encapsulation, and hydrophobic interaction. At first, photosensitizers are activated with light and as a result, cytotoxic reactive oxygen species are formed and persuade required cell death (Park et al. 2012). Assembly of oxygen molecules are influenced by the energy transfer process from excited UCNPs relying on the spectral overlapping of donor and acceptor impurities in host lattice. There are two factors that quantify photodynamic therapy, one is high energy transfer efficiency and the other is large production of singlet oxygen and these two phenomenons are related with high stability of originally loaded photosensitizers in UCNPs. For good energy transfer process, non-covalent form of UCNP-PS is not ideal as this provides low loading efficiency and finally small production of singlet oxygen. For that a covalent conjugating of UCNPs ($NaYF_4$, Yb^{3+}, Eb^{3+}), photosensitizer and target molecule strategy can be adopted. For singlet oxygen and high fluorescence intensity production, a multifunctional upconversion nanoplatform has been already expanded on the basis of a particular energy transfer from UCNPs to photosensitizer for synchronous imaging and therapy (Fig. 9) (Liu et al. 2012). The therapeutic consequences of UCNP-PS compound for animal studies and cellular level have been already reported(Park et al. 2012). It has been reported in literature that for in vitro cancer cell destroying, mesoporous silica-coated UCNP-PS combination shows an interesting and encouraging result (Guo et al. 2010). Gu's group has revealed the uses of UCNP-PS nanocomplexes in the field of in vivo PDT (Cui et al. 2012). For in vitro and in vivo tracking, UCNPs have been used to designate stem cells by different groups (Wang et al. 2012).

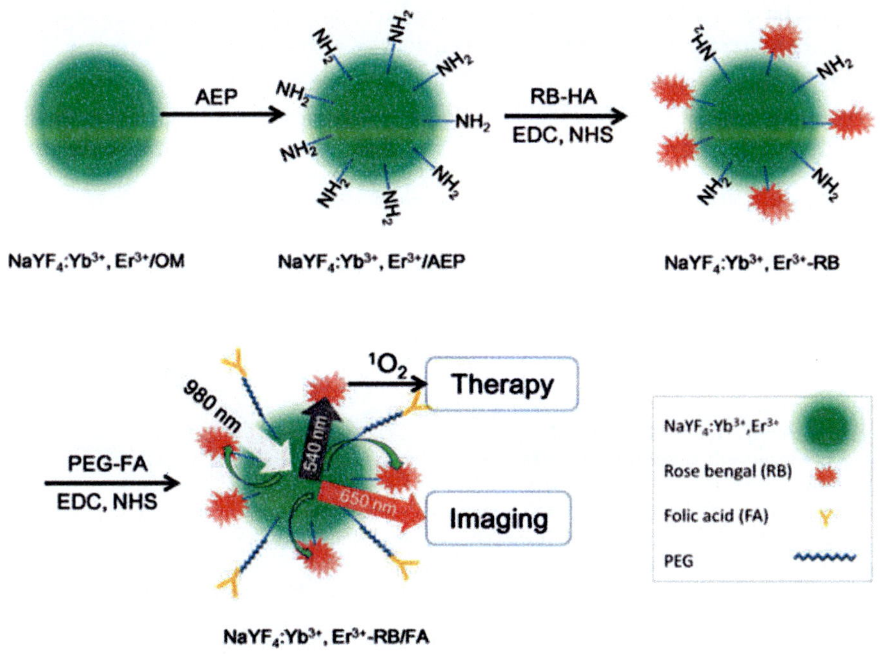

Fig. 9 Covalent conjugation of $NaYF_4$: Yb^{3+}, Er^{3+} UCNPs, photosensitizer (RB), target molecule (FA). Reproduced with permission (Liu et al. 2012). *Copyright* ACS Nano publication

7 Conclusions

This review explores recent developments of UCNPs along with a detailed survey and as application part, it focuses in bio-fields (imaging and therapy). Though there are tremendous amount of applications and a number of exciting aspects have been reported in literatures since past few years on this field but still many challenges are ahead especially in imaging and therapy.

With reducing the sizes of UCNPs, the quantum yield became lower which limits their use in various nano-probes. Later, several research groups modified the synthesis and production of UCNPs to increase their yield under continuous wave-excitation sources but still future efforts are required to make them useful for enormous number of in vivo applications. The long-term toxicity of RE-doped UCNPs is another perturbation. Though, bio-compatible coating-based UCNPs ($NaYF_4$) appeared to be safer to cells (with precise concentrations) compared to bared UCNPs, but, still other factors like their interactions with immune systems, interference with the reproductive system, whether the toxicity is affecting for next generation applications, that are still not known for which more number of systematic investigations are demanded. The effects of surface functionalization and sizes of UCNPs on in vivo behaviours are required to improve in reducing certain potential toxicity. Though imaging and therapies based on UCNPs and their different nanocomposites have been revealed in previous literatures, to achieve synergistic therapeutic effects and realize real-time scanning of treatment development, an improved design of UCNPs based on novel multifunctional agents is needed further for simultaneous medical diagnosis and cancer treatment.

References

Auzel F (2004) Upconversion and Anti-Stokes processes with f and d Ions in solids. Chem Rev 104:139–174

Barnes WL, Dereux A, Ebbesen TW (2003) Surface plasmon subwavelength optics. Nature 424:824–830

Beyazit S, Ambrosini S, Marchyk N (2014) Versatile synthetic strategy for coating Upconverting nanoparticles with polymer shells through localized photopolymerization by using the particles as internal light sources. Angew Chemie Int Ed 53:8919–8923

Boyer JC, Cuccia AL, Capobianco J (2007) Synthesis of colloidal upconverting NaYF 4: Er 3+/Yb 3+ and Tm 3+/Yb 3+ monodisperse nanocrystals. Nano Lett 7:847–852

Boyer JC, Manseau MP, Murray JI, van Veggel FCJM (2010) Surface modification of Upconverting $NaYF_4$ nanoparticles with PEG–Phosphate ligands for NIR (800 nm) biolabeling within the biological window. Langmuir 26:1157–1164

Boyer JC, Vetrone F, Cuccia LA, Capobianco JA (2006) Synthesis of colloidal Upconverting $NaYF_4$ nanocrystals doped with Er^{3+}, Yb^{3+} and Tm^{3+}, Yb^{3+} via thermal decomposition of lanthanide trifluoroacetate precursors. J Am Chem Soc 128:7444–7445

Chang IP, Hwang KC, Chiang CS (2008) Preparation of fluorescent magnetic nanodiamonds and cellular imaging. J Am Chem Soc 130:15476–15481

Chatterjee DK, Gnanasammandhan MK, Zhang Y (2010) Small Upconverting fluorescent nanoparticles for biomedical applications. Small 6:2781–2795
Chatterjee DK, Rufaihah AJ, Zhang Y (2008) Upconversion fluorescence imaging of cells and small animals using lanthanide doped nanocrystals. Biomaterials 29:937–943
Chatterjee DK, Yong Z (2008) Upconverting nanoparticles as nanotransducers for photodynamic therapy in cancer cells. Nanomedicine 3:73–82
Chen S, Weitemier AZ, Zeng X (2018) Near-infrared deep brain stimulation via Upconversion nanoparticle–mediated optogenetics. Science 359:679–684
Chen X, Sun J, Zhao H (2018) Theranostic system based on $NaY(Mn)F_4$:Yb/Er Upconversion nanoparticles with multi-drug resistance reversing ability. J Mater Chem B 6:3586–3599
Chen Z, Chen H, Hu H (2008) Versatile synthesis strategy for carboxylic acid−functionalized Upconverting nanophosphors as biological labels. J Am Chem Soc 130:3023–3029
Cheng L, Yang K, Li Y (2011) Facile preparation of multifunctional Upconversion nanoprobes for multimodal imaging and dual-targeted photothermal therapy. Angew Chemie Int Ed 50:7385–7390
Cheng L, Yang K, Shao M (2011) Multicolor in vivo imaging of Upconversion nanoparticles with emissions tuned by luminescence resonance energy transfer. J Phys Chem C 115:2686–2692
Cheng L, Yang K, Zhang S (2010) Highly-sensitive multiplexed in vivo imaging using pegylated Upconversion nanoparticles. Nano Res 3:722–732
Chien YH, Chou YL, Wang SW (2013) Near-infrared light photocontrolled targeting, bioimaging, and chemotherapy with caged Upconversion nanoparticles in vitro and in vivo. ACS Nano 7:8516–8528
Cooper ER, Andrews CD, Wheatley PS (2004) Ionic liquids and eutectic mixtures as solvent and template in synthesis of zeolite analogues. Nature 430:1012–1016
Cui S, Chen H, Zhu H (2012) Amphiphilic chitosan modified Upconversion nanoparticles for in vivo photodynamic therapy induced by near-infrared light. J Mater Chem 22:4861–4873
Dong A, Ye X, Chen J (2011) A generalized ligand-exchange strategy enabling sequential surface functionalization of colloidal nanocrystals. J Am Chem Soc 133:998–1006
Dou Q, Zhang Y (2011) Tuning of the structure and emission spectra of Upconversion nanocrystals by alkali ion doping. Langmuir 27:13236–13241
Doughan S, Han Y, Uddayasankar U, Krull UJ (2014) Solid-phase covalent immobilization of Upconverting nanoparticles for biosensing by luminescence resonance energy transfer. ACS Appl Mater Interfaces 6:14061–14068
Downing E, Hesselink L, Ralston J, Macfarlane R (2006) A three-color, solid-state, three-dimensional display. Science 273:1185–1189
Duhnen S, Haase M (2015) Study on the intermixing of core and shell in $NaEuF_4$/$NaGdF_4$ core/shell nanocrystals. Chem Mater 27:8375–8386
Ehlert O, Thomann R, Darbandi M, Nann T (2008) A four-color colloidal multiplexing nanoparticle system. ACS Nano 2:120–124
Eustis S, El-Sayed MA (2006) Why gold nanoparticles are more precious than pretty gold: Noble metal surface plasmon resonance and its enhancement of the radiative and nonradiative properties of nanocrystals of different shapes. Chem Soc Rev 35:209–217
Faklaris O, Garrot D, Joshi V (2008) Detection of single photoluminescent diamond nanoparticles in cells and study of the internalization pathway. Small 4:2236–2239
Feng X, Sayle DC, Wang ZL (2006) Converting Ceria polyhedral nanoparticles into single-crystal nanospheres. Science 312:1504–1508
Freitag M, Moller N, Ruhling A (2019) Photocatalysis in the dark: near-infrared light driven photoredox catalysis by an upconversion nanoparticle/photocatalyst system. Chem Photo Chem 3:24–27
Ghosh OSN (2014) Chitosan conjugation: A facile approach to enhance the cell viability of LaF3: Yb, Er Upconverting nanotransducers in human breast cancer cells. Carbohydr Polym 121:302–308

Grebenik EA, Nadort A, Generalova AN (2013) Feasibility study of the optical imaging of a breast cancer lesion labeled with Upconversion nanoparticle biocomplexes. J Biomed Opt 18:1–11

Guo H, Qian H, Idris NM, Zhang Y (2010) Singlet oxygen-induced apoptosis of cancer cells using Upconversion fluorescent nanoparticles as a carrier of photosensitizer. Nanomed Nanotechnol Biol Med 6:486–495

Hampl J, Hall M, Mufti NA (2001) Upconverting phosphor reporters in immunochromatographic assays. Anal Biochem 288:176–187

He M, Huang P, Zhang C (2011) Dual phase-controlled synthesis of uniform lanthanide-doped $NaGdF_4$ Upconversion nanocrystals via an OA/Ionic liquid two-phase system for in Vivo dual-modality imaging. Adv Funct Mater 21:4470–4477

Heer S, Kompe K, Gudel HU, Haase M (2004) Highly efficient multicolour Upconversion emission in transparent colloids of lanthanide-doped $NaYF_4$ Nanocrystals. Adv Mater 16:2102–2105

Idris NM, Gnanasammandhan MK, Zhang J (2012) In vivo photodynamic therapy using Upconversion nanoparticles as remote-controlled nanotransducers. Nat Med 18:1580–1585

Jackson SD (2004) Cross relaxation and energy transfer Upconversion processes relevant to the functioning of 2 μm Tm^{3+}-doped silica fibre lasers. Opt Commun 230:197–203

Jalani G, Naccache R, Rosenzweig DH (2016) Photocleavable hydrogel-coated Upconverting nanoparticles: a multifunctional theranostic platform for NIR imaging and on-demand macromolecular delivery. J Am Chem Soc 138:1078–1083

Jiang G, Pichaandi J, Johnson NJJ (2012) An effective polymer cross-linking strategy to obtain stable dispersions of upconverting $NaYF_4$ nanoparticles in buffers and biological growth media for biolabeling applications. Langmuir 28:3239–3247

Joubert MF (1999) Photon avalanche Upconversion in rare earth laser materials. Opt Mater (Amst) 11:181–203

Kamimura M, Miyamoto D, Saito Y (2008) Preparation of PEG and protein co-immobilized Upconversion nanophosphors as near-infrared biolabeling materials. J Photopolym Sci Technol 21:183–187

Kim WJ, Nyk M, Prasad PN (2009) Color-coded multilayer photopatterned microstructures using lanthanide (III) ion co-doped $NaYF_4$ nanoparticles with Upconversion luminescence for possible applications in security. Nanotechnology 20:185301

Kostiv U, Kotelnikov I, Proks V (2016) RGDS- and TAT-conjugated Upconversion of $NaYF_4$: Yb^{3+}/Er^{3+} & SiO_2 nanoparticles: in vitro human epithelioid cervix carcinoma cellular uptake, imaging, and targeting. ACS Appl Mater Interfaces 8:20422–20431

Kramer KW, Biner D, Frei G (2004) Hexagonal sodium yttrium fluoride based green and blue emitting upconversion phosphors. Chem Mater 16:1244–1251

Kubrakova IV, Toropchenova ES (2008) Microwave heating for enhancing efficiency of analytical operations (Review). Inorg Mater 44:1509–1519

Li C, Quan Z, Yang J (2007) Highly uniform and monodisperse β-$NaYF_4:Ln^{3+}$ (Ln = Eu, Tb, Yb/Er, and Yb/Tm) hexagonal microprism crystals: hydrothermal synthesis and luminescent properties. Inorg Chem 46:6329–6337

Li LL, Zhang R, Yin L (2012) Biomimetic surface engineering of lanthanide-doped upconversion nanoparticles as versatile bioprobes. Angew Chemie Int Ed 51:6121–6125

Liang ZQ, Zhao SL, Cui Y (2015) Phase transformation and morphology tuning of β-$NaYF_4$: Yb^{3+}, $E^{r3}+$ nanocrystals through K^+ ions codoping. Chinese Phys B 24:37801

Lim SF, Riehn R, Ryu WS (2006) In vivo and scanning electron microscopy imaging of Upconverting nanophosphors in caenorhabditis elegans. Nano Lett 6:169–174

Liu K, Liu X, Zeng Q (2012) Covalently assembled NIR nanoplatform for simultaneous fluorescence imaging and photodynamic therapy of cancer cells. ACS Nano 6:4054–4062

Liu Q, Li C, Yang T (2010) "Drawing" Upconversion nanophosphors into water through host–guest interaction. Chem Commun 46:5551–5553

Liu Y, Lu Y, Yang X (2017) Amplified stimulated emission in upconversion nanoparticles for super-resolution nanoscopy. Nature 543:229–233

Lu Q, Guo F, Sun L (2008) Silica-/titania-coated $Y2O_3$: Tm^{3+}, Yb^{3+} nanoparticles with improvement in Upconversion luminescence induced by different thickness shells. J Appl Phys 103:123533

Ma Y, Ji Y, You M (2016) Labeling and long-term tracking of bone marrow mesenchymal stem cells in vitro using $NaYF_4$: Yb^{3+}, E^{r3}+ Upconversion nanoparticles. Acta Biomater 42:199–208

Mader HS, Kele P, Saleh SM, Wolfbeis OS (2010) Upconverting luminescent nanoparticles for use in bioconjugation and bioimaging. Curr Opin Chem Biol 14:582–596

Mai HX, Zhang YW, Si R (2006) High-quality sodium rare-earth fluoride nanocrystals: controlled synthesis and optical properties. J Am Chem Soc 128:6426–6436

Mai HX, Zhang YW, Sun LD, Yan CH (2007) Highly efficient multicolor up-conversion emissions and their mechanisms of monodisperse $NaYF_4$: Yb, Er core and core/Shell-structured nanocrystals. J Phys Chem C 111:13721–13729

Mi C, Tian Z, Cao C (2011) Novel Microwave-assisted solvothermal synthesis of $NaYF_4$:Yb, Er Upconversion nanoparticles and their application in cancer cell imaging. Langmuir 27:14632–14637

Nichkova M, Dosev D, Gee SJ (2005) Microarray immunoassay for phenoxybenzoic acid using polymer encapsulated Eu: Gd_2O_3 nanoparticles as fluorescent labels. Anal Chem 77:6864–6873

Nyk M, Kumar R, Ohulchanskyy TY (2008) High contrast in vitro and in vivo photoluminescence bioimaging using near infrared to near infrared up-conversion in Tm^{3+} and Yb^{3+} doped fluoride nanophosphors. Nano Lett 8:3834–3838

Panda AB, Glaspell G, El-Shall MS (2007) Microwave Synthesis and optical properties of uniform nanorods and nanoplates of rare earth oxides. J Phys Chem C 111:1861–1864

Park Y, Kim HM, Kim JH (2012) Theranostic probe based on lanthanide-doped nanoparticles for simultaneous in vivo dual-modal imaging and photodynamic therapy. Adv Mater 24:5755–5761

Park YI, Kim JH, Lee KT (2009) Nonblinking and nonbleaching upconverting nanoparticles as an optical imaging nanoprobe and T1 magnetic resonance imaging contrast agent. Adv Mater 21:4467–4471

Rao L, Bu LL, Cai B (2016) Cancer cell membrane-coated upconversion nanoprobes for highly specific tumor imaging. Adv Mater 28:3460–3466

Rillings KW, Roberts JE (1974) A thermal study of the trifluoroacetates and pentafluoropropionates of praseodymium, samarium and erbium. Thermochim Acta 10:269–277

Russel C (1993) A pyrolytic route to fluoride glasses. Preparation and thermal decomposition of metal trifluoroacetates. J Non Cryst Solids 152:161–166

Schafer H, Ptacek P, Eickmeier H, Haase M (2009) Synthesis of hexagonal Yb^{3+}, E^{r3}+-doped NaY_F4 nanocrystals at low temperature. Adv Funct Mater 19:3091–3097

Scheps R (1996) Upconversion laser processes. Prog Quantum Electron 20:271–358

Schietinger S, Aichele T, Wang HQ (2010) Plasmon-enhanced upconversion in single $NaYF_4$: Yb^{3+}/Er^{3+} codoped nanocrystals. Nano Lett 10:134–138

Shao YZ, Liu LZ, Song S (2011) A novel one-step synthesis of Gd^{3+}-incorporated mesoporous SiO_2 nanoparticles for use as an efficient MRI contrast agent. Contrast Media Mol Imaging 6:110–118

Shikha S, Zheng X, Zhang Y (2017) Upconversion nanoparticles-encoded hydrogel microbeads-based multiplexed protein detection. Nano-Micro Lett 10:31

Sivakumar S, van Veggel FCJM, Raudsepp M (2005) Bright White light through up-Conversion of a single NIR source from sol−gel-derived thin film made with Ln^{3+}-Doped LaF_3 Nanoparticles. J Am Chem Soc 127:12464–12465

Son A, Dhirapong A, Dosev DK (2008) Rapid and quantitative DNA analysis of genetic mutations for polycystic kidney disease (PKD) using magnetic/luminescent nanoparticles. Anal Bioanal Chem 390:1829–1835

Sun Y, Chen Y, Tian L (2007) Controlled synthesis and morphology dependent upconversion luminescence of NaYF4: Yb Er nanocrystals. Nanotechnology 18:275609

Sun Y, Liu H, Wang X (2006) Optical spectroscopy and visible upconversion studies of YVO_4: Er^{3+} nanocrystals synthesized by a hydrothermal process. Chem Mater 18:2726–2732

Suyver JF, Aebischer A, Biner D (2005a) Novel materials doped with trivalent lanthanides and transition metal ions showing near-infrared to visible photon upconversion. Opt Mater (Amst) 27:1111–1130

Suyver JF, Grimm J, Kramer KW, Gudel HU (2005b) Highly efficient near-infrared to visible up-conversion process in $NaYF_4$: Er^{3+}, $Y^{b3}+$. J Lumin 114:53–59

Suyver JF, Grimm J, van Veen MK (2006) Upconversion spectroscopy and properties of $NaYF_4$ doped with Er^{3+}, Tm^{3+} and/or Yb^{3+}. J Lumin 117:1–12

Tsang MK, Ye W, Wang G (2016) Ultrasensitive detection of Ebola Virus oligonucleotide based on Upconversion nanoprobe/nanoporous membrane system. ACS Nano 10:598–605

van de Rijke F, Zijlmans H, Li S (2001) Up-converting phosphor reporters for nucleic acid microarrays. Nat Biotechnol 19:273–276

Vetrone F, Capobianco JA (2008) Lanthanide-doped fluoride nanoparticles: luminescence, upconversion, and biological applications. Int J Nanotechnol 5:1306–1339

Wang C, Cheng L, Xu H, Liu Z (2012) Towards whole-body imaging at the single cell level using ultra-sensitive stem cell labeling with oligo-arginine modified upconversion nanoparticles. Biomaterials 33:4872–4881

Wang F, Banerjee D, Liu Y (2010) Upconversion nanoparticles in biological labeling, imaging, and therapy. Analyst 135:1839–1854

Wang F, Deng R, Wang J (2011) Tuning upconversion through energy migration in core–shell nanoparticles. Nat Mater 10:968–973

Wang F, Han Y, Lim CS (2010) Simultaneous phase and size control of Upconversion nanocrystals through lanthanide doping. Nature 463:1061–1065

Wang F, Liu X (2008) Upconversion multicolor fine-tuning: visible to near-infrared emission from lanthanide-doped $NaYF_4$ Nanoparticles. J Am Chem Soc 130:5642–5643

Wang F, Liu X (2009) Recent advances in the chemistry of lanthanide-doped upconversion nanocrystals. Chem Soc Rev 38:976–989

Wang HQ, Nann T (2009) Monodisperse upconverting nanocrystals by microwave-assisted synthesis. ACS Nano 3:3804–3808

Wang L, Li P, Wang L (2009) Luminescent and hydrophilic LaF_3–polymer nanocomposite for DNA detection. Luminescence 24:39–44

Wang L, Li Y (2007) Controlled synthesis and luminescence of lanthanide doped $NaYF_4$ nanocrystals. Chem Mater 19:727–734

Wang L, Yan R, Huo Z (2005a) Fluorescence resonant energy transfer biosensor based on Upconversion-luminescent nanoparticles. Angew Chemie Int Ed 44:6054–6057

Wang X, Zhuang J, Peng Q, Li Y (2005b) A general strategy for nanocrystal synthesis. Nature 437:121–124

Wei Y, Lu F, Zhang X, Chen D (2006) Synthesis of oil-dispersible hexagonal-phase and hexagonal-Shaped $NaYF_4$:Yb, Er nanoplates. Chem Mater 18:5733–5737

Wisser MD, Chea M, Lin Y (2015) Strain-induced modification of optical selection rules in Lanthanide-Based Upconverting nanoparticles. Nano Lett 15:1891–1897

Xie X, Gao N, Deng R (2013) Mechanistic investigation of photon upconversion in Nd^{3+}-sensitized core–shell nanoparticles. J Am Chem Soc 135:12608–12611

Xie X, Liu X (2012) Upconversion goes broadband. Nat Mater 11:842–843

Xing H, Bu W, Zhang S (2012) Multifunctional nanoprobes for upconversion fluorescence, MR and CT trimodal imaging. Biomaterials 33:1079–1089

Xu H, Cheng L, Wang C (2011) Polymer encapsulated upconversion nanoparticle/iron oxide nanocomposites for multimodal imaging and magnetic targeted drug delivery. Biomaterials 32:9364–9373

Xu X, Wang Z, Lei P (2015) α-NaYb(Mn)F_4: Er^{3+}/Tm^{3+}@$NaYF_4$ UCNPs as "Band-Shape" luminescent nanothermometers over a wide temperature range. ACS Appl Mater Interfaces 7:20813–20819

Yang D, Dai Y, Liu J (2014) Ultra-small $BaGdF_5$-based upconversion nanoparticles as drug carriers and multimodal imaging probes. Biomaterials 35:2011–2023

Yang J, Li C, Cheng Z (2007) Size-tailored synthesis and luminescent properties of one-dimensional Gd_2O_3:Eu^{3+} nanorods and microrods. J Phys Chem C 111:18148–18154

Ye X, Collins JE, Kang Y (2010) Morphologically controlled synthesis of colloidal upconversion nanophosphors and their shape-directed self-assembly. Proc Natl Acad Sci 107: 22430–22435

Yi G, Lu H, Zhao S (2004) Synthesis, characterization, and biological application of size-controlled nanocrystalline $NaYF_4$: Yb, Er infrared-to-visible up-conversion phosphors. Nano Lett 4:2191–2196

Yi GS, Chow GM (2006) Synthesis of Hexagonal-Phase $NaYF_4$: Yb, Er and $NaYF_4$: Yb, Tm nanocrystals with efficient up-conversion fluorescence. Adv Funct Mater 16:2324–2329

Yi GS, Chow GM (2007) Water-soluble $NaYF_4$: Yb, Er(Tm)/NaY_F4/polymer core/shell/shell nanoparticles with significant enhancement of upconversion fluorescence. Chem Mater 19:341–343

Zeng JH, Su J, Li ZH (2005) Synthesis and upconversion luminescence of hexagonal-phase $NaYF_4$: Yb, Er^{3+} phosphors of controlled size and morphology. Adv Mater 17:2119–2123

Zhan Q, Liu H, Wang B (2017) Achieving high-efficiency emission depletion nanoscopy by employing cross relaxation in upconversion nanoparticles. Nat Commun 8:1058

Zhan Q, Qian J, Liang H (2011) Using 915 nm Laser Excited Tm^{3+}/Er^{3+}/Ho^{3+}-doped $NaYbF_4$ upconversion nanoparticles for in vitro and deeper in vivo bioimaging without overheating irradiation. ACS Nano 5:3744–3757

Zhang C, Chen J (2010) Facile EG/ionic liquid interfacial synthesis of uniform RE^{3+} doped $NaYF_4$ nanocubes. Chem Commun 46:592–594

Zhang H, Li Y, Ivanov IA (2010) Plasmonic modulation of the upconversion fluorescence in NaYF4:Yb/Tm hexaplate nanocrystals using gold nanoparticles or nanoshells. Angew Chemie Int Ed 49:2865–2868

Zhao J, Sun Y, Kong X (2008) Controlled synthesis, formation mechanism, and great enhancement of red upconversion luminescence of $NaYF_4$: Yb^{3+}, Er^{3+} nanocrystals/submicroplates at low doping level. J Phys Chem B 112:15666–15672

Zhu X, Zhou J, Chen M (2012) Core–shell Fe_3O_4@NaLuF4: Yb, Er/Tm nanostructure for MRI, CT and upconversion luminescence tri-modality imaging. Biomaterials 33:4618–4627

Zhu YJ, Wang WW, Qi RJ, Hu XL (2004) Microwave-assisted synthesis of single-crystalline Tellurium nanorods and nanowires in ionic liquids. Angew Chemie Int Ed 43:1410–1414

Zou W, Visser C, Maduro JA (2012) Broadband dye-sensitized upconversion of near-infrared light. Nat Photonics 6:560–564

Advances in Amphiphilic Assemblies and Its Immobilization in Room Temperature Supercooled Matrices

S. L. Gawali, S. B. Shelar, S. D. Kulkarni, and P. A. Hassan

Abstract Equilibrium nanostructures such as micelles are formed by the assembly of small amphiphilic molecules in appropriate solvents. Nature employs such a self-assembly process to create macroscopic objects with diverse micro/mesoscale structures. Micelles are one of the primitive self-assembling structures formed by amphiphiles in water. The interplay of various attractive and repulsive forces between amphiphiles leads to the formation of such equilibrium structures. The microstructure and interfacial characteristics of micelles can be conveniently tuned by the choice of amphiphiles, additives, etc. Micelles find extensive applications in a variety of products that we use in our everyday life. Control over the structure of these aggregates and macroscopic properties is important to fine-tune them for industrial applications. Owing to the colloidal nature, micelles undergo random Brownian motion in conventional solvents. Immobilization of such colloidal assemblies in selected solvents can be achieved by exploiting room temperature supercooled solvents as a matrix for self-assembly. This review summarizes the basic principles of self-assembly to form micelles and related structures, its application in diverse fields including drug delivery and recent developments in creating immobilized nanoscale assemblies.

Keywords Immobilization · Self-assembly · Colloids · Small angle X-ray scattering · Supercooled micelles

S. L. Gawali · S. B. Shelar · P. A. Hassan (✉)
Chemistry Division, Bhabha Atomic Research Centre, Trombay, Mumbai 400085, India
e-mail: hassan@barc.gov.in

S. L. Gawali · P. A. Hassan
Homi Bhabha National Institute, Training School Complex, Anushaktinagar, Mumbai 400094, India

S. D. Kulkarni
Chemistry Department, Sir Parashurambhau College, Pune 411030, India

A. Tripathi and J. S. Melo (eds.), *Immobilization Strategies*,
Gels Horizons: From Science to Smart Materials,
https://doi.org/10.1007/978-981-15-7998-1_10

Abbreviations

CMC	Critical micelle concentration
CPAs	Cryoprotective agents
CPP	Critical packing parameter
CTAB	Cetyltrimethylammonium bromide
DLS	Dynamic light scattering
DMSO	Dimethyl sulfoxide
HR-TEM	High-resolution transmission electron microscopy
SANS	Small-angle neutron scattering
SAXS	Small angle X-ray scattering
SDS	Sodium dodecyl sulfate
SLD	Scattering length density

1 Introduction

The name surfactant (also called amphiphile) arises as an acronym for 'surface-active agents' which exhibit interfacial activity (Hiemenz and Rajagopalan 1997; Myers 1999). They can stabilize the interface between different phases like liquid–liquid or liquid–gas or solid–liquid by reducing the surface tension of a single-phase or reducing the interfacial tension between two immiscible phases like oil in water (O/W) or water in oil (W/O) (Rosen 1989; Schramm 2000). Structurally, surfactants possess two distinct groups having opposing solubility tendencies. Typically, a water-soluble polar group (hydrophilic part) commonly called 'head group' and nonpolar part designated as 'tail group' (Myers 1991; Shaw 1992). The head group of the surfactant can be charged or uncharged, which contains heteroatoms such as O, S, P, or N, included in functional groups such as alcohol, thiol, ether, ester, acid, sulfonate, sulfate, phosphate, amine, amide, etc. On the other hand, the tail part is usually a hydrocarbon chain but may also be a fluorocarbon or siloxane chain of appropriate length. This hydrophilic–hydrophobic combination in surfactant molecules, often called as amphiphilic molecules (amphis = both) and (philia = love, friendship), makes it unique in its properties. Due to this dual solubility nature, amphiphilic molecules are associated with many useful interfacial phenomena and lead to spontaneous self-association in selected solvents giving rise to a variety of microstructures. One way of classifying surfactants is based on the nature of the hydrophilic group present on them (Myers 1991). Depending on the chemical structure of the hydrophilic head group, surfactants are basically classified as anionic, cationic, nonionic, and zwitterionic/amphoteric surfactant (see Fig. 1).

Anionic surfactants: The head group of anionic surfactants carries a negative charge, and this category includes the traditional long-chain carboxylate soaps and synthetic

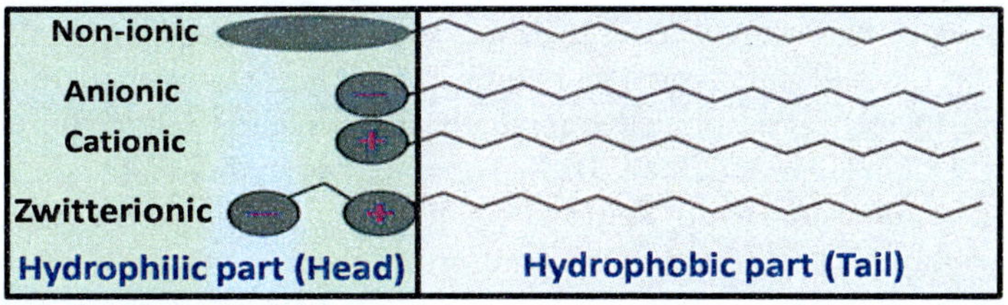

Fig. 1 Schematic illustration of different types of surfactants based on the type of hydrophilic head group

detergents, sulfates ($-OSO_3^-$) and sulfonates ($-SO_3^-$). These are excellent detergents and are extensively used in cleaning formulations.

Cationic surfactants: The surfactant molecule bears a positive charge on the head group, and they are usually quaternary ammonium salts ($R_4N^+X^-$), salt of long-chain amine ($RNH_3^+Cl^-$) or alkyl pyridinium compounds. Owing to the positive charge on the head group, they have a strong affinity to negatively charged fibers such as cotton and hair, and hence, they are used as hair conditioners and fabric softeners.

Nonionic surfactants: These surfactants do not carry any charge on the head group, and their water solubility is derived from polar groups like alkyl ethoxylates. This class of surfactants includes several semi-polar compounds such as amine oxides, sulphoxides, phosphine oxides, alkanolamides, pyrrolidones. Nonionic surfactants are regularly used as emulsifiers and in low-temperature detergency.

Zwitterionic surfactants: These surfactants possess both positively and negatively charged groups in their hydrophilic part, and hence, they can act as either anionic or cationic depending on the pH of the solution. This class of surfactant includes long-chain amino acid, betaines, sulfobetaines, naturally occurring surfactants of the class lecithin and phosphatidyl cholines, etc. These compounds are milder on the skin and have low eye-sting efficacy which leads to their use in toiletries and baby shampoos.

2 Modulating the Microstructure of Micelles

The widespread importance of surfactants in practical applications, and scientific interest in their nature and properties, has precipitated a wealth of published literature on the subject. Surfactants are employed in a variety of applications including detergents, dispersants, wetting agents, emulsifiers, foaming agents, corrosion inhibitors, bactericides, antistatic agents, etc. Owing to the amphiphilic nature, they have the ability to accumulate at the interface in a specific orientation as well as form self-assembled structures such as membrane, microemulsion, liquid crystal, liposome, micelles, vesicle, worm-like micelles, gel in selected solvents (Lombardo et al. 2015).

Such association phenomenon arises from the balance of various non-covalent forces acting among the molecules, such as hydrogen bonding, electrostatic interactions, hydrophobic interactions, solvophobic, steric and depletion interactions, van der Waals, π-π stacking (Ma and Zhao 2015). The extensive hydrogen-bonding network in water and its associated hydrophobic effect makes it an extensively studied solvent for self-assembly. Nature employs such a self-assembly phenomenon to create macroscopic objects. Association of lipids to form cell membranes, bile salt aggregation during digestion, stabilization of fat globules in milk, and attainment of the tertiary structure of proteins from linear strands of covalently linked amino acids, etc., are some of the well-known examples of self-assembly in nature (Patton and Carey 1979; Mezzenga et al. 2005; Capito et al. 2008). In this respect, the amphiphilic molecules become an attractive candidate and provide effective opportunities for designing novel materials for advanced applications in the biomedical field.

The self-assembled structure such as micellar aggregates is formed by the interplay of opposing forces: the attractive tail–tail hydrophobic interaction provides the driving force for the aggregation of the amphiphile molecules, while the electrostatic repulsion between the hydrophilic head groups limits the packing of the molecules in an aggregate. Micelle formation occurs only when the given concentrations of amphiphile molecules exceed above certain concentration called critical micelle concentration (CMC). Micelle is an aggregate formed by surfactant in equilibrium with the molecules or ions that contributes to micelles formation in a specific solvent above the CMC (Rana et al. 2017; de Gennaro 2019). Therefore, they are thermodynamically stable structures with well-defined aggregation number and size distribution. Micelle formation has been reported in a large variety of non-aqueous solvents. This has demonstrated that hydrophobic interaction is not unique to water, but a special case of a more general solvophobic effect (Ray 1971). Some of the non-aqueous solvents that have been explored as alternative media for micelle formation are glycerol, ethylene glycol, formamide, molten salts, ionic liquids, and deep eutectic solvents (Ray 1971; He et al. 2006). Micelles are used in diverse fields including drug delivery, oil recovery, extraction, etc., and to control order or porosity in inorganic materials (Li et al. 2011; Xiao et al. 2012). Typically, when water is the solvent, the nonpolar portions of the amphiphiles are in the interior and the polar portions at the exterior surface, exposed to water. However, if the solvent is relatively nonpolar, then the orientation of the hydrophilic groups in the micelle is reversed and hence called reverse micelle. The polar heads are at the interior, and the hydrophobic chains are protruding to the exterior. Cartoons of a micellar aggregate formation in water and nonpolar solvent (oil) are shown in Fig. 2.

Often, lyotropic phases are observed when the concentration of amphiphilic molecules in solvents is increased, as schematically shown in Fig. 3. Below the CMC, amphiphiles are molecularly dispersed in the solvent; but with increasing the concentrations, they exhibit a rich phase behavior with an entire gamut of structures such as micelles, which can be of the spherical, disk, ribbons, rod type, hollow capsules, and so on depending on the molecular structure. At even higher concentrations, these micellar aggregate changes to more ordered structures and can form hexagonal, cubic, or lamellar phases (Wärnheim and Jönsson 1988; Wolff and Klaussner

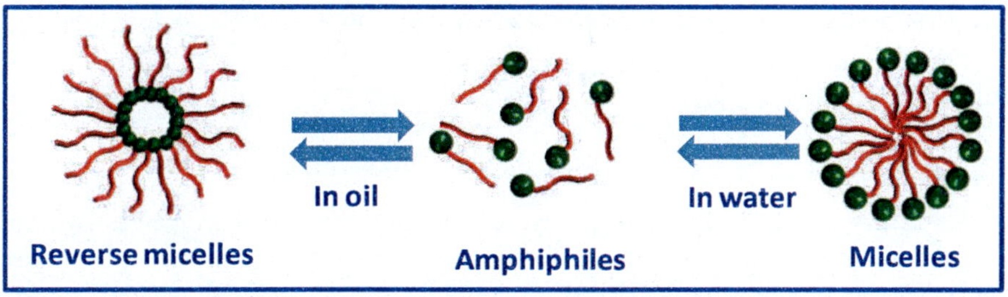

Fig. 2 Schematic illustration of micelle formation through intermolecular forces in the different solvent systems

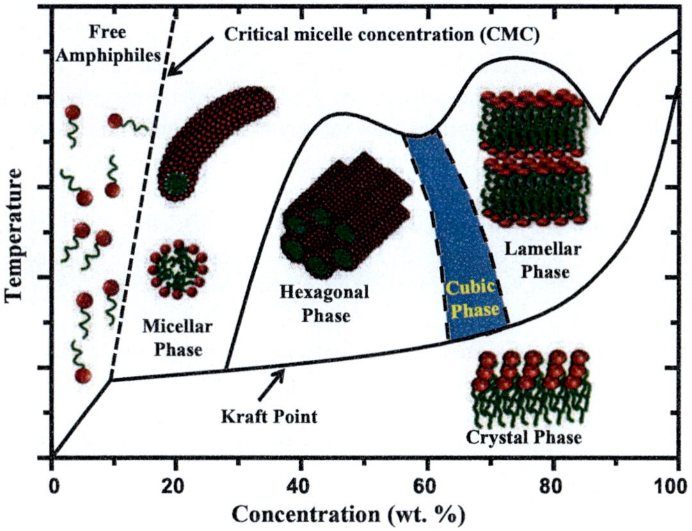

Fig. 3 Schematic illustrations of the concentration-dependent phase diagram of an amphiphilic surfactant in the aqueous solvent. It shows the formation of spherical or rod-like micelles above a critical concentration (CMC). Further increase in the surfactant concentrations leads to liquid crystalline phases, namely hexagonal and lamellar phases (Redrawn from Auvray et al. 1989)

1995). The observed phase diagrams can be quite complex and characteristics of these aggregates are not only depending on concentration, but also on solvent, pH, temperature, ionic strength, type and concentration of additives, and so on (Dierking and Al-Zangana 2017; Mayans et al. 2017; Kundu et al. 2017).

A theoretical description of the formation of equilibrium aggregates with finite aggregation number and its dependence on the molecular structure of amphiphiles has been given in the primary work of Tanford (Tanford 1973) and Israelachivli (Israelachvili et al. 1976). According to Tanford, the standard free energy change per molecule upon aggregation involves contribution from three terms, namely the transfer-free energy, interfacial free energy, and the repulsive interaction energy between the head groups (Tanford 1973; Nagarajan 2002).

$$\left(\frac{\Delta\mu_g^0}{kT}\right) = \left(\frac{\Delta\mu_g^0}{kT}\right)_{\text{Transfer}} + \left(\frac{\Delta\mu_g^0}{kT}\right)_{\text{Interface}} + \left(\frac{\Delta\mu_g^0}{kT}\right)_{\text{Head}} \tag{1}$$

The first term in the free energy expression accounts for the changes in free energy associated with the transfer of the hydrocarbon tail form an unfavorable contact with water molecules to that of a lipophilic environment in the micelles core. As it removes the unfavorable contact of hydrocarbon tail with water, this leads to a negative contribution to the free energy. The second term represents the interfacial region between the lipophilic core of the micelle and the polar solvent. This gives a positive free energy contribution due to the formation of an interface. The third term accounts for the repulsive interactions arising from the steric effects of solvated head groups which will tend to keep the hydrophilic part away from each other. In the case of ionic surfactants, there is an additional contribution from the electrostatic repulsion of the charged head groups. Again, these factors give rise to an increase in the free energy of the system. The optimum geometry of the aggregate can be obtained from the minimization of the total free energy per molecule, and this optimal packing of the molecules in the aggregate can be linked to the effective head group area of surfactant. Considering the transfer-free energy to be independent of the head group area of the surfactant, the optimum packing of the molecule is decided by the last two terms in the above equation, since the interfacial free energy per molecule is a product of interfacial tension (γ) and the area per molecule (a). Assuming the repulsive energy varies inversely with the area per molecule, one can write the above free energy Eq. 1 as,

$$\left(\frac{\Delta\mu_g^0}{kT}\right) = \left(\frac{\Delta\mu_g^0}{kT}\right)_{\text{Transfer}} + \left(\frac{\gamma}{kT}\right)a + \left(\frac{\alpha}{kT}\right)\frac{1}{a} \tag{2}$$

where α is the repulsion parameter, T is the temperature and$k$ is the Boltzmann constant.

Using Eq. 2, one can construct a free energy plot as a function of area per molecule. Obviously, such plot shows a minimum in the energy, corresponding to the optimum area per molecule (a_0), as shown in Fig. 4a. It may be noted that the position of the minimum can be varied by changing the strength of the interaction parameter (Fig. 4b). This clearly shows that the optimum area per molecule can be varied by maneuvering the intermolecular interaction which in turn affects the packing of the molecules in the aggregate.

In addition to the above factors, the contributions from the tails of the surfactant towards the packing of the molecules are also important. The optimal packing of molecules in the aggregate led to the development of a predictive geometrical parameter known as critical packing parameter (CPP) which is used as a guiding principle to forecast the geometry of aggregates. The packing parameter of an amphiphile can

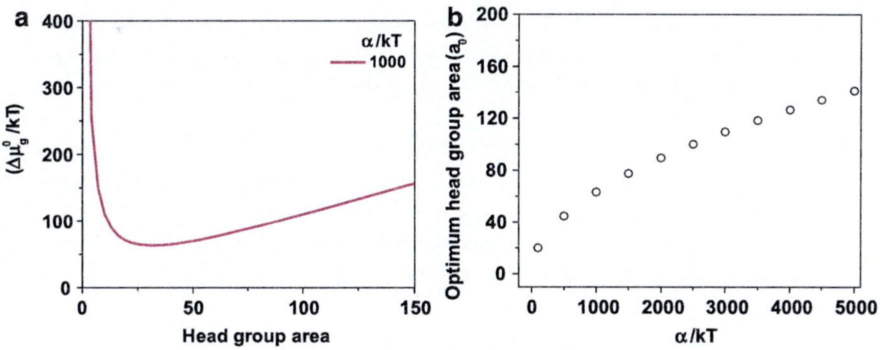

Fig. 4 **a** Simulated curves of change in the free energy per molecule in aggregate as a function of the head group area per amphiphile. The transfer-free energy is taken as zero, the interface energy term (γ/kT) as 1 units and the repulsive interaction term (α/kT) as 1000 units, **b** Variation of optimum head group area per molecule with a repulsive interaction term (α /kT) from 100 to 5000 units

be defined as:

$$C_{PP} = \frac{v}{a_0 l} \tag{3}$$

where v is the volume of the hydrophobic part, l is the length of the hydrophobic chain, and the a_0 is the effective head group area of the surfactant molecule. Schematic illustration of the corresponding changes in the packing of molecules to form structures such as spheres, rods, vesicles, and bilayers is shown in Fig. 5. From Eqs. 2 and 3, it is

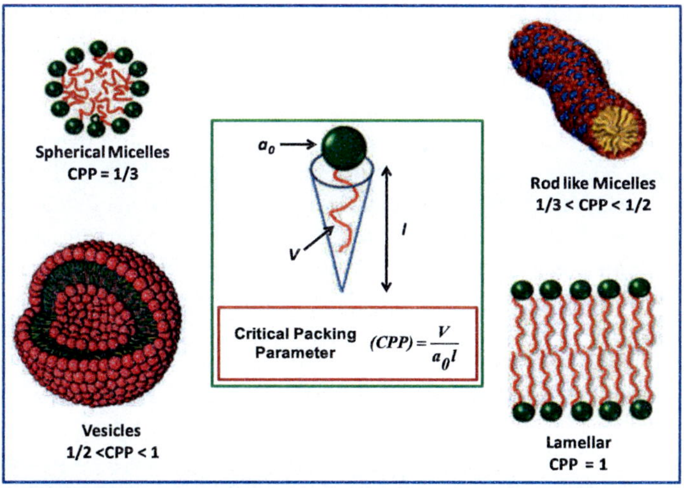

Fig. 5 Schematic of different shape aggregates formed by amphiphiles with different values of the critical packing parameter

obvious that one can modulate the equilibrium area and hence the CPP by changing the molecular repulsion parameter. Therefore, it is possible to tune the geometry of nanostructure architectures from spherical micelles ($C_{pp} \leq 1/3$) to rod/cylinder-like micelles ($1/3 \leq C_{pp} \leq 1/2$), vesicles ($1/2 \leq C_{pp} \leq 1$) and lamellar structures ($C_{pp} = 1$), by changing the experimental conditions (Degiorgio 1985). It is also possible to form aggregates with larger values of CPP (say $C_{pp} > 1$), in which case the amphiphiles will assemble into 'inverted' phases (Lombardo et al. 2015).

The use of the CPP model to explain morphological changes in micelles from spheres to rods, worms, and vesicles have been satisfactorily employed in a wide variety of systems. For example, nearly globular micelles can be formed by conventional surfactants like CTAB and SDS, at concentrations close to the CMC. However, these micelles transform into rod-shaped aggregates with an increase in surfactant concentration or addition of electrolytes (Hassan et al. 2002). This is due to the fact that in the absence of any additives, the presence of both steric and electrostatic repulsion of the head groups makes the effective area large and structures with low CPP are preferred near the CMC. The ionic strength of the medium increases with the addition of salts leading to a decrease in the Debye screening length and hence reduces the electrostatic repulsion. This in turn decreases the effective area of the head group a_0 and the micelle structure will be transformed into rods. The sphere-to-rod transition in micelles is also observed in ionic surfactants with an increase in surfactant concentrations. This arises mainly from an increase in the number density of counterions arising from the dissociation of amphiphiles. The micellar structure can also be changed by the addition of alcohols, phospholipids, etc. It may be noted that the importance of molecular packing in dictating the morphology is not limited to the concept of CPP. A comparison of the different models to predict the geometry of aggregates indicates that a direct correlation between the CPP model and other approaches such as hydrophile–lipophile balance or the spontaneous curvature model, used in predicting micro-emulsion phase behavior, can be arrived (Kunz et al. 2009). This approach of controlling the micelle morphology by molecular packing has been employed in a variety of systems including cationic–anionic pairs through changes in the pH of the medium (Lin et al. 2008; Ali et al. 2009).

The self-assembly-based approach offers various equilibrium nanostructures in the liquid phase. This is particularly useful in terms of the stability of the nanophase materials. Conventional nano-phase materials are kinetically stabilized. This is due to the inherent instability of nanophase materials due to the increased surface energy contributions. This has been addressed by electrostatic or steric stabilization of the nanoparticles to prevent agglomeration. Nanophase materials formed by the self-assembly of small molecules such as micelles and micro-emulsions are equilibrium structures and hence possess a potential energy minimum. However, one disadvantage of such aggregates is that they are dynamic in nature. The dynamic processes involve an exchange of monomers with bulk and Brownian motion mediated collision of the aggregates. Because of the dynamic nature, they are susceptible to dissociation upon changes in the external conditions. Lipids are also another class of amphiphiles that are capable of forming aggregates such as liposomes. Often, these types of aggregates are used in trapping drugs physically within the hydrophobic cores or linking drugs covalently to component molecules of the amphiphile. There are several examples of such systems being explored for drug delivery. They proved to be an excellent novel drug delivery system, at least in some of the cases, due to their compatibility and stability in physiological conditions, high loading capacity, and high accumulation of drugs at the target site. Ambisome, Doxil, etc., are some of the drug formulations currently available in the market. Quang et al. reported the formation of self-assembled amphiphilic micelles based on chitosan and polycaprolactone and are used as carriers of paclitaxel to improve its intestinal pharmacokinetic profile (Vu-Quang et al. 2016). Polymeric micelles composed of di-block copolymers mono-methoxy poly(ethylene glycol)-poly(lactide) were reported as a nanocarrier for solubilization and delivery of a promising anticancer drug Ethaselen (Li et al. 2009). Also, different self-assembled structures of amphiphiles are commonly used in the diagnostics and delivery of active ingredients (Langer 2001; Kabanov and Alakhov 2002; Salem et al. 2003).

3 Immobilization of Self-assemblies

As discussed earlier, colloidal materials undergo random Brownian motion under ambient conditions. One of the ways to stabilize these nanomaterials is to arrest the Brownian motion of the colloids to prevent collision mediated agglomeration. The study of the structural arrest of colloids has been the subject of intense investigation both theoretically (Gotze and Sjogren 1992; Angell 1997) and experimentally (Romer et al. 2000; Grandjean and Mourchid 2004). Colloids have been employed as a model system to understand the liquid to glass transition driven by jamming (also called immobilization). Immobilization refers to the structural arrest of the dynamic motion of colloidal particles in the solvent medium. Often jammed colloids are formed by increasing the volume fraction of the colloids. When the volume fraction of a colloidal suspension is increased beyond 0.5, its low-shear viscosity increases markedly, and beyond critical shear stress, the system often displays a shear-thickening or jamming transition (Smith et al. 2010). Jammed/immobilized colloidal particles are found in many industrial formulations which include toothpaste, paints, inks, cosmetics, etc. (Mattsson et al. 2009; Zheng et al. 2014; Holmqvist 2014).

There are various day-to-day life examples of immobilized colloidal particles. Since jamming depends on the degree to which particles block each other from rearranging inside a given volume, it does not require any particular particle size or type. There is no major structural transition associated with jamming: The system is amorphous on either side of the transition. These two aspects have made the study of colloidal jamming as a model for glassy behavior observed in a wide range of systems (Jaeger 2015). Jammed colloidal systems are often observed in many industrial processes. Milk to yogurt conversion, starch gelatinization, shaving foams, mayonnaise, etc., are examples of jammed colloids. Milk is an oil-in-water emulsion, with the fat globules dispersed in a continuous phase. The fat would rise and form a cream layer with jammed fat globules when raw milk was left to stand. Such coagulation of fat can be slowed down by homogenization. In the homogenization treatment, the fat globules in milk are passed through a tiny orifice under high pressure, which results in a decrease in the average size and an increase in number density and surface area, of the fat globules. This results in more stable colloidal suspension, which has a much-reduced tendency for the creaming of fat globules. Three factors contribute to this enhanced stability of homogenized milk: (1) a decrease in the mean diameter of the fat globules, (2) a decrease in the size distribution of the fat globules, and (3) an increase in number density of the globules (bringing them closer to the continuous phase). In addition, heat pasteurization breaks down the cryo-globulin complex, which tends to cluster fat globules causing them to rise. In the cake batter, starch is an essential component. Starch granules are typically on the order of microns to tens of microns in size. However, depending on the source of starch (potato, corn, wheat, etc.) the sizes and shapes can vary. In the cold water, there is no gelatinization of starch observed. However, if the mixture of water and starch is heated to the gelatinization temperature (typically around 60–70 °C), then the amylose is pushed out of the granules and loses its helices and get the dispersion of amylopectin granules surrounded by a solution of amylose in water. When the water is cooled back down and no stirring occurs, the amylose polymers link the amylopectin granules together, and we get a gel. In the gelatinization, there is a loss of crystalline structure of the starch granule, and essentially turning the starch molecules into a liquid. Once the heat is removed, the starch molecules start to solidify, but they are no longer organized in the same manner as they were before cooking. Amylose molecules can therefore sit together with forming tight junctions. If the starch is cooled very rapidly, there will be the formation of amorphous domains within the crystalline superstructure. Another example of jamming is aqueous foam, which is constituted by closely-packed polydisperse gas bubbles dispersed in a small volume of surfactant solution. Shaving foam, when sprayed on a surface, retains its shape and behaves like a solid, but can flow by the stick–slip motion of the bubbles when gently tapped with a finger (Weitz 1996). Suspensions, which are mixtures of two substances, one of which is finely divided and dispersed in the other (like, colloidal particles in water), and emulsions which are tiny droplets of one material suspended in another immiscible material (e.g., milk, mayonnaise, and paint) exhibit power-law shear-thinning due to flow-induced microstructural rearrangements (Barnes 1997).

4 Immobilization of Micelles Using Supercooled Matrix

Arresting the global motion of micelles or other self-assembled structures in suspension medium is of great relevance in controlling the rate of chemical reactions in micro-heterogenous media, retarding the growth of nanoparticles prepared via micellar route, etc. For example, the formation of nanoparticles in micelles or microemulsions occurs through the collision of droplets and the exchange of the ingredients in the droplets (Magno et al. 2011). In general, the Brownian motion of colloids such as polymer micelles, microgels, emulsion droplets, or proteins gets arrested by increasing the volume fraction (>0.58) of the colloid such that the particles are in contact with each other (Mason et al. 1995; Chen et al. 2003; Stradner et al. 2004). Though jamming of colloids can be achieved by increasing its volume fraction (Smith et al. 2010), in the case of micelles changing the concentration of aggregates has a remarkable effect on its microstructure and is pronounced in the case of assemblies with bulky or aromatic counterions (Hassan et al. 2002; Shi et al. 2013). At moderate temperatures, surfactants like CTAB and SDS in water forms globular micelles at low volume fraction (for CTAB ~0.2 and for SDS ~0.36) and transforms to cylindrical micelles with a hexagonal close-packed arrangement, and lamellar crystals at high concentrations (Kékicheff et al. 1989; Auvray et al. 1989; Zech and Kunz 2011). Thus, an alternate path to arrest the global motion of the micelles is to add other water-soluble ingredients to increase the total volume fraction of solids in the system. Jamming has also been reported in binary mixtures of colloids. Anomalous diffusion due to jammed state in a model binary colloid mixture has shown the small spheres diffuses in a slowly rearranging glassy matrix of the large sphere (Sentjabrskaja et al. 2016).

We demonstrated the jammed state of SDS micelles in a supersaturated solution of sucrose in water using dynamic light scattering (DLS) and small-angle X-ray scattering (SAXS). First, the SAXS measurements were carried out at different surfactant concentrations, in the absence of sucrose. Figure 6a shows the SAXS pattern of SDS micelles at different surfactant concentrations ranging from 5 to 20% at room temperature in the absence of sucrose. At low surfactant concentration, the SAXS pattern shows a broad peak in the q-range between 0.1–0.2 $Å^{-1}$ indicative of the core–shell morphology of micelles. With an increase in concentration, it clearly shows a correlation peak, at lower q value, characteristic of interacting colloids and the peak becomes prominent as concentration increases. However, in the presence of 0.4 M NaCl, the correlation peak appeared at lower q value becomes less prominent due to the screening of intermicellar interaction (Fig. 6b).

The simulated SAXS pattern of SDS micelles using a core–shell ellipsoidal(prolate) micelles model (see solid lines in Fig. 6) yields a hydrocarbon core of semimajor and semiminor radii as 18.4 Å and 13.4 Å, respectively (for 5% aqueous SDS micelles). The interparticle interaction is taken into account using screened Coulomb potential (Hayter and Penfold 1981). At low surfactant concentrations, the effect of intermicellar interactions can be minimized by the addition of electrolytes like NaCl. From measurements at different electrolyte concentrations, it was noted that the addition of 0.4 M NaCl is sufficient to suppress the correlation peak in 5% SDS. The structural parameters of SDS micelles obtained from model fitting the SAXS data with and without NaCl are summarized in Table 1a, b. It may be noted that the structural parameters of the micelles in the presence of NaCl are in the same range as that of micelles in water, with the exception that the charges on micelles in water are much higher than that in the presence of salt. More importantly, the micellar charge observed in the presence of NaCl is very low and hence can be treated as dilute enough to neglect interparticle interaction. This suggests that SDS micelles at a concentration of 5% or lower and in the presence of 0.4 M NaCl are

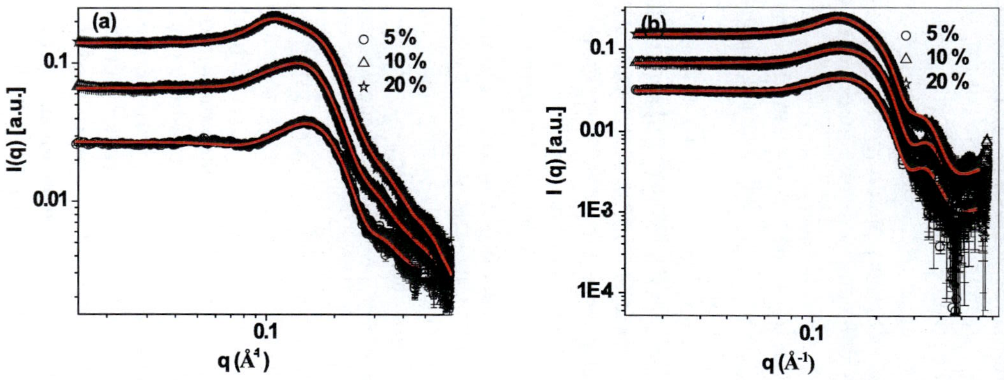

Fig. 6 SAXS pattern from aqueous SDS micelles (5, 10, and 20 wt.%) **a** in water and **b** in 0.4 M NaCl solution. Solid lines show the corresponding simulated scattering curve using core–shell ellipsoidal micelles

Table 1 Structural parameters of aqueous SDS micelles at different surfactant concentrations as obtained from SAXS data analysis using the prolate core–shell ellipsoid model. a. Structural parameters of SDS micelles in water at different concentrations (wt.%). b. Structural parameters of SDS micelles in 0.4 M NaCl solution at different concentrations (wt.%)

SDS (wt.%)	Semi-minor core (Å)	Semi-major core (Å)	Shell thickness (Å)	SLD of shell (10^{-6} Å^{-2})	Charge (e.u.)	Reduced χ^2
a						
5	13.4 ± 0.2	18.4 ± 0.2	12.8 ± 0.2	10.04 ± 0.2	26.8 ± 1.5	1.8
10	13.1 ± 0.2	18.2 ± 0.1	11.0 ± 0.1	10.14 ± 0.1	38.3 ± 2.5	3.7
20	14.9 ± 0.2	20.6 ± 0.2	10.5 ± 0.3	10.52 ± 0.2	40.5 ± 2.6	5.3
b						
5	13.5 ± 0.2	18.1 ± 0.2	12.5 ± 0.2	10.02 ± 0.2	2.8 ± 1.0	1.4
10	13.2 ± 0.3	18.4 ± 0.1	12.2 ± 0.2	10.10 ± 0.1	3.0 ± 1.0	2.3
20	13.5 ± 0.2	18.9 ± 0.2	12.1 ± 0.3	10.17 ± 0.2	4.2 ± 1.2	3.6

dilute enough to study jamming of soft colloids using DLS. Since the autocorrelation function of scattered intensity obtained from DLS is susceptible to interparticle interactions, it is necessary to include an appropriate amount of electrolyte in the system.

DLS is an appropriate tool to monitor the global diffusion of colloids. Jamming of colloids often leads to several orders of magnitude decrease in the diffusion coefficient. Figure 7 shows the normalized intensity autocorrelation functions of SDS micelles (5 wt% 0.4 M NaCl) at different concentrations of sucrose, ranging from 0 to 80% w/w. In the absence of sucrose, nearly single exponential decay in the intensity correlation function was observed with a relaxation time of 16 s. It was observed that with the addition of sucrose, the characteristic decay time increases and an additional

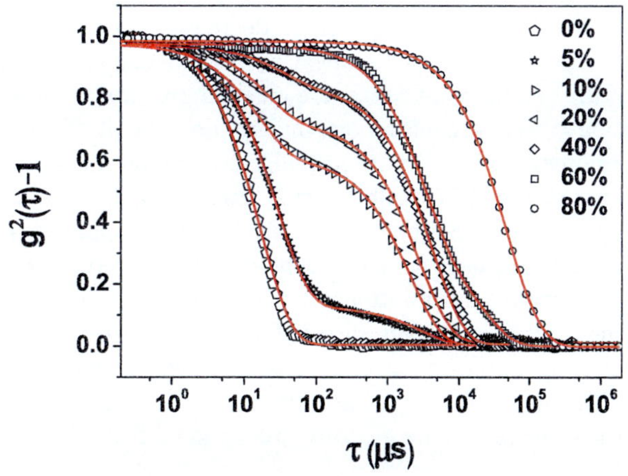

Fig. 7 Normalized autocorrelation function of scattered light from 5% SDS micelles (0.4 M NaCl) at different concentrations of added sucrose (Sucrose concentration from 0 to 80 (wt.%))

slow mode arises in the system. The amplitude of the slow mode becomes prominent at high concentrations of sucrose and the fast mode diminishes. In particular, the correlation function changes from a single exponential to a double exponential function (solid lines) as is observed in the case of arrested macromolecules in a crowded environment. Such double exponential behavior of the correlation function is observed experimentally and predicted from MD simulations in various colloidal glasses (Kwaśniewski et al. 2014; Augusto de Melo Marques et al. 2015; Sentjabrskaja et al. 2016). The fast mode is ascribed to the local motion of the particles. The fast mode is often represented by a single exponential decay while the slow mode is found to follow stretched exponential with the exponent varying with the volume fraction of colloids (Mattsson et al. 2009). It has been reported that the slow relaxation time is related to the structural relaxation of the material (Augusto de Melo Marques et al. 2015). From optical microscopic studies coupled with oscillatory rheometry, Pine and coworkers demonstrated that the dynamics of concentrated emulsions in jammed state captures the signature of the double exponential nature of the autocorrelation function (Knowlton et al. 2014). The walking confined diffusion of the particles in glassy or porous materials is another explanation given to the occurrence of slow mode (Daumas et al. 2003; Ochab-Marcinek et al. 2012). The magnitudes of the relaxation time due to fast and slow modes of the autocorrelation functions in the presence of sucrose can be obtained by fitting with a double exponential function of the form.

$$g^2(\tau) - 1 = A_1 \exp\left(-\frac{\tau}{\tau_1}\right) + A_2 \exp\left(-\frac{\tau}{\tau_2}\right) \tag{4}$$

where the parameters A_1 and A_2 are amplitudes of the two relaxation modes, and τ_1 and τ_2 are the corresponding relaxation times. The solid line in Fig. 7 represents the fit to the data using Eq. 4 Here, the amplitudes and relaxation times of the fast and slow modes are used as variables. The variation in the amplitudes of the fast (A_1) and slow (A_2) modes with sucrose concentration is shown in Fig. 8a. It can be seen that the amplitude of the fast mode decreases at the expense of the slow mode and the variation in it is almost linear at low sucrose concentration. A linear extrapolation of the amplitude of fast mode meets the x-axis as the sucrose concentration reaches about 55%. Previous reports on jamming of colloidal glasses comprising hard-sphere colloids indicate that the colloids undergo jamming transition when the volume fraction reaches 58% (Mattsson et al. 2009). So, the observed trend in the disappearance of the fast mode is consistent with colloidal jamming. The relaxation time due to the fast mode of the correlation function increases by almost six orders of magnitude, as the sucrose concentration reached 80%, clearly indicating jamming of the micelles. The corresponding variation in the micelle diffusion coefficient is depicted in Fig. 8b. As compared to micelles formed in water, a six order of magnitude decrease in the diffusion coefficient of the micelles is observed when the micelles are present in 20% water, the rest being glucose, urea, and surfactant.

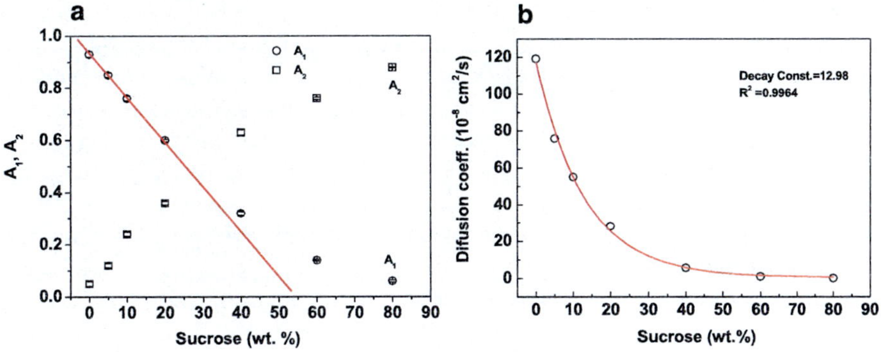

Fig. 8 Variation in the **a** amplitudes and **b** diffusion coefficient of 5% SDS micelle with 0.4 M NaCl of the double exponential decay of the correlation function as a function of sucrose concentration (wt.%)

To correlate the observed changes in the diffusion coefficient of micelles with bulk viscosity of the solution, we performed shear dependent viscosity measurements. According to the Stokes–Einstein equation, the diffusion coefficient of particles is inversely proportional to the viscosity of the medium. Rheological studies indicate that 5% aqueous SDS micelles show Newtonian behavior in water as well as in the presence of sucrose (Fig. 9a). Micelles in water have very low viscosity, being close to that of water. However, after the addition of sucrose in SDS micelles, the viscosity of the medium is increased drastically as indicated in Fig. 9b. The observed change in viscosity with sucrose concentration can be fitted to an exponential function. This exponential change in viscosity correlates well with the restricted motion of micelles.

Jamming transitions are observed in binary colloid mixtures of different particle sizes. There have been efforts to measure the diffusion of small particles in the presence of glasses formed by large particles. It has been observed that the anomalous

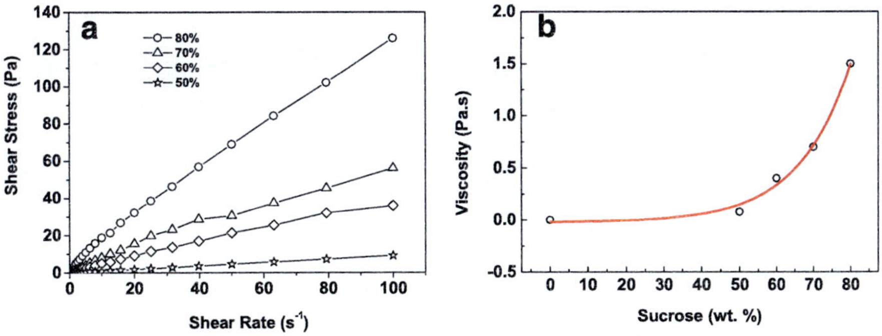

Fig. 9 **a** Variation of shear stress versus shear rate of 5% SDS micelles in water at different concentrations of added sucrose and **b** the corresponding zero shear viscosity of 5% SDS micelles as a function of sucrose

diffusion behavior of the small particles in a matrix of large particles depends on the size asymmetry and probing length scale. (Voigtmann and Horbach 2009). The present system of jammed micelles in sugar solution has an analogy to such colloid mixtures. Here, we monitored the arrested motion of large soft colloids in a crowded matrix of small molecules. Such crowding is observed in binary mixtures of colloids, colloid-polymer mixtures, colloidal particles in biological fluids, etc. (Daumas et al. 2003; Ochab-Marcinek and Hołyst 2011; Tuinier et al. 2015; Toyotama et al. 2016). Single-particle tracking of G-protein coupled receptor in fibroblast cells display such anomalous diffusion behavior due to depletion interaction (Daumas et al. 2003). Thus, the addition of water-soluble ingredients in a supersaturated state offers a convenient strategy to slow down translational diffusion of micelles and other self-assembled systems.

A close examination of the magnitude of the slow relaxation time of micelles in sucrose corresponds to a hydrodynamic correlation length of the order of hundreds of microns. Such slow motion can happen if the micelles undergo growth/phase separation in the presence of sucrose, leading to the formation of large droplets as in the case of clouding of nonionic micelles with heating. Therefore, SAXS measurements were carried out on these samples at different sucrose concentrations. Figure 10 shows the representative SAXS pattern of 5% SDS micelles (in 0.4 M NaCl) at two different sucrose concentrations (10 and 80%).

The SAXS pattern shows that even at 80% sucrose, the nanostructure is retained with a similar length scale as observed in water. The structural parameter of the micelles obtained from SAXS data fitting (solid lines) is summarized in Table 2. It is noted that the addition of sucrose leads to a slight increase in the micellar core dimensions. It may be noted that the structure model employed here need not be the only model that fits the data, as any elongation/swelling of the micelle beyond the resolution limit of the instrument cannot be captured here.

By combining the results from DLS and SAXS experiments, it is inferred that the global diffusion of SDS micelles can be arrested by the addition of sugars like

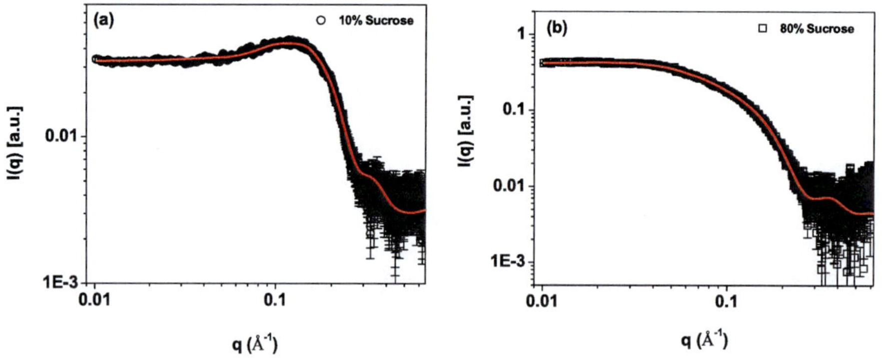

Fig. 10 SAXS pattern of 5% SDS micelles (in 0.4 M NaCl) in the presence of **a** 10% and **b** 80% sucrose, respectively. The solid lines are fit to the data using core–shell ellipsoid micelles

Table 2 Structural parameters of 5% aqueous SDS micelles (with 0.4 M NaCl) at different sucrose concentrations (wt.%) as obtained from SAXS data analysis using the prolate core–shell ellipsoid model

Sucrose (wt.%)	Semiminor core (Å)	Semimajor core (Å)	Shell thickness (Å)	SLD of shell (10^{-6} Å^{-2})	SLD of solvent (10^{-6} Å^{-2})	Charge (e.u.)	Reduced χ^2
0	13.5 ± 0.2	18.1 ± 0.2	12.5 ± 0.2	10.02 ± 0.2	9.44	2.8 ± 1.0	1.4
10	13.2 ± 0.3	20.6 ± 0.4	12.8 ± 0.1	10.22 ± 0.1	9.73	3.5 ± 1.2	1.3
80	14.0 ± 0.2	25.8 ± 0.2	13.1 ± 0.3	11.5 ± 0.2	11.8	3.8 ± 1.0	5.2

sucrose. Such behavior has been observed in other micellar systems like cationic micelles of CTAB in a supersaturated solution of sucrose (Gawali et al. 2017). It is worth mentioning that the micelles formed at high solid content are in a supersaturated metastable state, as the solubility of sucrose in water is about 67% at room temperature (Mathlouthi and Reiser 1995). However, the samples at high sucrose content can be solubilized by heating the sample in a water bath at 80 °C which remains in the supersaturated state even after cooling to room temperature (25 °C).

Such jammed micelles in supersaturated sucrose still contain a significant amount of water. To increase the kinetic stability of the metastable jammed micelles and increase further the viscosity of the matrix, it is desirable to remove water completely from the system. Our attempts to remove water completely from the above-jammed micelles by slow evaporation at 80 °C or under vacuum showed that the sugar crystallizes and micelles disrupt to form lamellar structures. Therefore, we attempted an alternative to form micelles without water. For this, we made use of the opposing behavior of sugar and urea on the micellization of conventional surfactants. Sugars such as glucose and sucrose are known to decrease the CMC of micelles, promote micelle growth, while, urea and its derivatives influence the structure and properties of micelles in the opposite way (Broecker and Keller 2013; Espinosa et al. 2018). For a variety of micellar systems, urea is found to increase the CMC of surfactants in water and decrease the micellar dimension (Sarkar and Bhattacharyya 1991; Kumar et al. 2004). Recent MD simulation studies on SDS micelles in the presence of urea indicate that up to 4 M urea solution; there is a systematic decrease in the aggregation number of SDS micelles (Espinosa et al. 2018). The action of urea on influencing micellization parameters in aqueous solutions is attributed to various contributions. One of them being, urea interacts with the polar head groups on the surface of micelles and changes the hydration layer. As a result, the surface area of head groups increases, causing strong steric repulsion between head groups and decrease in overall micellar parameters. It has also been suggested that the role of urea is merely an increase in the polar character of the solvent leading to increased dissociation of ionic surfactants (Dias et al. 2002). Comparison of the micelle properties of ionic surfactants in polar non-aqueous solvents such as formamide and N-methyl acetamide indicate that the observed decrease in the CMC in these non-aqueous solvents arises from a combined effect of the dielectric constant of the medium, the nature of hydrogen bonding, and the dispersion forces among the alkyl chains of the surfactant ions.

In aqueous micellar solution, urea can be solubilized even at concentrations as high as 10 M, without disrupting micelles. Thus, we postulated that sugar–urea mixtures will be an ideal matrix for immobilizing micelles.

To immobilize micelles, first, we added sugar–urea mixture, at appropriate composition, to an aqueous micellar solution followed by removal of the solvent by vacuum. In this process, we noticed that though most of the solvent can be removed by vacuum there is a significant amount of bound solvent molecules that could not be removed at ambient temperature. To eliminate any such possibility of bound water in the system, we later examined the formation of micelles in a sugar–urea melt, with the conjecture that successive cooling of this mixture can provide water free micelles trapped in a solid or glassy matrix. We found that micelles could be formed in such molten sugar–urea mixtures which can be supercooled to room temperature forming arrested micelles. These micelles were found to sustain over a wide range of temperatures; from as high as 90 °C to subzero degree temperatures (−25 °C), without cryogenic quenching (Gawali et al. 2018, 2019). Considering the ability of various sugar-additive combinations to form a glassy matrix at ambient conditions, this offers a unique way of forming frozen nanoscale assemblies in a glassy matrix.

One example of a system that forms arrested micelles comprises fructose, urea, and CTAB, at a weight ratio of 54:36:10. The micelles are prepared by dissolving the CTAB molecules in the fructose–urea mixture above its melting point. Once micelles are formed in the liquid state, further cooling of the melt leads to the formation of trapped micelles. The structure of micelles formed in the supercooled matrix is probed by SAXS. It was observed that up to 25% w/w of CTAB surfactant can be solubilized in the glucose–urea melt, without any phase separation upon cooling to 25 °C (Gawali et al. 2018). This is different from the phase behavior of CTAB in water (Auvray et al. 1989), where globular micelles exist up to a concentration of 20% w/w, then hexagonal phase from 20 to 62% w/w, and lamellar phase exists above ~80% w/w.

A schematic illustration of micelle formation in the melt upon heating and supercooling is shown in Fig. 11. Inset photographs of Fig. 11 show the crystalline powder before melting and the supercooled liquid that became optically transparent and

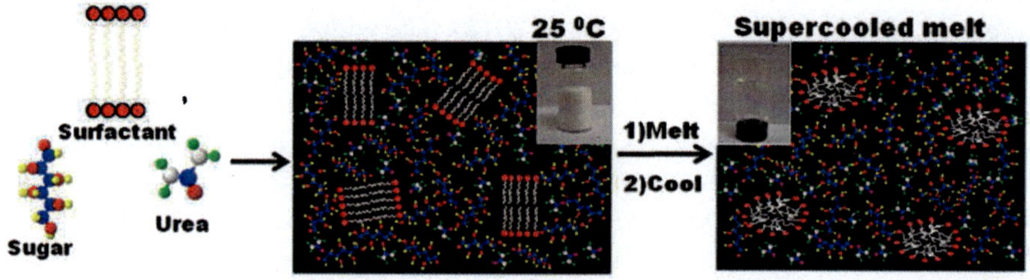

Fig. 11 Schematic illustration of the surfactant crystals dissolution in sugar–urea melt and subsequent trapping in a supercooled state. Inset photographs show the crystalline powder before melting and the supercooled liquid that became optically transparent and viscous enough to hold its own weight

viscous enough to hold its own weight (Gawali et al. 2018). The existence of micelles in the supercooled sugar–urea melt was also confirmed by using HR-TEM at ambient temperature. Direct images of micelles were previously observed only at cryogenic temperature, upon vitrification of suspensions. However, here TEM observations could be done at ambient temperature, as imaging of the micelles was possible due to the supercooled nature of the system.

Figure 12 shows the temperature-dependent SAXS pattern of 10% w/w CTAB surfactant in the fructose–urea mixture. At 25 °C, the SAXS pattern of the sample (fructose–urea-CTAB 54:36:10 w/w) shows characteristic lamellar peaks of CTAB. The peaks occur at scattering vector magnitude (q) values of 0.234 and 0.475 $Å^{-1}$. The ratio of the peak positions is close to 1:2, with a lattice parameter of 2.68 nm, as expected for lamellar crystals (Battaglia and Ryan 2005). At 60 °C, the contribution from the lamellar structure becomes negligible and the system mostly comprises micellar aggregates as shown by the broad peak at q value 0.05 $Å^{-1}$. At 80 °C, the lamellar structure completely disappears and the SAXS pattern is reminiscent of pure micelles. Moreover, upon cooling these preformed micelles to 15 °C (within a span of 30 s) the micelles SAXS pattern is retained, indicating kinetically trapped supercooled micelles. More importantly, it is observed that the SAXS pattern is almost identical even when the sample is cooled to −20 °C. This suggests that the aggregates are kinetically trapped and remain in the micellar form at least up to −20 °C. However, for CTAB micelles in water, where the surfactant starts crystallizing at ~15 °C, and diffraction peaks corresponding to CTAB crystals are found already at 15 °C (Fig. 13).

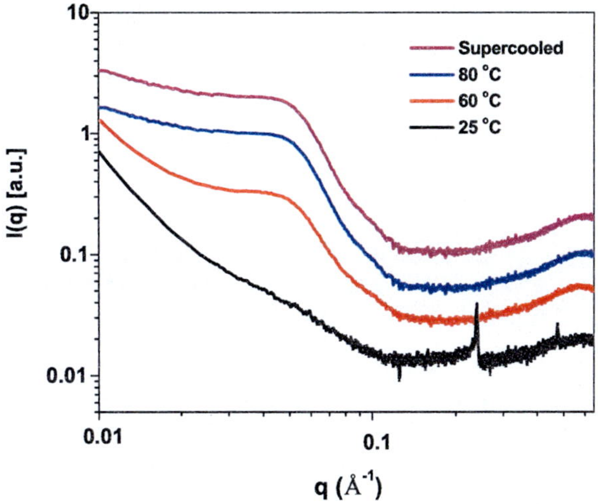

Fig. 12 Temperature-dependent SAXS intensity patterns of 10% CTAB in fructose–urea (3:2) mixture

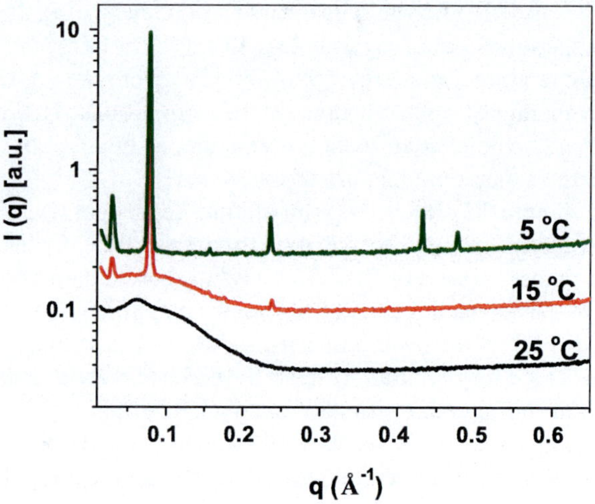

Fig. 13 Temperature-dependent SAXS patterns of 10% CTAB in water indicating crystallization of the surfactant below 15 °C

5 Supercooled Matrices for Cryopreservation of Biological Materials

Water plays a major role in the degradation of stored biological materials, providing conditions to promote the growth of spoilage organisms (Adams 2007). It is therefore necessary to immobilize or reduce the water content of stored samples, to improve its shelf life. One such methodology often used to stabilize vaccines, microorganisms, and other biological materials is by freeze-drying or lyophilization (Adams 1996; Gheorghiu et al. 1996). Lyophilization is a convenient method of dehydrating labile products by vacuum desiccation, in order to attain immobilized macromolecules. In the freeze-drying process, the material of interest is frozen in the presence of a cryoprotectant such as glucose and sucrose and the solvent is removed by sublimation. Freeze-drying process involves the following steps: (Adams 1996; Gheorghiu et al. 1996).

1. Cooling of the liquid sample to form an amorphous matrix comprising non-crystallizing solutes associated with unfrozen moisture.
2. Evaporation of water from the amorphous matrix by sublimation under vacuum.

Cryopreservation is also employed in long term storage of cells and other biological materials. However, when living cells are exposed to water under very low temperatures, intra- and extracellular ice crystal formation, osmotic shock, and membrane damage during freezing and thawing cycle may lead to cell death (Baust et al. 2001; Yang et al. 2003). Although bio-materials can be stored at very low temperatures for a long time without loss of their functionality, the cycles of cooling and then thawing for their use can be lethal to them. There are few criteria to be met for using the additives for cryopreservation like: (1) The material must be highly water-soluble and remain so at low temperatures in order to produce a profound depression of the freezing temperature; (2) it must be able to penetrate into the cells with low toxicity so that it can be used in the high concentrations that are required to produce these effects. Many biologically accepted compounds like glycerol, dimethyl sulfoxide, ethanediol, and propanediol have these properties and are used as cryoprotectants to avoid the intracellular freezing and extracellular ice crystallization. Cryopreservation has been of prime importance in biotechnology, regenerative medicine, and chemical biology along with clinical medicine and species conservation in plant and animal biology. Typically, there are two pathways for cryopreservation: slow freezing and vitrification (Elliott et al. 2017).

It is known that cells can be preserved for a long time at low temperature (−180 °C). by slow cooling at the rate of 1–5 °C/minute. During slow cooling of water containing solute, it is known that the ability of water for solute solvation is lost gradually; hence, solute particles are excluded from ice crystal structures. This results in an increase in the concentration of solute in the remaining fraction of water and responsible for colligative freezing point depression (Mazur 1984). In biological freezing, if cells are to be preserved in a medium containing water and other solutes, the same phenomenon takes place. As the temperatures are progressively decreased, solutes are excluded from the growing ice matrix. Thus, the concentration of solute in 'no ice' fraction works as a major damaging factor during the freezing of cells (Lovelock 1954). It is known that the increased salt concentration led at some point to irreversible membrane damage and subsequent cell lysis on thawing.

Vitrification is another pathway for cryopreservation of cells in which ice formation is prevented in intra- as well as extracellular medium hence bypassing the damages due to osmotic imbalances. In this process, the solidification of a formulation at low temperatures achieved with an extremely high viscosity without ice nucleation and crystallization (Wowk 2010). For vitrification, the glass transition temperature of the medium should be less than crystallization (freezing) temperature. This can be achieved by using highly viscous media to increase the glass transition temperature and ultrafast cooling rates (Armitage 2009). The choice of the cryopreservation method usually depends on the characteristics of the sample. For most suspended cells, slow-freezing looks promising as it is easy to perform. Although vitrification is tedious, it shows better results with respect to post-thaw survival rate, morphology and successful implantation of cryopreserved human embryos as compared to that of slow freezing (Armitage et al. 2002).

Cryoprotective agents (CPAs) are the materials when added to live cells in their medium and allow higher post-thaw recoveries. It plays a key role in the storage of cells at deep cryogenic temperatures and facilitates them to recovered at metabolic temperatures without losing the functionality (Carpenter and Crowe 1988; Doshi et al. 2013). The role of CPAs towards cryopreservation is attributed to two different mechanisms: (1) The CPAs action is linked to an ability to alter water properties because of hydrogen bonding between water and (2) CPA form a protective coating around the cells thereby protecting them by external shocks. Generally, CPAs are alcohols and their derivatives, sugars, polymers (of biological and synthetic origin), sulfoxides, amides, and amines. The most commonly employed material for cryopreservation is dimethyl sulfoxide (DMSO). Also, the inclusion of various non-toxic, non-penetrating additives such as hydroxyethyl starch, methylcellulose, polyvinylpyrrolidone, polyethylene glycol, and saccharides into cryoprotective medium are routinely used to achieve optimal cell viability (Fuller 2004; Petrenko et al. 2014). Sugars in combination with DMSO, glycerol, or 1,2 propanediol are known to be responsible for better survival rate after thawing of different cells (Rodriguez et al. 2005; Ma et al. 2006; Hu et al. 2009). The protective effect of sugars has also been shown in a number of model systems including liposomes, membranes, and proteins during both air and freeze-drying (Petrenko et al. 2014; Elliott et al. 2017).

Having noticed that sugars in combination with urea and other amide containing molecules are capable of sustaining self-assembly at sub-zero degree Celsius temperatures, we explored the possibility of using sugar-amino acid mixtures as CPAs for different cancer cell lines. The formulations were prepared by mixing fructose along with two different amino acids namely glycine and β-alanine. Two cancer cell lines A549 and MCF7 are used as model cell lines to study the effect of fructose-amino acid mixture as CPAs. Primarily, the efficacy of the formulation towards cryo-protection was measured by monitoring the effect of CPAs on cell morphology using optical microscopy. For comparison, cells were freeze-thawed in 10% DMSO and observed under the microscope.

After cryopreservation, the cells were thawed in an incubator. The optical microscopy revealed the effect of freezing on the cell morphology. Figures 14 and 15 show the cell morphology of A459 cell lines at 30% and 40% w/w of CPAs, respectively. It can be seen from Figs. 14 and 15 that, when 10% DMSO was used as CPAs, then cell membranes of cells were intact but they were slightly swollen. In the case of control which contains only culture medium, the cells were seen to be damaged completely due to ice formation. However, in the case of fructose-amino acid mixture as CPAs many cells were intact although some were seen to be affected. CPAs at

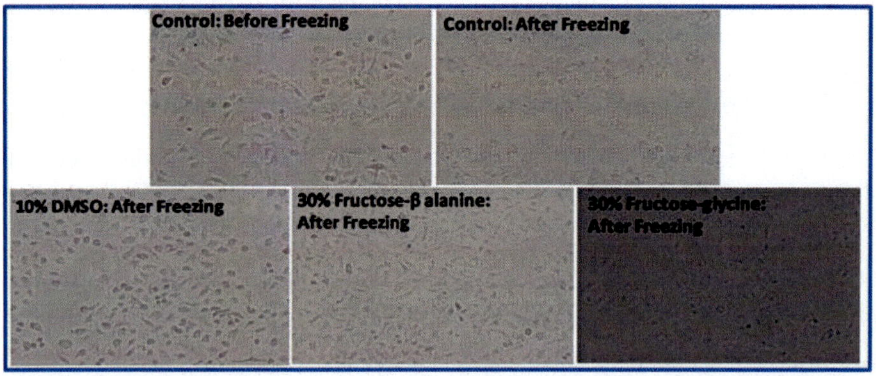

Fig. 14 Effect of fructose-amino acid mixture (30%) as cryoprotectants on A549 cell morphology after freezing at -25 °C and thawing after 24 h

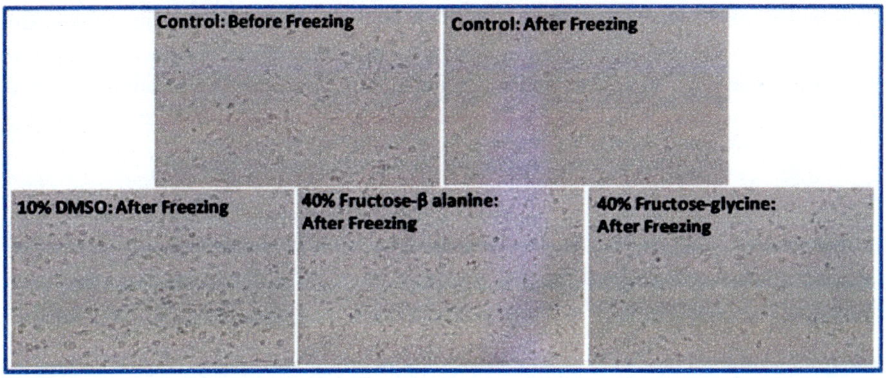

Fig. 15 Effect of fructose-amino acid mixture (40%w/w) as cryoprotectants on A459 cell line morphology after freezing at -25 °C and thawing after 24 h

40% concentration is found to more effective than 30% solutions, for cryopreservation, as cells were mostly intact and no change in morphology was seen. Comparing the cell survival efficiency using a trypan blue assay indicated that the fructose–alanine mixture is more effective than the fructose–glycine mixture. However, more work is needed in this direction to identify appropriate sugar–amino acid mixtures as mimics of natural cryoprotectants. It can be inferred that a proper combination of sugar–amino acid mixtures could be used as biocompatible cryoprotectants for the immobilization of biological materials.

6 Conclusion

In conclusion, this review focuses on the basic principles of self-assembling systems in common solvents and their immobilization in a room temperature supercooled solvent. The fundamental rules governing the microstructure (size, shape, etc.) of the assemblies formed by surfactants are discussed. It has now become evident that the intermolecular interactions between amphiphiles in a micelle can be conveniently tuned to obtain precise morphologies of the aggregates. More importantly, such an association phenomenon is sensitive to the solvent and a variety of solvents could be designed by using the appropriate combination of hydrophilic molecules. This offers a way of creating new supercooled solvents that are capable of inducing self-assembly of surfactants. In particular, a molten mixture of sugar and urea or other amide group-containing small molecules could be used as solvents for self-assembly. The intermolecular interaction between sugar and other additives could prevent the crystallization of the solvent below the melting points and remain in a supercooled state. The high viscosity and low mobility of supercooled solvents make it an attractive system to arrest the global diffusion of such assemblies or other biological macromolecules. This will have possible implications in controlling collision mediated agglomeration of nanoparticles, preservation of pharmaceutical products or biomolecules, etc. Preliminary studies indicate that suitable sugar-amino acid combinations could be used for the immobilization of cells without ice nucleation.

References

Adams G (2007) The principles of freeze-drying. In: Wolkers, Willem F, Oldenhof H (eds) Cryopreservation and freeze-drying protocols, 2nd edn, pp 15–38

Adams GDJ (1996) Lyophilization of vaccines BT—vaccine protocols. In: Robinson A, Farrar GH, Wiblin CN (eds). Humana Press, Totowa, NJ, pp 167–185

Ali M, Jha M, Das SK, Saha SK (2009) Hydrogen-bond-induced microstructural transition of ionic micelles in the presence of neutral naphthols: pH dependent morphology and location of surface activity. J Phys Chem B 113:15563–15571. https://doi.org/10.1021/jp907677x

Angell CA (1997) Entropy, fragility, "landscapes", and the glass transition BT—complex behaviour of glassy systems. In: Rubí M, Pérez-Vicente C (eds). Springer, Berlin, Heidelberg, pp 1–21

Armitage J (2009) Cryopreservation for corneal storage. Dev Ophthalmol 43:63–69. https://doi.org/10.1159/000223839

Armitage WJ, Hall SC, Routledge C (2002) Recovery of endothelial function after vitrification of cornea at −110 degrees C. Invest Ophthalmol Vis Sci 43:2160–2164

Auvray X, Petipas C, Anthore R, Rico I, Lattes A (1989) X-ray diffraction study of mesophases of cetyltrimethylammonium bromide in water, formamide, and glycerol. J Phys Chem 93:7458–7464. https://doi.org/10.1021/j100358a040

Barnes HA (1997) Thixotropy—a review. J Nonnewton Fluid Mech 70:1–33. https://doi.org/10.1016/S0377-0257(97)00004-9

Battaglia G, Ryan AJ (2005) The evolution of vesicles from bulk lamellar gels. Nat Mater 4:869–876. https://doi.org/10.1038/nmat1501

Baust JM, Vogel MJ, Van Buskirk R, Baust JG (2001) A molecular basis of cryopreservation failure and its modulation to improve cell survival. Cell Transplant 10:561–571

Broecker J, Keller S (2013) Impact of urea on detergent micelle properties. Langmuir 29:8502–8510. https://doi.org/10.1021/la4013747

Capito RM, Azevedo HS, Velichko YS, Mata A, Stupp SI (2008) Self-assembly of large and small molecules into hierarchically ordered sacs and membranes. Science 319:1812–1816. https://doi.org/10.1126/science.1154586

Carpenter JF, Crowe JH (1988) The mechanism of cryoprotection of proteins by solutes. Cryobiology 25:244–255. https://doi.org/10.1016/0011-2240(88)90032-6

Chen S-H, Chen W-R, Mallamace F (2003) The glass-to-glass transition and its end point in a copolymer micellar system. Science 300:619–622. https://doi.org/10.1126/science.1082364

Daumas F, Destainville N, Millot C, Lopez A, Dean D, Salome L (2003) Confined diffusion without fences of a g-protein-coupled receptor as revealed by single particle tracking. Biophys J 84:356–366. https://doi.org/10.1016/S0006-3495(03)74856-5

de Gennaro B (2019) Chapter 3—surface modification of zeolites for environmental applications. In: Mercurio M, Sarkar B, Langella ABT-MC and ZNM (eds) Micro and nano technologies. Elsevier, pp 57–85

de Melo A, Marques F, Angelini R, Zaccarelli E, Farago B, Ruta B, Ruocco G, Ruzicka B (2015) Structural and microscopic relaxations in a colloidal glass. Soft Matter 11:466–471. https://doi.org/10.1039/C4SM02010C

Degiorgio V (ed) (1985) Physics of amphiphiles: micelles, vesicles and microemulsions. Elsevier Science Publishers, New York

Dias LG, Florenzano FH, Reed WF, Baptista MS, Souza SMB, Alvarez EB, Chaimovich H, Cuccovia IM, Amaral CLC, Brasil CR, Romsted LS, Politi MJ (2002) Effect of urea on biomimetic systems: neither water 3-D structure rupture nor direct mechanism, simply a more "polar water." Langmuir 18:319–324. https://doi.org/10.1021/la010176u

Dierking I, Al-Zangana S (2017) Lyotropic liquid crystal phases from anisotropic nanomaterials. Nanomater 7

Shaw D (1992) Introduction to colloid and surface chemistry, 4th edn. Butterworth-Heinemann

Elliott GD, Wang S, Fuller BJ (2017) Cryoprotectants: a review of the actions and applications of cryoprotective solutes that modulate cell recovery from ultra-low temperatures. Cryobiology 76:74–91. https://doi.org/10.1016/j.cryobiol.2017.04.004

Espinosa YR, Grigera RJ, Ferrara CG (2018) Mechanisms associated with the effects of urea on the micellar structure of sodium dodecyl sulphate in aqueous solutions. Prog Biophys Mol Biol 140:117–123. https://doi.org/10.1016/j.pbiomolbio.2018.05.006

Fuller BJ (2004) Cryoprotectants: the essential antifreezes to protect life in the frozen state. Cryo Lett 25:375–388

Gawali SL, Kumar S, Hassan PA (2017) Jammed micelles in high solid content sucrose solutions. J Nanofluids 6 https://doi.org/10.1166/jon.2017.1352

Gawali SL, Zhang M, Kumar S, Aswal VK, Danino D, Hassan PA (2018) Dynamically arrested micelles in a supercooled sugar urea melt. Commun Chem 1:33. https://doi.org/10.1038/s42004-018-0032-0

Gawali SL, Zhang M, Kumar S, Ray D, Basu M, Aswal VK, Danino D, Hassan PA (2019) Discerning the structure factor of charged micelles in water and supercooled solvent by contrast variation X-ray scattering. Langmuir 35:9867–9877. https://doi.org/10.1021/acs.langmuir.9b00912

Gheorghiu M, Lagranderie M, Balazuc AM (1996) Stabilisation of BCG vaccines. Dev Biol Stand 87:251–261

Gotze W, Sjogren L (1992) Relaxation processes in supercooled liquids. Reports Prog Phys 55:241–376. https://doi.org/10.1088/0034-4885/55/3/001

Grandjean J, Mourchid A (2004) Re-entrant glass transition and logarithmic decay in a jammed micellar system. Rheol Dyn Invest Europhys Lett 65:712–718. https://doi.org/10.1209/epl/i2003-10172-5

Hassan PA, Raghavan SR, Kaler EW (2002) Microstructural changes in SDS micelles induced by hydrotropic salt. Langmuir 18:2543–2548. https://doi.org/10.1021/la011435i

Hayter JB, Penfold J (1981) An analytic structure factor for macroion solutions. Mol Phys 42:109–118. https://doi.org/10.1080/00268978100100091

He Y, Li Z, Simone P, Lodge TP (2006) Self-assembly of block copolymer micelles in an ionic liquid. J Am Chem Soc 128:2745–2750. https://doi.org/10.1021/ja058091t

Hiemenz PC, Rajagopalan R (1997) Principles of colloid and surface chemistry. CRC Press

Holmqvist P (2014) Short-time dynamic signature of the liquid-crystal-glass transition in a suspension of charged spherical colloids. Langmuir 30:6678–6683. https://doi.org/10.1021/la5010853

Hu J-H, Li Q-W, Jiang Z-L, Yang H, Zhang S-S, Zhao H-W (2009) The cryoprotective effect of trehalose supplementation on boar spermatozoa quality. Reprod Domest Anim 44:571–575. https://doi.org/10.1111/j.1439-0531.2007.00975.x

Israelachvili JN, Mitchell DJ, Ninham BW (1976) Theory of self-assembly of hydrocarbon amphiphiles into micelles and bilayers. J Chem Soc Faraday Trans 2 Mol Chem Phys 72:1525–1568. https://doi.org/10.1039/F29767201525

Doshi JC (2013) Ice recrystallization inhibitors: from biological antifreezes to small molecules. In: Wilson P (ed) Recent developments in the study of recrystallization. InTech, New York, pp 177–224

Jaeger HM (2015) Celebrating soft matter's 10th anniversary: toward jamming by design. Soft Matter 11:12–27. https://doi.org/10.1039/C4SM01923G

Kabanov AV, Alakhov VY (2002) Pluronic block copolymers in drug delivery: from micellar nanocontainers to biological response modifiers. Crit Rev Ther Drug Carrier Syst 19:1–72. https://doi.org/10.1615/critrevtherdrugcarriersyst.v19.i1.10

Kékicheff P, Grabielle-Madelmont C, Ollivon M (1989) Phase diagram of sodium dodecyl sulfate-water system: 1. A calorimetric study. J Colloid Interf Sci 131:112–132. https://doi.org/10.1016/0021-9797(89)90151-3

Knowlton ED, Pine DJ, Cipelletti L (2014) A microscopic view of the yielding transition in concentrated emulsions. Soft Matter 10:6931–6940. https://doi.org/10.1039/C4SM00531G

Kumar S, Parveen N, Kabir-ud-Din (2004) Effect of urea addition on micellization and the related phenomena. J Phys Chem B 108:9588–9592. https://doi.org/10.1021/jp036552w

Kundu A, Verma PK, Cho M (2017) Role of solvent water in the temperature-induced self-assembly of a triblock copolymer. J Phys Chem Lett 8:3040–3047. https://doi.org/10.1021/acs.jpclett.7b01008

Kunz W, Testard F, Zemb T (2009) Correspondence between curvature, packing parameter, and hydrophilic−lipophilic deviation scales around the phase-inversion temperature. Langmuir 25:112–115. https://doi.org/10.1021/la8028879

Kwaśniewski P, Fluerasu A, Madsen A (2014) Anomalous dynamics at the hard-sphere glass transition. Soft Matter 10:8698–8704. https://doi.org/10.1039/C4SM01671H

Langer R (2001) Drug delivery. Drugs on target. Science 293:58–59. https://doi.org/10.1126/science.1063273

Li W, Liu S, Deng R, Zhu J (2011) Encapsulation of nanoparticles in block copolymer micellar aggregates by directed supramolecular assembly. Angew Chemie Int Ed 50:5865–5868. https://doi.org/10.1002/anie.201008224

Li X, Yang Z, Yang K, Zhou Y, Chen X, Zhang Y, Wang F, Liu Y, Ren L (2009) Self-assembled polymeric micellar nanoparticles as nanocarriers for poorly soluble anticancer drug ethaselen. Nanoscale Res Lett 4:1502–1511. https://doi.org/10.1007/s11671-009-9427-2

Lin Y, Han X, Cheng X, Huang J, Liang D, Yu C (2008) pH-regulated molecular self-assemblies in a cationic−anionic surfactant system: from a “1−2” surfactant pair to a “1−1” surfactant pair. Langmuir 24:13918–13924. https://doi.org/10.1021/la802593n

Lombardo D, Kiselev M, magazù S, Calandra P (2015) Amphiphiles self-assembly: basic concepts and future perspectives of supramolecular approaches. Adv Condens Matter Phys 1–22. https://doi.org/10.1155/2015/151683

Lovelock JE (1954) The protective action of neutral solutes against haemolysis by freezing and thawing. Biochem J 56:265–270. https://doi.org/10.1042/bj0560265

Mathlouthi M, Reiser P (eds) (1995) Sucrose, properties and applications. Chapman & Hall, London

Ma W, O'Shaughnessy T, Chang E (2006) Cryopreservation of adherent neuronal networks. Neurosci Lett 403:84–89. https://doi.org/10.1016/j.neulet.2006.04.064

Ma X, Zhao Y (2015) Biomedical applications of supramolecular systems based on host-guest interactions. Chem Rev 115:7794–7839. https://doi.org/10.1021/cr500392w

Magno LM, Angelescu DG, Sigle W, Stubenrauch C (2011) Microemulsions as reaction media for the synthesis of Pt nanoparticles. Phys Chem Chem Phys 13:3048–3058. https://doi.org/10.1039/C0CP01085E

Mason B, Weitz (1995) Elasticity of compressed emulsions. Phys Rev Lett 75:2051–2054. https://doi.org/10.1103/PhysRevLett.75.2051

Mattsson J, Wyss HM, Fernandez-Nieves A, Miyazaki K, Hu Z, Reichman DR, Weitz DA (2009) Soft colloids make strong glasses. Nature 462:83–86. https://doi.org/10.1038/nature08457

Mayans E, Ballano G, Sendros J, Font-Bardia M, Campos JL, Puiggali J, Cativiela C, Aleman C (2017) Effect of solvent choice on the self-assembly properties of a diphenylalanine amphiphile stabilized by an ion pair. ChemPhysChem 18:1888–1896. https://doi.org/10.1002/cphc.201700180

Mazur P (1984) Freezing of living cells: mechanisms and implications. Am J Physiol 247:C125–C142. https://doi.org/10.1152/ajpcell.1984.247.3.C125

Mezzenga R, Schurtenberger P, Burbidge A, Michel M (2005) Understanding foods as soft materials. Nat Mater 4:729–740. https://doi.org/10.1038/nmat1496
Rosen MJ (1989) Surfactants and interfacial phenomena, 2nd edn. Wiley Inc., New York
Myers D (1991) Surfaces, interfaces, and colloids—principles and applications. VCH Publishers, New York, NY, United States
Myers D (1999) Surfaces, interfaces, and colloids, 2nd edn, Wiley Inc.
Nagarajan R (2002) Molecular packing parameter and surfactant self-assembly: the neglected role of the surfactant tail. Langmuir 18:31–38. https://doi.org/10.1021/la010831y
Ochab-Marcinek A, Hołyst R (2011) Scale-dependent diffusion of spheres in solutions of flexible and rigid polymers: mean square displacement and autocorrelation function for FCS and DLS measurements. Soft Matter 7:7366–7374. https://doi.org/10.1039/C1SM05217A
Ochab-Marcinek A, Wieczorek SA, Ziębacz N, Hołyst R (2012) The effect of depletion layer on diffusion of nanoparticles in solutions of flexible and polydisperse polymers. Soft Matter 8:11173–11179. https://doi.org/10.1039/C2SM25925G
Patton JS, Carey MC (1979) Watching fat digestion. Science (80–204):145 LP–148. https://doi.org/10.1126/science.432636
Petrenko YA, Rogulska OY, Mutsenko VV, Petrenko AY (2014) A sugar pretreatment as a new approach to the Me2SO- and xeno-free cryopreservation of human mesenchymal stromal cells. Cryo Lett 35:239–246
Rana S, Bhattacharjee J, Barick KC, Verma G, Hassan PA, Yakhmi J V (2017) Chapter 7—interfacial engineering of nanoparticles for cancer therapeutics. In: Ficai A, Grumezescu AMBT-N for CT (eds) Micro and nano technologies. Elsevier, pp 177–209
Ray A (1971) Solvophobic interactions and micelle formation in structure forming nonaqueous solvents. Nature 231:313–315. https://doi.org/10.1038/231313a0
Rodriguez L, Velasco B, Garcia J, Martin-Henao GA (2005) Evaluation of an automated cell processing device to reduce the dimethyl sulfoxide from hematopoietic grafts after thawing. Transfusion 45:1391–1397. https://doi.org/10.1111/j.1537-2995.2005.00213.x
Romer S, Scheffold F, Schurtenberger P (2000) Sol-gel transition of concentrated colloidal suspensions. Phys Rev Lett 85:4980–4983. https://doi.org/10.1103/PhysRevLett.85.4980
Sarkar N, Bhattacharyya K (1991) Effect of urea on micelles: fluorescence of p-toluidino naphthalene sulphonate. Chem Phys Lett 180:283–286. https://doi.org/10.1016/0009-2614(91)90320-9
Salem AK, Searson PC, Leong KW (2003) Multifunctional nanorods for gene delivery. Nat Mater 2:668–671. https://doi.org/10.1038/nmat974
Schramm LL (ed) (2000) Surfactants fundamentals and applications in the petroleum industry. Cambridge University Press, UK
Sentjabrskaja T, Zaccarelli E, De Michele C, Sciortino F, Tartaglia P, Voigtmann T, Egelhaaf SU, Laurati M (2016) Anomalous dynamics of intruders in a crowded environment of mobile obstacles. Nat Commun 7:11133. https://doi.org/10.1038/ncomms11133
Shi L, Wei Y, Sun N, Zheng L (2013) First observation of rich lamellar structures formed by a single-tailed amphiphilic ionic liquid in aqueous solutions. Chem Commun 49:11388–11390. https://doi.org/10.1039/C3CC45550E
Smith MI, Besseling R, Cates ME, Bertola V (2010) Dilatancy in the flow and fracture of stretched colloidal suspensions. Nat Commun 1:114. https://doi.org/10.1038/ncomms1119
Stradner A, Sedgwick H, Cardinaux F, Poon WCK, Egelhaaf SU, Schurtenberger P (2004) Equilibrium cluster formation in concentrated protein solutions and colloids. Nature 432:492–495. https://doi.org/10.1038/nature03109
Tanford C (1973) The hydrophobic effect. In: Formation of micelles and biological membranes. Wiley-Interscience, New York
Toyotama A, Okuzono T, Yamanaka J (2016) Spontaneous formation of eutectic crystal structures in binary and ternary charged colloids due to depletion attraction. Sci Rep 6:23292. https://doi.org/10.1038/srep23292

Tuinier R, Fan T-H, Taniguchi T (2015) Depletion and the dynamics in colloid–polymer mixtures. Curr Opin Colloid Interf Sci 20:66–70. https://doi.org/10.1016/j.cocis.2014.11.009

Voigtmann T, Horbach J (2009) Double transition scenario for anomalous diffusion in glass-forming mixtures. Phys Rev Lett 103:205901. https://doi.org/10.1103/PhysRevLett.103.205901

Vu-Quang H, Vinding MS, Nielsen T, Ullisch MG, Nielsen NC, Kjems J (2016) Theranostic tumor targeted nanoparticles combining drug delivery with dual near infrared and (19)F magnetic resonance imaging modalities. Nanomedicine 12:1873–1884. https://doi.org/10.1016/j.nano.2016.04.010

Wärnheim T, Jönsson A (1988) Phase diagrams of alkyltrimethylammonium surfactants in some polar solvents. J Colloid Interf Sci 125:627–633. https://doi.org/10.1016/0021-9797(88)90030-6

Weitz DA (1996) Foams flow by stick and slip. Nature 381:475–476. https://doi.org/10.1038/381475a0

Wolff T, Klaussner B (1995) Overlap of colloid chemistry and photochemistry in surfactant systems. Adv Colloid Interf Sci 59:31–94. https://doi.org/10.1016/0001-8686(95)80004-M

Wowk B (2010) Thermodynamic aspects of vitrification. Cryobiology 60:11–22. https://doi.org/10.1016/j.cryobiol.2009.05.007

Xiao C, Fujita N, Miyasaka K, Sakamoto Y, Terasaki O (2012) Dodecagonal tiling in mesoporous silica. Nature 487:349–353. https://doi.org/10.1038/nature11230

Yang H, Acker JP, Cabuhat M, McGann LE (2003) Effects of incubation temperature and time after thawing on viability assessment of peripheral hematopoietic progenitor cells cryopreserved for transplantation. Bone Marrow Transplant 32:1021–1026. https://doi.org/10.1038/sj.bmt.1704247

Zech O, Kunz W (2011) Conditions for and characteristics of nonaqueous micellar solutions and microemulsions with ionic liquids. Soft Matter 7:5507–5513. https://doi.org/10.1039/C0SM01103G

Zheng Z, Ni R, Wang F, Dijkstra M, Wang Y, Han Y (2014) Structural signatures of dynamic heterogeneities in monolayers of colloidal ellipsoids. Nat Commun 5:3829. https://doi.org/10.1038/ncomms4829

Immobilization of Enzymes onto Silica-Based Nanomaterials for Bioprocess Applications

Devendra Sillu, Yeshaswi Kaushik, and Shekhar Agnihotri

Abstract Enzyme-assisted catalysis have long been considered a risky venture for many industrial processes despite their unique advantages, specificity, and activity. The use of free enzymes often lack economic viability and commercial feasibility for industrial use since they are expensive, incapable to withstand harsh conditions during downstream processing, and are extremely difficult to recover. Developing novel interventions employing nanomaterials as an immobilization template has greatly improved in minimizing these obstacles, thereby promoting activity and recycling of enzymes. As a result, self-assembled enzymes over nanomaterials have demonstrated great success particularly in biomedical sector, using novel nanobiocatalysts (NBCs) for diagnostic and therapeutic purposes. However, the use of such hybrid system for the production of value-added biochemical products is yet another domain, which has not been much explored. This chapter thus provides some useful insights using enzyme-nanohybrids for bioprocessing applications, particularly in the field of carbohydrate hydrolysis, biotransformation, and biofuel production. Several nanocomposites were considered for enzyme immobilization owing to their extraordinary characteristics such as non-toxic, ease in modification, high porosity ,and extremely stable toward adverse conditions. All selected studies were critically evaluated in terms of enzyme used, immobilization strategies, process parameters (enzymes' stability, activity, loading, and recycling), and product specifications (yield, productivity, and purity). Relevant challenges encountered using nanomaterials, especially for the production of industrially relevant biochemical products, were also discussed at the end, which needs to be addressed in near future.

Keywords Nano-bioprocessing · Biochemical product · Biotransformation · Nanobiocatalyst · Enzyme immobilization · Silica-based material · Clay materials

---

D. Sillu · Y. Kaushik
Department of Biotechnology, Thapar Institute of Engineering & Technology, Bhadson Road, Patiala 147004, Punjab, India

S. Agnihotri (✉)
Department of Agriculture and Environmental Sciences, National Institute of Food Technology Entrepreneurship and Management, Kundli, Sonepat 131028, Haryana, India
e-mail: shekharagnihotri@gmail.com

A. Tripathi and J. S. Melo (eds.), *Immobilization Strategies*,
Gels Horizons: From Science to Smart Materials,
https://doi.org/10.1007/978-981-15-7998-1_11

Abbreviations

3-D	Three dimensional
°C	Degree Celsius
Al	Aluminium
APTMS	(3-Aminopropyl)trimethoxysilane
B	Boron
CaA	Calcium aluminosilicate
Cu	Copper
Fe	Iron
G	Gram
Ga	Galliu
H	Hour
HMF	5-Hydroxyfurfural
L	Liter
MCFs	Mesostructured silica foams
MCM	Mobil Composition of Matter
mg	Milligram
mL	Milliliter
MTS	Micelle-templated silica
NaA	Sodium forms of zeolites A
NaX	Sodium forms of zeolites X
NaY	Sodium forms of zeolites Y
NBC	Nanobiocatalyst
PDDA	Poly(diallyldimethylammonium chloride)
PEs	Polyelectrolytes
SBA	Santa Barbara Amorphous
Si	Silicon
TMAH	Tetramethyl ammonium hydroxide
v/v	Volume by volume
wt%	Weight percentage
ZIF	Zeolitic imidazolate framework
ZSM	Zeolite Socony Mobil

1 Introduction

The over exploitation of limited fossil feedstock as resource and environmentally damaging production processes for chemical production are alarming concerns, which motivated the scientific community to pursuit ways to replace these resources with renewable raw materials. The current environmental concerns and future sustainable development can be swiftly handled with the use of non-conventional sources for production of bio-based industrially relevant products. Lignocellulosic biomass, organic, and inorganic waste are widely applied as non-renewable natural resources for the production of biofuels and other biochemicals (Biermann et al. 2011; Calvo-Flores and Dobado 2010; Ramesh and Tharanathan 2003; Tyagi and Lo 2013). However, their production is often hampered by environmental hazardous by-product formed during the chemical conversion process in association with the limited productivity and associated high costs (de Vries 2016). The appropriate substitute for the chemical methods are biocatalysts or enzymes as they are considered as remarkable molecules for biotransformation/bioprocess technology (Jäger and Büchs 2012; Maitan-Alfenas et al. 2015). Some of the important enzymes in bioprocessing and their industrial applications are listed in Table 1.

Enzyme-based catalytic reactions have been employed widely for production of a vast range of products owing to their product specificity and low carbon footprint mechanism (Ojeda and Kafarov 2009; Sheldon 2016). The use of enzyme in pharmaceutical, food, leather, textile, baking, and detergent industry has witnessed a boom in the past few decades owing to exquisite activity and ease of operation. Enzymes represent a green solution for ever-increasing demand of industrial products; however, for industrial-scale application, modulation in enzyme properties is an essential step (Iyer and Ananthanarayan 2008; Sarrouh et al. 2012). Enzyme immobilization is considered as a cost-effective approach for formulating a comparatively robust combination of enzyme and a support matrix (Brena and Batista-Viera 2006; Sheldon 2007; Spahn and Minteer 2008). The stability of enzymes against varying process parameters also improves by the process; however, any change in the native structure of enzyme upon immobilization is a serious concern as it directly impacts the specific activity and product formation efficiency (Rodrigues et al. 2013; Secundo 2013). With current advances in biotechnological tools and techniques, the degree of change in intrinsic structure can be predicted, and best option can be chosen from a range of support matrices as well as immobilization approach. Some of the essential properties/features of support matrices for enzyme immobilization are shown in Fig. 1.

Materials with high surface area are preferred for enzyme immobilization for two basic reasons first, they provide favorable interaction between enzyme–substrate facilitating the reaction and second, high enzyme loading resulting in high conversion rate (Ahmad and Sardar 2015; Hartmann and Kostrov 2013; Magner 2013). The involvement of nanotechnology in various sectors intrigued researchers to extend its applications in biotechnology and biomedical area owing to the protection provided by nanomaterials (Agnihotri and Dhiman 2017; Agnihotri et al. 2012, 2019; Chauhan et al. 2019; Dhiman et al. 2019; Mukherji et al. 2012; Singh et al. 2016). Nanosized materials owing to their unique intrinsic properties have been studied in the past for immobilization-based applications (Agnihotri et al. 2015, 2013; Bharti et al. 2015; Gupta et al. 2011; Puri et al. 2013). The various routes for synthesis and functionalization impart feasibility to modify near surface environment (Agnihotri et al. 2018a, b; Agnihotri et al. 2014; Shekhar Agnihotri et al. 2018a, b). The influence of support matrix-substrate/product interaction on activity and stability of enzyme by support-enzyme interaction is also taken into consideration while selecting the support matrix(Santos et al. 2015). The impact of support on the substrate, product,

Table 1 List of industrial sectors and enzymes employed for bioprocessing applications

Industrial sector	Enzyme	Application
Leather industry	Proteases	To hydrolyze the protein fraction of dermatan sulfate
	Lipases	To hydrolyze fats, oils, and greases present in the hypoderm
	Keratinases	To hydrolyze the keratin present in hair and epidermis
Food and allied sectors	α-Amylases	Hydrolysis of α-1, 4 glycosidic bonds in polysaccharides
	Glucoamylases	Hydrolysis of polysaccharide starch from the non-reducing end
	Proteases	Hydrolysis of peptide bonds present in proteins and polypeptides
	Lactase	Hydrolysis of lactose
	Lipases	Hydrolysis of long-chain triglycerides
	Phospholipases	Break down phospholipids into fatty acids and other lipophilic substances
	Esterases	Facilitate the splitting of esters into acid and alcohol
	Lipoxygenases	Dioxygenation of polyunsaturated fatty acids in lipids
	Cellulases	Hydrolysis of β-1, 4 linkages to liberate glucose units
	Xylanases	Cleaves xylans, a major constituent of hemicellulose
	Pectinases	Hydrolysis of glycosidic bonds in pectic polymers
	Glucose oxidase	Convert glucose to gluconic acid
	Laccase	Oxidize phenolic compounds
	Catalase	Hydrogen peroxide decomposition
	Peroxidase	Reduction of peroxides and oxidation of a wide range of inorganic and organic compounds
Animal feed industry	Phytase	Improve the digestibility of phosphorous (P)
	Xylanase	Pre-bleaching of pulp improves digestibility
	β-glucanase	Digestion of cereal grains
Pulp and paper industry	Lipase	Pitch control
	Protease	Biofilm removal
	Amylase	drainage improvement

(continued)

Table 1 (continued)

Industrial sector	Enzyme	Application
	Xylanase	Bleach boosting
	Laccase	Non-chlorine bleaching, delignification
	Cellulase	Deinking, drainage improvement
Detergent industry	Amylase	Carbohydrate stain removal
	Lipase	Fat stain elimination
	Protease	Protein stain removal
	Cellulase	Color clarification
	Cutinase	Triglyceride removal
	Mannanase	Mannan spot removal
Cosmetics industry	Superoxide dismutase	Free radical scavenging, skin care
	Protease	Removal of dead skin
	Endoglycosidase	Teeth and gum tissue care
	Laccase	Hair dye
	Lipase	Skin care
Chemical industry	Nitrile hydratase	Synthesis of acrylamide, butyramide, nicotinamide
	Glycosyltranferase	Synthesis of oligosaccharides
	Lipase	Synthesis of pharmaceuticals, polymers, biodiesels, biosurfactants
	Glucose isomerase	Production of high-fructose corn syrup
	Acyltransferase	Synthesis of hydroxamic acids
	Laccase	Production of textile dyes, cosmetic pigments, flavor agents, and pesticides

and water essential for catalytic reactions to proceed can never be neglected while selecting the support matrix (De Castro et al. 1997; Margolin 1991). Immobilized enzymes can be used in bioreactors for continuous production; however, the cost of support matrix, impact of support matrix on intrinsic enzyme property, and reusability are some of the critical aspect to be taken into consideration. The economic viability of these systems majorly depends on the cost of matrix and their capability to anchor enzyme.

A vast array of materials (e.g., organic, inorganic, amorphous, crystalline, porous, non-porous) have been studied extensively in the past to fulfil the quest for suitable support matrix (Tripathi et al. 2010; Fang et al. 2011; Homaei et al. 2013; Zdarta et al. 2018; Zucca and Sanjust 2014). Porous materials have gained special attention for bioprocess applications owing to their active surface area, easy functionalization, and biocompatible nature (Fried et al. 2013; Tripathi and Melo 2019; Zhao et al. 2006). Specifically, porous silica materials represent a cost-effective substitute, as they are cheap and easily available. The high thermal/chemical resistance along with exceptional mechanical properties favors their employment for enzyme immobilization

Fig. 1 Main features of support materials used for enzyme immobilization (Reprinted with permission from Zdarta et al. 2018)

(Poorakbar et al. 2018). In addition, inherent high surface area and porous structure of silica materials results in good sorption characteristic allowing high enzyme anchoring with reduced diffusional limitations (Jesionowski et al. 2014). The presence of many forms of silica materials, i.e., sol–gel silica, fumed silica, colloidal silica nanoparticles, and silica gel, provides the diversity required for various applications where the existence of surface hydroxyl groups would favor an easy in functionalization. The functionalization strategies for silica materials have been widely explored in the past, which is an entirely different domain and is not included in the scope of this chapter. However, the impact of different functionalization of silica materials on the immobilized enzyme has been covered to a thorough extent.

One of the other silica-based materials, zeolites, has been also employed in the past for enzyme immobilization (Jesionowski et al. 2014; Mitchell and Pérez-Ramírez 2011). Zeolites are microporous crystalline solids with well-defined structures and shape-selective properties (Ghasemi et al. 2018; Misaelides 2011). The limited pore width redirects anchoring of enzyme on to the external surface and hence less protection of enzyme from the harsh environmental factors (Zhang et al. 2012). The uptake of enzyme can be increased with the use of hierarchical zeolites, i.e., zeolites with auxiliary mesoporosity (Holm et al. 2011). With the advancement in synthesis protocols, the new category of zeolites known as nanocrystallites or delaminated zeolites has also been introduced. The intrinsic properties of synthesized zeolites change with varying Si/Al ratio, and a higher Si content results in increased mesoporosity. The other associated surface properties of zeolites are easy to manipulate via alkaline treatment (Zhao et al. 2010).

In addition to these materials, inherent physical and chemical uniqueness of layered clay materials presents them as a potential support for enzyme immobilization (An et al. 2015; Zdarta et al. 2018). The two-dimensional structure and hydrous aluminosilicate nature of clay materials impart numerous advantages such as large surface area, swelling, and ion exchangeability to the table (Forano et al. 2006). The clay minerals are also reported to prevent enzyme from microbial contamination. The interaction between enzymes and clay material could be established through functional moieties present at clay material surface, i.e., Si–O–Si, Si–O–OH, and Al–OH (Boyd and Mortland 1990; Satyanarayana 2004; Sinegani et al. 2005). The hydroxyl group could be utilized to form hydrogen bond between matrix and enzyme, hence imparting some anchoring strength. Easy modifications of clay material using polymeric, organic, biological molecules, etc., enhance the applicability of clay material for different applications (Dai and Huang 1999; Jlassi et al. 2017; Mravčáková et al. 2006; Rogers and Glendinning 1996; Sarkar et al. 2019; Zulfiqar et al. 2009).Clay minerals such as bentonite (Sedaghat et al. 2009a), montmorillonite (Gopinath and Sugunan 2007), kaolinite (Ajayi et al. 2012), sepiolite (Sedaghat et al. 2009b), halloysite (Zhai et al. 2010), etc., have been explored for enzyme immobilization. Consequently, modified composite clay minerals have demonstrated excellent properties for enzyme immobilization, including increment in the loading amount, enhanced thermal and pH durability, and the superior biocompatibility (Bayramoglu et al. 2013; Chao et al. 2013; Songurtekin et al. 2013; Tzialla et al. 2009).

In industrial sector, the numbers of products that can be produced from biomass origin or via a bioprocessing route is amazingly large. The demand of chemical-based product has increased exponentially as the global population quadrupled in the past century. The number of studies reporting successful immobilization enzymes onto various support matrices has arisen exponentially, but practical application of these immobilized systems lacks behind, specifically in bioprocessing. This chapter focuses on the employment of immobilized enzymes for bioprocess applications.

2 Mesoporous Silica for Enzyme Immobilization

Improvement in stability and enzyme catalytic activity via confinement of enzyme in a specific space is a well-reported fact. A wide range of solid supports have been employed in the past for enzyme immobilization, which reduces the denaturation of protein by restricting the conformational change in protein 3-D structure. The quest for novel materials with high specific surface area, inherent porous nature, and easily modifiable properties has seen drastic increase. Use of silica materials specifically, mesoporous silica for enzyme immobilization, has attracted attention in last decade owing to two distinctive advantages over non-porous materials, first extremely high surface area, and seconds a wide pore size range (2–50 nm) (Zhou and Hartmann 2013). Depending on the ratio of pore and enzyme size, different mesoporous materials such as Mobil Composition of Matter (MCM), Santa Barbara Amorphous (SBA), Hollow Mesoporous Silica (HMS), etc., and their variants have been utilized in the recent past for enzyme immobilization.

SBA-15 (pore size up to 10 nm), a most commonly used mesoporous silica materials, has been investigated widely for immobilization of laccase (Hu et al. 2009, 2007), lipase (Cai et al. 2016; Gholamzadeh et al. 2017; Rios et al. 2016), papain (Dai et al. 2017), glucose oxidase (Majumdar et al. 2016), etc. Although the use of such porous materials, often deals with restriction related to the size of enzyme to be immobilized as this should be lesser then the pore size of matrix, slow loading rate. and nominal enzyme loading (<10 wt.%). In a study conducted by Han et al. (2002), they compared different porous materials (MCFs, SBA-15, SBA-16 and MCM-48) for enzyme immobilization. The authors clearly demonstrated that support matrixes having smaller pore size have limited loading and retained activity. They observed that mesostructured cellular foams (MCFs) obtained highest enzyme loading, owing to their large pore size. However, it has been proven scientifically that the use of support matrix with large pore size might result in high enzyme loading but imparts adverse effect on the enzyme catalytic activity (Weber et al. 2010). The best combination of both enzyme loading and retained activity has been achieved by matching the support matrix pore diameter with the enzyme size (Ikemoto et al. 2010; Sang and Coppens 2011). Surface volume becomes an important aspect while matching of pore diameter and enzyme size, as it has direct impact on loading of enzyme. The high surface volume results in enhanced loading and might further affect the catalytic reaction. Along with pore size and surface volume, the structure of materials also plays an important role with respect to the capacity and activity of the enzymes immobilized. The first generation of mesoporous silica materials had one-dimensional pore structure which resulted in enzyme blocking at pore mouth. The utilization of 3D interconnected pores and highest surface area results in better loading as well as catalytic activity of enzyme (Washmon-Kriel et al. 2000). Surface functionalization of materials having one-dimensional pore structure is an efficient approach to overcome these problems. For instance, Bautista et al. (2010) demonstrated the applicability of functionalized SBA-15 for immobilization of laccase. The immobilized laccase was employed for the degradation of polycyclic aromatic hydrocarbons, naphthalene in absence of any oxidation mediator system. The functionalization of SBA-15 using aminopropyl and aminobutyl functional groups resulted in increased specific surface area, hence improved loading of enzyme. Aminopropyl functionalized SBA-15 exhibited higher stability after multiple reuses than aminobutyl functionalized and non-functionalized SBA-15. The researchers also concluded that lower amino content is preferred to accomplish higher enzyme loading and corresponding catalytic activity, as higher amino content on silica surface limits the diffusion of enzyme molecules in pores of silica material. The plausible mechanism of biotransformation could be that the immobilized laccase system might act as a mediator and acts as a diffusible electron carrier or a co-oxidant that enables the laccase to accept electrons directly from the non-natural substrates (Alcalde et al. 2002). In another study, Mohammadi et al. (2018) demonstrated the immobilization of lipase on amine-functionalized silica nanoparticles (SBA-15) for enantioselective resolution of rac-ibuprofen. The systematic characterization of bionanoconjugate confirmed the high and efficient loading under extremely mild conditions. The effect of reaction temperature and time was also investigated; an increase in either reaction temperature or time resulted in decline in E-value of the reaction. The main cause of this impact was ascribed to increment in the rate of esterification reaction. The use of appropriate functionalization hence becomes a major concern while using these immobilized systems for industrial operations. The incorporation of organic moieties is yet another interesting approach to improve the biotransformation reaction (Hoffmann et al. 2006).

The general mechanism for enzyme loading on mesoporous silica materials is their adsorption in the pores; hence, possibilities are high that the loaded enzyme might desorb completely when the sample comes in contact with the solution. One of the possible solutions for this problem could be coating with nanoscale multilayer shells (Fig. 2) via the sequential deposition of oppositely charged polyelectrolytes (PEs) (Wang and Caruso 2004, 2005). However, with the advancement in synthesis and modification procedure in last few years, a number of mesoporous silica materials are being reported with excellent enzyme loading and improved stability under reaction conditions.

Synthesis procedures are also known to have impact on the immobilization efficiency and catalytic activity of enzyme as different synthesis method results in varying pore size and other surface related properties. In a previous study, Mureseanu et al. (2005) introduced a new mesoporous micelle-templated silica (MTS) route for lipases entrapment. A single pot reaction was carried out, where lipases was entrapped simultaneously with the matrix synthesis, overcoming the limitation associated with the entrapment procedure such as limited covalent bonding, leaching of entrapped enzyme, and limited enzyme–substrate interactions. The authors demonstrated that with use of natural surfactants, an intermediate between enzyme and silica matrix would preserve the enzyme activity along with improved diffusivity of substrate. The as synthesized entrapped lipase exhibited improved catalytic activity then commercially available encapsulated lipase, and was employed for ester hydrolysis. The emergence of hetero-functional supports in the past decade opened up new doors to combine the benefits of both adsorption and covalent binding, making the final product more robust and practical for industrial applications. Adsorption of enzyme on hydrophobic support are reported in the past to preserve the catalytic activity of immobilized enzyme as this method does not induce any kind of conformational change in enzyme structure. However, the enzyme leakage is a major concern in this method, which can be resolved by combining with covalent linkage using epoxy functionalities, a well-established method for enzyme immobilization. Garmroodi

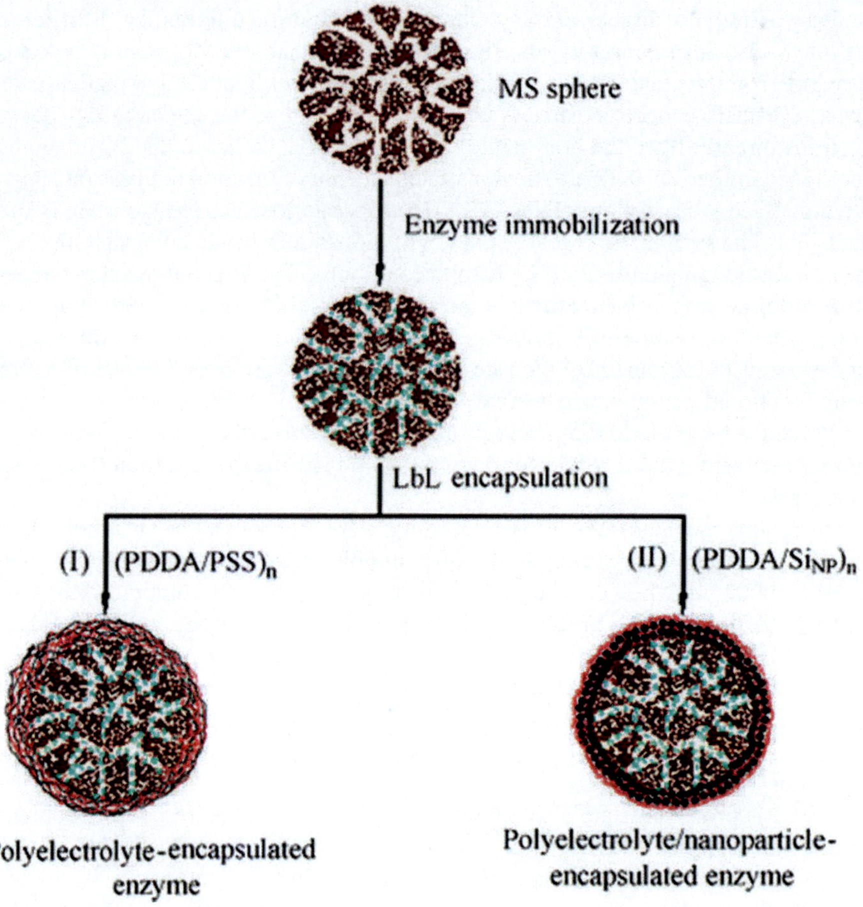

Fig. 2 Schematic representation of enzyme encapsulation using MS spheres as supports. The enzyme is first immobilized and subsequently encapsulated by a **a** PE or **b** PE/nanoparticle multilayer shell (Reprinted with permission from Wang and Caruso (2005) © 2005 American Chemical Society)

et al. (2016) employed such an approach on silica mesoporous nanoparticles with the help of two coupling linkers, one for hydrophobic interaction (octyl group) and other acting as covalent linker (epoxy group) between lipase and support. A twofold improvement in catalytic activity was observed with this approach as compared to the free enzyme. The increase in epoxy group, however, was reported to have negative impact on the specific activity of lipase. These findings might be helpful for further development of immobilized biocatalytic system for industrial applications.

The encapsulation of industrially relevant enzyme in hybrid silica matrix is an alternative approach to impart thermal stability and improving the substrate-enzyme interactions. In a recent study, Verri et al. (2016) encapsulated lipase enzyme in organic–inorganic nanospheres formed by a hybrid silica shell (with an internal liposomal core). The immobilized system was employed in the enantioselective esterification of (R, S)-ibuprofen for potential application in pharmaceutical sector. The use of surfactant in polymerization of silica shell enhanced the catalytic activity. The heterogeneous catalytic reaction carried out in presence of immobilized system resulted in a 16-fold increase in productivity as compared to the free lipase. The immobilized system could be reused for more than nine times with sufficient catalytic activity. Bayramoglu et al. (2015) employed biosilica-polymer particle for immobilization of lipase, and the resultant system was used for transesterification of algal oil. The immobilized enzyme system demonstrated improved conversion efficiency with 96.4% conversion rate as compared to free counterpart (85.7%). The reusability of the immobilized system was evaluated till six cycles which indicated higher operational stability with retaining 76% activity.

The efficiency of biotransformation can also be elevated using appropriate reactor technology. Over the last few years, a lot of scientific efforts have been invested in design of continuous flow bioreactors which could replace batch processes (Lindeque and Woodley 2019). Combination of enzyme immobilization and continuous flow reactors is an obligatory strategy in the current industrial scenario. Boros et al. (2013) immobilized *candida antarctica* lipase B on surface functionalized silica gels for kinetic resolutions of racemic 1-phenylethanol and racemic 1-phenylethanamine. The authors reported that silica gels modified with mixtures of phenyl and aminoalkyl silanes resulted in enhanced durability and a higher degree of enantiomer selectivity at elevated temperatures. They also concluded that higher productivity and selectivity at lower temperature can be achieved by adsorption of enzyme on functionalized silica support, while adsorption combined with cross-linked enzyme is essential requirement to impart these properties (productivity and selectivity) at elevated temperatures. The high surface area, well-ordered porous structure, and great hydrothermal resistance of mesoporous silica particles have gained them reputation as an excellent support matrix for enzyme immobilization. Specifically, the large pore size of silica particles provides a suitable environment for catalytic operation for efficient and continuous biocatalytic reactions. Piao et al. (2019) illustrated the use of laccase immobilized porous silica particles for biotransformation of bisphenol. They employed fluidized bed reactor, first of its kind in means of mesoporous silica, for the continuous remediation of polluted water. They investigated two different approaches of enzyme immobilization, i.e., direct adsorption and cross-linking, and concluded that cross-linked lacasse could be reused for three full cycles without any decrease in the catalytic efficiency. Even after ten complete uses, the immobilized system retained more than half of the initial biotransformation capacity. The degradation of 90% of 25 mg/L bisphenol was achieved in just 8 h of reaction time. This strategy might pave the path for use of mesoporous silica particles in various other biotransformation reactions in a continuous manner.

The enhancement in biocatalytic transformation can be achieved by taking into consideration the synergistic combination of intrinsic properties of enzyme with flow dynamic of microstructured reactors (Dhiman and Agnihotri 2020) . Microreactors provide a combination of high surface area-to-volume ratio, exquisite control of

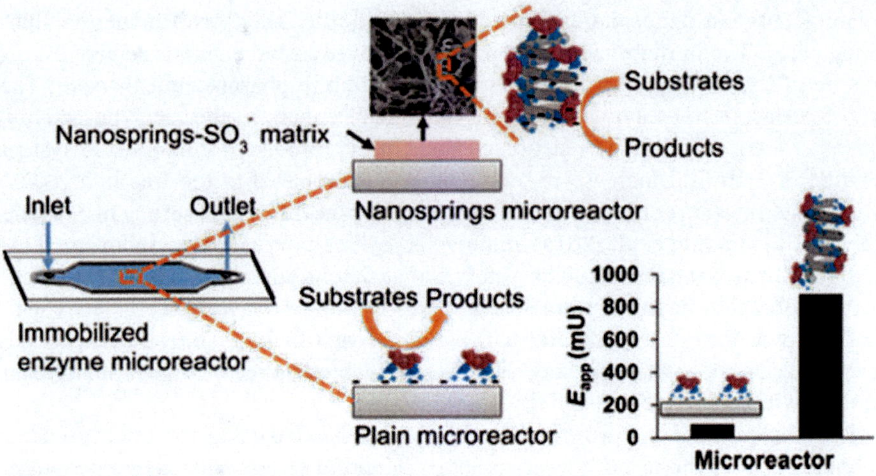

Fig. 3 A comparison of plain and nanospring microreactor for substrate conversion. (Reprinted with permission from Valikhani et al. (2017) © 2017 American Chemical Society.)

temperature, and high rates of mass transport that makes them particularly effective for enzyme-catalyzed reactions. Microfluidic reactors works well with continuous flow of substrate; however, the leaching of biocatalyst with flow poses a great threat to the operational capability for industrial applications. Strong anchoring of the enzyme to reactor wall could be employed as potentially viable solution, and silica materials owing to their high compatibility with solvents as well as enzyme have been employed widely for microfluidic applications (Fu et al. 2012; Schilke et al. 2010). Valikhani et al. (2017) recently introduced a novel concept of nanosprings microreactors with silica-binding modules for sucrose phosphorylase immobilization (Fig. 3). The porous silica nanospring coating in microchannel increased the specific surface area for enzyme immobilization in addition to low resistance to solvent flow. The specific activity of employed enzyme demonstrated 4.5 fold increases upon surface coating with nanosprings as compared to plain surface of microreactor. The introduction of functional group on nanosprings further improved the production of α-glucose 1-phosphate by tenfold. Even after 840 reactor cycles, the microreactor merely lost 15% of its initial conversion rate. The employed approach appears to be broadly applicable to different enzymes as well as microreactor designs at industrial scale.

In another recent study, Muderrisoglu et al. (2018) employed silica gel formulation in microfluidic platform for biotransformation of β-glucuronides. The potential application of microreactors in pharmaceutical industry is aimed in this study. The authors compared the conversion efficiency for batch and continuous operations, and concluded that the use of microfluidic reactor enhanced the yield by 50%. The improvement was ascribed to advanced mass transfer in a microfluidic platform. The impact of flow rate was also investigated; a high flow rate seems to have adverse effect on the catalytic reaction. The change in other process parameters would also have drastic impact on the biotransformation process; however, evaluation of all the parameters is not feasible under microfluidic reactor system. Mathematical modeling tool have been employed in the past to forecast the possible outcome while changing various process conditions in microreactors (Miložič et al. 2017). The models could be employed to optimize design and process condition parameters for high product yield.

The separation efficacy of immobilized system from the reaction solution is an important industrial aspect, from the recyclability for iterative uses and economical point of view. Although, the use of micro-sized inorganic matrices itself implies the easy separation form reaction mixture, but the same cannot be said about nanoscale inorganic matrices. The incorporation of magnetic properties within the support matrix is a very common approach for improving the separation efficiency or retrieval of immobilized system (Zhu et al., 2007). Immobilizing different enzyme on the same support matrix, a concept known as co-immobilization, has further facilitated the minimization of multistep approach for various bioprocess reactions. Lee et al. (2013) demonstrated the co-immobilization of cellulase and glucose isomerase on Fe_3O_4 loaded mesoporous silica nanoparticles for continuous cellulose to fructose conversion. The production efficiency of the continuous process was equivalent to the free enzyme process. Fructose can subsequently be converted into, 5-hydroxyfurfural (HMF), a well-established precursor for biofuels production. The stability and reusability of both the enzyme demonstrated improvement as compared to their free counterpart. Some of the recent studies are also listed in Table 2.

Despite the advancement made in mesoporous silica materials, the industrial applications are still limited because of several factors associated with the synthesis and practical application such as high cost and time duration required for synthesis, limited application due to pore sixe constrain, and limited capability to retain immobilized biomolecules.

3 Zeolites for Enzyme Immobilization

The inherent adsorption capability, unique electrostatic characteristic, wide range of framework arrangements, shape-selective properties, and presence of reactive silanol groups on surface are some of the exciting features which grabbed the attention of scientists for biomolecule anchoring. Enzyme-immobilized zeolites have been used in various other applications such as sensing (Naseri et al. 2018), bioremediation (Wang et al. 2019), biomolecule adsorption (Liu et al. 2017), antioxidant, and antimicrobial applications (Tegl et al. 2018). However, in comparison with ordered mesoporous solids (MCM-41, SBA-15, and FSM-16), zeolite as enzyme carriers for biotransformation has received little attention. Excellent water retention capability and capacity to adsorb free radicals in addition to high chemical and thermal stability favors the use of zeolites in different reaction environment.

The presence of water is mandatory in some enzyme-mediated reaction; hence, the selection of support matrix based on their water retention capability becomes a critical

Table 2 Immobilization of important bioprocess enzymes on mesoporous silica materials

Enzyme	Support matrix	Application	Outstanding feature	**References**
Lipases	Epoxy-functionalized SBA-15	Biodiesel production	20 cycles reuse with 94% retained activity	Babaki et al. (2016)
β-1,3-1,4-glucanase	Porous silica	Production of β-glucooligosaccharides	Reused till ten cycles with 42% retained activity	Cho et al. (2018)
β-galactosidase	Hexagonal ultra-large-pore SBA-15	Production of galacto-oligosaccharides (GOS)	High yield of GOS (20.2%) as compared to free enzyme (11.2%)	González-Delgado et al. (2018)
Protease	Functionalized silica nanoparticles	Whey protein hydrolysis	Efficiently reused up to six times with high degree of hydrolysis	Sinha & Khare (2015)
Glucoamylase	Mesostructured cellular foam silica	Tapioca starch hydrolysis	Dextrose equivalent (DE) values of 81.70% was obtained	Agustian & Hermida (2019)
Phospholipase D	Silica-coated magnetic NPs	Synthesis of functional phosphatidylserine	Eight times reused with 40% retained activity	Han et al., (2019)
Esterase	SBA-15	Hydrolysis and transesterification	Biocatalyst retained more than 20% of its activity even after ten cycles	Bonzom et al. (2018)
β-galactosidase	Silica/chitosan support	Lactose hydrolysis	200 h of continuous use in a fixed-bed reactor with 90% retained activity	Ricardi et al. (2018)
Haloacid dehalogenase	Aminosilane functionalized SBA-15	D-2-hydroxy alkyl acid production	Retained 60% of its original activity after seven cycles of reuse in batch operation Retained 100% of its original activity after ten cycles in the semi-continuous flow reactor	Jian et al. (2016)

(continued)

Table 2 (continued)

Enzyme	Support matrix	Application	Outstanding feature	**References**
Cyclodextrin glycosyltransferase	Pre-activated silica	Productionof α-, β-, or γ-cyclodextrins (CD)	Productions of 4.9 mg mL^{-1} of α-CD, 3.6 mg mL^{-1} of β-CD, and 3.5 mg mL^{-1} of γ-CD	da Natividade Schöffer et al. (2017)
Lipase	Electrodeposited silica films	Hydrolysis of 4-nitrophenyl butyrate	Conversion efficiencies of up to 100% after eight cycles in continuous fluidic bioreactor	Xiao et al. (2018)
Catalase	Silica particles	Decomposition rate of hydrogen peroxide	Reactor could operate continuously for 120 h with retaining 64% activity	Bayramoglu et al. (2016)
Lipase	Silica modified with protic ionic liquid	Ethyl esters from crude vegetable oil	Maximum conversion of ethyl esters47 $\pm$ 2.0% in 72 h	Santana et al. (2018)

factor specifically for hydrolytic enzymes acting in non-aqueous solvent. Zeolites are well known for their high water storage characteristic in intracrystalline void space, while external surface with high specific surface area is available for enzyme–substrate interaction. They can be utilized as an interface between aqueous phase present in zeolite, and organic medium on the outer surface of zeolite containing the substrates. The main mechanism for anchoring of biomolecule is via adsorption on external surface as microporous nature of zeolite limits the loading in internal matrix (Gonçalves et al. 1996). The hydroxyl group on the surface facilitates the adsorption process via hydrogen bonding between zeolite and biomolecules (Datta et al. 2013). The hydrophobicity of zeolites can be modified via thermal and chemical treatments, making them suitable for both organic and aqueous reaction medium. Investigation has been carried out in the past for the effect of Si/Al ratio on hydrophobicity. Lie and Molin (1991) systematically demonstrated that high Si/Al ratio results in nominal hydrophilicityas compared to low Si/Al ratio. The particular ratio of Si/Al could be employed for specific reaction as the authors concluded that high Si/Al ratio zeolite provides favorable environment for esterification, while low Si/Al ratio zeolite was better for hydrolysis reactions. Along with these features, zeolites support provided distinct advantage in respect to easy recovery and reuse. In a yet another study, Gonçalves et al. (1996) employed numerous zeolites (NaA, NaX, NaY, NaUSY, and modified forms) for immobilization of recombinant cutinase. They investigated the impact of structural and composition variations on catalytic activity of cutinase for hydrolysis of tricaprylin. Based on a thorough study focused on impact of water content, the authors concluded that zeolites could serve as excellent support matrices for immobilized enzymes under organic solvent medium. Therefore, the importance of framework composition is an important aspect while designing zeolite immobilized biocatalysts for bioprocess or biotransformation application. In a similar study, Serralha et al. (1998) immobilized recombinant cutinaseon various zeolites for alcoholysis reaction of butyl acetate with hexanol, in organic media (isooctane). They also compared the enzymatic stability as a factor of compositional changes in zeolite, water content of reaction, as well as different support matrices (polyamide accurel-PA6, silica and alumina). The zeolites with lower Si/Al ratio along with little acidity were found to be better support system for the catalytic reaction. Yagiz et al. (2007) examined the effect of four different types of zeolites on lipase immobilization for biodiesel production form waste oils. They achieved the enzyme loading of 9 mg/g, even though the loading was lower than hydrotalcite (13 mg/g). Waste cooking oil with methanol was catalyzed using immobilized lipase to produce methyl esters. The impact of various physicochemical parameters on stability and activity of adsorbed enzyme was examined, and the authors concluded that immobilized lipase can be reused seven times with significant retained catalytic activity.

Electrostatic characteristic of zeolites governs additional interaction aspect to biomolecules, and negatively charged aluminosilicate crystalline structures provide sites for cationic biomolecules. The additional property such as resistance to biodegradation and complete control on synthesis process to modify specific properties makes them a suitable support for enzyme immobilization (Xing et al. 2000). The increase in external surface area by use of nanocrystallites or delaminated zeolites often results in reduced processability. Utilization of zeolites with auxiliary mesoporosity is another approach for improvement in loading capacity for large carriers. Hierarchical zeolites have been examined to enhance the involvement of catalytic sites adsorbed in the micropores (Perez-Ramirez et al. 2008). The introduction of mesoporosity can be induced using various methods, i.e., by templating, controlled crystallization, or demetallation. The composition also has impact on the degree of mesoporosity of zeolites as high Si/Al ratio results in high mesoporosity and is commonly preferred over dealumination. The surface characteristic could be tuned via oragnic functionalization of mesopores, which would further improve the compatibility for enzyme immobilization (Mitchell and Pérez-Ramírez 2011).

Mitchell and Pérez-Ramírez (2011) compared pure microporous ZSM-5, functionalized hierarchical ZSM-5 zeolite, and mesoporous MCM-41 for lipase immobilization and corresponding catalytic activity. The studies performed indicated a correlation among mesoporous surface area and enzyme uptake, which in turn affected the catalytic activity. They further concluded that introduction of functional moieties on mesoporous wall is matter of prime importance as purely inorganic framework demonstrated rapid enzyme leaching. The lipase immobilized on thiol functionalized hierarchical ZSM-5 zeolite exhibited excellent catalytic activity as compared to other support matrices. The reason for high loading and catalytic activity of thiol functionalized hierarchical ZSM-5 was ascribed to increased mesoporous surface area. The disorder and interconnectivity of hierarchical zeolite facilitates the mesopores accessibility after surface functionalization.

The immobilization using a cross-linker is a common approach to minimize the leaching of enzyme in reaction medium. Glutaraldehyde has been used in numerous studies as cross-linker for immobilization on different kind of zeolites. In previous study, sodium aluminosilicate (NaA) and calcium aluminosilicate (CaA) were functionalized with glutaraldehyde for tyrosinase immobilization. The authors reported maximum of 85% loading in case of functionalized calcium aluminosilicate (CaA) zeolite. The repeated batch reactions till 48 h did not have any significant impact on the catalytic activity. A production rate of 31–36 mg/L/h was achieved for 7 h catalytic reaction for production of L-3,4-dihydroxyphenylalanine (Seetharam and Saville 2002). Denaturation of enzyme adsorbed on the external surface of zeolite is a serious concern while addressing industrial scope of these matrices. Starch hydrolyzing enzymes have a wide range of applications in textile, paper, detergent, and fermentation industries, and therefore, enhancement in their operation stability becomes priority, especially with respect to economical point of view. In a recent report, researchers investigated the application of nanopore zeolites for improving the stability of immobilized α-amylase. Glutaraldehyde was employed as a cross-linker between support matrix and amylase to enhance the loading and stability. The binding efficiency of enzyme was calculated to be more than 58% of the initial enzyme activity. The immobilized enzyme was tested under varying process parameters such as range of pH, temperature, reaction, and storage time (Talebi et al. 2016).

The recyclability of inorganic matrices has been improved using introduction of magnetic material, as discussed for mesoporous silica materials. In a recent study,

Wang et al. (2019) reported the synthesis of Fe_3O_4@ZIF-8 nanoparticles for laccase immobilization. The immobilized enzyme retained 46% of initial catalytic activity even after 6 h incubation at 80 °C, demonstrating the thermal resistant capability of immobilization matrix. The impact of storage time on activity of immobilized laccase was also investigated; it lost only about 13% of catalytic activity after ten days of storage time. Although the study is focused on organic pollutant remediation, this study has opened up the scope for investigating magnetic zeolite composites for immobilization of biocatalysts and their industrial applications in near future.

The use of zeolites with other materials could be a useful approach as the final composite will be benefited by the characteristic features of both the materials. One such approach could be utilized with metal organic frameworks, a well-known matrix for enzyme immobilization (Lyu et al. 2014; Wu et al. 2015). Zeolitic imidazolate frameworks (ZIFs) are one such material which is used in various applications, as well as for enzyme immobilization. A range of enzymes have been immobilized in the recent past on ZIF-8 such as glucose oxidase (Mylonas and Svergun 2007), horseradish peroxidase (Liang et al. 2015), Cyt c (Lyu et al. 2014), lipase (He et al. 2016), (R)-1-phenylethanol dehydrogenase (Pu et al. 2019), β-glucosidase (Shi et al. 2019), and cellulose (Xu et al. 2018) for different applications.

The immobilization of multiple enzymes is an interesting approach for enzyme-assisted biotransformation. In a two sub-sequential enzymatic reactions, the transfer of intermediate product from a production unit to other might come in contact with contamination which would have negative effect on the activity and specificity of enzyme employed later. The multi-enzyme system eliminates this risk by great folds. Other than that, sometimes the conversion of specific substrate or intermediate by-products prior to the other substrate conversion is important, specifically in case of biomass transformation into products. In a recent study, Care et al. (2017) co-immobilized three enzymes (β-Glucosidase, β-Xylanase, and β-mannanase) on zeolite support matrix for biomass hydrolysis. The authors employed linker technology to impart directionality and orientation to the immobilization of enzymes, which resulted in enhanced enzyme reusability. The strategy can be used for hydrolysis of a diverse lignin-based biomass sources. The latest published reports using zeolite/zeolite-based materials for enzyme immobilization are listed in Table 3.

Extraordinary advances have been made in the synthesis and functionalization of zeolites, further improving their loading efficiency as well as the protection capability for enzyme immobilization. Introduction of trivalent metal atoms such as Al, B, Ga, and Fe into the framework have been reported in the past to alter the acidity and other associated properties of zeolites (Yuan et al. 2002). Recently Marthala et al. (2017) fabricated hierarchical nanosheet assemblies of boron-containing MFI-type zeolites for lipase immobilization. The incorporation of heteroatoms specifically boron induces generation of high amount of defects (terminal silanol groups) around boron atom, resulting in easy functionalization and high enzyme loading. The optimum results for biotransformation of substrate into butyl propionate (82% yield in 48 h) were observed in case of Si/B = 50.

Table 3 Immobilization of important bioprocess enzymes on zeolites

Enzyme	Support matrix	Application	Outstanding feature	References
Alkaline protease	ZSM-5 zeolite	Milk coagulant for local dairy products	Retained 80% catalytic activity even after 12 cycles	Kumari et al. (2015)
Cellulase	Zeolitic imidazolate framework (ZIF-8)/polyvinylidene fluoride hybrid membranes	Catalytic reaction	Maintained 42% activity after seven cycles	Xu et al. (2018)
Glucose oxidase	ZIF-8-coated magnetic regenerated cellulose-coated nanoparticles	Stability of immobilized enzyme	24.2% enhanced enzyme relatively activity after immobilization	Cao et al. (2017)
Lipases	Nanozeolite NaX	Transesterification of palm oil to fatty acid ethyl esters	94% yield of fatty acid ethyl esters	de Vasconcellos et al. (2015)
Xylanase	Acid-activated zeolite matrix	Reducing sugar production	Reused six times with 56.77% retained activity	Na'imah et al. (2017)
α-amylase	Nanopore Zeolite	Improve enzyme's functions and stability	58.44% binding efficiency α-amylase	Talebi et al. (2016)
Lipases	Nanozeolites functionalized with APTMS	Ethanolysis transesterification of microalgae oil	Fatty acid ethyl esters yield more than 93% with stability up to five cycles	de Vasconcellos et al. (2018)
Lipase and Laccase	ZIF-8 and ZIF-zni	Electrochemical detection of the substrates	Immobilization on ZIF-zni results in better catalytic activity	Naseri et al. (2018)

4 Clay Materials for Enzyme Immobilization

Clay minerals, apart from mesoporous silicas and zeolites, owing to their inherent properties have also been investigated in the past for biomolecule loading/immobilization. Although zeolites and clay minerals (sepiolite, kaolinite, montmorillonite, smectite, bentonite, halloysite, etc.) are same in respect to their compositional aspect, however, the crystalline structures are different. Clay minerals are basically hydrous aluminium or magnesium phyllosilicates possessing a two-dimensional layered structured (Bergaya and Lagaly 2013). The distinguished layered structure feature results in large surface area, swelling, and ion exchangeability (Zhou et al. 2011; Zhou and Keeling 2013). The vast availability, biocompatible nature, and low cost provide strong feasibility for industrial-level applications. The hydrophobic functional group (Si–O–Si) present on the outer surface and hydrophilic groups (–OH) at broken edges of tetrahedral and octahedral sheets is available for enzyme anchoring. Furthermore, the introduction of additional functionalities is doable, as the surface modifications of clay minerals by organic, polymeric, biological means are simple and economical

(Chang and Juang 2007). The improvement in enzyme loading efficiency, nominal leaching, more protective nature against process conditions, and storage stability is reported to improve after additional functionalization. The application of various clay minerals for enzyme immobilization and their subsequent application in area of biosensing, bioremediation, etc., has experienced exponential growth in the past few years. However, the reports on application of enzyme immobilized clay minerals for biotransformation are limited, and proper attention needs to be diverted to this area.

The interactions between enzyme and clay minerals play major role in deciding the stability and catalytic activity of the composite system. The interactions can be classified majorly in two classes: non-covalent interaction and covalent interaction. Non-covalent interaction can be further branched into adsorption (Van der Waals forces, electrostatic interaction, hydrogen bonding, and hydrophobic interaction) and intercalation between layered structure of clay minerals. Non-covalent interactions preserves the conformational structure of enzyme while covalent interaction results in strong attachment of enzyme with the support matrix. The interaction depends on the type of clay minerals, enzymes, and functionalization employed for immobilization. Different clay minerals have different anchoring sites for covalent and non-covalent interactions, so does the type of functional group introduced after modifications. The different types of clay minerals as immobilization support of industrially important enzyme and latest advancement made in the area for transformation of these laboratory scale procedure to industrial level are summarized according to the literature published in recent past.

Yang et al. (2019) investigated different clay minerals such as hematite, kaolinite, and montmorillonite for β-glucosidase adsorption. The adsorbed enzyme activities were determined for transformation of two organic substrates (cellobiose and indican) under varying temperature. Hematite and kaolinite were better at adsorption as compared to montmorillonite. The immobilized system indicated 2-fourfold enhanced catalytic activity when indicant was employed as substrate, and 5-ninefold increased activity when substrate was cellobiose. These results indicated that choices of clay mineral as well as substrate employed have direct impact on immobilized enzyme activities; hence, in-depth modeling of the complex enzyme support complex is essential for further advancements. As demonstrated by the authors in the study report above, the classification of different clay minerals on their characteristic structure basis is also important as it plays important part in enzyme loading efficacy. Kaolinite, $Al_2O_3 \cdot 2SiO_2 \cdot 2H_2O$, is made up of a single tetrahedral sheet condensed with a single octahedral sheet as one unit. The narrow interlayer spacing restricts the intercalation of enzymes, leaving behind only the external (hydrophobic interaction) and edge (hydrophilic interaction) surfaces available as anchoring sites for enzyme immobilization. Montmorillonite, on the other hand, have high interlayer spacing which could be used in addition to hydrophobic force and hydrogen bonding interactions for enzyme immobilization. The negative electrostatic charge on layer surface is balanced by cations in the interlayer space. Halloysite was described as a dioctahedral 1:1 clay mineral of the kaolin group, and even though halloysite is chemically similar to kaolinite, there is significant structural difference, as in halloysite there is a water monolayer between the internal and external surface. The structural formula of hallysite is $Al_2(OH)_4Si_2O_5 \cdot nH_2O$. Halloysite have multilayer tube-like structure resulting from the infolding of mineral layers under favorable environmental circumstances. The unique tubular structure provides additional space to accommodate biomolecules, resulting in high enzyme loading efficacy. The internal surface consists positively charged aluminol (Al–OH) groups as compared to the negatively charged outer surface which is made up of siloxane group (Si–O–Si).

The adsorption (hydrophilic/hydrophobic, electrostatic, hydrogen bonding, etc.) interactions between enzyme and clay materials are studied extensively in the past. The hydrophobic interactions are reported to direct enzyme molecules toward a particular orientation and protecting the active state enzyme after immobilization (Wannerberger et al. 1997). Synthesis of industrially relevant lipophilic antioxidants has been reported by researchers using lipase B adsorbed on kaolin (Tanasković et al. 2017). The impact of pH and ionic strength on the adsorption of enzyme was investigated. The authors studied the adsorption kinetics and found that immobilization occurred via the Langmuir model, and electrostatic forces along with hydrophobic and Van der Waals forces were attributed as major interaction mechanism behind it. The negatively charged sites at edges of clay minerals also act as adsorption sites for positively charged enzymes via electrostatic interaction. Generation of such sites are directly correlated with the deprotonation of hydroxyl groups, which can be induced by changing the pH of the solution. Öztürk et al. (2016) used neat and organo-modified forms of sepiolite and montmorillonite for *Candida antarctica* lipase B (CALB) immobilization. The immobilized lipase was employed for polyester synthesis. The researchers evaluated the influence of solvent, enzyme loading, cross-linking, and type of clay support on immobilization efficiency and catalyst hydrolytic activity. The authors concluded that organo-modified clays demonstrated superior catalytic activity as compared to their neat counterpart. Golbaha et al. (2016) employed montmorillonite for immobilization of lipase and subsequently used them for hydrolysis of tributyrin. The immobilization of enzyme was performed via adsorption mechanism through hydrogen bonding of enzyme functionalities with the silanol groups. The hydrolysis activity of adsorbed enzyme was found to be superior to the free form. The reusability of the immobilized system was tested till five time reuse. Tully et al. (2016) used electrostatic interaction to load different enzymes (laccase, glucose oxidase, and lipase) in the positively charged hollow lumen. The enzymes were incubated with halloysite nanotubes above their isoelectric point. The authors showed that one-third of loaded enzyme leached out within 5–10 h, leaving behind the two-third of total enzyme loaded. The loaded enzyme demonstrated enhanced stability at acidic pH and higher temperature. In case of adsorption interaction between enzyme and matrix, the rapid leaching of enzyme creates a serious concern regarding industrial applications, which can be dealt with by using covalent interaction.

The presence of cationic moieties in interlayer space of clay materials have been utilized for enzyme loading in the past. These moieties can be replaced with positively charged enzyme molecules such as catalase (Alkan et al. 2005), laccase (Wang et al. 2013), β-glucosidase (Serefoglou et al. 2008) via cation exchange reaction (Johnston et al. 2011). However, cation exchange reactions are only feasible in case of small enzyme molecules, due to the limited interlayer space of clay minerals. The interlayer spacing can be modified via replacing the inorganic cations in the interlayer space with bulky organic cations and then replacing these organic cations with positively charged molecules (Fatimah and Huda 2013; Fujimori et al. 2014).

Acid activation, another aspect for modification of clay mineral, has been studied to improve enzyme immobilization and enhancing other inherent properties. Wu et al. (2019) fabricated Fe_3O_4-acid activated montmorillonite composite for cellulase enzyme immobilization. Adsorption and activity of cellulase were evaluated with respect to the change in structure of composite. It was found that size of Fe_3O_4 was affected by the amount of acid-activated montmorillonite, and content of Fe_3O_4 in the final composite had significant impact on adsorption of cellulase. The highest cellulase loading of 860 mg/g with 51.88% catalytic activity was obtained at mass ratio (Fe_3O_4/montmorillonite) 1:1. The increase in clay mineral mass adversely affected the adsorption and corresponding catalytic activity of cellulase. Higher fraction of α-helix and lower fraction of β-sheet was obtained when cellulase was immobilized on montmorillonite as compared to the enzyme adsorbed on composite.

The addition of polymeric molecules is a well-known approach for introduction of new functionalities in clay minerals. Chitosan has been widely employed to provide additional functional group (hydroxyl and amino groups) on surface of different clay materials (Dhiman and Agnihotri 2020). Researchers investigated the impact of various modifications, i.e., chitosan and copper, on sepiolite for laccase loading and corresponding catalytic activity (Olshansky et al. 2018). They observed a whooping increase by 250% in catalytic activity with Cu (II) adsorption on sepiolite and 500% increase in case of chitosan-sepiolite system. The combination of both chitosan and Cu (II) demonstrated an increase of 700% as compared to the native form. The retention capability of the as synthesized system was also found to be about 90% of initial loading. This study further confirms that addition of exchangeable cationic metals could improve the overall stability and activity of immobilized enzyme while using clay materials as inorganic solid support matrix. In a recent report, researchers developed a chitosan/clay nanocomposite film system for wine making application by immobilizing proteolytic enzyme pineapple stem bromelain (Benucci et al. 2018). They also investigated the impact of various clay materials unmodified and organically modified as well as their amount for superior catalytic activity. The different chemical composition, morphology, hydrophilicity/hydrophobicity, and surface charge gave rise to different interaction between support and enzyme, in turn affecting the enzyme loading efficiency. The introduction of chitosan enhanced the mechanical properties of the resulted films but adversely affected the catalytic activity of immobilized enzyme. Thus, the impact of incorporating additional moieties must beanalyzed properly. In another study, β-glucosidase was covalently immobilized on clay composed of montmorillonite and layer silicates with chitosan (Chang et al. 2008). The as prepared beads were tested for their stability and catalytic activity under varying reaction environment. Further, the catalytic activities of both dry and wet beads were evaluated, and it was found that use of the dried materials led to higher catalytic efficiency. In a recent study, magnetic halloysite nanotube functionalized with chitosan were utilized for covalent immobilization of laccase (Kadam et al. 2018). The synthesized system was able to achieve 92.47 mg/g of laccase loading with 92% of enzyme activity recovery. Higher pH and temperature stability was observed for immobilized laccase as compared to free counterpart. The biocatalytic system could be reused successfully for 11 cycles with 33% retained activity. The employment of catalytic films is becoming an important aspect for industrial sector. Recently, pineapple juice clarification was done using pectinase covalently immobilized on alginate-montmorillonite beads (Mohammadi et al. 2019). Authors did not observed any change in optimum temperature after immobilization, however, optimal pH changed from 5.5 to 5.0. The kinetic parameter of free and immobilized enzyme were also determined. The affinity of enzyme–substrate complex demonstrated an increase after immobilization, which resulted in improved bioconversion. The immobilized catalyst system retained more than 53% of initial activity even after six consecutive cycles of catalytic reaction. In another report, Sun et al. (2017) synthesized a halloysite nanotube-chitosan membrane for lipase immobilization. Loading of 3 wt.% lipase was achieved via electrostatic interaction. Long-term usability, enhanced pH, and temperature stability were displayed by catalytic films. The resultant system could retain more than 90% of initial activity at elevated temperature, i.e., 80 °C, and 85% activity even after ten cycles of use. The authors hypothesized the application of antimicrobial enzyme–lysozyme along with this system for anti-biofouling applications in industrial sector. The clay/chitosan composite make appropriate support for covalent immobilization of enzymes, with improved thermal and chemical resistance. Along with that, chitosan acts as a binder to hold together the clay particles as a unique structure. The composite system also provides the favorable micro-environment for the enzyme–substrate complex, resulting in enhanced catalytic activity as compared to their free counterpart. The application of clay/chitosan complex for food industry enzymes are well summarized by Cacciotti et al. (2019).

The composites of clay minerals have also been studied for enzyme immobilization. Wang et al. (2015) immobilized lysozyme on natural nanotube/layered double hydroxide composites with flower-like morphology to evaluate the immobilization potential of the matrix (Fig. 4).

The introduction of functional groups (amino, epoxy, carbonyl) on clay mineral is yet another strategy widely employed for covalent immobilization of enzyme on clay minerals. Recently Kołodziejczak-Radzimska and Jesionowski (2019) examined the impact of introduction of different functional groups on halloysite nanotubes for aminoacylase immobilization. Authors concluded that amino-functionalized halloysite nanotube performs better at varying process condition and have high

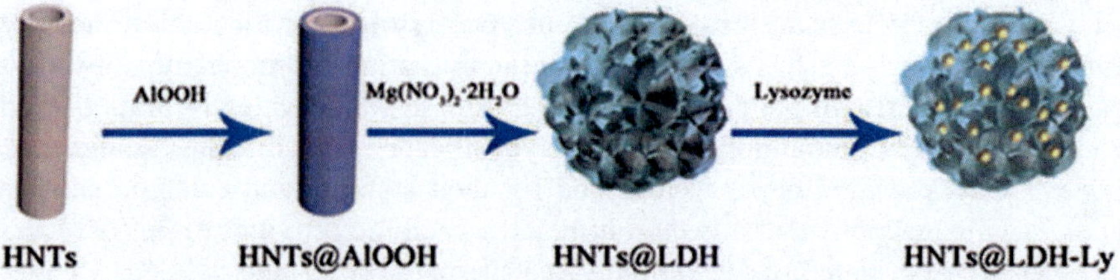

Fig. 4 Schematic representations of the overall preparation process of HNTs@LDH. Yellow spheres indicate lysozyme molecules (Reprinted with permission from Wang et al. (2015) © 2015 American Chemical Society)

reusable potential. In another study, researchers immobilized α-amylase on aminosilane functionalized halloysite nanotube (Pandey et al. 2017). The impact of time, pH, temperature, and metal ions on the enzymatic activity of immobilized α-amylase was also evaluated. The presence of metal ions seems to have less adverse effect on the immobilized enzyme. The optimum pH and temperature for best catalytic activity was found to be 7.4 and 40 °C. Kadam et al. (2017) employed a covalent immobilization for laccase immobilization on aminosilane functionalized halloysite nanotubes. Glutaraldehyde was employed as a cross-linker to form covalent interaction between functional moieties of laccase and support matrix. The separation efficiency of immobilized system was improved by using magnetically modified halloysite nanotube. They researchers achieved 84.26 mg/g of enzyme loading with 90% activity recovery. The stability and activity of immobilized laccase enhanced significantly under a range of pH and temperature. The immobilized system retained more than 90% of its initial activity after nine cycles of catalytic reactions. In a recent study, Sillu and Agnihotri (2020) used amino-functionalized magnetic halloysite nanotubes for covalent immobilization of cellulase (Fig. 5). The enzyme loading of 111.6 mg/g support with 93.5% retained catalytic activity was successfully achieved. The immobilized system demonstrated high stability against varying process parameters (pH, temperature, ionic liquid concentration). The as synthesized nanobiocatalyst was employed for hydrolysis of sugarcane bagasse extracted cellulose with high glucose yield. Study indicated the potential for of clay minerals for immobilization of different industrially relevant enzyme for industrial applications.

Some of the recent reports are listed in Table 4. As can be clearly summarized from the table, use of clay minerals for enzyme immobilization have been studied extensively; however, industrial biotransformation using these support matrix is lacking behind. The capability of these materials for enzyme immobilization is immense, and the exploration of these materials in industrial sector will certainly pave new path for sustainable and economic production of important bioproducts in near future.

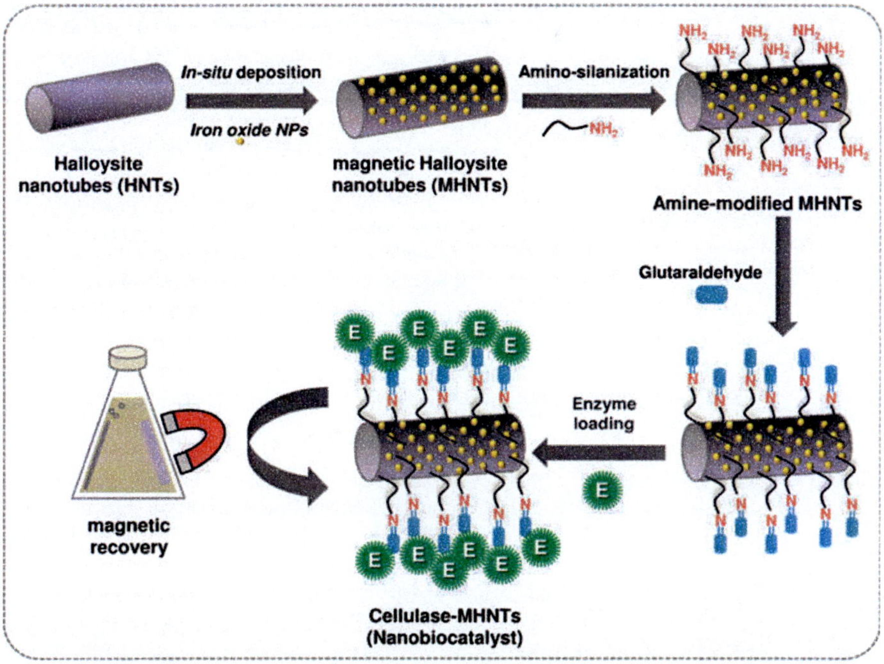

Fig. 5 Representative illustration for immobilizing cellulase enzyme on to halloysite nanotubes as a template/matrix using aminosilane as a cross-linker. Prior to this, iron oxide nanoparticles were synthesized over halloysite nanotubes in concurrent with their deposition, rendering the nanobiocatalyst recoverable using a magnet (Reprinted with permission from Sillu and Agnihotri (2020) © 2020 American Chemical Society)

5 Conclusions

Design of robust biocatalysts for industrial sector has emerged as a rapidly growing research and design filed, employing the fabrication of flexible yet strong economic support matrices and design of appropriate reaction reactors. The exciting challenges to do the same will pave the path for a bright future in nanobioprocessing sector. Recent advancements into synthesis and functionalization of various inorganic nanomaterials could possibly lead us to generate tuneable "support matrix," which can be employed for demand-specific biotransformation applications. These matrices could even be employed to a large group of bioprocessing applications, irrespective of enzyme immobilized, reaction condition, and sale of production. However, the interdisciplinary efforts in this direction would be crucial for two reasons. First, synthesis of an immobilized system and second implementation of as synthesized system at industrial scale with demanded physico-chemical parameters. The high cost of bioprocessed products could also be reduced by using a well-designed support matrix enhancing the stability as well as life cycles of the enzyme. The incorporation of more than one enzyme on the same immobilizing matrix is a promising yet

Table 4 Immobilization of important bioprocess enzymes on clay minerals

Enzyme	Support matrix	Application	Outstanding feature	References
Endoglucanase	Functionalized kaolin	Activity analysis	Retained 80% catalytic activity after 8 cycles	de Souza Lima et al. (2019)
Lysozyme	Nanotube/layered double hydroxide composites	To determine immobilization potential	High enzyme loading of 237.6 mg/g support	Wang et al. (2015)
Laccase	Chitosan–halloysite hybrid porous microspheres	Phenols removal	The removal efficiency could reach as high as 95.0%	Yao et al. (2015)
Lipase	Zirconium-pillared clay	Biodiesel production	High catalytic activity	Colín-Luna et al. (2018)
Catalase	Chitosan and chitosan–bentonite	To compare immobilization potential	Chitosan–bentonite complex retained 70% activity after 20 cycles	Kaushal et al. (2018)
Glucose Oxidase	Tetramethyl ammonium hydroxide (TMAH) modified bentonite	To determine capability to immobilize glucose oxidase	84.71% loading with 5% (v/v) TMAH	Chrisnasari et al. (2015)
Laccase	Poly(diallyldimethylammonium chloride) (PDDA)-modified halloysite nanotubes	Improve the activity and stability	Immobilized enzyme retained more than half of initial activity after 10 cycles	Chao et al. (2017a, b)
Lysozyme	Bentonite–tyrptophane microcomposite	Adsorption of enzyme	Enzyme adsorption of 365.16 mg/g support	Kalburcu et al. (2015)
Laccase	Porous polyvinyl alcohol/halloysite hybrid beads	To improve the stability and reusability of enzyme	81.17% of enzyme activity could be retained after storage at 4 °C for 5 weeks	Chao et al. (2017a, b)
Proteases	Alginate and alginate-bentonite microcapsules	To compare entrapment efficiency	100% retained activity of entrapped enzyme	Rodriguez et al. (2018)
Cellulase	Attapulgite nanoclay	To determine immobilization efficiency	High retained activity and reusable potential	Zang et al. (2015)
Phosphatase	Nanoclay-phosphatase	To determine adsorption capacity	Secondary conformational changes of protein asa result of adsorption	Paul et al. (2015)

(continued)

Table 4 (continued)

Enzyme	Support matrix	Application	Outstanding feature	References
Lipase	Bentonite	To evaluate electrostatic interaction	Retained 90% catalytic activity after 12 time use	Cao et al. (2016)

complex approach. The impact of product formation on the stability/activity of both enzymes needs to be taken into consideration. Employment of simulation tools to predict the product formation and inhibition kinetics could be a futuristic thing.

Acknowledgements This work was financially supported by Department of Science and Technology, India (DST-SERB), under the project Grant YSS/2015/001599, dated 23 March 2016 in Engineering Sciences.

References

Agnihotri S, Dhiman NK, Tripathi A (2018) Antimicrobial surface modification of polymeric biomaterials. Elsevier, pp 435–486

Agnihotri S, Bajaj G, Mukherji S, Mukherji S (2015) Arginine-assisted immobilization of silver nanoparticles on ZnO nanorods: an enhanced and reusable antibacterial substrate without human cell cytotoxicity. Nanoscale 7(16):7415–7429

Agnihotri S, Dhiman NK (2017) Development of nano-antimicrobial biomaterials for biomedical applications. In: advances in biomaterials for biomedical applications. Springer, pp 479–545

Agnihotri S, Mukherji S, Mukherji S (2012) Antimicrobial chitosan–PVA hydrogel as a nanoreactor and immobilizing matrix for silver nanoparticles. Appl Nanosci 2(3):179–188

Agnihotri S, Mukherji S, Mukherji S (2013) Immobilized silver nanoparticles enhance contact killing and show highest efficacy: elucidation of the mechanism of bactericidal action of silver. Nanoscale 5(16):7328–7340

Agnihotri S, Mukherji S, Mukherji S (2014) Size-controlled silver nanoparticles synthesized over the range 5–100 nm using the same protocol and their antibacterial efficacy. RSC Adv 4(8):3974–3983

Agnihotri S, Mukherji S, Mukherji S (2019) Impact of background water quality on disinfection performance and silver release of immobilized silver nanoparticles: modeling disinfection kinetics, bactericidal mechanism and aggregation behavior. Chem Eng J 372:684–696

Agnihotri S, Sillu D, Sharma G, Arya RK (2018) Photocatalytic and antibacterial potential of silver nanoparticles derived from pineapple waste: process optimization and modeling kinetics for dye removal. Appl Nanosci 8(8):2077–2092

Agustian J, Hermida L (2019) Capability of immobilised glucoamylase on mesostructured cellular foam silica to hydrolyse tapioca starch. In: Paper presented at the IOP conference series: materials science and engineering

Ahmad R, Sardar M (2015) Enzyme immobilization: an overview on nanoparticles as immobilization matrix. Biochem Anal Biochem 4(2):1

Ajayi OA, Nok AJ, Adefila SS (2012) Immobilization of cassava linamarase on kankara kaolinite clay. J Nat Sci Res 2:55–62

Alcalde M, Bulter T, Arnold FH (2002) Colorimetric assays for biodegradation of polycyclic aromatic hydrocarbons by fungal laccases. J Biomol Screen 7(6):547–553

Alkan S, Ceylan H, Arslan O (2005) Bentonite-supported catalase. J Serbian Chem Soc 70(5):721–726

An N, Zhou CH, Zhuang XY, Tong DS, Yu WH (2015) Immobilization of enzymes on clay minerals for biocatalysts and biosensors. Appl Clay Sci 114:283–296

Babaki M, Yousefi M, Habibi Z, Mohammadi M, Yousefi P, Mohammadi J, Brask J (2016) Enzymatic production of biodiesel using lipases immobilized on silica nanoparticles as highly reusable biocatalysts: effect of water, t-butanol and blue silica gel contents. Renew Energy 91:196–206

Bautista LF, Morales G, Sanz R (2010) Immobilization strategies for laccase from Trametes versicolor on mesostructured silica materials and the application to the degradation of naphthalene. Bioresour Technol 101(22):8541–8548

Bayramoglu G, Akbulut A, Ozalp VC, Arica MY (2015) Immobilized lipase on micro-porous biosilica for enzymatic transesterification of algal oil. Chem Eng Res Des 95:12–21

Bayramoglu G, Arica MY, Genc A, Ozalp VC, Ince A, Bicak N (2016) A facile and efficient method of enzyme immobilization on silica particles via Michael acceptor film coatings: immobilized catalase in a plug flow reactor. Bioprocess Biosyst Eng 39(6):871–881

Bayramoglu G, Senkal BF, Arica MY (2013) Preparation of clay–poly (glycidyl methacrylate) composite support for immobilization of cellulase. Appl Clay Sci 85:88–95

Benucci I, Liburdi K, Cacciotti I, Lombardelli C, Zappino M, Nanni F, Esti M (2018) Chitosan/clay nanocomposite films as supports for enzyme immobilization: an innovative green approach for winemaking applications. Food Hydrocoll 74:124–131

Bergaya F, Lagaly G (2013) Handbook of clay science, vol 5, Newnes

Bharti S, Agnihotri S, Mukherji S, Mukherji S (2015) Effectiveness of immobilized silver nanoparticles in inactivation of pathogenic bacteria. J Environ Res Dev 9(3A):849–856

Biermann U, Bornscheuer U, Meier MAR, Metzger JO, Schäfer HJ (2011) Oils and fats as renewable raw materials in chemistry. Angew Chem Int Ed 50(17):3854–3871

Bonzom C, Schild L, Gustafsson H, Olsson L (2018) Feruloyl esterase immobilization in mesoporous silica particles and characterization in hydrolysis and transesterification. BMC Biochem 19(1):1

Boros Z, Weiser D, Márkus M, Abaháziová E, Magyar Á, Tomin A, Koczka B, Kovács P, Poppe L (2013) Hydrophobic adsorption and covalent immobilization of *Candida antarctica* lipase B on mixed-function-grafted silica gel supports for continuous-flow biotransformations. Process Biochem 48(7):1039–1047

Boyd SA, Mortland MM (1990) Enzyme interactions with clay and clay-organic matter complexes. Soil Biochemistry 6, 1–128

Brena BM, Batista-Viera F (2006) Immobilization of enzymes. Immobilization of enzymes and cell. Springer, pp 15–30

Cacciotti I, Lombardelli C, Benucci I, Esti M (2019). Clay/chitosan biocomposite systems as novel green carriers for covalent immobilization of food enzymes. J Mater Res Technol

Cai C, Gao Y, Liu Y, Zhong N, Liu N (2016) Immobilization of *Candida antarctica* lipase B onto SBA-15 and their application in glycerolysis for diacylglycerols synthesis. Food Chem 212:205–212

Calvo-Flores FG, Dobado JA (2010) Lignin as renewable raw material. Chemsuschem 3(11):1227–1235

Cao J, Li Y, Tu Ni, Lv Y, Chen Q, Dong H (2016) Novel enzyme/exfoliated bentonite nanohybrids as highly efficient and recyclable biocatalysts in hydrolytic reaction. J Mol Catal B Enzym 132:41–46

Cao S-L, Xu H, Lai L-H, Gu W-M, Xu P, Xiong J, Yin H, Li X-H, Ma Y-Z, Zhou J (2017) Magnetic ZIF-8/cellulose/Fe_3O_4 nanocomposite: preparation, characterization, and enzyme immobilization. Bioresour Bioprocess 4(1):56

Care A, Petroll K, Gibson ESY, Bergquist PL, Sunna A (2017) Solid-binding peptides for immobilisation of thermostable enzymes to hydrolyse biomass polysaccharides. Biotechnol Biofuels 10(1):29

Chang M-Y, Juang R-S (2007) Use of chitosan–clay composite as immobilization support for improved activity and stability of β-glucosidase. Biochem Eng J 35(1):93–98

Chang M-Y, Kao H-C, Juang R-S (2008) Thermal inactivation and reactivity of β-glucosidase immobilized on chitosan–clay composite. Int J Biol Macromol 43(1):48–53

Chao C, Guan H, Zhang J, Liu Y, Zhao Y, Zhang B (2017) Immobilization of laccase onto porous polyvinyl alcohol/halloysite hybrid beads for dye removal. Water Sci Technol 77(3):809–818

Chao C, Liu J, Wang J, Zhang Y, Zhang B, Zhang Y, Xiang Xu, Chen R (2013) Surface modification of halloysite nanotubes with dopamine for enzyme immobilization. ACS Appl Mater Interfces 5(21):10559–10564

Chao C, Zhao Y, Guan H, Liu G, Hu Z, Zhang B (2017) Improved performance of immobilized laccase on poly (diallyldimethylammonium chloride) functionalized halloysite for 2, 4-dichlorophenol degradation. Environ Eng Sci 34(10):762–770

Chauhan A, Sillu D, Agnihotri S (2019) Removal of pharmaceutical contaminants in wastewater using nanomaterials: a comprehensive review. Curr Drug Metab 20(6):483–505

Cho HJ, Jang WJ, Moon SY, Lee JM, Kim J-H, Han H-S, Kim K-W, Lee B-J, Kong I-S (2018) Immobilization of β-1, 3–1, 4-glucanase from Bacillus sp. on porous silica for production of β-glucooligosaccharides. Enzyme Microb Technol 110:30–37

Chrisnasari R, Wuisan ZG, Budhyantoro A, Widi RK (2015) Glucose oxidase immobilization on TMAH-modified bentonite. Indones J Chem 15(1):22–28

Colín-Luna JA, Zamora-Rodea EG, González-Brambila MM, Barrera-Calva E, Rosas-Cedillo R, Medina-Mendoza AK, García-Martínez JC (2018) Biodiesel production using immobilized lipase supported on a zirconium-pillared clay. Effect of the immobilization method. Int J Chem React Eng 16(11)

Dai H, Ou S, Liu Z, Huang H (2017) Pineapple peel carboxymethyl cellulose/polyvinyl alcohol/mesoporous silica SBA-15 hydrogel composites for papain immobilization. Carbohydr Polym 169:504–514

Dai JC, Huang JT (1999) Surface modification of clays and clay–rubber composite. Appl Clay Sci 15(1–2):51–65

Datta S, Christena LR, Rajaram YRS (2013) Enzyme immobilization: an overview on techniques and support materials. 3 Biotech 3(1), 1–9

De Castro HF, De Oliveira PC, Pereira EB (1997) Evaluation of different approaches for lipase catalysed synthesis of citronellyl acetate. Biotechnol Lett 19(3):229–232

de Souza Lima J, Costa FN, Bastistella MA, de Araújo PHH, de Oliveira D (2019) Functionalized kaolin as support for endoglucanase immobilization. Bioprocess Biosyst Eng 1–9

de Vasconcellos A, Laurenti JB, Miller AH, da Silva DA, de Moraes FR, Aranda Donato AG, Nery JG (2015) Potential new biocatalysts for biofuel production: the fungal lipases of *Thermomyces lanuginosus* and *Rhizomucor miehei* immobilized on zeolitic supports ion exchanged with transition metals. Microporous Mesoporous Mater 214:166–180

de Vasconcellos A, Miller AH, Aranda DAG, Nery JG (2018) Biocatalysts based on nanozeolite-enzyme complexes: effects of alkoxysilane surface functionalization and biofuel production using microalgae lipids feedstock. Colloids Surf B 165:150–157

de Vries JG (2016) Catalytic conversion of renewable resources into bulk and fine chemicals. Chem Rec 16(6):2787–2800

Dhiman NK, Agnihotri S (2020) Hierarchically aligned nano silver/chitosan-PVA hydrogel for point-of-use water disinfection: Contact-active mechanism revealed. Environ Sci Nano 7:2337–2350

Dhiman NK, Agnihotri S, Shukla R (2019) Silver-based polymeric nanocomposites as antimicrobial coatings for biomedical applications. In: Nanotechnology in modern animal biotechnology. Springer, pp 115–171

Fang Y, Huang X-J, Chen P-C, Xu Z-K (2011) Polymer materials for enzyme immobilization and their application in bioreactors. BMB Rep 44(2):87–95

Fatimah Is, Huda T (2013) Preparation of cetyltrimethylammonium intercalated Indonesian montmorillonite for adsorption of toluene. Appl Clay Sci 74:115–120

Forano C, Vial S, Mousty C (2006) Nanohybrid enzymes-layered double hydroxides: potential applications. Curr Nanosci 2(3):283–294

Fried DI, Brieler FJ, Froeba M (2013) Designing inorganic porous materials for enzyme adsorption and applications in biocatalysis. ChemCatChem 5(4):862–884

Fu H, Dencic I, Tibhe J, Pedraza CAS, Wang Q, Noel T, Meuldijk J, de Croon MHJM, Hessel V, Weizenmann N (2012) Threonine aldolase immobilization on different supports for engineering of productive, cost-efficient enzymatic microreactors. Chem Eng J 207:564–576

Fujimori A, Arai S, Soutome Y, Hashimoto M (2014) Improvement of thermal stability of enzyme via immobilization on Langmuir-Blodgett films of organo-modified aluminosilicate with high coverage. Colloids Surf A 448:45–52

Garmroodi M, Mohammadi M, Ramazani A, Ashjari M, Mohammadi J, Sabour B, Yousefi M (2016) Covalent binding of hyper-activated Rhizomucor miehei lipase (RML) on hetero-functionalized siliceous supports. Int J Biol Macromol 86:208–215

Ghasemi Z, Sourinejad I, Kazemian H, Rohani S (2018) Application of zeolites in aquaculture industry: a review. Rev Aquac 10(1):75–95

Gholamzadeh P, Mohammadi Ziarani G, Badiei A (2017) Immobilization of lipases onto the SBA-15 mesoporous silica. Biocatal Biotransformation 35(3), 131–150

Golbaha N, Ramli Z, Endud S (2016) Immobilization of lipase onto mesoporous silica KIT-6 and montmorillonite K10 for enzymatic hydrolysis of tributyrin. Malay J Fundamental Appl Sci 12(1)

Gonçalves APV, Lopes JM, Lemos F, Ribeiro FR, Prazeres DMF, Cabral JMS, Aires-Barros MR (1996) Zeolites as supports for enzymatic hydrolysis reactions. Comparative study of several zeolites. J Mol Catal B Enzym 1(2), 53–60

González-Delgado I, Segura Y, Martín A, López-Muñoz M-J, Morales G (2018) β-Galactosidase covalent immobilization over large-pore mesoporous silica supports for the production of high galacto-oligosaccharides (GOS). Microporous Mesoporous Mater 257:51–61

Gopinath S, Sugunan S (2007) Enzymes immobilized on montmorillonite K 10: effect of adsorption and grafting on the surface properties and the enzyme activity. Appl Clay Sci 35(1–2):67–75

Gupta MN, Kaloti M, Kapoor M, Solanki K (2011) Nanomaterials as matrices for enzyme immobilization. Artif Cells Blood Subst Biotechnol 39(2):98–109

Han Q, Zhang H, Sun J, Liu Z, Huang W-C, Xue C, Mao X (2019) Immobilization of phospholipase D on silica-coated magnetic nanoparticles for the synthesis of functional phosphatidylserine. Catalysts 9(4):361

Han Y-J, Watson JT, Stucky GD, Butler A (2002) Catalytic activity of mesoporous silicate-immobilized chloroperoxidase. J Mol Catal B Enzym 17(1):1–8

Hartmann M, Kostrov X (2013) Immobilization of enzymes on porous silicas–benefits and challenges. Chem Soc Rev 42(15):6277–6289

He H, Han H, Shi H, Tian Y, Sun F, Song Y, Li Q, Zhu G (2016) Construction of thermophilic lipase-embedded metal–organic frameworks via biomimetic mineralization: a biocatalyst for ester hydrolysis and kinetic resolution. ACS Appl Mater Interfaces 8(37):24517–24524

Hoffmann F, Cornelius M, Morell J, Fröba M (2006) Silica-based mesoporous organic–inorganic hybrid materials. Angew Chem Int Ed 45(20):3216–3251

Holm MS, Taarning E, Egeblad K, Christensen CH (2011) Catalysis with hierarchical zeolites. Catal Today 168(1):3–16

Homaei AA, Sariri R, Vianello F, Stevanato R (2013) Enzyme immobilization: an update. J Chem Biol 6(4):185–205

Hu X, Wang P, Hwang H-M (2009) Oxidation of anthracene by immobilized laccase from Trametes versicolor. Bioresour Technol 100(21):4963–4968

Hu X, Zhao X, Hwang H-M (2007) Comparative study of immobilized Trametes versicolor laccase on nanoparticles and kaolinite. Chemosphere 66(9):1618–1626

Ikemoto H, Chi Q, Ulstrup J (2010) Stability and catalytic kinetics of horseradish peroxidase confined in nanoporous SBA-15. J Phys Chem C 114(39):16174–16180

Iyer PV, Ananthanarayan L (2008) Enzyme stability and stabilization—aqueous and non-aqueous environment. Process Biochem 43(10):1019–1032

Jäger G, Büchs J (2012) Biocatalytic conversion of lignocellulose to platform chemicals. Biotechnol J 7(9):1122–1136

Jesionowski T, Zdarta J, Krajewska B (2014) Enzyme immobilization by adsorption: a review. Adsorption 20(5–6):801–821
Jian H, Wang Y, Bai Y, Li R, Gao R (2016) Site-specific, covalent immobilization of dehalogenase ST2570 catalyzed by formylglycine-generating enzymes and its application in batch and semi-continuous flow reactors. Molecules 21(7):895
Jlassi K, Krupa I, Chehimi MM (2017) Overview: clay preparation, properties, modification. In: Clay-polymer nanocomposites. Elsevier, pp 1–28
Johnston CT, Premachandra GS, Szabo T, Lok J, Schoonheydt RA (2011) Interaction of biological molecules with clay minerals: a combined spectroscopic and sorption study of lysozyme on saponite. Langmuir 28(1):611–619
Kadam AA, Jang J, Jee SC, Sung J-S, Lee DS (2018) Chitosan-functionalized supermagnetic halloysite nanotubes for covalent laccase immobilization. Carbohydr Polym 194:208–216
Kadam AA, Jang J, Lee DS (2017) Supermagnetically tuned halloysite nanotubes functionalized with aminosilane for covalent laccase immobilization. ACS Appl Mater Interfaces 9(18):15492–15501
Kalburcu T, Tabak A, Ozturk N, Tuzmen N, Akgol S, Caglar B, Denizli A (2015) Adsorption of lysozyme from aqueous solutions by a novel bentonite–tyrptophane (Bent–Trp) microcomposite affinity sorbent. J Mol Struct 1083:156–162
Kaushal J, Singh G, Arya SK (2018) Immobilization of catalase onto chitosan and chitosan–bentonite complex: a comparative study. Biotechnol Rep 18:e00258
Kołodziejczak-Radzimska A, Jesionowski T (2019) Characterization of amino-, epoxy-and carbonyl-functionalized halloysite and its application in the immobilization of aminoacylase from Aspergillus melleus. Physicochem Probl Miner Process 55
Kumari A, Kaur B, Srivastava R, Sangwan RS (2015) Isolation and immobilization of alkaline protease on mesoporous silica and mesoporous ZSM-5 zeolite materials for improved catalytic properties. Biochem Biophys Rep 2:108–114
Lee Y-C, Chen C-T, Chiu Y-T, Wu K-W (2013) An effective cellulose-to-glucose-to-fructose conversion sequence by using enzyme immobilized Fe_3O_4-loaded mesoporous silica nanoparticles as recyclable biocatalysts. ChemCatChem 5(8):2153–2157
Liang K, Ricco R, Doherty CM, Styles MJ, Bell S, Kirby N, Mudie S, Haylock D, Hill AJ, Doonan CJ (2015) Biomimetic mineralization of metal-organic frameworks as protective coatings for biomacromolecules. Nat Commun 6:7240
Lie E, Molin G (1991) Hydrolysis and esterification with immobilized lipase on hydrophobic and hydrophilic zeolites. J Chem Technol Biotechnol 50(4):549–553
Lindeque RM, Woodley JM (2019) Reactor selection for effective continuous biocatalytic production of pharmaceuticals. Catalysts 9(3):262
Liu G, Xu Y, Han Y, Wu J, Xu J, Meng H, Zhang X (2017) Immobilization of lysozyme proteins on a hierarchical zeolitic imidazolate framework (ZIF-8). Dalton Trans 46(7):2114–2121
Lyu F, Zhang Y, Zare RN, Ge J, Liu Z (2014) One-pot synthesis of protein-embedded metal–organic frameworks with enhanced biological activities. Nano Lett 14(10):5761–5765
Magner E (2013) Immobilisation of enzymes on mesoporous silicate materials. Chem Soc Rev 42(15):6213–6222
Maitan-Alfenas GP, Visser EM, Guimarães VM (2015) Enzymatic hydrolysis of lignocellulosic biomass: converting food waste in valuable products. Curr Opin Food Sci 1:44–49
Majumdar P, Khan AY, Bandyopadhyaya R (2016) Diffusion, adsorption and reaction of glucose in glucose oxidase enzyme immobilized mesoporous silica (SBA-15) particles: experiments and modeling. Biochem Eng J 105:489–496
Margolin AL (1991) Enzymes: use them. ChemTech 21(3):160–167
Marthala VRR, Urmoneit L, Zhou Z, Machoke AGF, Schmiele M, Unruh T, Schwieger W, Hartmann M (2017) Boron-containing MFI-type zeolites with a hierarchical nanosheet assembly for lipase immobilization. Dalton Trans 46(13):4165–4169

Miložič N, Lubej M, Lakner M, Žnidaršič-Plazl P, Plazl I (2017) Theoretical and experimental study of enzyme kinetics in a microreactor system with surface-immobilized biocatalyst. Chem Eng J 313:374–381
Misaelides P (2011) Application of natural zeolites in environmental remediation: a short review. Microporous Mesoporous Mater 144(1–3):15–18
Mitchell S, Pérez-Ramírez J (2011) Mesoporous zeolites as enzyme carriers: synthesis, characterization, and application in biocatalysis. Catal Today 168(1):28–37
Mohammadi M, Heshmati MK, Sarabandi K, Fathi M, Lim L-T, Hamishehkar H (2019) Activated alginate-montmorillonite beads as an efficient carrier for pectinase immobilization. Int J Biol Macromol 137:253–260
Mohammadi M, Habibi Z, Gandomkar S, Yousefi M (2018) A novel approach for bioconjugation of Rhizomucor miehei lipase (RML) onto amine-functionalized supports; application for enantioselective resolution of rac-ibuprofen. Int J Biol Macromol 117:523–531
Mravčáková M, Boukerma K, Omastová M, Chehimi MM (2006) Montmorillonite/polypyrrole nanocomposites. The effect of organic modification of clay on the chemical and electrical properties. Mater Sci Eng C 26(2–3), 306–313
Muderrisoglu C, Sargin S, Yesil-Celiktas O (2018) Application of β-glucuronidase-immobilised silica gel formulation to microfluidic platform for biotransformation of β-glucuronides. Biotechnol Lett 40(5):773–780
Mukherji S, Ruparelia J, Agnihotri S (2012) Antimicrobial activity of silver and copper nanoparticles: variation in sensitivity across various strains of bacteria and fungi. In: Nano-antimicrobials. Springer pp 225–251
Mureseanu M, Galarneau A, Renard G, Fajula F (2005) A new mesoporous micelle-templated silica route for enzyme encapsulation. Langmuir 21(10):4648–4655
Mylonas E, Svergun DI (2007) Accuracy of molecular mass determination of proteins in solution by small-angle X-ray scattering. Appl Crystallogr 40(s1):s245–s249
Na'imah J, Prasetyawan S, Srihardyastutie A (2017) In vitro and in silico studies of immobilized xylanase on zeolite matrix activated with hydrochloric acid. J Pure Appl Chem Res 6(3):181
Naseri M, Pitzalis F, Carucci C, Medda L, Fotouhi L, Magner E, Salis A (2018) Lipase and laccase encapsulated on zeolite imidazolate framework: enzyme activity and stability from voltammetric measurements. ChemCatChem 10(23):5425–5433
Ojeda K, Kafarov V (2009) Exergy analysis of enzymatic hydrolysis reactors for transformation of lignocellulosic biomass to bioethanol. Chem Eng J 154(1–3):390–395
Olshansky Y, Masaphy S, Root RA, Rytwo G (2018) Immobilization of Rhus vernicifera laccase on sepiolite; effect of chitosan and copper modification on laccase adsorption and activity. Appl Clay Sci 152:143–147
Öztürk H, Pollet E, Phalip V, Güvenilir Y, Avérous L (2016) Nanoclays for lipase immobilization: biocatalyst characterization and activity in polyester synthesis. Polymers 8(12):416
Pandey G, Munguambe DM, Tharmavaram M, Rawtani D, Agrawal YK (2017) Halloysite nanotubes-An efficient 'nano-support'for the immobilization of α-amylase. Appl Clay Sci 136:184–191
Paul R, Datta SC, Manjaiah KM, Bhattacharyya R (2015) Characterization of nanoclay-phosphatase complex with IR spectroscopy and electron microscopy. Clay Res 34:99–109
Perez-Ramirez J, Christensen CH, Egeblad K, Christensen CH, Groen JC (2008) Hierarchical zeolites: enhanced utilisation of microporous crystals in catalysis by advances in materials design. Chem Soc Rev 37(11):2530–2542
Piao M, Zou D, Ren X, Gao S, Qin C, Piao Y (2019) High efficiency biotransformation of bisphenol A in a fluidized bed reactor using stabilized laccase in porous silica. Enzyme Microb Technol 126:1–8
Poorakbar E, Shafiee A, Saboury AA, Rad BL, Khoshnevisan K, Ma'mani L, Derakhshankhah H, Ganjali MR, Hosseini M (2018) Synthesis of magnetic gold mesoporous silica nanoparticles core shell for cellulase enzyme immobilization: Improvement of enzymatic activity and thermal stability. Process Biochem 71:92–100

Pu S, Zhang X, Yang C, Naseer S, Zhang X, Ouyang J, Li D, Yang J (2019) The effects of NaCl on enzyme encapsulation by zeolitic imidazolate frameworks-8. Enzyme Microb Technol 122:1–6
Puri M, Barrow CJ, Verma ML (2013) Enzyme immobilization on nanomaterials for biofuel production. Trends Biotechnol 31(4):215–216
Ramesh HP, Tharanathan RN (2003) Carbohydrates—the renewable raw materials of high biotechnological value. Crit Rev Biotechnol 23(2):149–173
Ricardi NC, de Menezes EW, Benvenutti EV, da Natividade Schöffer J, Hackenhaar CR, Hertz PF, Costa TMH (2018) Highly stable novel silica/chitosan support for β-galactosidase immobilization for application in dairy technology. Food Chem 246:343–350
Rios NS, Pinheiro MP, dos Santos JCS, Fonseca TS, Lima LD, de Mattos MC, Freire DMG, da Silva Júnior IJ, Rodríguez-Aguado E, Goncalves LRB (2016) Strategies of covalent immobilization of a recombinant *Candida antarctica* lipase B on pore-expanded SBA-15 and its application in the kinetic resolution of (R, S)-Phenylethyl acetate. J Mol Catal B Enzym 133:246–258
Rodrigues RC, Ortiz C, Berenguer-Murcia Á, Torres R, Fernández-Lafuente R (2013) Modifying enzyme activity and selectivity by immobilization. Chem Soc Rev 42(15):6290–6307
Rodriguez YE, Laitano MV, Pereira NA, López-Zavala AA, Haran NS, Fernández-Gimenez AV (2018) Exogenous enzymes in aquaculture: alginate and alginate-bentonite microcapsules for the intestinal delivery of shrimp proteases to Nile tilapia. Aquaculture 490:35–43
Rogers CDF, Glendinning S (1996) Modification of clay soils using lime. Lime stabilization. Thomas Telford, London, pp 99–112
Sang L-C, Coppens M-O (2011) Effects of surface curvature and surface chemistry on the structure and activity of proteins adsorbed in nanopores. Phys Chem Chem Phys 13(14):6689–6698
Santana JL, Oliveira JM, Carvalho NB, Osório NMFM, Mattedi S, Freitas LS, Cavalcanti EB, Lima ÁS, Soares CMF (2018) Analysis of the performance of a packed bed reactor to production ethyl esters from crude vegetable oil using lipase immobilized in silica modified with protic ionic liquid. Química Nova 41(8):891–898
Santos JCS, Barbosa O, Ortiz C, Berenguer-Murcia A, Rodrigues RC, Fernandez-Lafuente R (2015) Importance of the support properties for immobilization or purification of enzymes. ChemCatChem 7(16):2413–2432
Sarkar B, Rusmin R, Ugochukwu UC, Mukhopadhyay R, Manjaiah KM (2019) Modified clay minerals for environmental applications. In: Modified clay and zeolite nanocomposite materials. Elsevier, pp 113–127
Sarrouh B, Santos TM, Miyoshi A, Dias R, Azevedo V (2012) Up-to-date insight on industrial enzymes applications and global market. J Bioprocess Biotech S 4(002)
Satyanarayana KG (2004) Clay surfaces: fundamentals and applications, vol 1. Elsevier
Schilke KF, Wilson KL, Cantrell T, Corti G, McIlroy DN, Kelly C (2010) A novel enzymatic microreactor with Aspergillus oryzae β-galactosidase immobilized on silicon dioxide nanosprings. Biotechnol Prog 26(6):1597–1605
Schöffer N, Jéssie M, Roberta C, Charqueiro DS, de Menezes EW, Costa TMH, Benvenutti EV, Rodrigues RC, Hertz PF (2017) Effects of immobilization, pH and reaction time in the modulation of α-, β-or γ-cyclodextrins production by cyclodextrin glycosyltransferase: Batch and continuous process. Carbohydr Polym 169, 41–49
Secundo F (2013) Conformational changes of enzymes upon immobilisation. Chem Soc Rev 42(15):6250–6261
Sedaghat ME, Ghiaci M, Aghaei H, Soleimanian-Zad S (2009a) Enzyme immobilization. Part 3: immobilization of α-amylase on Na-bentonite and modified bentonite. Appl Clay Sci 46(2):125–130
Sedaghat ME, Ghiaci M, Aghaei H, Soleimanian-Zad S (2009b). Enzyme immobilization. Part 4. immobilization of alkaline phosphatase on Na-sepiolite and modified sepiolite. Appl Clay Sci 46(2):131–135
Seetharam G, Saville BA (2002) L-DOPA production from tyrosinase immobilized on zeolite. Enzyme Microb Technol 31(6):747–753

Serefoglou E, Litina K, Gournis D, Kalogeris E, Tzialla AA, Pavlidis IV, Stamatis H, Maccallini E, Lubomska M, Rudolf P (2008) Smectite clays as solid supports for immobilization of β-glucosidase: synthesis, characterization, and biochemical properties. Chem Mater 20(12):4106–4115

Serralha FN, Lopes JM, Lemos F, Prazeres DMF, Aires-Barros MR, Cabral JMS, Ribeiro FR (1998) Zeolites as supports for an enzymatic alcoholysis reaction. J Mol Catal B Enzym 4(5–6):303–311

Sheldon RA (2007) Enzyme immobilization: the quest for optimum performance. Adv Synth Catal 349(8–9):1289–1307

Sheldon RA (2016) Engineering a more sustainable world through catalysis and green chemistry. J R Soc Interf 13(116):20160087

Shi X, Xu J, Lu C, Wang Z, Xiao W, Zhao L (2019) Immobilization of high temperature-resistant GH3 β-glucosidase on a magnetic particle Fe_3O_4-SiO_2-NH_2-Cellu-ZIF8/zeolitic imidazolate framework. Enzyme Microb Technol 129:109347

Sillu D, Agnihotri S (2020) Cellulase immobilization on to magnetic halloysite nanotubes: enhanced enzyme activity and stability with high cellulose saccharification. ACS Sustain Chem Eng 8, 900−913

Sinegani AAS, Emtiazi G, Shariatmadari H (2005) Sorption and immobilization of cellulase on silicate clay minerals. J Colloid Interface Sci 290(1):39–44

Singh S, Tuteja SK, Sillu D, Deep A, Suri CR (2016) Gold nanoparticles-reduced graphene oxide based electrochemical immunosensor for the cardiac biomarker myoglobin. Microchim Acta 183(5):1729–1738

Sinha R, Khare SK (2015) Immobilization of halophilic *Bacillus* sp. EMB9 protease on functionalized silica nanoparticles and application in whey protein hydrolysis. Bioprocess Biosyst Eng 38(4):739–748

Songurtekin D, Yalcinkaya EE, Ag D, Seleci M, Demirkol DO, Timur S (2013) Histidine modified montmorillonite: Laccase immobilization and application to flow injection analysis of phenols. Appl Clay Sci 86:64–69

Spahn C, Minteer SD (2008) Enzyme immobilization in biotechnology. Recent Patents Eng 2(3):195–200

Sun J, Yendluri R, Liu K, Guo Y, Lvov Y, Yan X (2017) Enzyme-immobilized clay nanotube–chitosan membranes with sustainable biocatalytic activities. Phys Chem Chem Phys 19(1):562–567

Talebi M, Vaezifar S, Jafary F, Fazilati M, Motamedi S (2016) Stability improvement of immobilized α-amylase using nano pore zeolite. Iran J Biotechnol 14(1):33

Tanasković SJ, Jokić B, Grbavčić S, Drvenica I, Prlainović N, Luković N, Knežević-Jugović Z (2017) Immobilization of *Candida antarctica* lipase B on kaolin and its application in synthesis of lipophilic antioxidants. Appl Clay Sci 135:103–111

Tegl G, Stagl V, Mensah A, Huber D, Somitsch W, Grosse-Kracht S, Guebitz GM (2018) The chemo enzymatic functionalization of chitosan zeolite particles provides antioxidant and antimicrobial properties. Eng Life Sci 18(5):334–340

Tripathi A, Sami H, Jain SR, Viloria-Cols M, Zhuravleva N, Nilsson G, Jungvid H, Kumar A (2010) Improved bio-catalytic conversion by novel immobilization process using cryogel beads to increase solvent production. Enzyme Microb Technol 47:44–51

Tripathi A, Melo JS (2019) Self-assembled biogenic melanin modulated surface chemistry of biopolymers-colloidal silica composite porous matrix for the recovery of uranium. J Appl Polym Sci 136(5):46937

Tully J, Yendluri R, Lvov Y (2016) Halloysite clay nanotubes for enzyme immobilization. Biomacromolecules 17(2):615–621

Tyagi VK, Lo S-L (2013) Sludge: a waste or renewable source for energy and resources recovery? Renew Sustain Energy Rev 25:708–728

Tzialla AA, Kalogeris E, Enotiadis A, Taha AA, Gournis D, Stamatis H (2009) Effective immobilization of Candida antarctica lipase B in organic-modified clays: application for the epoxidation of terpenes. Mater Sci Eng, B 165(3):173–177

Valikhani D, Bolivar JM, Viefhues M, McIlroy DN, Vrouwe EX, Nidetzky B (2017) A spring in performance: silica nanosprings boost enzyme immobilization in microfluidic channels. ACS Appl Mater Interfaces 9(40):34641–34649

Verri F, Diaz U, Macario A, Corma A, Giordano G (2016) Optimized hybrid nanospheres immobilizing *Rhizomucor miehei* lipase for chiral biotransformation. Process Biochem 51(2):240–248

Wang J, Yu S, Feng F, Lu L (2019) Simultaneous purification and immobilization of laccase on magnetic zeolitic imidazolate frameworks: recyclable biocatalysts with enhanced stability for dye decolorization. Biochem Eng J 107285

Wang Q, Peng L, Li G, Zhang P, Li D, Huang F, Wei Q (2013) Activity of laccase immobilized on TiO_2-montmorillonite complexes. Int J Mol Sci 14(6):12520–12532

Wang Y, Caruso F (2004) Enzyme encapsulation in nanoporous silica spheres. Chem Commun 13:1528–1529

Wang Y, Caruso F (2005) Mesoporous silica spheres as supports for enzyme immobilization and encapsulation. Chem Mater 17(5):953–961

Wang Y, Liu C, Zhang Y, Zhang B, Liu J (2015) Facile fabrication of flowerlike natural nanotube/layered double hydroxide composites as effective carrier for lysozyme immobilization. ACS Sustain Chem Eng 3(6):1183–1189

Wannerberger K, Welin-Klintström S, Arnebrant T (1997) Activity and adsorption of lipase from *Humicola lanuginosa* on surfaces with different wettabilities. Langmuir 13(4):784–790

Washmon-Kriel L, Jimenez VL, Balkus Jr KJ (2000) Cytochrome c immobilization into mesoporous molecular sieves. J Mol Catal B Enzym 10(5):453–469

Weber E, Sirim D, Schreiber T, Thomas B, Pleiss J, Hunger M, Gläser R, Urlacher VB (2010) Immobilization of P450 BM-3 monooxygenase on mesoporous molecular sieves with different pore diameters. J Mol Catal B Enzym 64(1–2):29–37

Wu LM, Zeng QH, Ding R, Tu PP, Xia MZ (2019) Fe_3O_4/Acid activated montmorillonite/cellulase composites: Preparation, structure, and enzyme activity. Appl Clay Sci 179:105129

Wu X, Ge J, Yang C, Hou M, Liu Z (2015) Facile synthesis of multiple enzyme-containing metal–organic frameworks in a biomolecule-friendly environment. Chem Commun 51(69):13408–13411

Xiao X, Siepenkoetter T, Whelan R, Salaj-Kosla U, Magner E (2018) A continuous fluidic bioreactor utilising electrodeposited silica for lipase immobilisation onto nanoporous gold. J Electroanal Chem 812:180–185

Xing G-W, Li X-W, Tian G-L, Ye Y-H (2000) Enzymatic peptide synthesis in organic solvent with different zeolites as immobilization matrixes. Tetrahedron 56(22):3517–3522

Xu W, Sun Z, Meng H, Han Y, Wu J, Xu J, Xu Y, Zhang X (2018) Immobilization of cellulase proteins on zeolitic imidazolate framework (ZIF-8)/polyvinylidene fluoride hybrid membranes. New J Chem 42(21):17429–17438

Yagiz F, Kazan D, Akin AN (2007) Biodiesel production from waste oils by using lipase immobilized on hydrotalcite and zeolites. Chem Eng J 134(1–3), 262–267

Yang Z, Liao Y, Fu X, Zaporski J, Peters S, Jamison M, Liu Y, Wullschleger SD, Graham DE, Gu B (2019) Temperature sensitivity of mineral-enzyme interactions on the hydrolysis of cellobiose and indican by β-glucosidase. Sci Total Environ 686:1194–1201

Yao J, Wang Q, Wang Y, Zhang Y, Zhang B, Zhang H (2015) Immobilization of laccase on chitosan–halloysite hybrid porous microspheres for phenols removal. Desalination Water Treat 55(5):1293–1301

Yuan SP, Wang JG, Li YW, Jiao H (2002) Brønsted acidity of isomorphously substituted ZSM-5 by B, Al, Ga, and Fe. Density functional investigations. J Phys Chem A 106(35):8167–8172

Zang L, Qiu J, Mo H, Yang C, Sakai E (2015) 113 study on immobilization of cellulase on attapulgite nanoclay. In: Paper presented at the the proceedings of autumn conference of Tohoku Branch 2015, vol 51

Zdarta J, Meyer AS, Jesionowski T, Pinelo M (2018) A general overview of support materials for enzyme immobilization: characteristics, properties, practical utility. Catalysts 8(2):92

Zhai R, Zhang B, Liu L, Xie Y, Zhang H, Liu J (2010) Immobilization of enzyme biocatalyst on natural halloysite nanotubes. Catal Commun 12(4):259–263

Zhang B, Weng Y, Xu H, Mao Z (2012) Enzyme immobilization for biodiesel production. Appl Microbiol Biotechnol 93(1):61–70

Zhao L, Xu C, Gao S, Shen B (2010) Effects of concentration on the alkali-treatment of ZSM-5 zeolite: a study on dividing points. J Mater Sci 45(19):5406–5411

Zhao XS, Bao XY, Guo W, Lee FY (2006) Immobilizing catalysts on porous materials. Mater Today 9(3):32–39

Zhou C-H, Shen Z-F, Liu L-H, Liu S-M (2011) Preparation and functionality of clay-containing films. J Mater Chem 21(39):15132–15153

Zhou CH, Keeling J (2013) Fundamental and applied research on clay minerals: from climate and environment to nanotechnology. Appl Clay Sci 74:3–9

Zhou Z, Hartmann M (2013) Progress in enzyme immobilization in ordered mesoporous materials and related applications. Chem Soc Rev 42(9):3894–3912

Zhu Y, Kaskel S, Shi J, Wage T, van Pée K-H (2007) Immobilization of Trametes versicolor laccase on magnetically separable mesoporous silica spheres. Chem Mater 19(26):6408–6413

Zucca P, Sanjust E (2014) Inorganic materials as supports for covalent enzyme immobilization: methods and mechanisms. Molecules 19(9):14139–14194

Zulfiqar S, Ahmad Z, Ishaq M, Sarwar MI (2009) Aromatic–aliphatic polyamide/montmorillonite clay nanocomposite materials: Synthesis, nanostructure and properties. Mater Sci Eng A 525(1–2):30–36

Immobilization of Molecular Assemblies on 2D Nanomaterials for Electrochemical Biosensing Applications

Sheela Berchmans and T. Balamurugan

Abstract Immobilization or surface functionalization of molecular assemblies is of tremendous importance in the field of building functional micro/nanolevel devices for a wide range of applications starting from sensing, catalysis, photovoltaics, fuel cells and biomaterials to molecular electronics. Research in this area contributes as well to provide fundamental understanding of electron transfer, controlling reactions at surfaces, and tailoring surface properties. During the last few decades, we have witnessed significant advances in the area of electronics which has resulted in the development of highly compatible and simple electrical transducers from which we can process signals in a highly efficient manner than ever before. Biosensors are analytical tools whose applications stretch out in areas like health care, environment, agriculture and biotechnology. Electrochemical transducers are of highly advantageous in the fabrication of biosensors owing to their simplicity, amenability for field measurements and miniaturization, adaptability for wireless transmission and so on. The challenging task in the fabrication of biosensors is to effectively immobilize/functionalize the molecular recognition assemblies on to the electrode surface with their properties intact. High surface area materials with suitable functional groups that can be electrically wired to electrode surfaces efficiently for effective sensing are required for the fabrication of biosensors. Currently, two-dimensional (2D) materials have evolved as promising potential candidates in electrically transduced chemical sensing. They provide a large surface-to-volume ratio for substrate–analyte interactions when compared to their 0D, 1D and 3D analogues, which is essential for improving sensitivity and through the incorporation of known recognition centres like enzymes, or ligands, onto the surface of 2D materials selectivity is ensured. Apart from the well-studied 2D nanomaterial graphene, a diverse range of 2D materials like phosphorene, hexagonal boron nitride (h-BN), transition metal dichalcogenides (TMDCs), layered metal oxides, 2D metal–organic frameworks (MOFs), covalent organic frameworks (COFs) and other 2D compounds are being

S. Berchmans (✉) · T. Balamurugan
Academy of Scientific Innovative Research, Ghaziabad, Uttar Pradesh, India
e-mail: sheelaberchmans@yahoo.com

Electrodics and Electrocatalysis Division, CSIR-Central Electrochemical Research Institute, Karaikudi, Tamil Nadu, India

A. Tripathi and J. S. Melo (eds.), *Immobilization Strategies*,
Gels Horizons: From Science to Smart Materials,
https://doi.org/10.1007/978-981-15-7998-1_12

explored for biosensing recently. This book chapter will describe how the molecular recognition moieties like enzymes, ligands, ionophores, antigens/antibodies, host molecules, etc., would be immobilized on the 2D materials. Case studies from the literature and few examples from the authors' work would be highlighted.

Keywords Immobilization · Functionalization · Biosensors · Molecular recognition · Electrochemical transduction · 2D nanomaterials

Abbreviations

4-CN	4-chloro-1-naphthol
AAB	Antiapolipoprotein
AC	Alternate current
AFP	α-fetoprotein
ALD	Atomic layer deposition
AuNP	Gold nanoparticles
BP	Black Phosphorous
BSA	Bovine serum albumin
CA125	Cancer antigen 125
CEA	Carcinoembryonic antigen
Chox	Cholesterol oxidase
CPE	Carbon paste electrode
CS	Chitosan
CV	Cyclic voltammetry
CWE	Coated wire electrodes
DPV	Differential pulse voltammetry
E	Potential
EIS	Electrochemical impedance spectroscopy
FcMeOH	Ferrocene methanol
FET	Field effect transistor
GA	Glucoamylase
GC	Glassy carbon
GO	Graphene oxide
GOx	Glucose oxidase
Gr	Graphene
GTA	Glutaraldehyde
Hb	Haemoglobin
HCG	Human chorionic gonadotropin
HET	Heterogeneous electron transfer
HQ	Hydroquinone
HRP	Horseradish peroxidase
IgG	Immunoglobulin G
LACC	Laccase

LOD	Limit of detection
Mb	Myoglobin
MCH	Mercapto hexanol
MMP-2	Metalloproteinase-2
MOF	Metal–organic frameworks
MPA	Mercapto propionic acid
MXenes	Transition metal carbides/nitrides
NPV	Normal pulse voltammetry
NW	Nanowire
OTA	Ochratoxin-A
PAA	Polyacrylic acid
PEDOT	Poly(3,4-ethylenedioxythiophene)
PEI	Polyethyleneimine
PFIL	Polyethyleneimine functionalized ionic liquid
PLD	Pulsed laser deposition
PLL	Polylysine
PSA	Prostate-specific antigen
PSS	poly(styrenesulfonate)
PTCA	Perylene tetracarboxylic acid
RGO	Reduced graphene oxide
SWV	Square-wave voltammetry
TMDC	Transition metal dichalcogenides
TMOs	Transition metal oxides
TYR	Tyrosinase
V	Volt

1 Introduction

The term immobilization refers to the art of surface confining molecular assemblies (most often applicable to enzymes, microorganisms, cells, antigens, antibodies, etc.) onto some physical supports. The process of immobilization localizes the specific entities onto a defined region of space while retaining the maximum activity, reuse and stability of the immobilized entities. The supporting matrix allows the exchange of medium containing the reactant (analyte) molecules. The advantages of immobilization like increased stability under various pH and temperature and possibility of reuse make the process a favourable strategy for several applications in different fields such as food industry, health care and environmental analysis. The immobilization strategies are used in the development of biosensors and tremendous advances have been made over the past few decades. Ever since, the chemical sensors and biosensors have become essential elements of our present civilization (Velasco et al. 2003; Thakur and Ragavan 2013; Rodriguez-Mozaz et al. 2005; Su et al. 2018). Biosensors are analytical tools containing a unique union of biological molecular recognition unit such as enzymes, microbes, antigens and tissue that can specifically bind with the analyte and a transducer that can follow the chemical/physical changes in the reaction between the analyte and the recognition unit. Depending upon the nature/phase of the targeted analyte, we can have solid/liquid/gas detection systems possible (Sekhar et al. 2010). Further, the sensors can be subdivided based upon the nature of the transduction strategies employed (e.g. optical/thermometric/gravimetric, electrochemical and piezoelectric) (Banica 2002). The advantageous traits of electrochemical transducers are their simplicity, adaptability with wireless transmission, possibility of miniaturization, portability, possibility for uninterrupted monitoring, amenability for field measurements, etc. The inclusion of wireless sensing networks allows us to install the transducers in remote locations for continuous collection of data. The reaction of the target analyte with the molecular recognition unit can bring about physical or chemical changes in the systems which can be sensed by the respective transducers. Physical changes like temperature/pressure can be analysed to a greater precision due to the current progresses in materials chemistry by allowing us to manipulate their physical properties and by cleverly immobilizing these materials onto proper supports by integrating them to devices. Measuring chemical changes involve the additional complexity of tackling the chemical interface created at the junction of the sensing element and the analyte. The interface plays a key role in deciding the analytical parameters like the sensitivity, selectivity, stability and biocompatibility of chemical sensing devices. The receptor in the sensing element responds to the analyte either by covalent or non-covalent interactions leading to changes in electrical properties of the reaction system like conductivity, capacitance, potential, current or impedance which can be detected by appropriate electrochemical transducers. An ideal sensing material should possess maximum specific active spots along with large surface area to enable maximum material–analyte-specific interactions and should be able to detect the binding events into a suitable measurable output (Fig. 1). Also, it should possess good mechanical and processing properties. Many interesting developments have been demonstrated in electrically transduced chemical sensing with the help of new materials like carbon, boron-doped diamond, conductive polymers, fullerenes, carbon nanotubes, semiconductor nanoparticles, metal oxides and so on. Each material would have its own advantages and disadvantages. The best combination of surface and electronic properties of an ideal material could be achieved when its synthesis could be regulated with care with respect to its dimensionality, flexibility, mechanical stability and lattice matching with the base substrate. Further, it should specifically interact with the analyte to produce a measurable signal. Two-dimensional materials (2D) show a greater promise in this direction owing to their thickness ranging from few to tens of nanometers and their length extending up to several centimetres. They possess remarkable physical and chemical properties with possible applications in chemical sensing with electrochemical transduction. Graphene is now the well-studied 2D material consisting of a sheet of carbon atoms with 2D hexagonal crystal structure which has been demonstrated for a wide range of applications. A diverse range of 2D materials are being explored nowadays which include phosphorene, hexagonal boron nitride (h-BN), transition

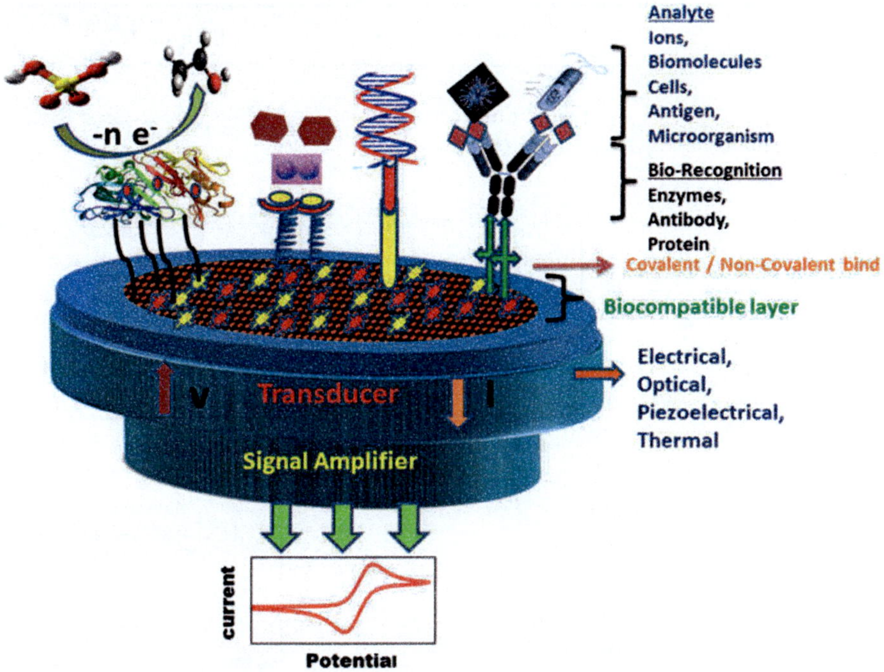

Fig. 1 Schematic representation of immobilization of various biomolecules for biosensor applications and various transduction techniques

metal dichalcogenides (TMDCs),layered metal oxides, 2D metal–organic frameworks (MOFs), covalent organic frameworks (COFs) and other 2D compounds (Siva Kumar et al. 2019; Wang et al. 2017; Zheng Meng et al. 2019). These materials are very good candidates for constituting the molecular recognition part of the biosensor owing to their advantages such as high surface-to-volume ratio, high aspect ratio, excellent stability, high density of reactive states, good thermal stability and compatibility with device miniaturization. Some of the disadvantages include lack of mass production of materials with large area, high and uniform quality, and lack of facile effective strategies for device integration and limited stability of some forms under ambient conditions. This chapter discusses how the 2D materials can be engineered to immobilize biological recognition elements for the development of biosensors.

2 Electrochemical Transduction

Among the various transduction strategies known today, electrochemical transducers represent the most popular transduction technologies adopted in the fabrication of chemical and biosensors (Fig. 2). When compared to optical, mass and thermal

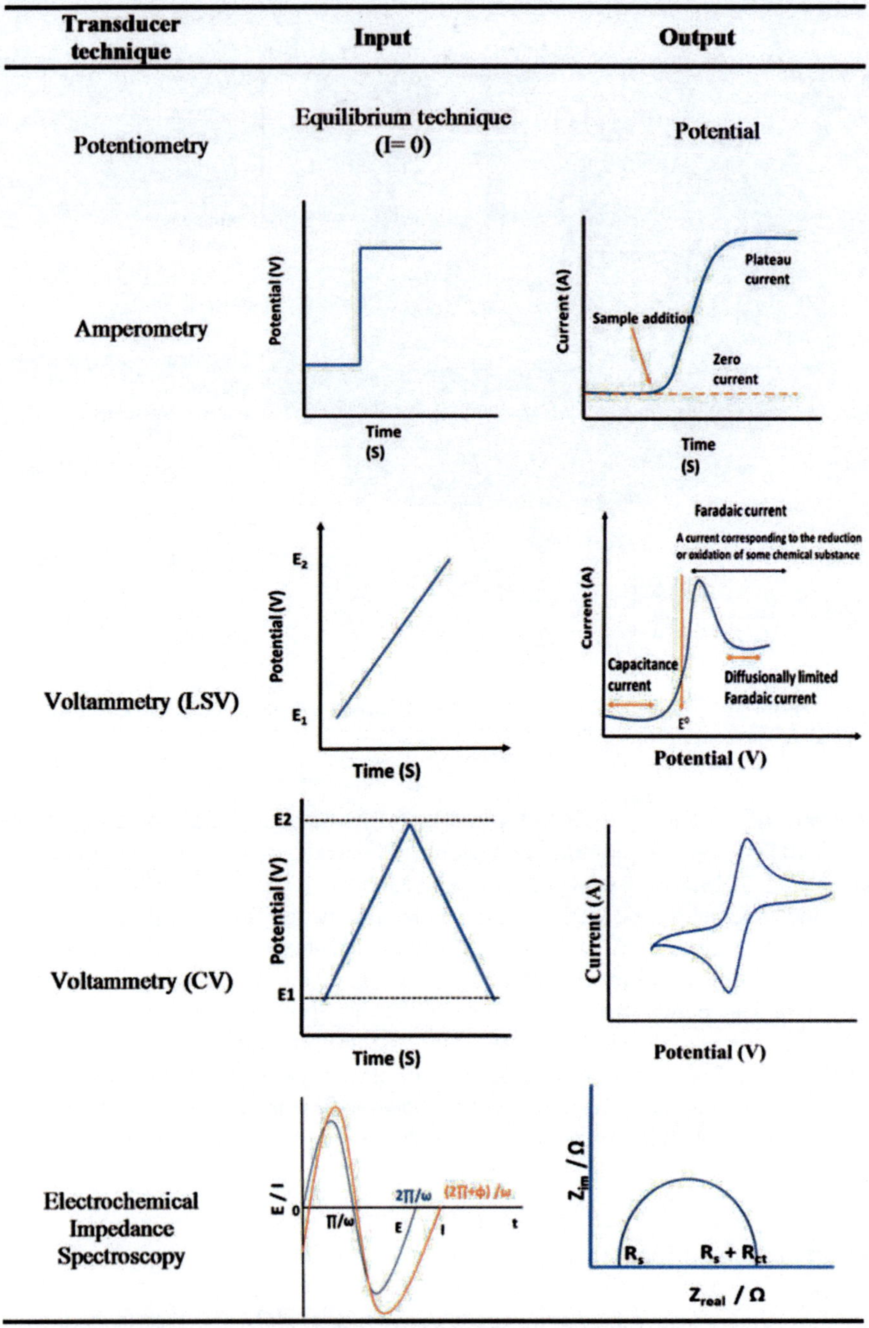

Fig. 2 Various types of electrochemical transduction technologies

sensors, electrochemical sensors are more advantageous owing to their ultra-low detection limits, simplicity and cost effectiveness. They have an edge over the other type of transduction strategies at the commercial level and have found a wide spectrum of significant applications in areas like clinical, industrial, environmental and agricultural analyses.

2.1 Potentiometric Sensors

Potentiometric sensors are devices which consist of a recognition element (ISE, ionophore, lipophilic groups) and an electrochemical transducer which determines the potential difference between two electrodes under zero current flow. The measured potential is correlated to estimate the concentration of the analyte in the solution. The potential E generated is a function of concentration as given by the Nernst equation. This equation predicts that the potential output of the sensor is directly proportional to the logarithm of a function of the activity of the ion in solution:

$$E = E^{\circ} + \frac{RT}{nF}(\ln a_i) \quad (1)$$

where E is the potential (V), R is the gas constant (8.314 J/K), F is the Faraday's constant (96,500 C/mol), n is the number of electrons and a_i is the activity of the analyte. E° is the standard potential. The sensor is an electrochemical cell consisting of the ion-selective and reference electrode whose potential output is E. A typical potentiometric sensor is shown in Fig. 3.

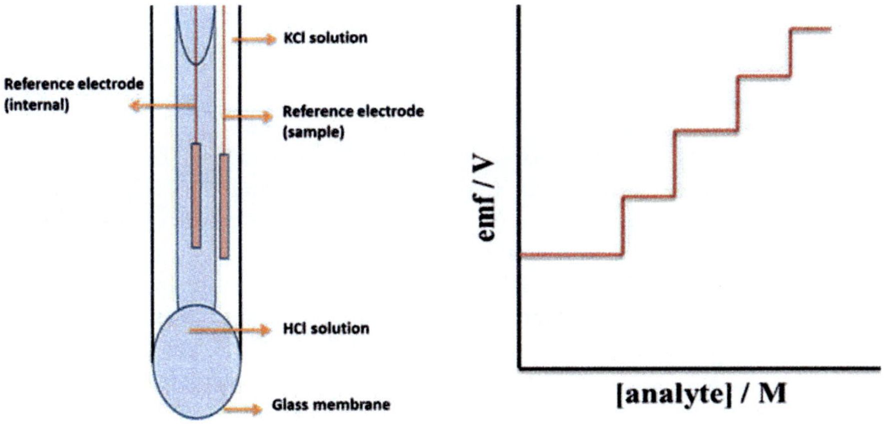

Fig. 3 Typical representation of potentiometric sensor and the output potential response for a target analyte

Typically, potentiometric sensors are predominantly membrane-based devices, consisting of ion-selective membrane, which isolates the sample and prevents contact with the inside of the electrode. The field of potentiometric sensors bridges fundamental host–guest chemistry, fundamental membrane science and engineering. Developing a selective and sensitive molecular/ion exchange membrane is a challenging process and advances in this area of research have led to infinite number of applications for various analytes. ISEs can be divided into three groups: glass, liquid or solid electrodes, respectively, based on the nature of the membrane. The pH electrode is the most popular potentiometric device made up of thin ion-selective glass membrane that is selective to hydrogen ion. The tremendous success of the glass electrode is due to its simple nature, feasibility for fast measurements, non-destructiveness, low cost, usability over a broad range of concentrations and its extraordinarily high specificity for hydrogen ions. Other monovalent cations, like sodium, lithium, ammonium and potassium, can also be sensed based on new glass compositions (Gadzekpo et al. 1985; Metzger 1987). Liquid-membrane-electrode ISEs, consisting of water-immiscible liquid substances integrated in a polymeric membrane, are employed for direct potentiometric detection of various polyvalent cations and some anions. A polymeric membrane will divide the test solution and the target ion solution in the inner compartment. The molecule/ion recognition at the membrane takes place through a liquid ion exchanger (Wang 1994) or by a neutral macrocyclic compound (Beer et al. 1995) with molecular-level cavities that can capture the target ions. In the solid contact ISEs, the latest version of sensors, the internal filling solution is absent and these solid contact ISEs are more robust and amenable for miniaturization than their analogues (Nikolskii and Materova 1985). The introduction of coated wire electrodes (CWEs) in 1970s can be considered as the birth of solid contact ISEs. However, the obstructed connection between the ion-selective membrane, ISM and the electrical conductor resulted in poor potential stability of the CWEs questioning their analytical applications. When good electroactive materials containing mixed ionic and electronic conductivities are interfaced with ISM, good enhancement in potential stability is observed. They function as ion-to-electron transducers when placed between the ISM and the underneath conductor (Hu et al. 2016). Electrochemically active conductive polymers such as (e.g. poly(3,4-ethylenedioxythiophene)-poly(styrenesulphonate) (PEDOT-PSS), poly(3-octyl)thiophene or polyaniline (PANI)), (Bobacka 2006; Migdalski et al. 1996) carbonaceous materials, (Hu et al. 2014; Lai et al. 2007; Crespo et al. 2008, 2009) and 2D nanomaterials (Ping et al. 2011; He et al. 2017; Hernández et al. 2012) offer great scope for providing good ion-to-electron transduction for solid contact ISEs. Ion-selective field effect transistors (ISFET) are another class of CWEs where the ion-selective membrane is directly attached to the gate region of a field effect transistor (FET). The FET is a solid–state device that can precisely measure the charge accumulated on the ion-sensing membrane and exhibits high input impedance and low-output impedance. FETS of smaller sizes can be made using the microelectronic chip technology and can be used for multianalyte sensing by including multiple gates (Kelly and Owen 1986). The miniaturized FET sensors can be easily employed for in vivo measurements. The advantages of potentiometric detection are the usage of just two electrodes allowing simplified device design, the possibility of taking measurements at thermodynamic equilibrium under zero current flow and the detection of analytes in the absence of redox processes, enabling the detection of redox-inert species. Some of the limitations of ISEs include the effect of other interfering constituents in the solution, the effect of ionic strength at high concentrations of the analyte causing deviations in the observed readings and the necessity for repeated sensor recalibration caused by potential drift.

2.2 *Voltammetric and Amperometric Sensors*

Voltammetric and amperometric transduction strategies are usually employed in the construction of electrochemical sensors owing to their simplicity, cost-effective instrumentation, great sensitivity, selectivity, miniaturization possibilities and amenability for field measurements (Bakker and Telting-Diaz 2002; Kimmel et al. 2012). Potential is the driving force which drives the electron transfer reactions which is applied as a linearly or triangularly varying potential input in the case of voltammetry and constant applied potential in the case of amperometry. The output is produced in the form of current which is a function of the concentrations of analytes in the solution. The redox conversions of the electroactive biologically important molecules occur at their characteristic potentials, and the cyclic voltammograms provide the fingerprinting necessary to detect the analyte of interest. In addition to cyclic voltammetry, several other voltammetric techniques are being used in electrochemical detection for additional improvement in the sensitivity and detection limit of the analysis. Differential pulse voltammetry (DPV), square-wave voltammetry (SWV) and normal pulse voltammetry (NPV) are the other pulse voltammetric techniques commonly employed in quantitative analysis (Shao et al. 2010; Zhang et al. 2008). In amperometry, a constant potential is applied to the electrode and the current is monitored with respect to time. The current recorded is in proportionate with the concentration of the analyte present. The response times, dynamic ranges and sensitivities of the amperometric sensors are akin to potentiometric sensors. The main limitation of amperometry is observed when any other interfering species gets oxidized or reduced under the applied bias potential. Hence, selective detection cannot be made based on the shape of recorded current (Hendry et al. 1990). Voltammetric and amperometric techniques usually employ a three-electrode cell assembly comprising of a working electrode, counter electrode and reference electrode. The potential difference between the reference and the working electrode indicates the applied potential, and the counter electrode is used to complete the current flowing circuit. The exterior surface of the working electrode is suitably functionalized to carry out the sensing reaction selectively to the target analyte. The functionalization of electrode surfaces with molecular assemblies is one of the foundations of molecular nanotechnology. There exists a number of ways by which we can design molecular assemblies on surfaces (Tanja and Chem 2011; Daniel et al. 2011; Justin and Chem 2011; Samia et al. 2011). The choice of the method depends on the substrate used for the investigations. Some of the methods commonly employed are physisorption, chemisorption (covalent modification using chemical reagents like carbodiimide, acetyl chloride, glutaraldehyde or electrografting), electrostatic adsorption and self-assembly. Electrodes can be functionalized by using smart materials such as electrochemically active polymers, ionic liquids, carbonaceous materials, metals and 2D nanomaterials. These materials act as supporting matrices for the immobilization of active molecular recognition species/catalysts for specific applications.

2.3 Impedance Sensors

Electrochemical impedance spectroscopy (EIS) is a steady state and a non-destructive technique that can be employed for examining the relaxation phenomena over a broad range of frequencies covering 10^6–10^{-4} Hz. EIS measurements are carried out with a three-electrode electrochemical cell for electroanalytical measurements in solutions. Impedance measurements are made with a frequency response analyser (FRA) that is used to apply a small-amplitude AC signal to the cell under a constant applied potential. The AC voltage and the corresponding current output of the cell are scrutinized by the FRA to estimate the resistive, capacitive and inductive constituents of the impedance of the cell at that particular frequency. Physicochemical processes happening within the electrochemical cell like heterogeneous electron transfer, ion transport, gas and solid-phase reactant transport in ion-selective membranes, etc., have distinct time constants and therefore are displayed at characteristic AC frequencies. When the experiments are carried out over a broad range of frequencies, impedance spectroscopy can be used to determine the impedance associated with these various processes. The ratio of the potential's amplitude and the current's amplitude provides the total impedance. The impedance measured is a complex number. This number can be articulated in the complex plane in polar coordinates by means of Z as the length of the vector and Φ as the angle. The real part of the impedance Z', and imaginary part is Z''. The value of the phase change is governed by the type of the electrolyte, diffusion process, electrode kinetics and chemical reactions that can happen in the electrochemical cell. Nyquist plot is obtained by plotting imaginary part of the impedance against the real part. Bode plot refers to the double logarithmic plot of magnitude and phase (ϕ) of the impedance as a function of the sinusoidal excitation frequency. Since EIS is a function of surface changes, it has become a proven analytical tool for sensor applications. In the case of 2D materials, EIS is a powerful technique for examining the properties and phenomena observed at each step of functionalization and incorporation of materials into device for sensing applications.

3 The 2D Layered Nanomaterials

Ever since its isolation in 2004, the distinctive properties of graphene such as its extraordinary electrical conductivity, large surface-to-volume ratio, unique optical/vibrational modes, exceptional mechanical strength and excellent biocompatibility have attracted the analytical chemists for biosensing applications (Novoselov et al. 2004; Lee et al. 2008; Liu et al. 2011; Hendry et al. 2010). These materials provide defect sites for catalysis and chemical functionalities for further functionalization for the development of biosensors. The success story of graphene has motivated researchers to search for similar materials leading to the synthesis of other 2D layered nanomaterials such as transition metal dichalcogenides (TMDs),transition metal oxides (TMOs), transition metal carbides/nitrides (MXenes) and hexagonal boron nitride (hBN) for biosensing applications (Wang et al. 2011; Li and Wong, 2017; Luo et al. 2018) (Fig. 4).

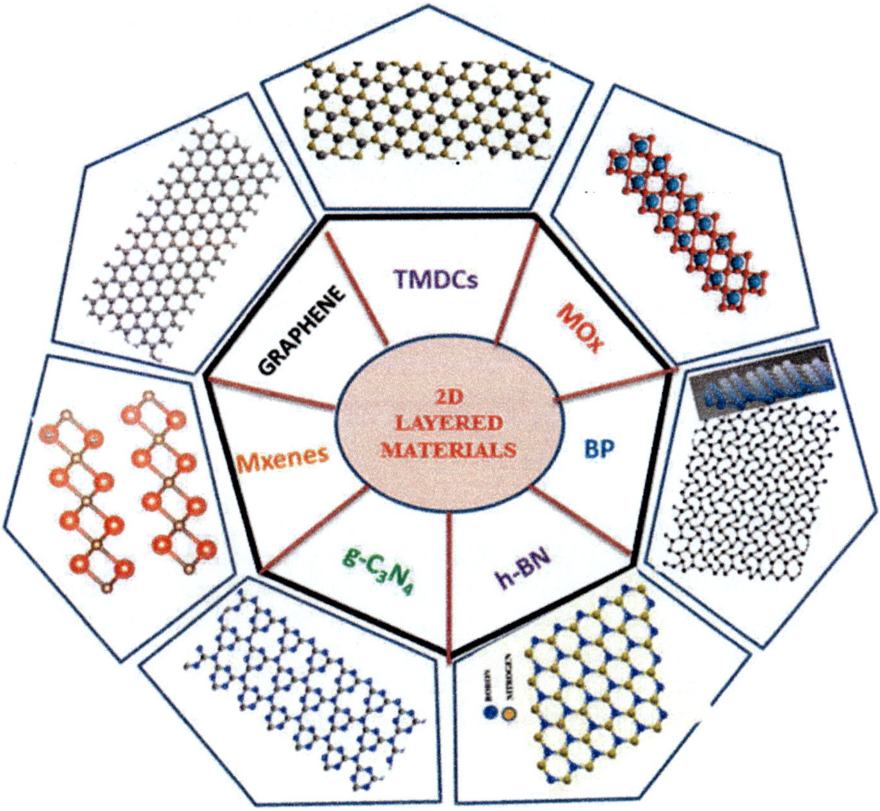

Fig. 4 Different types of 2D layered materials

Much progress has been made recently in the engineering of these materials by way of stacking, doping, functionalization and alloying enabling the construction of the required architecture for desired applications. The 2D nanomaterials enable improved mass transport, high surface area, high sensitivity and large signal-to-noise ratio. They also improve the heterogeneous electron transfer rate when modified on the electrode surface.

3.1 Graphene (Gr) and Graphene Oxide (GO)

The targeted analyte can interact with the graphene through π-π interactions, electrostatic interaction or by increasing the charge transfer rate of the analyte at the graphene/GO-modified electrode. The ratio of basal to edge plane content determines the sensing properties of graphene. The surface of the graphene can be suitably functionalized with molecular recognition units for sensing applications. The superior performance of graphene-based biosensors is due to high surface-to-volume ratio and high conductivity of graphene.

3.1.1 Immobilization of Glucose Oxidase on Graphene-Modified Electrodes for the Fabrication of Glucose Biosensors

The procedures adopted so far for the immobilization of glucose oxidase on graphene are (1) direct immobilization of GOx on graphene, (2) immobilization on graphene–polymer composite surfaces either by layer-by-layer assembly, physical adsorption or covalent linkage, (3) immobilization on graphene–metal nanocomposite surfaces and (4) immobilization on graphene–DNA surfaces.

Direct Immobilization on Graphene

Direct immobilization of glucose oxidase (GOx) on graphene surface has been achieved by drop casting 0.005 ml of 10 mg/ml enzyme solution on a glassy carbon (GC) surface already modified by dropcasting 6 μl of homogeneous dispersion of 1 mg graphene in 0.5 wt% chitosan. Thus, fabricated biosensor exhibited a linear range for the detection of glucose from 0.08 to 12 mM down to a level as low as 0.02 mM. The rate constant for electron transfer was $2.83 \pm 0.18\ s^{-1}$ (Kang et al. 2009). Similar immobilization strategy has been adopted using nitrogen-doped graphene in chitosan on GC electrode where GOx was modified by immersion of the nitrogen-doped graphene–chitosan electrode in 10 mg/ml enzyme solution at pH 7.0 for 24 h. Concentrations as low as 0.01 mM could be detected (Wang et al. 2010).

Immobilization on Graphene-Polymer Nanocomposite

Li et al. have shown how GOx–graphene conjugate immobilized on polypyrrole-modified GC electrode could sense glucose with a LOD of 0.003 mM. In this work, a precleaned GC electrode was modified by electropolymerization of pyrrole by cycling the potential from 0.8 to 1.3 V for 25 times using cyclic voltammetric technique. Over this electrode, the GOx–graphene conjugate prepared by mixing 2 mg GOx in pH 7.4 PBS buffer with 2 mg graphene in vortex mixture for 50 s was immobilized by dropcasting (Alwarappan et al. 2010). Zeng et al. followed a layer-by-layer assembly process for the immobilization of GOx and glucoamylase (GA) by using pyrene-grafted polyacrylic acid (PAA) and polyethyleneimine (PEI) polymers for biosensing glucose and maltose with LOD of 0.168 and 1.37 mM, respectively. PAA was first functionalized by 4.3% pyrene groups by DCC coupling between PAA and L-phenyl ethylamine hydrochloride in N-methyl pyrrolidone. Graphene prepared by Hummer's method is reduced in the presence of PAA–pyrene by hydrazine monohydrate in NaOH resulting in PAA–graphene. The anionic PAA–pyrene can be assembled by layer-by-layer assembly with cationic PEI. GOx and GA are negatively charged at pH 7. By alternate immersion in PEI and PAA–graphene solutions, multilayers of PEI-PAA graphene are prepared on GC electrode. Then, the electrode is further immersed in PEI before modification with either GOx or GA by immersing in a 2 mg/ml of the enzyme solution in 0.05 M PBS buffer (Zeng et al. 2010a, b). Shan et al. prepared the sensing platform by mixing PVP protected graphene that was prepared based on literature protocol with polyethylene imine functionalized ionic liquid (PFIL) to form a uniform dispersion which was then dropcasted on GC electrode. Then, the graphene–PFIL-modified GC electrode was immersed in 2 mg/mL GOx solution of pH 7.4, prepared using 0.05 M phosphate-buffered saline, for 24 h at 4 °C to obtain the graphene–GOx–PFIL-modified GC electrode. The counter-anions in the PFIL film freely exchange with negatively charged GOx at neutral pH and GOx (pI $\sim$ 4.5). This biosensor showed a linear response to glucose in the range 2–14 mM (Shan et al. 2009).

Immobilization on Graphene-Metal Nanocomposite

By incorporating metal nanoparticles onto graphene, further enhancement in glucose biosensing could be achieved. Chen et al. used GOx-Au-Gr biosensor for the estimation of blood sugar in human serum with an LOD of 8.9 μM and a linear range from 0.0436–0.2616 mM. The authors prepared Graphene–Au nanohybrid separately which was dropcasted on GC electrode. Glucose oxidase was subsequently immobilized through covalent linkage in the presence of EDC (1-ethyl3-(3 dimethyl1-amino propyl) carbodiimide hydrochloride) and N-hydroxy succinimide (Chen et al. 2011). It has been demonstrated by Wu et al. that the detection limit of glucose can be decreased to a level as low as 0.0006 mM through electrochemical deposition of Pt nanoparticles (PtNps) onto GO. Functional graphene sheets obtained by thermal exfoliation were mixed with chitosan solution to form a uniform dispersion and

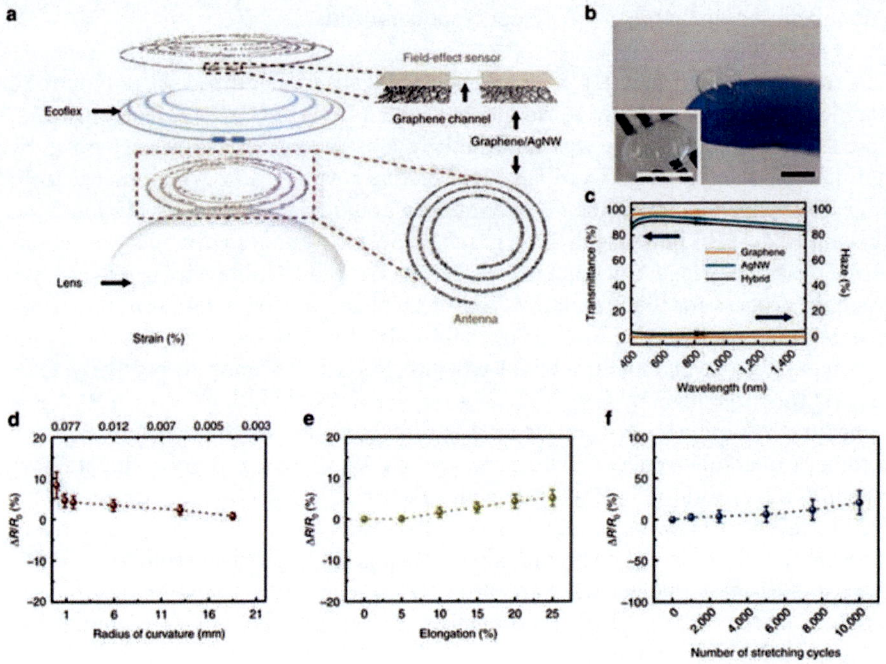

Fig. 5 **a** Schematic of the wearable contact lens sensor, integrating the glucose sensor and intraocular pressure sensor. **b** A photograph of the contact lens sensor. Scale bar, 1 cm. (Inset: close-up image of the antenna on the contact lens. Scale bar, 1 cm.) **c** Optical transmittance and haze spectra of the bare graphene, AgNWs film and their hybrid structures. **d** Relative changes in resistance as a function of outer radius of cylindrical supports. **e** Relative changes in resistance as a function of tensile strain. **f** Relative change in resistance of the graphene FET is for 10,000 cycles of stretching and relaxation. Each data point indicates the mean value for 20 samples, and error bars represent the s.d. (Adapted from Kim et al. 2017, Nature communications)

were drop casted on GC electrode. PtNPs were then electrochemically deposited at −0.25 V in 5 ml of 1 mM H_2PtCl_6 in 0.5 M H_2SO_4 · GOx solution in chitosan was drop-casted on to Graphene–PtNp surface for glucose biosensing (Wu et al. 2009). Park and co-workers fabricated a wireless FET device wherein the source and the drain electrodes are made up of graphene–Ag NW (Ag nanowires) composite. Graphene with covalently immobilized GOx using 1-pyrene butanoic acid and succinimidyl ester is used as the active material for sensing in the channel. Cr/Au function as interconnects and SU8 epoxy layer acts as a passivation layer. The FET device was incorporated on a soft contact lens and applied for the in vivo detection of glucose in tears of rabbit which is shown in Figs. 5 and 6 (Kim et al. 2017).

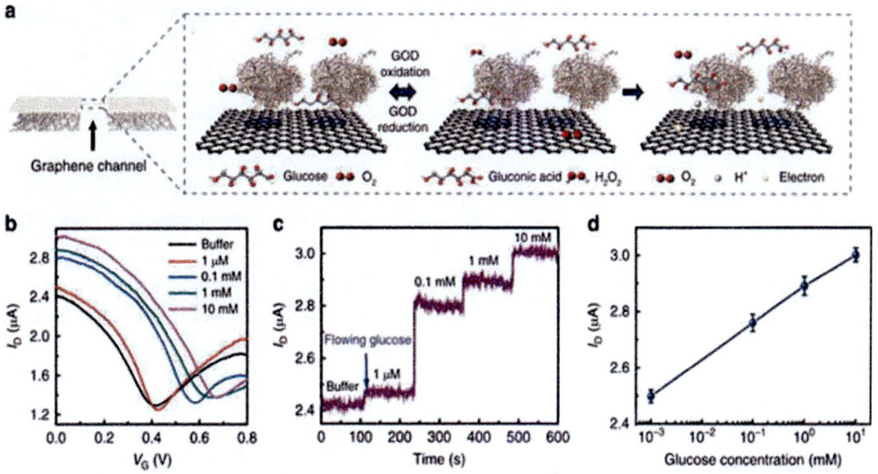

Fig. 6 **a** Schematic illustration and principle of glucose detection with the GOD-pyrene functionalized graphene. **b** Transfer (I_D–V_G) characteristics of the sensor at varied concentrations of glucose ($V_D = 0.1$ V). **c** Real-time continuous monitoring of glucose concentrations ($V_G = 0$ V). **d** The calibration curve generated by averaging current values and the glucose concentration from 0.001 to 10 mM. Each data point indicates the mean value for 10 samples, and error bars represent the s.d. (Adapted from Kim et al. 2017, Nature communications) (Fig. 7)

3.1.2 Immobilization of Horseradish Peroxidase, HRP

H_2O_2 sensing is of high significance as it is produced as a product in many of the enzymatic reactions and also has wide applications in food, pharmaceutical, clinical, industrial and environmental analysis. The enzymatic biosensing of H_2O_2 offers an enhanced method of sensing. Lu et al. demonstrated better sensitivity to H_2O_2 biosensing through immobilization of the heme-containing enzyme, horseradish peroxidase (HRP), onto the surface of exfoliated graphene sheets. The exfoliation was carried out by tetrasodium 1, 3, 6, 8-pyrenetetrasulphonic acid. The exfoliated graphene sheets could efficiently anchor HRP enzymes on the GC electrode surface with enhanced charge transfer. An over layer of nafion was also present on the modified electrode. The resulting biosensors could detect H_2O_2 up to a lower level of 0.106 μM and exhibited a linear range of 0.00063–0.0168 mM (Lu et al. 2010). Zeng et al. used layer-by-layer assembly approach to immobilize the enzyme HRP on the electrode surface. They could synthesize hierarchical enzyme–graphene nanocomposites through electrostatic self-assembly of enzymes and graphene nanosheets. HRP is (pI = 8.8) cationic in physiological media. Graphene prepared by Hummer's method is dispersed in an anionic surfactant to form well-dispersed, anionic graphene nanosheets solution which could be self-assembled with a cationic enzyme layer. This method led to a LOD of 0.0001 mM (Zeng et al. 2010a, b). Zhou et al. mixed reduced graphene oxide, RGO dispersed in chitosan with the microperoxidase-11 enzyme solution and dropcasted on Au electrode for the estimation of H_2O_2 by

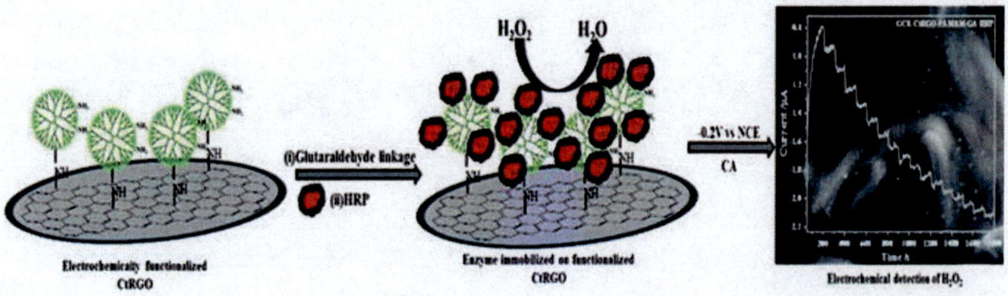

Fig. 7 Schematic of the electrochemical amination of graphene, glutaraldehyde linkage of horseradish peroxidase enzyme and sensor response (Adapted with permission from Manju et al. 2016, Royal chemical society)

amperometry in the linear range from 0.0025 to 0.135 mM (Zhou et al. 2010). In our laboratory, it has been shown that by electrografting PAMAM dendrimer molecules to graphene surface, HRP can be tethered easily on to the surface as depicted in Fig. 7 (Manju et al. 2016; Berchmans et al. 2018). Graphene oxide prepared by Hummer's method is reduced by carrot extract and designated as Ct-RGO. Before surface modification, glassy carbon electrode GC ($\phi = 3$ mm) was pretreated, by polishing by means of 0.05 μm α-Al_2O_3 and emery sheets followed by sonication in alcohol and double distilled water. Then, the cleaned GC was dried in N_2 stream. The PAMAM was electrografted on Ct-RGO by subjecting the GC/Ct-RGO to potential cycling between +0.0 and +1.0 V versus NCE at the scan rate of 20 mV s^{-1} for 5 times in an aqueous solution of 0.1 M $LiClO_4$ containing 0.020 mM PAMAM. The PAMAM anchored surface thus obtained was designated as GC/Ct-RGO–PAMAM. Then, HRP was immobilized onto the electrode by glutaraldehyde linkage. The sensitivity and linear range of detection obtained by this method are 28.96 mA mM^{-1} cm^{-2} and 50–800 mM. The method was validated by measuring H_2O_2 recovery in blood serum as shown in Fig. 8 (Manju et al. 2016; Berchmans et al. 2018).

3.1.3 Immobilization of Cholesterol Oxidase, ChOx

The point of care diagnostics of cholesterol in blood serum is of major importance as the extra build-up of cholesterol and its esters leads to cardiac problems. The surface/volume ratio for immobilization of cholesterol oxidase (ChOx) could be improved by using TiO2-Gr-Pt-Pd/Au Np/ChOx nanocomposite-modified electrodes. The modification was performed in several steps. Initially TiO_2–graphene composite was prepared which was then dispersed in PDDA solution and dropcasted on GC electrode. This was followed by the electrodeposition of Pd and Pt NPs on the electrode surface. Again a layer of TiO_2–graphene was applied on the electrode followed by a modification of Au monolayer from Au colloids by immersion and finally ChOx immobilization was carried out by dropcasting. Thus, constructed biosensor exhibited a broad linear range of response to cholesterol from 5×10^{-5} to 0.59 mM with a LOD of 1.7×10^{-5} mM and a response time of 7secs (Cao et al.

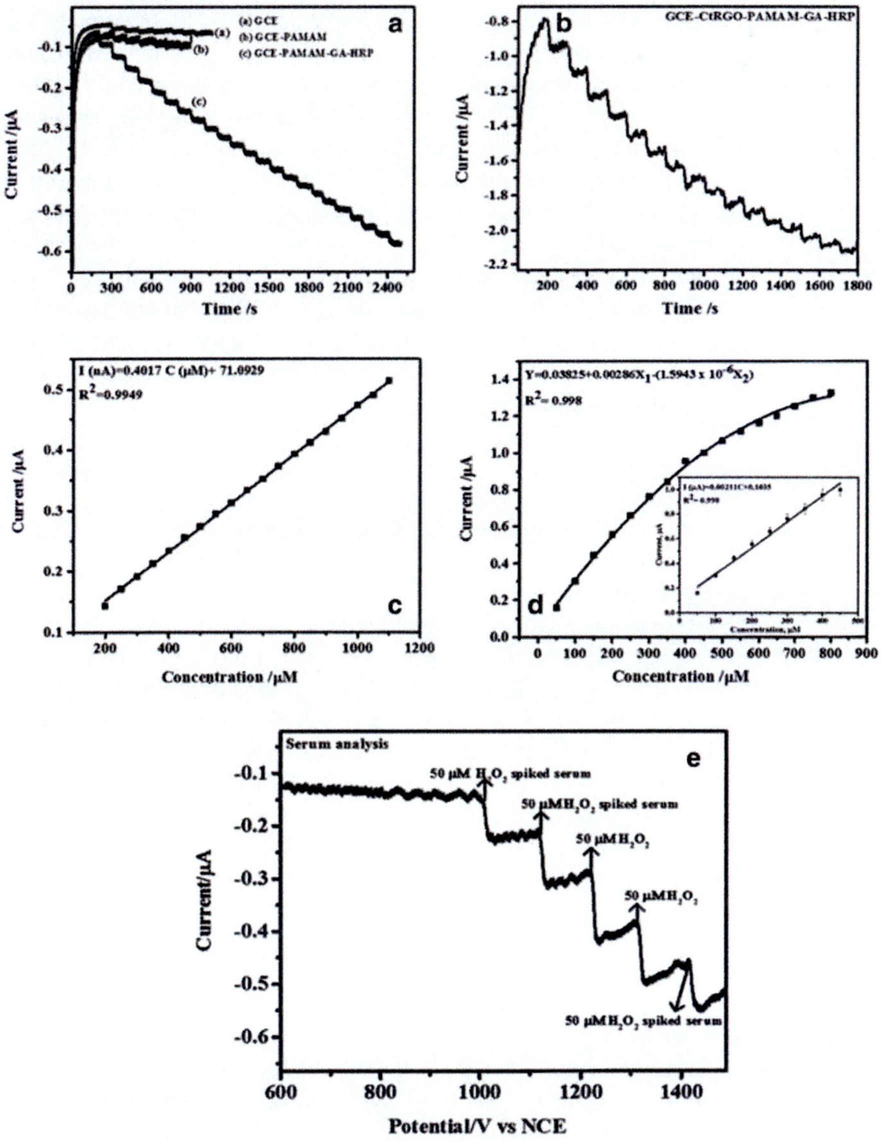

Fig. 8 Chronoamperograms recorded with GCE, GCE–PAMAM, GCE–PAMAM–HRP (**a**), GCE–CtRGO–PAMAM–HRP (**b**) at − 0.2 V in PBS pH-7.2 during successive addition of H_2O_2 and the corresponding calibration plots for GCE–PAMAM–GA–HRP (**c**) and GCE–CtRGO–PAMAM–GA–HRP and GCE–CtRGO–PAMAM–GA–HRP electrode on the addition of N_2 saturated serum, 50 µM of H_2O_2 spiked serum and 50 µM of H_2O_2 in PBS at pH-7 (**d**) (Adapted with permission from Manju et al. 2016, Royal chemical society)

2013). Raj et al. effectively immobilized cholesterol esterase and ChOx enzymes on the graphene nanosheets (GNS)–PtNPs-modified electrodes. By amperometric detection, cholesterol and its esters could be analysed with a LOD of 0.0002 mM. The bienzyme solution was prepared by dispersing cholesterol oxidase (10 mg) and cholesterol esterase (5 mg) in 10 μL of 1:1 isopropyl alcohol and PBS mixture and was kept at 4 °C. The enzyme mixture (3×10^{-3} mL) was dropcasted on the GNS-PtNPs electrode and dehydrated at 4 °C for 6 h to obtain the enzyme-modified electrode (Dey and Raj 2010). In our laboratory, we have shown that the use of cationic polymer overlayers on electrochemically formed alkaline NiOOH layers on GC electrode has facilitated the enzymatic detection of glucose and cholesterol at neutral pH and further it has been shown that the inclusion of graphene layer on GC electrode further decreased the overpotential for the amperometric detection. The enzymes are immobilized by dropcasting on the NiOOH electrode (Selvarani and Berchmans 2017).

3.1.4 Immobilization of Polyphenol Oxidase and Laccase

Oliveira et al. developed a bienzymatic biosensor using the enzymes laccase, LACC and tyrosinase, TYR for the detection of the common pesticide, carbamates. The base electrode of the biosensor was constructed by adding graphene during the fabrication of the carbon pate electrode (CPE). Then, a composite solution mixture was prepared by mixing Au nanoparticles, Au NPs, chitosan solution (CS) and the solutions of the two enzymes laccase (LACC) and tyrosinase (TYR). Several proportions of Au NPs and CS were evaluated in this work. The LACC-TYR-Au NPs-CS/GPE sensor was obtained by immersing the CPE electrode in the above solution and applying a constant potential of −1.5 V for 200 s. This sensor could detect carbamate residue in citrus fruits without any interferences (Oliveira et al. 2014).

3.1.5 Immobilization of DNA

This section will cover some examples from the literature where DNA-specific electroactive labels are immobilized on the electrode surface for sensing biomolecules. A highly sensitive electrochemical DNA biosensor was made by tethering a probe labelled gold nanoparticles (ssDNA–AuNP) with thiol group containing (GT) DNA strand (d(GT)29SH) to electrochemically reduced graphene oxide (ERGO)-modified electrode followed by the anchoring of horseradish peroxidase (HRP) immobilized on carbon sphere (CNS) as signal tag. Initially, graphene oxide film coated GC electrode was prepared which was subsequently allowed for electrochemical reduction by scanning between −1.5 and 0.1 V at 100 mV/s in 10 mM pH 5 PBS (K_2HPO_4/KH_2PO_4) to obtain electrochemically reduced graphene oxide-modified electrode (ERGO). The d(GT)29SH was then assembled on the electrode by the π-π stacking interaction between d(GT)29SH and ERGO for anchoring ssDNA–AuNPs. A triplex signal amplification strategy for DNA detection was attained by anchoring streptavidin

(SA)–horseradish peroxidase (HRP) functionalized carbon sphere (CNS) as a signal tag. The authors could attain a high sensitivity to DNA with the LOD as low as 5×10^{-3} mM and a linear response covering 5 decades of concentration from 1×10^{-14}–1×10^{-10} mM. This biosensor displayed very good selectivity for recognizing single-base mismatched and three-base mismatched DNA sequences (Dong et al. 2012). A highly sensitive and selective DNA biosensor AuNPs/RGO/GCE was reported by Benvidi et al. where no electrochemical label was attached to DNA. ERGO-modified GC electrode was used as the base electrode which was further functionalized by Au NPs by electrochemical deposition. Au NPs are useful for immobilization of thiolated bioreceptors. Then, a single-stranded DNA (ssDNA) probe for BRCA1 5382 insC mutation detection was immobilized on the modified electrode for a specific time by dropcasting and this electrode was named as ssDNA/AuNP/RGO/GCE. To circumvent the problem of non-specific adsorption of DNA at the AuNP/RGO/GCE surface that have not been covered by DNA molecules, the electrode was kept in the 6-mercapto 1-hexanol, MCH solution (1.0 mM) at ambient temperature for 30 min. This electrode is called as MCH/ssDNA/AuNP/RGO/GCE. The electrochemical response for DNA hybridization was monitored using electrochemical impedance spectroscopy (EIS) and cyclic voltammetry (CV) methods. Very low LODs of the order of 1.0×10^{-17} mM for target DNA could be achieved by this label-free electrochemical biosensor. The complementary and non-complementary sequences can be distinguished by two linear response ranges covering 13 decades of concentration (Benvidi et al. 2015).

3.1.6 Immobilization Procedures Used in Immunosensors

Generally, the antibodies are immobilized on to modified graphene surface by incubation in presence of Au NPs. In many of the reports, functionalized graphene (with Prussian blue, chitosan, etc.) along with Au NPs are used.

Alpha-Fetoprotein Detection

In one of the reports, graphene-doped chitosan solution (GDCS) was drop casted on a GC electrode modified with polythionine (Thi) through electropolymerization. The obtained electrode is dipped in gold colloids for 12 h at 4 °C. Subsequently, the Au NPs/GDCS/Thi-modified GC was then dipped into the HRP–anti-AFP solution and soaked for 12 h at 4 °C (Su et al. 2010). By this method, the detection of AFP became possible up to a limit of 0.7×10^{-3} mg/ml. In a slightly different approach, Wei et al. incorporated anti-AFP on GCE modified by Gr and thionine (Thi) through π-π interaction. Subsequently, AFP antibodies were covalently cross-linked with Thi (Wei et al. 2010). When the AFP encounters the electrode, electron transfer and mass transfer of Thi are reduced causing a decreased output. This immunosensor could estimate AFP as low as 5.77×10^{-9} mg/ml. The same immunoassay was validated in clinical human serum analysis. In another report, Prussian blue, PB/RGO nanocomposite and Au NPs-PEDOT are separately prepared and mixed together to form a uniform dispersion and 10×10^{-3} ml of the dispersion was drop casted on GC surface. The electrode was then rinsed with PBS to eliminate free AuNPs-PEDOT/PB-RGO and soaked in anti-AFP solution (11.4×10^{-3} mg/ml, 10×10^{-3} ml) at 37 °C for 2 h. After rinsing with PBS to get rid of excess anti-AFP, the electrode was immersed in 10 μl of 1% BSA to cover the leftover active sites. The composite electrode showed high specificity for the capture of the AFP antibody resulting in lower detection limits of 3.3×10^{-9} mg/ml. These methods were also validated by carrying out the analysis in serum samples (Yang et al. 2017).

Detection of Carcinoembryonic Antigen

An FET device has been reporter for carcinoembryonic antigen detection. A graphene film was transferred on the FET device (GFET). GFET was immersed in a solution of 5 mM 1-pyrenebutanoic acid succinimidyl ester (PYR-NHS) in dry dimethylformamide (DMF), for 2 h at room temperature, followed by washing with DMF. In the next step, anti-CEA was covalently immobilized on the graphene surface, by incubating the device in a solution of 2 mg/ml anti-CEA antibody suspended in 10 mM phosphate buffer solution (pH 7.2) overnight. After immobilizing anti-CEA and washing with blank PBS, the deactivation and blocking of the surplus active functionalities on the graphene surface were carried out by adding 100 mM ethanolamine onto the channel of the GFET for 1 h. Finally, the anti-CEA-modified GFETs were washed with deionized water and employed for the estimation of target molecules. The bifunctional molecule, 1-pyrenebutanoic acid succinimidyl ester, consisting of a pyrene and reactive succinimide ester functionalities, reacts with graphene non-covalently through π-π stacking. The succinimide ester group interacts with the amine group leading to antibody surface immobilization. The design of the constructed FET device allowed label-free estimation of this cancer biomarker (CEA) at 5×10^{-10} mM concentrations (Zhou et al. 2017). In another report, carcinoembryonic antigen was estimated by the formation of sandwich-type complex using two antibodies. A GC electrode was first modified by AuNPs by electrochemical deposition. Then, Ab1 antibodies were tethered to the surface of gold nanoparticles by means of the reaction between cysteine or NH_3^+-lysine residues of the proteins and gold nanoparticles. Blocking buffer was used to prevent non-specific adsorption at the leftover active sites. HRP–GO–Ab2 was used as molecular tags to form the sandwich-type immunocomplex. Detection antibodies and HRP molecules are linked by covalent linkage by using carbodiimide to the surface of graphene oxide nanosheets and were used as molecular tags. When H_2 O_2 was introduced in presence of 4-chloro-1-naphthol (4-CN), the immobilized HRP can catalyse the 4-CN substrate to produce an insoluble benzo-4-chlorohexadienone product, which precipitates on the surface of the impedimetric immunosensors, thus hampering the electron transfer process of the redox probes between the base electrode and the solution. At optimum conditions, the dynamic concentration range of the impedimetric immunosensors covers 1.0×10^{-9} to 80×10^{-6} mg/ml CEA with a detection limit

(LOD) of 6.4×10^{-10} mg/ml. Further, the method was validated in 10 clinical serum samples (Hou et al. 2013).

Detection of MMP

An electrochemical immunosensor based on sandwich immunocomplex formation with two antibodies was established for highly sensitive estimation of matrix metalloproteinase-2 (MMP-2), which is one of the main biomarkers in blood. In this method, a composite of Au NPs and nitrogen-doped graphenes is prepared and this composite facilitated sturdy immobilization of antibodies, improved electron transfer and displayed outstanding electrochemical activity. The secondary antibody was covalently conjugated to HRP through glutaraldehyde linkage and it was used for forming a hybrid with polydopamine functionalized GO. This innovative signal amplification method using a sandwich-type immunoreaction improved the sensitivity of estimation of biomarkers to a great extent. The proposed immunosensor showed outstanding demonstration in the estimation of MMP-2 in the linear range from 5×10^{-10} to 5.0×10^{-5} mg/ml, down to a level as low as 1.1×10^{-10} mg/ml and the method was validated in real samples (Yang et al. 2013a, b).

Detection of Low Density Lipoprotein

A mesoporous reduced graphene oxide consisting of few layers and functionalized with antiapolipoprotein B 100 (AAB) and nickel oxide (rGO-NiO) forming a nanocomposite was employed to demonstrate the estimation of low-density lipoprotein by a label-less, highly reproducible, sensitive and specific biosensor. With the help of electrophoretic technique using colloidal solution of rGO sheets and NiO nanoparticles, mesoporous rGO-NiO composite could be formed on indium tin oxide electrode. EDC-NHS coupling chemistry was used to immobilize the antibody AAB molecules on the rGO-NiO matrix. The antigen–antibody (LDL-AAB) binding kinetics and analytical parameters of this biosensor were determined by impedance method. This biosensor displays a very good sensitivity of 510 Ω $(\text{mg/dl})^{-1}$ cm^{-2} for estimation of LDL molecules. A very low detection limit of 5 mg/dl could be obtained in the linear range of 0–130 mg/dl (Ali et al. 2016).

3.1.7 Evaluation of Cell Apoptosis Using Caspase-3 Activity

A new method for the estimation of cell apoptosis by the electrochemical technique differential plus voltammetry (DPV) has been described. In this study, caspase-3 activity was employed for the measurement of cell apoptosis. An N-terminal blocked (acetylated) peptide Ac-Gly-Gly-His-Asp-Glu-Val-Asp-His-Gly-Gly-Gly-Cys is utilized in this method as the substrate which was covalently tethered to the gold electrode through Au–S bond. Due to the non-availability of free amine groups, it becomes impossible to tether the blocked peptide to graphene oxide (GO) through amide bonds. However, the cleavage of peptide which occurs in the presence of caspase-3 exposés new free N-terminal amine group after rinsing the acetylated pieces. Now GO can be tethered covalently to the electrode by the EDC/NHS chemistry. As the concentration of caspase-3 increases, additional amine functionalities are exposed for the EDC/NHS chemistry, resulting in the binding of more GO to the electrode. This increases the availability of electroactive methylene blue (MB) molecules modified on GO through π-π stacking and electrostatic interactions. MB acts as an electrochemical reporter. Caspase-3 could detect as low as 0.6×10^{-12} mg/ml by this highly sensitive method (Chen et al. 2015).

3.1.8 Detection of Cells

The 26-mer DNA aptamer, AS1411, is currently in phase II clinical trials for relapsed or refractory acute myeloid leukaemia and for renal cell carcinoma. Under the conditions when AS1411 exhibits stable G-quadruplex structure, it can bind to nucleolin with high affinity. Compared to normal cells, nucleolin is overexpressed on the plasma membrane of tumour cells. It is known from the literature that the nucleolin overexpressed on cancer cell surface that acts as a receptor for anticancer aptamer AS1411. The ability of nucleolin to get overexpressed in cancer cells allows AS1411 to differentiate cancer cells from the normal ones. Label-free cancer cell detection has been reported using aptamer AS1411, with the help of a reusable graphene-based electrochemical aptasensor. Initially, 1,3,4,9,10-perylene tetracarboxylic acid (PTCA), a water-soluble perylene derivative assembled on graphene through π-π stacking, is used to evade aggregation of graphene and to increase the number of negatively charged -COOH groups on graphene surface, without upsetting the conjugated π-system of graphene. Thus, the prepared PTCA-functionalized chemically converted graphene (PTCA/CCG) was linked to NH_2-modified aptamer strand using carbodiimide chemistry. The resulting aptamer–PTCA nanocomposite could successfully capture cells on electrode surface through the specific binding between cell surface nucleolin and the aptamer AS1411. The electrode can be made recyclable, when hybridized with the aptamer complementary DNA (cDNA), which causes the disruption of AS1411 quadruplex structure resulting in its inability to bind the overexpressed nucleolin on the plasma membrane of cancer cells. After slight rinsing (30 s in room temperature distilled water), the recycled electrode becomes reusable for cancer cell detection. Thus, fabricated electrochemical aptasensor could discriminate between cancer and normal cells to a low detection limit of 1000 cells (Feng et al. 2011). Yang et al. established a cytosensor by a layer-by-layer approach using polyethylene imine and carboxymethyl chitosan-modified graphene electrode functionalized with folic acid (FA). FA could be detected with high specificity by folate receptors overexpressed on the membrane of the tumour cells. These sensors could sense cancer cells as low as 500 cells/mL (Yang et al. 2013a, b). A sandwich immunocomplex formation using two antibodies has been reported for the detection of human prostate cancer cell line Du-145. A gold electrode was immersed in 3-mercapto propionic acid (3MPA) solution (0.5 mM, pH = 11) for 2 h at room temperature. Later, the 3MPA-modified electrode was rinsed with water and soaked in 0.1 mg/ml CD166 antibody solution and subjected to EDC/NHS chemistry (100 mM) for 24 h under an inert atmosphere. In the next step, the electrode was rinsed to remove excess reagents followed by soaking in various concentrations of cell solutions for 1 h at 37°C. The resulting Du-145 cell/antibody/gold electrode was rinsed and soaked with secondary antibody–HRP conjugate tethered to Au-NP graphene composite. HRP–antibody conjugate was tethered to graphene–Au NP composite through 3MPA via EDC-NHS chemistry. Selective detection of Du-145 cancer cells was achieved with a LOD of 20 cells/ml along with an extended linear range from 10^2 to 10^6 cells/ml (Yadegari et al. 2017). A biocompatible film composed of graphene oxide (GO) and poly-L-lysine (PLL) for better adhesion with leukaemia K562 cells was reported for the electrochemical impedance detection of leukaemia K562 cells. The biocompatible film provided better capability for immobilization and activity retention of immobilized cells. The immobilized K562 cells on the biocompatible film-modified electrode can be directly investigated by electrochemical impedance spectroscopy using $[Fe(CN)6]^{3-/4-}$ as redox probes. At optimum situations, the electron transfer resistance increase was proportional to the logarithmic value of concentration of K562 cells in the range 10^2–10^7 cells/ml, and a detection limit of 30 cells/ml (S/N = 3) could be achieved. Further, the described method was employed to measure the viability of cells and to assess the efficacy of antitumour drug Nilotinib on K562 cells (Zhang et al. 2013). Akhavan et al. could achieve improved sensitivity to leukaemia cells (LOD of 0.02 cell/ml) by depositing chemically exfoliated graphene oxide sheets on graphite rods in the presence of Mg salt by electrophoresis, which resulted in Mg^{2+}-doped positively charged spongy reduced graphene oxide composites (SGE). The SGEs were able to present two distinct signals (arising due to electrochemical oxidation of guanine) in differential pulse voltammetry (DPV) of leukaemic and normal blood cells, which is contrary to the response of glassy carbon electrodes giving only one overlapped peak. Hence, the SGEs could be employed for rapid (60 min) and highly sensitive detection of leukaemia (single abnormal cells in ~ 10^9 normal cells) in a blood serum (Akhavan et al. 2014). Acute myeloid leukaemia and acute lymphocytic leukaemia could be detected together by using dual aptamer-functionalized, graphene–Au nanoparticle electrodes as described by Wang et al. This biosensor showed a low detection level of ~350 cells per mL and a broad linear range from 5×10^2–1×10^7 cells/ml for both HL-60 and CEM cells, with negligible cross-reactivity and interference from non-targeting cells, such as non-leukaemia cancer cells, K562 (a chronic leukaemia cell line) and normal red blood cells. DPV was used to detect the cells by attaching the redox probes antharaquinone, AQ and thionine to SBA-15, a mesoporous nanostructure of silica, which generated the electrochemical signals for cyto-sensing. The redox tagged SBA-15 was further attached to HRP conjugated cells (Zheng et al. 2013). An aptamer-based electrochemical biosensor for early stage estimation of leukaemic cancer cells has been demonstrated using a thiolated sgc8c aptamer immobilized on gold nanoparticles-coated magnetic Fe_3O_4 nanoparticles (Apt-GMNPs). The estimation of leukaemic cancer cells was made by ethidium bromide (EB), intercalated into the stem of the aptamer hairpin. When leukaemic cancer cells encounters Apt-GMNPs, the hairpin structure of the aptamer is disturbed and the intercalator molecules are released, leading to a decrease of the electrochemical signal. The presence of nitrogen-doped reduced graphene oxide on screen-printed carbon electrode provides an excellent matrix for amplifying the read-out signal. At optimum conditions, the aptasensor showed a linear and broad dynamic range of leukaemia cancer cells from 10 to 1×10^6 cell ml-1. This method was validated in human blood plasma without any interferences (Khoshfetrat and Mehrgardi 2017). A gold nanoparticles-modified free-standing graphene paper electrode for detection of *Escherichia coli* O157:H7 (*E. coli* O157:H7) was reported using impedimetric detection. Graphene paper was made by chemically reducing graphene oxide paper prepared by vacuum filtration. The surface of graphene paper electrode was decorated by gold nanoparticles by single-step electrodeposition technique. The immobilization of anti-*E. Coli* O157:H7 antibodies on paper electrode was carried out through biotin–streptavidin system. Electrochemical impedance spectroscopy was employed to detect *E. coli* O157:H7 collected on the paper electrode. Thus, the described paper immunosensor exhibited improved sensing performance, over a broad linear range (1.5×10^2–1.5×10^7 cfu/ml), low detection limit (1.5×10^2 cfu/ml) and very high selectivity (Wang et al. 2013).

3.2 *Transition Metal Chalcogenides*

Transition metal dichalcogenides (TMDCs), generally represented as MX_2, belong to a class of semiconductors, where M is a transition metal atom (such as Mo or W) and X is a chalcogen atom (such as S, Se or Te) (Wilson and Yoffe 1969; Mattheiss 1973). The 2D structure of TMDCs coupled with their atomic-scale thickness, direct bandgap in the visible frequency range, strong spin–orbit coupling and favourable electronic and mechanical properties, make them suitable candidate materials for fundamental investigations and for applications related to energy, flexible electronics, DNA sequencing and electrochemistry. Due to the presence of different coordination spheres of the transition metal atoms in MX_2, TMDCs exist in numerous structural phases. The two typical common structural phases observed are trigonal prismatic (2H) or octahedral (1 T) coordination of metal atoms. These structural phases are also differentiated in terms of dissimilar stacking orders of the three atomic planes (chalcogen–metal–chalcogen) constituting the individual layers of these materials (Podberezskaya et al. 2001; Sajedeh and Ovchinnikov, 2017; Deep Jariwala et al. 2014). In electrochemistry, heterogeneous electron transfer (HET) capabilities of TMDCs towards electroactive molecules make them suitable for electrochemical devices. Some of the TMDCs are molybdenum disulphide (MoS_2, tungsten disulphide (WS_2, gallium arsenide (GaAs), gallium telluride (GaTe) and bismuth telluride (Bi_2Te_3), and they have been employed in aforementioned applications. In their thermodynamically stable 2H phase, MoS_2, $MoSe_2$, WS_2 and WSe_2 are semiconductors which make them suitable for electronic devices. MoS_2 and WS_2 occur in nature in the form a layered crystal (Lopez-Sanchez et al. 2013; Cheng et al. 2014; Pradhan et al. 2016).

Large-scale synthesis of 2D TMDCs are precisely controlled by their compositions, thicknesses, lateral sizes, crystal phases, doping, defects, strains, vacancies and surface properties. It is of great significance to find out the connection between their structural features and properties. Several synthesis methods have been used to prepare graphene and its analogue materials either in thin film or powder form. The synthesis methodologies basically include two categories such as top-down and bottom-up methods, and both physical and chemical approaches have been employed for this purpose. Micromechanical exfoliation generally known as the "Scotch tape" method is used for the preparation of graphene nanosheets. However, researchers are currently using this technique for the preparation of MoS_2 and other 2D layered nanosheets (Benameur et al. 2011; Novoselov et al. 2015; Late et al. 2012; Mahesh et al. 2015). Atomic layer deposition (ALD), a physical vapour deposition method (Valdivia et al. 2016; Thimsen et al. 2012; Kim et al. 2016) and pulsed laser deposition (PLD) (Serna et al. 2016; Mahjouri-Samani et al. 2014) were used for the preparation of atomically thin coatings of 2D materials. Chemical exfoliation methods have been also employed to exfoliate 2D bulk TMDCs in to single or ultra-thin layer using direct ultrasonic exfoliation, electrochemical exfoliation, shear exfoliation, ion intercalation exfoliation and anion exchange exfoliation. The solvents used in these techniques are N-methyl pyrrolidone (Brent et al. 2015), 1-methyl 2-pyrrolidone (Gopalakrishnan et al. 2014), dimethyl sulphoxide (Han et al. 2014), N–N dimethylformamide (Huang et al. 2014a, b, c, d, e) and aqueous solution of tetrabutylammonium hydroxide (Hu and Guo 2015). Electrochemical exfoliation is a fast and large-scale method in which exfoliation takes place under the applied electrochemical potential (Eda et al. 2011; Zeng et al. 2011; Kappera et al. 2014; Gordon et al. 2002). Liquid-phase exfoliation carried out in organic solvents is a scalable substitute to mechanical exfoliation and enables the preparation of solutions containing flakes with controllable thickness (Coleman et al 2011; Nicolosi et al. 2013). Highly practical and great quality TMDCs are prepared using the chemical vapour transport method (Wilson and Yoffe, 1979; Cai et al. 2018). Molecular beam epitaxy (MBE) can be considered as one of the scalable methods of synthesis (Joyce 1985).

3.2.1 MoX_2 and Its Composite-Based Electrochemical Biosensors

Among the TMDCs, MoS_2 is the most studied 2D material which is explored in electrochemical applications. Flexible Ebola biosensor has been realized using a MoS_2 thin film (6 μm thickness) prepared through restacking of the numerous flakes deposited on a flexible polyimide support from ethanol dispersion. MoS_2 sheets (with dimensions up to 50 μm) were prepared by a simple cathodic exfoliation approach in non-aqueous electrolyte to avoid surface oxidation and for maintaining the intact crystalline structure of the MoS_2 sheets. MoS_2 sheets provided a highly specific platform for the immobilization of VP40 antibodies. The immobilization was carried out using carboxylic acid terminated thiol, 11-mercaptoundecanoic acid (11-MUA)

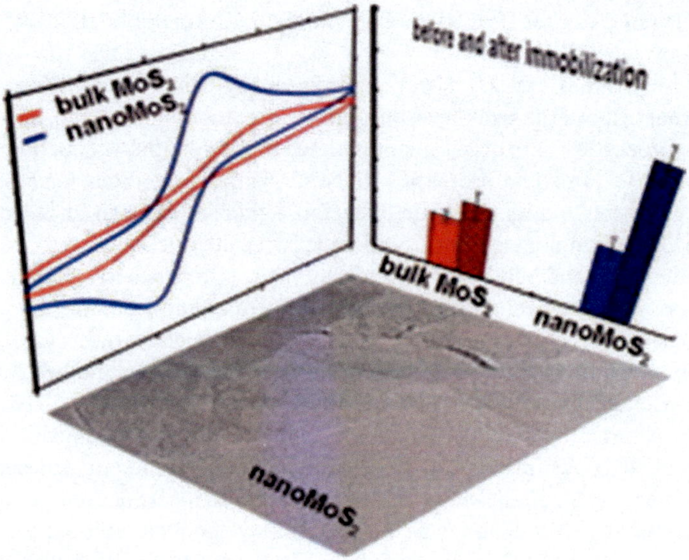

Fig. 9 Graphical representation of electrochemical detection of DNA using thin layer of MoS_2 nanosheet (Adapted with permission from Wang et al. 2015a, b, c; Elsevier)

as cross-linker, which self-assembles on the boundaries or sulphur vacancies and subsequently tethered to the antibody through carbodiimide chemistry. Non-specific adsorption was prevented by a further soaking in 0.5 mg/ml bovine serum albumin (BSA). The results obtained were compared with those found using *Staphylococcal enterotoxin B* (SEB) as analyte, which displays early signs like Ebola disease. A very low detection limit of the order of femtomolar–picomolar range could be achieved by this approach (Zhang et al. 2019).

A label-less highly sensitive electrochemical biosensor was demonstrated by Wang et al., using MoS_2 for the estimation of DNA (Fig. 9). The thin-layer MoS_2 nanosheets were prepared by the ultrasonic exfoliation with distinct attraction of MoS_2 for single-stranded DNA and double-stranded DNA (dsDNA), respectively. Immobilization of the single-stranded DNA molecules was deposited on to thin-layer MoS_2 nanosheets-modified carbon paste electrode (CPE). Thus, fabricated sensor can detect a target complimentary DNA sequence in the wide linear range from 1.0×10^{-13} to 1.0×10^{-7} mM to a level as low as 1.9×10^{-15} mM, with high selectivity and sensitivity. The simultaneous detection of adenine and guanine was also demonstrated in a similar manner (Wang et al. 2015a, b). Also, a similar electrochemical biosensor based on MoS_2/multiwalled carbon nanotubes has been fabricated wherein the active material was prepared by hydrothermal technique. A thiol-attached DNA probe on to the MoS_2/MWCNT composite and GOx/AuNPs-modifed electrode was assembled to develop the biosensor. The detection range of DNA biosensor was from 1×10^{-11} to $10^7 \times 10^{-12}$ mM with limit of detection up to 0.79×10^{-12} mM (Huang et al.

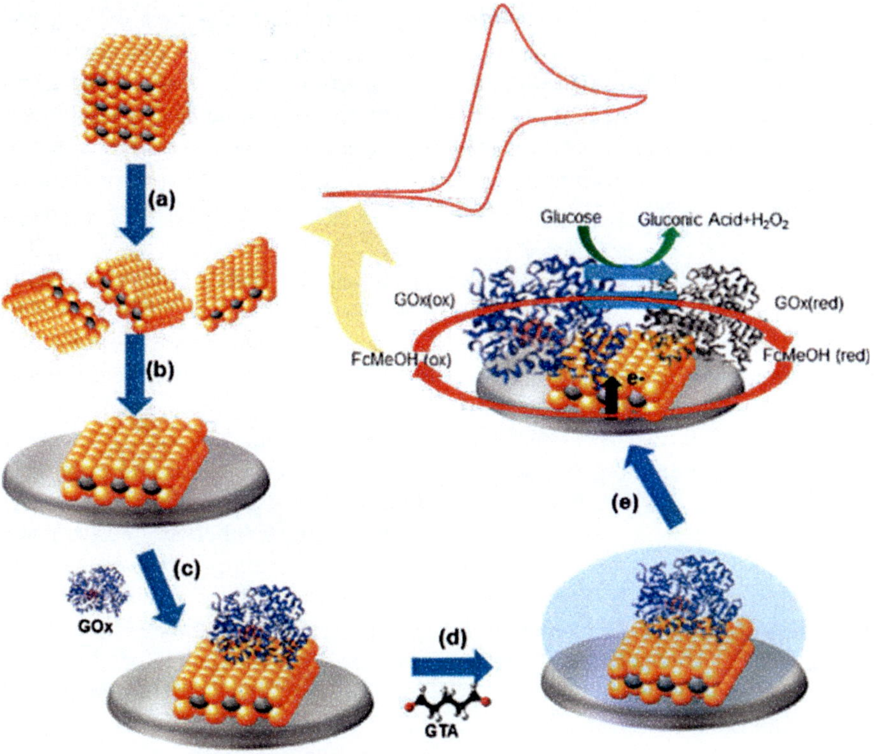

Fig. 10 Schematic representation illustrating **a** exfoliation of bulk TMDs employing t-BuLi as the intercalating agent; **b** construction of the biosensor electrode by drop casting exfoliated TMDs onto bare GC electrode, followed by **c** GOx, then **d** cross-linking using GTA; finally, **e** electrochemical detection mechanism of glucose (Adapted with permission from Toh et al. 2017, American Chemical Society)

2014a). A unique biosensor with amperometric detection was constructed by immobilizing catalase on a MoS_2 nanosheet–AuNR hybrid film to detect H_2O_2 released from SP2/0 cells. The electrochemical characterizations reveal that the structure and biological activity of the enzyme catalase remains intact after immobilization. The catalase–MoS_2–Au hybrid sensing platform provided highly active surface area to enhance the DET between catalase and the electrode materials (Shu et al. 2017). The negative charge of the layered MoS_2 facilitated the formation of Au-MoS_2/HRP hybrids by immobilizing IgG bound to HRP by electrostatic attraction. To accomplish this, bulk layer of MoS_2 was prepared in situ on a polymer-printed circuit board. When hydroquinone HQ was used as a mediator the linear range of detection for the reduction of H_2O_2 extended over a wide linear range. (Kim et al. 2015). Similarly, a cholesterol biosensor was developed using a ChOx immobilized on to MoS_2-AuNP material to detect cholesterol in egg yolk and pig liver. Hydrothermal and electrodeposition methods were used to synthesize MoS_2-AuNPs. AuNPs were self-assembled with sulphur atoms in the MoS_2 layer which increases surface area and provides high loading of cholesterol oxidase molecules. The synergized catalysis of MoS_2 and AuNPs led to adequate sensitivity, reproducibility and stability. The apparent Michaelis–Menten constant and sensitivity were calculated as 0.325 mM and 4.46 $mAmM^{-1}\ cm^{-2}$, respectively. The proposed mechanism of the oxidation of cholesterol to H_2O_2 and cholest-4-en-3-one in the presence of enzyme is given as below (Lin et al. 2016).

$$\text{Cholesterol} + O_2 \xrightarrow{\text{Chox}} \text{Cholest}-4-\text{en}-3-\text{one} + H_2O_2 \qquad (2)$$

2D TDMCs nanosheets and its composite embrace a great assurance for biomedical applications as these materials have desirable physical and chemical properties and can be directly dispersed in aqueous solution. Various biomolecular targets can be detected by electrochemical biosensor using antibodies and nucleic acid aptamers which are gradually recognized as potential candidate for the fabrication of biosensors. TMDCs also find applications in electrochemical immunosensor. A biosensor based on a dielectric layer-free MoS_2 using field effect transistor has been demonstrated by incorporating antiprostate-specific antigen (anti-PSA) antibody on the surface of 2D layered MoS_2 material. The hydrophobic environment of the MoS_2 facilitates the interaction between the anti-PSA antibodies and MoS_2 material. The experimental results revealed that the selective binding of negatively charged antiprostate-specific antigen (PSA) at near physiological pH to the immobilized antibodies led to detection level as low as 1×10^{-9} mg/ml (Lee et al. 2015). Similar approach was carried out by Wang et al. that fabricated a label-free multilayer MoS_2-based FET for the detection of PSA. The antibody–PSA interaction cause changes in drain current and is responsible for the recognition of cancer marker protein with the limit of detection (LOD) of 3.75×10^{-10} mM, as well as great selectivity in the presence of bovine serum albumin (Wang et al. 2014). Further, same author studied immunosensor based on carcinoembryonic antigen (CEA) in which CEA primary antibody, Ag nanoparticles, GOD and CEA secondary antibody have been immobilized on MoS_2-Au as solid support and thus fabricated biosensor for carcinoembryonic antigen showed a wide detection range from 1×10^{-10} mg/ml to 5×10^{-5} mg/ml with a LODs of 2.7×10^{-10} mg/ml in HSA sample. Besides this, immunoassay displayed least interference from human chorionic gonadotropin (HCG), α-fetoprotein (AFP), cancer antigen 125 (CA125) and PSA (Wang et al. 2015a, b, c).

3.2.2 The 2D WX_2 and Its Composite-Based Biosensor

Chemical exfoliation of WX_2 (WS_2 and WSe_2) are carried out using tert-butyllithium (t-BuLi) as organic intercalate for evolving a heterogeneous electron transfer (HET)-based second-generation electrochemical biosensor. Glucose oxidase enzyme was immobilized on WX_2 as immobilization matrix using glutaraldehyde (GTA) linkage.

Heterogeneous electron transfer (HET) investigations were carried out in phosphate-buffered saline solution (PBS, pH 7.2) containing ferrocenemethanol as mediator in the presence of glucose. The sensing principle in this second-generation biosensor is based on electron transfer of flavin redox centre (FAD) of GOx mediated by FcMeOH. The mediator acts as electron shuttle to regenerate FAD-GOx. The electron transfer to the electrode is facilitated by the small molecule FcMeOH which can electrically communicate with the deeply buried flavin redox centre. Redox behaviour of FAD cofactor mediated by ferrocene (FcMeOH)/ferrocenium ion ($Fc^{+}MeOH$) finally gives rise to an oxidation signal that can be correlated to glucose levels (Eq. 3). The presence of WX_2 helps in enhancing the signal at low overpotential. Three linear ranges are observed for glucose detection using this method, viz. 0.176–0.766 mM ($r = 0.997$), 1.3–7.7 mM ($r = 0.996$) and 7.7–22.3 mM ($r = 0.995$). The WSe_2-based biosensor showed two linear ranges, viz. 0.077–0.274 mM ($r = 0.999$) and 0.77–22.3 mM ($r = 0.998$) (Nasuha et al. 2017).

$$\text{Glucose} + \text{GOx} - \text{FAD} + O_2 \rightarrow \text{gluconic acid} + \text{GOx} - \text{FADH} + H_2O_2 \quad (3)$$

$$\text{GOx} - \text{FADH} + 2\text{Fc+MeOH} \rightarrow \text{GOx} - \text{FAD} + 2\text{FcMeOH} + 2\text{H} + \text{TMDs} \quad (4)$$

$$2\ \text{FcMeOH} \xrightarrow[100\ \text{mV}]{\text{TMDs}} 2\text{Fc+MeOH} + 2\ e^{-} \quad (5)$$

Same group established the beneficial role of t-BuLi in 2D TMDCs exfoliation. The 1 T-phase WS_2 for the construction of a heme-based H_2O_2 biosensor was compared to 1 T-phase MoS_2, $MoSe_2$ and WSe_2. With this objective, the GTA/Hb/1 T-WS_2/GC biosensor was fabricated based on a layer-by-layer deposition of 1 T-WS_2 and Hb, and entrapment by GTA. Chronoamperometry was used to analyse the reduction currents with successive additions of H_2O_2 into stirred degassed solution. At optimum conditions, the GTA/Hb/1 T-WS_2/GC biosensor exhibited high sensitivity and selectivity to H_2O_2 with a broad linear range for H_2O_2 concentration from 2×10^{-4} to 0.038 mM and 0.048 to 1.782 mM with detection limit of 36 nM. Consequently, author showed the capability of this biosensor for H_2O_2 detection in human serum (Toh et al. 2017) (Fig. 10). In addition, Martin Pumera et al. studied that heterogeneous electron transfer (HET) properties of 2D WX_2 (WS_2, WSe_2 and WTe_2). Nanohybrids with 2D TiO_2 in the presence of redox mediator were evaluated for biosensor application. The second-generation biosensor was established based on TiO_2@WX_2 towards H_2O_2 detection by using FcMeOH as a mediator. HRP enzyme was immobilized on TiO_2@WX_2 nanohybrids and their catalytic responses were evaluated by cyclic voltammetry. To investigate the effect of TiO_2 decoration on the HET properties of WX_2, cyclic voltammetric technique was employed using standard redox probe ferri/ferrocyanide $[Fe(CN)_6]^{3-/4-}$. Experimental results revealed the dependence on exfoliation method of WX_2 nanosheets in determining HET catalytic activity. TiO_2@WX_2 nanohybrids with t-BuLi-exfoliated WX_2 nanosheets displayed faster HET rates in comparison to their n-BuLi-treated counterparts. For instance,

TiO_2@t-BuLi WS_2 has the $k^0{}_{obs}$ value of 5.39 × 10^{-3} cm/s, which is marginally better than that of TiO_2@ n-BuLi WS_2 at 5.12 × 10^{-3} cm/s. In a similar fashion, TiO_2@t-BuLi WSe_2 and TiO_2@t-BuLi WTe_2 afforded faster HET rates 5.17 × 10^{-3} and 5.05 × 10^{-3} cm/s, respectively. For the biosensor application, HRP enzyme immobilized on TiO_2@WX_2 nanohybrids was electrochemically evaluated in presence of mediator. A reductive peak at low positive potential was obtained in the cathodic scan of cyclic voltammetry which is due to electroreduction of FcMeOH as mediator. The enzymatic reaction involves the reduction of H_2O_2 into water catalysed by HRP enzyme and the enzyme was reduced to its intermediate state. The HRP intermediate oxidized the FcMeOH (red) to FcMeOH (ox). Finally, the oxidized FcMeOH was regenerated by the improved electron transfer platform of TiO_2@WX_2 (Rahmanian et al. 2018). Huang and co-workers fabricated an AuNPs/chitosan/graphene-WS_2 composite for the covalent immobilization of ssDNA probe. Thus, fabricated biosensor can detect DNA hybridization using ferro/ferri as the electroactive redox probe. A unique 2D graphene–tungsten sulphide (WS_2–Gr) composite was reported to obtain excellent electrochemical properties for DNA biosensor. The 2D conductive graphene platform supported a high surface area of WS_2 nanocomposite. Incorporation of Au nanoparticles (AuNPs) on WS_2–Gr–chitosan composites facilitate the capturing of ssDNA sequences. Cyclic voltammetry and electrochemical impedance spectroscopy were employed for the analysis of modified electrodes. The fabricated biosensor exhibited a linear relationship between the current value and logarithm of the target DNA concentration in the range from 0.1 × 10^{-12} to 5 × 10^{-7} mM with detection as low as 2.3 × 10^{-12} mM. The DNA biosensor displayed superb capability to differentiate one-base mismatched DNA, three-base mismatched DNA and non-complementary DNA sequence (Huang et al. 2014b). The synergism between WS_2, graphene and Au led to substantial signal amplification of biosensor output. Huang et al. developed label-free electrochemical biosensor for the estimation of 17b-estradiol through the immobilization of the selective aptamers on the surface of WS_2 nanosheets decorated with Au nanoparticles (Huang et al. 2014c). The resulting biosensor responded linearly to 17b-estradiol in the linear range from 1.0 × 10^{-8} to 5.0 × 10^{-6} mM with a low detection limit of 2.0 × 10^{-9} mM. The same author developed a biosensing matrix for immunoglobulin E (IgE) detection based on Au nanoparticle and aptamer functionalized WS_2-graphene nanocomposites (Huang et al. 2014d).

3.3 *Black Phosphorous (BP)*

Black phosphorus is currently evolving as a new layered material. Its nanoparticles display a bigger bandgap when compared to bulk materials and they are usually prepared by exfoliation of bulk black phosphorus crystals. Cardiovascular disease marker myoglobin (Mb) could be detected using black phosphorous functionalized by anti-Mb-DNA aptamer prepared by SELEX approach. A dispersion of polylysine, PLL and 2D BP was prepared and was dropcasted on screen-printed electrode.

The exfoliated BP nanosheets were modified at the electrode surface with PLL to make it suitable for DNA aptamer functionalization for specific Mb sensing. Further, functionalization was carried out by dropcasting 1×10^{-3} mg of anti–Mb-DNA aptamer (0.2 mg/ml stock in 10 mM Tris buffer along with 150 mM NaCl and 5 mM $MgCl_2$, pH ~ 7.2) on PLL-BP-modified electrodes. The cyclic voltammograms of the aptasensor were recorded at different concentrations of Mb starting from 1×10^{-9} mg ml to 16×10^{-3} mg ml (Kumar et al. 2016). Similar procedure was adopted by Zhao et al. for the inclusion of haemoglobin by preparing Hb-PLL-BP electrode for detection of oxygen and H_2O_2. PLL-BP composite preparation was based on electrostatic and hydrophobic interactions. The H_2O_2 detection could be carried out over a linear range between 10×10^{-3} and 700×10^{-3} mM (Zhao et al. 2018). A recent report on FET type device for the detection of human immunoglobulin G (IgG) establishes the choice of using BP as the sensing channel in the FET biosensor for the estimation of biomolecules. The mechanically exfoliated BP nanosheets were coated on gold interdigitated electrodes made by electron beam lithography on a highly doped silicon wafer with a SiO_2 top layer. The device was then coated with an Al_2O_3 thin film for surface passivation by using atomic layer deposition, ALD. Anti-human IgG-conjugated gold nanoparticles were deposited on the surface of Al_2O_3 dielectric layer as the probes for antigen (IgG) detection by initially depositing gold NPs followed by conjugation of antibodies through glutaraldehyde linkage. When the IgG interacts with the sensor, it binds with the antibody resulting in a significant variation in the BP's electrical conductivity, which was recorded as the output. The BP-based FET sensor displayed good stability at room temperature for nearly one week with marginal changes in the electrical resistance and on–off current ratio. The sensor displayed a very good sensitivity (down to 10×10^{-6} mg/ml) to human IgG with a response time of the order of seconds (Chen et al. 2017). In a recent report by Mayorga-Martinez et al., electrochemically exfoliated BP has been used for rabbit immunoglobulin G (IgG) detection by making use of the hydrogen evolution properties of BP as the detection signal. Magnetic bead conjugated antirabbit IgG complex and BP NPs conjugated rabbit IgG were prepared separately and reacted together to form antigen–antibody interaction. The electrochemical detection of BP NPs captured through magneto-immunoassay was carried out through hydrogen evolution reaction, HER monitored by chronoamperometry. The electrochemical experiments were carried out with screen-printed electrodes by attaching a magnet on the reverse side of the working electrode. A potential of −0.88 V versus RHE was applied for 100 s, and chronoamperograms were recorded. A detection limit of 0.98 ng/ml could be achieved by this method (Mayorga-Martinez et al. 2016).

3.4 Metal Oxides

Nanostructured metal oxide films of Zn, Zr, Ce, Ti have emerged as novel 2D materials owing to their exciting optical, electrical and molecular properties for biosensor applications. Impedimetric cell analysis was demonstrated by 15 nm thickness of passive ZrO_2 coatings on gold interdigitated electrodes on PDMS biochips on glass covers. The sensor could be applied for the detection of human lung adenocarcinoma epithelial cell cultures (H441) and human dermal fibroblast cells (NHDF) (Sticker et al. 2015). Nanostructured zinc oxide film obtained on indium tin oxide, ITO by sol–gel process which contained preferred (002) plane and 10 nm crystallite size was found to be a good immobilization platform for cholesterol oxidase (ChOx). The immobilization was carried out by physisorption. With the help of cyclic voltammetry, the developed biosensor exhibited better detection range from 0.05–4 mg/ml, detection level as low as 0.005 mg/ml, response time of 10 s, sensitivity of the order of 5×10^{-6} mA/mg dl^{-1} cm^{-2} and Michaelis–Menten constant value of 9.8×10^{-3} mg/dl (Km) (Solanki et al. 2009b). The same research group used similar base electrode for sensing mycotoxin. An indium–tin–oxide (ITO) glass plate on which a nanostructured zinc oxide (Nano-ZnO) film was deposited was used for co-immobilization of rabbit-immunoglobin antibodies (r-IgGs) and bovine serum albumin (BSA) for ochratoxin-A (OTA) detection. An electrochemical impedance spectroscopy (EIS) technique was used to evaluate BSA/r-IgGs/Nano-ZnO/ITO immune electrode. This method could offer a linear range from 6×10^{-12} to 1×10^{-11} mM/ml, a detection limit as low as 6×10^{-12} mM/ml, a response time of 25 s and a sensitivity of the order of 189 Ω/nM/dm^3cm^{-2} (Ansari et al. 2010). A gold electrode modified by the electrodeposited nanostructured zirconium oxide, ZrO_2 film with a particle size of 35 nm, was used to immobilize 21-mer oligonucleotide probe ssDNA selective to *Mycobacterium tuberculosis* by the interaction between oxygen atom of phosphoric group and zirconium to fabricate DNA biosensor. This DNA-ZrO_2/Au bioelectrode can be used for timely and fast diagnosis of *M. tuberculosis* with detection limit as low as 65×10^{-6} mg/ml within 60 s (Das et al. 2010). Nanostructured ZrO_2 film deposited on ITO electrode by sol–gel process has proved to be good immobilization platform for 17 base single-stranded ssDNA identified from the 16 s rRNA coding region for *E. coli* for hybridization detection with complementary, non-complementary and genomic DNA of *E. coli*. This method showed very good sensitivity towards hybridization detection (10^{-15}–10^3 mM complementary DNA) using DPV technique (Solanki et al. 2009a). Balendhran et al. used 2D α MoO_3 nanoflakes of thickness ~1.4 nm obtained in high yield by liquid-phase exfoliation process to fabricate a FET-type device to detect the model protein BSA. The crystalline 2D α-MoO_3 nanoflakes were utilized for creating an electron conduction channel with high carrier mobility, in FET configurations on alumina substrates. An overlayer of Au of thickness 1–1.5 nm was deposited on MoO_3 film to facilitate charge transfer from the source and drain electrodes to the film and for the effective interaction with the protein. Thus, fabricated sensor exhibited a low detection limit of 250×10^{-3} mg/ml and a response time of <10 s. Small response times could be achieved due to the high permittivity nature of the MoO_3 nanoflakes, thus providing an extremely viable electronic-based biosensing platform (Balendhran et al. 2013).

3.5 MXenes

MXenes are a new type of 2D materials, with hydrophilic surface and metallic conductivity. Owing to their excellent mechanical, electrical, chemical and physical properties, MXenes' applications in the fields like energy storage, nanocomposite fabrication and chemical sensing are increasing at a rapid pace ever since its introduction in 2011. The 2D layered atomic structures of Mxenes along with their hydrophilic surface make them promising candidates for the fabrication of FET like devices for monitoring biological events. Ti_3C_2–MXene was synthesized by specific etching of the Al layer in precursor material Ti_3AlC_2 with 48% hydrofluoric acid (HF). The ultra-thin MXene micropatterns were created by microcontact printing (μCP), with the help of PDMS stamp and MXene ink prepared by dispersing MXenes in aqueous solution onto 3-aminopropyltriethoxysilane-modified glass cover slips. These MXene stripes afford an active surface to bind small biomolecules, which changes the conductivity of the MXene stripes. For the electrical characterization of MXene micropatterns, a pair of silver electrodes was deposited orthogonally to the strips with gap distances ranging from 2×10^6 to 1×10^7 nm. Initially, the MXene-micropattern-based FET device was used to detect a small biological molecule, dopamine, which is a typical neurotransmitter involved in regulating many brain functions. Since neurotransmitters are released as a result of action potential firing, the authors used the MXene-based FET device to evaluate real-time monitoring of spiking activity in cultured primary hippocampal neurons. The MXene material showed a very good biocompatibility with the culture samples (Xu et al. 2016). A glucose biosensor based on Au/MXene nanocomposite was reported wherein glucose oxidase (GOx) enzyme was immobilized on Au/MXene nanocomposite prepared in nafion over glassy carbon electrode (GCE). The Au nanoparticles facilitated electron transfer between the electroactive centre of GOx and the electrode. The GOx/Au/MXene/nafion/GCE-based amperometric biosensor displayed a linear range from 0.1 to 18 mM with a sensitivity of 4.2×10^{-3} mA mM^{-1} cm^{-2} and a LOD of 5.9×10^{-3} mM (S/N = 3) (Rakhi et al. 2016). A novel organ-like MXene-Ti_3C_2 nanomaterial was successfully prepared by etching Al from Ti_3AlC_2 in HF at room temperature and then used to immobilize haemoglobin (Hb) to fabricate a H_2O_2 biosensor with an oxidized surface. The enzyme electrode was prepared by dropcasting method. Firstly, 1 ml of 2 mg/ml MXene-Ti_3C_2 suspension and 0.5 ml of 10 mg/ml Hb were mixed and stirred for 30 min. This was followed by the addition of 0.5 ml of nafion (5%) to the mixture and then the mixture was subjected to stirring for 10 min. Lastly, 4×10^{-3} mL of the mixture was applied to the surface of a freshly polished GC electrode to prepare the nafion/Hb/Ti_3C_2/GC electrode. Thus, prepared biosensor exhibited a wide linear range of 0.1×10^{-3}–260×10^{-3} mM for H_2O_2 and a very low detection limit of 20×10^{-6} mM (Wang et al. 2015c). Similar electrode configuration has been used to detect nitrite in the range 0.5 to 11.8 mM, with an extremely low detection limit of 0.12×10^{-3} mM (Liu et al. 2015).

4 Conclusions and Future Scope

The 2D layered nanomaterials which are introduced into the scientific world quite recently have captured the attention of researchers to greater extents which could be witnessed by the numerous literature reports. The unique properties of these materials make them suitable candidates for biosensing applications. This chapter describes the different types of strategies that are adopted to immobilize biocatalysts on 2D nanomaterials for biosensing applications related to health care. The synergism between the unique properties of the 2D nanomaterials along with the electrochemical detection has resulted in very good improvements in the analytical performance of biosensors as discussed in this chapter. The atomic thickness of the materials provides easy access to the diffusion of targeted analytes, and the 2D basal planes provide numerous active sites for specific interaction with the targeted analyte. Biosensing is facilitated mainly by different approaches. Direct immobilization of biocatalysts/molecular recognition units on 2D materials is facilitated by the edge active sites which could be introduced by molecular engineering. The atomic-level dimensions of the materials provide high surface-to-volume ratio resulting in high sensitivity. The ultrashort diffusion paths and the active available in the atomic thickness of the 2D materials enhance the charge transfer properties of the materials. Further enhancement is achieved by introducing metal nanocomposites and conducting polymers onto the 2D materials. Au NPs incorporated on to the 2D materials provide anchoring sites for immobilization of antibodies through Au–S bonds. Doping of heteroatoms also facilitates charge transfer in 2D nanomaterials. Further research in the direction of scaled-up synthesis of 2D nanomaterials with tailored properties suitable for biosensing and new strategies that lead to the fabrication of robust functional devices are required to for analytical applications.

References

Akhavan O, Ghaderi E, Rahighi R, Abdolahad M (2014) Spongy graphene electrode in electrochemical detection of leukemia at single-cell levels. Carbon 79:654–663

Ali MA, Singh C, Mondal K, Srivastava S, Sharma A, Malhotra BD (2016) Mesoporous few-layer graphene platform for affinity biosensing application. ACS Appl Mater Interf 8:7646–7656

Alwarappan S, Liu C, Kumar A, Li CZ (2010) Enzyme-doped graphene nanosheets for enhanced glucose biosensing. J Phys Chem C 114:12920–12924

Ansari AA, Kaushik A, Solanki PR, Malhotra BD (2010) Nanostructured zinc oxide platform for mycotoxin detection. Bioelectrochemistry 77:75–81

Bakker E, Telting-Diaz M (2002) Electrochemical sensors. Anal Chem 74:2781–2800

Balendhran S, Walia S, Alsaif M, Nguyen E, Ou J, Zhuiykov S, Sriram S, Bhaskaran M, Kalantar-Zadeh K (2013) Field effect biosensing platform based on 2D α-MoO_3. ACS Nano 7:9753–9760

Banica FG (2002). Chemical Sens Biosens Fundamentals Appl. https://doi.org/10.1002/9781118354162

Beer PD, Mortimer RJ Stradiotto NR, Szemes F, Weightman JS (1995) Electrochemical and spectral recognition of chloride ions by novel acyclic ruthenium(II) bipyridyl receptor complexes. Anal Proc 32(12):485–527

Benameur MM et al (2011) Visibility of dichalcogenide nanolayers. Nanotechnology 22:125706

Benvidi A, Firouzabadi AD, Moshtaghiun SM, Mazloum-Ardakani M, Tezerjani MD (2015) Ultrasensitive DNA sensorbased on gold nanoparticles/reduced graphene oxide/glassycarbon electrode. Anal Biochem 484(1):24–30

Berchmans S, Venkatesan M, Vusa CSR, Arumugam P (2018) PAMAM dendrimer modified reduced graphene oxide postfunctionalized by horseradish peroxidase for biosensing H_2O_2. Methods Enzymol 609:143–170

Bobacka J (2006) Conduct Polym Solid-State Ion-Sel Electrodes Electroanal 18:7–18

Brent JR, Lewis DJ, Lorenz T, Lewis EA, Savjani N, Haigh SJ, Seifert G, Derby B, O'Brien P (2015) J. Am tin(II) sulfide (SnS) nanosheets by liquid-phase exfoliation of herzenbergite: IV-VI main group two-dimensional atomic crystals. Chem Soc 137:12689

Cao S, Zhang L, Chai Y, Yuan R (2013) An Integrated sensing system for detection of cholesterol based on TiO_2-graphene-Pt-Pd hybrid nanocomposites. Biosens Bioelectron 42:532–538

Chen H, Zhang J, Gao Y, Liu S, Koh K, Zhu X, Yin Y (2015) Sensitive cell apoptosis assay based on caspase-3 activity detection with graphene oxide-assisted electrochemical signal amplification. Biosens Bioelectron 68:777–782

Chen Y, Ren R, Pu H, Chang J, Mao S, Chen J (2017) Field-effect transistor biosensors with two-dimensional black phosphorus nanosheets. Biosens Bioelectron 89:505–510

Chen Y, Li Y, Sun D, Tian D, Zhang J, Zhu J (2011) Fabrication of gold nanoparticles on bilayer graphene for glucose electrochemical biosensing. J Mater Chem 21:7604

Cheng L, Huang W, Gong Q, Liu C, Liu Z, Li Y, Dai H (2014) Ultrathin WS_2 nanoflakes as a high-performance electrocatalyst for the hydrogen evolution reaction. Angewande Chemie Int Edn 53:7860

Coleman JN et al (2011) Two-dimensional nanosheets produced by liquid exfoliation of layered materials. Science 331(6017):568–571

Crespo GA, Macho S, Bobacka J, Rius FX (2009) Transduction mechanism of carbon nanotubes in solid-contact ion-selective electrodes. Anal Chem 81:676–681

Crespo GA, Macho S, Rius FX (2008) Ion-selective electrodes using carbon nanotubes as ion-to-electron transducers. Anal Chem 80:1316–1322

Daniel B, Chem JP (2011) A powerful method for surface modification electro grafting. Chem Soc Rev 40:3995–4048

Das M, Sumana G, Nagarajan R, Malhotra BD (2010) Zirconia based nucleic acid sensor for mycobacterium tuberculosis detection. Appl Phys Lett 96:133703

Deep Jariwala VK, Sangwan Lincoln J, Lauhon Tobin J, Marks, Mark C, Hersam (2014) Emerging device applications for semiconducting two-dimensional transition metal dichalcogenides. ACS Nano 8(2):1102–1120

Dey RS, Raj CR (2010) Development of an amperometric cholesterol biosensor based on graphene-pt nanoparticle hybrid material. J Phys Chem 114:21427–21433

Dong H, Zhu Z, Ju H, Yan F (2012) Triplex signal amplification for electrochemical DNA biosensing by coupling probe-gold nanoparticles-graphene modified electrode with enzyme functionalized carbon sphere as tracer. Biosens Bioelectron 33:228–232

Eda G et al (2011) Photoluminescence from chemically exfoliated MoS_2. Nano Letter 11:5111–5116

Feng L, Chen Y, Ren J, Qu X (2011) A graphene functionalized electrochemical aptasensor for selective label-free detection of cancer cells. Biomaterials 32:2930–2937

Gadzekpo VPY, Hungerford JM, Kadry AM, Ibrahim YA, Christian GD (1985) Lipophillic lithium ion carrier in a lithium ion selective electrode. Anal Chem 57(2):493–495

Gopalakrishnan D, Damien D, Shaijumon MM (2014) MoS_2 quantum dot-interspersed exfoliated MoS_2 nanosheets. ACS Nano 8:5297

Gordon RA, Yang D, Crozier ED, Jiang DT, Frindt RF (2002) Structures of exfoliated single layers of WS_2 MoS_2 and $MoSe_2$ in aqueous suspension. Phys Rev B 65:125407

Han JT, Jang JI, Kim H, Hwang JY, Yoo HK, Woo JS, Choi S, Kim HY, Jeong HJ, Jeong SY, Baeg KJ, Cho K, Lee GW, (2014) Extremely efficient liquid exfoliation and dispersion of layered materials by unusual acoustic cavitation. Sci Rep 4 Article number 5133

He Q, Das SR, Garland NT, Jing D, Hondred JA, Cargill AA, Ding S, Karunakaran C, Claussen JC (2017) Enabling inkjet printed graphene for ion selective electrodes with postprint thermal annealing. ACS Appl Mater Interf 9:12719–12727

Hendry E, Hale PJ, Moger J, Savchenko AK, Mikhailov SA (2010) Coherent nonlinear optical response of graphene. Phys Rev Lett 105(9):1–4

Hendry SP, Higgins IJ, Bannister JV (1990) Amperometric biosensors. J Biotechnol 15:229–238

Hernández R, Riu J, Bobacka J, Vallés C, Jiménez P, Benito AM, Maser WK, Rius FXC (2012) Reduced graphene oxide films as solid transducers in potentiometric all-solid-state ion-selective electrodes. J Phys Chem C 116:22570–22578

Hou L, Cui Y, Xu M, Gao Z, Huang J, Tang D (2013) Graphene oxide-labeled sandwich-type impedimetric immunoassay with sensitive enhancement based on enzymatic 4-chloro-1-naphthol oxidation. Biosens Bioelectron 47:149–156

Hu J, Stein A, Bühlmann P (2016) Rational design of all-solid-state ion-selective electrodes and reference electrodes. TrAC Trends Anal Chem 76:102–114

Hu J, Zou XU, Stein A, Buhlmann P (2014) Ion-selective electrodes with colloid-imprinted mesoporous carbon as solid contact. Anal Chem 86:7111–7118

Hu Y, Guo L (2015) Rapid preparation of perovskite lead niobate nanosheets by ultrasonic-assisted exfoliation for enhanced visible-light-driven photocatalytic hydrogen production. ChemCatChem 7:584–587

Huang KJ, Liu YJ, Wang HB, Wang YY, Liu YM (2014a) Sub-femto molar DNA detection based on layered molybdenum disulfide/multi-walled carbon nanotube composites Au nanoparticle and enzyme multiple signal amplification. Biosens Bioelectron 55:195–202

Huang KJ, Liu YJ, Wang HB, Gan T, Liu YM, Wang LL (2014b) Signal amplification for electrochemical DNA biosensor based on two-dimensional graphene analogue tungsten sulfide–graphene composites and gold nanoparticles. Sens Actuators B 191:828–836

Huang KJ, Liu YJ, Cao JT, Wang HB (2014c) An aptamer electrochemical assay for sensitive detection of immunoglobulin E based on tungsten disulfide—graphene composites and gold nanoparticles. RSC Adv 4:36742

Huang KJ, Liu YJ, Zhang JZ, Liu YM (2014d) A novel aptamer sensor based on layered tungsten disulfide nanosheets and Au nanoparticles amplification for 17β-estradiol detection. Anal Methods 6:8011–8017

Joyce BA (1985) Molecular beam epitaxy. Rep Prog Phys 48:1637–1697

Justin GJ, Chem SC (2011) The molecular level modification of surfaces from self-assembled monolayers to complex molecular assemblies. Chem Soc Rev 40:2704–2718

Kang X, Wang J, Wu H, Aksay IA, Liu J, Lin Y (2009) Glucose oxidase-graphene-chitosan modified electrode for direct electrochemistry and glucose sensing. Biosens Bioelectron 25:901–905

Kappera R, Voiry D, Yalcin S et al (2014) Phase-engineered low-resistance contacts for ultrathin MoS_2 transistors. Nat Mater 13:1128–1134

Kelly RG, Owen AE (1986) Solid-state Ion sensors. J Chem Soc Faraday Trans I 82:1195–1208

Khoshfetrat SM, Mehrgardi MA (2017) Amplified detection of leukemia cancer cells using an aptamer-conjugated gold-coated magnetic nanoparticles on a nitrogen-doped graphene modified electrode. Bioelectrochemistry 114:24–32

Kim Y, Song JG, Park YJ, Ryu GH, Lee SJ, Kim JS, Jeon PJ, Lee CW, Woo WJ, Choi T, Jung H, Lee HBR, Myoung JM, Im S, Lee Z, Ahn JH, Kim PJ (2016) Self-limiting layer synthesis of transition metal dichalcogenide. Sci Rep 6:18754

Kim HU, Kim H, Ahn C, Kulkarni A, Jeon M, Yeom GY, Lee MH, Kim T (2015) In situ synthesis of MoS_2 on a polymer based gold electrode platform and its application in electrochemical biosensing. RSC Adv 5(14):10134–10138

Kim J, Kim M, Lee MS, Kim K, Ji S, Kim YT, Park J, Na K, Bae KH, Kyun Kim H, Bien F, Young Lee C, Park JU (2017) Wearable smart sensor systems integrated on soft contact lenses for wireless ocular diagnostics. Nat Commun 8:14997

Kimmel DW, LeBlanc G, Meschievitz ME, Cliffel DE (2012) Electrochemical sensors and biosensors. Anal Chem 84:685–707

Kumar V, Brent JR, Shorie M, Kaur H, Chadha G, Thomas AG, Lewis EA, Rooney AP, Nguyen L, Zhong XL, Burke MG, Haigh SJ, Walton A, McNaughter PD, Tedstone AA, Savjani N, Muryn CA, O'Brien P, Ganguli AK, Lewis DJ, Sabherwal P (2016) Nanostructured aptamer-functionalized black phosphorus sensing platform for label-free detection of myoglobin a cardiovascular disease biomarker. ACS Appl Mater Interf 8:22860–22868

Lai CZ, Fierke MA, Stein A, Buhlmann P (2007) Ion-selective electrodes with three-dimensionally ordered macroporous carbon as the solid contact. Anal Chem 79:4621–4626

Late DJ, Liu B, Matte SR, Rao CNR, Dravid VP (2012) Adv Func Mater 22:1894

Lee J, Dak P, Lee Y, Park H, Choi W, Alam M, Kim S (2015) Two-dimensional layered MoS_2 biosensors enable highly sensitive detection of biomolecules. Sci Rep 4:7352

Lee C, Wei X, Kysar JW, Hone J (2008) Measurement of the elastic properties and intrinsic strength of monolayer graphene. Science 321(5887):385–388

Li Z, Wong SL (2017) Functionalization of 2D transition metal dichalcogenides for biomedical applications. Mater Sci Eng C 70(2):1095–1106

Lin X, Ni Y, Kokot S (2016) Electrochemical cholesterol sensor based on cholesterol oxidase and MoS_2-AuNPs modified glassy carbon electrode. Sens Actuators B 233:100–106

Liu H, Duan C, Yang C, Shen W, Wang F, Zhu Z (2015) A novel nitrite biosensor based on the direct electrochemistry of haemoglobin immobilized on MXene-Ti_3C_2. Sens Actuators 218:60–66

Liu K, Zhang JJ, Cheng FF, Zheng TT, Wang C, Zhu JJ (2011) Green and facile synthesis of highly biocompatible graphene nanosheets and its application for cellular imaging and drug delivery. J Mater Chem 21(32):12034

Lopez-Sanchez O, Lembke D, Kayci M, Radenovic A, Kis A (2013) Ultrasensitive photo detectors based on monolayer MoS_2. Nat Nanotechnol 8(7):497–501

Lu Q, Dong X, Li LJ, Hu X (2010) Direct electrochemistry-based hydrogen peroxide biosensor formed from single-layer graphene nano platelet-enzyme composite film. Talanta 82:1344–1348

Luo Y, Li Z, Zhu C, Cai X, Qu L, Du D, Lin Y (2018) Graphene-like metal-free 2D nanosheets for cancer imaging and theranostics. Trends Biotechnol 36(11):1145–1156

Mahesh KV, Rashada R, Kiran M, Peer Mohamed A, Ananthakumar S (2015) Shear induced micromechanical synthesis of Ti_3SiC_2 MAXene nanosheets for functional applications. RSC Adv 5:51242

Mahjouri-Samani M, Gresback R, Tian M, Wang K, Puretzky AA, Rouleau CM, Eres G, Ivanov IN, Xiao K, McGuire MA, Duscher G, Geohegan DB (2014) Pulsed laser deposition of photoresponsive two-dimensional GaSe nanosheet networks. Adv Func Mater 24:6365

Manju V, RaoVusa CS, Berchmans S, Arumugam P (2016) Electrochemical amination of graphene using nano sized PAMAM dendrimer for sensing applications. RSC Adv 6:33409–33418

Mattheiss LF (1973) Band structures of transition-metal dichalcogenide layer compounds. Phys Rev 8:3719–3740

Mayorga-Martinez CC, MohamadLatiff N, Eng AY, Sofer Z, Pumera M (2016) Black phosphorus nanoparticle labels for immunoassays via hydrogen evolution reaction mediation. Anal Chem 88:10074–10079

Metzger E, Dohner R, Simon W, Vonderschmitt, DJ, Gautschi K (1987) Lithium/sodium ion concentration ratio measurements in blood serum with lithium and sodium ion selective liquid membrane electrodes. Anal Chem 59(13):1600

Migdalski J, Blaz T, Lewenstam A (1996) Conducting polymer-based ion-selective electrodes. Anal Chim Acta 322:141–149

Nasuha R, Carmen C, Mayorga-Martinez, ZS, Pumera M (2017) 1T-phase transition metal dichalcogenides (MoS_2, $MoSe_2$, WS_2, and WSe_2) with fast heterogeneous electron transfer: application on second-generation enzyme-based biosensor. ACS Appl Mater Interf 9:40697—40706. https://doi.org/10.1021/acsami.7b13090

Nicolosi V, Chhowalla M, Kanatzidis M, G, Strano MS, Coleman JN (2013) Liquid exfoliation of layered materials. Science 340:1226419

Nikolskii BP, Materova EA (1985) Solid contact in membrane ion-selective electrodes. Ion-Sel Electrode Rev Elsevier Great Britain 7:3—39

Novoselov KS et al (2015) Two-dimensional atomic crystals. Proc Nat Acad Sci USA 102(30):10451–10453

Novoselov KS, Geim AK, Morozov SV, Jiang D, Zhang Y, Dubonos SV, Grigorieva IV, Firsov (2004) Electric field effect in atomically thin carbon films. Science 306(5696):666–669

Oliveira TM, Barroso MF, Morais S, Araujo M, Freire C, de Lima-Neto P, Correia AN, Oliveira MB, Delerue-Matos C (2014) Sensitive bi-enzymatic biosensor based on polyphenoloxidases-gold nanoparticles-chitosan hybrid film-graphene doped carbonpaste electrode for carbamates detection. Bioelectrochemistry 98:20–29

Ping J, Wang Y, Wu J, Ying Y (2011) Development of an all-solid-state potassium ion-selective electrode using graphene as the solid-contact transducer. Electrochem Commun 13:1529–1532

Podberezskaya NV, Magarill SA, Pervukhina NV, Borisov SV (2001) Crystal chemistry of dichalcogenides MX_2. J Struct Chem 42:654–681

Pradhan SK, Xiao B, Pradhan AK (2016) Enhanced photo-response in p-Si/MoS_2 heterojunction-based solar cells. Sol Energy Mater Sol Cells 144:117–127

Rahmanian E, Carmen C, Mayorga-Martinez RM, ZdenekSofer JL, Pumera M (2018) 1T-phase tungsten chalcogenides (WS_2 WSe_2 WTe_2) decorated with TiO_2 nanoplatelets with enhanced electron transfer activity for biosensing applications. ACS Appl Nano Mater 1:7006–7015

Rakhi RB, Nayak P, Xia C, Alshareef HN (2016) Novel amperometric glucose biosensor based on MXene nanocomposite. Sci Rep 6:36422

Rodriguez-Mozaz S, Alda M, Marco M, Barcelo D (2005) Biosensors for environmental monitoring a global perspective. Talanta 65:291–297

Toh RJ, Carmen C, Mayorga-Martinez ZS, Pumera M (2017) 1T-phase WS_2 protein-based biosensor. Adv Func Mater 27:1604923

Sajedeh M, Ovchinnikov D, Diego Pasquier Oleg V, Yazyev AK (2017) 2D transition metal dichalcogenides. Nat Rev Mater 2:17033 (1–15)

Samia MC, Gam-Derouich S, Mangeney C, Mohamed M, Chehimi C (2011) Aryl diazonium salts: a new class of coupling agents for bonding polymers, biomacromolecules and nanoparticles to surfaces. Chem Soc Rev 40:4143–4166

Sekhar PK, Brosha EL, Mukundan R, Garzon F (2010) Chemical sensors for environmental monitoring and homeland security. Electrochem Soc Inter 19:35–40

Selvarani K, Berchmans S (2017) Biosensing of cholesterol and glucose facilitated by cationic polymer overlayers on $Ni(OH)_{(2)}$/NiOOH at physiological pH. J Electrochem Soc 164(9):561–571

Serna MI, Yoo SH, Moreno S, Xi Y, Oviedo JP, Choi H, Alshareef HN, Kim MJ, Minary-Jolandan M, Quevedo-Lopez MA (2016) Large-area deposition of MoS_2 by pulsed laser deposition with in situ thickness control. ACS Nano 10:6054

Shan C, Yang H, Song J, Han D, Ivaska A (2009) Electrochemistry of glucose oxidase and biosensing for glucose based on graphene. Anal Chem 81:2378–2382

Shao Y, Wang J, Wu H, Liu J, Aksay IA, Lin Y (2010) Graphene based electrochemical sensors and biosensors. Rev Electroanal 22:1027–1036

Shu Y, Chen J, Xu Q, Wei Z, Liu F, Lu R, Xu S, Hu X (2017) MoS_2 nanosheet–au nanorod hybrids for highly sensitive amperometric detection of H_2O_2 in living cells. J Mater Chem B 5(7):1446–1453

Siva Kumar K, Singh E, Singh P, Meyyappan M, Nalwa HS (2019) A review on graphene-based nanocomposites for electrochemical and fluorescent biosensors. RSC Adv 9:8778

Solanki PR, Kaushik A, Chavhan PM, Maheshwari SN, Malhotra BD (2009) Nanostructured zirconium oxide based genosensor for escherichia coli detection. Electrochem Commun 11:2272–2277

Solanki PR, Kaushik A, Ansari AA, Malhotra BD (2009) Nanostructured zinc oxide platform for cholesterol sensor. Appl Phys Lett 94:143901

Sticker D, Rothbauer M, Charwat V, Steinkühler J, Bethge O, Bertagnolli E, Wanzenboeck H, D, Ertl P (2015) Zirconium dioxide nanolayer passivated impedimetric sensors for cell-based assays. Sens Actuators B 213:35−44

Su B, Tang J, Huang J, Yang H, Qiu B, Chen G, Tang D (2010) Graphene and nanogold-functionalized immunosensing interface with enhanced sensitivity for one-step electrochemical immunoassay of α-fetoprotein in human serum. Electroanalysis 22:2720–2728

Su S, Chen S, Fan C (2018) Recent advances in two-dimensional nanomaterials-based electrochemical sensors for environmental analysis. Green Energy Environ 3:97–106

Tanja N, Chem GN (2011) Strategies for '"wiring"' redox-active proteins to electrodes and applications in biosensors, bio fuel cells, and nanotechnology. Chem Soc Rev 40:3564–3576

Thakur MS, Ragavan KV (2013) Biosensors in food processing. J Food Sci Technol 50:625–641

Thimsen E, Riha SC, Baryshev SV, Martinson ABF, Pellin EJW (2012) Atomic layer deposition of the quaternary chalcogenide Cu_2ZnSnS_4. Chem Mater 24:3188

Valdivia A, Tweet DJ, Conley JF (2016) Atomic layer deposition of two dimensional MoS_2 on 150 mm substrates. J Vacuum Sci Technol 34:02151556

Velasco-G MN, Mottram T (2003) Biosensor technology: addressing agricultural problems. Biosyst Eng 84(1):1–12

Wang X, Nan F, Zhao J, Yang T, Ge T, Jiao K (2015a) A label-free ultrasensitive electrochemical DNA sensor based on thin-layer MoS_2 nanosheets with high electrochemical activity. Biosens Bioelectron 64:386–391

Wang X, Chu C, Shen L, Deng W, Yan M, Ge S, Yu J, Song X (2015b) An ultrasensitive electrochemical immunosensor based on the catalytical activity of MoS_2-Au composite using Ag nanospheres as labels. Sens Actuators 206:30–36

Wang F, Yang C, Duan C, Xiao D, Tang Y, Zhu J (2015c) An organ-like titanium carbide material (MXene) with multilayer structure encapsulating hemoglobin for a mediator-free biosensor. J Electrochem Soc 162:B16–B21

Wang J (1994) Analytical electrochemistry. VCH, New York

Wang L, Wang Y, Wong JI, Palacios T, Kong J, Yang H (2014) Functionalized MoS_2 nanosheet-based field-effect biosensor for label-free sensitive detection of cancer marker proteins in solution. Small 10:1101–1105

Wang L, Xiong Q, Xiao F, Duan H (2017) 2D nanomaterials based electrochemical biosensors for cancer diagnosis. Biosens Bioelectron 89:136–151

Wang Y, Li Z, Wang J, Li J, Lin Y (2011) Graphene and graphene oxide: biofunctionalization and applications in biotechnology. Trends Biotechnol 29(5):205–212

Wang Y, Ping J, Ye Z, Wu J, Ying Y (2013) Impedimetric immunosensor based on gold nanoparticles modified graphene paper for label-free detection of escherichia coli O157:H7. Biosens Bioelectron 49:492–498

Wang Y, Shao Y, Matson DW, Li J, Lin Y (2010) Nitrogen-doped graphene and its application in electrochemical biosensing. ACS Nano 4:1790–1798

Wei Q, Mao K, Wu D, Dai Y, Yang J, Du B, Yang M, Li H (2010) A novel label-free electrochemical immunosensor based on graphene and thionine nanocomposite. Sens Actuators 149:314–318

Wilson JA, Yoffe AD (1969) the Transition metal dichalcogenides discussion and interpretation of the observed optical electrical and structural properties. Adv Phys 18(73):193–335

Wilson JA, Yoffe AD (1979) The transition metal dichalcogenides discussion and interpretation of the observed optical electrical and structural properties. Adv Phys 18:193–335

Wu H, Wang J, Kang X, Wang C, Wang D, Liu J, Aksay IA, Lin Y (2009) Glucose biosensor based on immobilization of glucose oxidase in platinum nanoparticles/graphene/chitosan nanocomposite film. Talanta 80:403–406

Xu B, Zhu M, Zhang W, Zhen X, Pei Z, Xue Q, Zhi C, Shi P (2016) Ultrathin MXene-micropattern-based field-effect transistor for probing neural activity. Adv Mater 28:3333–3339

Yadegari A, Omidi M, Yazdian F, Zali H, Tayebi L (2017) An electrochemical cytosensor for ultrasensitive detection of cancer cells using modified graphene−gold nanostructures. RSC Adv 7:2365–2372

Yang G, Cao J, Li L, Rana RK, Zhu JJ (2013) Carboxymethyl chitosan-functionalized graphene for label-free electrochemical cytosensing. Carbon 51:124–133

Yang G, Li L, Rana RK, Zhu JJ (2013) Assembled gold nanoparticles on nitrogen-doped graphene for ultrasensitive electrochemical detection of matrix metalloproteinase-2. Carbon 61:357–366

Yang TT, Jia HY, Liu Z, Qiu XL, Gao YS, Xu JK, Lu LM, Yu YF (2017) Label-free electrochemical immunoassay for α-fetoprotein based on a redox matrix of prussian blue-reduced graphene oxide/gold nanoparticles-poly(3, 4-ethylenedioxythiophene) composite. J Electroanal Chem 799:625–633

Huang Y, Miao Y-E, Zhang L, Tjiu WW, Panb J, Liu T (2014) Synthesis of few-layered MoS_2 nanosheet-coated electrospun SnO_2 nanotube heterostructures for enhanced hydrogen evolution reaction. Nanoscale 6:10673–10679

Zeng Z et al (2011) Single-layer semiconducting nanosheets: high-yield preparation and device fabrication. Angewande Chemie Int Edn 50:11093–11097

Zeng G, Xing Y, Gao J, Wang Z, Zhang X (2010) Unconventional layer-by-layer assembly of graphene multilayerfilms for enzyme-based glucose and maltose biosensing. Langmuir 26:15022–15026

Zeng Q, Cheng J, Tang L, Liu X, Liu Y, Li J, Jiang J (2010) Self-assembled graphene-enzyme hierarchical nanostructures for electrochemical biosensing. Adv Func Mater 20:3366–3372

Zhang D, Zhang Y, Zheng L, Zhan Y, He L (2013) Graphene oxide/poly-L-lysine assembled layer for adhesion and electrochemical impedance detection of leukemia K562 cancer cells. Biosens Bioelectron 42:112–118

Zhang X, Ju H, Wang J (2008) Electrochemical sensors, biosensors and their biomedical applications. Academic Press, New York

Zhao Y, Zhang YH, Zhuge Z, Tang YH, Tao JW, Chen Y (2018) Synthesis of a poly-L-lysine/black phosphorus hybrid for biosensors. Anal Chem 90:3149–3155

Zheng T, Tan T, Zhang Q, Fu JJ, Wu JJ, Zhang K, Zhu JJ, Wang H (2013) Multiplex acute leukemia cytosensing using multifunctional hybrid electrochemical nanoprobes at a hierarchically nanoarchitectured electrode interface. Nanoscale 5:10360–10368

Meng Z, Stolz RM, Mendecki L, Mirica KA (2019) Electrically-transduced chemical sensors based on two-dimensional nanomaterials. Chem Rev 119:478–598

Cai Z, Liu B, Zou X, Cheng H-M (2018) Chemical vapor deposition growth and applications of two dimensional materials and their hetero structures. Chem Rev 118:6091–6133

Zhang P, Yang S, Pineda-Gómez R, Ibarlucea B, Ma J, Lohe MR, Akbar TF, Baraban L, Cuniberti G, Feng X (2019) Electrochemically exfoliated high-quality 2H-MoS_2 for multi flake thin film flexible biosensors. Small 15:1901265

Zhou L, Mao H, Wu C, Tang L, Wu Z, Sun H, Zhang H, Zhou H, Jia C, Jin Q, Chen X, Zhao J (2017) Label-free graphene biosensor targeting cancer molecules based on non-covalent modification. Biosens Bioelectron 87:701–707

Zhou Y, Liu S, Jiang HJ, Yang H, Chen HY (2010) Direct electrochemistry and bioelectrocatalysis of microperoxidase-11immobilized on chitosan-graphene nanocomposite. Electroanalysis 22:1323–1328

Advancement of Immobilization Techniques in Forensic Science

Akanksha Roberts, Deepshikha Shahdeo, and Sonu Gandhi

Abstract Immobilization techniques involve the confining/anchoring of a molecule (enzyme or cells) in or on an inert support/substrate matrix for stability, increased efficiency and easy recoverability of the molecule. The immobilization of molecules to a substrate is carried out on the basis of five major principle methods, i.e. adsorption, entrapment, membrane confinement (encapsulation), cross-linking (copolymerization) and covalent binding. Various different scientific processes make use of immobilization techniques to increase the efficiency of the process. In the upcoming field of forensic science, which involves the application of scientific methods to matters under investigation by a court of law, there has been advancement in the immobilization techniques being used to solve various cases. This involves both qualitative and quantitative analyses. Immunoassays, biosensors, adsorption and chromatographic methods make use of immobilization of molecules and are used for various forensic analyses such as narcotic drug, poison, toxin, heavy metal and pesticide detection, serological, latent fingerprinting analysis of biomolecules. In this chapter, we shall discuss in detail the different types of immobilization techniques, principle and their applications in forensic science.

Keywords Immobilization · Forensic science · Immunoassay · Biosensor · Chromatography · Adsorption

Abbreviations

2,4-D	2,4-Dichlorophenoxyacetic acid
Ab	Antibody
AChE	Acetylcholinesterase
Ag	Antigen
AgNPs	Silver nanoparticles

A. Roberts · D. Shahdeo · S. Gandhi (✉)
DBT-National Institute of Animal Biotechnology, Hyderabad, Telangana 500032, India
e-mail: gandhi@niab.org.in

A. Tripathi and J. S. Melo (eds.), *Immobilization Strategies*,
Gels Horizons: From Science to Smart Materials,
https://doi.org/10.1007/978-981-15-7998-1_13

Ag-Pd	Silver palladium
AuNPs	Gold nanoparticles
BSA	Bovine serum albumin
CM	*Carboxymethyl*
CNTs	Carbon nanotubes
CoPC	Cobalt phthalocyanine
CP	Carbon paste
CV	Cyclic voltammetry
DDT	Dichlorodiphenyltrichloroethane
DEAE	*Diethylaminoethyl*
DNA	Deoxyribonucleic acid
DPV	Differential pulse voltammetry
FET	Field-effect transistor
GCE	Classy carbon electrode
GLC	Gas–liquid chromatography
GOx	Glucose oxidase
GraFET	Graphene field-effect transistor
HPLC	High-performance liquid chromatography
HPTLC	High-performance thin layer chromatography
HRP	Horse radish peroxidase
IgG	Immunoglobulin G
LOD	Limit of detection
MAM	Monoacetyl morphine
MIPS	Molecularly imprinted polymers
MMD	Multimetal deposition
MS	Mass spectroscopy
MWCNTs	Multiwalled carbon nanotubes
NHSDA	National Household Survey on Drug Abuse
OP	Organophosphate
PCR	Polymerase chain reaction
PEDOT:PSS	Poly(3,4-Ethylenedioxythiophene) polystyrene sulphonate
PSA	Prostate-specific antigen
QDs	Quantum dots
RSID	Rapid stain identification
SPDP	Succinimidyl 3-(2-Pyridyldithio) propionate
SPR	Surface plasmon resonance
SWCNTs	Single-walled carbon nanotubes
SWV	Square wave voltammetry
TLC	Thin layer chromatography
TNT	Trinitrotoluene
UV	Ultraviolet
WHO	World Health Organization
ZnSNPs	Zinc sulphide nanoparticles

1 Introduction

Immobilization is a technique of fixing molecules to a substrate matrix in order to increase its stability, efficiency and functional reuse. This concept of immobilization has been used since the time of simple techniques such as paper chromatography and with the current advancements in this techniques, it has a variety of applications in food, drug, medicine, agricultural field, water treatment, etc. Immobilization method selection is based on the physiochemical characteristics of the molecule and substrate matrix. It is based on five different methods, i.e. adsorption, covalent binding, entrapment, cross-linking (copolymerization) and membrane confinement (encapsulation) of the molecule (Datta et al. 2013). Several modifications have been made, and there is continuous evolution to make immobilization technique compatible with new emerging applications. Immobilization techniques are economical as they are cost-effective since the immobilized substrate can be reused without losing its stability. It also minimizes the downstream processing and is highly sensitive (Homaei et al. 2013).

These exceptional properties of immobilization techniques have been made use of in the field of forensics. Forensic science is the application of various scientific techniques to matters under investigation to aid the legal system. Since the beginning of science, to prove or disapprove evidences at a crime scene, immobilization techniques such as paper chromatography for ink dye analysis (Zlotnick and Smith 1999), TLC for poison detection (Meade et al. 1972), etc., have been applied. Recent advancement in immobilization techniques applied to forensics could be of great value in providing criminal justice in analysing the evidences in a swift and more economical manner.

One of the major drawbacks of evidences that are analysed from a crime scene is the very minute quantity of samples usually available for analysis. For example, only one small blood splatter may be found at a crime scene and will have to be used for multiple forensic tests. While in the case of DNA, it is possible to multiply the sample available by PCR techniques, such is not always the case with other evidences as many of them cannot be multiplied. So when dealing with such small quantities of samples, it is essential that a minimum amount of sample is used up during each step of analysis. In this case, immobilization techniques provide a major advantage as immobilization helps fix one component, making the detection of the sample substance quicker and more sensitive. Hence, even small amounts of sample can be detected, and in many cases the sample can also be retrieved without losing its stability.

While the older immobilization techniques such as TLC for detection of poisons is still in use, the recent advances have made forensic testing more accurate, reliable, economical and quick.

Blood typing test using immunoassays and absorption–elution process are used to determine the blood group of an individual or a sample found at the crime scene (Bolton and Thorpe 1986). Even a piece of fibre with an old and dried blood stain can be immobilized on a cellulose–acetate strip, and the ABO blood group can be determined using antisera. Latent fingerprinting multimetal deposition (MMD) is another method where gold nanoparticles are immobilized on paper for enhancement of the visibility of latent fingerprints in forensic analysis (Sametband et al. 2007).

Application of immobilization is rapidly increasing in the field of toxicology (detection of poisons), pollution control (detection of pesticides/insecticides/herbicides Mahendra Wijaya et al. 2010; Ramansuri et al. 2009) and heavy metal content), food and water control (detection of microorganisms, heavy metals and pesticides/insecticides/herbicides Raman Suri et al. 2008; Sharma et al. 2010b; Talan et al. 2018), narcotic drug detection (Gandhi et al., 2008, 2015; Tey et al. 2010; Mishra et al. 2018), etc. Biosensor is a device which utilizes immobilization to detect trace amounts of biomolecules and is very useful in the field of forensics. Biological components such as serological fluids (blood, saliva, semen, vaginal fluid) can also be detected and distinguished at a crime scene. Detection of prostate-specific antigen (PSA) (Harbison and Fleming 2016), which is currently the most important serum marker for determining the presence of semen, can be made use of a biosensor to analyse serum samples in sexual assault cases.

Immobilization techniques provide the alternative of time-consuming, conventional methods. Especially with a large number of cases pending in forensic laboratories, it is necessary to find quicker methods of analysis. Also the components used in immobilization techniques can be retrieved, and reused, making this process very economical. In this chapter, we focus on the application of immobilization techniques in the field of forensics and its recent advancements.

2 Methods of Immobilization

Each immobilization technique is based on the physiochemical characteristics of the molecule and substrate matrix. The following are the major methods of immobilization:

- Adsorption
- Cross-linking (copolymerization)
- Covalent binding
- Entrapment
- Membrane confinement (encapsulation).

On the basis of physiochemical characteristics, these methods are classified into physical and chemical methods. Adsorption, encapsulation and entrapment fall under physical methods, where no chemical interaction between the molecule and substrate matrix exists, while covalent bonding and cross-linking fall under chemical methods. The different advantages and disadvantages of using these various immobilization techniques as well the types of carriers used in each technique have been discussed in Table 1.

Table 1 Overview of different immobilization techniques, their type of interaction and advantages/disadvantages, as well as the carriers used in each technique

Immobilization technique	Interaction	Advantage	Disadvantage	Carrier
Adsorption	Non-covalent interaction, i.e. hydrophobic, electrostatic and van der Walls forces	Cost-effective, rapid and simple	Unstable, very weak bond, susceptible to change in pH, temperature, etc.	Starch Cellulose Sepharose Alginate beads
Covalent bonding	Covalent bond	Stability, High binding strength	It may need prior modification, irreversible, expensive, may involve harsh and toxic chemical for modification	Collagen Gelatin Porous glass Polyacrylamide DEAE cellulose
Entrapment/Microencapsulation	Trapping or caging of molecules within polymers like chitosan, acrylamide	Stability	Leaching and reduction in binding efficiency	NA
Cross-linking	Molecules cross-linked by functional reactant	Prevents leakage, avoid desorption	Alteration in active site, loss of activity	NA

2.1 Physical Methods

2.1.1 Adsorption

It is one of the simplest methods of immobilization that is based on binding of molecule with the matrix. This method mainly involves low-energy bonds like hydrogen bonds, van der Waals' forces and ionic interactions. In this method, molecules get physically adsorbed onto the matrix surface (Mohamad et al. 2015). Activated carbon, porous glass, starch, dextrin, cellulose and chitosan are some of the examples of matrices used for the immobilization. Optimization of pH, temperature, ionic strength and concentration of the molecule to be immobilized is significant. It is also important to note that for significant surface bonding, the size of carrier particle size must be small, i.e. in the range of 500 A^0—1 mm diameter. Thin layer chromatography is one such technique which works on the method of adsorption. Adsorption of antibodies on the AuNPs is a physical interaction technique where positively charged groups in antibodies interact with the negatively charged surface of AuNPs which results in immobilization of antibodies onto the AuNPs as shown in Fig. 1.

When the molecules are adhered or attached to a solid support with non-covalent bonding like ionic or hydrophobic bonds, the molecule is said to have been adsorbed (Fig. 1). Ideal carrier for adsorption is insoluble matrices like ion-exchange matrices, glass fibre, porous carbon, polymeric aromatic resin and hydrous metal oxide. It is a reversible method of immobilization and comparatively weaker than other methods.

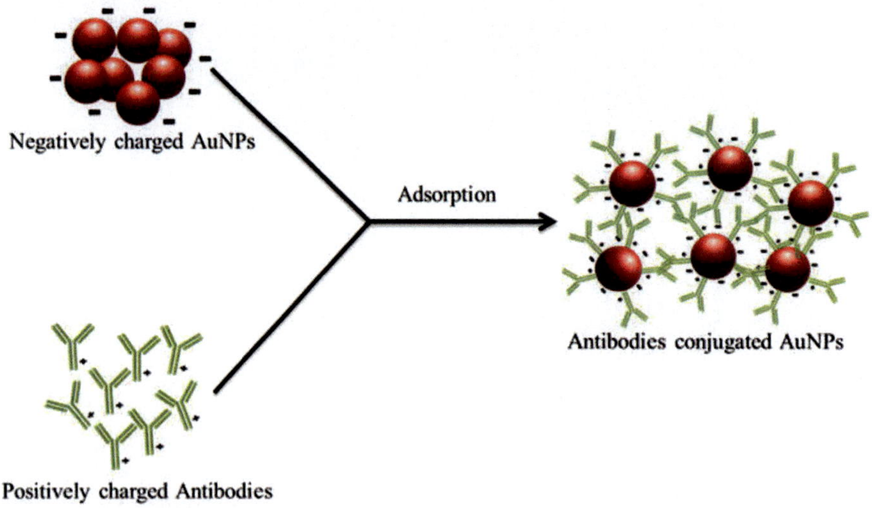

Fig. 1 Adsorption of antibodies on AuNPs. Negatively charged citrate-capped AuNPs interact with antibodies consisting of positively charged amide group as the functional group via ionic interaction

The carriers used for the adsorption-based techniques are further categorized into organic and inorganic support. Silica, titania and hydroxyapatite are the most common inorganic carriers, whereas chitin, chitosan, alginate, cellulose and the synthetic compounds, mainly polymers, are the organic carriers. The easy chemical modification of these matrices to increase the efficiency of a given enzyme is an added advantage (Jesionowski et al. 2014). Adsorption has both advantages and disadvantages. It is easy to handle, cost-effective and does not change the conformation of the molecules to be immobilized, but on the other hand, it gets easily affected by the change or fluctuation in temperature, pH, ionic strength, and is a slow process compared to the other methods.

Adsorption is used in the multimetal deposition technique for latent fingerprinting. In this technique, the gold particles are immobilized onto nitrocellulose paper to enhance the quality of latent fingerprints followed by addition of different developer.

2.1.2 Entrapment

It is an immobilization method for capturing the molecule on a fibre or within a gel. It consists of both covalent and non-covalent bonds. Matrix which surrounds the molecule is generally hydrophilic in nature, e.g. agar, gelatin, polyacrylamide, alginate matrices used for entrapment method (Hassan et al. 2016).

Entrapment by nanostructured supports like nanofibres, pristine materials and electrospun have revolutionized the immobilization of enzymes with their multiple applications in chemistry, biosensors, biofuels and biomedicine (Datta et al. 2013).

Immobilization techniques in which molecules are trapped inside a polymer are commonly called entrapment techniques. This method is lattice type where molecules bind to each other unlike other methods in which molecules get attached to the matrix. It is an irreversible process of immobilization and provides mechanical strength to the molecule. The molecules are trapped in the interstitial space between the cross-linked polymers. This is the main method of entrapment where the monomers are polymerized by cross-linking and form a three-dimensional network called lattice, and molecules can be entrapped in this lattice.

2.1.3 Encapsulation

This method includes enclosing of the molecules in a semipermeable polymer membrane (Nguyen 2017). It requires a well-controlled condition for synthesis of the microcapsule. In this entrapment method, molecules are free to move in a limited space as shown in Fig. 2 where the antibodies are trapped inside the capsule to increase their efficiency (Hemalatha et al. 2016).

Various materials that have been utilized to form microcapsules are in the size range of 10–100 μm in diameter, such as nylon and cellulose nitrate. The molecules enter the semi-permeable membrane through diffusion. One demerit of this method, however, is the possible rupturing of the encapsulation membrane due to diffusion of a.

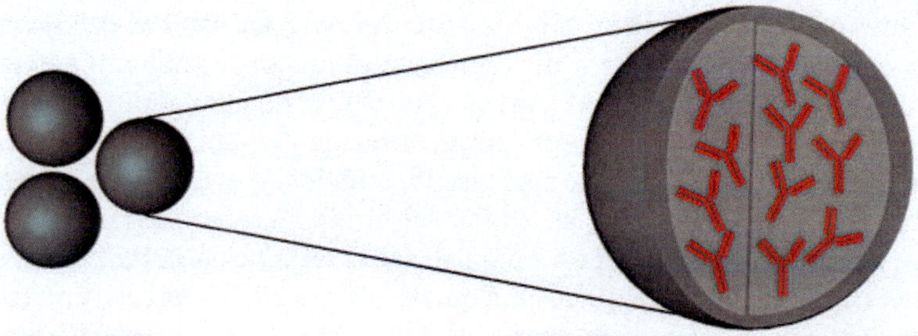

b.

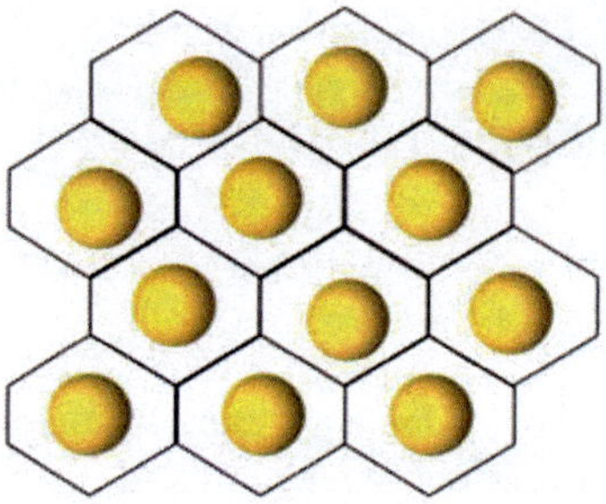

Fig. 2 There are two way to immobilize a molecule inside a structure: **a** Encapsulation of antibodies inside a microcapsule; **b** Entrapment of molecules inside a lattice structure

too many molecules into the membrane (more than the capacity) which may occur if the products from a reaction accumulate rapidly (Hassan et al. 2016).

2.2 *Chemical Method*

2.2.1 Covalent Attachment

In this method, the degree of association depends on the functional group of the molecules. Support material like mesoporous silica and chitosan increases the half-life and stability of the molecules to be immobilized. This method of immobilization is used in the industrial field of medicines and drugs because of the property of stability and reusability. This method has good stability, high binding strength and can be used over a long period of time (Hassan et al. 2016).

It is also significant that in this technique, the functional group of the support requires activation which results in modification of the support to increase the covalent binding efficiency as shown in Fig. 3 where the agarose bead (support) has been modified using aldehyde activation so that the amine ligand antibody may be immobilized onto it with increased efficiency. Hence, this method of immobilization

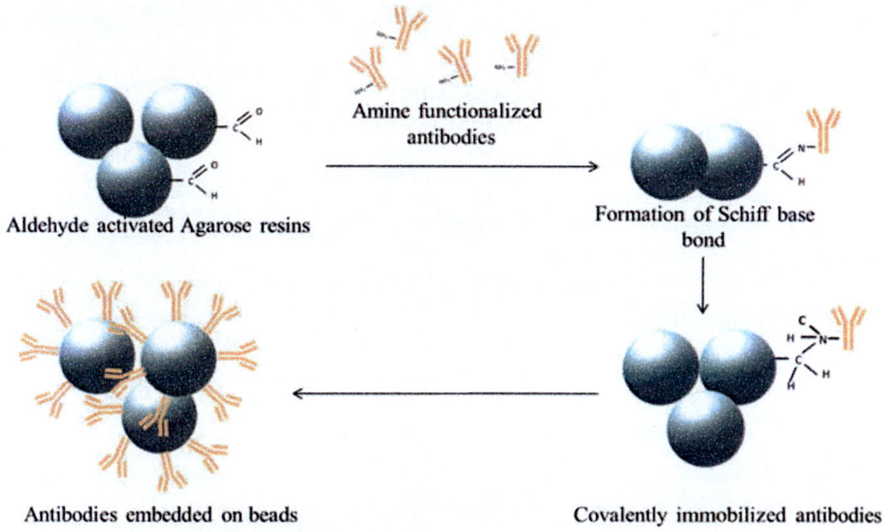

Fig. 3 Covalently attached antibody on aldehyde activated agarose resin. Agarose beads were activated by the aldehyde addition and interact with the amine group present in the antibody which results in covalent attachment of antibodies on the beads

requires chemical modification for fabrication. Matrices usually used in this type of interaction are cellulose, porous glass, etc.

2.2.2 Cross-Linking

In this method of immobilization, molecules are entrapped in a cross-linked matrix of a polymer by interacting with the molecules of the polymer. It is an irreversible technique of immobilization as the molecules form bonds within the polymer. This method of immobilization is also commonly called as copolymerization. It does not require any support matrix. This method involves the formation of intermolecular interaction between the molecules and the polymer with the help of bi and multifunctional reagents like glutaraldehyde (Nguyen 2017).

2.3 Support Material

Properties of the support matrix are essential for stabilization during the process of immobilization. Ideal support matrices for immobilization should be easily available, time saving, stable, reusable and economical as shown in Fig. 4. Ideal support material practically do not exist in nature; however, characteristics like hydrophilicity, insolubility in water, chemical stability, and protection against bacterial attacks can

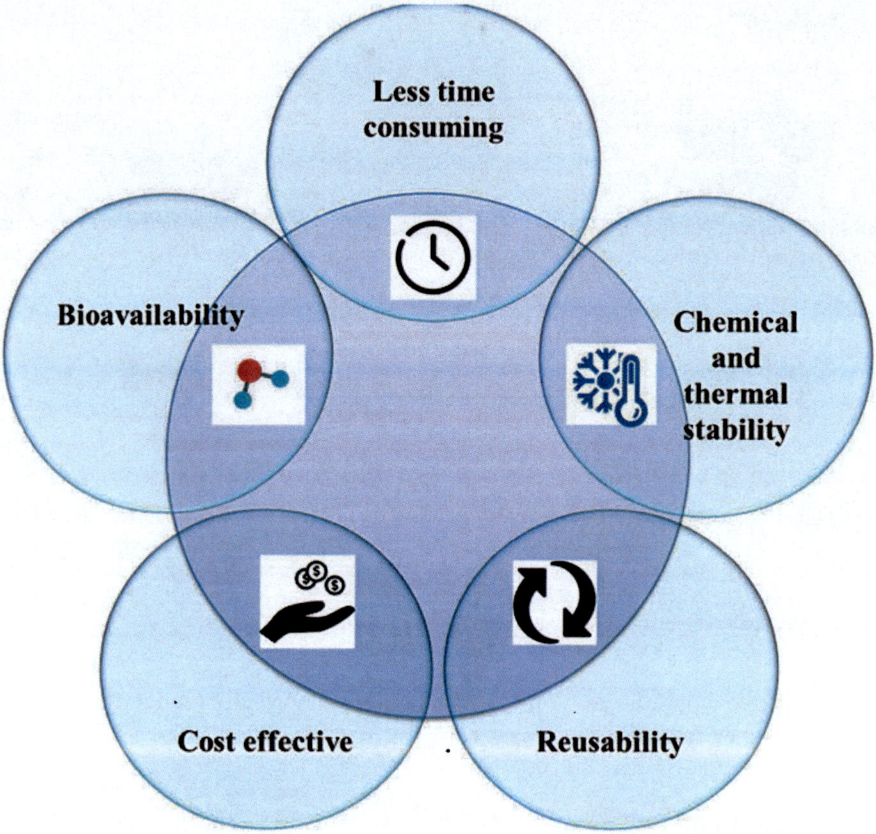

Fig. 4 Schematic representation of characteristics of an ideal support material. An ideal support material for efficient immobilization must be readily available, economical, reusable, time-saving and stable

be considered to deem a matrix as a good carrier (Hassan et al. 2016). The different criteria to be considered before choosing a particular carrier (mobile phase) for efficient immobilization has been shown in Table 2.

Table 2 Fundamental criteria during selection of carrier (mobile phase) for immobilization

Properties	Considerations
Physical	Strength, surface area per volume, porosity, permeability, density, flow rate and pressure
Chemical	Inertness towards immobilized molecules, hydrophilicity
Stability	Mechanical stability, regeneration of molecule activity
Resistance	Protection against fungal and bacterial attack

Table 3 Different examples of support material for the immobilization of molecules

Inorganic carriers	Organic carriers	Synthetic organic carriers
• SiO_2 • Porous glass • Silica • Celite • Centonite • Zeolites	• Cellulose derivatives: DEAE-cellulose • CM-cellulose • Starch • Chitosan • Agarose • Alginates • Dextran	• Acrylic polymers like Eupergit C (acrylic resin) • Polystyrene • Polyvinylacetate

Support can be divided into three categories according to their chemical composition, i.e. organic, inorganic and synthetic as shown in Table 3. It is important to know the nature and chemistry of the particular support matrix and molecules being used for any immobilization technique.

2.3.1 Most Common Natural Polymers as Used as Support

Alginate—Alginate is a polysaccharide found in brown algae cell wall and is used as an anionic polymer for immobilization of cells and molecules. This polymer has a vast range of applications in biomedical and food industry due to high biocompatibility, low toxicity and comparatively low cost. This polysaccharide is most commonly used for the entrapment of the molecules (Sharma et al. 2010a).

Chitosan and chitin—Chitosan and chitin are natural polymers. Poly (2-amine-2-deoxy-D-glucose) which is commonly known as chitosan is a deactylated form of chitin in alkaline medium derived from poly(N-acetyl-D-glucosamine) (Thirumavalavan and Lee 2015). Twice the amount of enzymes can be entrapped by chitosan in the form of beads as compared to chitin (Chang and Juang 2007).

Collagen—It is one of the most abundant proteins in a human body and plays an important role in fibre production in the body. Specific features of collagen like porous structure, thermal stability, mechanical strength, permeability and a large specific surface area makes it a good biocompatible matrix for immobilization (Hanachi et al. 2015).

3 Application of Immobilization Techniques

3.1 Forensic Finger Printing

Distinctive lines, loops, arches and swirls at the tip of our finger which are formed by ridge patterns are called fingerprints. These are unique for every individual and do not change throughout the person's life. Fingerprinting is a well-known biometric for criminal investigation, and it can be collected from any solid surface, porous or non-porous material, or fabric. In forensics, fingerprints are divided into patent or latent fingerprints. Visible prints are called patent fingerprints and are formed with ink, blood, paint, etc., whereas fingerprints which cannot be distinguished by the naked eye and require chemical and physical modification for the enhancement for visualization are called latent fingerprints (Guo and Tang 2013). Latent fingerprints are formed from the oil deposition or sweat secretion from the finger. Currently, the common method of latent fingerprinting is done through carbon powder, ninhydrin and cyanoacrylate, where carbon powder get adsorbed by the latent fingerprint, ninhydrin binds to the amino acids present in the sweat of the latent fingerprint and cyanoacrylate binds to the primary amines present in the sweat, for the visualization of fingerprints (Su et al. 2016). The basic process of latent fingerprinting consists of enrolment process and recognition process which has been elaborated in Fig. 5.

Fingerprints are obtained from the suspects by rolling the finger from one fingernail side to the other. It must be kept in mind that the latent fingerprint deposited on any surface changes with time due to several factors like evaporation, diffusion, exposure to light, etc. which may result in inaccurate analysis due to deformed, blurry image of latent fingerprint.

Integration of nanotechnology in forensic fingerprinting has vast applications. Nanoparticles improve latent fingerprinting techniques by enhancing the quality and

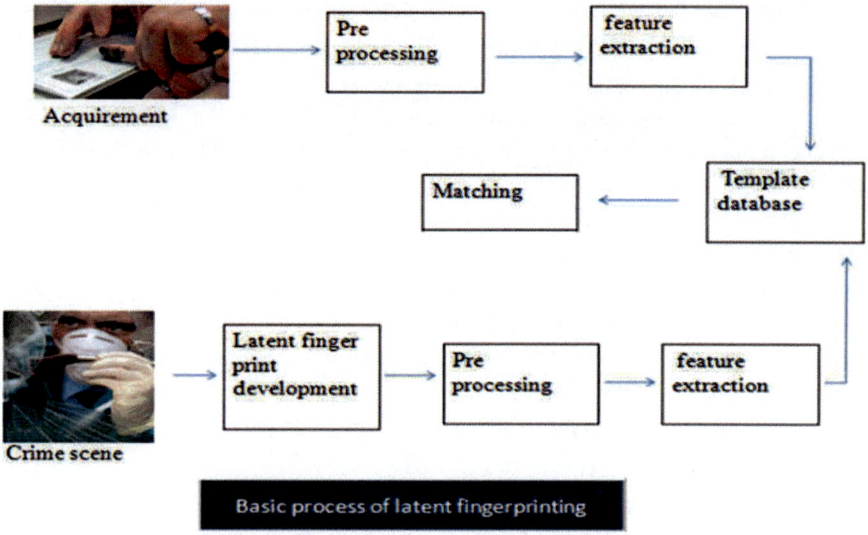

Fig. 5 Schematic representation of the process of fingerprinting in forensic science which involves acquiring the fingerprint from the individual by enrolment process to put in the database and then matching the latent fingerprint found at the crime scene with the database

sensitivity by making use of the concept of immobilization. Recently, various nanomaterials were designed to provide a decent contrast between the fingerprint and the background.

Clarity of the latent fingerprinting can be enhanced by the involvement of AuNPs followed by the treatment with various developers. This technique is based on the method of multi metal deposition (Hussain et al. 2010).

Kaushik et al. demonstrated the importance of gold particles for the visualization of latent fingerprinting by multimetal deposition method (Kaushik et al. 2017). First the author used citrate-capped negatively charged gold nanoparticle with positively charged components of fingerprints like primary amines of protein which further catalysed the precipitation of silver from silver-palladium (Ag-Pd) solution (Fig. 6). Also modified gold particles with some bifunctional reagents to interact with cellulose and form an invisible coating throughout the paper except on the sebaceous ridge of the fingerprint. Bifunctional reagent is bound with gold particles at the tail region, and their active head side has affinity towards the paper which results in attachment. Further it was treated with the Ag-Pd which resulted in darkening of the background coated with AuNPs due to precipitation of black silver particles. Deposition of silver particles significantly enhances the intensity of the latent fingerprint as a result of darkening of the background creating a contrast with the particle (Kaushik et al. 2017).

Another use of nanoparticles can be seen in the novel method of "intelligent" fingerprinting, i.e. the simultaneous identification of individuals and drug metabolites by using antibody-functionalized nanoparticles (Leggett et al. 2007). In this method, the antibody conjugated nanoparticles are used not only to visualize the latent fingerprint for identification but also to determine the presence of a particular drug metabolite in the sweat of the fingerprint which helps to provide a clue (lifestyle intelligence) about the individual fingerprint. Since sweat is the ultra-filtrate of blood plasma, it will contain drugs and its metabolites which are orally ingested and metabolized. So when a person touches a surface, he will leave behind trace amounts of the drug/metabolite along with the sweat in the fingerprint. By making a conjugate of a nanoparticle along with the antibody against a particular drug or metabolite, immobilization can be done with conjugate onto the fingerprint. A fluorescence-tagged secondary antibody is then attached to the primary antibody and visualized under the microscope. The nanoparticle in this case facilitates the immobilization of a larger number of antibodies by increasing the surface area and hence increasing the intensity of the fluorescent image. The image develops only if the targeted drug or metabolite is present and the clarity of the fluorescent image can be used to identify the fingerprint pattern as well for identification of the individual. This method can be further developed and used in the screening process of athletes. Using gold nanoparticles, the metabolite cotinine, which is a metabolite of nicotine, has been detected in the fingerprint of smokers (Leggett et al. 2007). In this, the conjugate was formed by immobilizing the anti-cotinine antibody onto the gold nanoparticle using SPDP bifunctional linkers and Protein A (cell-wall component of *Staphylococcus aureus*) as shown in Fig. 7. This conjugate was then immobilized onto the latent fingerprint by the antibody–antigen reaction between the anti-cotinine antibody and

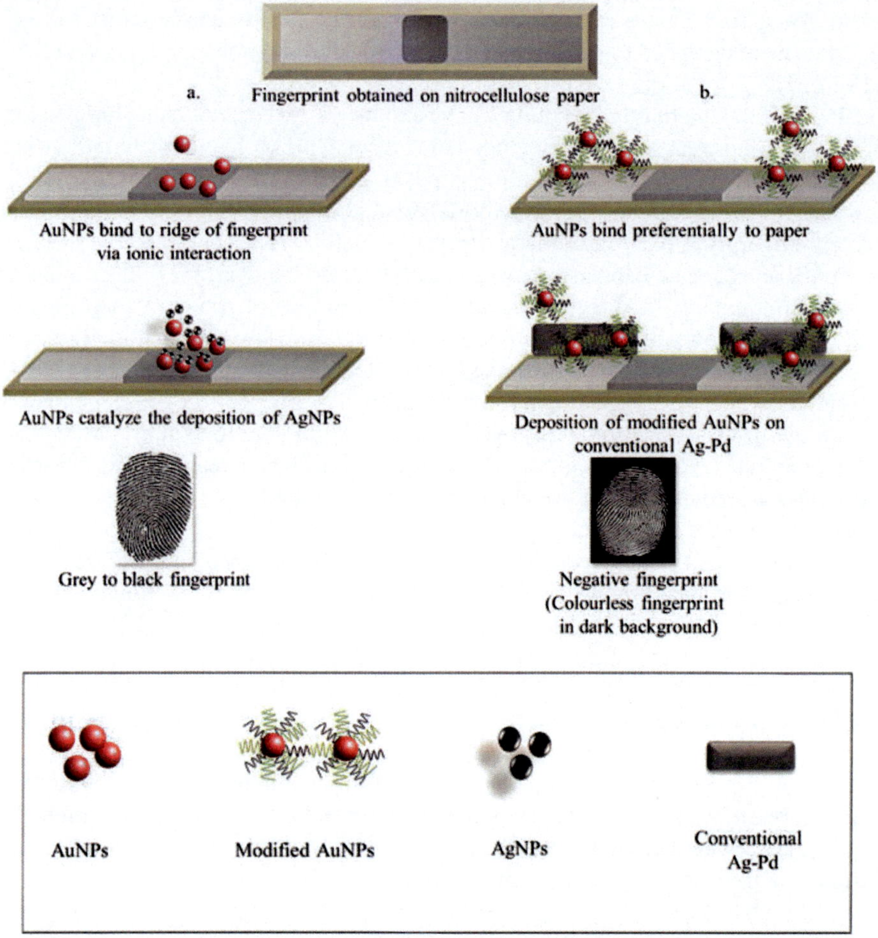

Fig. 6 Identification of latent fingerprinting on a wet paper using nanoparticles; **a** AuNPs binds to the fingerprint residue through ionic interaction and further silver from Ag-Pd solution precipitate on the surface; **b** Modified gold nanoparticles bind to paper by AuNPs-cellulose interaction to form an invisible coating on the surface and leave the finger mark residue

cotinine metabolite present in the sweat of the fingerprint. Using fluorescence tagged secondary antibody (Alexa Fluor), this immobilized conjugate could be visualized.

3.2 *Opiates Detection*

Opiates are drugs derived from opium which is naturally present in the poppy plant. Morphine is the main alkaloid of opium and clinically used as a potent analgesic in

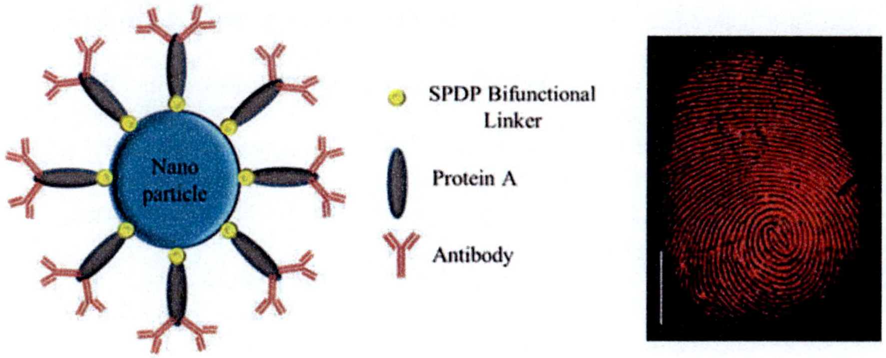

a. b.

Fig. 7 **a** Immobilization of anti-cotinine antibody (Ab) on AuNPs using linkers for the **b** Enhanced visualization of latent fingerprints by fluorescence

chronic pain in a moderate way so as to avoid the withdrawal symptoms. It suppresses the respiratory system and in high doses, can result in death. Heroin is one of the derivatives of morphine and has an additional two acetyl group. In a study, National Household Survey on Drug Abuse (NHSDA) concluded that the risks of starting alcohol use, illegal drug use, and tobacco chewing/smoking are largely found during adolescence and young adult years (Ali et al. 2011).

Widespread drug abuse among youngsters all over the world contributes towards the deterioration in health and also leads to involvement of people in illegal activities to earn quick money to buy illegal drugs. Keeping a check on drug use is important as it is used for medical screening, forensic testing and as a painkiller. Forensic analysis of drugs consists of forensic toxicology which tests for the drug on different body fluids from the suspect. Testing also determines whether the cause of death is drug overdose. According to the World Drug Report 2017 published by the United Nations Office on Drugs and Crime, opioids are the major drug with the potential to cause harmful health consequences even though they are not as widespread as cannabis (Liu et al. 2018).

Conventional analytical methods for monitoring of these drugs like gas–liquid chromatography (GLC), thin layer chromatography (TLC), mass spectroscopy (MS) and high-performance liquid chromatography (HPLC) are still used for opiates and other drug detection, but these analytical techniques are expensive, time-consuming and not applicable for on-site analysis (Singh et al. 2018). Out of these, TLC and HPTLC make use of immobilization adsorption method. The sample to be tested for the drug is put on the stationary phase which consists of the support, e.g. glass, aluminium/polyester sheet, and sorbents such as silica gel, aluminium oxide. Gypsum and starch are used as binders. The mobile phase (methanol, chloroform, etc.) carries the components over the stationary phase and based on the affinity between the components and the stationary phase, they travel different distances over the plate before the components of the drug immobilize by adsorption onto the stationary phase. The separated components can be detected by UV/Fluorescence to determine the presence of any particular drug when run with a control.

Development of immunosensors is one of the applications of immobilization, which analyses the signal during the formation of antigen–antibody complex. After ingestion, heroin gets deacetylated and converts into monoacetyl morphine (MAM) which further metabolizes to morphine. Detection of this heroin in the blood stream is very difficult since it has a short half-life. Immunosensors are developed by modifying the hapten (MAM) conjugated with carrier molecule to generate the specific antibody against an opiate molecule as shown in Fig. 8 (Singh et al. 2018).

One method to rapidly detect these opiates (e.g. morphine) can be done with the help of a lateral flow dipstick kit, which is based on the method of immobilization of antigen on a matrix, followed by interaction with antibodies (Gandhi et al. 2018, 2009).

Samples like oral fluids, blood, serum, hair, sweat can be used for detection. Heroin and its metabolite analogues last only for 9–38 h. Therefore, it is difficult to detect heroin in the blood sample after a prolonged period of time. M-6-M can be used to quantify the amount of heroin in oral fluids. Antibodies labelled with AuNPs have been used for rapid and sensitive detection of 6-MAM in a lateral flow assay via a colorimetric signal (Liu et al. 2018).

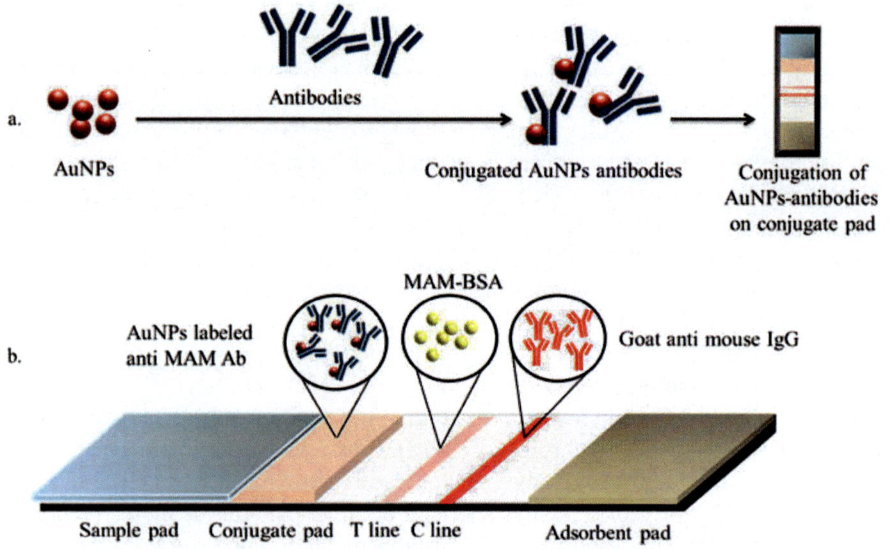

Fig. 8 **a** Immobilization of anti MAM antibody on the surface of AuNPs via ionic interaction; **b** Lateral-flow dipstick: Different components of lateral flow dipstick which consists of plastic backing case, a sponge adsorbent strip, sample pad, conjugated pad and nitrocellulose membrane. T & C lines are test and control lines, respectively. Conjugate pad consists of gold nanoparticles labelled anti-MAM antibody, test line consists of BSA-MAM as antigen, and control line is coated with goat anti-mouse IgG

Component of lateral flow dipstick: Lateral flow dipstick kit consists of plastic backing case, a sponge adsorbent strip, sample pad, conjugated pad and nitrocellulose membrane. The sponge adsorbent is for collection of fluid. AuNPs-antibody conjugate was fixed at conjugation pad, and nitrocellulose membrane part is coated with MAM-BSA conjugate at test line and goat anti-rabbit antibody at Control line.

In this competitive assay, when sample is loaded onto the sample pad, it flows towards the conjugation pad due to lateral flow and binds to the AuNPs labelled antibody. Excess of antibody continues to move along laterally by chromatography capillary effect and further binds to antigen on test line. This lateral flow technique has been demonstrated in Fig. 9. Intensity of signal/colour is based on the concentration of morphine in the sample, less morphine means more number of antibody flows toward the T line, and intensity of colour would be high and vice versa. Darker the colour, lower the morphine concentration in the sample.

There are different immobilization techniques which have been used to develop immunoassays based on various approaches such as electrochemical, optical, micromechanical, piezoelectric, aptameric, etc. for the detection of opiate drugs (Gandhi et al. 2015).

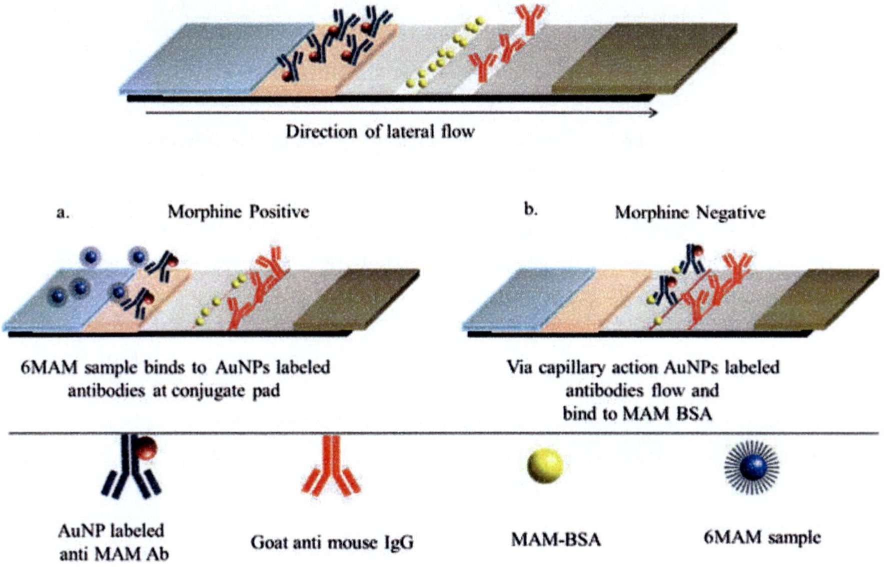

Fig. 9 Schematic representation of the lateral flow strip for 6-MAM detection; **a** In the presence of morphine, 6-MAM binds to the Ab-AuNPs restricting all of the Ab-AuNPs from flowing to the test line; **b** In the absence of Morphine, all the 6-MAM Ab-AuNPs binds to 6MAM-BSA on the test line

3.3 Pesticide Detection

Analytical toxicology is an important branch of forensic science which deals with analysis of poisons and other toxic substances. With fatal poisoning being the number one method of suicides and homicides, such cases received for toxicological analysis is constantly on the rise. With the increase in population resulting in increase of agricultural needs, there has been a rise in the use of insecticides, pesticides and other potentially poisonous substances in the last five years for the purpose of crop protection, but their misuse has led to suicidal, homicidal and/or accidental poisoning cases. The poison isolated from biological material in such cases is usually in microgram quantities. The conventional methods of chemical analysis are not feasible in such cases as they are time-consuming and less sensitive and so microchemical and instrumental techniques have to be developed for the specific, quick and sensitive detection, isolation and quantification of isolated poisons.

WHO concluded that two million suicidal poisonings and one million accidental poisonings occur due to insecticides worldwide every year, out of which 2,00,000 patients die mainly in developing countries. In India, the commonest pesticides such as organophosphates, organochlorates, and organocarbamates are used with suicidal and homicidal intent because of easy access (Singh and Sharma 2000; Siwach and Gupta 1995).

Josef et al. have prepared a WHO report of significant morbidity and mortality cases related to acute pesticide poisoning, especially in developing countries. Due to the lack of standardized case definition and unreported cases in rural areas since cause of death remains unknown, pesticide-related health effects cannot be estimated. The case definition should include all circumstances of poisoning such as occupational hazard, suicide, accidental exposure and homicide.

In developed countries, it has been shown that the acute poisoning cases in the agricultural sector annually are around 18.2 per 1,00,000 full-time workers (Calvert et al. 2004). Whereas in developing countries, where there is inadequate surveillance and information systems, insufficient regulation, less enforcement, lack of training and personal protective equipment, and large agriculture-dependant populations, the cases will be higher. Highly toxic pesticides, as classified by WHO, have been banned in industrialized countries. Improper storage and obsolete stockpiles increase risks of poisoning cases in developing countries.

Ultimately, it is shown that pesticides and insecticides are the most common form of poison used for suicide attempts as it is cheap and easily available to people from all walks of life. Its availability also makes it accessible for homicides, especially those made to look like suicides. Also, pesticide and insecticide poisoning cases occur due to accidental poisoning or long exposure to these toxic substances mainly due to occupational hazard. So the aim of a forensic scientist dealing with a pesticide/insecticide poisoning case is to extract, detect and quantify the pesticide from different sources such as water, food, human body, etc. There are several techniques used for the qualitative as well as quantitative analysis of pesticides/insecticides, ranging from conventional chromatographic methods (TLC, HPTLC) to the modern, biosensor detection, etc. methods, which make use of immobilization methods.

The conventional methods of analysis of pesticide/insecticide detection involved the basic method of immobilization, i.e. adsorption. First the extracted sample must be put on the stationary phase which will be the coated chromatography plate which consists of the support, e.g. glass, aluminium/polyester sheet, and sorbents such as silica gel, aluminium oxide, etc. Binders such as gypsum and starch are used. The mobile phase (methanol, chloroform, etc.) carries the components over the stationary phase and based on the affinity between the components and the stationary phase, they travel different distances over the plate before they immobilize by adsorption onto the stationary phase. The separated components can then be detected by UV/fluorescence to determine the presence of any particular pesticide/insecticide when run with a control.

The recent advances in pesticide detection involve detection using biosensors and paper-based devices. The biosensors have been divided into different types such as electrochemical, optical and piezoelectric based on the method of detection. The immobilized biorecognition elements have also been used as a basis to classify biosensors, i.e. whole cells, enzymes, antibodies and DNA. Recently, molecularly imprinted polymers (MIPs) and aptamers which are tailor-designed biomolecules, along with the incorporation of nanomaterials, are proving to be highly efficient in the detection of pesticides (Sassolas et al. 2012).

The detection of acetylcholinesterase (AChE) inhibition, i.e. the enzyme activity before and after exposure to pesticides, is the most common method of pesticide detection. Colorimetric Ellman assay is typically used to measure AChE activity. Hossain et al. fabricated a paper-based sensor where the AChE enzyme entrapped inside sol–gel-derived layers of silica was inkjet printed onto paper. The residual activity of AChE was measured on the paper-based sensor using Ellman's colorimetric assay to detect pesticides. Both lateral flow and dipstick were used in this detection which resulted in LOD of ~100 nM for paraoxon and 30 nM for aflatoxin B1, with a response time of less than 5 min. In a follow-up work, Hossain et al. immobilized the enzyme and the reagents onto paper. The AuNPs were co-entrapped along with the enzyme onto the paper for the AChE-catalysed enlargement of AuNPs (3 nm). The acetylthiocholine substrate and Au(III) salt were both spotted on the paper. Upon hydrolysis, thiocoline was generated by the enzyme substrate, which resulted in further reduction of Au(III) to AuNPs, leading to particle growth and increase in intensity of colour. Enzyme inhibition by pesticides was correlated with the colour intensity produced. 500 nM to 1 mM range was reported for paraoxon (Bülbül et al. 2015). Similarly, there are other enzyme-based biosensors which are used in dipstick & lateral-flow methods for detection of pesticides/insecticides. Various AChE-based electrochemical biosensors developed for pesticide detection have been listed in Table 4 along with the type of immobilization technique and material used, range, limit of detection, etc. (Sassolas et al 2012).

Detection of OPs is being carried out using aptamer-based biosensors, i.e. aptasensors (Fig. 10). Aptamer-modified AuNPs are immobilized on a biosensor to detect omethoate based on colorimetric assay. The aptamer has more affinity towards

Table 4 AChE-based electrochemical biosensors developed for pesticide detection along with the type of immobilization technique and material used, range, limit of detection, etc.

Target analyte	Detection technique	Enzyme immobilization technique	Electroactive materials	Linearity range (Molar)	Limit of detection (Molar)	References
Organophosphorus Insecticides						
Chlorpyrifos	CV	Covalent binding	Exfoliated graphite nanoplatelets	ND	1.58×10^{-10}	Ion et al. (2010)
Chlorpyrifos	SWV	Cross-linking	SWCNT	10^{-11}–10^{-6}	10^{-12}	Viswanathan et al. (2009)
Chlorpyrifos	Amperometry	Covalence	ZnS NPs	1.5×10^{-9}–4×10^{-8}	ND	Chauhan et al. (2011)
Chlorpyrifos	CV, DPV	Covalent binding	Flourine Doped Tin Oxide-AuNPs	1×10^{-15}–1×10^{-6}	10^{-15}	Talan et al. (2018)
Chlorpyrifos	Amperometry	Entrapment	[BMIM][BF_4]/MWCNT-CP	4×10^{-9}	4×10^{-9}	Zamfir et al. (2011)
Chlorpyrifos oxon	CV	Entrapment	PEDOT:PSS	ND	4×10^{-9}	Istamboulie et al. (2010)
Chlorpyrifos oxon	Amperometry	Entrapment	7,7,8,8-tetracyanoquinodemathane	6×10^{-9}–2.4×10^{-9}	6×10^{-9}	Hildebrandt et al. (2008a, 2008b)
Paraoxon	Amperometry	Affinity	MWCNT	3.6×10^{-14}–3.6×10^{-11}	5×10^{-15}	Ivanov et al. (2010)
Paraoxon	Amperometry	Entrapment	-	1.3×10^{-7}–5×10^{-6}	3.5×10^{-2}	Sinha et al. (2010)
Paraoxon	Amperometry	Adsorption	AuNPs, graphene oxide nanosheets	ND	10^{-13}	Wang et al. (2011)

(continued)

Table 4 (continued)

Target analyte	Detection technique	Enzyme immobilization technique	Electroactive materials	Linearity range (Molar)	Limit of detection (Molar)	References
Paraoxon	Amperometry	Cross-linking	CoPC-Prussian blue	7.3×10^{-9}–1.8×10^{-8}	7.3×10^{-9}	Arduini et al. (2006)
Methylparaoxon	Amperometry	Entrapment	CoPC	$2 \times 10^{-9} 4 \times 10^{-6}$	2.6×10^{-9}	Valdés-Ramírez et al. (2008a)
Methylparaoxon	Amperometry	Affinity	MWCNT	3.8×10^{-14}–3.8×10^{-11}	5.3×10^{-15}	Ivanov et al. (2010)
Triazophos	Amperometry	Adsorption	MWCNT	3×10^{-8}–7.8×10^{-6}	10^{-8}	Du et al. (2007c)
Dichlorvos	Amperometry	Adsorption	–	ND	10^{-10}	Vakurov et al. (2004)
Dichlorvos	Amperometry	Entrapment	CoPC	ND	7×10^{-12}	Valdés-Ramírez et al. (2008b)
Dichlorvos	Amperometry	Entrapment	CoPC	2×10^{-10}–10^{-8}	9.6×10^{-11}	Valdés-Ramírez et al. (2008a)
Dichlorvos	Amperometry	Adsorption	-	Up to 10^{-16}	10^{-17}	Sotiropoulou et al. (2005)
Dichlorvos	Amperometry	Cross-linking	Prussian blue	4.52×10^{-11}–4.52×10^{-8}	1.13×10^{-11}	Sun and Wang (2010)
Omethoate	Amperometry	Cross-linking	Prussian blue	2.34×10^{-10}–4.69×10^{-8}	7.04×10^{-11}	Sun and Wang (2010)

(continued)

Table 4 (continued)

Target analyte	Detection technique	Enzyme immobilization technique	Electroactive materials	Linearity range (Molar)	Limit of detection (Molar)	References
Trichlorfon	Amperometry	Cross-linking	Prussian blue	1.16×10^{-10}–1.94×10^{-8}	1.94×10^{-11}	Sun and Wang (2010)
Phoxim	Amperometry	Cross-linking	Prussian blue	1.68×10^{-10}–3.35×10^{-8}	3.35×10^{-11}	Sun and Wang (2010)
Trichlorfon	Amperometry	Adsorption	TiO_2 and PbO_2 particles	10^{-8}–2×10^{-5}	10^{-10}	Wei et al. (2009)
Monocrotophos	Amperometry	Adsorption	AuNPs	4.5×10^{-9}–4.5×10^{-6}	2.7×10^{-9}	Du et al. (2008)
Monocrotophos	Amperometry	Covalent binding	AuNPs-QDs	4.5×10^{-9}–4.5×10^{-6}	1.3×10^{-9}	Du et al. (2007a)
Acephate	FET	Affinity	CNT	ND	5.45×10^{-14}	Ishii et al. (2008)
Dimethoate	Amperometry	Adsorption	CNTs, Zirconia NPs, Au colloid coated Fe_3O_4 magnetic NPs, Prussian blue	4.4×10^{-6}–4.4×10^{-2}	2.4×10^{-6}	Gan et al. (2010)

(continued)

Table 4 (continued)

Target analyte	Detection technique	Enzyme immobilization technique	Electroactive materials	Linearity range (Molar)	Limit of detection (Molar)	References
Carbamate Insecticides						
Aldicarb	Amperometry	Cross-linking	CoPC-Prussian blue	6.3×10^{-8}–1.6×10^{-7}	1.3×10^{-7}	Arduini et al. (2006)
Carbaryl	Amperometry	Adsorption	–	2.5×10^{-8}–5×10^{-7}	1.5×10^{-8}	Du et al. (2007b)
Carbaryl	Amperometry	Covalent binding	QDs	5×10^{-9}–2.5×10^{-7}	3×10^{-9}	Du et al. (2008)
Carbaryl	Amperometry	Cross-linking	CoPC-Prussian blue	1.2×10^{-7}–4.9×10^{-7}	1.2×10^{-7}	Arduini et al. (2006)
Carbaryl	Amperometry	Adsorption	MWCNT	5×10^{-13}–5×10^{-10}	5×10^{-15}	Zhang et al. (2008)
Carbaryl	Amperometry	Entrapment	CoPC	9×10^{-8}–4×10^{-6}	1.6×10^{-7}	Valdés-Ramírez et al. (2008a)
Carbaryl	Amperometry	Entrapment	Prussian Blue-Chitosan-GCE	0.01×10^{-6}–0.4×10^{-6} and 10^{-6}–5×10^{-6}	3×10^{-9}	Song et al. (2011)
Carbofuran	DPV	Adsorption	CNTs-AuNPs	4.8×10^{-9}–0.9×10^{-8}	4×10^{-9}	Qu et al. (2010)
Carbofuran	Amperometry	Cross-linking	CoPC	10^{-10}–10^{-7}	4.9×10^{-10}	Laschi et al. (2007)
Carbofuran	Amperometry	Entrapment	CoPC	$4 \times 10^{-9} \times 8 \times 10^{-8}$	4.5×10^{-9}	Valdés-Ramírez et al. (2008a)

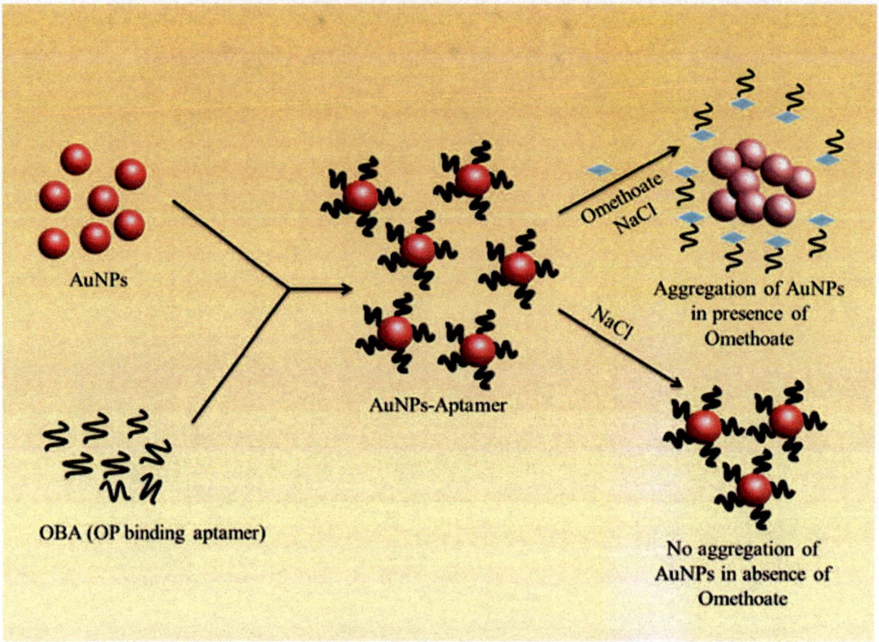

Fig. 10 OP binding aptamer was adsorbed (immobilized) on gold nanoparticles to form a stable probe. When omethoate is present, OBA detaches from the gold particles and binds to omethoate, causing aggregation of gold particles (the colour of shifts from red to blue). In contrast, AuNPs–aptamer is highly resistant to salt-induced aggregation (solution stays red)

omethoate and hence detaches from the AuNPs which successively aggregate. Using the methoate-binding aptamer and measuring the colour changes due to aggregation of AuNPs, this biosensor gave a linear range from 0.1 μM to 10 μM and LOD of 0.1 μM (Wang et al. 2016). A similar detection technique was developed for malathion by using a malathion-specific aptamer, cationic peptide and unmodified AuNPs. This biosensor gave a linear detection range of 0.01–0.75 nM and LOD 1.94 pM (Bala et al. 2017).

There are various other biosensors used for lateral-flow and dipstick assays using different components such as whole DNA, antigen and antibodies. To enhance the sensitivity of these biosensors, modification using nanoparticles has also been done. Chlorpyrifos can be detected in river water by using nanoparticles based on iridium oxide, modified with tyrosinase, on screen-printed carbon electrodes which acts as a disposable enzymatic biosensor (Mayorga-Martinez et al. 2014). In another research, methyl parathion was detected using an electrochemical biosensor which gave LOD at femtogram level (5 fg/mL). The biosensor was made up of a nanoporous carbon paste electrode with chitosan, Nafion (immobilized protective membrane) and AuNPs. The recognition element here was acetylcholinesterase. Methyl parathion showed high affinity to the substrate of acetylcholinesterase which resulted in quick response due to biocompatibility and good electron transfer rate of AuNPs and chitosan. In another study, chlorpyrifos was detected by immobilizing antichlorpyrifos antibodies onto graphene field-effect transistors (GraFET) with Si/SiO_2 substrate with a linear range from 1 fM to 1 μM and LOD up to 1.8 fM (Islam et al. 2019).

Optical immunosensors based on surface plasmon resonance are also being developed. Real-time monitoring is possible based on SPR, and labelled molecules are not required (Homola et al. 1999). Competitive detection of 2,4-D was carried out by Miura and coworkers using an SPR-based immunosensor (Gobi et al. 2007, 2005; Kim et al. 2008, 2007). Immobilized 2,4-D-ovalbumin conjugate on the Au surface of the sensor chip competed with free 2,4-D to bind selectively with monoclonal anti- 2,4-D. Avidin–biotin amplification interactions enhanced sensor sensitivity. A biotinylated secondary antibody against anti-2,4-D, avidin was used in successive incubations for amplification followed by biotin-BSA molecules. The amplification of the sensor signal was produced up to a factor of 10 (Kim et al. 2007). Different SPR-based biosensors have been shown in Table 5 along with the immobilization technique employed and the linearity range and the LOD (Sassolas et al. 2012).

3.4 Heavy Metal Detection

Heavy metals such as copper, zinc, mercury, lead, cadmium, arsenic, chromium and others are constantly released into the environment from various anthropogenic sources as a supplement to natural geochemical sources. They mainly find their way to water bodies and cause contamination of drinking water leading to heavy metal toxicity in living organisms, possibly even death. These heavy metals cannot be broken down or degraded naturally, i.e. they are non-biodegradable leading to bioaccumulation through the food chain. Various man-made activities and substance such as the burning of fossil fuels with heavy metal content, tetra-ethyl lead additive in gasoline, increased use of metals in industrial applications, e.g. dye and chemical manufacture, metal plating, tanning, mining, have made environmental pollution by heavy metals a major issue in today's world (Klaassen 1996). The most frequently found metal contaminants include lead, chromium, cadmium, copper, zinc and mercury.

Old immobilization techniques such as TLC have become outdated as they take up too much time and require laboratory equipment. Hence, the most recent immobilization techniques used in heavy metal detection are biosensors in the form of dipsticks or lateral-flow assays. Inhibition-based biosensors have been developed by immobilization of enzymes such as alkaline phosphatase, urease, acetylcholinesterase, GOx and HRP on electrodes for detection of Cd^{2+}, Cu^{2+}, Cr^{3+}, Zn^{2+}, Ni^{2+}, Pb^{2+}, Co^{2+}, Cr^{6+}, Mn^{2+} and Fe^{2+} (Rebollar-Pérez et al. 2016). There are also other biosensors available that are based on protein (Bontidean et al. 1998), whole cell (Corbisier et al. 1999), DNA (Li and Lu 2000), etc. Often nanoparticles are incorporated into the biosensors to enhance the sensitivity and/or specificity (Liu and Lu 2003). Different types of biosensors based on the type of molecule immobilized have been shown in Fig. 11 (Verma and Singh 2005).

Table 5 SPR-based immunosensor targets, immobilization techniques, linearity range and limit of detection

Target	Immobilization technique	Linearity range (Molar)	Limit of detection (Molar)	References
DDT	Covalent binding	1.7×10^{-10}–5×10^{-9}	5.64×10^{-11}	Mauriz et al. (2006)
2,4-D	Adsorption of 2,4-D-BSA conjugate	2.3×10^{-9}–4.5×10^{-6}	2.3×10^{-9}	Gobi et al. (2005)
2,4-D	Adsorption of 2,4-D-ovalbumin conjugate	4.5×10^{-10}–1.4×10^{-6}	4.5×10^{-10}	Gobi et al. (2005)
2,4-D	Covalent immobilization of 2,4-D-BSA conjugate	ND	3.6×10^{-11}	Kim et al. (2007)
2,4-D	Adsorption of 2,4-D-BSA conjugate	4.5×10^{-10}	4.5×10^{-10}	Kim et al. (2008)
Chlorpyrifos	Covalent binding	6.6×10^{-10}–1.4×10^{-7}	1.4×10^{-10}	Mauriz et al. (2006)
Carbaryl	Covalent binding	8.2×10^{-9}–7.3×10^{-8}	4.5×10^{-9}	Mauriz et al. (2006)
Isoproturon	Covalent binding	6.3×10^{-9}–7.9×10^{-8}	4.8×10^{-10}	Gouzy et al. (2009)
Atrazine	Covalent immobilization of Atrazine-BSA conjugate	1.3×10^{-10}–3.7×10^{-9}	9.3×10^{-11}	Farré et al. (2007)
Atrazine	Covalent binding	4.64×10^{-6}–4.64×10^{-4}	4.64×10^{-6}	Nakamura et al. (2003)
TNT	Adsorption of TNT-BSA conjugate	2.6×10^{-11}–1.3×10^{-6}	2.6×10^{-11}	Shankaran et al. (2006)
TNT	Covalent immobilization of AuNPs	ND	4.4×10^{-11}	Kawaguchi et al. (2008)
TNT	Covalent binding	3.5×10^{-11}–1.3×10^{-7}	ND	Kawaguchi et al. (2007)
TNT	Covalent binding	Up to 3.1×10^{-8}	4.8×10^{-10}	Singh et al. (2009)
2,4-dinitrotoluene	Covalent binding	4.4×10^{-9}–4.4×10^{-7}	8.8×10^{-11}	Nagatomo et al. (2009)
2,4-dichlorophenol	Affinity (Protein G)	ND	1.2×10^{-7}	Soh et al. (2003)

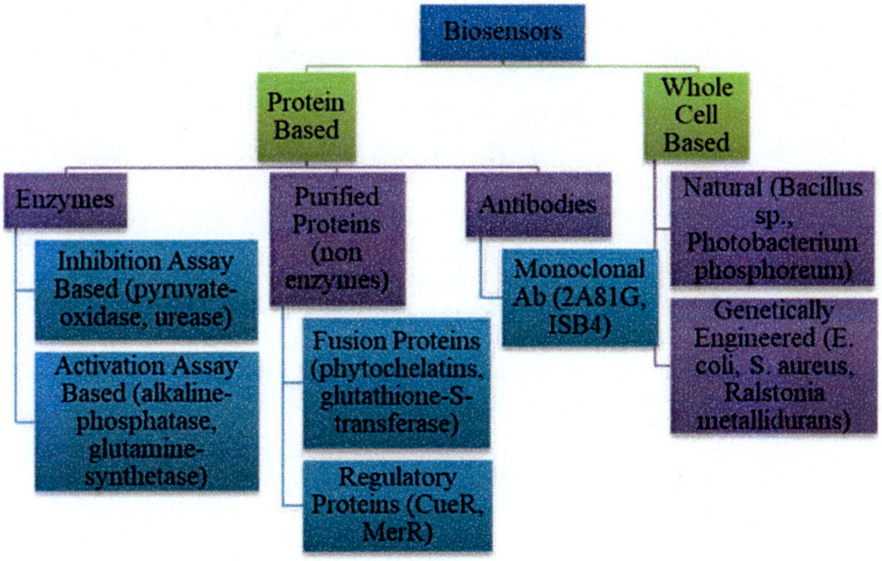

Fig. 11 Classification of biosensors for detection of heavy metals based on the type of molecule immobilized

A novel non-protein-based example of a biosensor has recently been introduced. It involves a colorimetric Pb(II) sensor based on DNAzyme-directed assembly of AuNPs. A DNAzyme or deoxyribozymes are DNA oligonucleotides which can perform a specific chemical reaction similar to other biological enzymes such as proteins or ribozymes (Breaker 1997). The heavy metal sensing element, i.e. the "8–17" DNAzyme, is highly selective towards Pb(II) and shows high activity. The aggregation of the AuNPs, DNAzyme and its substrate results in blue colour formation, but in case of presence of Pb(II), the AuNPs do not aggregate and a red colour forms. The sensor has a linear detection range of 100 nM to 4.0 1 M (Liu and Lu 2003).

3.5 Serological Analysis

Apart from the simple confirmatory tests such as luminol, Kastle-Meyer, PSA detection, salivary-amylase, etc. to detect the presence of body fluids at a crime scene, various immobilization techniques are also used out of which immunological tests are the most common.

The OneStepABAcard® HemaTrace test strip is an immunochromatographic test for human blood detection. If human haemoglobin is present, it combines with the mobile monoclonal antihuman haemoglobin Ab which has been immobilized onto the test strip. Any Ab–Ag complex formed migrates to the test line through the absorbent pad. The pink line is visible in the control area above the LOD 0.05 μg/ml. The results have shown the OneStepABAcard® HemaTrace to be a highly specific, convenient, quick and sensitive test for the detection of human blood both at the scene of crime and the laboratory (An et al. 2012).

Similarly, apart from the microscopic identification of sperm cells, the detection of prostate-specific antigen (PSA) is used to detect semen samples as it is produced in large amount by the male prostate gland and can even be detected in azoospermic male semen. One of the most commonly used commercial kits, the OneStepABAcard® PSA, uses the technique of immobilization of monoclonal antihuman PSA Ab on a test strip, which binds to human PSA and moves along the absorbent pad to the control line, forming a visible result line as I migrates. Similarly, the RSID™-saliva test is based on antihuman salivary anti-amylase Ab for the detection of saliva and RSID™-urine test for the detection of urine (Harbison and Fleming 2016).

4 Summary

From this chapter, it is evident that immobilization techniques in the laboratory are a vital part of forensic investigations. Most of the time with only a minute amount of sample retrieved from the crime scene, it is important that the sample is not lost and the testing of the evidence is done in an economical manner. The advantage of immobilization techniques is that the fixing of components to substrates enables faster reactions and also increases the sensitivity of the procedure along with making sure reagents and samples are not used excessively or get wasted unnecessarily. This results in an economical use of samples as well as reagents.

Most of the older immobilization techniques such as TLC and HPTLC are still predominantly being used in most forensic laboratories for a lot of scenarios such as toxin, poison and pesticide detection, ink analysis, heavy metal detection, etc. While these older methods are still reliable, they are less sensitive, less economical, time-consuming and require a huge laboratory set-up with trained personnel as well. So it has become important to move on to more economical and user-friendly immobilization techniques such as biosensors, dipsticks, lateral flow assays, etc. These sensors give a quick result and do not require elaborate laboratory set-ups or trained personnel and are especially excellent for preliminary tests. The incorporation of nanoparticles in these techniques is also an upcoming field which enhances the test results by increasing sensitivity. However, these new methods are not yet being widely used, and hence, it is important to continue to research and develop these immobilization techniques as the future of forensic science

Acknowledgements The support provided, to carry out this research work, in the form of research grants from the Department of Science and Technology (DST), New Delhi (DST/ECR/2016/000,075), and the Department of Biotechnology (DBT-Biocare), New Delhi (BT/PR18069/BIC/101/574/2016) is highly appreciated.

References

Ali S, Mouton CP, Jabeen S et al (2011) Early detection of illicit drug use in teenagers. Innov Clin Neurosci 8:24–28

An JH, Shin KJ, Yang WI, Lee HY (2012) Body fluid identification in forensics. BMB Rep 45(10):545–553

Arduini F, Ricci F, Tuta CS et al (2006) Detection of carbamic and organophosphorous pesticides in water samples using a cholinesterase biosensor based on Prussian Blue-modified screen-printed electrode. Anal Chim Acta 580:155–162

Bala R, Dhingra S, Kumar M et al (2017) Detection of organophosphorus pesticide—malathion in environmental samples using peptide and aptamer based nanoprobes. Chem Eng J 311:111–116

Bolton S, Thorpe JW (1986) Enzyme-linked immunosorbent assay for A and B water soluble blood group substances. J Forensic Sci 31(1):27–35

Bontidean I, Berggren C, Johansson G et al (1998) Detection of heavy metal ions at femtomolar levels using protein-based biosensors. Anal Chem 70(19):4162–4169

Breaker RR (1997) DNA enzymes. Nat Biotechnol 15:427–431

Bülbül G, Hayat A, Andreescu S (2015) Portable nanoparticle-based sensors for food safety assessment. Sensors 15:30736–30758

Calvert GM, Plate DK, Das R et al (2004) Acute occupational pesticide-related illness in the US, 1998–1999: Surveillance findings from the SENSOR-pesticides program. Am J Ind Med 45:14–23

Chang MY, Juang RS (2007) Use of chitosan–clay composite as immobilization support for improved activity and stability of β-glucosidase. Biochem Eng J 35:93–98

Chauhan N, Narang J, Pundir CS (2011) Immobilization of rat brain acetylcholinesterase on ZnS and poly (indole-5-carboxylic acid) modified Au electrode for detection of organophosphorus insecticides. Biosens Bioelectron 29:82–88

Corbisier P, van der Lelie D, Borremans B et al (1999) Whole cell- and protein-based biosensors for the detection of bioavailable heavy metals in environmental samples. Anal Chim Acta 387:235–244

Datta S, Christena LR, Rajaram YRS (2013) Enzyme immobilization: an overview on techniques and support materials. Biotech 3:1–9

Du D, Chen S, Cai J, Zhang A (2007) Immobilization of acetylcholinesterase on gold nanoparticles embedded in sol–gel film for amperometric detection of organophosphorous insecticide. Biosens Bioelectron 23:130–134

Du D, Chen S, Song D et al (2008) Development of acetylcholinesterase biosensor based on CdTe quantum dots/gold nanoparticles modified chitosan microspheres interface. Biosens Bioelectron 24:475–479

Du D, Ding J, Cai J, Zhang A (2007) Determination of carbaryl pesticide using amperometric acetylcholinesterase sensor formed by electrochemically deposited chitosan. Colloids Surf. B Biointerf. 58:145–150

Du D, Huang X, Cai J, Zhang A (2007) Amperometric detection of triazophos pesticide using acetylcholinesterase biosensor based on multiwall carbon nanotube–chitosan matrix. Sens. Actuators B Chem 127:531–535

Farré M, Martínez E, Ramón J et al (2007) Part per trillion determination of atrazine in natural water samples by a surface plasmon resonance immunosensor. Anal Bioanal Chem 388:207–214

Gan N, Yang X, Xie D et al (2010) A Disposable organophosphorus pesticides enzyme biosensor based on magnetic composite nano-particles modified screen printed carbon electrode. Sensors 10:625–638

Gandhi S, Banga I, Maurya PK, Eremin SA (2018) A gold nanoparticle-single-chain fragment variable antibody as an immunoprobe for rapid detection of morphine by dipstick. RSC Adv 8:1511–1518

Gandhi S, Caplash N, Sharma P, Raman Suri C (2009) Strip-based immunochromatographic assay using specific egg yolk antibodies for rapid detection of morphine in urine samples. Biosens Bioelectron 25:502–505

Gandhi S, Sharma P, Capalash N et al (2008) Group-selective antibodies based fluorescence immunoassay for monitoring opiate drugs. Anal Bioanal Chem 392:215–222

Gandhi S, Suman P, Kumar A et al (2015) Recent advances in immunosensor for narcotic drug detection. Bioimpacts 5(4):207–213

Gobi KV, Kim SJ, Tanaka H et al (2007) Novel surface plasmon resonance (SPR) immunosensor based on monomolecular layer of physically-adsorbed ovalbumin conjugate for detection of 2,4-dichlorophenoxyacetic acid and atomic force microscopy study. Sens Actuat B Chem 123:583–593

Gobi KV, Tanaka H, Shoyama Y, Miura N (2005) Highly sensitive regenerable immunosensor for label-free detection of 2,4-dichlorophenoxyacetic acid at ppb levels by using surface plasmon resonance imaging. Sens Actuat B Chem 111–112:562–571

Gouzy M-F, Kess M, Krämer PM (2009) A SPR-based immunosensor for the detection of isoproturon. Biosens Bioelectron 24:1563–1568

Guo W, Tang Y (2013) Latent fingerprint recognition: challenges and advances. CCBR 2013. Lecture Notes in Computer Science 8232. Springer, Cham, 208–215

Hanachi P, Jafary F, Jafary F, Motamedi S (2015) Immobilization of the alkaline phosphatase on collagen surface via cross-linking method. Iran J Biotechnol 13:32–38

Harbison S, Fleming R (2016) Forensic body fluid identification: state of the art. Res Rep Forensic Med Sci 11–23

Hassan ME, Tamer TM, Omer AM (2016) Methods of enzyme immobilization. Int J Curr Pharm Rev Res 7(6):385–392

Hemalatha V, Kalyani P, Chandana VK, Hemalatha KPJ (2016) Methods, applications of immobilized enzymes-a mini review . Int J Eng Sci Res Tech 11:523–526

Hildebrandt A, Bragós R, Lacorte S, Marty JL (2008) Performance of a portable biosensor for the analysis of organophosphorus and carbamate insecticides in water and food. Sens Actuat B Chem 133:195–201

Hildebrandt A, Ribas J, Bragós R et al (2008) Development of a portable biosensor for screening neurotoxic agents in water samples. Talanta 75:1208–1213

Homaei AA, Sariri R, Vianello F, Stevanato R (2013) Enzyme immobilization: an update. J Chem Biol 6:185–205

Homola J, Yee SS, Gauglitz G (1999) Surface plasmon resonance sensors: review. Sens Actuat B Chem 54:3–15

Hussain I, Hussain SZ, Rehman H et al (2010) In situ growth of gold nanoparticles on latent fingerprints—from forensic applications to inkjet printed nanoparticle patterns Nanoscale 2:2575–2578

Ion AC, Ion I, Culetu A et al (2010) Acetylcholinesterase voltammetric biosensors based on carbon nanostructure-chitosan composite material for organophosphate pesticides. Mater Sci Eng C 30:817–821

Ishii A, Takeda S, Hattori S et al (2008) Ultrasensitive detection of organophosphate insecticides by carbon nanotube field-effect transistor. Colloids Surfaces a Physicochem Eng Asp 313–314:456–460

Islam S, Shukla S, Bajpai VK et al (2019) Microfluidic-based graphene field effect transistor for femtomolar detection of chlorpyrifos. Sci Rep 9(276):1–7

Istamboulie G, Sikora T, Jubete E et al (2010) Screen-printed poly(3,4-ethylenedioxythiophene) (PEDOT): a new electrochemical mediator for acetylcholinesterase-based biosensors. Talanta 82:957–961

Ivanov Y, Marinov I, Gabrovska K et al (2010) Amperometric biosensor based on a site-specific immobilization of acetylcholinesterase via affinity bonds on a nanostructured polymer membrane with integrated multiwall carbon nanotubes. J Mol Catal B Enzym 63:141–148

Jesionowski T, Zdarta J, Krajewska B (2014) Enzyme immobilization by adsorption: a review. Adsorption 20:801–821

Kaushik M, Mahendru S, Chaudhary S, Kukreti S (2017) DNA fingerprints: advances in their forensic analysis using nanotechnology. J Forensic Biomech 8:1–4

Kawaguchi T, Shankaran DR, Kim SJ et al (2007) Fabrication of a novel immunosensor using functionalized self-assembled monolayer for trace level detection of TNT by surface plasmon resonance. Talanta 72:554–560

Kawaguchi T, Shankaran DR, Kim SJ et al (2008) Surface plasmon resonance immunosensor using Au nanoparticle for detection of TNT. Sens. Actuat. B Chem 133:467–472

Kim SJ, Gobi KV, Iwasaka H et al (2007) Novel miniature SPR immunosensor equipped with all-in-one multi-microchannel sensor chip for detecting low-molecular-weight analytes. Biosens Bioelectron 23:701–707

Kim SJ, Gobi KV, Tanaka H et al (2008) A simple and versatile self-assembled monolayer based surface plasmon resonance immunosensor for highly sensitive detection of 2,4-D from natural water resources. Sens Actuat B Chem 130:281–289

Klaassen G (1996) Acid rain and environmental degradation: the economics of emission trading

Laschi S, Ogończyk D, Palchetti I, Mascini M (2007) Evaluation of pesticide-induced acetylcholinesterase inhibition by means of disposable carbon-modified electrochemical biosensors. Enzyme Microb Technol 40:485–489

Leggett R, Lee-Smith EE, Jickells SM, Russell DA (2007) "Intelligent" fingerprinting: Simultaneous identification of drug metabolites and individuals by using antibody-functionalized nanoparticles. Angew Chemie Int Ed 46(22):4100–4103

Li J, Lu Y (2000) A highly sensitive and selective catalytic DNA biosensor for lead ions. J Am Chem Soc 122(42):10466–10467

Liu J, Hu X, Cao F et al (2018) A lateral flow strip based on gold nanoparticles to detect 6-monoacetylmorphine in oral fluid R. Soc Open Sci 5(6):180288 (1–11).

Liu J, Lu Y (2003) A colorimetric lead biosensor using DNAzyme-directed assembly of gold nanoparticles. J Am Chem Soc 125:6642–6643

Mahendra Wijaya IP, Nie TJ, Gandhi S et al (2010) Femtomolar detection of2,4-dichlorophenoxyacetic acidherbicidesvia competitive immunoassays using microfluidic based carbon nanotube liquid gated transistor. Lab Chip 10:634–638

Mauriz E, Calle A, Montoya A, Lechuga LM (2006) Determination of environmental organic pollutants with a portable optical immunosensor. Talanta 69:359–364

Mayorga-Martinez CC, Pino F, Kurbanoglu S et al (2014) Iridium oxide nanoparticle induced dual catalytic/inhibition based detection of phenol and pesticide compounds. J Mater Chem B 2:2233–2239

Meade BW, Higgins G, Widdop B et al (1972) Simple tests to detect poisons. Ann Clin Biochem An Int J Biochem Lab Med 9:35–46

Mishra P, Banga I, Tyagi R et al (2018) An immunochromatographic dipstick as an alternate for monitoring of heroin metabolites in urine samples. RSC Adv 8:23163–23170

Mohamad NR, Marzuki NHC, Buang NA et al (2015) An overview of technologies for immobilization of enzymes and surface analysis techniques for immobilized enzymes. Biotechnol Biotechnol Equip 29(2):205–220

Nagatomo K, Kawaguchi T, Miura N et al (2009) Development of a sensitive surface plasmon resonance immunosensor for detection of 2,4-dinitrotoluene with a novel oligo (ethylene glycol)-based sensor surface. Talanta 79:1142–1148

Nakamura C, Hasegawa M, Nakamura N, Miyake J (2003) Rapid and specific detection of herbicides using a self-assembled photosynthetic reaction center from purple bacterium on an SPR chip. Biosens Bioelectron 18:599–603

Nguyen HHM (2017) An overview of techniques in enzyme immobilization. Appl Sci Converg Technol 26:157–163

Qu Y, Sun Q, Xiao F et al (2010) Layer-by-Layer self-assembled acetylcholinesterase/PAMAM-Au on CNTs modified electrode for sensing pesticides. Bioelectrochemistry 77:139–144

Ramansuri C, Kaur J, Gandhi S, Shekhawat GS (2008) Label-free ultra-sensitive detection of atrazine based on nanomechanics. Nanotechnology 19(23):235502(1–6)
Ramansuri C, Boro R, Nangia Y et al (2009) Immunoanalytical techniques for analyzing pesticides in the environment. TrAC Trends Anal Chem 28:29–39
Rebollar Pérez G, Campos Terán J, Ornelas Soto N et al (2016) Biosensors based on oxidative enzymes for detection of environmental pollutants. Biocatalysis 1(1):118–129
Sametband M, Shweky I, Banin U et al (2007) Application of nanoparticles for the enhancement of latent fingerprints. Chem Commun 11:1142–1144
Sassolas A, Prieto Simón B, Marty JL (2012) Biosensors for pesticide detection: new trends. Am. J Anal Chem 3:210–232
Shankaran DR, Matsumoto K, Toko K, Miura N (2006) Development and comparison of two immunoassays for the detection of 2,4,6-trinitrotoluene (TNT) based on surface plasmon resonance. Sens Actuat B Chem 114:71–79
Sharma J, Mahajan R, Gupta V (2010) Comparison and suitability of gel matrix for entrapping higher content of enzymes for commercial applications. Indian J Pharm Sci 72(2):223–228
Sharma P, Gandhi S, Chopra A et al (2010) Fluoroimmunoassay based on suppression of fluorescence self-quenching for ultra-sensitive detection of herbicide diuron. Anal Chim Acta 676:87–92
Singh P, Onodera T, Mizuta Y et al (2009) Dendrimer modified biochip for detection of 2,4,6 trinitrotoluene on SPR immunosensor: Fabrication and advantages. Sens Actuat B Chem 137:403–409
Singh S, Mishra P, Banga I et al (2018) Chemiluminescence based immunoassay for the detection of heroin and its metabolites. Bioimpacts 8:53–58
Singh S, Sharma N (2000) Neurological syndromes following organophosphate poisoning. Neurol. India 48:308–313
Sinha R, Ganesana M, Andreescu S, Stanciu L (2010) AChE biosensor based on zinc oxide sol–gel for the detection of pesticides. Anal. Chim. Acta. 661:195–199
Siwach SB, Gupta A (1995) The profile of acute poisonings in Harayana-Rohtak Study. J Assoc Phys India 43:756–759
Soh N, Tokuda T, Watanabe T et al (2003) A surface plasmon resonance immunosensor for detecting a dioxin precursor using a gold binding polypeptide. Talanta 60:733–745
Song Y, Zhang M, Wang L et al (2011) A novel biosensor based on acetylecholinesterase/prussian blue–chitosan modified electrode for detection of carbaryl pesticides. Electrochim Acta 56:7267–7271
Sotiropoulou S, Fournier D, Chaniotakis NA (2005) Genetically engineered acetylcholinesterase-based biosensor for attomolar detection of dichlorvos. Biosens Bioelectron 20:2347–2352
Su CH, Yu CC, Cheng FY (2016) Rapid visualization of latent fingermarks using gold seed-mediated enhancement. J Nanobiotechnol 14(75):1–11
Sun X, Wang X (2010) Acetylcholinesterase biosensor based on prussian blue-modified electrode for detecting organophosphorous pesticides. Biosens Bioelectron 25:2611–2614
Talan A, Mishra A, Eremin SA et al (2018) Ultrasensitive electrochemical immuno-sensing platform based on gold nanoparticles triggering chlorpyrifos detection in fruits and vegetables. Biosens Bioelectron 105:14–21
Tey JN, Gandhi S, Wijaya IPM et al (2010) Direct detection of heroin metabolites using a competitive immunoassay based on a carbon-nanotube liquid-gated field-effect transistor. Small 6:993–998
Thirumavalavan M, Lee JF (2015) A short review on chitosan membrane for biomolecules immobilization. J Mol Genet Med 9:1–5
Vakurov A, Simpson CE, Daly CL et al (2004) Acetylcholinesterase-based biosensor electrodes for organophosphate pesticide detection: I modification of carbon surface for immobilization of acetylcholinesterase. Biosens Bioelectron 20:1118–1125
Valdés Ramírez G, Cortina M, Ramírez Silva MT, Marty JL (2008a) Acetylcholinesterase-based biosensors for quantification of carbofuran, carbaryl, methylparaoxon, and dichlorvos in 5% acetonitrile. Anal Bioanal Chem 392:699–707

Valdés Ramírez G, Fournier D, Ramírez Silva MT, Marty JL (2008b) Sensitive amperometric biosensor for dichlorovos quantification: Application to detection of residues on apple skin. Talanta 74:741–746

Verma N, Singh M (2005) Biosensors for heavy metals. Biometals 18:121–129

Viswanathan S, Radecka H, Radecki J (2009) Electrochemical biosensor for pesticides based on acetylcholinesterase immobilized on polyaniline deposited on vertically assembled carbon nanotubes wrapped with ssDNA. Biosens Bioelectron 24:2772–2777

Wang P, Wan Y, Ali A et al (2016) Aptamer-wrapped gold nanoparticles for the colorimetric detection of omethoate. Sci China Chem 59:237–242

Wang Y, Zhang S, Du D et al (2011) Self assembly of acetylcholinesterase on a gold nanoparticles–graphene nanosheet hybrid for organophosphate pesticide detection using polyelectrolyte as a linker. J Mater Chem 21:5319–5325

Wei Y, Li Y, Qu Y et al (2009) A novel biosensor based on photoelectro-synergistic catalysis for flow-injection analysis system/amperometric detection of organophosphorous pesticides. Anal Chim Acta 643:13–18

Zamfir LG, Rotariu L, Bala C (2011) A novel, sensitive, reusable and low potential acetylcholinesterase biosensor for chlorpyrifos based on 1-butyl-3-methylimidazolium tetrafluoroborate/multiwalled carbon nanotubes gel. Biosens Bioelectron 26:3692–3695

Zhang SP, Shan LG, Tian ZR et al (2008) Study of enzyme biosensor based on carbon nanotubes modified electrode for detection of pesticides residue. Chinese Chem Lett 19:592–594

Zlotnick J, Smith F (1999) Chromatographic and electrophoretic approaches in ink analysis. J Chromatogr B Biomed Sci Appl 733:265–272

Functionalization, Immobilization and Stabilization of Biomolecules in Microfluidic Devices

Sandeep Kumar Jha, Amrita Soni, Rishi Raj, Smriti Bala, Komal Sharma, Shweta Panwar, and Harpreet Singh

Abstract Since the advent of biosensors for rapid analytical applications and use of micro fabrication technology for their miniaturization, science has come a long way in addressing the requirements for multiplexing, reusability and automation required for multianalyte biosensing. Though, microfluidic device fabrication has become more or less a standard technique nowadays, it is often tricky to immobilize the bio recognition element onto them after device assembly. For example glass-PDMS based microfluidic devices needs to be bonded using oxygen plasma or UV-ozone. Therefore, biomolecules, such as enzymes, antibodies etc. cannot be immobilized prior to device bonding. In this chapter, we shall put special emphasis on such critical immobilization procedures, while we shall also discuss the methods of their stabilization for the sake of enhanced shelf-life and reusability. Our discussions shall be based also on lateral flow immuno-devices and other paper microfluidic devices apart from glass-glass, glass-PDMS, PMMA-PMMA, Si-wafers and other combinations.

Keywords Microfluidic chips · Film deposition techniques · Microfluidic substrates · Immobilization · Nanomaterials

Abbreviations

PDMS	Polydimethylsiloxane
Micro-TAS	Micro-total analysis system
PMMA	Polymethyl methacrylate
BSA	Bovine serum albumin
APTES	3-Aminopropyl triethoxysilane
GOPS	3-Glycidoxypropyltrimethoxysilane
PVD	Physical vapour deposition

S. K. Jha (✉) · A. Soni · R. Raj · S. Bala · K. Sharma · S. Panwar · H. Singh
Centre for Biomedical Engineering, Indian Institute of Technology, New Delhi, Delhi 110016, India
e-mail: sandeepjha@iitd.ac.in

A. Tripathi and J. S. Melo (eds.), *Immobilization Strategies*,
Gels Horizons: From Science to Smart Materials,
https://doi.org/10.1007/978-981-15-7998-1_14

CVD	Chemical vapour deposition
EPD	Electrophoretic deposition
MWCNT	Multiwalled carbon nanotube
SWCNTs	Single wall carbon nanotube
NMs	Nanomaterials
GO	Graphene oxide
LOD	Limit of Detection
CEA	Carcinoembryonic antigen
OTFT	Organic thin film transistor

1 Introduction

Microfluidics is the science of fluid dynamics at the microscale. It allows for precise control and opportunity to manipulate fluid (in the range of microlitres to picolitres) in channels with 10 to 100 s of microns in dimension (Abgrall and Gue 2007; Bein et al. 2018; Huang et al. 2018; Yu et al. 2018). Originating in the early 1990s (Merkel et al. 2000) microfluidics as a discipline has been vastly explored in the fields of science and technology, ranging from semiconductor industry (from its inception) to the field of life-science and engineering (recently), for its tremendous potential to offer highly robust miniaturized systems (Alzahid et al. 2018; Plummer and Griffin 2001). Its ability to perform multiplex assays in high-throughput fashion with improved precision of experiments and lower limits of detection along with decreased sample-reagent consumption, reduced time of experiment and hence the overall cost of application, presents it a promising alternative to conventionally used centralized laboratory techniques. The introduction of soft-lithography techniques in microfabrication have enabled patterning of microstructures on the surface of polymers such as PDMS (polydimethylsiloxane), one of the most widely used polymers for microchip fabrication(Campbell 2001; Friend and Yeo 2010; McDonald and Whitesides 2002), is optically transparent in the range of 240–1100 nm (Piruska et al. 2005), has low autofluorescence (Christen and Andreou 2007), gas permeable (Flueckiger et al. 2011; Goddard and Erickson 2009; Zhu et al. 2004) and hydrophobic (Trantidou et al. 2017) in nature by virtue of which it has facilitated applications such as on-chip PCR (Jha et al. 2012), droplet microfluidics: water-in-oil droplet system for robust compartmentalization of reactions that has enabled a wide range of applications such as diagnostics, single cell entrapment and visualization, droplet PCR etc. (Ng et al. 2002).

Further miniaturization was possible by integrating multiple such modules performing independent and/or interdependent tasks and hence termed as Lab-on-a-Chip (LOC) or Micro-Total Analysis System (micro-TAS) (Reyes et al. 2002). Given the potential of microTAS to perform high-throughput, multiplex lab-on-chip assays at the point-of-care, the discipline has garnered a huge interest from researchers in the biomedical and environmental research community worldwide. However, the execution of chemical and biological assays inside microchannels for biomedical and environmental analysis demands for additional surface modification techniques in order to prime the internal surface of microchannels to incorporate desired functional groups or linker moieties to either impart fouling or non-fouling properties to the channel, as per the requirements for a particular application (Vilkner et al. 2004).

Functionalization and immobilization of the channel surfaces can be achieved by two broad ways: adsorption or covalent modifications. Functional groups such as hydroxyl, amino, sulfhydryl, carboxyl and epoxy groups etc., results in significantly impacting the inherent chemical and physical properties of the surface once covalently introduced. Covalent linkage ensures the amenability of microchannel surface for further immobilization of biomolecules of interest. Depending upon the chemical and physical properties of the substrate chosen for microchip fabrication (e.g.: glass, PDMS, paper etc.), one may choose to either functionalize and/or immobilize the channel surface with biomolecule of interest to accomplish the end-goal. Immobilization can be done either directly by anchoring the biomolecule of interest via adsorption or covalent interaction with the channel walls or it can be done indirectly using a linker molecule which then affixes the biomolecule to the channel walls with the help of spacers. Affinity binding and physical entrapment are some of the other immobilization methods that can also be utilized, but are seldom used. Though these two methods are hugely popular in macro biosensors development, they are not favoured due to high leaching rates normally seen in microfluidic conditions. Sol–gel or compatible techniques are needed in addition to affinity and entrapment methods to effectively retain biomaterials within a microfluidic channel. However, for immobilization of whole cells, especially anchorage dependent ones, surface patterning at micro and nanoscale to promote cellular adhesion inside microfluidic devices is usually carried out by adsorption method (Chen et al. 2015; Hu et al. 2014; Wang et al. 2017).

Nanomaterials on the other hand serve a two-fold purpose of anchoring biomolecule onto the substrate and enhancing the overall assay sensitivity by virtue of their peculiar surface properties (Lee et al. 2018). As particular examples, nanomaterials are immobilized inside microchannels in order to study antibiotic properties of silver nanoparticles and enhanced sensitivity for optical detection using gold nanoparticles in localized surface Plasmon based studies (Yun'an Qing et al. 2018). In addition to nanomaterials, thin film deposition techniques, such as physical vapour deposition and chemical vapour deposition are also popular to achieve efficient on-chip detection and immobilization of analyte of interest. Among metals, deposited thin films of gold is most commonly used to form electrodes for on-chip detection of biomolecules. Furthermore, its ability to interact with thiol group provides a reactive surface to anchor analyte of interest (Bürgi 2015; Xue et al. 2014). For instance, thiolated antibodies and thiolated oligonucleotides can be immobilized on gold coated microchannels for on-chip detection of antigen, protein, cancer cells, bacterial and viral pathogens etc. using one of physical vapour deposition techniques such as thermal evaporation and sputtering, etc.

This chapter hence, provides an overview on functionalization and immobilization techniques of biomolecules and nanomaterials inside microfluidic chips made from diverse range of substrates for on-chip detection and analysis of biomedically and environmentally desired analytes.

2 Strategies of Immobilization

Immobilization of bioanalytes is the principal constituent in patterning the biorecognition part of biosensor. There are different types of techniques for the immobilization of biomolecules and can be classified majorly as covalent, adsorption, entrapment, cross-linking and affinity techniques (Fig. 1). In some cases, the protocols for immobilization of bioanalyte include mixture of different immobilization methods (Cao 2005). For example, the pre-immobilization of a bioanalyte on beads by covalence, adsorption or affinity method prior to entrapment in a porous polymer. The nature of bioanalyte, transducer and the detection mode also necessitate to choose the most suitable and sagacious technique. The finest method of bioanalyte immobilization varies which depends on the application of biosensor whether it requires greatest sensitivity or stability. The basic immobilization techniques thus include:

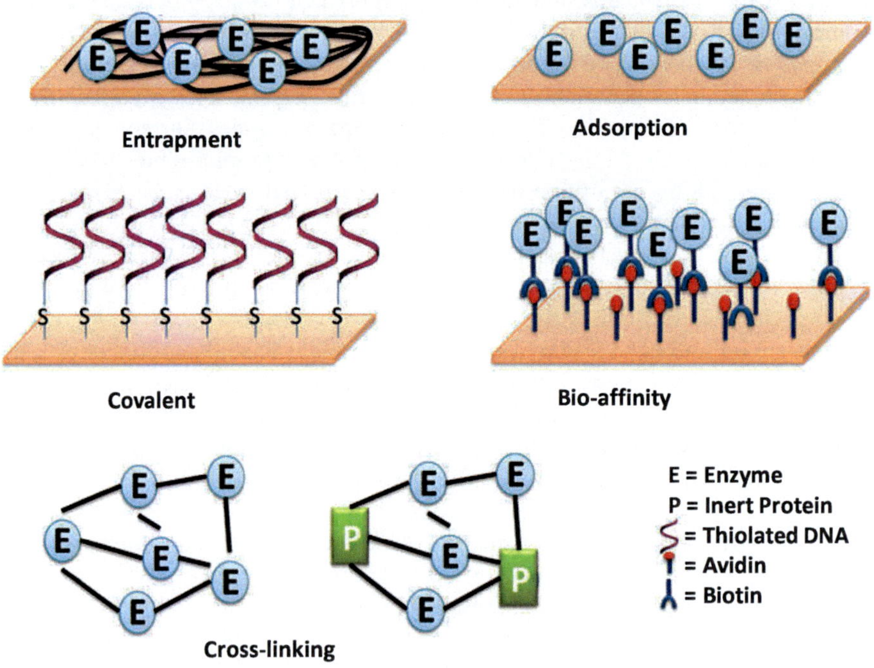

Fig. 1 Schematic of various types of immobilization techniques

2.1 Entrapment

There are different ways to immobilize biomolecules in three-dimensional frameworks, for example the electropolymerized film, a photopolymer, carbon paste, silica gel or polysaccharide on amphiphilic network of PDMS. This type of immobilization can be easily carried out. Consequently, the deposition of mediators, additives and enzymes are performed on the same detection surface. With this process of immobilization, there is no modification of the biorecognition element so that there is slight alteration in the activity of enzyme during immobilization and that too due to diffusional barriers rather than enzyme degradation or misconformation. To increase the storage and operational stability, generally physical entrapment method is used for bioanalytes-based biosensors. Though, there are some limitations such as leaching of the biological elements and conceivable diffusion barriers can reduce the performance of the biosensor.

In a few practical examples of use of immobilization techniques with particular needs, Sakai-Kato et al. explained a sol–gel which was prepared from tetramethoxysilane in HCL and water, further it has been hydrolysed to produce $SiOH_4$-n(OMe)n. After that trypsin was added to form trypsin-trapped sol–gel inside the PMMA chip (Sakai-Kato et al. 2003). The chip has been used for the digestion of amino acids and proteins. Therefore, PMMA microfluidic bioanalyte reactor based on silica sol–gel encapsulation of trypsin was done. This was one of the feasible ways to carry out sol–gel based immobilization inside microchannel without altering microchannel wall's properties. In contrast, Wu et al. reported a sandwiched PDMS microfluidic device with alumina sol-gels and titania in order to improve the antibody shelf-life and the faster digestion time in comparison to the homogenous solution phase reaction. First, they silanized the PDMS surface by treating with oxygen plasma. Afterwards, the tetrabutyl titanate and aluminium isopropoxide were used as solvent to prepare titania and alumina sol. Then, the trypsin was added to the solution. The PDMS microfluidic channel was poured with the solution. Subsequently, the condensation reaction took place with covalent bond formation between the silanol group of PDMS and hydroxyl group of the solution Finally, a stable sol–gel was formed on the PDMS surface with trypsin entrapped within the solution (Wu et al. 2004). With this method, microchannel wall became hydrophilic and thus, can retain sol–gel material within microchannel more efficiently.

2.2 Adsorption

The physical adsorption is the simplest method of immobilization of biomolecules onto the solid supports (Choi 2004). After preparing the solution, the solid support and the biomolecule solution is kept close for a period. The mechanism of adsorption depends on weak bonds i.e. electrostatic force, hydrophobic interactions or Van der Waal's forces. This technique does not affect the activity of bioanalyte and there is no involvement of the functionalization of support. Recently, Sia et al. presented an approach on a microfluidic device which includes the polystyrene lid coupled with PDMS surface designed with microchannels (Sia et al. 2004). In another method, Xiang et al. presented glass covered PDMS chip with "H"-channels. It was used to detect the E.*coli* antigen which was adsorbed on the PDMS surface and detection was performed using sandwich immunoassay technique in which primary and secondary antibodies was utilized (Xiang et al. 2006). However, physical adsorption is more suitable when leaching is intended. For example, if buffer salts and other proteinic ingredients are needed to be placed onto chip, they are simply adsorbed and gets washed out during flow of medium in microfluidic conditions. Besides, in electrochemical measurements, especially with faradic impedance spectroscopy and mediator based amperometry, adsorption does not guarantee full availability of active site as orientation of biomolecule on the substrate cannot be controlled. Adsorption often forms multi layered structure and not suitable for methods requiring formation of self-assembled monolayers (SAM) as in case of sandwich immunoassays.

2.3 *Cross-Linking*

Another technique for immobilization involves crosslinking of biomolecules with matrix, polymer sieve or within themselves. The inert proteins and their cross-linking with one another can be achieved using bifunctional agents e.g. glutaraldehyde, hexamethylenediamine or glyoxal. This method is mainly used in the development of macro-biosensors because of its simplicity and strong chemical bonding between biological elements. However, the main disadvantage of this technique is loss of enzyme activity because of the distortion of enzyme conformation and chemical changes to the active site during the reaction. In an example of using this technique, a conductometric biosensor was fabricated by carrying out enzyme immobilization in a gel, resulted from cross-linking of glutaraldehyde with bovine serum albumin (BSA) for the detection of various elements example, heavy metals (Berezhetskyy et al. 2008), pollutants (Gogol et al 2000), nitrite (Zhang et al. 2009) etc. Seong and Crooks reported a PDMS microfluidic device on which the enzyme packed functionalized PS beads were deposited with a masonry structure (Seong and Crooks 2002).

2.4 *Covalent immobilization*

The chemical immobilization method is mainly used for the covalent binding of bioanalytes to the surface of support materials. For this reason, biocatalysts contain functional groups which helps them to bind with the surface and are not crucial for their catalytic activity. The multifunctional reagents (example, glutaraldehyde, carbodiimide) have been used for the initial activation of the surface by immobilization of the biomolecules to the activated support, followed by its coupling to the support and then expulsion of unbound and excessive biomolecules. The support can either be a natural (i.e. cellulose), inorganic material (e.g. controlled pore glass), synthetic polymer (e.g. nylon) or microchip substrate itself, e.g. Si wafer, glass, polymethyl methacrylate (PMMA) etc. Pre-activated membranes have also been used. The enzymes can either be directly immobilized onto the transducer or onto nanoparticles to be conjugated latter onto the transducer.

As examples, Reginer et al. have used the glass surface and its aminosilanization was done with APTES [(3-aminopropyl) triethoxysilane] as it can bind with the -OH group of the glass surface and another end holds the amine functional group which facilitate the covalent linkage with proteins. (Xiong and Regnier 2001). To improve the specific binding and for increasing the ability of antibody binding, Yu et al. in comparison to original PDMS surface incorporated an intermediate layer of PVA on PDMS to decrease the non-specific adsorption of proteins. Salinization of oxygen-plasma-treated PDMS surface with was performed with APTES in which amino group of APTES was activated by glutaraldehyde to covalently attach with PVA to PDMS surface. After that, glutaraldehyde was used to link covalently with proteins like BSA and antibodies to the intermediate layer (Yu et al. 2007).

2.5 *Affinity Binding*

For the orientation and site-specific immobilization of enzymes, antibody, aptamers, etc., affinity-based interactions are commonly used. This interaction produces highly specific and stronger immobilization which results in minimizing the leaching of bioanalytes. Moreover, these interactions are unaffected by minor changes in pH, heat treatment or chemical treatments, and for the sake of reuse of biorecognition element or to regenerate the active surface, sequential changes in pH, heat, salinity or chemical strength is needed. For example, an aptamer immobilized covalently on glass surface can be regenerated from its ligand/analyte by passage of 1–3 M NaCl in a microfluidic channel. Other examples of bioaffinity immobilization is the interaction between streptavidin–biotin and metal-chelator. Several genetic engineering methods like protein fusion technology, post-transcriptional modification and site-specific mutagenesis have been used to attach bioaffinity tags such as biotin to the specific protein sequence (Andreescu and Marty 2006).

As example, Wang et al. reported a microfluidic based electrochemical thrombin biosensor in which aptamers were immobilized (Wang et al. 2010). In the PMMA microfluidic chip, three interspersed positive layers of Polydiallyl dimethylammonium chloride (PDADMAC) and a negative layer of gold nanoparticles were prepared by electrostatic interactions. To encapsulate thrombin, the thiolized aptamers was coupled with gold nanoparticles. Thrombin, and then subsequently alanine phosphatase functionalized aptamers, were injected in the microchannel. Finally, the amperometric detection was used to detect p-aminophenol which has been produced after the enzymatic conversion of 4-aminophenyl phosphate. In another example, Seong and Crook prepared a system to analyse the mixing efficiency of the enzyme substrate with PDMS microfluidic bead-packed channels. The streptavidin-coated PS beads were immobilized with biotinylated glucose oxidase and peroxidase (Seong and Crooks 2002).

3 Substrate Used in Microfluidic Devices

Microfluidic devices are fabricated on substrates. The selection of an appropriate substrate depends on numerous factors such as physical properties, chemical properties, ease of integration of the device, biocompatibility etc. Substrates aid to benefits like a precise and rapid response, low cost, portability and non-requirement of specialized equipment. This section deals with the list of materials which can be used as substrate for microfluidic devices and previously discussed methods of immobilization are suitable to these.

3.1 Glass

Glass is widely used as a substrate material in microfluidics owing to its transparency, robustness and hydrophilic properties. It is mostly non-reactive to chemicals and biological samples. Glass based microfluidic devices are primarily made through patterning and photolithography, metal deposition for electrical contacts followed by etching of unwanted layer. Borosilicate, Pyrex glass and Soda lime glass are the commonly used glass substrates. These glasses can bond easily with PDMS and borosilicate itself (Pandey et al. 2018). The credibility of a substrate depends upon the fact that how easily it can hold the biomolecules in case of biosensors and reactants in case of chemical sensors to its surface.

Glass provides enough flexibility for adsorption and/or covalently binding. Commonly used methods to functionalize the glass surface are cyanogen bromide treatment to glass surface with -OH groups, diazonium derivatives supports having aromatic amino groups to activate for enzyme immobilization, use of condensing agents like Woodward's reagent K to condense carboxyl or amino groups as supports and amino acid residues to yield peptide linkages, diazo coupling, and alkylation. Some other common reactions used for covalent binding are the Carbodiimide method by utilizing acyl groups via treatment of hydrazides with nitrous acid, derivatizing glass with amino alkyl triethoxy silane. For example, Wang et al. developed a microfluidic chip-based biosensor on a singular T-Microfluidic chip of dimensions (50 μm $\times$ 10 μm $\times$ 15 mm) using glass as substrate for sequence-specific identification of DNA in patients having oral cancer by using their saliva samples (Wang et al. 2013). Another instance of glass based microfluidic device can be found in work presented by Liu et al. which demonstrated trapping of single mouse embryos and its immobilization. The device showed hydrodynamic trapping which is caused due to vacuum generation at the micro-holes on glass substrate (Liu and Sun 2013).

3.2 *Silicon*

Originally microfluidics started on the silicon substrates, later glass and polymers became the popular choices. Silicon is preferred because of its high resistance to organic solvents. Moreover, it is easy to deposit metal on its wafer, it has high thermal conductivity and induced mobility of liquid under external potential is also stable. Silicon has (Si–OH) silanol group on its surface which can be functionalized to adhere and increase its specificity to bind to targeted biomolecules.

Microfluidic devices developed on silicon has many advantages like low sample volume requirement due to miniaturized size, guided flow of fluid and shorter execution time. Silicon bonds well even at lower temperatures making it suitable for multilayer device fabrication. Surface modification and derivatization is required to stabilize the biomolecules on the surface. Most common methods include surface modification using 3-aminopropyltriethoxysilane (APTES) (Fig. 2).

For instance, Yakovleva et al. had developed a microfluidic chip for enzyme immunoassay to detect atrazine. Surface modification of silicon was done with three different polymers viz. (APTES), 3-glycidoxypropyltrimethoxysilane (GOPS), linear or branched polyethyleneimine. The study showed that GL and LPEI coupling helped immobilized antibody to achieve highest stability (Yakovleva et al. 2003). The immunosensor thus produced contained silicon microchannels which were formed using screen printing technology, electronic circuitry and battery as power source. This immunosensor was used to detect *E. coli* in blood samples (RoyChaudhuri et al. 2011; Yakovleva et al. 2002). Chandrasekaran et al. developed a silicon based microfluidic device to carry out immunoassay using fluorescence as biosensing mechanism. The microfluidic device was integrated with an on-chip spectrophotometer. Alexafluor-647 tagged antibody and sheep IgG were the bio-detection elements, an example of antigen–antibody interaction where antigen adsorbs on the channel wall. The experiment showed that silicon is compatible, cost-effective, having good surface affinity to biomolecules, easily disposable and a good source for bulk manufacturing (Chandrasekaran et al. 2007).

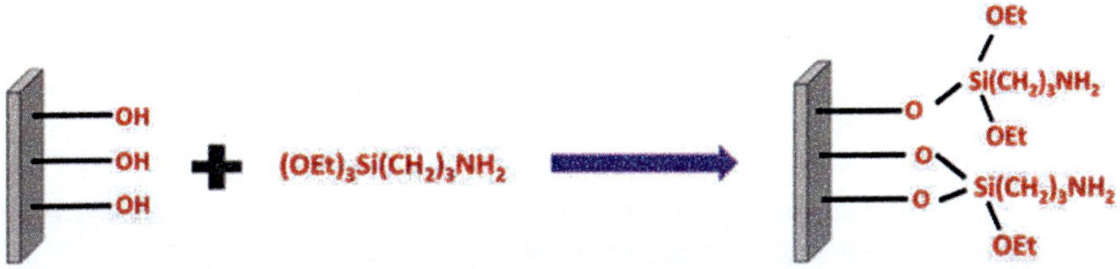

Fig. 2 Surface modification of silicon substrate using APTES (Concept from: Juvaste et al. 1999)

3.3 Polymer

PMMA and PDMS are the popular elastomers and thermoplastics respectively which are used in chip fabrication. These are cheap and can be easily fabricated. These polymers are chemically and electrically inert, transparent and can be used as a substitute to glass. The modulus of elasticity of PDMS is very low which is favourable to construct microvalves and micropumps (Pandey et al. 2018). The intrinsic structure of PDMS is hydrophobic which sometimes causes problem as some hydrophobic molecules from the flowing liquid might get absorbed leading to non-specificity in molecular adsorption and permeation. Chemical modifications are required on the PDMS surface to make it more analyte specific. On the other hand, it may help to remove unwanted hydrophobic molecules in flow, for example: on-chip PCR reaction wherein cell debris may bind to PDMS walls (Jha et al. 2012).

As example, Ko et al. fabricated PMMA and PDMS based microfluidic device for immunosensing, in which the top surface of the chip was made using PDMS by polymer casting technique, while the bottom surface of PMMA was obtained by hot embossing. Inlet ports and air vent were casted for sample loading and fluid velocity control respectively. Electrodes made up of gold were patterned on the bottom surface containing PMMA (Ko et al. 2003). Bai et al. presented a novel method to treat surface using poly ethyleneimine, to enhance binding of antibodies on PMMA surface, the work showed that the activity of antibodies were retained 10 times more compared to untreated surface (Bai et al. 2006).

3.4 Paper

Paper is a low-cost, flexible, light-weight and easily disposable material. It is made up of cellulose which is bio-compatible and recyclable. The flow of fluid is maintained by the internal porous structure of the paper which facilitates capillary based movement of fluid and does not require any external power supply or pumps (Pandey et al. 2018). The use of paper is restricted due to its weak mechanical properties (Ren et al. 2013). The main concern with paper-microfluidics is the creation of hydrophobic barriers to restrict the movement of fluid in the desired channel.

In order to create hydrophobic micro-channels the methods popularly used are photolithography on paper, plotting PDMS on paper, Wax printing, and Inkjet printing. The inks of different types of material from polymers to biomolecules have been utilized for biosensing applications (Fig. 3). For example, Rahbar et al. presented the use of anion-exchange filter paper to immobilize anionic reagents on detection channels which is of opposite charge containing protonated tertiary amine groups. Ion-Exchange paper showed strong adhesion to anionic reagents such as carboxylic and sulfonic as compared to cellulose paper. Chromogenic detection was possible when the fluid was allowed to travel the channel length so as to detect calcium and total acidity in serum and wine samples (Rahbar et al. 2019). Shangguan et al.

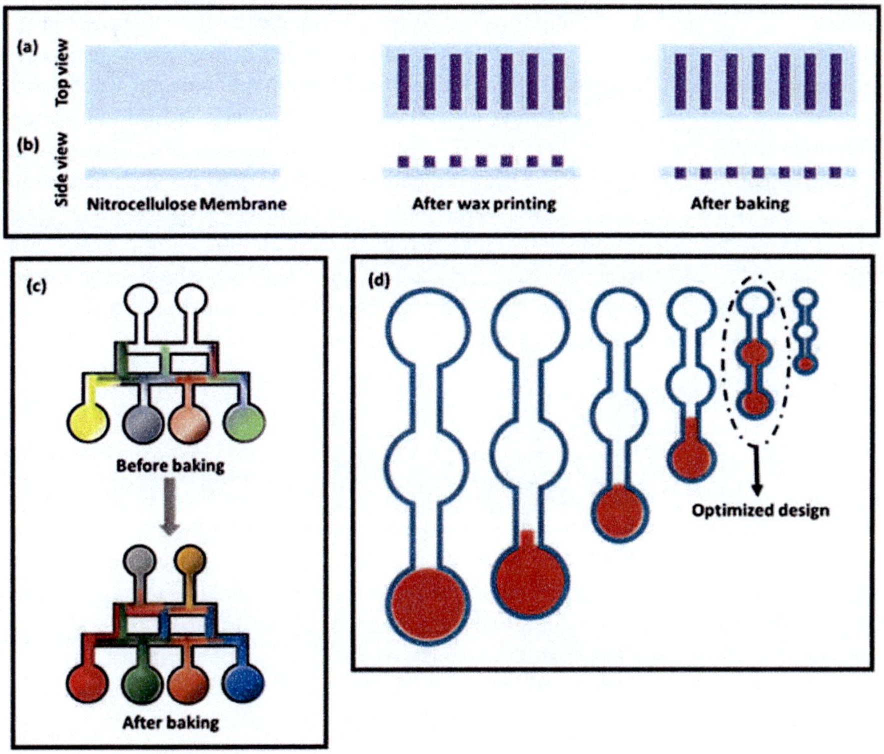

Fig. 3 Schematics of paper-based microfluidics **a–b** absorption of wax onto paper to form microchannel **c** fluid flow testing using dyes **d** optimization for the DNA microarray study (Concept from: Prasad et al. 2019)

fabricated microfluidic channels using PDMS on paper. The patterning of channels was done such that PDMS was adsorbed and percolated into the porous structure of paper forming a hydrophobic barrier. This design was used for biosensing of liver function biomarkers and serum protein (Shangguan et al. 2017).

4 Film Deposition Techniques on Substrates

Thin films are considered integral part in the modern sensor technology, especially for fabrication of microelectrodes or surface for immobilization. Thin film deposited sensors have been applied to various fields such as health care (Raposo and Ribeiro 2019), environmental (Wu et al. 2019), telecommunication (Derickson et al. 2003), energy storage devices (Lv et al. 2017) and so on. Thin film is defined by the thickness of the material, which can vary from nanometers to micrometers on the substrate it is deposited. The quality of the deposited film will depend on the substrate material, type of the material being deposited (amorphous, crystalline or polycrystalline) (Richardson et al. 2017), deposition technique or deposition methods and total deposition time of the film (Richardson et al. 2017). There are various film deposition techniques which include physical vapour deposition techniques (PVD), chemical vapour deposition techniques (CVD), and electrophoretic deposition techniques. In PVD, vapour which consists of the atoms/molecules from the target materials get strike and condense on the substrate which results in the thin film, but the deposition of the thin film in CVD occurs due to chemical reaction of vapour which is generated by the precursor gas, flowing into a heated chamber with the substrate on to which film has to be deposited (Pierson 1999).

4.1 Physical Vapour Deposition (PVD)

In this thin film deposition process, vapour of the atoms or molecules gets condensed at the surface of the substrate under vacuum condition in a plasma environment. By using PVD methods one can deposit thin films from a few nanometers to micrometer. Every PVD method includes some common steps such as formation of vapour of atoms/molecule from the target material in the deposition chamber via evaporation, sputtering, or electron beam bombardment; movement of the ejected ions, atoms, or molecules to the substrate and then condensation or nucleation at the substrate resulting of film formation (Bhat 1999; Mattox 2010). Major PVD techniques include Thermal evaporation, sputtering and electron beam deposition.

Thermal evaporation is the easiest and cost-effective method for the formation of thin films among these. In this method, target materials change their phase from solid to vapour, by heating of the source in ultra-high vacuum condition, and at the surface of substrate, vapour gets condensed which results in formation of thin film. Quality and thickness of the deposited film depends on some parameters like range of vacuum as high vacuum minimizes the chances of gaseous contamination in film formation, rate and duration of deposition (Bhat 1999). While chrome-gold or Ti-Au are preferred metals to be deposited, other combinations involving inert metal on top are also used. While an inert metal is choice of making microelectrodes, it generally has poor adhesion on glass or silicon and therefore requires roughly 10–20 nm thick layer of Cr or Ti metal to be deposited first followed by the active metal layer to 40–00 nm thickness. Standard photolithographic techniques are needed in combination of positive photoresist patterning for finer structure formation or shadow masks can be used for patterning these electrodes of larger geometries (>100 microns). Once this metal layer is formed in the shape of microelectrodes, next step involves surface modification by either adsorption, APTES or silanization based covalent linkage or using self-assembly via thiols (Silva et al. 2010). In particular, gold has excellent affinity for thiol groups and hence preferred choice of immobilization within the microchannel, in which biorecognition elements such as antibody, protein or aptamer is thiolated at one end (Sasso et al. 2014). In one such application, another application,

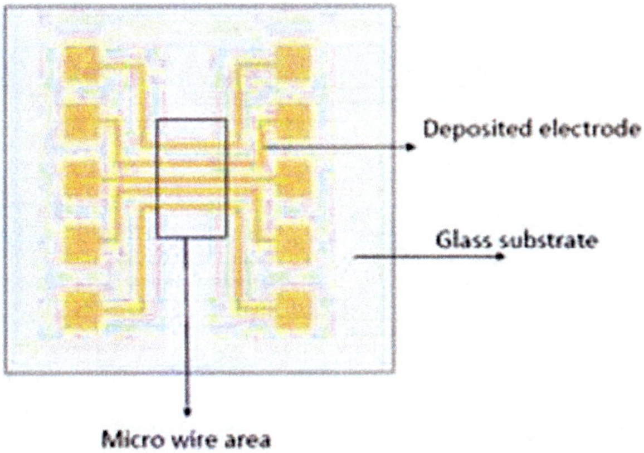

Fig. 4 Microelectrode fabricated from thermal evaporation (Concept from: Ginet et al. 2011)

gold microwires were fabricated over glass substrate using thermal evaporation for studying cell culture in a microfluidic devices (Fig. 4) (Ginet et al. 2011). In another example, microelectrodes fabricated using thermal evaporation in 64-working electrodes array form along with an Ag/AgCl as a reference electrode and Pt as a counter electrode has been successfully used for studying electrophysiological and in vitro electrochemical recording on live neurons (Song et al. 2012). In another study, 12 identical microelectrodes array was fabricated for DNA hybridization and glucose sensing using cyclic voltammetry and impedance spectroscopy techniques (Dimaki et al. 2014). Besides metal thin films, thermal evaporation can be used for deposition of organic molecules at low temperature as well. For example, Kim et al. fabricated an organic thin film transistor (OTFT) by depositing pentacene on gate region of TFT already having two gold microelectrodes. The OTFT was used to study DNA hybridization and subsequently for detection of pathogens (Kim et al. 2011).

Another way of thin film deposition is via sputtering, which is a non-thermal process for thin film deposition. In this, atoms get to escape from the surface of the target material because of atomic collisions which are carried out by the bombardment of ions of high energy to a metal target. The simplest form of the sputtering system consists of a vacuum chamber that contains anode and cathode of metal, to form a glow discharge from the gas residue which are present in the gas chamber (Bhat 1999; Song et al. 2012). Likewise, electron beam evaporation too is non-thermal method for thin film deposition in which materials get heated and evaporated by a beam of electrons that are generated from an electron source (which may be metal) in vacuum. Film deposited using this technique is of very good quality, however, these two are slightly expensive techniques in comparison to thermal evaporation.

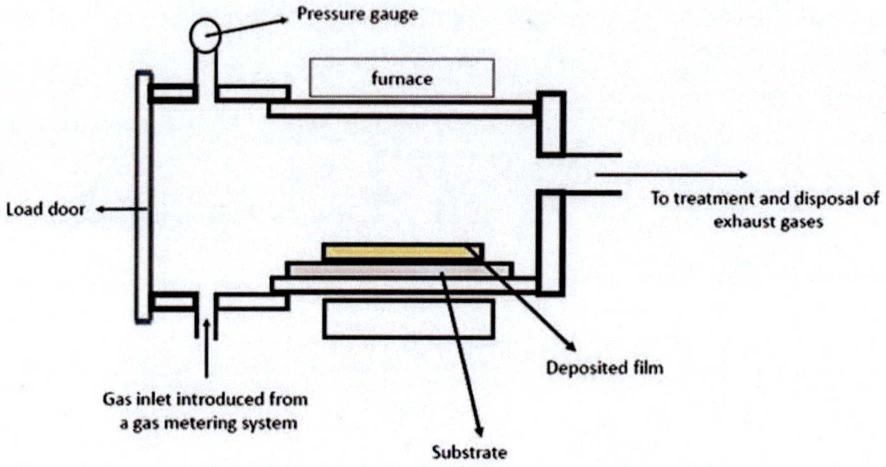

Fig. 5 CVD method for thin film deposition (Concept from: Amin et al. 2016)

4.2 Chemical Vapour Deposition (CVD)

CVD is the process of deposition of gaseous molecules by means of chemical reaction on the substrate (Amin et al. 2016). A typical process flow diagram for CVD method has been shown below in Fig. 5. As a result of chemical reaction, thin and uniform films are formed and there may not be a requirement to generate vacuum. However, carrier gas and other vapour exhaust lines may be necessary. As examples of use of CVD, contamination free graphene film and boron doped microelectrode were deposited (Asai and Einaga 2019; Lisi et al. 2017) for use in electrochemical sensors.

4.3 Electrophoretic Deposition

Electrophoretic deposition technique (EPD) is the process of deposition of thin film by using two-electrode method: one is anode and another is cathode. Its schematic has been shown in the Fig. 6 (Boccaccini et al. 2010). A low power (typically up to 10 V) is needed to achieve constant current or constant voltage based electrodeposition. The advantage of electrodeposition over CVD and PVD is that very thick films can be achieved by this process. For example, the electrochemical biosensors that need up to 200 nm thick gold microelectrode should be fabricated using electrodeposition rather than PVD to save on the cost of metal. Besides the metals, other organic molecules and substances which can be oxidized or reduced can be used with this technique inside a microchannel even when the microchip has been bonded. Methods other than EPD are not suitable for in-situ immobilization or thin film formation.

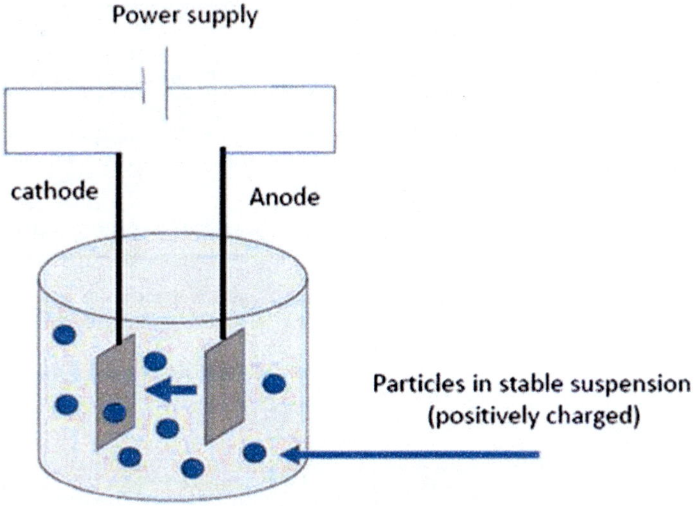

Fig. 6 EPD method for film deposition (Concept from: Boccaccini et al. 2010)

As example, MWCNT has been deposited via EPD onto ITO electrode for the detection of nucleic acid through the immobilization of DNA and subsequent hybridization on the surface (Ghrera et al. 2018). In another study, graphene films were deposited through EPD for electrochemical sensor application. Graphene nanostructured as electrode provides high surface area, good chemical stability and catalytic activity for electrochemical detection, such as TNT explosive (Tang et al. 2010). Reduced graphene oxide using hydrazine reduction or without reduction were deposited using EPD and both materials have been tested for the bacterial toxicity (Chavez-Valdez et al. 2013). A large scale gold nanoparticle array has been fabricated through laser induced lithography and EPD which have been tested for Rhodamine 6G through SERS (Hwang and Yang 2018).

5 Nanomaterials for Surface Modification of Microfluidic Devices

With the advancement in technology, the requirement of system that detect analytes quickly and efficiently is in great need. Therefore, the ideal systems should be portable (Lakshmipriya et al. 2016), stable (Kim et al. 2016), highly sensitive (Fenzl et al. 2016) and capable of multi-analyte detection (Li et al. 2016). The microfluidic biosensors are excellent choice to achieve aforementioned functionalities (Fig. 7) (Arrigan 2015).

To perform multiplex functionality and to improve sensitivity of device by keeping ultra-low concentration of analytes, the surface modification of detection electrodes

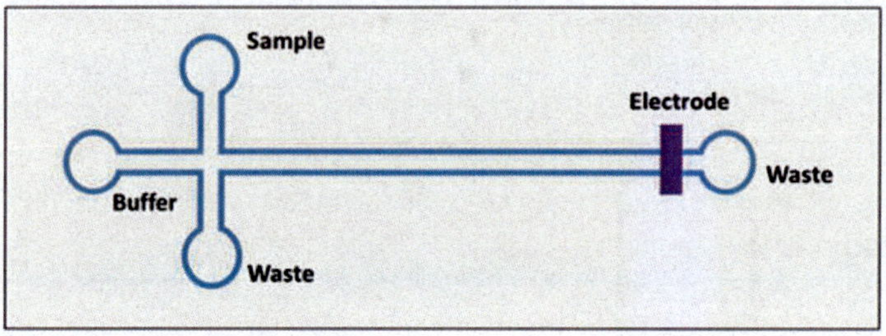

Fig. 7 Schematic representation of microchip design for biosensor applications in which samples are pre-treated, mixed with reagents and detection can be performed in one go either sequentially or simultaneously (Concept from: Arrigan 2015)

is required. Nanomaterials (NMs) are great choice for this purpose because of their small size, highly active surface area and compatibility with biological entities such as proteins, enzymes, DNA and RNA oligonucleotides. These materials contain at least one dimension in the range of less than 100 nm therefore, the physiochemical properties of NMs such as optical, electrical, mechanical properties show different behaviour than their bulk counterparts. These properties can be easily tailored by changing their size, shape and composition of NMs.

Noble metals such as Gold (Au), Silver (Ag), Platinum, Palladium (Pd), Titanium (Ti), Rhodium (Rh) and other metal nanomaterials are the most popular and extensively studied because of their biocompatibility, tuneable size, shape dependent optical and electronic properties and their relatively simple production and modification. Gold nanoparticles are centre of attention nowadays to enhance sensitivity of microfluidic device for detection of disease specific antigens because of its unique surface chemistry, chemical inertness, high electron densities, good electrical and optical properties (Altintas et al. 2014). For example BB Nunna et al. fabricated microchannel for the detection of ovarian cancer by immobilizing gold nanoparticle functionalized CA-125 antibodies on gold electrode surface (Fig. 8). The CA-125 antibody-antigen conjunction was studied by capacitance measurements which are function of physiochemical properties of an individual protein. The modified electrode with gold nanoparticles showed 2.5 times higher capacitive response than unmodified electrodes. Because of high surface area of gold nanoparticles, more number of antibodies could be immobilized on their surface to achieve higher antigen–antibody conjugation. The gold nanoparticles also provided orientational freedom to antibodies which enhanced the sensitivity of the microfluidic device (Nunna et al. 2019).

Carbon-based materials are also getting more interest in electrode surface modification in microfluidic biosensors. Carbon-based NMs such as graphene oxide (GO), carbon nanotubes (CNTs), carbon nanospheres (CNSs) and graphene nanoribbons (GNRs) are known for their high surface area, ultra-lightweight, excellent electrical and thermal conductivity, high mechanical strength and highly ordered structure which made them suitable candidate for biosensor applications. These carbon materials are either directly immobilized on electrode surface or functionalized with

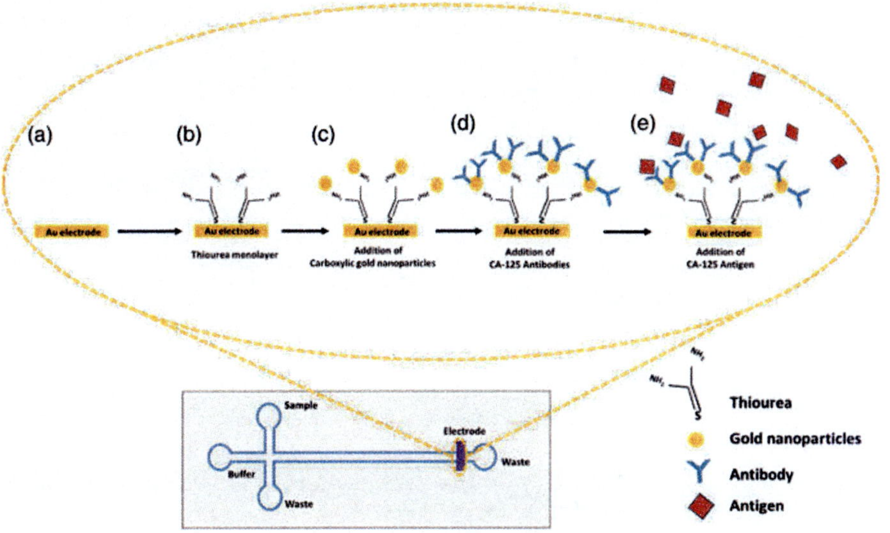

Fig. 8 Schematic showing various stages of bioelectrode fabrication: **a** Bare Au electrode **b** Thiourea monolayer on bare Au electrode **c** immobilization of gold nano particles on monolayer **d** imobilization of antibody on the electrode **e** antibody-antigen conjugation on the electrode (Concept from: Nunna et al. 2019)

biomolecules to enhance the selectivity of device (Reverté et al. 2016). For example Varadan et al. performed detection of sugar by fabricating field effect transistor (FET) consisting of single wall carbon nanotube (SWCNTs) channel in microfluidic device (Fig. 9a). The surface of SWCNTs was functionalized with boronic acid to achieve a Limit of Detection (LOD) of 5 mM (Varadan et al. 2008). In another study, Okuno et al. fabricated label-free immunosensor with modified platinum microelectrodes using SWCNTs for the detection of total prostate specific antigen (T-PSA) (Fig. 9b). Afterwards, surface of SWCNTs was modified with monoclonal antibodies (T-PSA-mAb) for antibody-antigen complexation with T-PSA. The SWCNTs provided better electron transfer and a LOD of 0.25 ng mL^{-1} could be achieved as compared to non-modified electrode (Okuno et al. 2007).

Apart from improving the detection sensitivity of electrochemical systems, the NMs can also use as a fluorescent biomarkers. The quantum dot materials completely fit into this category because of their stable, size dependent fluorescence properties. For example, Ang et al. (Liu et al. 2005) fabricated microfluidic channel for immunofluorescent detection of Iridovirus using quantum dots as fluorescent biomarker. The quantum dots were functionalized with polyclonal and monoclonal antibodies and results have compared with conventional detection methods. The detection sensitivity of Iridovirus could be improved from 360 down to 22 ng ml^{-1}. The detection time also improved form >3.25 to $\leq$0.5 h.

Next class of NMs consists of metal oxide magnetic nanoparticles. These nanoparticles are area of interest because they are biocompatible, less toxic, and exhibit

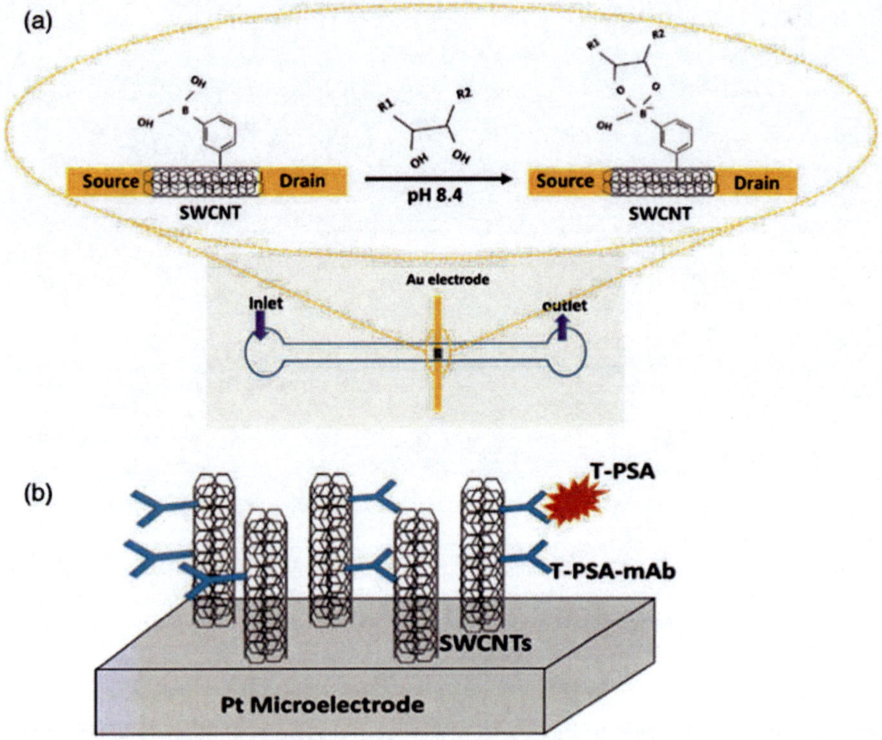

Fig. 9 Schematic representation of **a** Binding of glucose to boronic acid-functionalized carbon nanotube FET in microfluidic channel (Varadan et al. 2008) and **b** Label-free immunosensor: modified platinum microelectrodes with T-PSA-mAb functionalized SWCNTs (Concept from: Okuno et al. 2007)

different magnetic behaviour than their bulk counterpart because of phenomena known as superparamagnetic behaviour. Superparamagnetism is the property of the magnetic nanoparticles which consists of tiny domain magnets aligned randomly in the absence of external magnetic field. On applying external magnetic field these domains flip into same direction and start to exhibit magnetism in average zero time (Fig. 10) (Zhao et al. 2014).

Based on above principle Chen et al. (2011) fabricated microfluidic chip for sensitive and selective detection of carcinoembryonic antigen (CEA) by use of magnetic sandwich-type immunoassay (Fig. 11). The silica coated magnetic nanoparticles (IDA-SCMNPs) modified using iminodiacetic acid were synthesized directly in microfluidic channel by self-assembly to work as solid phase channel. The Monoclonal anti-CEA antibody (Ab2) modified Pb nanoparticles were then used for IDA-SCMNP, CEA and Ab2-Pb complexation and provided LOD of 0.058 $\mu g\ L^{-1}$.

The magnetic separation is also used for preconcentrating the sample. For example, low LOD could be achieved for detection of *Salmonella typhi* pathogen

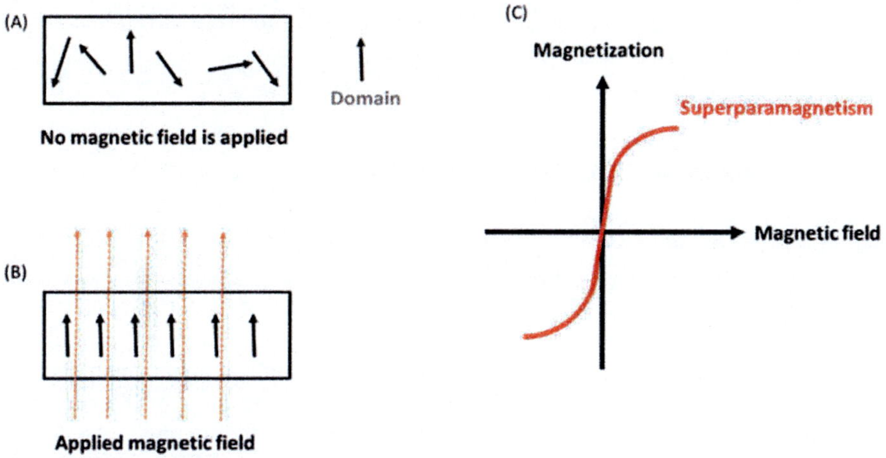

Fig. 10 Schematic representation of superparamagnetism: **a** in the absence of external magnetic field, the domains are aligned in random directions, **b** in the presence of external magnetic field, all the domains aligned in direction of magnetic field, **c** showing the magnetization curve of superparamegnetic materials (Concept from: Zhao et al. 2014)

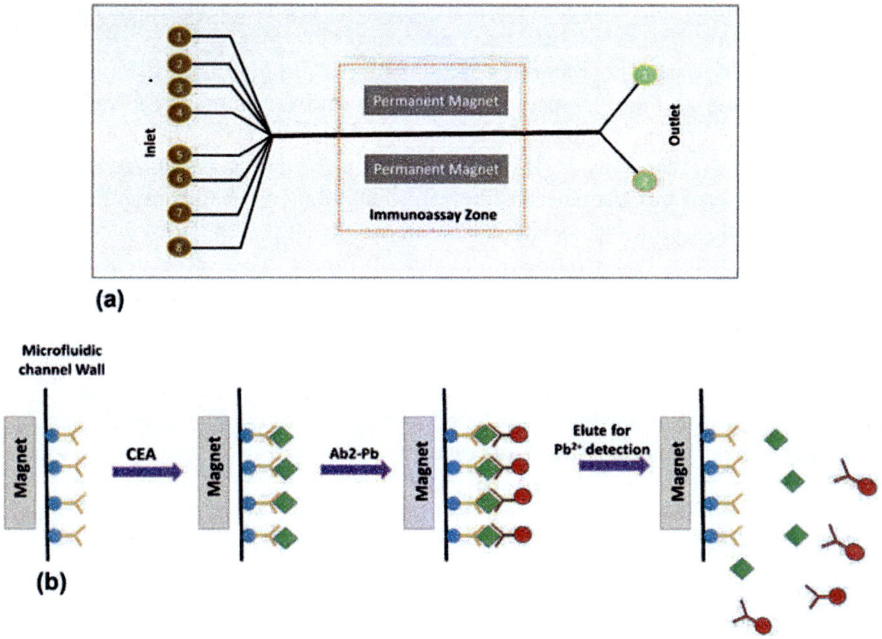

Fig. 11 **a** Schematic representation of microfluidic chip for magnetic immunoassay **b** step wise sandwich-type magnetic immunoassay for the determination of CEA (Concept from: Chen et al. 2011)

in milk and water using loop mediated thermal amplification (LAMP) based assay (Kaur et al. 2018a, b).

6 Conclusions and Future Perspectives

Immobilization of biomolecules in microfluidic channels is very critical step as it has to ensure high density packing of molecules in a very small area, while still retaining their activity. Careful and rational selection of suitable attachment techniques and platforms are vital to retain the activity of biomolecules as the surface inside microchannels are utilized for transportation of sample, reagent, analytes and substrate. Fundamentally, the main aim of using appropriate immobilization techniques is to improve the sensitivity, specificity, repeatability, throughput and to minimize the non-specific adsorption of biomolecules on biosensing platform. However, in some cases, the use of physical adsorption, bioaffinity interaction and covalent bonding technique alone is not able to achieve this goal. Therefore, combinations of these techniques are often used to get desirable results.

Various substrates are used for the fabrication of microfluidic devices and therefore, choice of immobilization technique depends on the functional group or surface roughness present on these substrates. For example, formation of Si–OH groups on the surface of glass and silicon is usual practice to immobilize the bioanalytes. Further, to minimize the non-specific interactions and to maintain the activity of bioanalytes, the surface coatings of various nanomaterials using different deposition techniques are carried out. In this regard, integration of nanomaterials with microfluidic system has become an interesting alternative for the immobilization and stabilization of biomolecules. NMs based lab-on-chip technology have added advantages in relation to its sensitivity and miniaturization irrespective of design and application of the device. Moreover, NMs can be used as microchannel modifiers, filters, actuators and for biomolecules capturing.

Microfluidic based systems are thus powerful platforms in measuring accurate intramolecular response in variety of experimental purposes. These systems are among the devices that has been developed in response to the alarming challenges in the field of healthcare, environment, security, food and agriculture or even space which sometimes necessitate detection of analytes at genomic to cellular level in even point-of-care form factor, or to be deployed for field applications for environmental monitoring. Such a diverse application in non-standard conditions require multiplexing, and suitably stabilized biorecognition elements. From commercial point of view, the real challenges is to choose right combination of substrate, immobilization and stabilization techniques and optimize fabrication of low-cost but sensitive detection systems. To address these challenges, miniaturization of the detection systems may be considered as reasonable solution since they are easily deployed and utilized for analysis with or without a particular laboratory setting.

References

Abgrall P, Gue A (2007) Lab-on-chip technologies: making a microfluidic network and coupling it into a complete microsystem—a review. J Micromech Microeng 17(5):R15

Altintas Z, Kallempudi SS, Gurbuz Y (2014) Gold nanoparticle modified capacitive sensor platform for multiple marker detection. Talanta 118:270–276

Alzahid YA, Mostaghimi P, Gerami A, Singh A, Privat K, Amirian T, Armstrong RT (2018) Functionalisation of polydimethylsiloxane (PDMS)-microfluidic devices coated with rock minerals. Sci Rep 8(1):15518

Amin SK, Abdallah H, Roushdy M, El-Sherbiny S (2016) An overview of production and development of ceramic membranes. Int J Appl Eng Res 11(12):7708–7721

Andreescu S, Marty J-L (2006) Twenty years research in cholinesterase biosensors: from basic research to practical applications. Biomol Eng 23(1):1–15

Arrigan DW (2015) Electrochemical strategies in detection science. Royal Society of Chemistry

Asai K, Einaga Y (2019) Fabrication of an all-diamond microelectrode using a chromium mask. Chem Commun 55(7):897–900

Bai Y, Koh CG, Boreman M, Juang Y-J, Tang I-C, Lee LJ, Yang S-T (2006) Surface modification for enhancing antibody binding on polymer-based microfluidic device for enzyme-linked immunosorbent assay. Langmuir 22(22):9458–9467

Bein A, Shin W, Jalili-Firoozinezhad S, Park MH, Sontheimer-Phelps A, Tovaglieri A, Chalkiadaki A, Kim HJ, Ingber DE (2018) Microfluidic organ-on-a-chip models of human intestine. Cell Molecul Gastroenterol Hepatol 5(4):659–668

Berezhetskyy A, Sosovska O, Durrieu C, Chovelon J-M, Dzyadevych S, Tran-Minh C (2008) Alkaline phosphatase conductometric biosensor for heavy-metal ions determination. Irbm 29(2–3):136–140

Bhat DG (1999) In Donal M (ed) A review of: handbook of physical vapor deposition (PVD) processing flim formation, adhesion, surface preparation and contamination control. Taylor & Francis, Matiox

Boccaccini A, Keim S, Ma R, Li Y, Zhitomirsky I (2010) Electrophoretic deposition of biomaterials. J Royal Soc Interf 7(suppl_5):S581–S613

Bürgi T (2015) Properties of the gold–sulphur interface: from self-assembled monolayers to clusters. Nanoscale 7(38):15553–15567

Campbell SA (2001) The science and engineering of microelectronic fabrication. The Oxford Series in Electrical and Computer Engineering

Cao L (2005) Immobilised enzymes: science or art? Curr Opin Chem Biol 9(2):217–226

Chandrasekaran A, Acharya A, You JL, Soo KY, Packirisamy M, Stiharu I, Darveau A (2007) Hybrid integrated silicon microfluidic platform for fluorescence based biodetection. Sensors 7(9):1901–1915

Chavez-Valdez A, Shaffer MS, Boccaccini AR (2013) Applications of graphene electrophoretic deposition. A review. J Phys Chem B 117(6):1502–1515

Chen B, Hu B, Jiang P, He M, Peng H, Zhang X (2011) Nanoparticle labelling-based magnetic immunoassay on chip combined with electrothermal vaporization-inductively coupled plasma mass spectrometry for the determination of carcinoembryonic antigen in human serum. Analyst 136(19):3934–3942

Chen S, Lu X, Hu Y, Lu Q (2015) Biomimetic honeycomb-patterned surface as the tunable cell adhesion scaffold. Biomater Sci 3(1):85–93

Choi MM (2004) Progress in enzyme-based biosensors using optical transducers. Microchim Acta 148(3–4):107–132

Christen JB, Andreou AG (2007) Design, fabrication, and testing of a hybrid CMOS/PDMS microsystem for cell culture and incubation. IEEE Trans Biomed Circuits Syst 1(1):3–18

Derickson DJ, Fortenberry R, Scobey M, Sommer R, Stokes L (2003) Advancements in thin film filters for telecommunications applications. In International society for optics and photonics, pp 595–607

Dimaki M, Vergani M, Heiskanen A, Kwasny D, Sasso L, Carminati M, Gerrard J, Emneus J, Svendsen W (2014) A compact microelectrode array chip with multiple measuring sites for electrochemical applications. Sensors 14(6):9505–9521

Fenzl C, Hirsch T, Baeumner AJ (2016) Nanomaterials as versatile tools for signal amplification in (bio) analytical applications. TrAC, Trends Anal Chem 79:306–316

Flueckiger J, Bazargan V, Stoeber B, Cheung KC (2011) Characterization of postfabricated parylene C coatings inside PDMS microdevices. Sensors Actuators B Chem 160(1):864–874

Friend J, Yeo L (2010) Fabrication of microfluidic devices using polydimethylsiloxane. Biomicrofluidics 4(2):026502

Ghrera AS, Pandey CM, Malhotra BD (2018) Multiwalled carbon nanotube modified microfluidic-based biosensor chip for nucleic acid detection. Sensors Actuators B Chem 266:329–336

Ginet P, Montagne K, Akiyama S, Rajabpour A, Taniguchi A, Fujii T, Sakai Y, Kim B, Fourmy D, Volz S (2011) Towards single cell heat shock response by accurate control on thermal confinement with an on-chip microwire electrode. Lab Chip 11(8):1513–1520

Goddard JM, Erickson D (2009) Bioconjugation techniques for microfluidic biosensors. Anal Bioanal Chem 394(2):469

Gogol E, Evtugyn G, Marty J-L, Budnikov H, Winter V (2000) Amperometric biosensors based on nafion coated screen-printed electrodes for the determination of cholinesterase inhibitors. Talanta 53(2):379–389

Hu J, Hardy C, Chen C-M, Yang S, Voloshin AS, Liu Y (2014) Enhanced cell adhesion and alignment on micro-wavy patterned surfaces. PLoS ONE 9(8):e104502

Huang Y-X, He C-L, Wang P, Pan Y-T, Tuo W-W, Yao C-C (2018) Versatile on-stage microfluidic system for long term cell culture, micromanipulation and time lapse assays. Biosens Bioelectron 101:66–74

Hwang JS, Yang M (2018) Sensitive and reproducible gold SERS sensor based on interference lithography and electrophoretic deposition. Sensors 18(11):4076

Jha SK, Chand R, Han D, Jang Y-C, Ra G-S, Kim JS, Nahm B-H, Kim Y-S (2012) An integrated PCR microfluidic chip incorporating aseptic electrochemical cell lysis and capillary electrophoresis amperometric DNA detection for rapid and quantitative genetic analysis. Lab Chip 12(21):4455–4464

Juvaste H, Iiskola EI, Pakkanen TT (1999) Preparation of new modified catalyst carriers. J Mol Catal A Chem 150(1–2):1–9

Kaur A, Das R, Nigam MR, Elangovan R, Pandya D, Jha S, Kalyanasundaram D (2018) Rapid Detection Device for Salmonella typhi in Milk, Juice, Water and Calf Serum. Indian J Microbiol 58(3):381–392

Kaur A, Kapil A, Elangovan R, Jha S, Kalyanasundaram D (2018) Highly-sensitive detection of Salmonella typhi in clinical blood samples by magnetic nanoparticle-based enrichment and in-situ measurement of isothermal amplification of nucleic acids. PLoS ONE 13(3):e0194817

Kim J-M, Jha SK, Chand R, D.-H. Lee, Kim Y-S (2011) Flexible pentacene thin film transistors as DNA hybridization sensor. Kaohsiung, Taiwan, pp 421–424

Kim YS, Raston NHA, Gu MB (2016) Aptamer-based nanobiosensors. Biosens Bioelectron 76:2–19

Ko JS, Yoon HC, Yang H, Pyo H-B, Chung KH, Kim SJ, Kim YT (2003) A polymer-based microfluidic device for immunosensing biochips. Lab Chip 3(2):106–113

Lakshmipriya T, Hashim U, Gopinath SC, Azizah N (2016) Microfluidic-based biosensor: signal enhancement by gold nanoparticle. Microsyst Technol 22(10):2389–2395

Lee J-H, Cho H-Y, Choi HK, Lee J-Y, Choi J-W (2018) Application of gold nanoparticle to plasmonic biosensors. Int J Mol Sci 19(7):2021.

Li J, Zhu Z, Zhu B, Ma Y, Lin B, Liu R, Song Y, Lin H, Tu S, Yang C (2016) Surface-enhanced Raman scattering active plasmonic nanoparticles with ultrasmall interior nanogap for multiplex quantitative detection and cancer cell imaging. Anal Chem 88(15):7828–7836

Lisi N, Dikonimos T, Buonocore F, Pittori M, Mazzaro R, Rizzoli R, Marras S, Capasso A (2017) Contamination-free graphene by chemical vapor deposition in quartz furnaces. Sci Rep 7(1):9927

Liu W-T, Zhu L, Qin Q-W, Zhang Q, Feng H, Ang S (2005) Microfluidic device as a new platform for immunofluorescent detection of viruses. Lab Chip 5(11):1327–1330

Liu X, Sun Y (2013) Microfluidic devices for single-cell trapping and automated micro-robotic injection. In: Microfluidic devices for biomedical applications. Elsevier, 351–365e

Lv Q, Chi K, Zhang Y, Xiao F, Xiao J, Wang S, Loh KP (2017) Ultrafast charge/discharge solid-state thin-film supercapacitors via regulating the microstructure of transition-metal-oxide. J Mater Chem A 5(6):2759–2767

Mattox DM (2010) Handbook of physical vapor deposition (PVD) processing. William Andrew

McDonald JC, Whitesides GM (2002) Poly (dimethylsiloxane) as a material for fabricating microfluidic devices. Acc Chem Res 35(7):491–499

Merkel T, Bondar V, Nagai K, Freeman BD, Pinnau I (2000) Gas sorption, diffusion, and permeation in poly (dimethylsiloxane). J Polym Sci Part B Polym Phys 38(3):415–434

Ng JM, Gitlin I, Stroock AD, Whitesides GM (2002) Components for integrated poly (dimethylsiloxane) microfluidic systems. Electrophoresis 23(20):3461–3473

Nunna BB, Mandal D, Lee JU, Singh H, Zhuang S, Misra D, Bhuyian MNU, Lee ES (2019) Detection of cancer antigens (CA-125) using gold nano particles on interdigitated electrode-based microfluidic biosensor. Nano Convergence 6(1):3

Okuno J, Maehashi K, Kerman K, Takamura Y, Matsumoto K, Tamiya E (2007) Label-free immunosensor for prostate-specific antigen based on single-walled carbon nanotube array-modified microelectrodes. Biosens Bioelectron 22(9–10):2377–2381

Pandey CM, Augustine S, Kumar S, Kumar S, Nara S, Srivastava S, Malhotra BD (2018) Microfluidics based point-of-care diagnostics. Biotechnol J 13(1):1700047

Pierson HO (1999) Handbook of chemical vapor deposition: principles, technology and applications. William Andrew

Piruska A, Nikcevic I, Lee SH, Ahn C, Heineman WR, Limbach PA, Seliskar CJ (2005) The autofluorescence of plastic materials and chips measured under laser irradiation. Lab Chip 5(12):1348–1354

Plummer JD, Griffin PB (2001) Material and process limits in silicon VLSI technology. Proc IEEE 89(3):240–258

Prasad A, Hasan SMA, Grouchy S, Gartia MR (2019) DNA microarray analysis using a smartphone to detect the BRCA-1 gene. Analyst 144(1):197–205

Rahbar M, Wheeler AR, Paull B, Macka M (2019) Ion-exchange based immobilization of chromogenic reagents on microfluidic paper analytical devices. Anal Chem 91(14):8756–8761

Raposo M, Ribeiro PA (2019) Thin film-based sensors. Multidisciplinary Digital Publishing Institute.

Ren K, Zhou J, Wu H (2013) Materials for microfluidic chip fabrication. Acc Chem Res 46(11):2396–2406

Reverté L, Prieto-Simón B, Campàs M (2016) New advances in electrochemical biosensors for the detection of toxins: Nanomaterials, magnetic beads and microfluidics systems. A Rev Anal Chimica Acta 908:8–21

Reyes DR, Iossifidis D, Auroux P-A, Manz A (2002) Micro total analysis systems. 1. Introduction, theory, and technology. Anal Chem 74(12):2623–2636

Richardson C, Devine-Stoneman J, Divitini G, Vickers ME, Chang C-Z, Amado M, Moodera J, Robinson J (2017) Structural properties of thin-film ferromagnetic topological insulators. Sci Rep 7(1):12061

RoyChaudhuri C, Das RD, Dey S, Das S (2011) Functionalised silicon microchannel immunosensor with portable electronic readout for bacteria detection in blood. IEEE. pp 323–326

Sakai-Kato K, Kato M, Toyo'oka T (2003) Creation of an on-chip enzyme reactor by encapsulating trypsin in sol—gel on a plastic microchip. Anal Chem 75(3):388–393

Sasso L, Suei S, Domigan L, Healy J, Nock V, Williams M, Gerrard J (2014) Versatile multi-functionalization of protein nanofibrils for biosensor applications. Nanoscale 6(3):1629–1634

Seong GH, Crooks RM (2002) Efficient mixing and reactions within microfluidic channels using microbead-supported catalysts. J Am Chem Soc 124(45):13360–13361

Shangguan J-W, Liu Y, Pan J-B, Xu B-Y, Xu J-J, Chen H-Y (2017) Microfluidic PDMS on paper (POP) devices. Lab Chip 17(1):120–127

Sia SK, Linder V, Parviz BA, Siegel A, Whitesides GM (2004) An integrated approach to a portable and low-cost immunoassay for resource-poor settings. Angew Chem Int Ed 43(4):498–502

Silva MM, Cavalcanti IT, Barroso MF, Sales MGF, Dutra RF (2010) Gold electrode modified by self-assembled monolayers of thiols to determine DNA sequences hybridization. J Chem Sci 122(6):911–917

Song Y, Lin N, Liu C, Jiang H, Xing G, Cai X (2012) A novel dual mode microelectrode array for neuroelectrical and neurochemical recording in vitro. Biosens Bioelectron 38(1):416–420

Tang L, Feng H, Cheng J, Li J (2010) Uniform and rich-wrinkled electrophoretic deposited graphene film: a robust electrochemical platform for TNT sensing. Chem Commun 46(32):5882–5884

Trantidou T, Elani Y, Parsons E, Ces O (2017) Hydrophilic surface modification of PDMS for droplet microfluidics using a simple, quick, and robust method via PVA deposition. Microsyst Nanoeng 3:16091

Varadan VK, Chen L, Xie J (2008) Nanomedicine: design and applications of magnetic nanomaterials, nanosensors and nanosystems. Wiley

Vilkner T, Janasek D, Manz A (2004) Micro total analysis systems. Rec Develop Anal Chem 76(12):3373–3386

Wang H, Liu Y, Liu C, Huang J, Yang P, Liu B (2010) Microfluidic chip-based aptasensor for amplified electrochemical detection of human thrombin. Electrochem Commun 12(2):258–261

Wang Y, Yu Z, Mei D, Xue D (2017) Fabrication of micro-wavy patterned surfaces for enhanced cell culturing. Procedia CIRP 65:279–283

Wang Z, Chen J, Fan Y, Wang W, Fu F (2013) A novel micro-fluidic biosensor for the rapid and sequence-specific detection of DNA with electrophoretic driving mode and laser-induced fluorescence detector. Microfluid Nanofluid 14(1–2):145–152

Wu H, Tian Y, Liu B, Lu H, Wang X, Zhai J, Jin H, Yang P, Xu Y, Wang H (2004) Titania and alumina sol−gel-derived microfluidics enzymatic-reactors for peptide mapping: design, characterization, and performance. J Proteome Res 3(6):1201–1209

Wu M-R, Li W-Z, Tung C-Y, Huang C-Y, Chiang Y-H, Liu P-L, Horng R-H (2019) NO gas sensor based on ZnGa 2 O 4 epilayer grown by metalorganic chemical vapor deposition. Sci Rep 9(1):7459

Xiang Q, Hu G, Gao Y, Li D (2006) Miniaturized immunoassay microfluidic system with electrokinetic control. Biosen Bioelectron 21(10)

Xiong L, Regnier FE (2001) Channel-specific coatings on microfabricated chips. J Chromatogr A 924(1–2):165–176

Xue Y, Li X, Li H, Zhang W (2014) Quantifying thiol–gold interactions towards the efficient strength control. Nat Commun 5(1):1–9

Yakovleva J, Davidsson R, Bengtsson M, Laurell T, Emnéus J (2003) Microfluidic enzyme immunosensors with immobilised protein A and G using chemiluminescence detection. Biosens Bioelectron 19(1):21–34

Yakovleva J, Davidsson R, Lobanova A, Bengtsson M, Eremin S, Laurell T, Emnéus J (2002) Microfluidic enzyme immunoassay using silicon microchip with immobilized antibodies and chemiluminescence detection. Anal Chem 74(13):2994–3004

Yu F, Nivasini D, Kumar OS, Choudhury D, Foo LC, Ng SH (2018) Microfluidic platforms for modeling biological barriers in the circulatory system. Drug Discovery Today 23(4):815–829

Yu L, Li CM, Zhou Q, Luong JH (2007) Poly (vinyl alcohol) functionalized poly (dimethylsiloxane) solid surface for immunoassay. Bioconjug Chem 18(2):281–284

Yun'an Qing LC, Li R, Liu G, Zhang Y, Tang X, Wang J, Liu H, Qin Y (2018) Potential antibacterial mechanism of silver nanoparticles and the optimization of orthopedic implants by advanced modification technologies. Int J Nanomed 13:3311

Zhang Z, Xia S, Leonard D, Jaffrezic-Renault N, Zhang J, Bessueille F, Goepfert Y, Wang X, Chen L, Zhu Z (2009) A novel nitrite biosensor based on conductometric electrode modified with cytochrome c nitrite reductase composite membrane. Biosens Bioelectron 24(6):1574–1579

Zhao X, Zhao H, Chen Z, Lan M (2014) Ultrasmall superparamagnetic iron oxide nanoparticles for magnetic resonance imaging contrast agent. J Nanosci Nanotechnol 14(1):210–220
Zhu X, Chu LY, Chueh B-h, Shen M, Hazarika B, Phadke N, Takayama S (2004) Arrays of horizontally-oriented mini-reservoirs generate steady microfluidic flows for continuous perfusion cell culture and gradient generation. Analyst 129(11):1026–1031

Biofilms: Naturally Immobilized Microbial Cell Factories

Sudhir K. Shukla, T. Manobala, and T. Subba Rao

Abstract Biocatalyst, including bacteria, fungi and their enzymes are an effective and environmentally preferred tools for industrial biotechnological applications, and for emerging wastewater treatment technologies. Due to low waste generation and energy requirements, biocatalyst-mediated processes have an edge over many established chemical processes and are also more economical. In most of the cases biocatalysts, especially whole cells need to be immobilized to make them reusable and cut down the downstream costs. Biofilms, which consists of surface-attached microbial cells, have natural immobilization due to the secretion of biopolymers by its own cells. Biofilms are identified as potential industrial workhorses for their sustainable production of chemicals, due to their high resistance to reactants and toxic chemicals, easy separation from the bulk liquid and long-sustaining enzymatic activity, which all facilitate continuous processing. Despite these favorable characteristics, utilizing biofilms as live biocatalysts has not yet broadly implemented concept in bioprocess development due to few bottlenecks. Of late, new prospective opportunities using these biofilm-based biocatalysts are showing an ascending application and seem to be feasible with the better understanding of mechanisms of biofilm biology. This chapter summarizes the recent developments in biofilm research with special emphasis on the advantages and disadvantages of biofilm-mediated processes for viable applications.

Keywords Biofilms · Biocatalysts · Biosorption · Bioremediation · Immobilization · Heavy metals

S. K. Shukla · T. S. Rao (✉)
Biofouling & Thermal Ecology Section, Water & Steam Chemistry Division, BARC Facilities, Kalpakkam 603102, India
e-mail: subbarao@igcar.gov.in

T. Manobala
Department of Applied Science and Technology, Anna University, Chennai, Tamil Nadu 600025, India

S. K. Shukla · T. S. Rao
Homi Bhabha National Institute, Mumbai 400094, India

A. Tripathi and J. S. Melo (eds.), *Immobilization Strategies*,
Gels Horizons: From Science to Smart Materials,
https://doi.org/10.1007/978-981-15-7998-1_15

Abbreviations

AI	Autoinducers
DNA	Deoxyribose nucleic acid
eDNA	Extracellular DNA
EPS	Extracellular polymeric substances (EPS)
MABR	Membrane aerated biofilm reactor
PAH	Polycyclic aromatic hydrocarbons
PCS	Plastic composite support
PPS	Polypropylene support
QS	Quorum sensing
RNA	Ribose nucleic acid
SEM	Scanning electron microscope
TBR	Tubular biofilm reactor

1 Introduction

The immobilization concept has greatly influenced the production of many important industrial products. Through the technique, the entire production process has become faster and ease of continuous operation has become possible. Alginates, agars and other polymeric substances have been used extensively in immobilization applications. As basic concepts of immobilization and methods have been already described in previous chapters, we refrain from discussing these topics again. Studies pertaining to immobilization of planktonic cells have been used for enzyme and other secondary metabolites production. These types of techniques facilitate easy downstream processing and separation. This artificial cellular immobilization process mimics a natural phenomenon of microbes known as biofilm formation (Flemming et al. 2007). Bacterial cells often form communities or aggregations entangled in a polymeric matrix often called the extracellular polymeric substances (EPS) which is similar to the artificial matrixes used for immobilization (Flemming and Wingender 2010). Microbial biofilms are ubiquitous in nature and they are present almost in all the ecological sites. In this chapter, we describe about the basics of biofilm formation, its composition, advantages and limitations of using biofilms as naturally immobilized biocatalysts when compared to planktonic cells. The prospective commercial scale application and underlying challenges are also described.

2 Biofilm as Natural Immobilizers

In nature, bacteria rarely grow as planktonic cultures. For the decades, microbes have been studied as individual bacterial species grown as planktonic cells in laboratory conditions for various purposes. However, in nature, they predominantly exist in the form of surface-attached moieties (biotic or abiotic) and as communities known as "microbial biofilms," a term coined by Costerton (Costerton et al. 1978). Definition of microbial biofilm is ambiguous pertaining to vast structural and compositional difference among of the microbial biofilms, from one environmental niche to another. Most of the biofilms have at least three characteristics, which qualifies them to be categorized as "microbial biofilms" as follows:

1. Cells that are irreversibly attached to the substratum or interfaces or to each other.
2. Cells that are embedded in a matrix of self-produced extracellular polymeric substances.
3. An altered phenotype as compared to that of their free-living or planktonic mode of growth with respect to growth rate and gene transcription.

Microbial biofilms are a heterogeneous complex structure enclosing cluster of microorganisms adhered to a abiotic substratum (e.g., glass, rocks, metals, plastic) or biotic surfaces (e.g., skin, mucosa) (Hall-Stoodley et al. 2004). The nature of biofilms is highly dynamic which depends on the bacterial species, various environmental cues and biological factors (e.g., quorum sensing), therefore, biofilms are still being studied to understand the underlying complexity even after the decades of its first discovery (Costerton et al. 1978).

3 The Biofilm Matrix: Natural Immobilization Matrix

Extracellular polymeric substances (commonly known as EPS), soluble microbial products, and live and inert biomass altogether form the biofilm matrix, which act as a natural immobilizer matrix for microbial cells. The chemical composition of biofilms is markedly influenced by the metabolism of the biofilm microbiota and their growth substrates (Mangwani et al. 2016b). Major components of the biofilm matrix and their respective function have been summarized in Table 1. Major components of the biofilm matrix are detailed as follows:

Table 1 Composition of biofilm matrix (Flemming and Wingender 2010)

Components	% of biofilm matrix
Water	Up to 93–97%
Microbial cells	2–5%
Polysaccharides	1–2% (neutral and polyanionic)
Proteins including enzymes	<1–2%
Nucleic acids (DNA and RNA)	1–2%
Divalent ions	<0.1% (bound and free)

3.1 *Extracellular Polysaccharides*

The major component of the biofilm matrix is exopolysaccharides (Wingender et al. 2001). Earlier "extracellular polysaccharides" alone were believed to be responsible for the cell attachment and biofilm maturation. With the growing understanding of biofilms, it was learned that biofilm matrices also contain proteins, nucleic acids, lipids, etc. (Flemming et al. 2007). The extracellular polysaccharides/exopolysaccharides also determine the biofilm structure and stability and it also affects its biodegradation abilities (Chen and Stewart 2002; Mangwani et al. 2016b, 2014). Exopolysaccharides have been isolated and extensively characterized from various bacterial species and diverse environments. Exopolysaccharides are long, linear or branched molecules with high-molecular-weight ranging from 500 to 2000 kDa (Shukla et al. 2014). These exopolysaccharides can be homo-polysaccharides or hetero-polysaccharides consist of neutral as well as charged sugar residues, interlinked with each other with the help of various ionic, hydrophilic, hydrophobic and Van der Waals forces (Wingender et al. 2001). Alginate, the most studied matrix used in the artificial immobilization methods is an unbranched heteropolymer exopolysaccharide with high-molecular mass produced in *Pseudomonas aeruginosa* biofilms (Wingender et al. 2001).

3.2 *Extracellular Proteins*

Cell surface-associated proteins and extracellular proteins are the second most dominant extracellular polymeric substance present in the biofilm matrix. Many reports have shown their critical role in primary adhesion, intercellular adhesion, and in the maturity of the biofilm. In the last decade, several studies established that biofilm matrix consists of significant amounts of proteins and play critical role in biofilm architecture and stability (Shukla and Rao 2013a, b). Some studies have even identified surface proteins as a potential therapeutic target to get rid of biofilm-mediated infections (Sharma et al. 2019; Shukla and Rao 2017; Vuotto and Donelli 2019). Some of these proteins are able to establish biofilm formation even in the absence of exopolysaccharides. Apart from extracellular enzymes and structural proteins, amyloids are the second most found proteinaceous components of the biofilm matrix. The amyloids have been detected ubiquitously in various types of environmental biofilms such as freshwater lakes, brackish water, drinking-water reservoirs and wastewater treatment plants (Otzen and Nielsen 2008).

3.3 Extracellular DNA

The importance of extracellular DNA (eDNA) in biofilms was recognized recently by Whitchurch et al (2002), which was thought to be present in isolated EPS as the product of lysed cells of biofilms. Growing pieces of evidences show that eDNA comprises an essential component of the biofilm matrix. eDNA has been shown to play a critical role in the stabilization and self-organization of bacterial biofilms (Gloag et al. 2013; Molin and Tolker-Nielsen 2003). eDNA comprises fragments of high-molecular weight up to 30 Kb (Tetz et al. 2009). Few studies have even indicated that eDNA contributes to higher antibiotic resistance in biofilms (Mulcahy et al. 2008; Tetz et al. 2009). eDNA carries a negative charge on phosphorus backbones and interlinks with each other via the crosslinking of divalent ions such as Ca^{2+}, thus stabilizes the biofilm matrix and entangles microbial cells within (Chen and Stewart 2002).

3.4 Lipids and Biosurfactants

Extracellular polymeric substances also have hydrophobic properties due to the substantial presence of hydrophobic molecules such as phenol soluble modulins and rhamnolipids. The surface-active molecules, also known as biosurfactants, interfere with the surface tension at the air–water interface and in turn aid in the exchange of gases (Leck and Bigg 2005). These biosurfactants aid in the maturation of the biofilms by formation of water channels, thus, affecting the diffusion mass transfer. Phenol soluble modulins and rhamnolipids help biofilms in acquiring a characteristic architecture and dispersal (Dusane et al. 2010; Periasamy et al. 2012). The hydrophobic character of the EPS is attributed to lipids and alkyl group-linked polysaccharides, such as methyl and acetyl groups (Busalmen et al. 2002; Neu et al. 1992). Secretion of hydrophobic lipopolysaccharides molecules has been shown to be very critical for adhesion in *Thiobacillus ferrooxidans* (Sand and Gehrke 2006) and surface-active extracellular lipids in *Serratia marcescens* (Matsuyama and Nakagawa 1996).

3.5 Quorum Sensing

Quorum sensing (QS) is one of the regulatory factors involved in coordinating the biofilm formation with varying roles from the initial attachment to maturation of biofilm (Vidal et al. 2013). QS is a cell density-dependent phenomenon that acts as a mechanism for inter-cell communication within the biofilm, and mediated by specific signaling molecules known as *autoinducers* (AI). These QS molecules bind to receptors of new bacteria and regulate the transcription of various genes. QS molecules do not comprise any significant portion of the matrix, but it plays the most important role in deciding the fate of biofilm matrix by regulating a number of genes which are responsible for phenotypic and genotypic changes during the biofilm initiation, maturation and its dispersion (Boles et al. 2004). QS regulates the biofilm development process (Abee et al. 2011) and also plays an important role in enabling communications between interspecies and intra-species communities present in the biofilm.

In certain species of *Aeromonas*, three-QS system collectively regulates vital functions like formation of biofilm, cell motility and induction of virulence (Talagrand-Reboul et al. 2017). Secretion of matrix lysing enzyme such as autolysin and DNase are regulated by the QS and in turn the stability of biofilm matrix (Mangwani et al. 2012). Biofilm formation and various phases in the biofilm development are tightly regulated by QS-related mechanisms (Li and Tian 2012). Studies on different QS mutant strains have revealed the involvement of array of quorum regulatory units in biofilm formation and maturation (Kjelleberg and Molin 2002). The QS system inside the biofilm architecture aids in the synthesis of degradative enzymes to protect it from deleterious environment (Cáp et al. 2012). QS is also involved in regulating the expression of catalase and superoxide dismutase genes, whose products are involved in quenching the free radicals generated due to activity of biocides (Hassett et al. 1999). Any deficiency in QS leads to formation of thinner biofilm with lesser EPS production and increased susceptibility to biocides (Shih and Huang 2002). This sensing system induces or represses the overall community activities which have intense effects on biofilm structure. This QS signaling helps in enabling communications during food shortage, environmental stress conditions like antibiotics and disinfectants attack and protection from harmful bacterial species attack.

Autoinducers are synthesized by the bacterial cells and are exported across the membranes by specific transporters. For a specific gene expression, a threshold concentration of AI is needed. Their concentration is critical for proper gene expression in the cells (Charlton et al. 2000). QS mechanism is found in both gram-positive as well as gram-negative bacteria. However, there is a difference in the QS molecules in the gram-positive and gram-negative cells. There are mainly three classes of quorum sensing system, namely: (1) acyl-homoserine lactones (AHLs) molecules in gram-negative bacteria, (2) oligopeptide-type QS in gram-positive bacteria and (3) LuxS-encoded autoinducer 2 (AI-2) that are expressed both in gram-negative and gram-positive bacteria (Li and Tian 2012). The QS system specific for intra-species signaling are AHL and peptide mediated. AI-2 is termed as the universal language as it helps in interspecies communications and cooperation among across the bacterial species in the biofilm. LuxS homologs are very prevalent among bacterial species (Federle and Bassler 2003; Von Bodman et al. 2008). This QS machinery apart from its vital role in biofilm formation has role in exopolysaccharide synthesis, motility and chemotaxis which are crucial for bacteria during degradation or remediation of any pollutant (Mangwani et al. 2016a).

4 Steps in Microbial Biofilm Formation

The development of biofilm is a complex and multistage process that is dependent on a variety of variables including the type of microorganism, the surface of attachment and environmental factors. There are specific gene expression pattern changes that guide the biofilm formation. The bacterial cells undergo various changes during this complete biofilm formation process. In a biofilm, the bacteria are closely associated and clustered within a matrix. It is a survival strategy that provides equilibrium and stability to the bacterial communities to form aggregates, mature and disperse in a highly dynamic environment. Biofilm growth and maturation is a stepwise process, guided by physical, chemical and biological interactions. There are five major steps involved in a complete biofilm life cycle: (1) initial reversible attachment, (2) irreversible attachment, (3) micro-colony formation, (4) maturation stage, (5) dispersion (Stoodley et al. 2002).

The first stage involves the reversible attachment of the bacterial cells on to any suitable surface. The cells are in free-living planktonic stage and prefer to switch to a sessile biofilm form, because of the advantages in the biofilm mode of dwelling. The cells first sense their proximity to the adhering surface or any interface (Costerton 1999). For best attachment the roughness, hydrophilicity and the surface coating of biofilm are the important factors. The critical level of flow velocity, favorable temperature condition and availability of nutrients are crucial factors which determine this initial attachment (Donlan and Costerton 2002). In the first stage of biofilm formation, the planktonic cells adhere to a suitable surface by physical forces or by using bacterial appendages such as pili or flagella. The surface attachment of these planktonic cells are not permanent. This process may be active or passive, depending on whether the bacteria are motile or transported by the surrounding aqueous phase (Annous et al. 2009) (Fig. 1).

In the second stage, most of the reversibly attached cells remain immobilized to the surface and get irreversibly adhered. This attachment stage is mediated by the steady increase in the expression of quorum sensing molecules and the production of extracellular polymeric substances. Cell surface hydrophobicity plays a crucial role in biofilm formation depending on the nature of surface on which the microbial cells have to be attached. Slowly, micro-colony formation takes place on to the attached surface since binding becomes stronger. These two initial stages of biofilm formation are succeeded by specific strong association (Garrett et al. 2008; Hall-Stoodley et al. 2004).

In the third step, irreversible attachment is followed by multiplication of cells and secretion of extracellular polymeric substances that attach the cells to the surface and also act as a glue that binds microbial cells with each other. Multiplication of cells starts with the input from the chemical signals. These chemical signals and increased gene expression pattern aid in the multiplication of cells. In the maturation stage of biofilm formation, biofilm starts to acquire specific architecture and topography that depends on various biological and physiological factors such as the presence of O_2, nutritional factors, shear forces and bulk fluid flow. (Davies et al. 1998). The biofilm

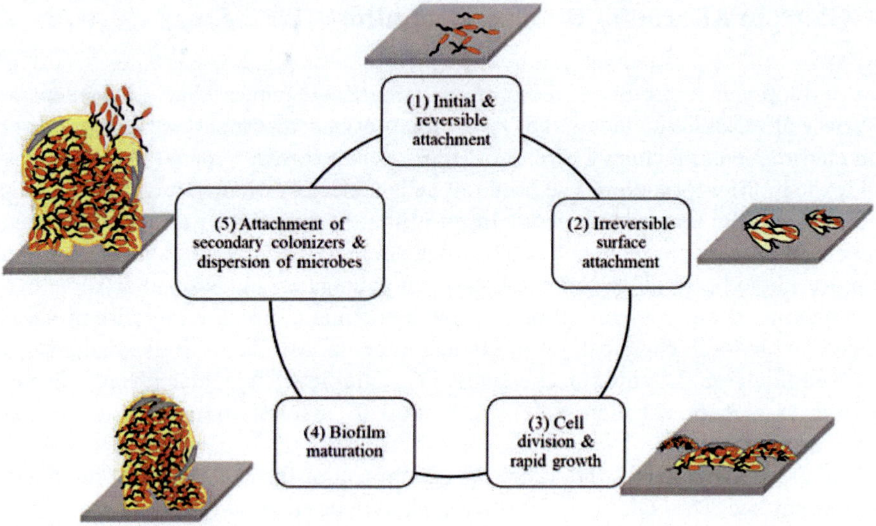

Fig. 1 Schematic illustration of the life cycle of microbial biofilm formation, depicting the different stages in the biofilm development process

starts to acquire a three-dimensional structure with the cells packed in clusters. There is formation of channels within and between the cell clusters that aid in the transport of water, nutrients and waste removal. These water channels act as circulatory system which aid in distributing the different nutrients and removing waste materials from the bacterial communities. The biofilm becomes multi-layered and their thickness also increases up to 10 μm. A shift in the gene expression levels also happens. Many biofilm-specific genes start expressing, whereas those related to motile lifestyle of bacteria start to down-regulate. In a biofilm mode of dwelling, there is enhanced substrate exchange, better distribution of metabolic products and removal of toxic end products (Davey and O'toole 2000). In this maturation stage, biofilm becomes adopted with the external condition by manipulating its structure, physiology and metabolism (Costerton et al. 1999; Parsek and Singh 2003).

The final stage of biofilm formation marks the shredding of biofilm under unfavorable conditions or to propagate and colonizes at new ecological niches. Initiation of new biofilm from the detachment and dispersion of cells from the preformed biofilm is observed. Slowly, the sessile cells transform into the motile form. These dispersed cells from the biofilm are morphologically similar to the planktonic cells than to the mature biofilm cells. The microbial community in the biofilm produces different matrix degrading enzymes that help in breaking the biofilm matrix that holds microbial cells together and releases bacteria to colonize on a new surface. Biofilm being a highly dynamic structure, with the increase in time, there is slow detachment of cells from the matrix. This dispersion process is a highly regulated process. This complete dispersion process is regulated by both intercellular and intracellular signals by the bacteria in the biofilm. There is also possibility of cell dispersion due to changes in environmental conditions like nutrition limitation, pH and external pressure. Other conditions for biofilm dispersal include fluid shear force and abrasion. In some cases, there is a decrease in the production of EPS by the bacteria, which causes dispersion of bacteria from the attachment surface (Kaplan 2010).

5 Biofilms as a Living Catalyst for Bioprocesses

Biocatalysis or biotransformation has been identified as an effective and environmentally safer tool for the industrial production of many chemicals (Schmid et al. 2001; Straathof et al. 2002). Biocatalysts are isolated enzymes or whole living cells that include bacteria, fungi, and their enzymes. Whole cells are preferred over the isolated enzyme-mediated biocatalysis as it reduces the steps required to establish a biotransformation process thus are more economic. Biocatalytic processes perform the chemical reactions with high specificity and turn over, under mild physicochemical conditions. Biocatalysts are also biodegradable and involve biodegradable substrate and products, therefore, waste generation and energy requirements are often considerably lesser when compared with chemical processes (Liese et al. 2006). Isolated enzymes do have a shorter active life as compared to the whole cells as they tend to lose enzymatic activity in an exposed environment. In addition, the enzyme purification step increases the overall operating cost significantly, giving a strong reason to choose whole-cell catalysts over enzyme-mediated catalysis.

Many commercially established productive biocatalysis using whole cells are limited to batch and fed-batch processes in which microbial cells are often grown suspended in a liquid medium and are not reused. Immobilization methods are used to make them reusable. The major techniques used for immobilization include: flocculation, surface adsorption, crosslinking of cells, covalent bonding to carriers, entrapment, encapsulation, and nanocoating (Martins et al. 2013). Majority of the viable immobilization techniques for microbial cells are only feasible at the laboratory scale. Apart from reusability, immobilization offers other advantages such as cell retention, higher catalyst stability, self-generated expensive cofactors, and reactions involving multi-component assembly of membrane-bound proteins (Woodley 2006).

Immobilizing cells to any mechanical carrier avoids the problem of cell leaching out. Immobilization of cells provides increased mechanical, thermal and chemical resistance depending on the methods and materials employed for the entrapment process. This immobilization process aids in multiple usages of the same bacterial cells without much loss in the biological activity of these immobilized cells (Ying et al. 2007). In immobilized whole-cell processes, continuous operations are possible, wherein microbial cells establish a steady-state condition with constant product qualities (Halan et al. 2012). However, despite the above-mentioned potential features, the number of commercial whole-cell continuous processes are very few due to the various reasons, a few prominent reasons are discussed below. Most common methods used at industrial level immobilization biocatalysis are achieved either by fixing whole cells to a carrier material using chemicals or by physically

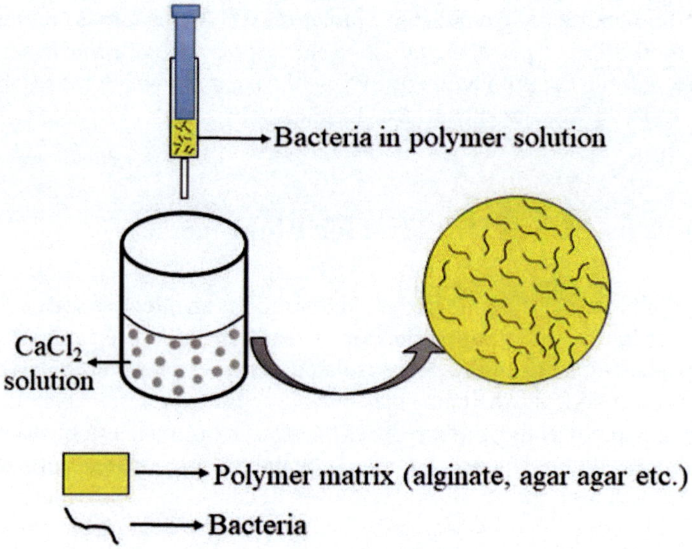

Fig. 2 Conventional immobilization technique of bead formation using polymer solution

retaining them by entrapment or encapsulation in different matrices (Liese et al. 2006; Rosche et al. 2009). For whole cells, physical entrapment in a polymeric matrix is the most preferred technique, whereas chemical binding is mostly used for enzymes (see Fig. 2). A few earlier reports have identified major drawbacks in immobilization procedures, (Buchholz et al. 2012; Hekmat et al. 2007) and are listed as below:

i. Enzyme immobilization is subjected in reduced absolute activity and reduced viability of the entrapped organisms.
ii. The physical barrier used in entrapment methods interfere with diffusion mass transfer of substrates and oxygen lowering the reaction rates.
iii. Immobilization processes do not allow self-regeneration of the biocatalyst thus making long-term use of the immobilized biocatalyst as a limiting factor.
iv. Attachment of the bacterial cells to any carrier will limit the mass transfer process, which might interfere with the substrate uptake and extracellular secretion process.
v. There are many factors which affect the immobilized bacterial cells.

    a. *Support*: Roughness factor, porosity, hydrophobicity, presence of functional groups
    b. *Environmental factors*: pH, temperature, presence of oxygen, nutrient availability, flow velocity, external forces, presence of cations/anions
    c. *Microbial cell*: Age of cell, EPS, hydrophobicity, surface proteins

vi. There is no universally applicable immobilization methods, which could be used for all biocatalyst.

vii. Development and optimization of intricate preparation steps for individual biocatalyst's immobilization that add additional costs decreasing it economically viability.

Microbial biofilms, a naturally immobilized systems are an elegant, powerful and cheap solution to above-mentioned shortcomings of immobilization methods. In biofilms, microbial cells are immobilized in their self-produced polymers (polysaccharides, proteins, extracellular DNA, etc.), without the necessity of any added polymers or chemicals. Biofilm-forming microorganisms are self-sustainable, where cells are in tune with each other, spatially and metabolically and are in general tolerant to toxic substrates and/or products that would be present in a bioprocess. Bacterial cells assembled in the form of biofilms provide benefits to the cells. Bacterial cells entangled in a biofilm lowers the chances of cell leaching out, it enhances cell–cell communications and aids better biochemical resistances with enhances substrate utilization (Ławniczak et al. 2011). Biofilms have great potential as industrial catalyst for the sustainable production of chemicals because of their inherent characteristics of self-immobilization, high resistance to reactants and long-term activity, which all facilitate a continuous processing (Rosche et al. 2009). Moreover, a number of sequential reactions can be catalyzed within a biofilm matrix of multispecies biofilms where microbes with the ability to catalyze different reactions and thus costly intermediate work-up steps can be circumvented. Their overall robustness makes them attractive living catalysts for organic syntheses, which can be harnessed for technical applications. The bacterial cells present in the biofilms are metabolically and functionally integrated consortia. It is a very dynamic system, where the cells are localized and can adapt to spatial configurations. There is enhanced molecular diffusion generated in the biofilm through the chemical gradients generated by various electron acceptors (such as O_2, CO_2, etc.), waste products produced in the biofilm (Berlanga and Guerrero 2016). With developing fundamental understanding of biofilm development, and literature on biofilm-mediated biotransformation and bioremediation/biodegradation of organic pollutants, biofilms are looked upon as potential living biocatalysts for performing challenging conversions in controlled environments. These biofilms can be successfully used for the production of value added products and potentially useful industrially relevant substances.

6 Immobilization Versus Biofilm Formation by Bacteria

Biofilms are a natural case of whole-cell catalysis in which microbial cells are immobilized by polymeric substances. Strong stability is conferred to the bacterial cells residing in the biofilm, which aids in an efficient cell to cell communication and also facilitates horizontal gene transfer. The cells are physiologically and metabolically benefited in the biofilm (Berlanga and Guerrero 2016). Figure 3 shows scanning electron microscopic (SEM) images of bacterial cells immobilized as a biofilm and

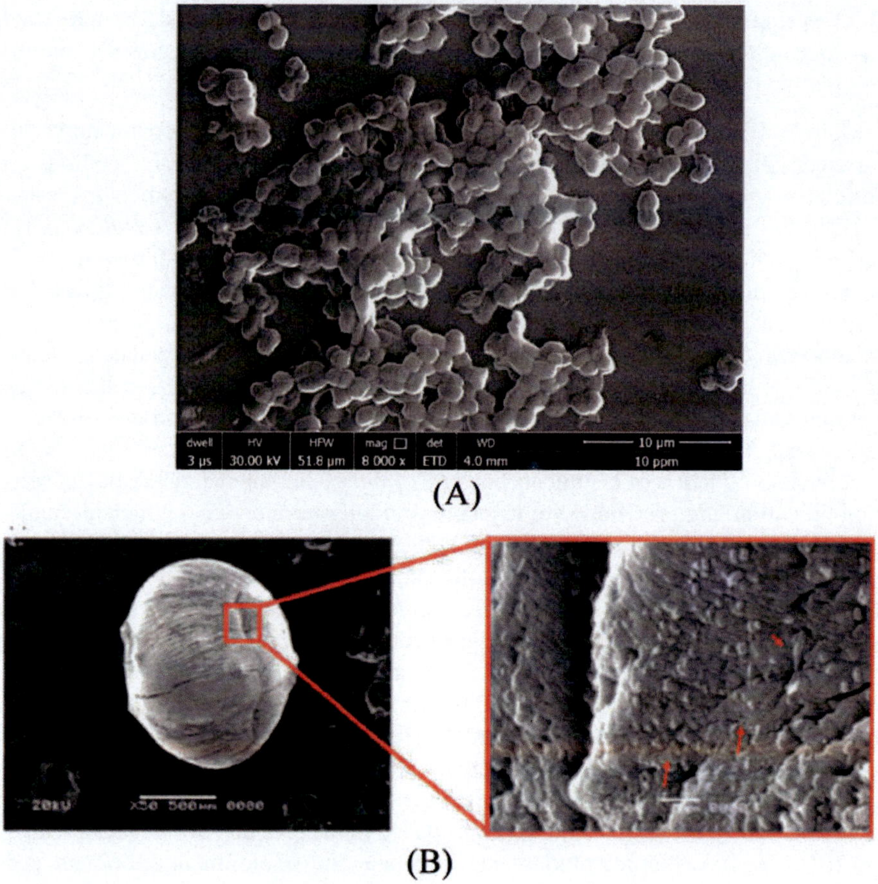

Fig. 3 Scanning electron microscopic images of **a** bacterial biofilm (Manobala et al. 2019a) and **b** calcium alginate-inulin beads (Atia et al. 2016)

bacterial cells immobilized in calcium alginate bead (Atia et al. 2016; Manobala et al. 2019a).

Table 2 lists out major differences between immobilization between naturally entrapped bacterial biofilm and immobilization by chemical means. A major shortcoming in the artificial immobilization process is decreased stability of biocatalyst, in terms of enzyme activity or the metabolic activity and viability of organism that is entrapped.

For example, in a comparative study of immobilization, *Saccharomyces cerevisiae* cells were entrapped in alginate beads or Lentikat® disks and compared for their bioethanol production efficiency (Mathew et al. 2013). The bioethanol production by *S. cerevisiae* cells entrapped in alginate beads and Lentikat® disks were 24% and 28%, respectively, higher as compared to the free-living planktonic cells. Bioethanol yield was significantly higher in the immobilized alginate beads, but its

Table 2 Comparison of immobilization of microbial cells in biofilm versus by polymer matrix

Immobilization in biofilm	Immobilization in polymer matrix
Bacterial cells can be kept alive aiding in enzymatic reaction apart from adsorption	It involves mainly adsorption phenomenon
Biofilm typically consists of mixture of growing, dormant and persister cells	Steps followed in immobilization normally comprise of cell that belong to same growth phase
Biofilm is not very stable and depends on the bacterial strains used	Comparatively more stable than biofilm
Possibility of disintegration, but regeneration is possible under favorable conditions	Disintegration is observed after repeated treatments
Based on application, biofilm composition can be altered based on the bacterial strain used	Size of the beads, bacterial strain used can be altered based on the application
Separation after treatment process can be difficult as there is an equilibrium between biofilm and planktonic phase	Easy physical separation is possible

instability makes calcium alginate unsuitable for commercial applications. It has been a major challenge in applying a bioprocess reaction and the main reason why most bioprocesses are run in either batch or fed-batch mode. Biofilm catalysts have shown long-term activity and enhanced tolerance to toxic reactants, and these characteristics might help to overcome the issues of limited biocatalyst stability and thus allow continuous processing.

7 Biofilms as Biocatalysts When Compared to Planktonic Cells

Bacteria residing in biofilms have several ecological and physiological advantages over their planktonic counterparts. Bacterial cells switch over to sessile form of growth from a free-swimming planktonic phase, they express phenotypic traits that are often distinct from those that are expressed during planktonic growth (Stoodley et al. 2002). Biofilm mode not only confers a protective physical barrier to antimicrobial and toxic substrates as well as products but also makes them less susceptible to all environmental assaults due to owing lower metabolic activity and by forming a subpopulation of cells, known as persister cells (Burmølle et al. 2014; Lee et al. 2014). As the metabolic requirements in a biofilm mode and planktonic lifestyle are different, biofilm cells adapt to new environmental milieu by either varying gene expression patterns or physiological function. Different concentrations of nutrients, availability of oxygen and signaling molecules inside the matrix result in a gradient of various niches (Cogan et al. 2005). As a net result of these changes in phenotype and gene expression of bacteria, biocatalysis ability of bacteria might increase or decrease to carry out a particular reaction.

From an industrial viewpoint, an ideal biofilm catalysis should have few desired characteristics as follows:

i. It should form a good biofilm on any given substratum of interest. Biofilm ability should not drastically change on different substratum.
ii. EPS produced by the bacteria should be sufficient for high robustness but should not reduce the mass transfer.
iii. Genetically engineered biofilm-grown cells should be stable.
iv. The biofilm should be highly porous for diffusion of growth substrates, biotransformation substrates and products.
v. It should attain a steady state under industrial sustaining diverse and adverse environmental conditions (e.g., toxicity and shear forces) without loss of catalytic activity.
vi. It should also be able to self-regenerate having a balance between growth and biocatalysts activity.

Keeping these requirements in mind, we will discuss the challenges and existing bottlenecks to achieve a commercially viable biofilm-based process for industrial applications. Despite having many utilizing biofilms as living catalysts, it is not yet a broadly implemented concept in bioprocess development. Here, first, we discuss, the recent developments in biofilm-based biocatalysis or bioremediation as compared to the planktonic cells.

Table 3 lists a compilation of few studies where a direct comparison was made between biofilm-mediated productivity and planktonic cells mediated productivity. *Zymomonas mobilis* biofilm-based bioreactors showed a great difference in the productivity of ethanol production when compared to the productivity of ethanol production using planktonic cells, in three separate studies using different support

Table 3 Productivity of synthesis of products by biofilm reactors in comparison with the planktonic cells (Berlanga and Guerrero 2016)

Species	Product	Productivity of biofilm (g/l/h)	Productivity of planktonic cells (g/l/h)	References
Zymomonas mobilis	Ethanol	105	<4	Bland et al. (1982)
Zymomonas mobilis	Ethanol	536	5	Kunduru and Pometto (1996)
Zymomonas mobilis	Ethanol	13.40	0.43	Todhanakasem et al. (2014)
Saccharomyces cerevisiae	Ethanol	76	5	Kunduru and Pometto (1996)
Clostridium acetobutylicum	Butanol	1.53	~ 0.22	Förberg and Häggström (1985)
Actinobacillus succinogenes	Succinic acid	8.8	7	Urbance et al. (2004)

for the biofilm growth (Bland et al. 1982; Kunduru and Pometto 1996; Todhanakasem et al. 2014). Especially, the study by Kunduru and Pometto (1996) showed an increase in productivity by more than 100 times as compared to that of planktonic cells. Similarly, butanol production by *Clostridium acetobutylicum* biofilm was also reported to be enhanced by sevenfold (Förberg and Häggström 1985) and succinic acid by *Actinobacillus succinogenes* biofilm was found to be increased by 25% (Urbance et al. 2004), when compared to their respective planktonic cells.

These studies using *Zymomonas mobilis* biofilm bioreactors for ethanol production, a great difference among the productivities was observed with different support material for biofilm growth, which highlights the critical role played by the support for biofilm growth and emphasizes its importance when developing a biofilm-based biocatalyst process at commercial level (Bland et al. 1982; Kunduru and Pometto 1996; Todhanakasem et al. 2014). In a similar study by Mathew et al. (2013) showed a significant difference in bioethanol yield (g bioethanol per kg oilseed rape straw) when *Saccharomyces cerevisiae* biofilm was grown on different supports. The bioethanol production yield by *S. cerevisiae* biofilm grown on spent grains, reticulated foam, and Leca was found to be 101.38, 88.4 and 65.97 g bioethanol kg^{-1}, respectively. Earlier studies using *S. cerevisiae* ATCC 24,859 as a continuous bioethanol production system in the form of biofilm, supported using plastic composite supports (PCS: made of agriculture materials such as soybean hulls, oat hulls, complex nutrients, and polypropylene) and polypropylene alone support (PPS) has proved that PCS (two to ten times more bioethanol production) was more suitable than PPS (Demirci et al. 1997).

In biofilm-based bioremediation studies of toxic polycyclic aromatic hydrocarbons (PAH), biofilm-mediated degradation has been shown to be more efficient as compared to bioremediation by planktonic cells (Shukla et al. 2019, 2014). Recent studies in our laboratory have shown that biofilm formation, its architecture and extracellular polymeric substance composition greatly affect the rate of PAHs degradation (Mangwani et al. 2016b). In one study, *Stenotrophomonas acidaminiphila* biofilm culture efficiently degraded ~71 and ~40% of phenanthrene and pyrene with an initial concentration of 100 mg L^{-1}, whereas planktonic culture could degrade ~38 and ~29% of phenanthrene and pyrene, respectively. Sometime, cofactors such as Ca^{2+} might modulate the biofilm architecture by enhanced EPS production and significantly alter the PAHs degradation efficiency (Mangwani et al. 2014).

It is not always the case that biofilm increases biodegradation or biocatalysis. Few studies indicated reduced biodegradation efficiency in biofilm mode as compared to planktonic cells. For example, *Pseudomonas fluorescens* DSM 8341 planktonic cells grown in chemostat reactor showed higher biodegradation rate of fluorinated xenobiotics when compared to that of biofilm cells grown in a tubular biofilm reactor (TBR) with fluoroacetate as the sole carbon source. This study proposed that with enrichment techniques *P. fluorescens* biofilm could outweigh the planktonic cells in xenobiotics utilization (Heffernan et al. 2009a). A similar study by Heffernan et al. (2009a) showed complete mineralization of fluoroacetate in the membrane aerated biofilm reactor (MABR) designed using pure culture of *P. fluorescens*. In MABR, biofilm naturally immobilizes in the form of an oxygen-permeable membrane. Because of which high biofilm thickness and complete oxygen utilization rate could be achieved that helped in the fluoroacetate degradation up to 50 mM concentrations (Heffernan et al. 2009b). These two studies highlight that biofilm-forming ability alone is not enough to achieve better biocatalysts, but biofilm reactor design also plays a very important role in a biocatalysis process to be efficient and commercially viable.

8 Challenges in Industrial Applications of Biofilm-Based Biocatalysis

One of the major drawbacks that limit the operational stability of biofilm immobilization is cell dispersion. Biofilms are dynamic microbial systems where biofilm and planktonic cells both are simultaneously present in the bioreactor system. In batch systems, operational conditions involving mixing causes shear force disturbance, which may lead to biofilm dispersal. Development of a smoother biofilm that leads to lesser mass transfer limitation can have better efficiency (Picioreanu et al. 2006). On the other hand, in continuous process, the planktonic cell dilution rate (due to influx of the bulk liquid) is higher than that of the growth rate of the bacteria species involved, the net result would be the washout of suspended bacterial cells and surface-attached biofilm. Thus, the robustness and feasibility of a biofilm process depend on the biology of the biofilm-forming microbes, which pose potential as well as challenges for their application in industrial biocatalysis. In a study by Mathew et al. (2013), lower ethanol production yield in biofilm-based immobilization as compared to the immobilization in calcium alginate was attributed to higher desorption rate of the *S. cerevisiae* biofilm cells that leached from all the three support surfaces used in the study.

Higher EPS production is desired to minimize the cell detachment-related problem in biofilm-based biocatalysts. EPS binds microbial cells together and to the surface thus provide higher robustness to shear forces and lower desorption rate (Manobala et al. 2019a; Shukla et al. 2020). High EPS production is mostly useful where biosorption is a major mechanism of choice, such as in heavy metal bioremediation (Manobala et al. 2019b; Shukla et al., 2017). However, the accumulation of unproductive EPS reduces the availability of active cells and hamper the productive catalytic biofilm process (Rosche et al. 2009). Therefore, there has to be a trade-off between the desired level of robustness and stability of the biofilm on one hand and biofilm-based biocatalytic process on the other hand for optimum production of the EPS. To achieve this, some optimization steps need to be established.

A biofilm generally comprises growing cells as well as dormant and persister cells. These cells have different physiological states and thus have variation in metabolic activity. These cells in various growth phases might not contribute to be a part of productive biomass. This also results in the presence of a significant amount of unproductive biomass which adds up to the diffusion barrier and prevents the free diffusion of substrate and products. Synchronization of all the cells present in biomass to be in the biocatalytic phase is one of the biggest challenges in biofilm-based bioprocess development.

The surface properties also determine the stability of biofilm and its architecture (Garrett et al. 2008). We discussed earlier how the different support materials resulted in diverse bioethanol production yield by *S. cerevisiae* biofilm (Mathew et al. 2013). Presence of some cofactors such as Ca^{2+} increases biofilm stability on a hydrophilic surface (Dixon et al. 2018). In a few studies, the presence of calcium stabilized the biofilm by increasing the EPS production, which in turn also significantly enhanced the bioremediation of heavy metals (Manobala et al. 2019a; Shukla and Subba 2015) and PAHs degradation (Mangwani et al. 2014).

Quorum sensing, the cell to cell signaling, is intricately involved in the regulation of biofilm maturation, determining its architecture and formation of water channels and voids via rhamnolipids and biosurfactants production or triggering biofilm dispersal. Thus, quorum sensing plays a very critical role in the overall biocatalytic activity of biofilms. Water channels and biofilm topography are important factors in deciding overall mass diffusion rates of substrate and products. Shear force and unfavorable condition have been shown to be responsible for biofilm dispersal by depleting the QS molecule (Irie and Parsek 2008). Currently, the understanding of the relation between QS and biofilm dispersal in industrial settings is limited and needs to be further expanded by carrying out in-depth studies.

9 Conclusion

Tolerance of bacterial biomass to harsh environmental conditions as well as to various toxic substrates and products makes biofilm as unique and natural immobilized whole-cell biocatalyst. These biocatalysts have higher stability, self-regeneration and long-term bioactivity. These characteristics are the most important features regarding their industrial applications. Significant progress in the understanding of biofilm biology has been achieved in the last two decades of the research. This expanding knowledge would be helpful in improving the technological applications of biofilm at the commercial scale. Though most of the studies have been carried out using single pure culture bacteria, efforts are also being made to use consortia of bacteria to evaluate their biocatalytic process. Social engineering aspects of ecological interactions, such as competition, predation and commensalism, must be evaluated for their efficiency when a bacterial consortium is used.

References

Abee T, Kovács ÁT, Kuipers OP, Van der Veen S (2011) Biofilm formation and dispersal in Gram-positive bacteria. Curr Opin Biotechnol 22:172–179

Annous BA, Fratamico PM, Smith JL (2009) Scientific status summary: quorum sensing in biofilms: why bacteria behave the way they do. J Food Sci 74:R24–R37

Atia A, Gomaa A, Fliss I, Beyssac E, Garrait G, Subirade M (2016) A prebiotic matrix for encapsulation of probiotics: physicochemical and microbiological study. J Microencapsul 33:89–101

Berlanga M, Guerrero R (2016) Living together in biofilms: the microbial cell factory and its biotechnological implications. Microb Cell Fact 15:165

Bland R, Chen H, Jewell W, Bellamy W, Zall R (1982) Continuous high rate production of ethanol by Zymomonas mobilis in an attached film expanded bed fermentor. Biotech Lett 4:323–328

Boles BR, Thoendel M, Singh PK (2004) Self-generated diversity produces "insurance effects" in biofilm communities. Proc Natl Acad Sci USA 101:16630–16635

Buchholz K, Kasche V, Bornscheuer UT (2012) Biocatalysts and enzyme technology. Wiley.

Burmølle M, Ren D, Bjarnsholt T, Sørensen SJ (2014) Interactions in multispecies biofilms: do they actually matter? Trends Microbiol 22:84–91

Busalmen J, Vazquez M, De Sanchez S (2002) New evidences on the catalase mechanism of microbial corrosion. Electrochim Acta 47:1857–1865

Cáp M, Váchová L, Palková Z (2012) Reactive oxygen species in the signaling and adaptation of multicellular microbial communities. Oxid Med Cell Longev 2012:976753–976753

Charlton TS, De Nys R, Netting A, Kumar N, Hentzer M, Givskov M, Kjelleberg S (2000) A novel and sensitive method for the quantification of N-3-oxoacyl homoserine lactones using gas chromatography–mass spectrometry: application to a model bacterial biofilm. Environ Microbiol 2:530–541

Chen X, Stewart PS (2002) Role of electrostatic interactions in cohesion of bacterial biofilms. Appl Microbiol Biotechnol 59:718–720

Cogan NG, Cortez R, Fauci L (2005) Modeling physiological resistance in bacterial biofilms. Bull Math Biol 67:831–853

Costerton JW (1999) Introduction to biofilm. Int J Antimicrob Agents 11:217–221

Costerton JW, Geesey G, Cheng K (1978) How bacteria stick. Sci Am 238:86–95

Costerton JW, Stewart PS, Greenberg EP (1999) Bacterial biofilms: a common cause of persistent infections. Science 284:1318–1322

ME Davey GA O'toole (2000) Microbial biofilms: from ecology to molecular genetics Microbiol Mol Biol Rev 64:847–867

Davies DG, Parsek MR, Pearson JP, Iglewski BH, Costerton JW, Greenberg EP (1998) The involvement of cell-to-cell signals in the development of a bacterial biofilm. Science 280:295–298

Demirci A, Pometto A III, Ho KG (1997) Ethanol production by *Saccharomyces cerevisiae* in biofilm reactors. J Ind Microbiol Biotechnol 19:299–304

Dixon M, Flint S, Palmer J, Love R, Chabas C, Beuger A (2018) The effect of calcium on biofilm formation in dairy wastewater. Water Pract Technol 13:400–409

Donlan RM, Costerton JW (2002) Biofilms: survival mechanisms of clinically relevant microorganisms. Clin Microbiol Rev 15:167–193

Dusane DH, Nancharaiah YV, Zinjarde SS, Venugopalan VP (2010) Rhamnolipid mediated disruption of marine *Bacillus pumilus* biofilms. Colloids Surf, B 81:242–248

Federle MJ, Bassler BL (2003) Interspecies communication in bacteria. J Clin Investig 112:1291–1299

Flemming H-C, Neu TR, Wozniak DJ (2007) The EPS matrix: the house of biofilm cells. J Bacteriol 189:7945–7947

Flemming H-C, Wingender J (2010) The biofilm matrix. Nat Rev Microbiol 8:623

Förberg C, Häggström L (1985) Control of cell adhesion and activity during continuous production of acetone and butanol with adsorbed cells. Enzyme Microbial Technol 7:230–234

Garrett TR, Bhakoo M, Zhang Z (2008) Bacterial adhesion and biofilms on surfaces. Prog Nat Sci 18:1049–1056

Gloag ES, Turnbull L, Huang A, Vallotton P, Wang H, Nolan LM, Mililli L, Hunt C, Lu J, Osvath SR, Monahan LG, Cavaliere R, Charles IG, Wand MP, Gee ML, Prabhakar R, Whitchurch CB (2013) Self-organization of bacterial biofilms is facilitated by extracellular DNA. Proc Natl Acad Sci USA 110:11541–11546

Halan B, Buehler K, Schmid A (2012) Biofilms as living catalysts in continuous chemical syntheses. Trends Biotechnol 30:453–465

Hall-Stoodley L, Costerton JW, Stoodley P (2004) Bacterial biofilms: from the natural environment to infectious diseases. Nat Rev Microbiol 2:95–108

Hassett DJ, Ma JF, Elkins JG, McDermott TR, Ochsner UA, West SE, Huang CT, Fredericks J, Burnett S, Stewart PS, McFeters G, Passador L, Iglewski BH (1999) Quorum sensing in *Pseudomonas aeruginosa* controls expression of catalase and superoxide dismutase genes and mediates biofilm susceptibility to hydrogen peroxide. Mol Microbiol 34:1082–1093

Heffernan B, Murphy CD, Casey E (2009) Comparison of planktonic and biofilm cultures of *Pseudomonas fluorescens* DSM 8341 cells grown on fluoroacetate. Appl Environ Microbiol 75:2899–2907

Heffernan B, Murphy CD, Syron E, Casey E (2009) Treatment of fluoroacetate by a *Pseudomonas fluorescens* biofilm grown in membrane aerated biofilm reactor. Environ Sci Technol 43:6776–6785

Hekmat D, Bauer R, Neff V (2007) Optimization of the microbial synthesis of dihydroxyacetone in a semi-continuous repeated-fed-batch process by in situ immobilization of *Gluconobacter oxydans*. Process Biochem 42:71–76

Irie Y, Parsek MR (2008) Quorum sensing and microbial biofilms. Curr Top Microbiol Immunol 322:67–84

Kaplan Já (2010) Biofilm dispersal: mechanisms, clinical implications, and potential therapeutic uses. J Dent Res 89:205–218

Kjelleberg S, Molin S (2002) Is there a role for quorum sensing signals in bacterial biofilms? Curr Opin Microbiol 5:254–258

Kunduru MR, Pometto A (1996) Continuous ethanol production byZymomonas mobilis and *Saccharomyces cerevisiae* in biofilm reactors. J Ind Microbiol 16:249–256

Ławniczak Ł, Kaczorek E, Olszanowski A (2011) The influence of cell immobilization by biofilm forming on the biodegradation capabilities of bacterial consortia. World J Microbiol Biotechnol 27:1183–1188

Leck C, Bigg EK (2005) Biogenic particles in the surface microlayer and overlaying atmosphere in the central Arctic Ocean during summer. Tellus B 57:305–316

Lee KWK, Periasamy S, Mukherjee M, Xie C, Kjelleberg S, Rice SA (2014) Biofilm development and enhanced stress resistance of a model, mixed-species community biofilm. ISME J 8:894

Li Y-H, Tian X (2012) Quorum sensing and bacterial social interactions in biofilms. Sensors 12:2519–2538

Liese A, Seelbach K, Wandrey C (2006) Industrial biotransformations. Wiley.

Mangwani N, Dash HR, Chauhan A, Das S (2012) Bacterial quorum sensing: functional features and potential applications in biotechnology. J Mol Microbiol Biotechnol 22:215–227

Mangwani N, Kumari S, Das S (2016) Bacterial biofilms and quorum sensing: fidelity in bioremediation technology. Biotechnol Genet Eng Rev 32:43–73

Mangwani N, Shukla SK, Kumari S, Das S, Rao TS (2016) Effect of biofilm parameters and extracellular polymeric substance composition on polycyclic aromatic hydrocarbon degradation. RSC Advances 6:57540–57551

Mangwani N, Shukla SK, Rao TS, Das S (2014) Calcium-mediated modulation of *Pseudomonas mendocina* NR802 biofilm influences the phenanthrene degradation. Colloids Surf, B 114:301–309

Manobala T, Shukla SK, Rao TS, Kumar MD (2019a) A new uranium bioremediation approach using radio-tolerant *Deinococcus radiodurans* biofilm. J Biosci 44:122

Manobala T, Shukla SK, Rao TS, Kumar MD (2019b) Uranium sequestration by biofilm-forming bacteria isolated from marine sediment collected from Southern coastal region of India. Int Biodeterior Biodegradation 145:104809
Martins SCS, Martins CM, Fiúza LMCG, Santaella ST (2013) Immobilization of microbial cells: a promising tool for treatment of toxic pollutants in industrial wastewater. Afr J Biotech 12:4412–4418
Mathew AK, Crook M, Chaney K, Humphries AC (2013) Comparison of entrapment and biofilm mode of immobilisation for bioethanol production from oilseed rape straw using *Saccharomyces cerevisiae* cells. Biomass Bioenerg 52:1–7
Matsuyama T, Nakagawa Y (1996) Surface-active exolipids: analysis of absolute chemical structures and biological functions. J Microbiol Methods 25:165–175
Molin S, Tolker-Nielsen T (2003) Gene transfer occurs with enhanced efficiency in biofilms and induces enhanced stabilisation of the biofilm structure. Curr Opin Biotechnol 14:255–261
Mulcahy H, Charron-Mazenod L, Lewenza S (2008) Extracellular DNA chelates cations and induces antibiotic resistance in *Pseudomonas aeruginosa* biofilms. PLoS Pathogen 4:e1000213
Neu TR, Dengler T, Jann B, Poralla K (1992) Structural studies of an emulsion-stabilizing exopolysaccharide produced by an adhesive, hydrophobic *Rhodococcus* strain. J Gen Microbiol 138:2531–2537
Otzen D, Nielsen PH (2008) We find them here, we find them there: functional bacterial amyloid. Cell Mol Life Sci 65:910–927
Parsek MR, Singh PK (2003) Bacterial biofilms: an emerging link to disease pathogenesis. Ann Rev Microbiol 57:677–701
Periasamy S, Joo H-S, Duong AC, Bach T-HL, Tan VY, Chatterjee SS, Cheung GY, Otto M (2012) How *Staphylococcus aureus* biofilms develop their characteristic structure. Proc Natl Acad Sci USA 109:1281–1286
Picioreanu C, Rittmann B, Van Loosdrecht M (2006) Mathematical modelling of biofilms: IWA Publishing, London, UK
Rosche B, Li XZ, Hauer B, Schmid A, Buehler K (2009) Microbial biofilms: a concept for industrial catalysis? Trends Biotechnol 27:636–643
Sand W, Gehrke T (2006) Extracellular polymeric substances mediate bioleaching/biocorrosion via interfacial processes involving iron (III) ions and acidophilic bacteria. Res Microbiol 157:49–56
Schmid A, Dordick J, Hauer B, Kiener A, Wubbolts M, Witholt B (2001) Industrial biocatalysis today and tomorrow. Nature 409:258
Sharma R, Bajpai P, Sayyed U, Ahmad IZ (2019) Approaches towards microbial biofilm disruption by natural bioactive agents. In: Biofilms in human diseases: treatment and control. Springer, pp 233–261
Shih PC, Huang CT (2002) Effects of quorum-sensing deficiency on *Pseudomonas aeruginosa* biofilm formation and antibiotic resistance. J Antimicrobial Chemotherapy 49:309–314
Shukla SK, Hariharan S, Rao TS (2020) Uranium bioremediation by acid phosphatase activity of *Staphylococcus aureus* biofilms: Can a foe turn a friend? J Hazard Mater 384:121316
Shukla SK, Mangwani N, Karley D, Rao TS (2017) 19-bacterial biofilms and genetic regulation for metal detoxification. In: Handbook of metal-microbe interactions and bioremediation, 317
Shukla SK, Mangwani N, Rao TS (2019) Bioremediation approaches for persistent organic pollutants using microbial biofilms. In: Microbial biofilms in bioremediation and wastewater treatment, 179
Shukla SK, Mangwani N, Rao TS, Das S (2014) 8–Biofilm-mediated bioremediation of polycyclic aromatic hydrocarbons. Microbial biodegradation and bioremediation, 203–232
Shukla SK, Rao TS (2013a) Dispersal of Bap-mediated *Staphylococcus aureus* biofilm by proteinase K. J Antibiotics 66:55–60
Shukla SK, Rao TS (2013b) Effect of calcium on *Staphylococcus aureus* biofilm architecture: A confocal laser scanning microscopic study. Colloids Surf B 103:448–454
Shukla SK, Rao TS (2017) *Staphylococcus aureus* biofilm removal by targeting biofilm-associated extracellular proteins. Indian J Med Res 146:S1

Shukla SK, Subba R (2015) Heavy metals-bioremediation by highly radioresistant Deinococcus radiodurans biofilm prospective use in nuclear reactor decontamination. In: Proceedings of the symposium on water chemistry and corrosion in nuclear power plants in Asia-2015

Stoodley P, Sauer K, Davies DG, Costerton JW (2002) Biofilms as complex differentiated communities. Ann Rev Microbiol 56:187–209

Straathof AJ, Panke S, Schmid A (2002) The production of fine chemicals by biotransformations. Curr Opin Biotechnol 13:548–556

Talagrand-Reboul E, Jumas-Bilak E, Lamy B (2017) The social life of *Aeromonas* through biofilm and quorum sensing systems. Front Microbiol 8:37

Tetz GV, Artemenko NK, Tetz VV (2009) Effect of DNase and antibiotics on biofilm characteristics. Antimicrob Agents Chemother 53:1204–1209

Todhanakasem T, Sangsutthiseree A, Areerat K, Young GM, Thanonkeo P (2014) Biofilm production by *Zymomonas mobilis* enhances ethanol production and tolerance to toxic inhibitors from rice bran hydrolysate. New Biotechnol 31:451–459

Urbance SE, Pometto AL, DiSpirito AA, Denli Y (2004) Evaluation of succinic acid continuous and repeat-batch biofilm fermentation by *Actinobacillus succinogenes* using plastic composite support bioreactors. Appl Microbiol Biotechnol 65:664–670

Vidal JE, Howery KE, Ludewick HP, Nava P, Klugman KP (2013) Quorum-sensing systems LuxS/autoinducer-2 and *Com* regulate *Streptococcus pneumoniae* biofilms in a bioreactor with living cultures of human respiratory cells. Infect Immun 81:1341–1353

Von Bodman SB, Willey JM, Diggle SP (2008) Cell-cell communication in bacteria: united we stand. Am Soc Microbiol

Vuotto C, Donelli G (2019) Novel treatment strategies for biofilm-based infections. Drugs, 1–21

Whitchurch CB, Tolker-Nielsen T, Ragas PC, Mattick JS (2002) Extracellular DNA required for bacterial biofilm formation. Science 295:1487

Wingender J, Strathmann M, Rode A, Leis A, Flemming HC (2001) Isolation and biochemical characterization of extracellular polymeric substances from *Pseudomonas aeruginosa.* Methods Enzymol 336:302–314

Woodley JM (2006) Microbial biocatalytic processes and their development. Adv Appl Microbiol 60:1–15

Ying W, Ye T, Bin H, H Zhao, J Bi, B-l Cai (2007) Biodegradation of phenol by free and immobilized *Acinetobacter sp.* strain PD12. J Environ Sci 19:222–225

Bioremediation of Industrial Effluents by Aerobic Bacterial Granules

Kisan M. Kodam, Sunil S. Adav, Viresh R. Thamke, and Ashvini U. Chaudhari

Abstract Aerobic bacterial granules are a special kind of biofilm composed of the self-immobilized cell without support material and used in the various wastewater treatment processes due to their efficiency and stability. Their compact structure, wide microbial diversity, and good settling ability offer better separation of treated water. Other advantages of these granules are high bioactivity and toxicity tolerance where bacterial extracellular polymeric substances provide protection to the residing microbial community. They are potentially valuable in terms of investment cost, energy consumption, and efficiency; their stability in highly toxic pollutants and high strength organic load remains the crucial factor for their industrial commercial application. This chapter exploits state-of-the-art aerobic granulation technology, effective reactor operating conditions, and potential applications of aerobic granules in the bioremediation of different pollutants from industry as well as in the environment. An attempt has been made to correlate the long-term stability of aerobic granules with a spatial selection of microbial community structure, substrate concentration, and feeding cycles. The factors that can help to improve the efficiency, stability, and reuse of aerobic granules were also discussed.

Keywords Aggregation · Aerobic granules · Immobilization · Extracellular polymeric substances

K. M. Kodam (✉) · V. R. Thamke · A. U. Chaudhari
Biochemistry Division, Department of Chemistry, Savitribai Phule Pune University, Pune 411007, India
e-mail: kodam@chem.unipune.ac.in

S. S. Adav (✉)
Singapore Phenome Centre, Lee Kong Chian School of Medicine, Nanyang Technological University, 59 Nanyang Drive, Singapore 636921, Singapore
e-mail: ssadav@ntu.edu.sg

A. Tripathi and J. S. Melo (eds.), *Immobilization Strategies*,
Gels Horizons: From Science to Smart Materials,
https://doi.org/10.1007/978-981-15-7998-1_16

Abbreviations

ABGs	Aerobic bacterial granules
AHL	N-acyl-homoserine lactone
AI-2	Autoinducer-2
ATP	Adenosine triphosphate
CLSM	Confocal laser scanning microscopy
COD	Chemical oxygen demand
DGGE	Denaturing gradient gel electrophoresis
EPS	Extracellular polymeric substances
FISH	Fluorescence in situ hybridization
FITC	Fluorescein isothiocyanate
HRT	Hydraulic retention time
KI	Inhibitory constant
K_m	Michaelis–Menten constant
OLR	Organic loading rate
P	Phosphorus
PCR	Polymerase chain reaction
PNP	P-nitrophenol
QS	Quorum sensing
SBR	Sequencing batch reactor
V_{max}	Maximum velocity
VSS	Volatile suspended solids

1 Introduction

Aerobic bacterial granules are self-immobilized highly dense microbial consortium used in biological wastewater treatment. The granular sludge was first observed in the late 1970s in the up-flow anaerobic sludge blanket reactor (Lettinga et al. 1980). However, anaerobic granulation technology has numerous drawbacks such as the long start-up period, a relatively high operating temperature, and less efficiency in removal of nitrogen and phosphorus from wastewater. The aerobic granular sludge is first reported in the aerobic up-flow sludge blanket reactor (Mishima and Nakamura 1991). Morgenroth et al. (1997) cultivated aerobic granular sludge in the sequencing batch reactor (SBR). Aerobic granules are densely packed millions of microbial species per gram of biomass (Liu and Tay 2004). Gao et al. (2011a) called it aggregates of microbial origin, which does not coagulate under reduced hydrodynamic shear, and settle significantly faster than activated sludge flocs. They possess compact structure, diverse microbial community, good settling ability, and high biomass retention.

Microbial granules can be cultivated in SBR using properly chosen selection factors such as appropriate settling time from 2 to 10 min, high aeration rate ensuring superficial up-flow air velocity in a column SBR (Adav et al. 2007c), and feeding time. The aerobic granules cultivated using optimized SBR operational parameters results in the formation of compact structured, biologically efficient aerobic granules with wide diverse microbial species and excellent settling capabilities. The main purpose of the cultivation of compact structured aerobic microbial granules and using them in industrial and municipal wastewater treatment is to avoid the construction of secondary settling tanks or to diminish their size. Other advantages are their higher resistance to toxic pollutants and potential to withstand high organic loading. Hence, such granules have been successfully used in treating industrial wastewater (Adav et al. 2009c; Schwarzenbeck et al. 2005).

Microbial granules have been exploited in biological wastewater treatment since the microbial community is organized and can remove biodegradable organic matter, nitrogen, phosphorus, and other different pollutants. The microbial community structure plays a major role in granulation process, granule stability as well as impacts removal efficiencies of environmental pollutants. The microbial community has been studied by using techniques like polymerase chain reaction (PCR) (Lv et al. 2014), denaturing gradient gel electrophoresis (DGGE), fluorescence in situ hybridization (FISH), and other biological assays. Recently microbial community genomics also been employed to understand uncultivable microbial species, genes, and naturally occurring genomic variability (DeLong 2005). The microbial community structure of aerobic granules remains dynamic and changes with variations in substrate composition (Adav et al. 2010b; Weissbrodt et al. 2014) and temperature (Ebrahimi et al. 2010). The microbial community and structure of aerobic granules are strongly impacted by feeding strategy, aeration intensities (Adav et al. 2007c), settling time (Adav et al. 2009a), organic loading rate (Adav et al. 2010a; Wang et al. 2009), reactor design, exchange ratio and shear force (Adav et al. 2008a), and many other operating and environmental factors.

The literature review indicates that the long-term SBR operation results in loss of stability of aerobic granules that limits its full-scale application. To increase the life span of aerobic granules with its full functional potential, several operational parameters need to be optimized. In fact, Adav et al. (2007e) stored the granules in different media and at different temperatures and evaluated their activity. Based on these evaluations, storing granules at low temperatures like −20 °C may serve the purpose. This chapter depicts the recent development in aerobic biogranulation technology, SBR operational parameters and their impact on biogranulation, and their applications in industrial wastewater treatment. The possibilities of storage of the granules and their reuse are discussed in detail (Lee et al. 2010). The role of extracellular polymeric substances (EPS) in the stability of granules and protection to residing microbial communities is explored. Further, the microbial community structure and their role have been delineated in this chapter.

2 Granulation Process and Granule Formation

The basic mechanism of aerobic microbial granulation involves cell-to-cell interactions that contain biological, physical, and chemical phenomena. In a selective state or condition, microorganisms attach to each other, aggregate and through self-immobilization granules are formed. The sequencing batch reactor and the formed granules are shown in Fig. 1. It involves biotic and abiotic interactions between microorganisms and particles from the seed, which leads to the formation of aggregates.

In biology, the microbial aggregation has been hypothesized which occurs thorough intra-, inter-, and multi-generic cell-to-cell attachment where cell surface receptor plays a major role. Based on the localization and distribution of EPS, proteins, lipids, total cells, and dead cells that been explored by using confocal laser scanning microscopy (CLSM) revealed that the microbial aggregation serves as an initial step in granule formation, where microbial cells are self-immobilized in EPS matrix (Adav et al. 2008b, 2010d; Liu and Tay 2004). The process of aerobic granules involves several steps including cell-to-cell contacts, aggregation of the cells with EPS, multiplication of microbial cells, and further maturation of aggregates leading to formation of stable granules. It has been proposed that the filamentous fungi act as a backbone providing more surface area where bacterial species can attach (Weber et al. 2007). On the contrary, single-strain granulation using *Acinetobacter calcoaceticus* has also been documented (Adav and Lee 2008b). This suggests

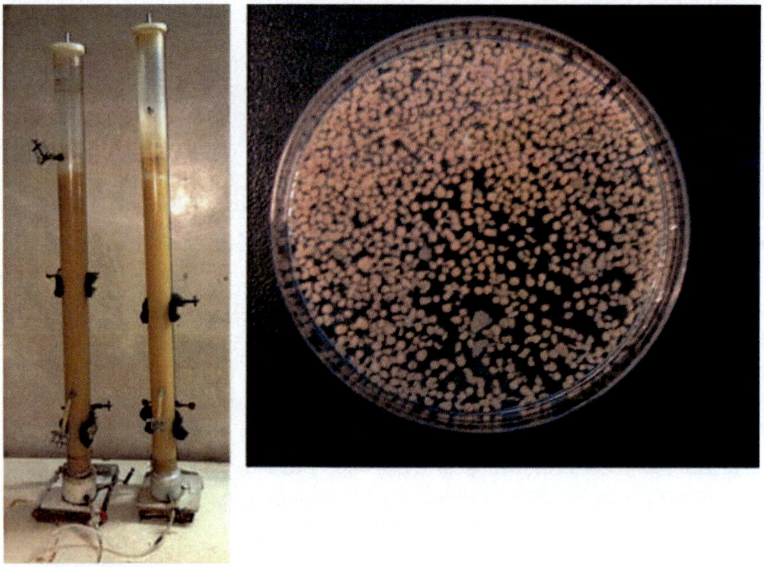

Fig. 1 Sequencing batch reactor and cultivated granules

that the process of microbial granulation is complex process governed by different physical, chemical, and cellular processes.

2.1 Cell–Cell Contact and Microaggregation

Cell-to-cell contact and aggregate formation are important and mostly determined by cellular mechanisms and their physical and chemical characteristics. Cell surface hydrophobicity plays major role and is attributed to protein/polysaccharide ratio. The higher protein/polysaccharide ratio means lower negative surface charge of granular floc leading to reduced electrostatic repulsion between bacterial cells. The low electrostatic repulsion could lead to enhanced granulation with more stable granules with excellent settling ability (Yuan et al. 2018). The negatively charged bacteria may also attach to positively charged inorganic such as calcium and phosphate, aggregate along with EPS, and enhance granulation process (Wan et al. 2015). Thus, cell surface hydrophobicity, electrostatic forces, and inorganic ions also have their role in the initiation of granule formation and maintaining structural stability.

2.2 EPS Production and Granulation

Based on the published literature survey, it is been well-established that EPS plays a major role in granulation process and long-term stability of the formed granules. Microbes aggregate to form the biofilms through the network of cells and EPS. The abbreviation "EPS" has often been defined as extracellular polysaccharides or exopolysaccharides. However, the detailed composition of EPS includes a matrix of polymers such as proteins, polysaccharides, nucleic acids, and humic acids (Adav and Lee 2008c; Adav et al. 2008c). The EPS stained using different chromophores (Fig. 2), and EPS is considered as biological glue that forms a gel-like network keeping bacteria together and provides protection to residing microbial community against noxious environmental conditions (Flemming 2016). Earlier we studied the roles of individual components of EPS on the structural stability of phenol-fed granules (Adav et al. 2008c) and we found that selective enzymatic hydrolysis of extracellular proteins, lipids, and α-polysaccharides had little effect, whereas granule disintegration was noted when β-polysaccharides were hydrolyzed. Seviour et al. (2009) noted exopolysaccharides or glycosides as a significant gelling agent in aerobic granules than in activated sludge. The alginate-like exopolysaccharides were extracted from aerobic granules, which showed gel-forming properties in the presence of calcium chloride (Lin et al. 2010). In summary, EPS acts as a biological glue and play major role in granule formation and structural stability.

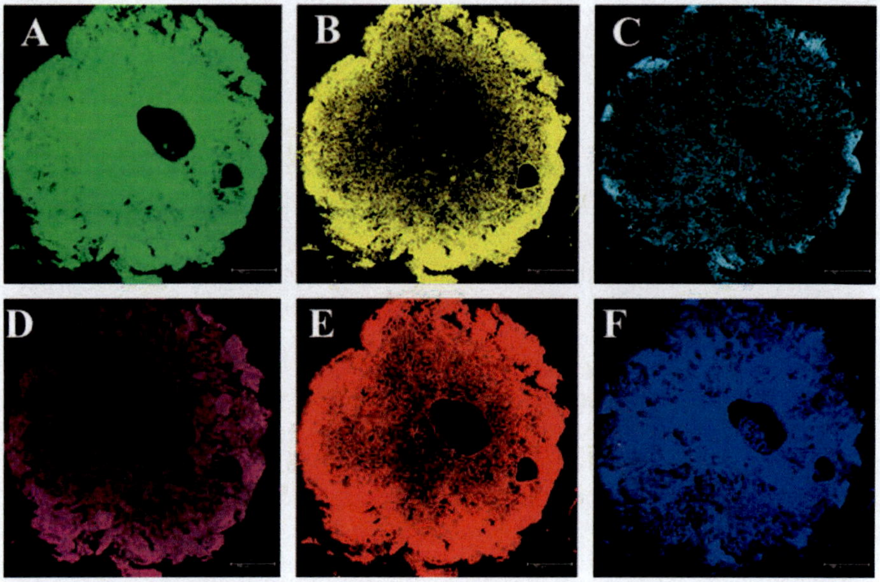

Fig. 2 CLSM images of granules showing stained EPS. **a** proteins (green): Fluorescein isothiocyanate (FITC); **b** lipids (yellow): Nile red; **c** α-polysaccharide (light blue): Con A rhodamine; **d** dead cells (violet): Sytox blue; **e** total cells (red): Syto 63; and **f** β-polysaccharide (blue): calcofluor white

2.3 Cell-To-Cell Communication

Bacteria communicate with one another using chemical signal molecules. Such chemical communication involves production, release, detection, and response to small hormone-like molecules termed autoinducers, while the process is called quorum sensing (QS). In a complex microbial consortium, bacteria employ this sophisticated communication mechanism wherein signaling molecules are released into the extracellular environment. When signal molecules reach the threshold concentration in proportion to a certain quorum, it binds to receptor protein and induces the particular mechanism. By using such mechanism, bacteria also modulate gene expression to compete with other bacteria. Thus, the information provided by these molecules remains critical in monitoring the environment, synchronizing the cellular activities of the cells, and altering behavior on microbial community. Bacteria use different types of QS signaling mechanisms such as N-acyl-homoserine lactone (AHL) signals of Gram-negative bacteria, peptide signals of Gram-positive bacteria, and autoinducer-2 (AI-2) molecules of interspecies signals. N-acyl homoserine lactone (AHL)-based QS has been recognized to play an important role in the formation of biofilm and aerobic granules (Li and Zhu 2014). The QS in aerobic granulation has been mainly focused on the AI-2 (Zhang et al. 2011). According to researchers (Xiong and Liu 2010), AI-2 remains important in the maturation of the aerobic granules and it is ATP-dependent. Further, addition of boron promotes the activity of AI-2-based QS and plays a major role in formation and maturation of aerobic granules (Zhang et al. 2011). In activated sludge or in wastewater treatment, several bacteria synthesize and release AHL-like molecules (Chong et al. 2012). The production of QS chemical signaling molecules from aerobic granules, their role in initial cell adherence, and granule formation have been established (Ren et al. 2013). However, the quantitative relationship between QS chemical signaling molecules, granule formation, and maturation need to be established. The inhibition of ATP synthesis results in reduction of AHLs and EPS production that negatively influenced the granule formation (Jiang and Liu 2012). A higher QS activity during granulation has been correlated with a higher production of gel-forming EPS, with higher hydrophobicity, involved in increased aggregation and stability of granules (Li and Zhu 2014). The positive correlation between the aerobic granulation and AHL-based QS was established, that emphasized the importance of AHL-based QS in the aerobic granulation (Li and Zhu 2014). These authors provided the experimental evidence that AHL-based QS play a major role in granulation and inhibition of AHL-based QS weakened the process of aerobic granule formation (Tan et al. 2014). They used a microbial granulation system such as SBR as a model ecosystem for complex microbial community assembly and demonstrated a strong positive correlation of EPS production, community composition changes, and AHL concentrations (Tan et al. 2014). The addition of chemically synthetic N-acyl-homoserine lactones (AHLs) into bioreactors enhanced the growth and attachment of nitrifying microorganisms, improved the granulation process with increase in extracellular protein and autotropic biomass, and contributed significantly to nitrifying sludge granulation (Wu et al. 2017).

3 Factors Affecting Aerobic Granulation

3.1 Seed Sludge

In the granulation process, mostly, there are two types of seed sludge, i.e., traditional activated sludge taken from municipal wastewater treatment plant and anaerobic granular sludge that have been used to cultivate granules. The granules cultivated using activated sludge and the microbial community are shown in Fig. 3. The microbial population differs with kind of sludge. A different group of bacterial species or strain has different potential to agglomerate due to their cellular activities (Bos et al. 1999). The hydrophobic bacteria attach firmly to sludge floc than hydrophilic and hence seed with higher abundances of hydrophobic bacteria that result in faster and stable aerobic granules formation with excellent settling velocity. Anaerobic granular sludge has also been used to cultivate aerobic granules. It is important to note that the cultivation of stable anaerobic granules needs three months at least.

Recently Song et al. (2009) used activated sludge from municipal wastewater treatment plants and from beer wastewater treatment plants as a seed to cultivate

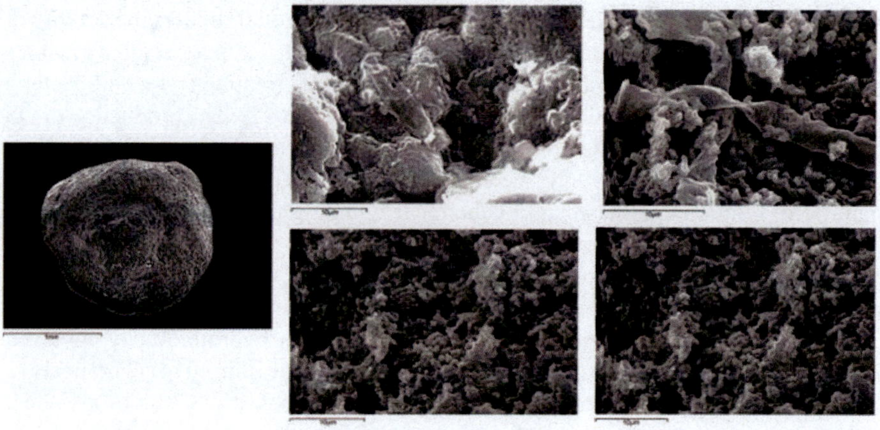

Fig. 3 Scanning electron microscopic image of aerobic granules cultivated using activated sludge and the microbial structural diversity

aerobic granules and found that the sludge of beer wastewater treatment plants more suitable as a seed inoculum. The authors concluded that better the hydrophobicity of seed sludge, the faster the aerobic granulation with excellent settling ability. Kolekar et al. (2012) used bacterial isolates that were isolated from soil/sludge contaminated with textile dye industrial wastewater as a seed to cultivate the aerobic granules. The authors successfully cultivated the aerobic granules and tested their application in decolorization and biodegradation of azo dye. Adav and Lee (2008a, b) used single pure bacterial strain *Bacillus thuringiensis* or *Acinetobacter calcoaceticus* as a seed to cultivate granules. Aerobic mature granule as inoculum with the intention to improve the start-up period has also been tested; however, no clear differences were noted compared to a reactor inoculated with activated sludge. To reduce the long start-up time, the mixture of floccular and crushed aerobic granules in proportion of 50:50 was used as an inoculum (Pijuan et al. 2011). The pellet obtained from activated sludge can also be used as a seed inoculum, which may shorten start-up time compared with that obtained using activated sludge as inoculum. All these studies with different types of seed inoculum were attributed to the different bacterial populations and characteristics of formed aerobic granules including their settling velocity. In summary, the properties and characteristics of the seed inoculum play a major role in the formation of stable aerobic granules.

3.2 Substrate Composition

At the beginning of the aerobic granulation era, most studies on aerobic granulation have been performed in laboratory-scale SBR using readily biodegradable synthetic substrates containing mainly acetate and glucose as carbon sources. It is also been considered as an innovative technology to simultaneously remove nitrogen, phosphorus, and carbon from wastewater. Since this technology was quickly adopted for the treatment of domestic and some industrial wastewaters, the need for testing different carbon compounds on the granulation process was crucial. Moreover, aerobic granules have been cultivated using a wide variety of organic substrates in SBR, including glucose, acetate (Adav et al. 2010c; Tay et al. 2001), peptone, propionate, ethanol (De Kreuk and van Loosdrecht 2006), phenol (Adav et al. 2007a, b), molasses, particulate organic matter-rich wastewater (Wagner et al. 2015), and real municipal wastewater. In addition to these substrates, other organic carbon source and ammonia have also been used in the cultivation of aerobic bacterial granules (ABGs) (Azari et al. 2017; Kent 2019). When ammonia and inorganic carbon source were used as substrates, ABGs were dominated with ammonia-oxidizing and nitrite-oxidizing bacteria. Similarly, when acetate and glucose were employed as carbon and nitrate as a nitrogen source, granules were dominated by the *Epistylis*, *Poterioochromonas*, *Geotrichum*, and *Geotrichum klebahnii* bacterial genus (Williams and De los Reyes 2006). Phenol and phenolic compounds are a major pollutant in industrial wastewater, and its removal by using aerobic microbial granules has been tested (Adav et al. 2007a, b; Jiang et al. 2004). When phenol was used as a substrate, dominancy of β- or γ-*Proteobacteria* group was noted (Jiang et al. 2004). Autoaggregation of the microbe has been noted when exposed to higher concentration of phenol (Farrell and Quilty 2002). Such response of the microbes may be due to phenol toxicity and considered as an adaptive response which is noted with other toxic substances or at high organic loading. When propionate was used as a substrate, granules were dominated by *Zoogloea* (40%), *Acidovorax*, and *Thiothrix*, whereas acetate-fed granules were dominated mainly by *Thiothrix* (60%) (Gonzalez-Gil and Holliger 2011). In addition to the microbial community structure, substrate composition has significant effect on structure and physical properties of the ABGs, e.g., acetate-fed ABGs had a very compact structure with excellent settling and strength properties than glucose-fed granules at identical organic loading (Moy et al. 2002). In aerobic granules, diverse microbial community exists which displays a diverse functional role, i.e., some group of microorganisms mainly degrade the substrates while other groups help to maintain the granules' structural stability and integrity. In short, optimized substrate concentration may act as a selection pressure in stable granule formation which also display optimal activity and pollutant removal.

Some studies reported successful granule cultivation with real domestic wastewater (Wagner and da Costa 2013). The laboratory-scale studies also successfully cultivated aerobic granulation while treating a variety of industrial wastewaters such as petrochemical (Caluwe et al. 2017), malting (Schwarzenbeck et al. 2004), and potato industry (Dobbeleers et al. 2017) which evident the great potential of this technology for industrial applications. Aerobic granule has been successfully cultivated in an SBR while treating dairy wastewater (Schwarzenbeck et al. 2005), soybean-processing wastewater (Su and Yu 2005), brewery (Wang et al. 2007), and many more. It is important to note that the brewery wastewater has a high organic matter content due to the presence of starch, sugar, volatile fatty acids, etc., which are easily biodegradable compounds and suitable for aerobic as well as anaerobic granule formation. The major pollutants in industrial wastewater removal are preferred by biological methods because of lower costs, the possibility of complete mineralization, and environmentally friendly than physical or chemical methods. However, microbes are very sensitive to type and strength of the pollutants and microbial growth gets inhibited at higher concentration of the substrates and pollutants. Although in aerobic microbial granules, microbes get protection by EPS but high organic loading during granule formation or treatment of high strength wastewater, failure of granular system and disintegration of granules have been noted. In summary, wide variety of substrate was used to cultivate the aerobic microbial granules since bacterial community has potential to degrade different pollutants. The optimized substrate concentration acts as a selection pressure. The substrate concentration beyond certain limit leads to granule disintegration and thus treating high-strength wastewater may cause failure of ABGs system.

3.3 Substrate Loading Rate

The substrate loading rate influences the bacterial population and granulation characteristics. The successful granulation has been achieved in a chemical oxygen demand (COD) loading range between 1 and 15 kg m^{-3} d^{-1} (Li et al. 2008). Li et al. (2008) used three SBRs, with COD loading rates of 1.5 (R1), 3.0 (R2) and 4.5 (R3) kg m^{-3} d^{-1}, respectively, and noted that the different organic loading results in different rates of granulation and altered morphological features of the granules. Interestingly, the lowest microbial species diversity was noted in the high loading rate, while the reactor with the lowest substrate loading rate had the highest species diversity. The dominancy of the species of β- and γ-*Proteobacteria* and *Flavobacterium* within the granule communities at different substrate loading rate as determined by PCR and DGGE technique clearly indicates that few common bacterial species play an important role in the formation of ABGs as well as in maintaining structural integrity of the granules. The researcher used acetate and glucose to test the formation of aerobic granules at an organic loading rate (OLR) of 6–15 kg COD m^{-3} d^{-1} (Moy et al. 2002). During aerobic granulation, it was indicated that glucose and peptone-fed granules can only sustain the structural integrity up to 4 kg COD m^{-3} d^{-1}, whereas acetate-fed granules can sustain up to 9 kg COD m^{-3} d^{-1} (Tay et al. 2004b). When organic loading was increased to 8 kg COD m^{-3} d^{-1}, aerobic granules were not stable (Tay et al. 2004b). At 6–15 kg COD m^{-3} d^{-1}, stable granulation can be achieved but need to provide a sufficient air velocity (Chen et al. 2008).

Some studies have examined the effects of different OLR on the microbial community and the structure of cultivated granules. The species diversity was found to be decreased with an increase in OLR when glucose was used as the organic source (Li et al. 2008). Interestingly, Adav et al. (2009b) cultivated aerobic granules at low OLR (9.0 kg COD m^{-3} d^{-1}) and then challenged them with stepwise increase in OLR to 12.6 and 16.7 kg COD m^{-3} d^{-1}. The functional microbial consortium that tolerated high OLR effectively was isolated using an organic shock-loading-to-extinction approach. These authors noted that the strain *Zoogloea resiniphila* and at least two uncultured strains, *Acinetobacter* sp. clone JT2 and bacterium clone P1D1-516, formed the functional consortium of the aerobic granules under a high OLR. The loss of these uncultured strains caused protein leakage from granules, thereby destabilizing the granules. The microbial diversity decreased as the OLR serially increased, finally leaving four viable strains in granules—*Z. resiniphila*, *Chryseobacterium gleum* strain 2-1-9a, *Rhodocyclales* bacterium TP411, and *Acinetobacter* sp. TDWCW6, with *Zoogloea* sp. as the predominant strain. The proposed organic shock-loading-to-extinction approach is effective in isolating the functional consortium from aerobic granules under high OLR 19.5 −21.3 kg COD m^{-3} d^{-1}. To investigate the major cause of structural stability, the acetate-fed granule that was cultivated at high OLR was further studied for microbial community (Adav et al. 2010a). The isolated microbial isolates were tested for the autoaggregation indices and the capability of isolates to secrete extracellular polymeric substances and found correlation between that the reduced protein quantity secreted by isolates and its impact on structural stability of the granules. In summary, the substrate concentration plays a major role in aerobic granulation and it exerts a selection pressure in bacterial granulation process. The stepwise increase in substrate concentration and studying microbial community structure may help to demonstrate the function of the microbes and also isolate functional consortium for treating high-strength wastewater or particular pollutants.

3.4 Hydrodynamic Shear Force

Several methods and mechanisms of aerobic granulation have been proposed. Based on the methods and mechanism, the granulation process is complex and the ecological mechanisms are largely unknown. It is generally thought that granules can be obtained by (1) applying high hydrodynamic shear forces, (2) feast–famine alternation, and (3) washing-out of the non-granulated biomass (Adav et al. 2008b; Show et al. 2012). In aerobic granulation, hydrodynamic shear force results from liquid and/or airflow or particle-to-particle attrition, and it plays a major role in the formation of granules. It is believed that the high shear forces stimulate bacteria to increase the production of extracellular polymers with a higher polysaccharide/protein ratio, which benefits in the formation of compact and denser aerobic granules. According to Adav et al. (2007c), high aeration rate accelerates granule formation with higher quantities of EPS, particularly, with higher protein/polysaccharide ratio when phenol acts as a carbon source.

It is important to state that different researchers experimented and established various factors as selection pressures for aerobic granulation. The settling velocity is a major hydraulic selection pressure for the formation of aerobic granules and three key parameters, namely settling time, volume exchange ratio, and discharge time, need to be considered since they influence settling velocity (Bindhu and Madhu

2015). Hydrodynamic shear force may influence substrate transfer, oxygen diffusion, microbial selection, and granule structural integrity. The dense surface layer and compact interior core would generate mass transfer resistant to all nutrients, metabolites oxygen, etc. By adopting technique like CLSM and microelectrode, researcher demonstrated the layered structure in granules (Chen et al. 2007). As per this model, aerobic bacteria remain on the surface of the granules while anaerobic bacteria in the core of the granules. The apparent oxygen diffusivity in acetate and phenol-fed granules was determined by probe (Chiu et al. 2006). With enough flow velocity across the granule to minimize the effects of external mass transfer resistance, the diffusivity coefficients of the two types of granules were estimated and concluded that the oxygen diffusivity declined with decreasing granule diameter (Chiu et al. 2006). The oxygen diffusivity depends on the carbon source (Chiu et al. 2006). Further, Adav et al. (2008a) investigated the interior structure of aerobic granules and mass exchange rates between intra- and inter-granular void spaces using size-exclusion chromatography. One-dimensional convection–dispersion of tracer in intra- and inter-granular void spaces was modeled, from which the effective diffusivities of various tracers, the intra-granular permeability, and convection and diffusion transit time inside granules were examined.

3.5 Feast–Famine Regime

The microbial community of the aerobic granules in the bioreactor or industrial wastewater experiences rapid changing environments including changes in dissolved oxygen, substrate concentration, pH, temperature, etc., depending on the operation parameters set to operate the reactor. The aerobic granules cultivation in the SBR involves a sequencing cycle of feeding, aeration, settling, and discharging of supernatant. In fact, the aeration period mainly consists of two phases: a degradation phase in which the substrate is depleted to a minimum, followed by an aerobic starvation phase in which the external substrate is no longer available. Thus, the feed is only active during a short period of time called "feast" where a desired maximal concentration is reached. The feeding phase is followed by a period of no feed referred as "famine" in which the cells take up the substrate. Thus, the microbial community residing in the aerobic granules is subjected to a periodic feast and famine regime.

Feast–famine regime affects the microbial community and leads to decrease in product and biomass yield compared to steady-state conditions (de Jonge et al. 2014). The decrease in biomass yield of *S. cerevisiae* of up to 25% was noted and was dependant on the frequency and the magnitude of the substrate fluctuations (van Kleeff et al. 1996). It is important to note that under rapid dynamic conditions strong metabolic flux changes are expected which exclusively depend on the extracellular substrate concentration. Suarez-Mendez et al. (2014) noted slight reduction in biomass yield under the feast–famine regime. Several researchers (Suarez-Mendez et al. 2014; van Kleeff et al. 1996) designed experiments that produce regimes of high and low substrate availability. The floc-forming microbes might have relatively high substrate uptake kinetics since it aggregates a large number of bacteria. Then, the concept of intermittent feeding came into existence. To study the impact of intermittent feeding, McSwain et al. (2004a) operated three parallel SBRs with the same operating parameters like settling time, aeration rate, and volumetric loading rate, but introduced intermittent feeding to study the effect of feast–famine on the formation and structure of granules. These authors noted that intermittent feeding significantly affects the microbial community and helps in the selection floc-forming and filamentous organisms, while the substrate and oxygen consumption rates were comparable. According to Suarez-Mendez et al. (2014), after a few cycles of the feast–famine regime, a stable and repetitive pattern with a reproducible metabolic response can be obtained. Thus, such feast–famine regime can act as a robust platform for studying the microorganism's physiology under dynamic conditions.

Bacteria in the natural environment are often exposed to starvation or under nutrient-limiting conditions. Upon exposed to such conditions, microbes developed starvation–survival strategies, such as spore-forming, which facilitate them to persist in the environment until conditions become favorable. Non-spore-forming bacteria tend to aggregate. Applying suitable feast–famine regime may prompt bacteria to aggregate. Bacterial aggregation is the initiation stage of the microbial granulation. Thus, the periodic feast–famine regime in SBR can be regarded as a kind of microbial selection pressure for granulation process. The feast–famine regime affects the surface properties of the microbial cells, and thus, cell hydrophobicity and the content of extracellular polysaccharides also get affected.

3.6 *Settling Time*

Short settling time is always preferred in aerobic microbial granulation since it helps to select fast settling bacteria. It also helps with a washout of poor settleable flocs and suspended microorganisms. Several authors (Adav et al. 2007a, b; McSwain et al. 2004b; Qin et al. 2004) reported that aerobic granules were successfully cultivated and became dominant only in the SBR operated at a settling time of 5 min. Qin et al. (2004) operated SBR at different settling time like 20, 15, 10, and 5 min and found that SBR operated at 5 min revealed successful cultivation of aerobic granules, while a mixture of granules and suspended sludge was observed in other reactors that were operated at settling times of 20, 15, and 10 min. During the comparison of different strategies to enhance aerobic granule formation, Gao et al. (2011b) found the fastest granulation by shortening the settling time from 15 to 5 min in 11 days. Similarly, fast granulation was achieved by shortening the settling time to 2 min at OLR of 8 kg m^{-3} day^{-1} (Liu et al. 2016). Further, extracellular polysaccharides production was stimulated, and the cell surface hydrophobicity was also improved significantly at 5 min settling time. Shortening settling time gradually from 20 min to 8, 6, 5, and 4 min, the aerobic granular sludge became dominant in the reactor (Gao 2007). To achieve successful granulation at long settling times, higher shear force is required (Chen and Lee 2015; Zhou et al. 2014). The settling time helps to select the specific microbial community since the specific genera get washed out proportionally to their relative abundance on the floc particle (Szabo et al. 2017). It is well known that the biomass gets proportionally washed out until granules emerged. Once the granules emerged, microorganisms located on the granular surface will get washed out from the reactors due to erosion of the granules. In short, the settling time acts as a major hydraulic selection pressure on microbial community and hence it is an important SBR operational parameters that affects the granule formation as well as structure and properties of the granules.

3.7 *Other Factors That Affect Aerobic Granulation*

In addition to the above discussed factors, operational parameters like hydraulic retention time, volumetric exchange ratio, dissolved oxygen, feeding strategy, reactor configuration, and presence of ions in the feed also affect the granulation process. Hydraulic retention time (HRT) is another important parameter that affects granule formation, properties, and eventually reactor performance. HRT impacts the hydraulic conditions and contact time among different reactants within the reactor. The contact time between substrates and microorganisms profoundly influences the wastewater treatment efficiency as well as economic feasibility of a bioprocess since HRT is closely related to the amount of substrate that can be handled per unit time. The short HRT although good for aerobic microbial granulation, very short cycle time may suppress the growth due to frequent washout of the suspended material. Pan et al. (2004) investigated the impact of HRT on the development of aerobic microbial granules and found that HRTs between 2 and 12 h favors the formation and maintenance of stable aerobic granules with good settleability and activity. The volumetric exchange ratio is the volume of the mixed liquor removed after every cycle of the SBR. The effluent is removed from the port located at the height of the column type SBR reactor. An exchange ratio of 40, 50, or 80% can be maintained.

Environmental factors like pH and temperature play major role in formation and development of maturation of ABGs. The pH of the reactor medium affects overall microbial growth and is a decisive factor during aerobic granulation and degradation. The growth of fungi is highly dominated at low pH (Yang et al. 2008) while bacteria grow well at near-neutral or alkaline pH (Fu et al. 2010). Lowering pH causes the filamentous bulking of aerobic granules (Ji et al. 2011). In fact, the pH value inside the granules found to be lower than the outer liquid. It is noteworthy that the dominancy of bacterial and fungal populations depends on the type of substrates and pH controls the microbial community structure. Temperature impacts the growth of bacteria, and hence changes in temperature influence the performance of aerobic granular systems. The high temperature (42 ± 2 °C) and low temperature (26 ± 2 °C) can lead to decrease in biomass affecting granulation process (Hailei et al. 2006). However, effect of temperature in the range 29–38 °C on granulation is not very significant. Song et al. (2009) evaluated cultivation of aerobic granules in sequencing batch airlift reactors at 25, 30, and 35 °C, respectively, and found that 30 °C is an optimum temperature for biogranulation, that can lead formation of granules with compact structure, better settling ability, and higher biomass activity.

4 Applications of Aerobic Bacterial Granulation Technology

4.1 Treating Toxic Organic Wastewaters

The application of aerobic granular technology is regarded as one of the promising biotechnologies in biological wastewater treatment. The excellent settleability of the aerobic granules remains an outstanding feature which is a prerequisite to handling high liquid flows. Moy et al. (2002) applied aerobic granules for the treatment of high-strength organic wastewater. Upon the stabilization of the granular system and COD removal efficiency, the OLR was stepwise increased from 6 to 15 kg COD m^{-3} d^{-1} to evaluate the potential of granular process. Adav et al. (2009b) applied "organic shock-loading-to-extinction" concept to test the potential of aerobic granules and also to isolate the functional consortium from aerobic granules cultivated at OLR up to 21.3 kg m^{-3} d^{-1}. In another study, Adav et al. (2010a) tested the granule stability and high COD degradation rates at an OLR of 19.5 kg m^{-3} day^{-1} and noted loss of structural integrity at OLR of 21.3 kg m^{-3} day^{-1}. The aerobic granules subjected to high organic loading exhibited poor stability and disintegration. According to study conducted by Tay et al. (2004b), aerobic granules can be cultivated at 8 kg COD m^{-3} d^{-1} but then disintegrated following after 2 weeks of SBR operation and formation. The existence of proteolytic bacteria belonging to genera *Pseudomonas*, *Raoultella*, *Acinetobacter*, *Pandoraea*, *Klebsiella*, *Bacillus*, and uncultured bacterium inside aerobic granules were reported (Adav et al. 2009d) and experimentally it was demonstrated that these bacteria were mainly responsible for granule core deterioration, resulting in granule disintegration. However, limited research explored the biological causes of granule failure under high OLR. To investigate the biological cause of structural stability of aerobic granules, Adav et al. (2010a) isolated 54 isolates from the aerobic granules cultivated at high OLR of 21.3 kg m^{-3} day^{-1} and tested their autoaggregation at different OLR. These authors applied critical COD regime that correlated strongly with the OLR range in which granules started disintegrating. Reduced protein quantity secreted by isolates was associated with the noted poor granule integrity under high OLR.

Ammonium wastewaters containing phenolic compounds are typical from diverse industries, including petroleum refinement, coal tar processing industries, petrochemicals manufacturing plants, paints, and resins producing industries (Veeresh et al. 2005). The phenolic compounds are toxic and add an odor to drinking and food-processing water. Such toxicants also have a potential inhibitory effect over conventional biological treatment like activated sludge processes (Gao et al. 2017). Hence, activated sludge system is not useful to treat such wastewater, but ABGs system overcomes such toxic and inhibitory effects due to compact and dense structure of granules. Aerobic granules have been tested for the phenol removal from wastewater (Adav et al. 2007a, b). The granules could degrade phenol at a specific rate exceeding 1 g phenol g^{-1} volatile suspended solids (VSS) d^{-1} at 500 mg L^{-1} of phenol (Tay et al. 2004a), while Adav et al. (2007b) found phenol degradation rate 1.18 g phenol g^{-1} VSS d 1. The difference in the phenol degradation rate in two studies could be due to differences in the microbial community and mass transfer barrier provided by a granule matrix. Thus, the microbial community residing in the dense granules provides a platform for simultaneous removal of ammonium and phenolic compounds. Further, pyridine degradation by aerobic granules at different concentrations of 200–2500 mg L^{-1} was reported (Adav et al. 2007d). These authors found that the aerobic granules degrade pyridine at initial pyridine concentrations of 250 and 500 mg L^{-1}, with degradation kinetics followed closely a zero-order kinetics with no time delay. The specific degradation rate of pyridine was 73.0 and 66.8 mg pyridine g VSS^{-1} h^{-1} at 250 and 500 mg L^{-1} of pyridine, respectively. Adav et al. (2007d) tested biodegradation of phenol and pyridine and found a competitive inhibition pattern with Michaelis–Menten kinetics with corresponding parameters V_{max}, K_m, and KI of 63.7 mg L^{-1} h^{-1}, 827.8 and 1388.9 mg L^{-1}, respectively. In biological treatment of wastewater containing p-nitrophenol (PNP), aerobic granules have been evaluated and it was noted that the specific PNP degradation rates increase with PNP concentration from 0 to 40.1 mg of PNP L^{-1}, peaked at 19.3 mg of PNP/(g of VSS) h^{-1} (Yi et al. 2006). The efficiency and performance of aerobic granules have been evaluated in the presence of bisphenol-A, where authors noted inhibitory effect of bisphenol-A on heterotrophic and nitrifying microorganisms of aerobic granules (Li et al. 2015).

4.2 Decolorization and Biodegradation of Dyes

Dyes are synthetic organic compounds widely used in various industries such as textile, leather, plastic, paints manufacturing, and pharmaceutical industries. Approximately, 0.7 million tons of organic synthetic dyes are produced yearly, worldwide (Jonstrup et al. 2011). Dyes are grouped based on their applications and chemical structures. It is composed of chromophores, a group responsible for dye color [azo (–N = N–), carbonyl (–C = O), ethylene(–C = C–), carbon–nitrogen group (–C = NH–), nitro (–NO_2–),and quinoid groups], and auxochromes which are electron-withdrawing or electron-donating substituents. The most important auxochromes are amines (–NH_2), carboxyl (–COOH), sulfonate (–SO_3H), and hydroxyl (–OH). Based on the method of application, they are further categorized as reactive, acid, direct, basic, mordant, disperse, sulfur, and vat dyes (Popli and Patel 2015). The color in textile wastewater impairs light penetration and compromise ecosystems. The damage by textile industry wastewater to the environment is due to discharge of untreated effluents into water bodies. The dyes or their products are carcinogenic, mutagenic, or damaged DNA (Popli and Patel 2015). Thus, an important issue related to discharge of big volumes of highly polluted wastewater of textile, leather, plastic, and paint industries needs to be resolved. The dyes are soluble organic compounds, and their extreme solubility in water makes it difficult to remove them by conventional methods. Bioremediation by microorganisms could be the best option for the removal, reduction, or destruction of these harmful dyes by microorganisms such as bacteria, filamentous fungi, and yeasts.

The microbial biomasses are good biosorbent materials for the bioremediation of textile dyes (Tan et al. 2010). However, the main disadvantages of such microbial biosorption are the adsorption capacity of the biomass and again the final disposal of such biomass. In fact, microbes like *Aeromonas punctata* and *Shewanella putrefaciens* have the potential to degrade azo textile dyes Acid Red 88, Reactive Black 5, Direct Red 81, and Disperse Orange 3 (Khalid et al. 2008). *Aeromonas hydrophila* can biotransform the crystal violet dye into phenol, 2, 6-bis (1,1-dimethylethyl), 2′,6′-dihydroxyacetophenone and benzene (Bharagava et al. 2018). Complete mineralization of Disperse Blue 79 and Acid Orange 10 at the concentration of 1.5 g l^{-1} was observed by *Bacillus fusiformis* (Kolekar et al. 2008). *Alishewanella* sp. was reported to decolorize and degrade reactive blue 59 (Kolekar and Kodam 2012) as well as mixture of different dyes (Kolekar et al. 2013). *Lysinibacillus* sp. was found to simultaneously decolorize reactive Orange M2R dye and reduce chromate (Chaudhari et al. 2017). The co-cultures or hybrid bacterial cultures can lead to a higher level of biodegradation. The consortium that includes *Ochrobactrum* sp., *Pseudomonas aeruginosa*, and *Providencia vermicola* remains useful to treat a textile effluent of diverse composition (Lellis et al. 2019). Aerobic granulation technology has been successfully applied to treat industrial wastewaters containing synthetic dyes. Kolekar et al. (2012) cultivated aerobic granules using textile wastewater sludge and tested their potential to degrade reactive blue and found that these granules could degrade reactive blue 59 at dye loading of up to 5.0 g L^{-1}. The biotransformation of anthraquinone dye reactive blue 4 was carried out by using aerobic bacterial granules. Metagenomics study revealed the involvement of *Clostridia*, *Actinobacteria*, and *Proteobacterial* members in biotransformation and tolerance of high concentrations of reactive blue 4 dye (Chaudhari et al. 2017). Ma et al. (2014) cultivated aerobic granules using activated sludge as inoculum that can remove Congo red. The sequential or two-stage anaerobic–aerobic processes are one of the most powerful and efficient technologies for bioremediation of azo dye containing wastewaters. Franca et al. (2019) adopted liquid chromatography coupled with electrospray ionization tandem mass spectrometry technique and identified nineteen degradation products of an azo dye (Acid Red 14) during aerobic degradation cycle, while azo bond reduction and formation of the aromatic amine 4-amino-naphthalene-1-sulfonic acid was noted in anaerobic phase. The color removal of azo dyes was noted under anaerobic conditions but the efficiency of decolorization decreased with increase in salt content and nitrate ion concentration in the reactor (Assadi et al. 2017). Mostly, aromatic amines can be mineralized under aerobic phase through hydroxylation and ring fission of aromatic compounds. If azo dyes are not reduced and cleaved in the anaerobic stage, they most probably leave the aerobic stage intact. To alleviate such issue an anaerobic–aerobic treatment system is useful since the dead-end toxic by-products generated under anoxic conditions can be broken down aerobically.

4.3 Treating Dairy and Oil Refinery Wastewater

The real wastewaters show significantly different behavior in biological degradability from that of synthetic wastewaters. In real wastewaters, different substrates may present not only in the dissolved form but also in the colloidal form or as a particulate matter. Therefore, the overall degradation efficiency in terms of COD_{total} may not be an accurate determination, because the effluent can contain significant portions of biomass. The dense compact structured granules can be obtained for organic loading rates of up to 15–21 kg COD m^{-3} day^{-1}. Bumbac et al. (2015) adopted continuous flow airlift reactor seeded with aerobic granules but fed with dairy wastewater and evaluated its performance to treat dairy wastewater. These authors noted removal efficiencies of 81–93% and 85–94% for COD and biochemical oxygen demand, respectively, with simultaneous nitrification and denitrification. The removal efficiency of total nitrogen was between 52 and 80% and phosphate removal ranged between 65 and above 99%. Schwarzenbeck et al. (2005) successfully cultivated aerobic granules and that demonstrated the removal efficiencies of 90% COD_{total}, 80% N_{total}, and 67% P_{total} at a volumetric exchange ratio of 50% and a cycle duration of 8 h.

Effluents from the chemical and petroleum industries contain many hazardous chemicals and particularly rich in aromatic organic compounds such as polyaromatic hydrocarbons and phenolic substances. Iraqi water resources experienced a significant amount of nutrients in the effluent discharge from existing wastewater treatment plants. Al-Hashimi and colleagues (Al-Hashimi et al. 2017) tested the feasibility and flexibility of aerobic granular sludge technology in treating real wastewaters from oil refinery and dairy industries. These authors noted the COD and NH_4 removal efficiencies for oil refinery wastewater 86% and 92%, respectively, while the corresponding values were for dairy wastewater were 80% and 82%, (Al-Hashimi et al. 2017).

4.4 Other Applications

Aerobic granules have been successfully used in the nitrogen and phosphorus removal. The complete nitrogen removal takes place through nitrification and denitrification. Adav et al. (2010c) cultivated aerobic granules at high OLR that effectively converted 100–700 mg L^{-1} nitrite to nitrogen gas when SBR was supplemented with acetate as a carbon source. The denitrifying microbial community structure of these granules was investigated by 16S rRNA gene sequences which revealed 66 clone sequences belonging to phylum Proteobacteria (64.7%), 34 clones to Bacteroidetes

(33.3%), and only two clones affiliated to Firmicutes phyla (2.0%). The simultaneous removal of organics and nitrogen by aerobic granules has also been investigated, and the use of aerobic granules for nitrification and denitrification was also reported (Beun et al. 2001). Oxygen concentration in the reactor plays a major role in simultaneous nitrification and denitrification processes. Lim et al. (2003) adopted SBR and developed phosphorus (P)-accumulating microbial granules at different substrate P/COD ratios in the range of 1/100–10/100. The P uptake by granules was within the range of 1.9–9.3% by weight. These authors concluded that low aerobic respirometric activity favors P uptake by granules. The microbial biomass can also be used as a platform for heavy metal ion removal, which has several advantages over chemical precipitation methods. The bioaccumulation process development for bioextractive applications including the removal and recovery of heavy metal ions for downstream purification and refining, rather than disposal has been well reviewed (Diep et al. 2018).

5 Summary and Perspectives

Aerobic granulation technology is a promising technique for treating industrial wastewater. The major advantages of aerobic granules are their compact and strong structure, good settling ability, and wide microbial diverse structure that can withstand a high OLR and degradation potential of microbes. The aerobic granules have the potential to degrade or biotransform the pollutant, and hence it is very useful for treating wide variety of industrial wastewater. However, majority of the research work is done using SBRs and this technique is adopted and operated batchwise. Therefore, much more research is needed for testing the feasibility of developing aerobic granulation in continuous culture systems. When compared with anaerobic granules, aerobic granules are less stable, and more investigation is required to improve the stability of aerobic granules. The current aerobic granulation technology is applied mostly at room temperature. However, the temperature of industrial effluent remains higher than the normal temperature, therefore some research work is needed on thermophilic microbial granulation. The granules without pathogenic microbial community will be very useful in reducing health risk and public health.

References

Adav SS, Chen MY, Lee DJ, Ren NQ (2007a) Degradation of phenol by aerobic granules and isolated yeast *Candida tropicalis*. Biotechnol Bioeng 96(5):844–852

Adav SS, Chen MY, Lee DJ, Ren NQ (2007b) Degradation of phenol by *Acinetobacter* strain isolated from aerobic granules. Chemosphere 67(8):1566–1572

Adav SS, Lee DJ, Lai JY (2007c) Effects of aeration intensity on formation of phenol-fed aerobic granules and extracellular polymeric substances. Appl Microbiol Biotechnol 77(1):175–182

Adav SS, Lee DJ, Ren N (2007d) Biodegradation of pyridine using aerobic granules in the presence of phenol. Water Res 41(13):2903–2910

Adav SS, Lee DJ, Tay JH (2007e) Activity and structure of stored aerobic granules. Environ Technol 28(11):1227–1235

Adav SS, Lee DJ (2008a) Aerobic granulation of pure bacterial strain *Bacillus thuringiensis*. Front Environ Sci Eng China 2(4):461–467

Adav SS, Lee DJ (2008b) Single-culture aerobic granules with *Acinetobacter calcoaceticus*. Appl Microbiol Biotechnol 78(3):551–557

Adav SS, Lee DJ (2008c) Extraction of extracellular polymeric substances from aerobic granule with compact interior structure. J Hazard Mater 154(1–3):1120–1126

Adav SS, Chang CH, Lee DJ (2008a) Hydraulic characteristics of aerobic granules using size exclusion chromatography. Biotechnol Bioeng 99(4):791–799

Adav SS, Lee DJ, Show KY, Tay JH (2008b) Aerobic granular sludge: recent advances. Biotechnol Adv 26(5):411–423

Adav SS, Lee DJ, Tay JH (2008c) Extracellular polymeric substances and structural stability of aerobic granule. Water Res 42(6–7):1644–1650

Adav SS, Lee DJ, Lai JY (2009a) Aerobic granulation in sequencing batch reactors at different settling times. Biores Technol 100(21):5359–5361

Adav SS, Lee DJ, Lai JY (2009b) Functional consortium from aerobic granules under high organic loading rates. Biores Technol 100(14):3465–3470

Adav SS, Lee DJ, Lai JY (2009c) Treating chemical industries influent using aerobic granular sludge: recent development. J Taiwan Inst Chem Eng 40(3):333–336

Adav SS, Lee DJ, Wang A, Ren N (2009d) Functional consortium for hydrogen production from cellobiose: concentration-to-extinction approach. Biores Technol 100(9):2546–2550

Adav SS, Lee DJ, Lai JY (2010a) Potential cause of aerobic granular sludge breakdown at high organic loading rates. Appl Microbiol Biotechnol 85(5):1601–1610

Adav SS, Lee DJ, Lai J (2010b) Microbial community of acetate utilizing denitrifiers in aerobic granules. Appl Microbiol Biotechnol 85(3):753–762

Adav SS, Lin JC-T, Yang Z, Whiteley CG, Lee D-J, Peng X-F, Zhang Z-P (2010c) Stereological assessment of extracellular polymeric substances, exo-enzymes, and specific bacterial strains in bioaggregates using fluorescence experiments. Biotechnol Adv 28(2):255–280

Al-Hashimi M, Abbas T, Jumaha G (2017) Aerobic granular sludge: An advanced technology to treat oil refinery and dairy wastewaters. Eng Technol J 35(3 Part (A) Engineering): 216–221

Assadi A, Naderi M, Mehrasbi MR (2017) Anaerobic–aerobic sequencing batch reactor treating azo dye containing wastewater: effect of high nitrate ions and salt. J Water Reuse Desalination 8(2):251–261

Azari M, Lübken M, Denecke M (2017) Simulation of simultaneous anammox and denitrification for kinetic and physiological characterization of microbial community in a granular biofilm system. Biochem Eng J 127:206–216

Beun J, Heijnen J, Van Loosdrecht M (2001) N-removal in a granular sludge sequencing batch airlift reactor. Biotechnol Bioeng 75(1):82–92

Bharagava RN, Mani S, Mulla SI, Saratale GD (2018) Degradation and decolourization potential of an ligninolytic enzyme producing *Aeromonas hydrophila* for crystal violet dye and its phytotoxicity evaluation. Ecotoxicol Environ Saf 156:166–175

Bindhu B, Madhu G (2015) Influence of three selection pressures on aerobic granulation in sequencing batch reactor.

Bos R, van der Mei HC, Busscher HJ (1999) Physico-chemistry of initial microbial adhesive interactions—its mechanisms and methods for study. FEMS Microbiol Rev 23(2):179–230

Bumbac C, Ionescu IA, Tiron O, Badescu VR (2015) Continuous flow aerobic granular sludge reactor for dairy wastewater treatment. Water Sci Technol 71(3):440–445

Caluwe M, Dobbeleers T, D'Aes J, Miele S, Akkermans V, Daens D, Geuens L, Kiekens F, Blust R, Dries J (2017) Formation of aerobic granular sludge during the treatment of petrochemical wastewater. Biores Technol 238:559–567

Chaudhari AU, Paul D, Dhotre D, Kodam KM (2017) Effective biotransformation and detoxification of anthraquinone dye reactive blue 4 by using aerobic bacterial granules. Water Res 122:603–613

Chen MY, Lee DJ, Tay JH (2007) Distribution of extracellular polymeric substances in aerobic granules. Appl Microbiol Biotechnol 73(6):1463–1469
Chen Y-Y, Lee DJ (2015) Effective aerobic granulation: role of seed sludge. J Taiwan Inst Chem Eng 52:118–119
Chen Y, Jiang W, Liang DT, Tay JH (2008) Aerobic granulation under the combined hydraulic and loading selection pressures. Biores Technol 99(16):7444–7449
Chiu ZC, Chen MY, Lee DJ, Tay ST, Tay JH, Show KY (2006) Diffusivity of oxygen in aerobic granules. Biotechnol Bioeng 94(3):505–513
Chong G, Kimyon O, Rice SA, Kjelleberg S, Manefield M (2012) The presence and role of bacterial quorum sensing in activated sludge. Microb Biotechnol 5(5):621–633
De Jonge L, Buijs NA, Heijnen JJ, van Gulik WM, Abate A, Wahl SA (2014) Flux response of glycolysis and storage metabolism during rapid feast/famine conditions in *Penicillium chrysogenum* using dynamic (13)C labeling. Biotechnol J 9(3):372–385
De Kreuk MK, van Loosdrecht MC (2006) Formation of aerobic granules with domestic sewage. J Environ Eng 132(6):694–697
DeLong EF (2005) Microbial community genomics in the ocean. Nat Rev Microbiol 3(6):459–469
Diep P, Mahadevan R, Yakunin AF (2018) Heavy metal removal by bioaccumulation using genetically engineered microorganisms. Front Bioeng Biotechnol 6(157)
Dobbeleers T, Daens D, Miele S, D'aes J, Caluwé M, Geuens L, Dries J (2017) Performance of aerobic nitrite granules treating an anaerobic pre-treated wastewater originating from the potato industry. Biores Technol 226:211–219
Ebrahimi S, Gabus S, Rohrbach-Brandt E, Hosseini M, Rossi P, Maillard J, Holliger C (2010) Performance and microbial community composition dynamics of aerobic granular sludge from sequencing batch bubble column reactors operated at 20 °C, 30 °C, and 35 °C. Appl Microbiol Biotechnol 87(4):1555–1568
Farrell A, Quilty B (2002) Substrate-dependent autoaggregation of *Pseudomonas putida* CP1 during the degradation of mono-chlorophenols and phenol. J Ind Microbiol Biotechnol 28(6):316–324
Flemming HC (2016) EPS-then and now. Microorganisms 4(4):41
Franca RDG, Oliveira MC, Pinheiro HM, Lourenço ND (2019) Biodegradation products of a sulfonated azo dye in aerobic granular sludge sequencing batch reactors treating simulated textile wastewater. ACS Sustain Chem Eng 7(17):14697–14706
Fu J, Tang Y, Wang H, Ma X, You K (2010) The formation process and characteristics of aerobic granular sludge under different pH conditions. J Shenyang Jianzhu Univ Nat Sci 4:734–738
Gao D, Liu L, Liang H, Wu W-M (2011a) Aerobic granular sludge: characterization, mechanism of granulation and application to wastewater treatment. Crit Rev Biotechnol 31(2):137–152
Gao D, Liu L, Liang H, Wu WM (2011b) Comparison of four enhancement strategies for aerobic granulation in sequencing batch reactors. J Hazard Mater 186(1):320–327
Gao JF (2007) Effects of settling time and biofilm on the cultivation of nitrifying aerobic granular sludge. Huan Jing Ke Xue 28(6):1245–1251
Gao M, Diao MH, Yuan S, Wang YK, Xu H, Wang XH (2017) Effects of phenol on physicochemical properties and treatment performances of partial nitrifying granules in sequencing batch reactors. Biotechnol Rep 13:13–18
Gonzalez-Gil G, Holliger C (2011) Dynamics of microbial community structure of and enhanced biological phosphorus removal by aerobic granules cultivated on propionate or acetate. Appl Environ Microbiol 77(22):8041–8051
Hailei W, Guangli Y, Guosheng L, Feng P (2006) A new way to cultivate aerobic granules in the process of papermaking wastewater treatment. Biochem Eng J 28(1):99–103
Ji SL, Li JT, Qin ZP, Liu ZP, Cui DH (2011) Bacteria composition of aerobic granular sludge under filamentous bulking and control method of filamentous bulking. J Beijing Univ Technol 10
Jiang B, Liu Y (2012) Roles of ATP-dependent N-acylhomoserine lactones (AHLs) and extracellular polymeric substances (EPSs) in aerobic granulation. Chemosphere 88(9):1058–1064

Jiang HL, Tay JH, Maszenan AM, Tay STL (2004) Bacterial diversity and function of aerobic granules engineered in a sequencing batch reactor for phenol degradation. Appl Environ Microbiol 70(11):6767–6775

Jonstrup M, Kumar N, Murto M, Mattiasson B (2011) Sequential anaerobic–aerobic treatment of azo dyes: decolourisation and amine degradability. Desalination 280(1–3):339–346

Kent TR (2019) Mechanistic understanding of the NOB suppression by free ammonia inhibition in continuous flow aerobic granulation bioreactors, Virginia Tech

Khalid A, Arshad M, Crowley DE (2008) Accelerated decolorization of structurally different azo dyes by newly isolated bacterial strains. Appl Microbiol Biotechnol 78(2):361–369

Kolekar YM, Kodam KM (2012) Decolorization of textile dyes by *Alishewanella* sp. KMK6. Appl Microbiol Biotechnol 95(2): 521–529

Kolekar YM, Konde PD, Markad VL, Kulkarni SV, Chaudhari AU, Kodam KM (2013) Effective bioremoval and detoxification of textile dye mixture by *Alishewanella* sp. KMK6. Appl Microbiol Biotechnol 97(2): 881–889

Kolekar YM, Nemade HN, Markad VL, Adav SS, Patole MS, Kodam KM (2012) Decolorization and biodegradation of azo dye, reactive blue 59 by aerobic granules. Biores Technol 104:818–822

Kolekar YM, Pawar SP, Gawai KR, Lokhande PD, Shouche YS, Kodam KM (2008) Decolorization and degradation of Disperse Blue 79 and Acid Orange 10, by *Bacillus fusiformis* KMK5 isolated from the textile dye contaminated soil. Biores Technol 99(18):8999–9003

Lee DJ, Chen YY, Show KY, Whiteley CG, Tay JH (2010) Advances in aerobic granule formation and granule stability in the course of storage and reactor operation. Biotechnol Adv 28(6):919–934

Lellis B, Fávaro-Polonio CZ, Pamphile JA, Polonio JC (2019) Effects of textile dyes on health and the environment and bioremediation potential of living organisms. Biotechnol Res Innov 3(2):275–290

Lettinga G, van Velsen AFM, Hobma SW, de Zeeuw W, Klapwijk A (1980) Use of the upflow sludge blanket (USB) reactor concept for biological wastewater treatment, especially for anaerobic treatment. Biotechnol Bioeng 22(4):699–734

Li AJ, Yang SF, Li XY, Gu JD (2008) Microbial population dynamics during aerobic sludge granulation at different organic loading rates. Water Res 42(13):3552–3560

Li K, Wei D, Zhang G, Shi L, Wang Y, Wang B, Wang X, Du B, Wei Q (2015) Toxicity of bisphenol A to aerobic granular sludge in sequencing batch reactors. J Mol Liq 209:284–288

Li YC, Zhu JR (2014) Role of N-acyl homoserine lactone (AHL)-based quorum sensing (QS) in aerobic sludge granulation. Appl Microbiol Biotechnol 98(17):7623–7632

Lin Y, de Kreuk M, van Loosdrecht MC, Adin A (2010) Characterization of alginate-like exopolysaccharides isolated from aerobic granular sludge in pilot-plant. Water Res 44(11):3355–3364

Lin YM, Liu Y, Tay JH (2003) Development and characteristics of phosphorus-accumulating microbial granules in sequencing batch reactors. Appl Microbiol Biotechnol 62(4):430–435

Liu Y, Tay JH (2004) State of the art of biogranulation technology for wastewater treatment. Biotechnol Adv 22(7):533–563

Liu YQ, Zhang X, Zhang R, Liu WT, Tay JH (2016) Effects of hydraulic retention time on aerobic granulation and granule growth kinetics at steady state with a fast start-up strategy. Appl Microbiol Biotechnol 100(1):469–477

Lv Y, Wan C, Lee D-J, Liu X, Tay JH (2014) Microbial communities of aerobic granules: granulation mechanisms. Biores Technol 169:344–351

Ma DY, Wang XH, Song C, Wang SG (2014) Biodegradation of congo red by aerobic granules in a sequencing batch reactor. Adv Mater Res 955–959:656–662

McSwain BS, Irvine RL, Wilderer PA (2004a) The effect of intermittent feeding on aerobic granule structure. Water Sci Technol 49(11–12):19–25

McSwain BS, Irvine RL, Wilderer PA (2004b) The influence of settling time on the formation of aerobic granules. Water Sci Technol 50(10):195–202

Mishima K, Nakamura M (1991) Self-immobilization of aerobic activated sludge–a pilot study of the aerobic upflow sludge blanket process in municipal sewage treatment. Water Sci Technol 23(4–6):981–990

Morgenroth E, Sherden T, Van Loosdrecht MCM, Heijnen JJ, Wilderer PA (1997) Aerobic granular sludge in a sequencing batch reactor. Water Res 31(12):3191–3194
Moy BP, Tay JH, Toh SK, Liu Y, Tay SL (2002) High organic loading influences the physical characteristics of aerobic sludge granules. Lett Appl Microbiol 34(6):407–412
Pan S, Tay JH, He YX, Tay SL (2004) The effect of hydraulic retention time on the stability of aerobically grown microbial granules. Lett Appl Microbiol 38(2):158–163
Pijuan M, Werner U, Yuan Z (2011) Reducing the startup time of aerobic granular sludge reactors through seeding floccular sludge with crushed aerobic granules. Water Res 45(16):5075–5083
Popli S, Patel UD (2015) Destruction of azo dyes by anaerobic–aerobic sequential biological treatment: a review. Int J Environ Sci Technol 12(1):405–420
Qin L, Liu Y, Tay J-H (2004) Effect of settling time on aerobic granulation in sequencing batch reactor. Biochem Eng J 21(1):47–52
Ren TT, Li XY, Yu HQ (2013) Effect of N-acy-l-homoserine lactones-like molecules from aerobic granules on biofilm formation by *Escherichia coli* K12. Biores Technol 129:655–658
Schwarzenbeck N, Erley R, Mc Swain BS, Wilderer PA, Irvine RL (2004) Treatment of malting wastewater in a granular sludge sequencing batch reactor (SBR). Acta Hydrochim Hydrobiol 32(1):16–24
Schwarzenbeck N, Borges JM, Wilderer PA (2005) Treatment of dairy effluents in an aerobic granular sludge sequencing batch reactor. Appl Microbiol Biotechnol 66(6):711–718
Seviour T, Pijuan M, Nicholson T, Keller J, Yuan Z (2009) Gel-forming exopolysaccharides explain basic differences between structures of aerobic sludge granules and floccular sludges. Water Res 43(18):4469–4478
Show KY, Lee DJ, Tay JH (2012) Aerobic granulation: Advances and challenges. Appl Microbiol Biotechnol 167(6):1622–1640
Song Z, Ren N, Zhang K, Tong L (2009) Influence of temperature on the characteristics of aerobic granulation in sequencing batch airlift reactors. J Environ Sci 21(3):273–278
Su KZ, Yu HQ (2005) Formation and characterization of aerobic granules in a sequencing batch reactor treating soybean-processing wastewater. Environ Sci Technol 39(8):2818–2827
Suarez-Mendez CA, Sousa A, Heijnen JJ, Wahl A (2014) Fast "feast/famine" cycles for studying microbial physiology under dynamic conditions: a case study with *Saccharomyces cerevisiae.* Metabolites 4(2):347–372
Szabo E, Liebana R, Hermansson M, Modin O, Persson F, Wilen BM (2017) Comparison of the bacterial community composition in the granular and the suspended phase of sequencing batch reactors. AMB Express 7(1):168
Tan CY, Li G, Lu X-Q, Chen Zl (2010) Biosorption of basic orange using dried *A. filiculoides*. Ecol Eng 36(10): 1333–1340
Tan CH, Koh KS, Xie C, Tay M, Zhou Y, Williams R, Ng WJ, Rice SA, Kjelleberg S (2014) The role of quorum sensing signalling in EPS production and the assembly of a sludge community into aerobic granules. ISME J 8(6):1186–1197
Tay JH, Liu QS, Liu Y (2001) Microscopic observation of aerobic granulation in sequential aerobic sludge blanket reactor. J Appl Microbiol 91(1):168–175
Tay JH, Jiang HL, Tay STL (2004) High-rate biodegradation of phenol by aerobically grown microbial granules. J Environ Eng 130(12):1415–1423
Tay JH, Pan S, He Y, Tay STL (2004b) Effect of organic loading rate on aerobic granulation. II: characteristics of aerobic granules. J Environ Eng 130(10): 1102–1109
van Kleeff BH, Kuenen JG, Heijnen JJ (1996) Heat flux measurements for the fast monitoring of dynamic responses to glucose additions by yeasts that were subjected to different feeding regimes in continuous culture. Biotechnol Prog 12(4):510–518
Veeresh GS, Kumar P, Mehrotra I (2005) Treatment of phenol and cresols in upflow anaerobic sludge blanket (UASB) process: a review. Water Res 39(1):154–170
Wagner J, da Costa RHR (2013) Aerobic granulation in a sequencing batch reactor using real domestic wastewater. J Environ Eng 139(11):1391–1396

Wagner J, Weissbrodt DG, Manguin V, Ribeiro da Costa RH, Morgenroth E, Derlon N (2015) Effect of particulate organic substrate on aerobic granulation and operating conditions of sequencing batch reactors. Water Res 85:158–166

Wan C, Lee DJ, Yang X, Wang Y, Wang X, Liu X (2015) Calcium precipitate induced aerobic granulation. Biores Technol 176:32–37

Wang SG, Liu XW, Gong WX, Gao BY, Zhang DH, Yu HQ (2007) Aerobic granulation with brewery wastewater in a sequencing batch reactor. Biores Technol 98(11):2142–2147

Wang SG, Gai LH, Zhao LJ, Fan MH, Gong WX, Gao BY, Ma Y (2009) Aerobic granules for low-strength wastewater treatment: formation, structure, and microbial community. J Chem Technol Biotechnol Int Res Process Environ Clean Technol 84(7):1015–1020

Weber SD, Ludwig W, Schleifer KH, Fried J (2007) Microbial composition and structure of aerobic granular sewage biofilms. Appl Environ Microbiol 73(19):6233–6240

Weissbrodt DG, Shani N, Holliger C (2014) Linking bacterial population dynamics and nutrient removal in the granular sludge biofilm ecosystem engineered for wastewater treatment. FEMS Microbiol Ecol 88(3):579–595

Williams JC, De los Reyes F (2006) Microbial community structure of activated sludge during aerobic granulation in an annular gap bioreactor. Water Sci Technol 54(1):139–146

Wu LJ, Li AJ, Hou B-l, Li MX (2017) Exogenous addition of cellular extract N-acyl-homoserine-lactones accelerated the granulation of autotrophic nitrifying sludge. Int Biodeterior Biodegradation 118:119–125

Xiong Y, Liu Y (2010) Involvement of ATP and autoinducer-2 in aerobic granulation. Biotechnol Bioeng 105(1):51–58

Yang S, Li X, Yu H (2008) Formation and characterisation of fungal and bacterial granules under different feeding alkalinity and pH conditions. Process Biochem 43(1):8–14

Yi S, Zhuang WQ, Wu B, Tay ST, Tay JH (2006) Biodegradation of p-nitrophenol by aerobic granules in a sequencing batch reactor. Environ Sci Technol 40(7):2396–2401

Yuan S, Gao M, Ma H, Afzal MZ, Wang YK, Wang M, Xu H, Wang SG, Wang XH (2018) Qualitatively and quantitatively assessing the aggregation ability of sludge during aerobic granulation process combined XDLVO theory with physicochemical properties. J Environ Sci 67:154–160

Zhang SH, Yu X, Guo F, Wu ZY (2011) Effect of interspecies quorum sensing on the formation of aerobic granular sludge. Water Sci Technol 64(6):1284–1290

Zhou D, Niu S, Xiong Y, Yang Y, Dong S (2014) Microbial selection pressure is not a prerequisite for granulation: Dynamic granulation and microbial community study in a complete mixing bioreactor. Biores Technol 161:102–108

Application of Immobilization Techniques in Heavy Metal and Metalloid Remediation

Sudhakar Srivastava and Bunty Gupta

Abstract Heavy metal contamination of the environment is a serious problem affecting groundwater and soil quality throughout the world. This in turn leads to contamination of drinking water and food and impacts human health. The research on metal remediation from contaminated matrices has been a topic of intensive research for the past several decades and a number of potential physical, chemical and biological methods have been developed. However, bioremediation has been considered the most optimal of all the methods owing to its cost-effectiveness and feasibility of application even in remote areas without requirement of any electricity and sophisticated machinery. There are several approaches through which microorganisms-mediated remediation of metals can be achieved either in situ or ex situ. The immobilization methods offer great hope in this aspect due to their demonstrated success in diverse conditions and media (soil/water) and their cost-effectiveness. The technique may be immobilization of metal itself in the medium (soil/water) and/or the use of immobilized microbes for metal removal. Microbe-mediated metal immobilization in soil is applicable in large contaminated areas whether it is more beneficial to immobilize metal and reduce its bioavailability. The microbial immobilization allows utilization of microbes for multiple times that results into lowered cost of the process. This is particularly suitable for aqueous solutions, wastewater and effluents. The choice of microbe, the carrier material, adsorbent, optimization of conditions, etc., is the topic of research to apply immobilization based techniques in metal remediation. The present review summarizes the problem of metal contamination and its control through various approaches focusing on immobilization techniques.

Keywords Bacteria · Bioprecipitation · Immobilization · Fungi · Metals · Remediation

S. Srivastava (✉) · B. Gupta
Plant Stress Biology Laboratory, Institute of Environment and Sustainable Development, Banaras Hindu University, Varanasi 221005, India
e-mail: sudhakar.iesd@bhu.ac.in; sudhakar.srivastava@gmail.com

A. Tripathi and J. S. Melo (eds.), *Immobilization Strategies*,
Gels Horizons: From Science to Smart Materials,
https://doi.org/10.1007/978-981-15-7998-1_17

Abbreviations

Ag	Silver
Aqp	Aquaporin
As(III)	Arsenite
As(V)	Arsenate
As	Arsenic
ATSDR	Agency for Toxic Substances and Disease Registry
Au	Gold
B	Boron
BLMs	Bulk liquid membranes
BTC	1,3,5-Benzenetricarboxylate
Cd	Cadmium
Cr	Chromium
Cu	Copper
ELMs	Emulsion liquid membranes
EPS	Extracellular polymeric substances
GSH	Glutathione
Hg	Mercury
HMAs	Heavy metal ATPases
IONs	Ion oxide nanoparticles
LMs	Liquid membranes
MICB	Magnetic nanoparticles impregnated chitosan beads
MICP	Microbiologically induced carbonates precipitation
MNPs	Magnetic nanoparticles
MOFs	Metal-organic frameworks
MTPs	Metal-tolerance proteins ()
MTs	Metallothioneins
nHAp	Nano-hydroxyapatite particles
Ni	Nickel
NIPs	Nodulin 26-type intrinsic proteins
NPs	Nanoparticles
NRAMPs	Natural resistance-associated macrophage proteins
nZVI	Nano zerovalent iron
Pb	Lead
PCs	Phytochelatins
PIMs	Polymer inclusion membranes
Sb	Antimony
SLMs	Supported liquid membranes
Sn	Tin
Zn	Zinc

1 Introduction

Heavy metals, marked by their high relative atomic densities (>4–5 g/cm^3), are non-decomposable inorganic pollutants that are toxic to a wide variety of living biological organisms. For biological systems, several metals play essential roles and must be present in an optimum concentration range. However, at high concentrations, even the essential metals can act in a detrimental manner by blocking essential functional groups, displacing other metal ions or altering the active conformation of biological molecules. The non-essential metals are toxic even at very low concentrations. Metals like arsenic (As), lead (Pb), mercury (Hg) and cadmium (Cd) have been included by US Agency for Toxic Substances and Disease Registry (ATSDR) in the list of "Top 20 Hazardous Substances" (Zawierucha et al. 2016). Through various anthropogenic activities like industrial emissions, domestic wastes, sewage, acid mine drain, power generation liquid waste streams, our environment receives a number of heavy metals (Rahman and Singh 2019). Agricultural runoff also has negative impacts on the water quality as some metals may be contributed through pesticides, herbicies, insecticides and fertilizers used in crop production. Apart from anthropogenic sources, natural biogeochemical processes and volcanic eruptions contribute to metal contamination of the environment (Awasthi et al. 2017).

The entire ecosystem gets adversely affected when these metals reach to aquatic bodies and damage aquatic flora and fauna due to their toxicity, persistence and accumulative behavior. Toxic metals enter the human body mainly through food and drinking water. \query{Please check the edits made in the sentence 'Apart from this, exposure to metal-laden fine particles from air and absorption...'.}Apart from this, exposure to metal-laden fine particles from air and absorption via skin particularly as occupational exposure is other source. The toxicity of heavy metals may result into several ailments including cardiovascular, neurological and dermal effects and damage to lungs, kidneys, liver and other vital organs. Prolonged exposure may result into development of serious disorders and cancers such as arsenicosis in case of As (Zawierucha et al. 2016; Upadhyay et al. 2019).

The extent of spread of metals in environment today is of serious concern, and there are more and more applications of metals budding up with time. The health impacts of metals on humans have become increasingly visible, which is due to worldwide import and export of crop produce. The greater knowledge of metals' toxic effects also emanates from technological advancements in detection of metals in low levels (up to parts per trillion). The current scenario demands cleanup of contaminated water and soil in an effective, sustainable, feasible and cost-effective manner (Shukla et al. 2018). Metal-containing wastes, both solid and liquid, have been matter of interest as an extremely valuable resource and focus has been to extract metals from them. It is important to note that metal resources are invaluable and irreplaceable, and hence, their reuse and recycling are essential for sustainable development (Nancharaiah et al. 2016). Hence, the aim of metal removal from contaminated wastes, water and soil has also extended to recover valuable metals for reuse. Therefore, alternate sources, earlier considered as waste, like computers, smart phones, batteries, wastewater, mine tailings, etc., need to be utilized judiciously. Microorganisms offer hope in this aspect as an effective and yet low-cost tool. A number of physical and chemical methods have been developed and applied till date. These include extraction, chemical precipitation, membrane separation, ion exchange, etc. Of these techniques, sorption and membrane-based metal extraction techniques utilizing a variety of immobilized materials, enzymes and proteins, cells are easy to use, effective, have low-cost and do not generate secondary waste. The adsorptive materials can be of biological origin (microbial cells, enzymes, proteins, plant biomass, crop residues, etc.) or of physical origin (fly ash, red mud, kaolinite, bentonite, montmorillonite, etc.) (Demirbas et al. 2009). Biomining is performed at industrial level to extract copper (Cu) (about 20% of world production) and Au (about 5% of world production) from sulfidic ores (Johnson 2014; Nancharaiah and Lens 2015). Biomining is also performed for U (Williamson et al. 2014; Campbell et al. 2015).

2 Adsorption Processes for Metal Removal

Adsorption processes can be operated with high efficiency from very low to very high metal concentrations. The handling of adsorption processes is very easy. In addition, adsorption processes can be extremely sensitive too (Abdolali et al. 2014). The success of adsorption processes, nevertheless, depends on the type of material used and its life. The feasibility and wide-scale application demand that the material should be affordable too. The commonly used adsorbents like activated carbon, alumina, silica, etc., are expensive, and their removal from treated wastewater can itself be a problem. On the other hand, the use of traditional agricultural by-products and crop residues is very cost effective. The ideal material should have high internal volume for effective, rapid and more metal removal, should be durable and should have as much as possible functional groups on the surface so as to have greater life. The modification of solid surface of adsorbent by addition of active functional groups increases the adsorption capacity and removal efficiency of the adsorbent (Marahel et al. 2009). The impregnated chemically modified resins are used to recover the specific metal from the industrial effluents for reuse and for remediation of contaminated sites.

The mechanism of adsorption is passive uptake of metal by dead or inactive biological material by various physico-chemical processes. These include physical adsorption, ion exchange, chelation, complexation, and micro-precipitation (Abdolali et al. 2014). The commonly used biosorbents include algae, bacteria, waste products of industries and agriculture. Various biosorbents have been found to be extremely effective in removal of a variety of metals like Cd, chromium (Cr), nickel (Ni), Pb, zinc (Zn), As, etc. The immobilization of microbial biomass is performed for their application in biosorption, and this has yield high biosorption efficiency of microbial biomass. However, there are a few issues like swelling of biomass, the problem of solid–liquid separation and difficulties in using material for several times so as to make it cost affordable. A number of immobilization techniques have been developed to sort out such issues. The use of polymeric matrix as bio-career for immobilization of microbial biomass has been performed to enhance resistance, allow regeneration and separation from solution. The common polymeric matrix includes sodium or calcium alginate, polysulfone, polyacrylamide, polyurethane and silica (Vijayaraghavan and Yun 2008).

3 Immobilization Techniques for Metal Removal

Several types of materials have been used through immobilization for the removal of metals from wastewater and for cleanup of surface and ground water. Such immobilized materials include impregnated resins, bio-carriers with immobilized biomass and immobilized membranes.

Calixarenes are important macrocyclic compounds, which act as highly selective carriers for heavy metal and transition ions. These can be easily synthesized and functionalized with various functional groups. Calixarene-based ionophores may be sorbed onto synthetic resins, which are selective, thermally stable and have regeneration ability (Jain and Kanaiya 2011). The use of calixarene-based resin for Pb removal from aqueous effluents has been demonstrated (Adhikari et al. 2011). The *p-tert*-butylcalix[8]areneoctamide impregnated Amberlite XAD-4 resin was found to show high sorption of hexavalent chromium [Cr(VI)] and arsenate [As(V)] (Qureshi et al. 2009, 2010). The resin was found to be effective for As(V) removal in both laboratory and field conditions with removal efficiency higher than 90%. In addition, calix[4]arene impregnated resins can also remove arsenite [As(III)] effectively (Qureshi et al. 2011). Zawierucha et al. (2014) optimized the use of carboxypheny[l]resorcinarene impregnated resin for sorption of Pb from aqueous solutions and found Pb sorption to be as high as 107.6 mg g^{-1} which was about 99.7% of the initial concentration in the solution. Hu et al. (2010) studied thiacalix[4]arene-loaded resins for Cu, Pb and Cd removal and found that removal efficiency of three heavy metal ions increased with an increase in pH from 2 to 7. The removal of Pb from aqueous solutions by gallocyanine-impregnated XAD-16 was shown by Hosseini-Bandegharaei et al. (2014) who also observed the process of Pb removal to be pH dependent with the optimum removal occurring at pH 7–7.5.

Wang et al. (2014) used magnetic nanoparticles impregnated chitosan beads (MICB) for the As removal and found the maximum adsorption to be 35.7 mg g^{-1} for As(V) and 35.3 mg g^{-1} for As(III) at pH 6.8. The magnetic property of adsorbent allows easy removal from solution through application of magenic field. The MICB was studied for five cycles, and it was found that about 88.2 and 76.02% of the original As(V) and As(III) adsorption could be achieved. The MICB was found to be equally effective for treatment of real contaminated groundwater sample and could decrease the concentration of As from 0.103 mg L^{-1} to 0.01 mg L^{-1} in 5 h. Lee et al. (2012) studied nano-sized-iron oxide-immobilized-sand (INS) for the removal of Cu, Cd and Pb from wastewater. They found it to be effective in both batch mode and column mode and suggested its application in the remediation of wastewater contaminated with heavy metal. Papain was used for immobilization on activated charcoal by Dutta et al. (2009) for the removal of Hg from wastewater. The observed efficiency of Hg removal was 99.4% from initial concentration of 20 mg L^{-1} in 2 min at CIP mass of 0.03 g.

4 Metal-Organic Framewords (MOFs)

Metal–organic frameworks (MOFs) are materials that have large surface areas and highly porous structure that are commonly used for biosorption (Yang and Yin 2017). MOFs are developed by coordination of metal ion (e.g., iron (Fe), Cu, calcium (Ca), Zn, Cd, etc.) and organic ligand (e.g., p-phthalic acid, imidazole, benzoic acid, amines, sulfonates, carboxylates and phosphates, etc.) (Wu et al. 2019; Li et al. 2018). Highly efficient Cr(VI) removal has been noticed with the nano zerovalent iron (nZVI) materials like graphene-supported nZVI (162 mg/g) (Jabeen et al. 2011) and cellulose-supported nZVI (562.8 mg/g) (Sharma et al. 2015). MOFs have also been developed for extremely toxic and widespread contaminant, As. An Fe-BTC (BTC: 1,3,5-benzenetricarboxylate) polymer has been found to possess about sixfold more As adsorption efficiency than iron oxide nanoparticles (NPs) and 36-fold higher efficiency than commercial iron oxide powders (Li et al. 2018). The sorption of As(V) on to other MOFs has also been found to be good, e.g., MIL-53(Al) (105.6 mg/g) (Li et al. 2014) and MIL-100(Fe) (100.0 mg/g) (Cai et al. 2016). Mesoporous CoFe2O4@MIL-100(Fe) hybrid magnetic nanoparticles (MNPs) have been found to show As(III) adsorption up to 143.6 mg/g within 2 min (Yang and Yin 2017). Zhu et al. (2009) have demonstrated good potential of activated carbon loaded nZVI (nZVI/AC) for As(III) (about 18.2 mg/g) and As(V) (about 12.0 mg/g) removal. Nano-hydroxyapatite particles (nHAp) [Ca10(PO4)6(OH)2] are another class of potential adsorbent for metal(loid) removal from water and soil (Silva et al. 2017; Cai et al. 2019). nHAp have been found to successful remove metals like Pb, Cd, Cu, Zn, Co (Zhang et al. 2010; Sheha et al. 2016). \query{Both the terms 'ion oxide nanoparticles' and 'iron oxide nanoparticles' have been used in this chapter. Please suggest which one is to be followed throughout.}Magnetite (Fe3O4) and maghemite (c-Fe_2O_3) are common superparamagnetic ion oxide nanoparticles (IONs) (Mahmoudi et al. 2011).

5 Microbial Immobilization Techniques for Metal Removal

The negative charge in Gram-positive bacteria is attributable mainly to carboxyl and phosphate groups in teichoic acids, while to phospholipids and lipopolysaccharides in cell membrane of Gram-negative bacteria. Bacteria naturally carry a net negative charge that makes them excellent adsorbent for metal cations (Gadd 2009,

2018). Bacteria may also utilize extracellular polymeric substances (EPS; polyanionic polymers including proteins, polysachharides, nucleic acids, humic acids) to adsorb metals (Tiquia-Arashiro 2018). The role of EPS in Pb adsorption is demonstrated in several bacteria like *Klebsiella michiganensis*, *Providencia rettgeri*, *Raoultella planticola*, and *Serratia sp.* (Bowman et al. 2018). EPS play a major role in metal adsorption in bacterial biofilms and activated sludge (Gupta and Diwan 2017). In a biotechnological approach, a lead binding protein (PbrR) and lead-specific promoter (pbr) were cloned from *Cupriavidus metallidurans* and expressed in *E. coli* by using cell surface display technology. The engineered cells showed selective removal of lead from multi-metal solution with improved efficiency (Kuroda and Ueda 2010). The other effective method to enhance metal adsorption is to express a metal-specific peptide. A polypeptide was expressed as a fusion with an outer membrane protein of *E. coli* (OmpA), and it was found to increase *E. coli* survival to toxic metals (Mejare et al. 1998).

Apart from biological materials, whole bacterial cells have also been used to developed adsorbents. Microbial immobilization is an important that has been studied extensively for metal removal. Immobilization of whole cells is more beneficial than the use of cells as free suspension. As in immobilized form, higher cell density and more biomass of microbial cells may be used. Further, the restoration of biosorbent for reuse for a few cycles is possible (Jacob et al. 2018). A number of base substrates have been used for whole cell immobilization, like sodium alginate, cellulose and polyacrylamide (Dhanarani et al. 2016).

Podder and Majumder (2016) immobilized cells of *Bacillus arsenicus* strain MTCC4380 on sawdust/$MnFe_2O_4$ composite bed and studied As removal from industrial wastewaters. They observed a maximum capacity of about 87 and 89 mg g^{-1} adsorption for As(III) and As(V), respectively. The efficiency of As removal was dependent of biosorbent bed height. The biomass of *Trichoderma viride* was immobilized in a calcium alginate packed-bed column by Kumar et al. (2011). It was found that column performed well for sorption of Cr(VI), Ni and Zn ions from electroplating effluents as well as laboratory-made solutions. The sorption efficiency increased with an increase in bed height, flow rate and initial metal concentration. Ahmad et al. (2014) utilized immobilized *Bacillus subtilis* beads for Cd removal from water. They observed Cd biosorption of IBSB to reach 251.91 mg g^{-1} at equilibrium state achieved at 3 h at an initial Cd concentration of 496.23 mg L^{-1}. The important observation was regeneration potential of biomass for five times. Mahmoudi et al. (2011) developed a biosorbent with the use of heat-inactivated *Aspergillus ustus* and silicon dioxide nano-powder and evaluated its Cd removal potential from tap and seawater. The removal of Cd from both tap and seawater was about 97%. The dead biomass of *Bacillus drentensis* was immobilized in polysulfone polymer by Kim et al. (2014), and its performance was evaluated in laboratory- and pilot-scale experiments. In batch experiments, about 92% of Pb and Cu was removed from 10 mg L^{-1} of initial concentration with the use of 2 g polysulfone and 5% microbial biomass. In column experiment, up to 1.553 g of Pb per g of biosorbent was removed. The system was also tested for cleanup of groundwater, and it was found that more than

93% of Cu, Zn, Cd and Fe could be removed by one kg of biomass from 1098 L of groundwater.

6 Algal Utilization in Immobilization-Based Metal Removal

In addition to bacteria, algal biosorbents have been developed in recent years by using algal cells as such or after genetically modifying those (Cheng et al. 2019). The major approach has been the over-expression of metal binding peptides and moieties like metallothioneins (MTs), glutathione (GSH) and phytochelatins (PCs) at the cell surface by cell surface display technology (Grill et al. 2006). Glutathione, MTs and PCs have cysteine in their structure that allows binding of metal with –SH group via mercaptide bonds. MTs have been identified in several algae and even their genes have been identified such as *Synechococcus sp., Spirulina sp., Microcystis aeruginosa, Anabaena flos aquae, Nostoc sp., Fischerella*, etc. Phytochelatins have been naturally detected in algae like *Stichococcus bacillaris,* (Pawlik-Skowrońska et al. 2004), and *Scenedesmus acutus* (Torricelli et al. 2004). In addition, there are several metal-tolerance proteins (MTPs) which play a role in metal sequestration. In *Chlamydomonas reinhardtii*, five MTP genes have been identified which play variable role in tolerance to different metals like CrMTP4 for Cd (Ibuot et al. 2017). There are several studies where an algal or bacterial genes has been expressed within algal cell and enhanced tolerance to metals has been noticed. However, the over-expression of proteins or peptides at cell surface might be more advantageous owing to the fact that metals can adsorb on to cell surface in metabolically independent manner, and hence, dead biomass can be utilized. Further, cell surface-adsorbed metal may be easily recovered enhancing the overall benefits from the biosorption technology. In a study (Shen et al. 2017), it was found that microalgae engineered for gold biosorption showed self-flocculation that offers advantage of easy and efficient gold recovery.

Algal cell wall has several functional groups of polysaccharides and lipids like –COOH, –OH, SO_4^{2-}, –SH, etc., that participate in metal adsorption. These groups belong to cellulose skeleton of xylans or mannans and to embedding matrix made up of alginate, alginic acid and sulfated galactans. Such a wide array of functional groups and small size of microalgae lead to rapid adsorption of a number of metal species. However, specificity may be lost (Yang et al. 2015). To further enhance adsorptive potential of algal cells, cell wall pretreatment with alkali solution, hydrolysis or acid may be employed to free metal-binding sites (Gupta et al. 2010; Aravindhan et al. 2004). Yacou et al. (2018) treated *Turbinaria turbinate* to acid treatment and prepared activated carbon from the treated algae to achieve improved Cr(VI) adsorption (about 246% more adsorption). The pretreatment of algal cells with ethylenediamine in case of *Caulerpa serrulata* also resulted in improved efficiency for adsorption of Cu and Pb (Mwangi and Ngila 2012).

Further, several adsorbents have been developed with the use of biopolymers. Lim et al. (2008) synthesized a magnetite encapsulating biopolymer-based adsorbent that was found to remove heavy metals of both cationic and anionic categories, i.e., Cu and As. Sodium alginate gel was cross-linked with zirconium(IV) and prepared with magnetic beads to remove the Pb ion by Li et al. (2013a, b). In other experiments also, marine algae have been used to develop biochar with magnetic properties for metal remediation (Son et al. 2018). In a study, cross-linking with polyethylenimine (PEI) was performed with *Fucus vesiculosus* (Demey et al. 2018) that demonstrated significant potential for multi-metal removal including Cu (1.44 mmol g^{-1}), Pb (1.09 mmol g^{-1}), Ni (1.03 mmol g^{-1}), Zn (1.07 mmol g^{-1}) and Cd (0.87 mmol g^{-1}).

The performance of algae may be reduced for metal removal after alginate extraction (Bertagnolli et al. 2014). However, in the study of Cardoso et al. (2017), the alginate extraction from *Sargassum filipendula* was found to have no effect on adsorption efficiency of algae for Ag, Cr, Cu, Ni, Pb and Zn. However, Cd adsorption was decreased. Thus, different groups of algal cell walls and alginate may play specific roles in adsorption of different metals. Papageorgiou et al. (2006) extracted alginate from *Laminaria digitate* and demonstrated that biosorption efficiency of algae extracted alginate extract was greater than that of commercially available alginate beads for Cu^{2+} , Cd^{2+}, and Pb^{2+}. It has been suggested that a high mannuronate (M block) to gulutonate (G block) ratio plays major role in adsorption by algae. Another way of utilizing algae is to synthesize nanoparticles (NPs) from their biomass. NPs have very small size and provide a large surface area for adsorption. Dotto et al. (2012) developed *Spirulina platensis* NPs for Cr(VI) adsorption and observed up to 99.1% Cr(VI) removal at initial Cr(VI) level of 250 mg/l.

7 Membrane-Dependent Separation Processes for Metal Removal

Membrane-dependent metal removal processes have been widely used and offer a number of advantages. The membrane technology offers the feasibility of continuous operation and the application in combination with other technologies. There are several classes of liquid membranes like bulk liquid membranes (BLMs), emulsion liquid membranes (ELMs) and supported liquid membranes (SLMs) (Nghiem et al. 2006). Polymer inclusion membranes (PIMs) are a class of LMs, which are prepared by the physical immobilization of a receptor and an indicator in the plasticized polymer matrix. The base polymer provides mechanical strength and increases the membrane stability. It also does not hinder the metal transport within membrane. The base polymers like cellulose acetate propionate (CAP), cellulose tributyrate (CTB) and cellulose triacetate (CTA) have been used. Plasticizers are added in PIMs to generate more flexible and less brittle membranes. The individual molecular chains in a PIM are connected together by various types of attractive forces. Plasticizers added to PIMs also increase the flux of metals and softness of PIM. Di-octylphthalate (DOP), 2-nitrophenyl-octyl ether (2-NPOPE), di-octyl terephthalate (DOT), di-octyl phthalate (DOP) and tris-(2-ethylhexyl)phosphate (T2EHP) are commonly used plasticizers (Scindia et al. 2005). The preparation of the final membrane is result of an optimum combination of analyte-specific components: polymer matrix, the extractant and the plasticizer with a desirable texture and thickness. PIM-dependent metal separation processes from wastewater are efficient, and selective, easily operable with minimum use of toxic chemicals.

Membrane technology is a highly efficient and reliable technology for the desalination of water. The development of "aquaporin biomimetic membranes" is a novel and unique addition to a number of membranes. The aquaporin membranes however offer the advantage of natural selectivity. Aquaporin is a water-selective protein that allows movement of water molecules across cellular membranes. The aquaporins are embedded in an amphiphilic structure similar to cell membrane. In this regard, the use of electrokinetic approach has been found to be successful for stabilization of aquaporin membranes on a solid support in the presence of electric field (Fuwad et al. 2019). Such membranes have been found to be so selective for water molecules that a rejection of up to about 97–98% salt has been observed (Fuwad et al. 2019). Sun et al. (2013) fabricated a nanofiltration membrane with the immobilization of an aquaporin (Aqp) protein, AqpZ reconstituted liposome on polydopamine coated membrane. The membrane showed improved potential for salt (NaCl and $MgCl_2$) rejection for desalination. To this end, it is important to note that a subgroup of aquaporins named nodulin 26-type intrinsic proteins (NIPs) is known to be involved in transport of specific metalloids such as antimony (Sb), As, tin (Sn) and boron (B). (Bienert et al. 2008; Awasthi et al. 2017). Similarly, there are several metal-specific transporter proteins like natural resistance-associated macrophage proteins (NRAMPs) and heavy metal ATPases (HMAs). Although not yet tested, the approach of utilizing aquaporins of NIPs and other such metal-specific transporter proteins through their immobilization in solid support to develop selective membranes for metal removal would be an effective approach.

8 Microbe-Mediated In Situ Metal Immobilization

The process of biomineralization may be used for the bioremediation of soil and water and biomineral production. Diverse processes contribute to the immobilization of metal(loid)s such as biosorption (on cell walls, EPS), bioaccumulation, bioprecipitation (release of phosphate, sulfide, and oxalate) and bioreduction (through enzyme). Biominerals may be metallic (Ag(0), Au(0)), metal oxide (Fe_3O_4), carbonate ($CaCO_3$), phosphate ($HUPO_4$) and chalcogenide (CdS, CdSe). The prevailing conditions of the medium, the source of mineral and microorganisms utilized control the shape and size of biominerals as well as its composition.

Microbe-mediated metal immobilization occurs with a change in the physical and/or chemical state leading to reduced metal mobility. This is achieved through change in pH, redox processes, complexation and adsorption (Tamayo-Figueroa et al. 2019). The mechanisms of metal immobilization include four possible processes: biosorption, ion exchange, micro-precipitation and coordination complex formation (Jing and Kjellerup 2018). An important technique for immobilizing metal(loid) contaminants is microbiologically induced carbonates precipitation (MICP) (Tamayo-Figueroa et al. 2019). MCIP is a process of biomineralization where microorganisms deposit calcium carbonate extracellularly (Krajewska 2018). A number of metals like Cd, Co, Cu, Ni, Pb, etc., may be immobilized through MICP. *Serratia marcescens* has been demonstrated to remove 98% Cd at 5 mg/L while only 65% at 15 mg/L (Bhattacharya et al. 2018). This is due to lower availability of adsorption sites in microbial cells as Cd concentration increased. Thus, MICP-mediated metal immobilization can be a concentration-dependent phenomenon. Kumari et al. (2014) showed 90% Cd removal with *Exiguobacterium undae*. Achal et al. (2012a) found that *Kocuria flava* could enhance formation of lead carbonates from 5.9 to 35.1 mg/kg in soil augmented with 100 mg/kg Pb. Mwandira et al. (2017) studied Pb remediation potential of *Pararhodobacter sp.* with simulated Pb-containing soil and achieved 100% Pb removal in 6 h. This was done through repeated injections of microbial mix. In the case of As, its two species, As(V) and As(III), bind differently to natural calcite, As(V) being more dominant (Catelani et al. 2018). Arsenite bioremediation was studied Achal et al. (2012b) with the use of *Sporosarina ginsengisoli.* It was found that formation of arsenic carbonates increased from about 15% to about 22% with bacterial-mediated biomineralization. In addition to MICP-based immobilization of single metal, there are studies where multi-metal mixtures have been subjected to MICP (Li et al. 2013a, b, 2016; Mugwar and Harbottle 2016). Kang et al. (2016) used a bacterial consortium for metal mixture and observed increased Cu (5.6%), Pb (98.3%) and Cd (85.4%) immobilization as compared to that achieved by individual bacteria.

Apart from bacteria-mediated MICP, fungi can also perform the task equally effectively. With the use of fungi, higher biomass and thus greater extracellular enzyme production can be achieved allowing improved biomineralization (Dhami et al. 2017). Li et al. (2015) demonstrated biomineralization of Sr by fungi *Pestalotiopsis* sp. and *Myrothecium gramineum. Aspergillus sp.* strain UF3 and *Fusarium oxysporum* strain UF8 have been found to precipitate Pb (34% and 54%, respectively) and Sr (48% and 31%, respectively) (Dhami et al. 2017).

References

Abdolali A, Guo WS, Ngo HH, Chen SS, Nguyen NC, Tung KL (2014) Typical lignocellulosic wastes and by-products for biosorption process in water and wastewater treatment: a critical review. Bioresour Technol 160:57–66

Achal V, Pan X, Zhang D, Fu Q (2012a) Bioremediation of Pb-contaminated soil based on microbially induced calcite precipitation. J Microbiol Biotechnol 22:244–247

Achal V, Pan X, Fu Q, Zhang D (2012b) Biomineralization based remediation of As(III) contaminated soil by Sporosarcina ginsengisoli. J Hazard Mater 201–202:178–184

Adhikari BB, Kanemitsu M, Kawakita H, Jumina, Ohto K (2011). Synthesis and application of a highly efficient polyvinylcalix[4]arene tetraacetic acid resin for adsorptive removal of lead from aqueous solutions. Chem Eng J 172: 341–353

Ahmad MF, Haydar S, Bhatti AA, Bari AJ (2014) Application of artificial neural network for the prediction of biosorption capacity of immobilized Bacillus subtilis for the removal of cadmium ions from aqueous solution. Biochem Eng J 84:83–90

Aravindhan R, Madhan B, Rao JR, Nair BU (2004) Recovery and reuse of chromium from tannery wastewaters using *Turbinaria ornata* seaweed. J Chem Technol Biotechnol 79:1251–1258

Awasthi S, Chauhan R, Srivastava S, Tripathi RD (2017) The journey of arsenic from soil to grain in rice. Front Plant Sci 8:1007

Bertagnolli C, da Silva MGC, Guibal E (2014) Chromium biosorption using the residue of alginate extraction from *Sargassum filipendula.* Chem Eng J 237:362–371

Bhattacharya A, Naik SN, Khare SK (2018) Harnessing the bio-mineralization ability of urease producing for remediation of heavy metal cadmium(II). J Environ Manage 215:143–152

Bienert GP, Thorsen M, Schüssler MD, Nilsson HR, Wagner A, Tamás MJ et al (2008) A subgroup of plant aquaporins facilitate the bi-directional diffusion of As (OH)3 and Sb (OH)3 across membranes. BMC Biol 6:26

Bowman N, Patel D, Sanchez A, XuW AA, Tiquia-Arashiro SM (2018) Lead-resistant bacteria from Saint Clair River sediments and Pb removal in aqueous solutions. Appl Microbiol Biotechnol 102:2391–2398

Cai J, Wang X, Zhou Y, Jiang L, Wang C (2016) Selective adsorption of arsenate and the reversible structure transformation of the mesoporous metal-organic framework MIL-100(Fe). Phys Chem Chem Phys 18:10864–10867

Cai C, Zhao M, Yu Z, Rong H, Zhang C (2019) Utilization of nanomaterials for in-situ remediation of heavy metal(loid) contaminated sediments: A review. Sci Total Environ 662:205–217

Campbell KM, Gallegos TJ, Landa ER (2015) Biogeochemical aspects of uranium mineralization, mining, milling, and remediation. Appl Geochem 57:206–235

Cardoso SL, Costa CSD, Nishikawa E, da Silva MGC, Viera MGA (2017) Biosorption of toxic metals using the alginate extraction residue from the brown algae Sargassum filipendula as a natural ion-exchanger. J Clean Prod 165:491–499

Catelani T, Perito B, Bellucci F, Lee SS, Fenter P, Newville M, Rimondi V, Pratesi G, Costagliola P (2018) Arsenic uptake in bacterial calcite. Geochim Cosmochim Acta 222:642–654

Cheng SY, Show PL, Lau BF, Chang JS, Ling TC (2019) New prospects for modified algae in heavy metal adsorption Trends Biotechnol. in Press, https://doi.org/10.1016/j.tibtech.2019.04.007

Demey H, Vincent T, Guibal E (2018) A novel algal-based sorbent for heavy metal removal. Chem Eng J 332:582–595

Demirbas E, Dizge N, Sulak MT, Kobya M (2009) Adsorption kinetics and equilibrium of copper from aqueous solutions using hazelnut shell activated carbon. Chem Eng J 148:480–487

Dhami NK, Quirin MEC, Mukherjee A (2017). Carbonate biomineralization and heavy metal remediation by calcifying fungi isolated from karstic caves. Ecol Eng 103 (Part A): 106–117

Dhanarani S, Viswanathan E, Piruthiviraj P, Arivalagan P, Kaliannan T (2016) Comparative study on the biosorption of aluminum by free and immobilized cells of *Bacillus safensis* KTSMBNL 26 isolated from explosive contaminated soil. J Taiwan Inst Chem Eng 69:61–67

Dotto GL, Cadaval TRS, Pinto LAA (2012) Preparation of bionanoparticles derived from Spirulina platensis and its application for Cr (VI) removal from aqueous solutions. J Ind Eng Chem 18:1925–1930

Dutta S, Bhattacharyya A, De P, Ray P, Basu S (2009) Removal of mercury from its aqueous solution using charcoal-immobilized papain (CIP). J Hazard Mater 172:888–896

Fuwad A, Ryu H, Lee JH, Kim D, Yoo YE, Kim YR, Kim SM, Jeon TJ (2019) An electrokinetic approach to fabricating aquaporin biomimetic membranes for water purification. Desalination 452:9–16

Gadd GM (2009) Biosorption: critical review of scientific rationale, environmental importance and significance for pollution treatment. J Chem Technol Biotechnol 84:13–28

Grill E, Mishra S, Srivastava S, Tripathi RD (2006) Role of phytochelatins in phytoremediation of heavy metals. In: Singh SN, Tripathi RD (eds) Environmental bioremediation technologies. Springer, Heidelberg, pp 101–145

Gupta P, Diwan B (2017) Bacterial exopolysaccharide mediated heavy metal removal: a review on biosynthesis, mechanism and remediation strategies. Biotechnol Rep 13:58–71

Gupta VK, Rastogi A, Nayak A (2010) Biosorption of nickel onto treated alga (*Oedogonium hatei*): application of isotherm and kinetic models. J. Colloid Interface Sci. 342:533–539

Hosseini-Bandegharaei A, Karimzadeh M, Sarwghadi M, Heydarbeigi A, Hosseini SH, Nedaie M, Shoghi H (2014) Use of a selective extractant-impregnated resin for removal of Pb ion from waters and wastewaters: Kinetics, equilibrium and thermodynamic study. Chem Eng Res Des 92:581–591

Hu X, Li Y, Wang Y, Li X, Li H, Liu X, Zhang P (2010) Adsorption kinetics, thermodynamics and isotherm of thiacalix[4]arene-loaded resin to heavy metal ions. Desalination 259:76–83

Ibuot A, Dean AP, McIntosh OA, Pittman JK (2017) Metal bioremediation by *CrMTP4* over-expressing *Chlamydomonas reinhardtii* in comparison to natural wastewater-tolerant microalgae strains. Algal Res 24:89–96

Jabeen H, Chandra V, Jung S, Lee J, Kim K, Kim S (2011) Enhanced Cr(VI) removal using iron nanoparticle decorated graphene. Nanoscale 3:3583–3585

Jacob JM, Karthik C, Saratale RG, Kumar SS, Prabakar D, Kadirvelu K, Pugazhendhi A (2018) Biological approaches to tackle heavy metal pollution: A survey of literature. J Environ Manage 217:56–70

Jain VK, Kanaiya PH (2011) Chemistry of calix[4]resorcinarenes. Russ Chem Rev 80:75–102

Jing R, Kjellerup BV (2018) Biogeochemical cycling of metals impacting by microbial mobilization and immobilization. J Environ Sci 66:146–154

Johnson DB (2014) Biomining-biotechnologies for extracting and recovering metals from ores and waste materials. Curr Opin Biotechnol 30:24–31

Kang C-H, Kwon Y-J, So J-S (2016) Bioremediation of heavy metals by using bacterial mixtures. Ecol Eng 89:64–69

Kim I, Lee M, Wang S (2014) Heavy metal removal in groundwater originating from acid mine drainage using dead *Bacillus drentensis* sp. immobilized in polysulfone polymer. J Environ Manage 146:568–574

Krajewska B (2018) Urease-aided calcium carbonate mineralization for engineering applications: a review. J Adv Res 13:59–67

Kumar R, Bhatia D, Singh R, Rani S, Bishnoi NR (2011) Sorption of heavy metals from electroplating effluent using immobilized biomass Trichoderma viride in a continuous packed-bed column. Int Biodeterior Biodegrad 65:1133–1139

Rahman Z, Singh VK (2019) The relative impact of toxic heavy metals (THMs) (arsenic (As), cadmium (Cd), chromium (Cr)(VI), mercury (Hg), and lead (Pb)) on the total environment: an overview. Environ Monit Assess 191:419

Kumari D, Pan X, Lee D-J (2014) Immobilization of cadmium in soil by microbially induced carbonate precipitation with Exiguobacterium undae at low temperature. Int Biodeterior Biodegrad 94:98–102

Kuroda K, Ueda M (2010) Engineering of microorganisms towards recovery of rare metal ions. Appl Microbiol Biotechnol 87:53–60

Lee SM, Laldawngliana C, Tiwari D (2012) Iron oxide nano-particles-immobilized-sand material in the treatment of Cu, Cd and Pb contaminated waste waters. Chem Eng J 195–196:103–111

Li Q, Cheng X, Guo H (2013a) Heavy metal removal by biomineralization of urease producing bacteria isolated from soil. Int Biodeterior Biodegrad 76:81–85

Li X, Qi Y, Li Y, Zhang Y, He X, Wang Y (2013b) Novel magnetic beads based on sodium alginate gel crosslinked by zirconium (IV) and their effective removal for Pb^{2+} in aqueous solutions by using a batch and continuous systems. Bioresour Technol 142:611–619

Li J, Wu Y, Li Z, Zhu M, Li F (2014) Characteristics of arsenate removal from water by metal-organic frameworks (MOFs). Water Sci Technol 70:1391–1397

Li Q, Csetenyi L, Paton GI, Gadd GM (2015) CaCO3 and SrCO3 bioprecipitation by fungi isolated from calcareous soil. Environ Microbiol 17:3082–3097

Li M, Cheng X, Guo H, Yang Z (2016) Biomineralization of carbonate by Terrabacter tumescens for heavy metal removal and biogrouting applications. J Environ Eng 142:C4015005

Li J, Wang X, Zhao G, Chen C, Chai Z, Alsaedi A et al (2018) Metal-organic framework-based materials: superior adsorbents for the capture of toxic and radioactive metal ions. Chem Soc Rev 47:2322–2356

Lim S-F, Zheng YM, Zou SW, Chen JP (2008) Characterization of copper adsorption onto an alginate encapsulated magnetic sorbent by a combined FT-IR, XPS, and mathematical modeling study. Environ Sci Technol 42:2551–2556

Mahmoudi M, Sant S, Wang B, Laurent S, Sen T (2011) Superparamagnetic iron oxide nanoparticles (SPIONs): development, surface modification and applications in chemotherapy. Adv Drug Deliv Rev 63:24–46

Marahel F, Ghaedi M, Shokrollahi A, Montazerozohori M, Davoodi S (2009) Sodium dodecyl sulfate coated poly (vinyl) chloride: an alternative support for solid phase extraction of some transition and heavy metals. Chemosphere 74:583–589

Mejare M, Ljung S, Bulow L (1998) Selection of cadmium specific hexapeptides and their expression as OmpA fusion proteins in *Escherichia coli*. Protein Eng 11:489–494

Mugwar A, Harbottle M (2016) Toxicity efects on metals sequestration by microbially-induced carbonate precipitation. J Hazard Mater 314:237–248

Mwandira W, Nakashima K, Kawasaki S (2017) Bioremediation of lead-contaminated mine waste by *Para rhodobacter* sp. based on the microbially induced calcium carbonate precipitation technique and its effects on strength of coarse and fine grained sand. Ecol Eng 109:57–64

Mwangi IW, Ngila JC (2012) Removal of heavy metals from contaminated water using ethylenediamine-modified green seaweed (*Caulerpa serrulata*). Phys Chem Earth A/B/C 50(52):111–120

Nancharaiah YV, Lens PN (2015) Selenium biomineralization for biotechnological applications. Trends Biotechnol 33:323–330

Nancharaiah YV, Venkata Mohan S, Lens PNL (2016) Biological and bioelectrochemical recovery of critical and scarce metals. Trends Biotechnol 34:137–155

Nghiem LD, Mornane P, Potter ID, Perera JM, Cattrall RW, Kolev SD (2006) Extraction and transport of metal ions and small organic compounds using polymer inclusion membranes (PIMs). J Membr Sci 281:7–41

Papageorgiou SK, Katsaros FK, Kouvelos EP, Nolan JW, Le Deit H, Kanellopoulos NK (2006) Heavy metal sorption by calcium alginate beads from *Laminaria digitata*. J Hazard Mater 137:1765–1772

Pawlik-Skowrońska B, Pirszel J, Kalinowska R, Skowroński T (2004) Arsenic availability, toxicity and direct role of GSH and phytochelatins in As detoxification in the green alga *Stichococcus bacillaris*. Aquat Toxicol 70:201–212

Podder MS, Majumder CB (2016) Fixed-bed column study for As(III) and As(V) removal and recovery by bacterial cells immobilized on Sawdust/MnFe2O4 composite. Biochem Eng J 105:114–135

Qureshi I, Memon S, Yilmaz M (2009) Estimation of chromium(VI) sorption efficiency of novel regenerable *p-tert*-butylcalix[8]areneoctamide impregnated Amberlite resin. J Hazard Mater 164:675–682

Qureshi I, Memon S, Yilmaz M (2010) An excellent arsenic(V) sorption behavior of p-tertbutylcalix[8]areneoctamide impregnated resin. C R Chimie 13:1416–1423

Qureshi I, Qazi MA, Bhatti AA, Memon S, Sirajuddin YM (2011) An efficient calix[4]arene appended resin for the removal of arsenic. Desalination 278:98–104

Scindia YM, Pandey AK, Reddy AVR (2005) Coupled-diffusion transport of Cr(VI) across anion-exchange membranes prepared by physical and chemical immobilization methods. J Membr Sci 249:143–152

Sharma A, Kumar R, Mittal S, Hussain S, Arora M, Sharma R, Babu J (2015) In situ reductive regeneration of zerovalent iron nanoparticles immobilized on cellulose for atom efficient Cr(VI) adsorption. RSC Adv 5:89441–89446

Sheha RR, Moussa SI, Attia MA, Sadeek SA, Someda HH (2016) Novel substituted hydroxyapatite nanoparticles as a solid phase for removal of Co(II) and Eu(III) ions from aqueous solutions. J Environ Chem Eng 4:4808–4816

Shen N, Birungi ZS, Chirwa EMN (2017) Selective biosorption of precious metals by cell-surface engineered microalgae. Chem Eng Trans 61:25–30

Shukla A, Srivastava S, D'Souza SF (2018) An integrative approach towards biosensing and bioremediation of metals and metalloids. Int J Environ Sci Technol 15:2701–2712

Silva MM, Pérez DV, Wasserman JC, Santos-Oliveira R, Wasserman MAV (2017) The effect of nanohydroxyapatite on the behavior of metals in a microcosm simulating a lentic environment. Environ Nanotechnol Monit Manage 8:219–227

Son E-B, Poo KM, Chang JS, Chae KJ (2018) Heavy metal removal from aqueous solutions using engineered magnetic biochars derived from waste marine macro-algal biomass. Sci Total Environ 615:161–168

Sun G, Chung TS, Jeyaseelan K, Armugam A (2013) Stabilization and immobilization of aquaporin reconstituted lipid vesicles for water purification. Colloids Surf B Biointeraces 102:466–471

Tamayo-Figueroa DP, Castillo E, Brandao PFB (2019) Metal and metalloid immobilization by microbiologically induced carbonates precipitation. World J Microbiol Biotechnol 35:58

Tiquia-Arashiro SM (2018) Lead absorption mechanisms in bacteria as strategies for lead bioremediation. Appl Microbiol Biotechnol 102:5437–5444

Torricelli E, Gorbi G, Pawlik-Skowronska B, Di Toppi LS, Corradi MG (2004) Cadmium tolerance, cysteine and thiol peptide levels in wild type and chromium-tolerant strains of *Scenedesmus acutus* (Chlorophyceae). Aquat Toxicol 68:315–323

Upadhyay MK, Shukla A, Yadav P, Srivastava S (2019) A review of arsenic in crops, vegetables, animals and food products. Food Chem 276:608–618

Vijayaraghavan K, Yun YS (2008). Bacterial biosorbents and biosorption. Biotechnol Adv 26: 266–291

Wang J, Hong WX, Liang C, Jiu HX, Huai LJ (2014) Preparation and evaluation of magnetic nanoparticles impregnated chitosan beads for arsenic removal from water. Chem Eng J 251:25–34

Williamson AL, Caron F, Spiers G (2014) Radionuclide release from simulated waste material after biogeochemical leaching of uraniferous mineral samples. J Environ Radioactiv 138:308–314

Wu Y, Pang H, Liu Y, Wang X, Yu S, Fu D et al (2019) Environmental remediation of heavy metal ions by novel-nanomaterials: A review. Environ Pollut 246:608–620

Yacou C, Altenor S, Carene B, Gaspard S (2018) Chemical structure investigation of tropical Turbinaria turbinata seaweeds and its derived carbon sorbents applied for the removal of hexavalent chromium in water. Algal Res 34:25–36

Yang J, Yin X (2017) CoFe2O4@MIL-100(Fe) hybrid magnetic nanoparticles exhibit fast and selective adsorption of arsenic with high adsorption capacity. Sci Rep 7:40955

Yang T, Chen ML, Wang JH (2015) Genetic and chemical modification of cells for selective separation and analysis of heavy metals of biological or environmental significance. Trends Anal Chem 66:90–102

Zawierucha I, Kozlowska J, Kozlowski C, Trochimczuk A (2014) Sorption of Pb, Cd and Zn performed with the use of carboxyphenylresorcinarene-impregnated Amberlite XAD-4 resin. Desalin Water Treat 52:314–323

Zawierucha I, Kozlowski C, Malina G (2016) Immobilized materials for removal of toxic metal ions from surface/groundwaters and aqueous waste streams. Environ Sci Process Impacts 18:429–444

Zhang Z, Li M, Chen W, Zhu S, Liu N, Zhu L (2010) Immobilization of lead and cadmium from aqueous solution and contaminated sediment using nanohydroxyapatite. Environ Pollut 158:514–519

Zhu H, Jia Y, Xing W, He W (2009) Removal of arsenic from water by supported nano zero-valent iron on activated carbon. J Hazard Mater 172:1591–1596

Strategies, Challenges, and Advancement in Immobilizing Silver Nanomaterials

Sushrut Bhanushali and Murali Sastry

Abstract Silver nanoparticles have been widely studied for their distinct optoelectronic properties as well as their potent antimicrobial behavior. While these properties lend themselves to a host of applications ranging from sensors to biomedical applications to water purification, an important challenge is in developing experimental methods for the synthesis, assembly, and immobilization of silver nanoparticles over a range of sizes, shapes, and surface chemistry on different surfaces. This chapter will provide an overview on physical, chemical, as well as biological methods to produce silver nanoparticles and their immobilization on different surfaces such as Langmuir–Blodgett films, self-assembled monolayers, and renewable materials, to name a few. This chapter will discuss in some detail potential applications of immobilized silver nanoparticles.

Keywords Silver nanoparticles · Immobilization · Lithography · Self-assembly · Water purification

List of Abbreviations

$AgNO_3$	Silver nitrate
SDS	Sodium dodecyl sulfate
PVP	Polyvinylpyrrolidone
DMF	Dimethyl formamide
THF	Tetrahydrofuran
DMSO	Dimethylsulfoxide
CTAB	Cetyl trimethyl ammonium bromide

S. Bhanushali · M. Sastry
Department of Materials Science and Engineering, Monash University, Clayton, VIC 3800, Australia

M. Sastry (✉)
IITB-Monash Research Academy, IIT Bombay, Powai Mumbai 400076, India
e-mail: murali.sastry@iitbmonash.org

A. Tripathi and J. S. Melo (eds.), *Immobilization Strategies*,
Gels Horizons: From Science to Smart Materials,
https://doi.org/10.1007/978-981-15-7998-1_18

AOT	Sodium di2-ethylsulfosuccinate
CMC	Carboxymethyl cellulose
PVA	Polyvinyl alcohol
$NaNO_3$	Sodium nitrate
CNT	Carbon nanotube
$NaBH_4$	Sodium borohydride
PDMS	Polydimethoxysilane
PET	Poly(ethylene terephthalate)
MPEG	Methoxy polyethylene glycol
PEI	Poly(ethylene imine)
PAH	Poly(allylamine hydrochloride)
MPTMS	3-mercaptopropyltrimethoxysilane
APTES	3-aminopropyltriethoxysilane
ITO	Indium tin oxide
PAA	Porous anodic alumina
DNA	Deoxyribonucleic acid
HCP	Hexagonal close packing
FCC	Face-centered cubic
LSPR	Localized surface plasmon resonance
Nd:YAG	Neodymium-doped yttrium aluminum garnet ($Nd:Y_3Al_5O_{12}$)
EDL	Electric double layer
VdW	Van der Waals forces
DLVO	Derjaguin, Landau, Verwey, and Overbeek theory
A_H	Hamacker constant
κ	Inverse Debye screening length
ε_0	Permittivity of free space
ε_r	Dielectric constant
k_B	Boltzman constant
T	Temperature
EBL	Electron beam lithography
FIB	Focused ion beam lithography
NIL	Nanoimprint lithography
SERS	Surface-enhanced Raman spectroscopy
LED	Light-emitting diode
RFID	Radiofrequency identification

1 Introduction

Silver, among other coinage metals, has been historically an important element and has been used as jewelry, décor, currency, and wealth reserve. Apart from its ornamental and monetary value, silver, in various forms, has also been used as a functional material. For e.g., silverware was historically known to preserve food and water, achieved better wound healing and was used during ancient Greek, Roman, Persian, and Egyptian civilizations, long before the knowledge of microbial origin of infection (History of the Medical Use of Silver 2009). With unprecedented scientific progress in the nineteenth and twentieth centuries, silver found widespread use in technological applications such as photography (Bensley 1950), reflective coatings (Stenhouse 1843), electroplating (Barclay 1921), treatment of burns (Kissmeyer 1936), dental fillings (Auerbach 1953), and biological staining (Foot 1888; Taft and Ludlum 1930). Nanoforms of silver, conventionally referred to as colloids, have been investigated and used for more than a 100 years (Nowack et al. 2011). With the advent of nanotechnology in the past few decades, nanoparticulate silver has been one of the key materials subject to intense fundamental as well as applied research. The academic as well as industrial interest in silver as an important material arises from its unique properties in bulk form, for e.g., Ag has the highest electrical and thermal conductivity among all metals, as well as properties evidenced by Ag in form of nanomaterials. For e.g., metallic Ag nanoparticles exhibit localized surface plasmon resonance (LSPR), a unique light matter interaction resulting from the collective oscillation of free electrons (Jain et al. 2007). Since the resonant energy corresponds to the visible light wavelength range of the electromagnetic spectrum, dispersions of Ag nanomaterials exhibits striking colors, distinct from that of bulk silver metal (Sun and Xia 2003). This LSPR absorbance profile can be tuned precisely by tailoring the size, morphology, and chemical environment of the nanoparticles (Mock et al. 2002; Jackson and Halas 2001) and has been abundantly investigated for applications in optoelectronics (Rossouw et al. 2011) plasmonics (Maier et al. 2001), sensing (McFarland and Van Duyne 2003), imaging (Lee and El-Sayed 2006), theranostics (Mukherjee et al. 2014), etc. The size and morphology of the nanoparticles also have other important consequences, such as surface-to-volume ratio and consequently the specific surface area, which have direct implications in catalysis (Xu et al. 2006), antimicrobial surfaces (Pal et al. 2007), and water purification (Helmlinger et al. 2016). In many of these applications, the Ag nanomaterials need to be immobilized either as free-standing entities or on/within scaffolds for the desired applicability, efficiency and reusability. The assembly and interactions between these building blocks is not trivial, and thus, it is imperative to have a control over their size, morphology, and surface physico-chemical functionality. Here, we first briefly describe the established methods and protocols for synthesis of Ag nanomaterials in conjunction with their characterization. We then discuss approaches to assembly and immobilization of Ag nanoparticles with specific examples. Further, we highlight applications of immobilized Ag nanomaterials such as optoelectronic applications, biomedical applications, catalysis, and low-cost water purification.

2 Synthesis of Ag Nanomaterials

Nanomaterials are typically synthesized by either top-down or bottom-up approaches, the latter being more predominant in the case of metallic nanomaterials owing to the low material and energy costs, better control on size and size distribution, and viability to control the shape and surface functionality in a single step. The methods can be further classified broadly into physical, chemical, and biological methods, depending on the agents that are employed in converting the precursor into their nanoparticulate form. While there have been some comprehensive reviews on various aspects on synthesis of metallic nanomaterials (Xia et al. 2009; Cuenya 2010; Sau and Rogach 2010; Xia et al. 2015; Zhang et al. 2012; Fan and Zhang 2016; Chen et al. 2018), here, we focus on well-established synthesis methods relevant to metallic silver nanoparticles.

2.1 *Physical Methods*

2.1.1 Laser Ablation Method

Metallic nanoparticles can be produced by irradiating a metal target with a laser with sufficient energy leading to ablation of the metal. The metal atoms and clusters thus released aggregate into nanoparticles, which need to be capped suitably to prevent formation of larger aggregates. Ag nanoparticles were synthesized by Mafuné et al. by irradiating a silver metal plate in aqueous solution of sodium dodecylsulfate (SDS) using pulsed Nd:YAG laser (532 nm, 10 Hz) with a power of 20–100 mJ/pulse (Mafuné et al. 2000a). The resultant Ag nanoparticles were found to have spherical morphology with diameters of 8–20 nm. The particle size was found to be inversely proportional to laser power as well as the SDS concentration. The same group demonstrated the particle size and stability of the nanoparticles as a function of the alkyl chain length ($n = 8, 10, 12, 16$) of the capping surfactant (Mafuné et al. 2000b). In a similar system, Tsuji and co-workers demonstrated the dependence of Ag nanoparticle size on the wavelength of the laser source (Tsuji et al. 2002). The mean diameter was found to decrease with the wavelength, and particle sizes of 29 $\pm$ 13 nm, 26 $\pm$ 11 nm, and 12 $\pm$ 8 nm were reported for $\lambda = 1064$ nm, 532 nm, and 355 nm, respectively. Ag nanoparticles prepared by laser ablation in aqueous solutions of polyvinylpyrrolidone (PVP) were found to have mean diameter in the range of 10–20 nm, decreasing with increasing concentration of PVP (Tsuji et al. 2008). Tilaki and co-workers demonstrated the effect of solvent on the particle size of Ag nanoparticles prepared by ablation using Nd:YAG laser (1064 nm), where the mean diameter of 5, 13, and 22 nm was reported for the media acetone, water, and ethanol, respectively (Tilaki et al. 2006). Ablation-based synthesis has also been reported in other solvents such as acetonitrile, dimethylformamide (DMF), tetrahydrofuran (THF), and dimethylsulfoxide (DMSO) (Amendola et al. 2007). A much smaller particle size (2–5 nm) and a narrow size distribution was reported by Suzuki and co-workers by using higher laser power (340 mJ) and smaller spot sizes (0.6 mm). The size and polydispersity of the particles was found to increase with increasing the spot sizes (Pyatenko et al. 2004).

2.1.2 Arc Discharge Method

Arc discharge is a high-pressure discharge (1–100 atm) formed between two electrodes due to thermionic or field emission of electrons and characterized by high currents (1–100 A) and low voltages (1-50 V) (Wüthrich and Abou Ziki 2015). In a typical arc discharge setup, the anode and cathode are assembled and contained in a stainless steel chamber. The arc discharge is sustained between the closely placed electrodes by a constant power supply, and can be used to produce nanoparticles in gas as well as liquid phase (Hutchison et al. 2001). While this method is more relevant to materials such as carbon nanotubes (Shi et al. 1999) and graphene (Subrahmanyam et al. 2009), there have been a few reports on production of metallic nanoparticles including Ag using this method. For e.g., Ag nanoparticles were synthesized using 15A discharge current between titanium electrodes (99% pure, wire, $d = 2$ mm) in aqueous $AgNO_3$ solutions, with sodium citrate as stabilizer. The nanoparticles thus formed were found to have sizes in the range 27 ± 14 nm and were rather polydisperse (Ashkarran 2010). Ag nanoparticles were also produced without any surfactant in deionized water at a peak current of 6.4 A between silver wire electrodes (Tien et al. 2008). Ag nanoparticles were synthesized by Lo et al. in water ($d = 25$ nm) and ethylene glycol ($d = 13$ nm)(Lo et al. 2007). The effect of arc current on the particle size of Ag nanoparticles formed by arc discharge in aqueous medium was reported by Ashkarran and co-workers, where the particle size was found to marginally increase with current in the range of 10 to 20A. Effect of various surfactants such as cetyl trimethylammonium bromide (CTAB), polyvinylpyrrolidone (PVP), sodium citrate, sodium dodecyl sulfate (SDS), sodium di2-ethylsulfosuccinate (AOT), and carboxymethyl cellulose (CMC) was also reported (Ashkarran et al. 2009). The effect of temperature on the size of Ag nanoparticles synthesized in polyvinylalcohol (PVA) solution by arc discharge was reported, where the particle size increased from 10 nm to 30 nm when temperature was increased from 130 to 200 °C (Zhou et al. 1999a). Ag nanowires with diameter of 80 nm and lengths up to 14 μm were reported to form by arc discharge between Ag electrodes in $NaNO_3$ solution, showing the possibility of shape control using this method (Zhou et al. 1999b).

2.1.3 Vapor Deposition Method

In this technique, nanoparticles are grown by deposition of vapor phase of the precursor on to a suitable substrate under elevated temperatures (Cai et al. 2018). This method has been successfully used for materials such as silicon (oxide, nitride, carbide) (Dal Negro et al. 2003) and carbon (CNTs, graphene, nanodiamonds) (Chen et al. 2011), while very few reports exist for Ag nanomaterials. Thermally evaporated silver was deposited on thin layer of hafnium oxide (HfO_2) where Ag nanoparticles with size 5–20 nm were observed to be formed (Yan et al. 2002). Nepijko and co-workers reported formation of Ag nanorods, when gas phase Ag clusters were deposited on (100) plane of NaCl crystal (Nepijko et al. 2000). The nanorods with lengths up to 70 nm and aspect ratio (AR) of up to 10 exhibited penta-twinned symmetry along the (110) orientation. While these examples provide the proof of concept of formation of Ag nanomaterials by vapor deposition, the products have a rather low yield and high polydispersity, making it a method with limited practical applicability.

2.2 *Chemical Methods*

Chemical synthesis of metallic nanoparticles typically involves reduction of solvated metal ions to their zero-valent form using a reducing agent. This complex process is initiated by nucleation, which is the creation of a distinct thermodynamic phase throughout the homogenous medium (LaMer and Dinegar 1950). The reduced metal ions act as adatoms and deposit on the fluctuating nucleated structures leading to the growth of the crystal. This process is reversible, and smaller particles, due to their high surface energy and solubility, can redissolve and redeposit on larger particles, leading to size-focusing effect also known as Ostwald ripening (Thanh et al. 2014). In order to arrest the growth of the crystal within nanodimensions, capping agents are generally employed which bind to the surface of the nanoparticle and prevent their uncontrolled growth and coalescence. These capping agents are typically amphiphilic molecules such as surfactants, polymers, and biomolecules (amino acids, polysaccharides) (Campisi et al. 2016). Capping agents play a crucial role in nanomaterial synthesis, as they can be used to direct the shape of the growing nanoparticle (Zeng et al. 2010), render desired surface chemistry to the particle (Stankus et al. 2011; Niu and Li 2014), control the interaction of the nanoparticle with surrounding medium (Bhattacharya et al. 2006), and act as a linker moiety for the conjugation and assembly of nanoparticles (Zhou et al. 2014). The characteristics of nanoparticles, both as a single particle and as a population, are highly sensitive to reaction conditions such as temperature, pH, solvent medium, concentration of reactants, and presence of impurities, and thus, a precise control on these parameters is crucial to control the thermodynamics and kinetics of particle growth. Here, we describe the key methods to synthesize silver nanomaterials using wet chemical methods.

2.2.1 Chemical Reduction in Solution

Wet chemical synthesis of silver nanoparticles typically involves reduction of solvated Ag^+ ions from silver salt such as $AgNO_3$ with reducing agents such as hydrogen (Merga et al. 2007), hydrazine (Nickel et al. 2000), citrate (Henglein and Giersig 1999), ascorbate (Sondi et al. 2003), and sodium borohydride ($NaBH_4$) (Shirtcliffe et al. 1999). Use of a strong reducing agent such as $NABH_4$ has been reported to form nanoparticles with smaller sizes and narrow size distribution (Schneider et al. 1994; Shirtcliffe et al. 1999), while using a weaker reducing agent such as citrate leads to a slower reduction rate producing nanoparticles with rather poor monodispersity. In this method, citrate serves as both the reducing as well as capping agent and remains a popular approach to produce Ag nanoparticles with relative ease (Henglein and Giersig 1999). By altering the pH of the solution using NaOH in the citrate-based approach, Murphy et al. demonstrated the synthesis of Ag nanowires, where the shape control was attributed to the deprotonated states of citrate and consequently their binding with Ag^+ ions (Caswell et al. 2003). Dong et al. showed that by varying the pH of the solution in the range of 5–11, triangular Ag plates could be obtained at lower pH (5.7) while Ag nanorods were the predominant product in the higher pH range (11.1) (Dong et al. 2009) Ag nanorods and nanowires can be synthesized with a much better control on size and aspect ratio by using sodium dodecylsulfonate (SDS) as a shape-directing capping agent in addition to citrate (Hu et al. 2004). Apart from the reducing agents discussed above, alcohols too possess the capability to reduce Ag ions in solution. This polyol method for synthesizing Ag nanomaterials is a versatile process and can be used to produce a variety of particle shapes and sizes, where a polyol such as ethylene glycol, propylene glycol, and pentanediol acts as a solvent as well as a reducing agent at elevated temperatures (Fiévet et al. 1989, 2018). Xia and co-workers have investigated the mechanism of polyol process, where they reported that the growing Ag nanoseeds emerge as one of (i) single-crystalline, (ii) single-twinned, or (iii) multiply twinned structures. The single-crystalline structures grow into quasispherical or cuboctahedron morphology, and eventually evolve to form nanocubes or truncated nanocubes with further addition of Ag atoms (Wiley et al. 2004). On the contrary, the multiply twinned seeds grow into nanorods and nanowires with a fivefold symmetry. The growth occurs preferably at twin defects along the (110) axis, as it is associated with no strain and thus thermodynamically favored (as against the growth perpendicular to the axis, which has a large associated strain) leading to breaking the symmetry of the growing seed. The polyol synthesis methods typically use polyvinylpyrrolidone (PVP) as a capping agent, which selectively binds to (100) facets of Ag (Sun et al. 2003). Since all three types of seeds mentioned above can be formed during nucleation in the polyol method, the resulting PVP capped Ag particles are a mixture of different shapes such as nanocubes, nanorods, nanowires, and nanobipyramids. Xia et al. have successfully demonstrated that the reaction could be forced along one of the three pathways by introducing oxidative etchants in the system (Wiley et al. 2007). For e.g., twinned seeds are more susceptible to oxidative etching due to high intrinsic strain and thus can be selectively dissolved by introducing trace etchants such as Cl^-. The Cl^-/O_2 system selectively dissolves singly twinned and multiply twinned seeds and leaves behind single-crystalline seeds which eventually grow into Ag nanocubes (Wiley et al. 2004). Further, they also demonstrated that by carefully selecting the oxidative etchants, e.g., Br^-/O_2 (which is weaker than Cl^-/O_2), selectivity between singly twinned and multiply twinned seeds can be achieved as well (Wiley et al. 2006).

2.2.2 Seed-Mediated Synthesis

A related synthetic approach to generating nanocrystals is the use of preformed nanocrystals as seeds for subsequent growth of desired nanostructures. In this method, the nucleation and growth steps are separated, thus allowing a better control on the final size, shape, and aspect ratio of the nanoparticles (Xia et al. 2017). Typically, the seeds are synthesized in a short burst of nucleation using a strong reducing agent, and then allowed to age. These seeds are introduced in a growth solution, which contains the metal ions and a weak reducing agent (Pietrobon et al. 2009). The neutral adatoms deposit on the seed epitaxially, depending on the lattice parameter, leading to growth of the nanocrystal. The morphology of the resultant nanostructure is governed by the relative growth rates of different crystallographic faces, making it a complex yet versatile approach to growing a wide variety of nanostructures. For e.g., Aherne and co-workers synthesized triangular Ag nanoprisms using Ag seeds produced by borohydride reduction. A set quantity of the Ag seeds was introduced in a growth solution containing $AgNO_3$ and a mild reducing agent, ascorbic acid. The edge lengths of the Ag nanoprisms could be precisely varied from ~40 nm to 100 nm by simply varying the quantity of the seeds introduced (Aherne et al. 2008). Ag nanocubes were synthesized by Zhang and co-workers by introducing single-crystalline Ag seeds in $AgNO_3$ growth solution, where the edge lengths of the cubes were readily tuned between 30 nm to 200 nm by controlling the relative concentration of Ag seeds and $AgNO_3$ (Zhang et al. 2010). Pentagonal Ag nanorods with a narrow size distribution were synthesized by using decahedral Ag seeds in aqueous citrate solutions, where the diameter of the rods could be varied from 50 nm to 2 μm by tuning the size of the decahedral seeds (Pietrobon et al. 2009). Ag nanostructures can also be grown by heteroepitaxial seeded growth, where the seeds and the nanostructure overgrowth are chemically different. The lattice mismatch between the two elements plays a critical role in whether the growth would take place epitaxially or non-epitaxially (For e.g., Au and Ag have 0.25% lattice mismatch, while Pt and Ag have 4.15% lattice mismatch). In case of a large lattice mismatch, isotropic growth is unfavorable due to associated high strain, resulting in anisotropic growth. For e.g., Ag nanowires with uniform diameters (30–40 nm) and lengths up to 50 μm were synthesized with a high yield by seeded growth using Pt seeds in ethylene glycol solution containing PVP as capping agent (Sun et al. 2002). It should be noted that the experimental conditions such as temperature, rate of addition, mixing, age of the seeds, and impurities can drastically affect the product, and thus, these need to be well controlled. Heterogeneous nucleation can result from using contaminated and scratched glassware, stirrer bars, etc., and due care needs to be taken for achieving a high yield, high phase purity product.

2.2.3 Template-Directed Synthesis

Template-based synthesis of nanomaterials relies on growing the nanostructures on or within a scaffold, where the size and morphology of the generated structures complement the template. Template-based approaches can broadly be divided into hard templates and soft templates. Porous anodic alumina (PAA) with rigid channels is probably the best example of hard template in nanomaterial synthesis. PAA can be prepared with nanodimensional channels ranging from 5 to 500 nm in diameter, 10–500 μm in thickness, and high pore densities of up to 10^{11} channels/cm^2 (Riveros et al. 2006). For e.g., Ag nanowires with tuneable diameters ranging from 10 to 40 nm were synthesized by AC electrodeposition of silver in PAA (Zong et al. 2004). Similarly, mesoporous silica (SBA-15) can be used to grow Ag nanowires by electroless deposition (Han et al. 2000). The hard templates provide an opportunity to prepare multicomponent nanomaterials, such as nanowires with different metals along the length, which is difficult to achieve by other methods. On the other hand, the hard templates need to be etched out or dissolved at the end of the synthesis to liberate the nanostructures, limiting its applicability. It should also be noted that nanomaterials formed within templates are often polycrystalline.

Soft templates are microemulsions (micelles and reverse micelles) which form spontaneously in solution on introduction of their components (surfactants) above the critical micellar concentration (CMC) (Petit et al. 1993). The size and shape of the micellar structure (e.g., rod-shaped micelles) direct the dimensions and morphology of the growing nanostructure (Kijima et al. 2004; Capek 2004). Ag nanoparticles were synthesized by Pileni and co-workers in AOT reverse micelles, leading to highly monodisperse spherical particles with 2–4 nm diameter (Taleb et al. 1997). The nanoparticles were found to spontaneously self-organize in a 2D hexagonal close-packing structure. Highly shape-selective Ag nanoprisms with 24 nm thickness and 68 nm edge length were synthesized in CTAB micelles, which were found to self-assemble into chain-like structures along the (111) basal plane (Chen and Carroll 2002). Other soft templates such as DNA, peptides, and dendrimers can serve as soft templates for nanoparticle synthesis as well. For e.g., Braun and co-workers demonstrated electrically conductive Ag nanowires (12 μm length, 100 nm diameter) grown using DNA as a template toward application in nanocircuit interconnects (Braun et al. 1998). The shape-directed synthesis is a result of binding of Ag ions to DNA, which on reduction take the shape of the template. The soft-templated synthesis has a number of advantages, such as precise control on morphology, surface functionality, and prevention of coalescence of formed nanostructures due to steric repulsion.

2.2.4 Light-Mediated Synthesis

The sensitivity of silver to light is a well-known phenomenon and has been utilized thoroughly in photography on commercial scales. Light excitation has been demonstrated to have a significant effect on Ag nanomaterials pre as well as post-synthesis. In their seminal paper, Mirkin et al. demonstrated photoinduced conversion of Ag nanospheres to Ag nanoparisms using visible light (Jin et al. 2001). Further, the same group demonstrated the edge length of the nanoprisms could be precisely controlled depending on the wavelength of light used for photoexcitation. The resulting nanoprisms were monodisperse, and the edge length could be tuned from 30 to 120 nm (Jin et al. 2003). The proposed mechanism suggested that the LSPR excitation leads to the photooxidation of the surface bound citrate molecules. The decarboxylation of the oxidized citrate results in electron transfer to the Ag particle, which in turn increases the rate of Ag^{+} ion reduction and deposition. Similar photoexcitation-based size- and shape-controlled synthesis of other nanostructures such as Ag nanocubes (Zhou et al. 2016), bipyramids (Zhang et al. 2009), and alloy nanoshells (Aslam and Linic 2016), has been reported. It should be noted that the radiant power of the light used for plasmon excitation is typically close to ambient; as against the reports on high-power pulsed lasers causing reshaping of metallic nanoparticles due to photothermal effects (Link et al. 1999).

2.3 Biological Methods

Biological synthesis of metallic nanoparticles broadly refers to reduction of metal ions by exposure to biological entities such as microorganisms, e.g., bacteria, yeast, fungi, etc., and plant material, e.g., leaf, root, bark, fruit, and flower extracts (Thakkar et al. 2010). The microorganisms reduce the metal ions as a defense mechanism against the heavy metal ion cytotoxicity. This is achieved either by extracellular (biomineralization, biosorption, precipitation, etc.) or intracellular (e.g,. bioaccumulation) modes (Narayanan and Sakthivel 2010). Similarly, plants are rich in phytochemicals such as polyphenols, alkaloids, flavonoids, terpenoids, glycolipids, and glycoproteins that have the capability to reduce metal ions to their neutral form (Mittal et al. 2013). Biological methods provide an exciting pathway to generate metallic nanostructures because (a) the synthesis typically occurs at ambient conditions of temperature, pressure, and pH; (b) the biochemicals involved act as reducing as well as capping agents, and synthesis is typically performed in aqueous media; (c) complex particle morphologies can be achieved in a relatively straightforward manner, and (d) the biosynthesized nanoparticles are biocompatible, though this claim is highly debated.

2.3.1 Biosynthesis Using Plants

The Sastry group has reported the synthesis of Ag and Ag-Au alloy nanoparticles using the leaf broth of Neem plant (*Azadirachta indica*), where the flavanone and terpenoid constituents were found to be responsible for the particle synthesis (Shankar et al. 2004). The quasispherical nanoparticles formed were polydisperse with sizes ranging from 5 to 35 nm. Silver nanoparticle synthesis with size distribution of 15.2 ± 4.2 nm was reported by the same group using *Aloe vera* extract (Chandran et al. 2006). Aqueous $AgNO_3$ reduction performed using Geranium (*Pelargonium graveolens*) leaf extract resulted in crystalline Ag nanoparticles with 15–40 nm diameter, and was found to assemble in quasilinear structures in solution (Shankar et al.

2003). Huang and co-workers reported the synthesis of Ag nanoparticles with sizes ranging from 55–80 nm, using sun-dried biomass of *Cinnamomum camphora* leaf, where polyols were found to be the reducing and capping agent (Huang et al. 2007). Similar synthesis of Ag nanoparticles using plant extracts of *Acalypha indica* has been reported by Krishnaraj et al. (Krishnaraj et al. 2010). For more examples, the review by Mittal et al. comprehensively covers plant-based synthesis of Ag nanomaterials (Mittal et al. 2013). Biosynthesis using plant materials is relatively straightforward, can be carried out at high concentrations, and occurs much rapidly as compared to microbial synthesis. However, it suffers considerably from polydispersity of the nanoparticle population. The seasonal, geographical, and environmental variability of phytochemical makeup of a given plant also severely affects the reproducibility of the synthesis, and thus, precise chemical entities responsible for the synthesis need to be elucidated, isolated, and well characterized from the phytochemical cocktail, for scalable and reproducible use of this method.

2.3.2 Biosynthesis Using Microbes

Biosynthesis of metallic nanoparticles by microbes primarily occurs as a defense mechanism against heavy metal ions, which are reduced by the organism to its less toxic metallic form, typically of the order of nanodimensions (Ramanathan et al. 2013). This phenomena can be harnessed to our advantage as a green chemical pathway to nanoparticle synthesis. For e.g., Gurunathan et al. demonstrated the synthesis of Ag nanoparticles using the bacterium *Bacillus licheniformis*, where the enzyme nitrate reductase was speculated to be responsible for Ag^+ reduction (Kalimuthu et al. 2008). Ramanathan et al. reported the synthesis of Ag nanoplates using *Morganella psychrotolerans*, a silver-resistant psychrophilic bacterium. This was particularly remarkable because it was the first report that demonstrated the shape anisotropy in Ag nanoparticles by controlling the bacterial growth kinetics (Ramanathan et al. 2011). Sastry group reported the intracellular synthesis of Ag nanoparticles using the fungus *Verticillium* sp. (Mukherjee et al. 2001). The nanoparticles with particle size 25 ± 12 nm were formed beneath the cell wall possibly reduced by cell membrane enzymes. The fungal cells were viable and continued to grow post-synthesis. The same group reported the extracellular synthesis of Ag nanoparticles using the fungus *Fusarium oxysporum* with size range of 5–15 nm (Ahmad et al. 2003). Similar extracellular synthesis of Ag nanoparticles (5–25 nm) was demonstrated using *Aspergillus fumigatus* (Bhainsa and D'Souza 2006). It is important to note that extracellular synthesis is advantageous as a method of synthesis as the nanoparticles can be recovered by a relatively simple way. On the contrary, intracellular synthesis requires cell lysis, and the particle extraction in this case is both cumbersome and expensive. We would like to direct the reader to comprehensive reviews by Narayanan and Sakthivel (2010) and Thakkar et al. (2010) for more examples on microbial synthesis of metallic nanoparticles.

2.4 Comparison of Synthesis Methods

Here, we present an overview and general comments on the various synthesis methods briefly reviewed above. The physical methods are relatively expensive due to the cost of the specialized equipment, energy input, and hazards (high voltages and currents, high intensity lasers, high temperatures), making them less suitable for practical applications. Limited control over particle size, morphology, crystallinity, and a high polydispersity remains a concern for most of the physical methods. The chemical pathways are more versatile, low cost, and offer a precise control on the particle size, morphology, and surface functionality. The thermodynamic and kinetic parameters, while complicating the experiment, serve as suitable handles to precisely tailor the above aspects of particle genesis. While size distribution remains an issue in many protocols, with the wealth of knowledge available in the published literature, there are numerous reproducible methods of Ag nanoparticle synthesis, perfected over time by multiple research groups that result in nanoparticles with precise size, morphology, monodispersity, high yield, and at low cost. The seed-mediated methods give the chemical synthesis method a robust and powerful edge with the high degree of precision and accuracy that can be achieved in controlling the size, morphology, and crystallographic aspects of the growing nanostructures. Hard template-directed methods, though attractive due to the imposed size and morphology by design, suffer due to the polycrystallinity of the nanostructures. The template needs to be removed by dissolution or etching at the end of the synthesis adding to the cost, effort, and hazard. Also, there are a limited number of morphologies such as nanorods and nanowires, for which templates can be conveniently prepared. Conversely, soft templates are a versatile option, due to the precision they offer for tailoring the particle size, morphology, and surface functionality. As seen in examples above, they offer nanoparticles with high size selectivity, monodispersity, and dispersion stability. On the downside, however, the soft templates (micelles and reverse micelles) are composed of surfactants, such as AOT, CTAB, and SDS, and organic solvents, often in large quantity, which are toxic to the environment. The light excitation directed synthesis is a distinct method particularly applicable for Ag nanoparticles and offers a green pathway to generate uniform anisotropic particles simply by shining low-to-medium intensity light on the reactant solution. This method continues to garner a significant interest among researchers and has a high probability of deployment in real-world applications (e.g., sunlight-driven reactions). Biological methods provide a cost-effective and near environmentally benign route to nanoparticle synthesis due to the use of bioderived reducing and capping agents, mild reaction conditions (often room temperature and pressure), non-toxic solvent (most syntheses reported are in aqueous solutions), and no requirement of specialized equipment. However, a significant downside they suffer is that the syntheses often lead to a cocktail of particle sizes and morphology, which might be of limited use in certain applications such as self-assembly. The quantitative yield and time required for completion of syntheses is another limiting factor. While plant extracts require few hours to a day for synthesis, microbial systems typically require several days, limiting their practical use as viable nanomaterial production methods. The intracellular microbial synthesis suffers from a further disadvantage due to the additional steps required for recovery of the nanoparticles from inside the cells. Perhaps, the biggest limitation faced by biosynthesis methods is the seasonal, geographical variability, and the variation within the biological species for reproducible production of nanoparticles. The choice of synthesis method, however, would ultimately depend upon the envisaged application. In subsequent sections, various aspects of assembly and immobilization of Ag nanomaterials and their applications would be discussed.

3 Strategies and Mechanisms for Immobilization of Ag Nanomaterials

The initial research in nanoscience and nanotechnology focused primarily on the synthesis of nanoparticles and a wealth of synthetic protocols is now available to fabricate nanomaterials with control over their size, shape, crystal structures, and surface chemistry. A myriad of nanostructures such as spheres, polyhedra, plates, cubes, rods, wires, cages, and core-shell structures can be synthesized with a high yield and monodispersity (Burda et al. 2005). Many of applications developed based on these nanostructures such as heterogeneous catalysis (Zhu and Xu 2016), biosensors (Yan et al. 2009), waveguides (Zou and Schatz 2006), optoelectronic components (Maier et al. 2003a), stretchable soft electronics (Jung et al. 2019), and transparent flexible electrodes (Hu et al. 2010) are not based on individual nanoparticles but their ensembles, where these nanoparticle act as building blocks to organize and mutually interact in purposeful ways. Thus, the research focus in nanosciences has gradually evolved from synthesizing these individual components toward their controlled assembly into higher order structures and manipulating their collective properties in an efficient, precise, and robust manner.

3.1 Nanoscale Forces and Interactions

In order to elucidate the mechanistic pathways leading to these assemblies, it is important to better understand the nanoscale forces and the spatial and temporal scales they operate in and how the nanobuilding blocks interact mutually under the influence of these forces. Though the fundamental interactions between atoms/molecules and between macroscopic objects are well understood, extrapolating these laws to the intermediate nanoscale regimes is not trivial. Here, we briefly discus these forces and interactions that operate at the nanoscale, which can be creatively used for assembly of nanoparticles.

3.2 Van der Waals Interactions

Van der Waals forces are short range forces between two bodies such as atoms, molecules, and colloids, which arise from electromagnetic fluctuations caused due to incessant oscillations of charges within these bodies (Dzyaloshinskii et al. 1961). The Van der Waals forces are the most ubiquitous form of interactions which are generally attractive in nature and rapidly vanish with distance between the interacting bodies (Israelachvili 2011b). The Van der Waals interaction (u) between two bodies can be expressed as

$$u_{\text{vdW}}(r) = \frac{-C_{\text{vdW}}}{r^6} \tag{1}$$

where r is the distance between the two bodies, and C_{vdW} is a constant encompassing three distinct interactions viz (a) permanent dipole—dipole/multipole—multipole interactions also known as Keesom interaction, (b) permanent dipole—induced dipole interaction, also known as Debye interaction, and (c) fluctuating dipole—induced dipole interactions, also known as London dispersion forces; (Hamaker 1937) all of which scale inverse to the sixth power of the interbody distance (r). Since Van der Waals forces are omnipresent and their collective magnitude can be significant, they can cause agglomeration of nanoparticles and thus are often considered unfavorable. However, by modifying the nanoparticles with suitable capping agents and solvents, Van der Waals interactions can be used as a powerful tool to steer nanoparticles to self-assemble. The Van der Waals force-driven self-assembly of nanoparticles typically results in closed pack assemblies such as 2D hexagonal arrays and 3D hexagonal close-packed (HCP) or face-centered cubic (FCC) crystals. It is, however, complicated to gauge the contribution of Van der Waals forces to the self-assembly, since such effect can also be observed in case of entropy-driven assembly when particle volume fraction crosses a threshold value.

3.3 Electrostatic Interactions

Electrostatic interactions (also known as Coulombic interactions) arise from the charge acquired by a particle in solution, either from adsorption of ionic species on the surface or from the dissociation of ionizable surface chemical groups, also known as the single electrode potential. Electrostatic interactions can either be attractive or repulsive depending on the similar or opposite charge on the particles (Israelachvili 2011a). The excess charge on the particle leads to the formation of electric double layer, which consists of strongly bound layer of oppositely charged ions, known as Stern layer and a loosely bound layer of counter ions known as the diffuse layer (see Fig. 1) The length scale and magnitude of these interactions can be tailored by use of suitable ionic species (size, valency, concentration) as well as the solvent (dielectric

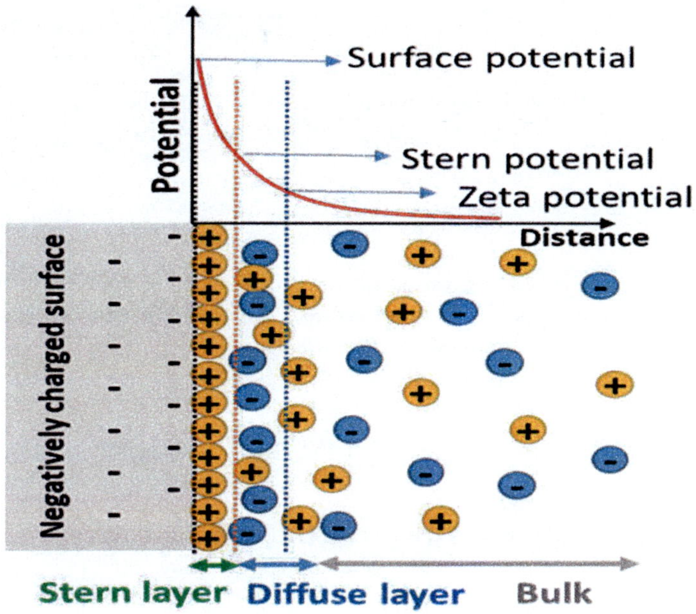

Fig. 1 Schematic of structure of electric double layer (EDL). The orange dotted line represents the Helmholtz plane, while the blue dotted line represents the slipping plane

constant), and thus can be used for electrostatic stabilization of nanoparticles in solution as well as for their assembly into superstructures.

The long-range attractive forces and short-range repulsive forces arising from the electric double layer (EDL) and the Van der Waals (VdW) interactions are summed up by the Derjaguin, Landau, Verwey, and Overbeek (DLVO) theory to estimate the interaction potential between colloidal particles (Verwey 1947; Derjaguin and Landau 1993). In brief, The DLVO interaction can be modeled as the summation of the attractive (VdW) and repulsive (double layer) interactions as:

$$V_{\mathrm{DLVO}} = V_A + V_R \tag{2}$$

For identical charged spheres, with radius r and inter-particle separation d, the attractive VdW forces can be expressed as

$$V_A = -\frac{A_H r}{12d} \tag{3}$$

where A_H is Hamacker constant, which depends on the polarizability and the density of the material. When particles are dispersed in a medium, the composite Hamaker constant can be estimated by taking the geometric mean of the particle $A_{H(\mathrm{p})}$ and that of the medium $A_{H(\mathrm{m})}$ with respect to their values in vacuum as (Cosgrove and Cosgrove 2010).

$$A_H = \left(\sqrt{A_{H(\mathrm{p})}} - \sqrt{A_{H(\mathrm{m})}}\right)^2 \tag{4}$$

As a result of the double layer forces arising between the two particles with nonzero surface potential (ϕ_0), the repulsive interactions are expressed as

$$V_R = \frac{64\pi k_\mathrm{B} T r C^* \Gamma_0^2 \mathrm{e}^{-\kappa d}}{\kappa^2} \tag{5}$$

where C^* is the number concentration of the ionic species in the bulk of the solution, k_B is Boltzman constant, T is temperature, Γ_0 is the reduced surface potential, expressed as a hyperbolic tangent of the ratio of electrostatic energy to random thermal energy and for symmetric electrolytes that can be expressed as

$$\Gamma_0 = \tanh\left(\frac{z q_e \phi_0}{4 k_\mathrm{B} T}\right) \tag{6}$$

where z is the ionic valency. κ is the inverse Debye screening length, which is the thickness of the diffuse electric double layer, and is given by

$$\kappa^{-1} = \left(\frac{\varepsilon_0 \varepsilon_r k_\mathrm{B} T}{2 q_e^2 C^*}\right)^{\frac{1}{2}} \tag{7}$$

where ε_0 is the permittivity of free space and ε_r is the dielectric constant. The Debye screening length is typically of the order of one to few tenths of a nanometer (Israelachvili 2011c). The DLVO theory has important implications in colloidal stability. For a detailed treatise on the DLVO theory, please refer to the works by Derjaguin, Churaev, and Muller (Derjaguin et al. 1987) and Hansen and Löwen (Hansen and Löwen 2000).

Figure 2 shows the plot of total potential as a function of the inter-particle distance. When two particles with similar charge start approaching each other, the EDLs interact in a repulsive manner due to the long-range electrostatic forces leading to a potential energy barrier maximum. This potential energy barrier can be crossed either by thermal energy or be lowered by introduction of suitable electrolytes, leading to charge screening. As the particles come closer, the short-range attractive VdW forces become dominant leading to a potential energy minimum, known as the primary minimum. At this stage, the particles irreversibly aggregate, and it is nearly impossible to redisperse them in solution. Hypothetically, if the inter-particle distance is further reduced, the particles experience a steep increase in inter-particle potential as a result of the Pauli exclusion principle (hard sphere repulsion). Thus, when a sufficient potential barrier as a result of surface charge exists, the particles will be repulsive and stay dispersed in the solution. Interestingly, at intermediate inter-particle separation lengths, the exponential repulsion vanishes more rapidly than the attraction resulting in a shallow minimum in the DLVO interaction profile as a result of net attraction, known as the secondary minima. The secondary minima

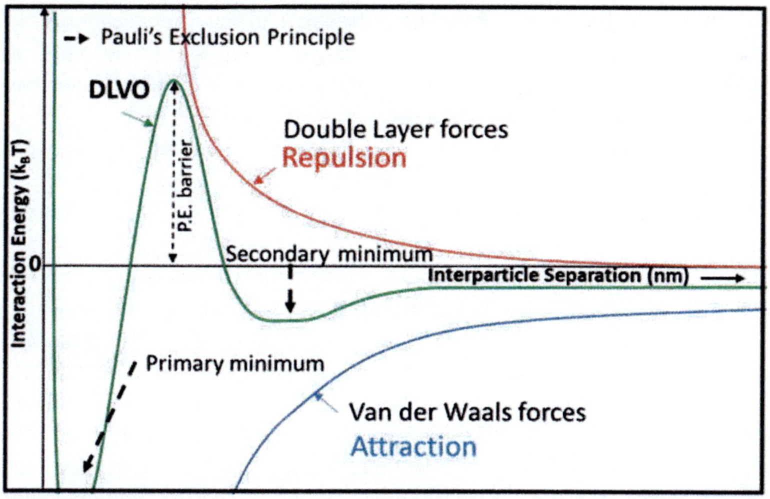

Fig. 2 Interaction energy profile as a function of inter-particle separation distance. The red curve represents repulsive interaction due to the electric double layer, and the blue curve represents attractive interactions due to Van der Waals forces. The green curve represents the net interaction energy profile (DLVO theory)

can be sufficiently deep for carefully designed colloidal systems and can be used to drive self-assembly of nanoparticles (Behrens and Borkovec 2000; Behrens et al. 1998).

3.4 *Molecular Interactions*

Apart from the intermolecular forces discussed above, other interactions collectively summed up as non-DLVO interactions play a significant role in stability and assembly of colloids (Grasso et al. 2002).

3.4.1 Steric Interactions

Typically, nanoparticles are capped with a variety of capping agents, as discussed in Sect. 2.2. These capping agents which arrest the growth of a particle to nanoscale also assist in stabilizing the colloid by preventing particle aggregation by steric interactions. The steric interactions arise from spatial interaction of the nanoparticle "corona" formed by the surface ligands, more specifically due to their overlapping electron clouds leading to repulsion. The magnitude of these interactions depend on the grafting density of these ligands on the particle surface and its interaction with the dispersion medium (de Gennes 1987). The steric interactions can be enhanced by using moieties such as hyper-branched polymers and polymer brushes (Hao et al. 2019). The stability of a colloidal dispersion can be also improved by using charged surface ligands (e.g., Tween-80) (Li et al. 2012), leading to a combined electro-steric stabilization (Pelley and Tufenkji 2008; McClements 2005).

3.4.2 Cross-Linking

Use of large polymers/macromolecules as capping agents can lead to aggregation where the polymer chain binds to more than one nanoparticle, but does not sufficiently cover the surface of each particle, at low concentrations (Biggs et al. 2000). This property can, however, be creatively used to assemble nanoparticles by using appropriate "linking" ligands. For e.g., Ag nanoparticles can be cross-linked using bidentate ligands (1–6, hexane dithiol) (Ahonen et al. 2006). These linkages can also be created by forming covalent bonds between the ligands, e.g., furan functionalized Ag nanoparticles can be cross-linked by Diels-Alder type reaction with bismaleimide (Liu et al. 2017). Further, the linking can be reversed simply by controlling the temperature. Nanoparticles can also be cross-linked using the classical streptavidin biotin conjugation (Aslan et al. 2004).

3.4.3 Hydrogen Bonding

Hydrogen bonding arises from the partial positive charge on the hydrogen atom covalently bonded to a strong electronegative atom such as oxygen or nitrogen, leading to formation of proton-mediated electrostatic attraction of the larger atoms (Kollman and Allen 1972; Jeffrey 2003). Hydrogen bond strengths typically range from 5 to 50 kJ mol^{-1} and are typically intermediate between covalent and non-covalent bond strengths. Hydrogen bonding can induce aggregation in metallic nanoparticles functionalized with ligands such as –OH, –COOH, and $–NH_2$ where the aggregation and ordering was found to depend on the individual hydrogen bond strength (Kinge et al. 2007; Johnson et al. 1998). Hydrogen bonding can be achieved for anisotropic assembly of nanoparticles as well, for e.g., Au nanorods could be assembled in end-to-end manner by selectively capping the nanorods ends with hydrogen bonding functionalities (Thomas et al. 2004). When these moieties were carboxylic groups, the pH of the solution could be used to rearrange the assemblies, at low pH the nanorods assembled end to end due to hydrogen bonding, while at higher pH they reassemble in side-by-side manner due to Van der Waals forces. Thus, different inter-particle interactions can be potentially used in unison toward self-assembled structures. Figure 3 schematically represents the various interactions between nanoparticles arising from molecular interactions.

The cumulative magnitude of the surface molecular forces is typically a product of the effective bond strength of individual bonds and the number of bonds formed between the functionalized surfaces.

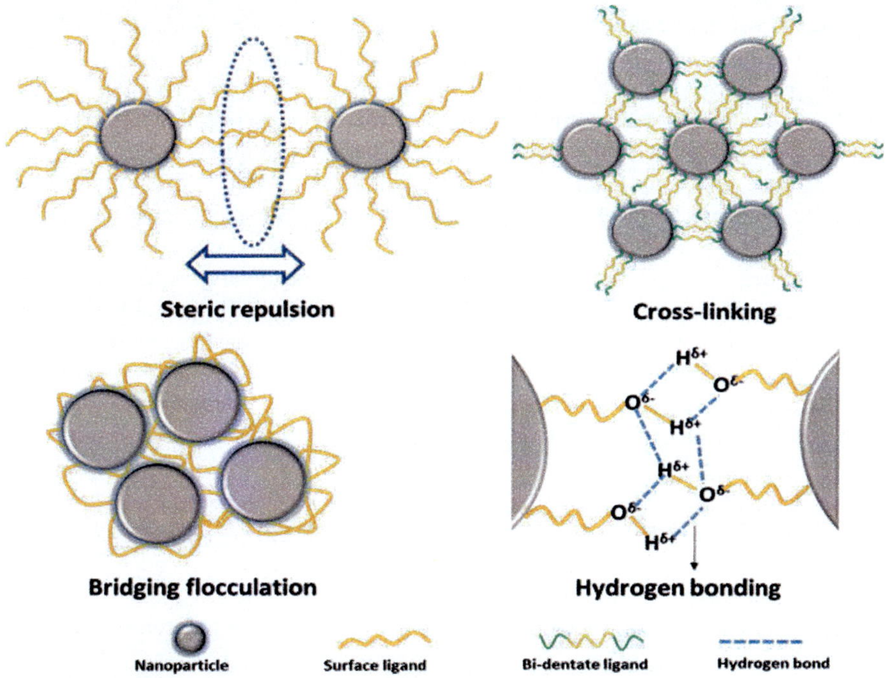

Fig. 3 Schematic representation of nanoscale molecular interactions

3.5 Entropic Interactions

3.5.1 Attractive Depletion Forces

In absence of other attractive forces, colloids can still assemble due to depletion forces or confinement. This effect is purely entropic and arises when non-interacting macromolecules such as surfactants or polymers are present in a stable colloidal solutions. When the inter-particle distance is smaller than the size of the macromolecules (depletant) present in the solution, a zone of exclusion is created around each particle, from where the macromolecules are excluded, leading to a phase of pure solvent (depleted of macromolecules). A force equivalent to the osmotic pressure of the macromolecular solution is generated due to this imbalance, pushing the particles closer together (Asakura and Oosawa 1954). This force scales with the excluded volume and can be greatly enhanced by using charged macromolecules as depletants (Asakura and Oosawa 1958; Piech and Walz 2000). Since the depletion forces are attractive in nature, they can be used for driving self-assembly of nanoparticles, the magnitude and range of which can be controlled by the molecular structure and concentration of the depletant (Ray et al. 2015; Cao et al. 2013). Since the depletion interactions depend on the inter-particle contact area, they can be used for shape-selective flocculation and separation of nanoparticles of distinct morphologies, even for particles with nominally similar mass by simply tuning the concentration of the surfactant (Park et al. 2010). The depletion forces can also be used in a directional (anisotropic) manner, where the shape of the nanoparticle building block directs the symmetry of the superstructure in order to maximize the system's entropy, while the surfactant charge determines the inter-particle spacing (Young et al. 2013). Similarly, flat-plate structures such as nanoprisms can be assembled in one-dimensional lamellar crystals by fine-tuning the balance between attractive depletion and repulsive electrostatic interactions (Young et al. 2012). These examples demonstrate the entropic effects in dilute regimes. In concentrated regimes, particles can transition into an ordered phase in order to maximize their entropy, while minimizing free energy. This effect is particularly pronounced in rod-shaped particles in concentrated systems, where the rods are translationally as well as rotationally constrained. The total entropy reaches a maxima when the rods orient themselves in nematic phases from the isotropic phase (Bishop et al. 2009).

3.5.2 Solvation Forces

The hydration forces (or generally, solvation forces) are a result of the water (or solvent) molecules getting ordered on the particle surface. The ordered structure of the solvent molecule can extend beyond the first layer, depending on the magnitude of interaction leading to a solid-like or crystalline interface (Israelachvili and Gourdon 2001). Under such conditions, the DLVO predictions deviate, where the Van der Waals attraction continuum is replaced by oscillatory solvation forces along the distance from the particle surface with a periodicity corresponding to the size of the immobilized solvent molecules (Christenson 1988; Israelachvili and Gourdon 2001) (Fig. 4).

3.5.3 Hydrophobic Interactions

When particles with hydrophobic surfaces are introduced in strongly hydrogen-bonded medium such as water, the hydrogen bonds reorganize such that the particle is excluded out of the spatial hydrogen-bonded network of water molecules (Rodnikova 2007). The bulk elasticity of the spatial network is of the order of 10^9 Pa, and is claimed to be the driving force behind these interactions (Rodnikova 2007). These non-specific interactions are responsible for the formation of micellar structures, as well as in biological systems such as vesicles, cell membranes, and protein folding and thus are essential for life (Blokzijl and Engberts 1993). The hydrophobic force is unique in context of nanoparticle assembly, as the particle organization arises from the repulsion from the solvent, as against attraction between the nanoparticles (Tanford 1978). This interaction is believed to be of long range (than a typical covalent bond), decaying exponentially with distance (Israelachvili and Pashley 1982). However, the "pure" hydrophobic effect arising from hydrogen bond structure reorganization beyond the very short length scale (~10 Å) is still a topic of academic

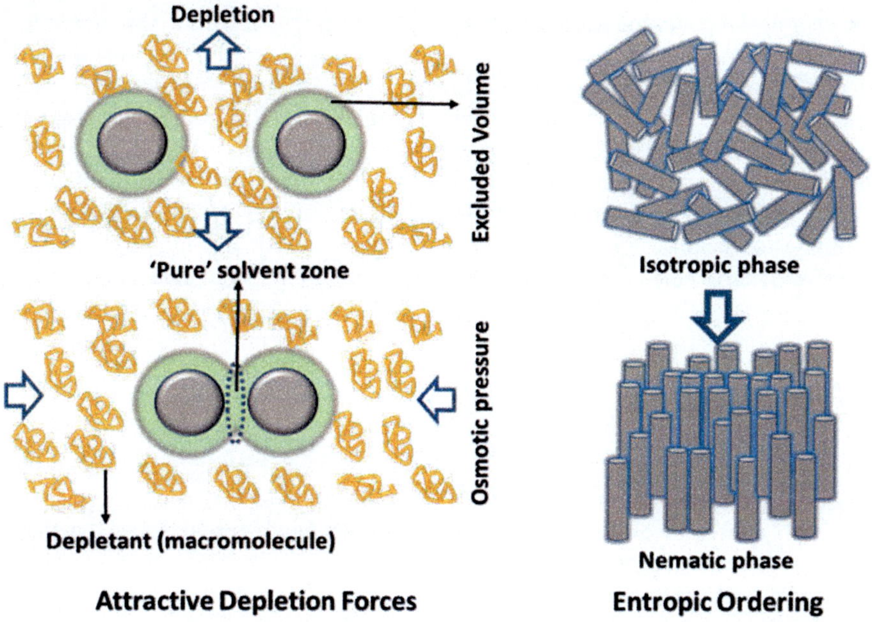

Fig. 4 Schematic representation of entropic interactions

debate (Hammer et al. 2010; Meyer et al. 2006). The estimation of hydrophobic interactions remains challenging in case of nanoparticles due to their incessant Brownian motion (Sánchez-Iglesias et al. 2012). However, these interaction are invoked for qualitatively explaining the aggregation and self-assembly of nanoparticles coated with hydrophobic functionalities (Sánchez-Iglesias et al. 2012; Wang et al. 2012; Nie et al. 2008).

3.6 Extrinsic Forces and Interactions

The nanoscale forces and interactions in context of colloidal dispersions discussed so far are broadly intrinsic in nature, i.e., they arise from the particle material, morphology, surface charge, surface ligands, solvent, etc. However, nanoparticle assembly can be directed by external forces as well. For e.g., capillary forces arise at fluid–solid–air interfaces from the Laplace pressure at the meniscus of fluid bridges around the particles, to minimize the surface energy of the interface (Kralchevsky and Nagayama 1994). Capillary forces can be used to assemble nanoparticles into one-, two-, or three-dimensional structures by controlled solvent evaporation from the colloidal solution (Nikoobakht et al. 2000). The lateral hydrodynamic drag generated during solvent evaporation drives the particles to form close-packed structures on flat substrates (Kralchevsky and Denkov 2001). The capillary forces can be combined with chemically patterned substrates, for e.g., hydrophilic domains to form complex 2D arrays of assembled particles (Aizenberg et al. 2000). Similarly, externally applied forces such as electric, magnetic, and flow fields can be employed for maneuvring particles. For e.g., flow fields resulting from external convective forces can be used to drive assembly of nanoparticles at high volume fractions (Prevo et al. 2007). Particles can be deposited in desired patterns by focusing them on substrates using electric fields by means of 'nanoscopic electrostatic lenses' (Kim et al. 2006). 2D and 3D assemblies of nanorods oriented perpendicular to the substrate could be formed by using DC electric field combined with solvent evaporation (Ryan et al. 2006).

4 Methods for Immobilization of Ag Nanomaterials

Nanomaterials can be immobilized in a wide variety of ways taking advantage of the nanoscale interactions discussed thus far. Immobilization involves adhering nanostructures to a substrate or confining or "jamming" them into assemblies such that their relative movements are restricted or minimized. This can be achieved in a number of ways such as using covalent and/or non-covalent interactions for attaching the nanoparticles to a substrate or template. This type of immobilization can either be stochastic, or can be ordered in close-packed structures, 2D and 3D superstructures, etc., or structures with intermediate orders such as nematic or smectic phases. These structures can be achieved using self-assembly techniques which creatively use the force balance between various inter-particle interactions and extrinsic fields. A variety of hierarchical structures can be achieved by combining the self-assembly approach with patterned templates. The templates can confine nanoparticles in ordered domains, and exotic nanoassemblies can be achieved with a high degree of control. Nanomaterials can also be grown as features by controlled deposition techniques such as vapor deposition, electrodeposition, nanolithography, etc., on various surfaces. Similar to the earlier case, using patterned templates in combination with deposition techniques can lead to uniform and tightly controlled arrays of nanoscale features. Another important aspect of nanoparticle immobilization is having a control over the inter-particle spacing, as this has important consequences in optical, electronic, mechanical, catalytic applications. Here, we describe methods to achieve immobilized silver nanostructures using various approaches.

4.1 Nanolithograpy

The term "nanolithography" literally means writing on a surface at nanoscale. This technique has been used on a commercial scale in microelectronics industry for manufacturing of integrated circuits, microprocessors, high-density memory devices, interconnects, etc. (Ito and Okazaki 2000). While there exist a wide variety of nanolithographic techniques, such as e-beam lithography (EBL)(Chen 2015), focused ion beam (FIB) lithography (WATT et al. 2005), optical lithography (Wu and Kumar 2007), and scanning probe lithography, here, we describe the techniques most relevant in case of silver.

4.1.1 E-Beam and Focused Ion Beam Lithography

EBL and FIB techniques rely on polymer resists as masks for etching or deposition of the material. For e.g., various morphologies of Ag nanomaterials were synthesized using e-beam irradiation (0.2–3 MV, 0.06–0.24 mA) on spin-coated substrates of silver dodecylthiolate (Kim et al. 2010). Dense silver nanorods arrays could be created using EBL and ion milling by impinging argon ions on thick silver films (Si et al. 2013). Other morphologies such as silver nanodisk assemblies can also be fabricated (Cinel et al. 2012). FIB milling was recently used to create silver nanocircuits using gallium ion beam (30 kV, 49 pA) on silver plates (Schörner et al. 2019). While they can achieve well-defined structures with a high degree of precision, these techniques are cumbersome, require specialized equipment, and are expensive for most applications.

4.1.2 Nanosphere Lithography

As against the EB and FIB techniques, there are low-cost, high-throughput lithographic techniques such as nanosphere lithography (Deckman and Dunsmuir 1982; Wood 2007). In this method, monodisperse colloidal spheres (typically polystyrene) are assembled in hexagonal close-packed 2D monolayers on suitable substrates, by capillary forces during solvent evaporation. This close-packed assembly serves as the mask on which silver is deposited by vapor deposition techniques. The silver forms a film on the spheres as well as on the substrate through the interstitial voids between the HCP assembly. The spheres are then removed by sonication, leaving behind uniform motifs of silver nanostructures on the substrate (Deckman and Dunsmuir 1983). The in-plane size of the silver particles can be controlled by tuning the diameter of the colloidal nanospheres, while the out-of-plane size can be tuned by the thickness of the deposited silver overlayer. Shape control can be achieved by using a double-layered nanosphere mask, which leaves behind hexagonal particles (Hulteen et al. 1999). Similarly, nanopyramids of silver (350–400 nm height) can be formed on glass substrates by depositing thicker overlayers of silver (Tabatabaei et al. 2013). Other morphologies such as silver nanoring arrays with tunable diameters and pitch can be formed by modifying the nanosphere assembly and deposition methodology (Halpern and Corn 2013). Complementary to the aforementioned structures that form on the substrate, large area silver nanobowl arrays can be obtained by peeling off the thermally evaporated silver on the polystyrene spheres using a sticky tape (Halpern and Corn 2013).

4.1.3 Soft Lithography

Soft lithographic methods refer to printing methods that are used to create nanoscale-patterned assemblies. Soft lithography overcomes many disadvantages of photolithographic processes such as optical diffraction limit, high energy beams, limited tolerance to variation of materials, and protocols apart from a high equipment and operational cost (Xia and Whitesides 1998). Soft lithography is a collection of techniques that use a stamp or a mold to create micro- or nanostructured patterns using wide variety of precursors and on a wide variety of substrates (Wisser et al. 2015). Such structured surfaces are indispensable components in optoelectronics, plasmonics, data storage, displays, photovoltaics, etc. (Wisser et al. 2015; Rogers et al. 2001). Microcontact printing is one of these methods, which uses a polydimethoxysilane (PDMS) stamp to transfer the nanoscale patterns to the substrates as first demonstrated by Whitesides group (Kumar et al. 1994). Silver nanoparticle ink can be printed on to PET substrates to form conducting grid patterns (Xin et al. 2017). The characteristics of the ink (viscoelasticity, solvent evaporation rate, adhesion) stamp (pitch, surface treatment) as well as the substrate (wettability) are critical for microcontact printing. Nanoimprint lithography (NIL) is another non-projection technique which uses compression molding using hard mold to create a pattern by mechanical deformation of an imprint resist (polymer/monomer) coated on a substrate (Gates et al. 2005; Chou et al. 1996). This process is akin to hot embossing since it involves heating the polymer above the glass transition temperature followed by pressure-induced transfer of topographic pattern from a hard mold to a soft polymer layer (Gates et al. 2005). This methodology can be used to create highly uniform nanostructures such as silver nanowire arrays (Schumm et al. 2013). For e.g., polymeric silver precursor consist of ethylene glycol, citric acid, and silver citrate spin coated on a substrate. At elevated temperature (80 °C), a PDMS stamp was imprinted into the viscous precursor film, to generate longitudinal groove patterns. The polymer was cured at 120 °C, followed by removal of the PDMS stamp. Finally, the conductive silver nanowire arrays were formed by thermal treatment of the patterned precursor. This method is versatile (can be used with a wide variety of substrates, precursors, and molds), low cost, and highly scalable. For e.g., flexible conducting silver nanomesh electrodes were formed on polyethylene terephthalate (PET) films using NIL in a roll-to-roll process (Yi et al. 2017).

4.1.4 Dip-Pen Lithography

Dip-pen lithography is a scanning probe technique where a nanoscale probe is used to create patterns on a wide variety of substrates typically at a sub-100 nm scale (Ginger et al. 2004). Since this technique comprises physical flow of materials (ink) from the probe to the substrate, and the physico-chemical interaction of the ink with the substrate, many parameters such as those of the (a) ink: ink composition, visco-elastic characteristics, solvent evaporation, flow rate, temperature, humidity; (b) substrate: composition, roughness, hydrophobicity/hydrophilicity, etc., and (c) scanning probe:

tip material, sharpness, contact angle, scanning rate, etc., play a critical role (Brown et al. 2014). A detailed treatise on these parameters is reviewed by Mirkin et al (Ginger et al. 2004; Brown et al. 2014). Pearton and co-workers demonstrated writing of conducting silver nanoparticle ink traces at sub-micron resolution on silicon oxide substrates using commercial silver inks (Wang et al. 2008). The same group further demonstrated the versatility of this approach by extending it to a variety of substrates and writing speeds with patterning at higher resolution (500 nm) and improved electrical conductivity of the silver ink traces, close to that of bulk silver (Hung et al. 2010).

4.2 *Self-Assembly and Templated Assembly*

Self-assembly refers to spontaneous organization of the nanoobject building blocks into ordered ensembles due to mutual interactions (inter-particle forces) or interactions with the environment (external fields or constraints). Self-assembled mesostructures are typically characterized by thermodynamic equilibrium to minimize the free energy of the system (Whitesides and Grzybowski 2002). Self-assembly is a powerful tool to organize nanomaterials with minimal external intervention. The building blocks (composition, morphology), surface ligands, particle environment, templates, and external fields are some of the key parameters that control the characteristics of self-assembled structures (Grzelczak et al. 2010; Panigrahi et al. 2006). Self-assembly induced by capillary forces is one of the most straightforward ways to assemble particles. For e.g., dodecanethiol-capped silver nanospheres could be assembled into long-range close-packed arrays on carbon and mica substrates by solvent evaporation (Korgel et al. 1998). The periodicity of the array was determined by the size of the soft corona leading to steric stabilization and counterbalancing the attractive Van der Waals forces. Similarly, the shape of the building blocks can direct the morphology of the assembly. For e.g., silver nanowire arrays were formed by self-assembly of dodecanethiol-capped prolate Ag nanocrystals, where the morphology and periodicity of the arrays was determined by the shape of the building block. This template-free self-assembly was directed by entropic interactions, leading to formation of free-standing close-packed assemblies (Korgel and Fitzmaurice 1998). CTAB-capped silver nanodiscs could spontaneously self-assemble into chain-like structures, owing to the self-assembled monolayer of CTAB on the basal plane of the nanodisc, which prevent particle aggregation (Chen et al. 2002). The morphology of the assembly can also be directed by the particle environment. For e.g., when silver nanoparticles were assembled in the polymer methoxy polyethylene glycol (MPEG), the linear orientation of the polymer acting as a soft template led to assembly of Ag nanoparticles into chain-like 1D structures (Mallick et al. 2005). Similarly, hard templates can be patterned and combined with solvent evaporation to form immobilized Ag nanoparticles into complementary patterns such as square grids (Layani and Magdassi 2011). These grid-patterned particles can further be sintered to form conducting substrates, making this process akin to lithography. Uniform silver nanorods readily self-assemble into densely packed arrays by solvent evaporation from a concentrated nanoparticle suspension (Pietrobon et al. 2009). Interestingly, the nanorods are assembled into hexagonal close-packed structures, despite their pentagonal symmetry. The assembly of 2D close-packed arrays can be extended to 3D structures by layer-by-layer assembly of the Ag nanoparticles (Cassagneau and Fendler 1999). PVP-functionalized silver nanowires ($d = 70$ nm, AR = 80-100) could be self-assembled on silicon substrates, and further a layer-by-layer approach led to formation of perpendicularly crossed arrays (Gao et al. 2003). While solvent evaporation strategy is quite straightforward to implement, it faces certain drawbacks. The solvent evaporation-driven assembly can be challenging to control due to the intensity of the solvation and capillary forces and their spatial fluctuations, leading to a poor long-range order (Rabani et al. 2003). The Yang group demonstrated sedimentation as a strategy, accompanied by depletion forces to assemble monodisperse silver polyhedra into long-range, densely packed superlattices. The driving force resulting from gradual sedimentation is comparatively gentle and homogenous as against solvent evaporation according to the authors (Henzie et al. 2012). Another limitation of solvent evaporation approach is that the nanoparticles adhere to the substrate by non-specific interactions such as Van der Waals forces; while this can be advantageous for transferring the assembly to other substrates, it can be a shortcoming when the nanoparticles need to be adhered strongly to the substrate (e.g., SERS substrate). This can be overcome by tethering the nanoparticles to the substrate by using strong attractive electrostatic interactions or by bond formation using linker moieties. For e.g., dip coating of substrates functionalized with charged polymers such as poly(ethylene imine) (PEI) and poly(allylamine hydrochloride) (PAH) in silver nanoparticle dispersions can form immobilized monolayers (Oćwieja et al. 2011). The surface coverage can be controlled by the ionic strength of the polyelectrolyte (tuned by pH) and the deposition time. Silver nanoparticles could be immobilized on glass substrates functionalized with 3-mercaptopropyltrimethoxysilane (MPTMS) (Fan and Brolo 2009). The silane functionality grafts to the glass surface by forming Si–O bonds, while exposing the thiol functionality. When this functionalized substrate is immersed in silver nanoparticle dispersion, the particles adhere to the exposed thiol groups. The substrate was washed repeatedly to remove the non-bonded particles. Ag nanoparticles and films can also be immobilized on such thiol top functionalized monolayer, by first immobilizing Ag^+ ions by chemisorption, followed by wet chemical reduction (Maoz et al. 2000). These surface bound ions could also be reduced by electrochemical reduction using tip probe lithographic method. The surface ligands of the nanoparticles can also be engineered to achieve programmed assembly of the particles, by rendering them with complementary functionality. When these particles interact in solution, they recognize and selectively bind to each other by bond formation (Fullam et al. 2002). Another ingenious approach to form uniform close-packed arrays is by assembling the nanoparticles at the air–water interface such as Langmuir monolayers. This nanoparticle assembly can be lifted off as a monolayer on solid substrates as Langmuir–Blodgett monolayer (Sastry et al. 2002). Sastry and co-workers demonstrated highly ordered, large domains of monodisperse silver nanoparticles by spontaneous reduction of silver ions on 3-pentadecylphenol

Langmuir monolayers (Swami et al. 2004). Similarly, octadecylamine-capped silver nanoparticles could be assembled into Langmuir monolayers at air–water interface and could be transferred on to various substrates as Langmuir–Blodgett films with hexagonal closed packed arrangement (Selvakannan et al. 2004). As seen from these examples, self-assembly is a versatile technique to immobilize a wide variety of nanomaterials in a scalable manner with relative ease. Common to all examples is the requirement of high monodispersity of the particles, the characteristics of the particle corona, and the physico-chemical environment of the system, to achieve a defect free, long-range ordering of nanomaterials.

5 Applications of Immobilized Silver Nanomaterials

Here, we highlight various applications of immobilized silver nanomaterials with their salient features. These are representative illustrations and are by no means exhaustive.

5.1 Optical Applications

As discussed earlier, silver, at nanoscale, exhibits remarkable optical and optoelectronic properties such as the localized surface plasmon resonance (LSPR). The optical excitation of these LSPR modes gives rise to intense local electromagnetic fields, which can be enhanced further by coupling of the plasmon modes by engineering the gaps between assembled nanoparticles (Hao and Schatz 2004). These properties can be modulated by the size, morphology, chemical environment, and assembly of the nanoparticles (Kelly et al. 2003). The E-field enhancement as a result of plasmonic excitation has been harnessed for imaging, sensing, theranostics, etc. (Jain et al. 2008). Surface-enhanced Raman spectroscopy (SERS) is a remarkable sensing technique, in which the Raman scattering by adsorbed molecules on the plasmonic substrates is enhanced by a factor of 10^{10}–10^{11} (Moskovits 2005). Silver nanomaterials are the most exploited systems in SERS owing to the very high ability to enhance the signal, where it is possible to detect single molecule (Nie and Emory 1997). Real-time optical sensors with zeptomolar sensitivity have been developed using silver nanoparticles as SERS substrates (McFarland and Van Duyne 2003). Arrays of closely spaced silver nanomaterials can also serve as optical waveguides, where the coupled plasmon modes between adjacent particles give rise to coherent energy propagation along the superstructure (Maier et al. 2001; Quinten et al. 1998). Confinement and propagation of light below the diffraction limit can be achieved by such closely spaced nanoparticles that convert optical modes to non-radiating surface plasmons, as demonstrated using arrays of silver nanorods (Maier et al. 2003b). The light confinement can also be used for sub-diffraction limit optical imaging by excitation of surface plasmons, as demonstrated using silver lens, with 60 nm half pitch resolution, a sixth of the illuminating wavelength (Fang et al. 2005). Silver trapezoidal arrays, as metal–insulator–metal stack, are capable of broadband resonant absorption of entire visible spectrum of light (400–700 nm), acting as superabsorbers with applicability in photovoltaic and thermophotovoltaic cells (Aydin et al. 2011).

5.2 *Electronic Applications*

Silver has the highest electrical conductivity and thermal conductivity among all metals, making it an element of choice for electronic applications such as circuit interconnects, sensor and actuator elements, and electronic packaging materials. Silver nanoparticle dispersions with a suitable capping such as gum arabic can be deposited on substrates and then sintered into conducting thin films (Balantrapu and Goia 2009). Concentrated dispersions of long-chain alkylamine-capped silver nanoparticles could be spin coated on to substrates to form thin films and elements with high conductivity (2$-$4 $\times$ 104 S cm^{-1}) (Li et al. 2005). Further, these elements used as source/drain electrodes in organic thin film transistors demonstrated field-effect transistor properties similar to those of vacuum-deposited electrodes. Silver nanoparticles can be formulated into conductive inks that can be used in printed electronics using drop-on-demand inkjet printing. For e.g., dodecanoic acid-capped silver nanoparticles (d = 7 nm) at 33 weight % loading in toluene could be printed into circuits on photopaper substrates using piezoelectric drop-on-demand inkjet head (Lee et al. 2006). Similarly, silver nanoparticle inks can be printed as highly conductive tracks with very low resistivity of 3.7 $\mu\Omega$ cm on to paper and PET substrates using a commercial printer (Shen et al. 2014). These conductive interconnects were used for connecting LEDs on the PET substrates to demonstrate their applicability in flexible electronics. The printed electronics that make use of spherical nanoparticles suffer degradation in performance and ultimately failure on stretching or bending due to cracks and defects in the conducting layer. This issue can be alleviated by using high aspect ratio silver nanowires which form a percolation network at much lower loading fractions and are also known to possess high mechanical strength and flexibility. For e.g., silver nanowire circuits printed on substrates such as paper and PET could retain their high conductivity even after 10,000 rolling cycles and 50 folding tests (Huang et al. 2014). These flexible nanowire electronics can be applied to a range of substrates as foldable circuits, radiofrequency identification (RFID) tags, and LED arrays (Yang et al. 2011). Another emerging application of these silver nanowire networks are transparent flexible electrodes in electronic diplays, which are a potent replacement for the brittle, rigid, and expensive indium tin oxide (ITO)-based displays (Zhang and Engholm 2018). Silver nanowires can be coated and welded on flexible transparent flexible substrates such as PET and PDMS films to form transparent conducting electrodes, which retain conductivity under tensile strain up to 75% and 100 + bending cycles (Miller et al. 2013). The out-of-plane transmittance and in-plane electrical conductivity comparable to ITO have been achieved by tuning the size, morphology, coverage, and immobilization methods for silver nanowire-based electrodes (Nam et al. 2014). Such transparent flexible conductors also find potential applications in organic solar cells (Kang et al. 2010), touch panels (Lee et al. 2012), etc. Electronic packaging is another important application of immobilized silver nanomaterials, owing to the high electrical and thermal conductivity of silver. For e.g., highly stretchable and transparent electromagnetic interference shielding films can be formed by percolation networks of silver nanowires (Jung et al. 2017). Silver nanoparticles can be dispersed in electronically conductive adhesives to fill the gaps and asperities between larger silver nanoflakes to improve their conductivity (Chen et al. 2009; Alshehri et al. 2012). Silver nanoparticles can also be used in electronic packaging as thermal interface materials rendering them with high thermal conductivity while still retaining their mechanical compliance (Lee et al. 2012). Silver nanowires filled in epoxy composites achieve the low thermal interfacial resistance and high compliance at much lower loadings owing to their ability to form percolation networks (Chen et al. 2014). Such thermal interface materials are vital for heat dissipation from electronic components to heat sinks and ensure their sustained performance.

5.3 Catalysis

Metallic nanoparticles including silver have been used widely in organic chemistry typically as heterogeneous catalysts (Tao et al. 2014). The catalytic properties are remarkable in case of nanostructured catalysts owing to their high surface-to-volume ratio, high number of surface defects, and a variety of crystallographic facets which serve as active sites for progression of a chemical reaction (Rodrigues et al. 2019). Additionally, the ability to modify a metal particle for its size, shape, surface chemistry, and crystal structure serve as important tools to enhance and modulate the catalytic activity, and even make it selective (Narayanan and El-Sayed 2005). Silver nanoparticles have been one of the important redox catalysts, for e.g., silver can reduce aromatic nitro compounds such as 2-nitrophenol, 4-nitrophenol, and 4-nitroaniline (Pradhan et al. 2002). However, recovery and repeated use of the catalyst is indispensable to achieve the desired efficiency and economy. In this regard, immobilizing the nanoparticles on suitable substrates is advantageous, as it combines their catalytic ability with recyclability. For e.g., silver nanoparticles can be immobilized on silica spheres for degradation of dyes such as eosin, methylene blue, and rose bengal (Jiang et al. 2005). Silver nanodendrites exhibited an improved catalytic activity toward reduction of p-nitrophenol and p-aminonphenol over silver nanospheres, highlighting the effect of shape and surface area (Rashid and Mandal 2007). Silver nanoparticles were generated within porous polymer fibers of poly[3-(trimethoxysilyl)propyl methacrylate], by electrospinning the polymer solution containing silver nitrate, followed by calcination. These polymer fiber mats with 10 wt% silver loading showed degradation of methylene blue dye and were recoverable (Patel et al. 2007). Silver nanoparticles tethered on amine-functionalized nanosilica could be used repeatedly up to ten times for reduction of 4-nitrophenol and 2-nitroaniline (Dong et al. 2014). Similar catalytic activity was reported for silver nanoparticles immobilized on thiol-terminated surface of halloycite nanotubes, an aluminosilicate clay (Liu and Zhao 2009). Redox activity was also reported for silver shell immobilized on cationic polystyrene beads, acting as a solid-phase reusable catalyst (Jana et al. 2006). Hydrotalcite-supported silver nanoparticles were shown to catalyze dehydrogenation of a wide variety of alcohols without loss of activity or selectivity (Mitsudome et al. 2008). The plasmonic properties of silver nanostructures can also be invoked for catalytic applications. For e.g., Linic and co-workers demonstrated catalytic oxidation reactions such as ethylene epoxidation, CO oxidation, and NH_3 oxidation by plasmonic silver nanostructures driven by low-intensity visible light, opening avenues to harvest solar energy and drive green chemical processes (Christopher et al. 2011). Photocatalytic performance of TiO_2 was shown to be significantly enhanced due to plasmonic excitation of immobilized silver nanoparticles (Awazu et al. 2008). Apart from E-field enhancements as a result of plasmon excitation, photothermal effects have also been shown to enhance the reaction rates catalyzed by silver nanoparticles immobilized on electrospun polyacrilonitrile microfiber network (Gao et al. 2016).

5.4 Biomedical Applications

The antibacterial and antiinflammatory properties of silver have been known since ancient times and have been the most important antimicrobial agent available before the discovery of antibiotics (History of the Medical Use of Silver 2009). With progress of nanotechnology in the past few decades and rise of the threat of multidrug resistant bacteria, they have shot silver nanomaterials in the spotlight, and they have been investigated extensively for their antibacterial properties (Rai et al. 2012). In addition to the broad spectrum antibacterial effects, the antiinflammatory, wound-healing, and burn treatment using silver nanoparticles has also been of immense academic interest (Chaloupka et al. 2010; Arvizo et al. 2012). The antibacterial effect has been variably attributed to (a) direct contact of bacterial cell wall with the nanoparticles leading to the cell wall's disruption (Morones et al. 2005; Sondi and Salopek-Sondi 2004), (b) release of Ag + ions from the nanoparticles that inhibit enzymes and interfere with DNA replication (Feng et al. 2000), and (c) generation of free radicals and reactive oxygen species which damage the cell membranes (Kim et al. 2007). Since the silver nanoparticles have a potential to cause cytotoxicity to healthy human cells, it would be advantageous to adhere these particles on suitable substrates for contact disinfection and treatments, and have been demonstrated as prophylactic coatings on various biomedical devices, implants, textiles, etc. (Chen and Schluesener 2008; Monteiro et al. 2009). For e.g., silver nanoparticles ($d = 5$–7 nm) coated on cotton as well as polyester fabrics show antibacterial activity against *E. coli*, *S. aureus*, and *K. pneumoniae,* and the silver particles persisted on the fabric even after 20 cycles of laundering (Lee et al. 2003). Surgical facemasks coated with silver nanoparticles showed bactericidal activity against *E. coli* and *S. aureus*, where no skin irritation was experienced by volunteers (Li et al. 2006). Silver bromide nanoparticles immobilized on the cationic polymer poly(4-vinyl-N-hexylpyridinium bromide) showed antibacterial activity against gram-positive *Bacillus cereus* and *Staphylococcus aureus* and gram-negative *Escherichia coli* and *Pseudomonas aeruginosa* (Sambhy et al. 2006). These composite materials could form coatings on surfaces and were found to be active against airborne as well as waterborne bacteria, and prevented biofilm formation. The release of silver ions from nanosilver surfaces can be systematically controlled by various competing surface chemical processes, as shown by Hurt and co-workers (Liu et al. 2010). For e.g., the release of Ag+ ions could be accelerated by peroxidation or particle size reduction, while it could be systematically retarded by surface passivation using thiol and citrate ligands, sulfidic coating or scavenging of peroxy intermediates (Liu et al. 2010).

Silver nanoparticles have also been shown to aid in wound healing, when incorporated in wound dressings (Maneerung et al. 2008). For e.g., electrospun gelatin microfiber mats incorporated with silver nanoparticles showed antibacterial activity against *seudomonas aeroginosa, Staphylococcus aureus, Escherichia coli*, and methicillin-resistant *S. aureus* in simulated body fluid (pH = 7.4) at physiological temperature (37 °C) (Rujitanaroj et al. 2008). Similarly, composite microfibrous mats of chitosan, polyvinyl alcohol, and silver nanoparticles cross-linked with glutaraldehyde showed synergistic bactericidal effect between chitosan and silver nanoparticles (Abdelgawad et al. 2014). More recently, hydrogels consisting of Ag/graphene composites and acrylic acid and *N,N′*-methylene bisacrylamide were shown to accelerate the healing rate of wounds in vivo in rats (Fan et al. 2014). The composite aids in preventing wound infection as well as retains appropriate level of moisture around the wound.

5.5 *Water Purification*

Since silver nanoparticles exhibit remarkable catalytic, photocatalytic, as well as antimicrobial activity (see Sects. 5.3 and 5.4), they can be employed for water disinfection and purification. However, similar to the earlier cases, they need to be immobilized on suitable substrates for preventing their aggregation and repeated usage. Another aspect of immobilization of silver for water purification is that it not only disinfects water by direct contact and slow release of ions, but also prevents silver nanoparticles from entering the human body and environment. Silver nanoparticles could be immobilized on common polyurethane foam and showed a complete removal of *E. coli* from water with bacterial load of 10^5 CFU/mL at a flow rate of 0.5 L/min (Jain and Pradeep 2005). Silver nanoparticles tethered to porous ceramic substrates via 3-aminopropyltriethoxysilane (APTES) linker showed a similar performance at 0.01 L/min (Lv et al. 2009). Silver nanoparticles anchored on methacrylic acid polymer beads, via carboxylate groups exhibited antibacterial activity against both gram-positive and gram-negative bacteria (Gangadharan et al.

2010). Silver nanoparticles impregnated on cellulosic paper exhibited up to log 6 reduction in bacterial load for water contaminated with *Escherichia coli* and *Enterococcus faecalis* (Dankovich and Gray 2011). Silver nanoparticles also exhibit antiviral activity as demonstrated by photocatalytic inactivation of Bacteriophage MS2 in aqueous media by silver-impregnated titania (Liga et al. 2011). Mercaptosuccinic acid-capped silver nanoparticles immobilized on alumina supports were shown to remove heavy metal ions such as mercury, with a high removal ability of 0.8 g mercury per gram of silver (Sumesh et al. 2011). A wide variety of porous substrates such as organic sponges (Deng et al. 2017), cryogels (Loo et al. 2013), and rice husk ash (He et al. 2013) have been demonstrated as suitable supports for nanosilver-based water disinfection. Bioderived silica is an important renewable material as a substrate, owing to its abundance, high porosity, surface area, and the versatility to functionalize its surface with variety of chemical groups (Parandhaman et al. 2015). Silver nanoparticles immobilized on nanosilica were shown to degrade organic dyes Congo red, eosin yellow, bromophenol blue, and brilliant blue, as well as antimicrobial activity against *Rhizopus oryzae*, *Escherichia coli*, and *Pseudomonas aeruginosa* (Das et al. 2013). Immobilized silver nanoparticles on biosourced supports such as silica and rice husk ash have been commercialized into low water purifiers, based on contact decontamination under gravity-driven percolating flow. Such point of use water purifiers are low cost and do not require electricity, potentially toxic chemicals or pressurized water flow, vis-a-vis ultraviolet—reverse osmosis (UV-RO)-based systems, and thus can be of immense benefit for communities deprived of safe drinking water.

6 Conclusion and Outlook

In this chapter, we have provided an overview of synthesis and assembly of silver nanomaterials and their immobilization on various substrates. The various forces and interactions operating at the nanoscale and their interplay are highlighted using a variety of examples. As these illustrative examples demonstrate, spectacular progress has been achieved in engineering the size, morphology, and arrangement of silver nanomaterials. While the "low-hanging" fruits on the application fronts such as point of use in water purification and antibacterial textiles, are already commercially deployed, applications such as e-skins, sensors, and flexible displays are in advanced development stages and can be expected for deployment in near future. The therapeutic applications of silver, such as alternative to antibiotics, are promising therapies in light of multidrug resistant infections; however, extensive in vivo testing and trials with a focus on nanoparticle toxicity, bioaccumulation, non-specific interactions, and long-term effects would be required before nanosilver-based therapies become a reality. Silver nanomaterials also hold promise in plasmonic catalysis toward solar-driven chemical transformations and solar fuels. Central to these applications is the immobilization of silver into assemblies and superstructures with desired functionality, while being economical and environmentally benign and continues to be a topic of intense research interest.

References

Abdelgawad AM, Hudson SM, Rojas OJ (2014) Antimicrobial wound dressing nanofiber mats from multicomponent (chitosan/silver-NPs/polyvinyl alcohol) systems. Carbohydr Polyme 100:166–178. https://doi.org/10.1016/j.carbpol.2012.12.043

Aherne D, Ledwith DM, Gara M, Kelly JM (2008) Optical properties and growth aspects of silver nanoprisms produced by a highly reproducible and rapid synthesis at room temperature. Adv Functional Mater 18(14):2005–2016. https://doi.org/10.1002/adfm.200800233

Ahmad A, Mukherjee P, Senapati S, Mandal D, Khan MI, Kumar R, Sastry M (2003) Extracellular biosynthesis of silver nanoparticles using the fungus Fusarium oxysporum. Colloids Surf B: Biointerf 28(4):313–318. https://doi.org/10.1016/S0927-7765(02)00174-1

Ahonen P, Laaksonen T, Nykänen A, Ruokolainen J, Kontturi K (2006) Formation of stable ag-nanoparticle aggregates induced by dithiol cross-linking. J Phys Chem B 110(26):12954–12958. https://doi.org/10.1021/jp0604692

Aizenberg J, Braun PV, Wiltzius P (2000) Patterned colloidal deposition controlled by electrostatic and capillary forces. Phys Rev Lett 84(13):2997–3000. https://doi.org/10.1103/physrevlett.84.2997

Alshehri AH, Jakubowska M, Młożniak A, Horaczek M, Rudka D, Free C, Carey JD (2012) Enhanced electrical conductivity of silver nanoparticles for high frequency electronic applications. ACS Appl Mater Interf 4(12):7007–7010. https://doi.org/10.1021/am3022569

Amendola V, Polizzi S, Meneghetti M (2007) Free silver nanoparticles synthesized by laser ablation in organic solvents and their easy functionalization. Langmuir 23(12):6766–6770. https://doi.org/10.1021/la0637061

Arvizo RR, Bhattacharyya S, Kudgus RA, Giri K, Bhattacharya R, Mukherjee P (2012) Intrinsic therapeutic applications of noble metal nanoparticles: past, present and future. Chem Soc Rev 41(7):2943–2970. https://doi.org/10.1039/c2cs15355f

Asakura S, Oosawa F (1954) On Interaction between Two Bodies Immersed in a Solution of Macromolecules. J Chem Phys 22(7):1255–1256. https://doi.org/10.1063/1.1740347

Asakura S, Oosawa F (1958) Interaction between particles suspended in solutions of macromolecules. J Polym Sci 33(126):183–192. https://doi.org/10.1002/pol.1958.1203312618

Ashkarran AA (2010) A novel method for synthesis of colloidal silver nanoparticles by arc discharge in liquid. Current Appl Phys 10(6):1442–1447. https://doi.org/10.1016/j.cap.2010.05.010

Ashkarran AA, Iraji zad A, Ahadian MM, Hormozi Nezhad MR (2009) Stability, size and optical properties of colloidal silver nanoparticles prepared by electrical arc discharge in water. Eur Phys J Appl Phys 48(1):10601

Aslam U, Linic S (2016) Kinetic trapping of immiscible metal atoms into bimetallic nanoparticles through plasmonic visible light-mediated reduction of a bimetallic oxide precursor: case study of Ag–Pt nanoparticle synthesis. Chem Mater 28(22):8289–8295. https://doi.org/10.1021/acs.chemmater.6b03381

Aslan K, Luhrs CC, Pérez-Luna VH (2004) Controlled and reversible aggregation of biotinylated gold nanoparticles with streptavidin. J Phys Chem B 108(40):15631–15639. https://doi.org/10.1021/jp036089n

Auerbach MB (1953) Filling the root canals of molar teeth with silver wires. J Am Dental Assoc 46(3):270–274. https://doi.org/10.14219/jada.archive.1953.0043

Awazu K, Fujimaki M, Rockstuhl C, Tominaga J, Murakami H, Ohki Y, Yoshida N, Watanabe T (2008) A plasmonic photocatalyst consisting of silver nanoparticles embedded in titanium dioxide. J Am Chem Soc 130(5):1676–1680. https://doi.org/10.1021/ja076503n

Aydin K, Ferry VE, Briggs RM, Atwater HA (2011) Broadband polarization-independent resonant light absorption using ultrathin plasmonic super absorbers. Nat Commun 2(1):517. https://doi.org/10.1038/ncomms1528

Balantrapu K, Goia DV (2009) Silver nanoparticles for printable electronics and biological applications. J Mater Res 24(9):2828–2836. https://doi.org/10.1557/jmr.2009.0336

Barclay WR (1921) Electro-silver plating. Part I. Early technical history. Part II. The author's investigation. Trans Farad Soc 16:515–523. https://doi.org/10.1039/tf9211600515

Behrens SH, Borkovec M (2000) Influence of the secondary interaction energy minimum on the early stages of colloidal aggregation. J Colloid Interf Sci 225(2):460–465. https://doi.org/10.1006/jcis.2000.6780

Behrens SH, Borkovec M, Schurtenberger P (1998) Aggregation in charge-stabilized colloidal suspensions revisited. Langmuir 14(8):1951–1954. https://doi.org/10.1021/la971237k

Bensley RD (1950) Natural color photography in colloidal silver. Science 112(2915):553–557. https://doi.org/10.1126/science.112.2915.553

Bhainsa KC, D'Souza SF (2006) Extracellular biosynthesis of silver nanoparticles using the fungus *Aspergillus fumigatus*. Colloids Surf B: Biointerf 47(2):160–164. https://doi.org/10.1016/j.colsurfb.2005.11.026

Bhattacharya S, Srivastava A, Pal A (2006) Modulation of viscoelastic properties of physical gels by nanoparticle doping: influence of the nanoparticle capping agent. Angew Chem Int Ed 45(18):2934–2937. https://doi.org/10.1002/anie.200504461

Biggs S, Habgood M, Jameson GJ, Yan Y-d (2000) Aggregate structures formed via a bridging flocculation mechanism. Chem Eng J 80(1):13–22. https://doi.org/10.1016/S1383-5866(00)00072-1

Bishop KJM, Wilmer CE, Soh S, Grzybowski BA (2009) Nanoscale forces and their uses in self-assembly. Small 5(14):1600–1630. https://doi.org/10.1002/smll.200900358

Blokzijl W, Engberts JBFN (1993) Hydrophobic effects. Opinions and facts. Angewandte Chemie Int Edn in Eng 32(11):1545–1579. https://doi.org/10.1002/anie.199315451

Braun E, Eichen Y, Sivan U, Ben-Yoseph G (1998) DNA-templated assembly and electrode attachment of a conducting silver wire. Nature 391(6669):775–778. https://doi.org/10.1038/35826

Brown KA, Eichelsdoerfer DJ, Liao X, He S, Mirkin CA (2014) Material transport in dip-pen nanolithography. Front Phys 9(3):385–397. https://doi.org/10.1007/s11467-013-0381-1

Burda C, Chen X, Narayanan R, El-Sayed MA (2005) Chemistry and properties of nanocrystals of different shapes. Chem Rev 105(4):1025–1102. https://doi.org/10.1021/cr030063a

Cai Z, Liu B, Zou X, Cheng H-M (2018) Chemical vapor deposition growth and applications of two-dimensional materials and their heterostructures. Chem Rev 118(13):6091–6133. https://doi.org/10.1021/acs.chemrev.7b00536

Campisi S, Schiavoni M, Chan-Thaw CE, Villa A (2016) Untangling the role of the capping agent in nanocatalysis: recent advances and perspectives. Catalysts 6(12). https://doi.org/10.3390/catal6120185

Cao X-Z, Merlitz H, Wu C-X, Egorov SA, Sommer J-U (2013) Effective pair potentials between nanoparticles induced by single monomers and polymer chains. Soft Matter 9(25):5916–5926. https://doi.org/10.1039/c3sm50495f

Capek I (2004) Preparation of metal nanoparticles in water-in-oil (w/o) microemulsions. Adv Colloid Interf Sci 110(1):49–74. https://doi.org/10.1016/j.cis.2004.02.003

Cassagneau T, Fendler JH (1999) Preparation and layer-by-layer self-assembly of silver nanoparticles capped by graphite oxide nanosheets. J Phys Chem B 103(11):1789–1793. https://doi.org/10.1021/jp984690t

Caswell KK, Bender CM, Murphy CJ (2003) Seedless, surfactantless wet chemical synthesis of silver nanowires. Nano Lett 3(5):667–669. https://doi.org/10.1021/nl0341178

Chaloupka K, Malam Y, Seifalian AM (2010) Nanosilver as a new generation of nanoproduct in biomedical applications. Trends Biotechnol 28(11):580–588. https://doi.org/10.1016/j.tibtech.2010.07.006

Chandran SP, Chaudhary M, Pasricha R, Ahmad A, Sastry M (2006) Synthesis of gold nanotriangles and silver nanoparticles using aloevera plant extract. Biotechnol Progr 22(2):577–583. https://doi.org/10.1021/bp0501423

Chen Y (2015) Nanofabrication by electron beam lithography and its applications: a review. Microelectr Eng 135:57–72. https://doi.org/10.1016/j.mee.2015.02.042

Chen S, Carroll DL (2002) Synthesis and characterization of truncated triangular silver nanoplates. Nano Lett 2(9):1003–1007. https://doi.org/10.1021/nl025674h

Chen X, Schluesener HJ (2008) Nanosilver: a nanoproduct in medical application. Toxicol Lett 176(1):1–12. https://doi.org/10.1016/j.toxlet.2007.10.004

Chen S, Fan Z, Carroll DL (2002) Silver nanodisks: synthesis, characterization, and self-assembly. J Phys Chem B 106(42):10777–10781. https://doi.org/10.1021/jp026376b

Chen D, Qiao X, Qiu X, Chen J (2009) Synthesis and electrical properties of uniform silver nanoparticles for electronic applications. J Mater Sci 44(4):1076–1081. https://doi.org/10.1007/s10853-008-3204-y

Chen Z, Ren W, Gao L, Liu B, Pei S, Cheng H-M (2011) Three-dimensional flexible and conductive interconnected graphene networks grown by chemical vapour deposition. Nat Mater 10(6):424–428. https://doi.org/10.1038/nmat3001

Chen C, Tang Y, Ye YS, Xue Z, Xue Y, Xie X, Mai Y-W (2014) High-performance epoxy/silica coated silver nanowire composites as underfill material for electronic packaging. Compos Sci Technol 105:80–85. https://doi.org/10.1016/j.compscitech.2014.10.002

Chen Y, Fan Z, Zhang Z, Niu W, Li C, Yang N, Chen B, Zhang H (2018) Two-dimensional metal nanomaterials: synthesis, properties, and applications. Chem Rev 118(13):6409–6455. https://doi.org/10.1021/acs.chemrev.7b00727

Chou SY, Krauss PR, Renstrom PJ (1996) Imprint lithography with 25-nanometer resolution. Science 272(5258):85–87. https://doi.org/10.1126/science.272.5258.85

Christenson HK (1988) Non-DLVO forces between surfaces -solvation, hydration and capillary effects. J Dispers Sci Technol 9(2):171–206. https://doi.org/10.1080/01932698808943983

Christopher P, Xin H, Linic S (2011) Visible-light-enhanced catalytic oxidation reactions on plasmonic silver nanostructures. Nat Chem 3(6):467–472. https://doi.org/10.1038/nchem.1032

Cinel NA, Bütün S, Özbay E (2012) Electron beam lithography designed silver nano-disks used as label free nano-biosensors based on localized surface plasmon resonance. Opt Expr 20(3):2587–2597. https://doi.org/10.1364/oe.20.002587

Cosgrove T, Cosgrove PT (2010) Colloid science: principles, methods and applications. Wiley, Hoboken

Cuenya BR (2010) Synthesis and catalytic properties of metal nanoparticles: size, shape, support, composition, and oxidation state effects. Thin Solid Films 518(12):3127–3150. https://doi.org/10.1016/j.tsf.2010.01.018

Dal Negro L, Cazzanelli M, Daldosso N, Gaburro Z, Pavesi L, Priolo F, Pacifici D, Franzò G, Iacona F (2003) Stimulated emission in plasma-enhanced chemical vapour deposited silicon nanocrystals. Physica E: Low-Dimens Syst Nanostr 16(3):297–308. https://doi.org/10.1016/S1386-9477(02)00605-7

Dankovich TA, Gray DG (2011) Bactericidal paper impregnated with silver nanoparticles for point-of-use water treatment. Environ Sci Technol 45(5):1992–1998. https://doi.org/10.1021/es103302t

Das SK, Khan MMR, Parandhaman T, Laffir F, Guha AK, Sekaran G, Mandal AB (2013) Nano-silica fabricated with silver nanoparticles: antifouling adsorbent for efficient dye removal, effective water disinfection and biofouling control. Nanoscale 5(12):5549–5560. https://doi.org/10.1039/c3nr00856h

de Gennes PG (1987) Polymers at an interface; a simplified view. Adv Colloid Interf Sci 27(3):189–209. https://doi.org/10.1016/0001-8686(87)85003-0

Deckman HW, Dunsmuir JH (1982) Natural lithography. Appl Phys Lett 41(4):377–379. https://doi.org/10.1063/1.93501

Deckman HW, Dunsmuir JH (1983) Applications of surface textures produced with natural lithography. J Vac Sci Technol B: Microelectron Process Phenom 1(4):1109–1112. https://doi.org/10.1116/1.582644

Deng C-H, Gong J-L, Zhang P, Zeng G-M, Song B, Liu H-Y (2017) Preparation of melamine sponge decorated with silver nanoparticles-modified graphene for water disinfection. J Colloid Interf Sci 488:26–38. https://doi.org/10.1016/j.jcis.2016.10.078

Derjaguin B, Landau L (1993) Theory of the stability of strongly charged lyophobic sols and of the adhesion of strongly charged particles in solutions of electrolytes. Progr Surf Sci 43(1):30–59. https://doi.org/10.1016/0079-6816(93)90013-L

Derjaguin BV, Churaev NV, Muller VM (1987) The Derjaguin—Landau—Verwey—Overbeek (DLVO) theory of stability of lyophobic colloids. In: Surface forces. Springer US, Boston, pp 293–310

Dong X, Ji X, Wu H, Zhao L, Li J, Yang W (2009) Shape control of silver nanoparticles by stepwise citrate reduction. J Phys Chem C 113(16):6573–6576. https://doi.org/10.1021/jp900775b

Dong Z, Le X, Li X, Zhang W, Dong C, Ma J (2014) Silver nanoparticles immobilized on fibrous nano-silica as highly efficient and recyclable heterogeneous catalyst for reduction of 4-nitrophenol and 2-nitroaniline. Appl Catal B: Environ 158–159:129–135. https://doi.org/10.1016/j.apcatb.2014.04.015

Dzyaloshinskii IE, Lifshitz EM, Pitaevskii LP (1961) The general theory of van der Waals forces. Adv Phys 10(38):165–209. https://doi.org/10.1080/00018736100101281

Fan M, Brolo AG (2009) Silver nanoparticles self assembly as SERS substrates with near single molecule detection limit. Phys Chem Chem Phys 11(34):7381–7389. https://doi.org/10.1039/b904744a

Fan Z, Zhang H (2016) Crystal phase-controlled synthesis, properties and applications of noble metal nanomaterials. Chem Soc Rev 45(1):63–82. https://doi.org/10.1039/c5cs00467e

Fan Z, Liu B, Wang J, Zhang S, Lin Q, Gong P, Ma L, Yang S (2014) A novel wound dressing based on Ag/graphene polymer hydrogel: effectively kill bacteria and accelerate wound healing. Adv Func Mater 24(25):3933–3943. https://doi.org/10.1002/adfm.201304202

Fang N, Lee H, Sun C, Zhang X (2005) Sub–diffraction-limited optical imaging with a silver superlens. Science 308(5721):534–537. https://doi.org/10.1126/science.1108759

Feng QL, Wu J, Chen GQ, Cui FZ, Kim TN, Kim JO (2000) A mechanistic study of the antibacterial effect of silver ions on Escherichia coli and Staphylococcus aureus. J Biomed Mater Res 52(4):662–668. https://doi.org/10.1002/1097-4636(20001215)52:4%3c662:aid-jbm10%3e3.0.co;2-3

Fiévet F, Lagier JP, Figlarz M (1989) Preparing monodisperse metal powders in micrometer and submicrometer sizes by the polyol process. MRS Bull 14(12):29–34. https://doi.org/10.1557/s0883769400060930

Fiévet F, Ammar-Merah S, Brayner R, Chau F, Giraud M, Mammeri F, Peron J, Piquemal JY, Sicard L, Viau G (2018) The polyol process: a unique method for easy access to metal nanoparticles with tailored sizes, shapes and compositions. Chem Soc Rev 47(14):5187–5233. https://doi.org/10.1039/c7cs00777a

Foot AW (1888) Note on silver-staining (Argyria). Trans R Acad Med Ireland 6(1):7. https://doi.org/10.1007/bf03171106

Fullam S, Rensmo H, Rao SN, Fitzmaurice D (2002) Noncovalent self-assembly of silver and gold nanocrystal aggregates in solution. Chem Mater 14(9):3643–3650. https://doi.org/10.1021/cm011182e

Gangadharan D, Harshvardan K, Gnanasekar G, Dixit D, Popat KM, Anand PS (2010) Polymeric microspheres containing silver nanoparticles as a bactericidal agent for water disinfection. Water Res 44(18):5481–5487. https://doi.org/10.1016/j.watres.2010.06.057

Gao Y, Jiang P, Liu DF, Yuan HJ, Yan XQ, Zhou ZP, Wang JX, Song L, Liu LF, Zhou WY, Wang G, Wang CY, Xie SS (2003) Synthesis, characterization and self-assembly of silver nanowires. Chem Phys Lett 380(1):146–149. https://doi.org/10.1016/j.cplett.2003.08.074

Gao S, Zhang Z, Liu K, Dong B (2016) Direct evidence of plasmonic enhancement on catalytic reduction of 4-nitrophenol over silver nanoparticles supported on flexible fibrous networks. Appl Catal B: Environ 188:245–252. https://doi.org/10.1016/j.apcatb.2016.01.074

Gates BD, Xu Q, Stewart M, Ryan D, Willson CG, Whitesides GM (2005) New approaches to nanofabrication: molding, printing, and other techniques. Chem Rev 105(4):1171–1196. https://doi.org/10.1021/cr030076o

Ginger DS, Zhang H, Mirkin CA (2004) The evolution of dip-pen nanolithography. Angew Chem Int Ed 43(1):30–45. https://doi.org/10.1002/anie.200300608

Grasso D, Subramaniam K, Butkus M, Strevett K, Bergendahl J (2002) A review of non-DLVO interactions in environmental colloidal systems. Rev Environ Sci Biotechnol 1(1):17–38. https://doi.org/10.1023/a:1015146710500

Grzelczak M, Vermant J, Furst EM, Liz-Marzán LM (2010) Directed self-assembly of nanoparticles. ACS Nano 4(7):3591–3605. https://doi.org/10.1021/nn100869j

Halpern AR, Corn RM (2013) Lithographically patterned electrodeposition of gold, silver, and nickel nanoring arrays with widely tunable near-infrared plasmonic resonances. ACS Nano 7(2):1755–1762. https://doi.org/10.1021/nn3058505

Hamaker HC (1937) The London—van der Waals attraction between spherical particles. Physica 4(10):1058–1072. https://doi.org/10.1016/S0031-8914(37)80203-7

Hammer MU, Anderson TH, Chaimovich A, Shell MS, Israelachvili J (2010) The search for the hydrophobic force law. Faraday Discuss 146:299–308. https://doi.org/10.1039/b926184b

Han Y-J, Kim JM, Stucky GD (2000) Preparation of noble metal nanowires using hexagonal mesoporous silica SBA-15. Chem Mater 12(8):2068–2069. https://doi.org/10.1021/cm0010553

Hansen J-P, Löwen H (2000) Effective interactions between electric double layers. Ann Rev Phys Chem 51(1):209–242. https://doi.org/10.1146/annurev.physchem.51.1.209

Hao E, Schatz GC (2004) Electromagnetic fields around silver nanoparticles and dimers. J Chem Phys 120(1):357–366. https://doi.org/10.1063/1.1629280

Hao Y, Gao J, Xu Z, Zhang N, Luo J, Liu X (2019) Preparation of silver nanoparticles with hyperbranched polymers as a stabilizer for inkjet printing of flexible circuits. New J Chem 43(6):2797–2803. https://doi.org/10.1039/c8nj05639k

He D, Ikeda-Ohno A, Boland DD, Waite TD (2013) Synthesis and characterization of antibacterial silver nanoparticle-impregnated rice husks and rice husk ash. Environ Sci Technol 47(10):5276–5284. https://doi.org/10.1021/es303890y

Helmlinger J, Sengstock C, Groß-Heitfeld C, Mayer C, Schildhauer TA, Köller M, Epple M (2016) Silver nanoparticles with different size and shape: equal cytotoxicity, but different antibacterial effects. RSC Adv 6(22):18490–18501. https://doi.org/10.1039/c5ra27836h

Henglein A, Giersig M (1999) Formation of colloidal silver nanoparticles: capping action of citrate. J Phys Chem B 103(44):9533–9539. https://doi.org/10.1021/jp9925334

Henzie J, Grünwald M, Widmer-Cooper A, Geissler PL, Yang P (2012) Self-assembly of uniform polyhedral silver nanocrystals into densest packings and exotic superlattices. Nat Mater 11(2):131–137. https://doi.org/10.1038/nmat3178

History of the Medical Use of Silver (2009) Surg Infect 10(3):289–292. https://doi.org/10.1089/sur.2008.9941

Hu J-Q, Chen Q, Xie Z-X, Han G-B, Wang R-H, Ren B, Zhang Y, Yang Z-L, Tian Z-Q (2004) A simple and effective route for the synthesis of crystalline silver nanorods and nanowires. Adv Funct Mater 14(2):183–189. https://doi.org/10.1002/adfm.200304421

Hu L, Kim HS, Lee J-Y, Peumans P, Cui Y (2010) Scalable coating and properties of transparent, flexible, silver nanowire electrodes. ACS Nano 4(5):2955–2963. https://doi.org/10.1021/nn1005232

Huang J, Li Q, Sun D, Lu Y, Su Y, Yang X, Wang H, Wang Y, Shao W, He N, Hong J, Chen C (2007) Biosynthesis of silver and gold nanoparticles by novel sundried Cinnamomum camphoraleaf. Nanotechnology 18(10):105104. https://doi.org/10.1088/0957-4484/18/10/105104

Huang G-W, Xiao H-M, Fu S-Y (2014) Paper-based silver-nanowire electronic circuits with outstanding electrical conductivity and extreme bending stability. Nanoscale 6(15):8495–8502. https://doi.org/10.1039/c4nr00846d

Hulteen JC, Treichel DA, Smith MT, Duval ML, Jensen TR, Van Duyne RP (1999) Nanosphere lithography: size-tunable silver nanoparticle and surface cluster arrays. J Phys Chem B 103(19):3854–3863. https://doi.org/10.1021/jp9904771

Hung S-C, Nafday OA, Haaheim JR, Ren F, Chi GC, Pearton SJ (2010) Dip pen nanolithography of conductive silver traces. J Phys Chem C 114(21):9672–9677. https://doi.org/10.1021/jp101505k

Hutchison JL, Kiselev NA, Krinichnaya EP, Krestinin AV, Loutfy RO, Morawsky AP, Muradyan VE, Obraztsova ED, Sloan J, Terekhov SV, Zakharov DN (2001) Double-walled carbon nanotubes fabricated by a hydrogen arc discharge method. Carbon 39(5):761–770. https://doi.org/10.1016/S0008-6223(00)00187-1

Israelachvili JN (2011a) 14-electrostatic forces between surfaces in liquids. In: Israelachvili JN (ed) Intermolecular and surface forces, 3rd edn. Academic Press, Boston, pp 291–340

Israelachvili JN (2011b) Chapter 13—Van der Waals Forces between particles and surfaces. In: Israelachvili JN (ed) Intermolecular and surface forces, 3rd edn. Academic Press, Boston, pp 253–289

Israelachvili JN (2011c) Chapter 14—electrostatic forces between surfaces in liquids. In: Israelachvili JN (ed) Intermolecular and surface forces, 3rd edn. Academic Press, San Diego, pp 291–340

Israelachvili J, Gourdon D (2001) Putting liquids under molecular-scale confinement. Science 292(5518):867–868. https://doi.org/10.1126/science.1061206

Israelachvili J, Pashley R (1982) The hydrophobic interaction is long range, decaying exponentially with distance. Nature 300(5890):341–342. https://doi.org/10.1038/300341a0

Ito T, Okazaki S (2000) Pushing the limits of lithography. Nature 406(6799):1027–1031. https://doi.org/10.1038/35023233

Jackson JB, Halas NJ (2001) Silver nanoshells: variations in morphologies and optical properties. J Phys Chem B 105(14):2743–2746. https://doi.org/10.1021/jp003868k

Jain P, Pradeep T (2005) Potential of silver nanoparticle-coated polyurethane foam as an antibacterial water filter. Biotechnol Bioeng 90(1):59–63. https://doi.org/10.1002/bit.20368

Jain PK, Huang X, El-Sayed IH, El-Sayed MA (2007) Review of some interesting surface plasmon resonance-enhanced properties of noble metal nanoparticles and their applications to biosystems. Plasmonics 2(3):107–118. https://doi.org/10.1007/s11468-007-9031-1

Jain PK, Huang X, El-Sayed IH, El-Sayed MA (2008) Noble metals on the nanoscale: optical and photothermal properties and some applications in imaging, sensing, biology, and medicine. Acc Chem Res 41(12):1578–1586. https://doi.org/10.1021/ar7002804

Jana S, Ghosh SK, Nath S, Pande S, Praharaj S, Panigrahi S, Basu S, Endo T, Pal T (2006) Synthesis of silver nanoshell-coated cationic polystyrene beads: a solid phase catalyst for the reduction of 4-nitrophenol. Appl Catal A: Gen 313(1):41–48. https://doi.org/10.1016/j.apcata.2006.07.007

Jeffrey GA (2003) Hydrogen-bonding: an update. Crystallogr Rev 9(2–3):135–176. https://doi.org/10.1080/08893110310001621754

Jiang Z-J, Liu C-Y, Sun L-W (2005) Catalytic properties of silver nanoparticles supported on silica spheres. J Phys Chem B 109(5):1730–1735. https://doi.org/10.1021/jp046032g

Jin R, Cao Y, Mirkin CA, Kelly KL, Schatz GC, Zheng JG (2001) Photoinduced conversion of silver nanospheres to nanoprisms. Science 294(5548):1901–1903. https://doi.org/10.1126/science.1066541

Jin R, Charles Cao Y, Hao E, Métraux GS, Schatz GC, Mirkin CA (2003) Controlling anisotropic nanoparticle growth through plasmon excitation. Nature 425(6957):487–490. https://doi.org/10.1038/nature02020

Johnson SR, Evans SD, Brydson R (1998) Influence of a terminal functionality on the physical properties of surfactant-stabilized gold nanoparticles. Langmuir 14(23):6639–6647. https://doi.org/10.1021/la9711342

Jung J, Lee H, Ha I, Cho H, Kim KK, Kwon J, Won P, Hong S, Ko SH (2017) Highly stretchable and transparent electromagnetic interference shielding film based on silver nanowire percolation network for wearable electronics applications. ACS Appl Mater Interf 9(51):44609–44616. https://doi.org/10.1021/acsami.7b14626

Jung J, Cho H, Yuksel R, Kim D, Lee H, Kwon J, Lee P, Yeo J, Hong S, Unalan HE, Han S, Ko SH (2019) Stretchable/flexible silver nanowire electrodes for energy device applications. Nanoscale 11(43):20356–20378. https://doi.org/10.1039/c9nr04193a

Kalimuthu K, Suresh Babu R, Venkataraman D, Bilal M, Gurunathan S (2008) Biosynthesis of silver nanocrystals by *Bacillus licheniformis*. Colloids Surf B: Biointerf 65(1):150–153. https://doi.org/10.1016/j.colsurfb.2008.02.018

Kang M-G, Xu T, Park HJ, Luo X, Guo LJ (2010) Efficiency enhancement of organic solar cells using transparent plasmonic Ag nanowire electrodes. Adv Mater 22(39):4378–4383. https://doi.org/10.1002/adma.201001395

Kelly KL, Coronado E, Zhao LL, Schatz GC (2003) The optical properties of metal nanoparticles: the influence of size, shape, and dielectric environment. J Phys Chem B 107(3):668–677. https://doi.org/10.1021/jp026731y

Kijima T, Yoshimura T, Uota M, Ikeda T, Fujikawa D, Mouri S, Uoyama S (2004) Noble-metal nanotubes (Pt, Pd, Ag) from lyotropic mixed-surfactant liquid-crystal templates. Angew Chem Int Ed 43(2):228–232. https://doi.org/10.1002/anie.200352630

Kim H, Kim J, Yang H, Suh J, Kim T, Han B, Kim S, Kim DS, Pikhitsa PV, Choi M (2006) Parallel patterning of nanoparticles via electrodynamic focusing of charged aerosols. Nat Nanotechnol 1(2):117–121. https://doi.org/10.1038/nnano.2006.94

Kim S-E, Han Y-H, Lee BC, Lee J-C (2010) One-pot fabrication of various silver nanostructures on substrates using electron beam irradiation. Nanotechnology 21(7):075302. https://doi.org/10.1088/0957-4484/21/7/075302

Kinge SS, Crego-Calama M, Reinhoudt DN (2007) Gold nanoparticle assemblies through hydrogen-bonded supramolecular mediators. Langmuir 23(17):8772–8777. https://doi.org/10.1021/la700514u

Kissmeyer A (1936) Treatment of burns and scalds with a silver nitrate ointment. The Lancet 228(5904):985. https://doi.org/10.1016/S0140-6736(00)47951-1

Kollman PA, Allen LC (1972) Theory of the hydrogen bond. Chem Rev 72(3):283–303. https://doi.org/10.1021/cr60277a004

Korgel BA, Fitzmaurice D (1998) Self-assembly of silver nanocrystals into two-dimensional nanowire arrays. Adv Mater 10(9):661–665. https://doi.org/10.1002/(sici)1521-4095(199806)10:9%3c661:aid-adma661%3e3.0.co;2-l

Korgel BA, Fullam S, Connolly S, Fitzmaurice D (1998) Assembly and self-organization of silver nanocrystal superlattices: ordered "soft spheres". J Phys Chem B 102(43):8379–8388. https://doi.org/10.1021/jp981598o

Kralchevsky PA, Denkov ND (2001) Capillary forces and structuring in layers of colloid particles. Curr Opin Colloid Interface Sci 6(4):383–401. https://doi.org/10.1016/S1359-0294(01)00105-4

Kralchevsky PA, Nagayama K (1994) Capillary forces between colloidal particles. Langmuir 10(1):23–36. https://doi.org/10.1021/la00013a004

Krishnaraj C, Jagan EG, Rajasekar S, Selvakumar P, Kalaichelvan PT, Mohan N (2010) Synthesis of silver nanoparticles using Acalypha indica leaf extracts and its antibacterial activity against water borne pathogens. Colloids Surf, B 76(1):50–56. https://doi.org/10.1016/j.colsurfb.2009.10.008

Kumar A, Biebuyck HA, Whitesides GM (1994) Patterning self-assembled monolayers: applications in materials science. Langmuir 10(5):1498–1511. https://doi.org/10.1021/la00017a030

LaMer VK, Dinegar RH (1950) Theory, production and mechanism of formation of monodispersed hydrosols. J Am Chem Soc 72(11):4847–4854. https://doi.org/10.1021/ja01167a001

Layani M, Magdassi S (2011) Flexible transparent conductive coatings by combining self-assembly with sintering of silver nanoparticles performed at room temperature. J Mater Chem 21(39):15378–15382. https://doi.org/10.1039/c1jm13174e

Kim JS, Kuk E, Yu KN, Kim J-H, Park SJ, Lee HJ, Kim SH, Park YK, Park YH, Hwang C-Y, Kim Y-K, Lee, Y-S, Jeong DH, Cho M-H (2007) Antimicrobial effects of silver nanoparticles. Nanomed Nanotechnol Biol Med 3(1):95–101. https://doi.org/10.1016/j.nano.2006.12.001

Lee K-S, El-Sayed MA (2006) Gold and silver nanoparticles in sensing and imaging: sensitivity of plasmon response to size, shape, and metal composition. J Phys Chem B 110(39):19220–19225. https://doi.org/10.1021/jp062536y

Lee HJ, Yeo SY, Jeong SH (2003) Antibacterial effect of nanosized silver colloidal solution on textile fabrics. J Mater Sci 38(10):2199–2204. https://doi.org/10.1023/a:1023736416361

Lee KJ, Jun BH, Kim TH, Joung J (2006) Direct synthesis and inkjetting of silver nanocrystals toward printed electronics. Nanotechnology 17(9):2424–2428. https://doi.org/10.1088/0957-4484/17/9/060

Lee J, Lee P, Lee H, Lee D, Lee SS, Ko SH (2012) Very long Ag nanowire synthesis and its application in a highly transparent, conductive and flexible metal electrode touch panel. Nanoscale 4(20):6408–6414. https://doi.org/10.1039/c2nr31254a

Li Y, Wu Y, Ong BS (2005) Facile synthesis of silver nanoparticles useful for fabrication of high-conductivity elements for printed electronics. J Am Chem Soc 127(10):3266–3267. https://doi.org/10.1021/ja043425k

Li Y, Leung P, Yao L, Song QW, Newton E (2006) Antimicrobial effect of surgical masks coated with nanoparticles. J Hosp Infect 62(1):58–63. https://doi.org/10.1016/j.jhin.2005.04.015

Li H-J, Zhang A-Q, Hu Y, Sui L, Qian D-J, Chen M (2012) Large-scale synthesis and self-organization of silver nanoparticles with Tween 80 as a reductant and stabilizer. Nanoscale Res Lett 7(1):612–612. https://doi.org/10.1186/1556-276x-7-612

Liga MV, Bryant EL, Colvin VL, Li Q (2011) Virus inactivation by silver doped titanium dioxide nanoparticles for drinking water treatment. Water Res 45(2):535–544. https://doi.org/10.1016/j.watres.2010.09.012

Link S, Burda C, Mohamed MB, Nikoobakht B, El-Sayed MA (1999) Laser photothermal melting and fragmentation of gold nanorods: energy and laser pulse-width dependence. J Phys Chem A 103(9):1165–1170. https://doi.org/10.1021/jp983141k

Liu P, Zhao M (2009) Silver nanoparticle supported on halloysite nanotubes catalyzed reduction of 4-nitrophenol (4-NP). Appl Surf Sci 255(7):3989–3993. https://doi.org/10.1016/j.apsusc.2008.10.094

Liu J, Sonshine DA, Shervani S, Hurt RH (2010) Controlled release of biologically active silver from nanosilver surfaces. ACS Nano 4(11):6903–6913. https://doi.org/10.1021/nn102272n

Liu L, Yang P, Li J, Zhang Z, Yu X, Lu L (2017) Temperature-controlled cross-linking of silver nanoparticles with diels-alder reaction and its application on antibacterial property. Appl Surf Sci 403:435–440. https://doi.org/10.1016/j.apsusc.2017.01.162

Lo C-H, Tsung T-T, Lin H-M (2007) Preparation of silver nanofluid by the submerged arc nanoparticle synthesis system (SANSS). J Alloy Compd 434–435:659–662. https://doi.org/10.1016/j.jallcom.2006.08.217

Loo S-L, Fane AG, Lim T-T, Krantz WB, Liang Y-N, Liu X, Hu X (2013) Superabsorbent cryogels decorated with silver nanoparticles as a novel water technology for point-of-use disinfection. Environ Sci Technol 47(16):9363–9371. https://doi.org/10.1021/es401219s

Lv Y, Liu H, Wang Z, Liu S, Hao L, Sang Y, Liu D, Wang J, Boughton RI (2009) Silver nanoparticle-decorated porous ceramic composite for water treatment. J Membr Sci 331(1):50–56. https://doi.org/10.1016/j.memsci.2009.01.007

Mafuné F, Kohno J-Y, Takeda Y, Kondow T, Sawabe H (2000a) Formation and size control of silver nanoparticles by laser ablation in aqueous solution. J Phys Chem B 104(39):9111–9117. https://doi.org/10.1021/jp001336y

Mafuné F, Kohno J-Y, Takeda Y, Kondow T, Sawabe H (2000b) Structure and stability of silver nanoparticles in aqueous solution produced by laser ablation. J Phys Chem B 104(35):8333–8337. https://doi.org/10.1021/jp001803b

Maier SA, Brongersma ML, Kik PG, Meltzer S, Requicha AAG, Atwater HA (2001) Plasmonics—a route to nanoscale optical devices. Adv Mater 13(19):1501–1505. https://doi.org/10.1002/1521-4095(200110)13:19%3c1501:aid-adma1501%3e3.0.co;2-z

Maier SA, Kik PG, Atwater HA (2003a) Optical pulse propagation in metal nanoparticle chain waveguides. Phys Rev B 67(20):205402. https://doi.org/10.1103/physrevb.67.205402

Maier SA, Kik PG, Atwater HA, Meltzer S, Harel E, Koel BE, Requicha AAG (2003b) Local detection of electromagnetic energy transport below the diffraction limit in metal nanoparticle plasmon waveguides. Nat Mater 2(4):229–232. https://doi.org/10.1038/nmat852

Mallick K, Witcomb MJ, Scurrell MS (2005) Self-assembly of silver nanoparticles in a polymer solvent: formation of a nanochain through nanoscale soldering. Mater Chem Phys 90(2):221–224. https://doi.org/10.1016/j.matchemphys.2004.10.030

Maneerung T, Tokura S, Rujiravanit R (2008) Impregnation of silver nanoparticles into bacterial cellulose for antimicrobial wound dressing. Carbohyd Polym 72(1):43–51. https://doi.org/10.1016/j.carbpol.2007.07.025

Maoz R, Frydman E, Cohen SR, Sagiv J (2000) Constructive nanolithography: site-defined silver self-assembly on nanoelectrochemically patterned monolayer templates. Adv Mater 12(6):424–429. https://doi.org/10.1002/(sici)1521-4095(200003)12:6%3c424:aid-adma424%3e3.0.co;2-s

McClements DJ (2005) Theoretical analysis of factors affecting the formation and stability of multilayered colloidal dispersions. Langmuir 21(21):9777–9785. https://doi.org/10.1021/la0512603

McFarland AD, Van Duyne RP (2003) Single silver nanoparticles as real-time optical sensors with zeptomole sensitivity. Nano Lett 3(8):1057–1062. https://doi.org/10.1021/nl034372s

Merga G, Wilson R, Lynn G, Milosavljevic BH, Meisel D (2007) Redox catalysis on "naked" silver nanoparticles. J Phys Chem C 111(33):12220–12226. https://doi.org/10.1021/jp074257w

Meyer EE, Rosenberg KJ, Israelachvili J (2006) Recent progress in understanding hydrophobic interactions. Proc Natl Acad Sci 103(43):15739–15746. https://doi.org/10.1073/pnas.0606422103

Miller MS, O'Kane JC, Niec A, Carmichael RS, Carmichael TB (2013) Silver nanowire/optical adhesive coatings as transparent electrodes for flexible electronics. ACS Appl Mater Interf 5(20):10165–10172. https://doi.org/10.1021/am402847y

Mitsudome T, Mikami Y, Funai H, Mizugaki T, Jitsukawa K, Kaneda K (2008) Oxidant-free alcohol dehydrogenation using a reusable hydrotalcite-supported silver nanoparticle catalyst. Angew Chem Int Ed 47(1):138–141. https://doi.org/10.1002/anie.200703161

Mittal AK, Chisti Y, Banerjee UC (2013) Synthesis of metallic nanoparticles using plant extracts. Biotechnol Adv 31(2):346–356. https://doi.org/10.1016/j.biotechadv.2013.01.003

Mock JJ, Barbic M, Smith DR, Schultz DA, Schultz S (2002) Shape effects in plasmon resonance of individual colloidal silver nanoparticles. J Chem Phys 116(15):6755–6759. https://doi.org/10.1063/1.1462610

Monteiro DR, Gorup LF, Takamiya AS, Ruvollo-Filho AC, Camargo ER, Barbosa DB (2009) The growing importance of materials that prevent microbial adhesion: antimicrobial effect of medical devices containing silver. Int J Antimicrob Agents 34(2):103–110. https://doi.org/10.1016/j.ijantimicag.2009.01.017

Morones JR, Elechiguerra JL, Camacho A, Holt K, Kouri JB, Ramírez JT, Yacaman MJ (2005) The bactericidal effect of silver nanoparticles. Nanotechnology 16(10):2346–2353. https://doi.org/10.1088/0957-4484/16/10/059

Moskovits M (2005) Surface-enhanced Raman spectroscopy: a brief retrospective. J Raman Spectrosc 36(6–7):485–496. https://doi.org/10.1002/jrs.1362

Mukherjee P, Ahmad A, Mandal D, Senapati S, Sainkar SR, Khan MI, Parishcha R, Ajaykumar PV, Alam M, Kumar R, Sastry M (2001) Fungus-mediated synthesis of silver nanoparticles and their immobilization in the mycelial matrix: a novel biological approach to nanoparticle synthesis. Nano Lett 1(10):515–519. https://doi.org/10.1021/nl0155274

Mukherjee S, Chowdhury D, Kotcherlakota R, Patra SB, Bhadra V, Sreedhar MP, Patra CR (2014) Potential theranostics application of bio-synthesized silver nanoparticles (4-in-1 System). Theranostics 4(3):316–335. https://doi.org/10.7150/thno.7819

Nam S, Song M, Kim D-H, Cho B, Lee HM, Kwon J-D, Park S-G, Nam K-S, Jeong Y, Kwon S-H, Park YC, Jin S-H, Kang J-W, Jo S, Kim CS (2014) Ultrasmooth, extremely deformable and shape recoverable Ag nanowire embedded transparent electrode. Sci Rep 4(1):4788. https://doi.org/10.1038/srep04788

Narayanan R, El-Sayed MA (2005) Catalysis with transition metal nanoparticles in colloidal solution: nanoparticle shape dependence and stability. J Phys Chem B 109(26):12663–12676. https://doi.org/10.1021/jp051066p

Narayanan KB, Sakthivel N (2010) Biological synthesis of metal nanoparticles by microbes. Adv Coll Interface Sci 156(1):1–13. https://doi.org/10.1016/j.cis.2010.02.001

Nepijko SA, Ievlev DN, Schulze W, Urban J, Ertl G (2000) Growth of rodlike silver nanoparticles by vapor deposition of small clusters. Chem Phys Chem 1(3):140–142. https://doi.org/10.1002/1439-7641(20001103)1:3%3c140:aid-cphc140%3e3.0.co;2-l

Nickel U, Castell A, Pöppl K, Schneider S (2000) A silver colloid produced by reduction with hydrazine as support for highly sensitive surface-enhanced raman spectroscopy. Langmuir 16(23):9087–9091. https://doi.org/10.1021/la000536y

Nie S, Emory SR (1997) Probing single molecules and single nanoparticles by surface-enhanced Raman scattering. Science 275(5303):1102–1106. https://doi.org/10.1126/science.275.5303.1102

Nie Z, Fava D, Rubinstein M, Kumacheva E (2008) "Supramolecular" assembly of gold nanorods end-terminated with polymer "pom-poms": effect of pom-pom structure on the association modes. J Am Chem Soc 130(11):3683–3689. https://doi.org/10.1021/ja711150k

Nikoobakht B, Wang ZL, El-Sayed MA (2000) Self-assembly of gold nanorods. J Phys Chem B 104(36):8635–8640. https://doi.org/10.1021/jp001287p

Niu Z, Li Y (2014) Removal and utilization of capping agents in nanocatalysis. Chem Mater 26(1):72–83. https://doi.org/10.1021/cm4022479

Nowack B, Krug HF, Height M (2011) 120 years of nanosilver history: implications for policy makers. Environ Sci Technol 45(4):1177–1183. https://doi.org/10.1021/es103316q

Oćwieja M, Adamczyk Z, Morga M, Michna A (2011) High density silver nanoparticle monolayers produced by colloid self-assembly on polyelectrolyte supporting layers. J Colloid Interf Sci 364(1):39–48. https://doi.org/10.1016/j.jcis.2011.07.059

Pal S, Tak YK, Song JM (2007) Does the antibacterial activity of silver nanoparticles depend on the shape of the nanoparticle? A study of the gram-negative bacterium *Escherichia coli*. Appl Environ Microbiol 73(6):1712–1720. https://doi.org/10.1128/aem.02218-06

Panigrahi S, Praharaj S, Basu S, Ghosh SK, Jana S, Pande S, Vo-Dinh T, Jiang H, Pal T (2006) Self-assembly of silver nanoparticles: synthesis, stabilization, optical properties, and application in surface-enhanced raman scattering. J Phys Chem B 110(27):13436–13444. https://doi.org/10.1021/jp062119l

Parandhaman T, Das A, Ramalingam B, Samanta D, Sastry TP, Mandal AB, Das SK (2015) Antimicrobial behavior of biosynthesized silica–silver nanocomposite for water disinfection: A mechanistic perspective. J Hazard Mater 290:117–126. https://doi.org/10.1016/j.jhazmat.2015.02.061

Park K, Koerner H, Vaia RA (2010) Depletion-induced shape and size selection of gold nanoparticles. Nano Lett 10(4):1433–1439. https://doi.org/10.1021/nl100345u

Patel AC, Li S, Wang C, Zhang W, Wei Y (2007) Electrospinning of porous silica nanofibers containing silver nanoparticles for catalytic applications. Chem Mater 19(6):1231–1238. https://doi.org/10.1021/cm061331z

Pelley AJ, Tufenkji N (2008) Effect of particle size and natural organic matter on the migration of nano- and microscale latex particles in saturated porous media. J Colloid Interface Sci 321(1):74–83. https://doi.org/10.1016/j.jcis.2008.01.046

Petit C, Lixon P, Pileni MP (1993) In situ synthesis of silver nanocluster in AOT reverse micelles. J Phys Chem 97(49):12974–12983. https://doi.org/10.1021/j100151a054

Piech M, Walz JY (2000) Analytical expressions for calculating the depletion interaction produced by charged spheres and spheroids. Langmuir 16(21):7895–7899. https://doi.org/10.1021/la000764s

Pietrobon B, McEachran M, Kitaev V (2009) Synthesis of size-controlled faceted pentagonal silver nanorods with tunable plasmonic properties and self-assembly of these nanorods. ACS Nano 3(1):21–26. https://doi.org/10.1021/nn800591y

Pradhan N, Pal A, Pal T (2002) Silver nanoparticle catalyzed reduction of aromatic nitro compounds. Colloids Surf, A 196(2):247–257. https://doi.org/10.1016/S0927-7757(01)01040-8

Prevo BG, Kuncicky DM, Velev OD (2007) Engineered deposition of coatings from nano- and micro-particles: A brief review of convective assembly at high volume fraction. Colloids Surf, A 311(1):2–10. https://doi.org/10.1016/j.colsurfa.2007.08.030

Pyatenko A, Shimokawa K, Yamaguchi M, Nishimura O, Suzuki M (2004) Synthesis of silver nanoparticles by laser ablation in pure water. Appl Phys A 79(4):803–806. https://doi.org/10.1007/s00339-004-2841-5

Quinten M, Leitner A, Krenn JR, Aussenegg FR (1998) Electromagnetic energy transport via linear chains of silver nanoparticles. Opt Lett 23(17):1331–1333. https://doi.org/10.1364/ol.23.001331

Rabani E, Reichman DR, Geissler PL, Brus LE (2003) Drying-mediated self-assembly of nanoparticles. Nature 426(6964):271–274. https://doi.org/10.1038/nature02087

Rai MK, Deshmukh SD, Ingle AP, Gade AK (2012) Silver nanoparticles: the powerful nanoweapon against multidrug-resistant bacteria. J Appl Microbiol 112(5):841–852. https://doi.org/10.1111/j.1365-2672.2012.05253.x

Ramanathan R, O'Mullane AP, Parikh RY, Smooker PM, Bhargava SK, Bansal V (2011) Bacterial kinetics-controlled shape-directed biosynthesis of silver nanoplates using morganella psychrotolerans. Langmuir 27(2):714–719. https://doi.org/10.1021/la1036162

Ramanathan R, Field MR, O'Mullane AP, Smooker PM, Bhargava SK, Bansal V (2013) Aqueous phase synthesis of copper nanoparticles: a link between heavy metal resistance and nanoparticle synthesis ability in bacterial systems. Nanoscale 5(6):2300–2306. https://doi.org/10.1039/c2nr32887a

Rashid MH, Mandal TK (2007) Synthesis and catalytic application of nanostructured silver dendrites. J Phys Chem C 111(45):16750–16760. https://doi.org/10.1021/jp074963x

Ray D, Aswal VK, Kohlbrecher J (2015) Micelle-induced depletion interaction and resultant structure in charged colloidal nanoparticle system. J Appl Phys 117(16):164310. https://doi.org/10.1063/1.4919359

Riveros G, Green S, Cortes A, Gómez H, Marotti RE, Dalchiele EA (2006) Silver nanowire arrays electrochemically grown into nanoporous anodic alumina templates. Nanotechnology 17(2):561–570. https://doi.org/10.1088/0957-4484/17/2/037

Rodnikova MN (2007) A new approach to the mechanism of solvophobic interactions. J Mol Liq 136(3):211–213. https://doi.org/10.1016/j.molliq.2007.08.003

Rodrigues TS, da Silva AGM, Camargo PHC (2019) Nanocatalysis by noble metal nanoparticles: controlled synthesis for the optimization and understanding of activities. J Mater Chem A 7(11):5857–5874. https://doi.org/10.1039/c9ta00074g

Rogers JA, Bao Z, Baldwin K, Dodabalapur A, Crone B, Raju VR, Kuck V, Katz H, Amundson K, Ewing J, Drzaic P (2001) Paper-like electronic displays: Large-area rubber-stamped plastic sheets of electronics and microencapsulated electrophoretic inks. Proc Natl Acad Sci 98(9):4835–4840. https://doi.org/10.1073/pnas.091588098

Rossouw D, Couillard M, Vickery J, Kumacheva E, Botton GA (2011) Multipolar plasmonic resonances in silver nanowire antennas imaged with a subnanometer electron probe. Nano Lett 11(4):1499–1504. https://doi.org/10.1021/nl200634w

Rujitanaroj P-O, Pimpha N, Supaphol P (2008) Wound-dressing materials with antibacterial activity from electrospun gelatin fiber mats containing silver nanoparticles. Polymer 49(21):4723–4732. https://doi.org/10.1016/j.polymer.2008.08.021

Ryan KM, Mastroianni A, Stancil KA, Liu H, Alivisatos AP (2006) Electric-field-assisted assembly of perpendicularly oriented nanorod superlattices. Nano Lett 6(7):1479–1482. https://doi.org/10.1021/nl060866o

Sambhy V, MacBride MM, Peterson BR, Sen A (2006) Silver bromide nanoparticle/polymer composites: dual action tunable antimicrobial materials. J Am Chem Soc 128(30):9798–9808. https://doi.org/10.1021/ja061442z

Sánchez-Iglesias A, Grzelczak M, Altantzis T, Goris B, Pérez-Juste J, Bals S, Van Tendeloo G, Donaldson SH, Chmelka BF, Israelachvili JN, Liz-Marzán LM (2012) Hydrophobic interactions modulate self-assembly of nanoparticles. ACS Nano 6(12):11059–11065. https://doi.org/10.1021/nn3047605

Sastry M, Rao M, Ganesh KN (2002) Electrostatic assembly of nanoparticles and biomacromolecules. Acc Chem Res 35(10):847–855. https://doi.org/10.1021/ar010094x

Sau TK, Rogach AL (2010) Nonspherical noble metal nanoparticles: colloid-chemical synthesis and morphology control. Adv Mater 22(16):1781–1804. https://doi.org/10.1002/adma.200901271

Schneider S, Halbig P, Grau H, Nickel U (1994) Reproducible preparation of silver sols with uniform particle size for application in surface-enhanced raman spectroscopy. Photochem Photobiol 60(6):605–610. https://doi.org/10.1111/j.1751-1097.1994.tb05156.x

Schörner C, Adhikari S, Lippitz M (2019) A single-crystalline silver plasmonic circuit for visible quantum emitters. Nano Lett 19(5):3238–3243. https://doi.org/10.1021/acs.nanolett.9b00773

Schumm B, Wisser FM, Mondin G, Hippauf F, Fritsch J, Grothe J, Kaskel S (2013) Semi-transparent silver electrodes for flexible electronic devices prepared by nanoimprint lithography. J Mater Chem C 1(4):638–645. https://doi.org/10.1039/c2tc00247g

Selvakannan PR, Swami A, Srisathiyanarayanan D, Shirude PS, Pasricha R, Mandale AB, Sastry M (2004) Synthesis of aqueous au core–ag shell nanoparticles using tyrosine as a ph-dependent reducing agent and assembling phase-transferred silver nanoparticles at the air–water interface. Langmuir 20(18):7825–7836. https://doi.org/10.1021/la049258j

Shankar SS, Ahmad A, Sastry M (2003) Geranium leaf assisted biosynthesis of silver nanoparticles. Biotechnol Prog 19(6):1627–1631. https://doi.org/10.1021/bp034070w

Shankar SS, Rai A, Ahmad A, Sastry M (2004) Rapid synthesis of Au, Ag, and bimetallic Au core–Ag shell nanoparticles using Neem (Azadirachta indica) leaf broth. J Colloid Interface Sci 275(2):496–502. https://doi.org/10.1016/j.jcis.2004.03.003

Shen W, Zhang X, Huang Q, Xu Q, Song W (2014) Preparation of solid silver nanoparticles for inkjet printed flexible electronics with high conductivity. Nanoscale 6(3):1622–1628. https://doi.org/10.1039/c3nr05479a

Shi Z, Lian Y, Zhou X, Gu Z, Zhang Y, Iijima S, Zhou L, Yue KT, Zhang S (1999) Mass-production of single-wall carbon nanotubes by arc discharge method 11. This work was supported by the National Natural Science Foundation of China, No. 29671030. Carbon 37(9):1449–1453. https://doi.org/10.1016/S0008-6223(99)00007-X

Shirtcliffe N, Nickel U, Schneider S (1999) Reproducible preparation of silver sols with small particle size using borohydride reduction: for use as nuclei for preparation of larger particles. J Colloid Interf Sci 211(1):122–129. https://doi.org/10.1006/jcis.1998.5980

Si G, Zhao Y, Lv J, Lu M, Wang F, Liu H, Xiang N, Huang TJ, Danner AJ, Teng J, Liu YJ (2013) Reflective plasmonic color filters based on lithographically patterned silver nanorod arrays. Nanoscale 5(14):6243–6248. https://doi.org/10.1039/c3nr01419c

Sondi I, Salopek-Sondi B (2004) Silver nanoparticles as antimicrobial agent: a case study on *E. coli* as a model for Gram-negative bacteria. J Colloid Interf Sci 275(1):177–182. https://doi.org/10.1016/j.jcis.2004.02.012

Sondi I, Goia DV, Matijević E (2003) Preparation of highly concentrated stable dispersions of uniform silver nanoparticles. J Colloid Interface Sci 260(1):75–81. https://doi.org/10.1016/S0021-9797(02)00205-9

Stankus DP, Lohse SE, Hutchison JE, Nason JA (2011) Interactions between natural organic matter and gold nanoparticles stabilized with different organic capping agents. Environ Sci Technol 45(8):3238–3244. https://doi.org/10.1021/es102603p

Stenhouse J (1843) CXXIII. On some of the substances which reduce oxide of silver and precipitate it on glass in the form of a metallic mirror. Mem Proc Chem Soc 2:242–244. https://doi.org/10.1039/mp8430200242

Subrahmanyam KS, Panchakarla LS, Govindaraj A, Rao CNR (2009) Simple method of preparing graphene flakes by an arc-discharge method. J Phys Chem C 113(11):4257–4259. https://doi.org/10.1021/jp900791y

Sumesh E, Bootharaju MS, Anshup Pradeep T (2011) A practical silver nanoparticle-based adsorbent for the removal of Hg2 + from water. J Hazardous Mater 189(1):450–457. https://doi.org/10.1016/j.jhazmat.2011.02.061

Sun Y, Xia Y (2003) Gold and silver nanoparticles: a class of chromophores with colors tunable in the range from 400 to 750 nm. Analyst 128(6):686–691. https://doi.org/10.1039/b212437h

Sun Y, Yin Y, Mayers BT, Herricks T, Xia Y (2002) Uniform silver nanowires synthesis by reducing $AgNO_3$ with ethylene glycol in the presence of seeds and Poly(Vinyl Pyrrolidone). Chem Mater 14(11):4736–4745. https://doi.org/10.1021/cm020587b

Sun Y, Mayers B, Herricks T, Xia Y (2003) Polyol synthesis of uniform silver nanowires: a plausible growth mechanism and the supporting evidence. Nano Lett 3(7):955–960. https://doi.org/10.1021/nl034312m

Swami A, Selvakannan PR, Pasricha R, Sastry M (2004) One-step synthesis of ordered two-dimensional assemblies of silver nanoparticles by the spontaneous reduction of silver ions by pentadecylphenol langmuir monolayers. J Phys Chem B 108(50):19269–19275. https://doi.org/10.1021/jp0465581

Tabatabaei M, Sangar A, Kazemi-Zanjani N, Torchio P, Merlen A, Lagugné-Labarthet F (2013) Optical properties of silver and gold tetrahedral nanopyramid arrays prepared by nanosphere lithography. J Phys Chem C 117(28):14778–14786. https://doi.org/10.1021/jp405125c

Taft AE, Ludlum SD (1930) A method for staining unfixed brain tissue with silver. Proc Soc Exp Biol Med 27(6):582–585. https://doi.org/10.3181/00379727-27-4866

Taleb A, Petit C, Pileni MP (1997) Synthesis of highly monodisperse silver nanoparticles from AOT reverse micelles: a way to 2D and 3D self-organization. Chem Mater 9(4):950–959. https://doi.org/10.1021/cm960513y

Tanford C (1978) The hydrophobic effect and the organization of living matter. Science 200(4345):1012–1018. https://doi.org/10.1126/science.653353

Tao F, Nguyen L, Zhang S (2014) Chapter 1 Introduction: Synthesis and catalysis on metal nanoparticles. In: Metal nanoparticles for catalysis: advances and applications. The Royal Society of Chemistry, pp 1–5

Thakkar KN, Mhatre SS, Parikh RY (2010) Biological synthesis of metallic nanoparticles. Nanomed Nanotechnol Biol Med 6(2):257–262. https://doi.org/10.1016/j.nano.2009.07.002

Thanh NTK, Maclean N, Mahiddine S (2014) Mechanisms of nucleation and growth of nanoparticles in solution. Chem Rev 114(15):7610–7630. https://doi.org/10.1021/cr400544s

Thomas KG, Barazzouk S, Ipe BI, Joseph STS, Kamat PV (2004) Uniaxial plasmon coupling through longitudinal self-assembly of gold nanorods. J Phys Chem B 108(35):13066–13068. https://doi.org/10.1021/jp049167v

Tien D-C, Tseng K-H, Liao C-Y, Huang J-C, Tsung T-T (2008) Discovery of ionic silver in silver nanoparticle suspension fabricated by arc discharge method. J Alloy Compd 463(1):408–411. https://doi.org/10.1016/j.jallcom.2007.09.048

Tilaki RM, Irajizad A, Mahdavi SM (2006) Stability, size and optical properties of silver nanoparticles prepared by laser ablation in different carrier media. Appl Phys A 84(1):215–219. https://doi.org/10.1007/s00339-006-3604-2

Tsuji T, Iryo K, Watanabe N, Tsuji M (2002) Preparation of silver nanoparticles by laser ablation in solution: influence of laser wavelength on particle size. Appl Surf Sci 202(1):80–85. https://doi.org/10.1016/S0169-4332(02)00936-4

Tsuji T, Thang DH, Okazaki Y, Nakanishi M, Tsuboi Y, Tsuji M (2008) Preparation of silver nanoparticles by laser ablation in polyvinylpyrrolidone solutions. Appl Surf Sci 254(16):5224–5230. https://doi.org/10.1016/j.apsusc.2008.02.048

Verwey EJW (1947) Theory of the Stability of Lyophobic Colloids. J Phys Colloid Chem 51(3):631–636. https://doi.org/10.1021/j150453a001

Wang H-T, Nafday OA, Haaheim JR, Tevaarwerk E, Amro NA, Sanedrin RG, Chang C-Y, Ren F, Pearton SJ (2008) Toward conductive traces: Dip Pen Nanolithography® of silver nanoparticle-based inks. Appl Phys Lett 93(14):143105. https://doi.org/10.1063/1.2995859

Wang H, Chen L, Shen X, Zhu L, He J, Chen H (2012) Unconventional chain-growth mode in the assembly of colloidal gold nanoparticles. Angew Chem Int Ed 51(32):8021–8025. https://doi.org/10.1002/anie.201203088

Watt F, Bettiol AA, Kan JAV, Teo EJ, Breese MBH (2005) Ion beam lithography and nanofabrication: a review. Int J Nanosci 04(03):269–286. https://doi.org/10.1142/s0219581x05003139

Whitesides GM, Grzybowski B (2002) Self-assembly at all scales. Science 295(5564):2418–2421. https://doi.org/10.1126/science.1070821

Wiley B, Herricks T, Sun Y, Xia Y (2004) Polyol synthesis of silver nanoparticles: use of chloride and oxygen to promote the formation of single-crystal, truncated cubes and tetrahedrons. Nano Lett 4(9):1733–1739. https://doi.org/10.1021/nl048912c

Wiley BJ, Xiong Y, Li Z-Y, Yin Y, Xia Y (2006) Right bipyramids of silver: a new shape derived from single twinned seeds. Nano Lett 6(4):765–768. https://doi.org/10.1021/nl060069q

Wiley B, Sun Y, Xia Y (2007) Synthesis of silver nanostructures with controlled shapes and properties. Acc Chem Res 40(10):1067–1076. https://doi.org/10.1021/ar7000974

Wisser FM, Schumm B, Mondin G, Grothe J, Kaskel S (2015) Precursor strategies for metallic nano- and micropatterns using soft lithography. J Mater Chem C 3(12):2717–2731. https://doi.org/10.1039/c4tc02418d

Wood MA (2007) Colloidal lithography and current fabrication techniques producing in-plane nanotopography for biological applications. J R Soc Interface 4(12):1–17. https://doi.org/10.1098/rsif.2006.0149

Wu B, Kumar A (2007) Extreme ultraviolet lithography: a review. J Vac Sci Technol B: Microelectr Nanometer Struct Proces Measurement Phenom 25(6):1743–1761. https://doi.org/10.1116/1.2794048

Wüthrich R, Abou Ziki JD (2015) Chapter 2—Historical overview of electrochemical discharges. In: Wüthrich R, Abou Ziki JD (eds) Micromachining using electrochemical discharge phenomenon, 2nd edn. William Andrew Publishing, Boston, pp 13–33

Xia Y, Whitesides GM (1998) Soft lithography. Angew Chem Int Ed 37(5):550–575. https://doi.org/10.1002/(sici)1521-3773(19980316)37:5%3c550:aid-anie550%3e3.0.co;2-g

Xia Y, Xiong Y, Lim B, Skrabalak SE (2009) Shape-controlled synthesis of metal nanocrystals: simple chemistry meets complex physics? Angew Chem Int Ed 48(1):60–103. https://doi.org/10.1002/anie.200802248

Xia Y, Xia X, Peng H-C (2015) Shape-controlled synthesis of colloidal metal nanocrystals: thermodynamic versus kinetic products. J Am Chem Soc 137(25):7947–7966. https://doi.org/10.1021/jacs.5b04641

Xia Y, Gilroy KD, Peng H-C, Xia X (2017) Seed-mediated growth of colloidal metal nanocrystals. Angew Chem Int Ed 56(1):60–95. https://doi.org/10.1002/anie.201604731

Xin Z, Liu Y, Li X, Liu S, Fang Y, Deng Y, Bao C, Li L (2017) Conductive grid patterns prepared by microcontact printing silver nanoparticles ink. Mater Res Expr 4(1):015021. https://doi.org/10.1088/2053-1591/aa5713

Xu R, Wang D, Zhang J, Li Y (2006) Shape-dependent catalytic activity of silver nanoparticles for the oxidation of styrene. Chem Asian J 1(6):888–893. https://doi.org/10.1002/asia.200600260

Yan X-M, Ni J, Robbins M, Park HJ, Zhao W, White JM (2002) Silver nanoparticles synthesized by vapor deposition onto an ice matrix. J Nanopart Res 4(6):525–533. https://doi.org/10.1023/a:1022884127552

Yan B, Thubagere A, Premasiri WR, Ziegler LD, Dal Negro L, Reinhard BM (2009) Engineered SERS substrates with multiscale signal enhancement: nanoparticle cluster arrays. ACS Nano 3(5):1190–1202. https://doi.org/10.1021/nn800836f

Yang C, Gu H, Lin W, Yuen MM, Wong CP, Xiong M, Gao B (2011) Silver nanowires: from scalable synthesis to recyclable foldable electronics. Adv Mater 23(27):3052–3056. https://doi.org/10.1002/adma.201100530

Yi P, Zhang C, Peng L, Lai X (2017) Flexible silver-mesh electrodes with moth-eye nanostructures for transmittance enhancement by double-sided roll-to-roll nanoimprint lithography. RSC Adv 7(77):48835–48840. https://doi.org/10.1039/c7ra09149d

Young KL, Jones MR, Zhang J, Macfarlane RJ, Esquivel-Sirvent R, Nap RJ, Wu J, Schatz GC, Lee B, Mirkin CA (2012) Assembly of reconfigurable one-dimensional colloidal superlattices due to a synergy of fundamental nanoscale forces. Proc Natl Acad Sci 109(7):2240–2245. https://doi.org/10.1073/pnas.1119301109

Young KL, Personick ML, Engel M, Damasceno PF, Barnaby SN, Bleher R, Li T, Glotzer SC, Lee B, Mirkin CA (2013) A directional entropic force approach to assemble anisotropic nanoparticles into superlattices. Angew Chem Int Ed 52(52):13980–13984. https://doi.org/10.1002/anie.201306009

Zeng J, Zheng Y, Rycenga M, Tao J, Li Z-Y, Zhang Q, Zhu Y, Xia Y (2010) Controlling the shapes of silver nanocrystals with different capping agents. J Am Chem Soc 132(25):8552–8553. https://doi.org/10.1021/ja103655f

Zhang R, Engholm M (2018) Recent progress on the fabrication and properties of silver nanowire-based transparent electrodes. Nanomaterials 8(8):628

Zhang J, Li S, Wu J, Schatz GC, Mirkin CA (2009) Plasmon-mediated synthesis of silver triangular bipyramids. Angew Chem Int Ed 48(42):7787–7791. https://doi.org/10.1002/anie.200903380

Zhang Q, Li W, Moran C, Zeng J, Chen J, Wen L-P, Xia Y (2010) Seed-mediated synthesis of Ag nanocubes with controllable edge lengths in the range of 30 − 200 nm and comparison of their optical properties. J Am Chem Soc 132(32):11372–11378. https://doi.org/10.1021/ja104931h

Zhang L, Niu W, Xu G (2012) Synthesis and applications of noble metal nanocrystals with high-energy facets. Nano Today 7(6):586–605. https://doi.org/10.1016/j.nantod.2012.10.005

Zhou Y, Liu HJ, Yu SH, Chen ZY, Zhu YR, Jiang WQ (1999a) Preparation of nanocrystalline silver by the method of liquid-solid arc discharge combined with hydrothermal treatment. Mater Res Bull 34(10):1683–1688. https://doi.org/10.1016/S0025-5408(99)00169-5

Zhou Y, Yu SH, Cui XP, Wang CY, Chen ZY (1999b) Formation of silver nanowires by a novel solid–liquid phase arc discharge method. Chem Mater 11(3):545–546. https://doi.org/10.1021/cm981122h

Zhou H, Zheng L, Jia H (2014) Facile control of the self-assembly of gold nanoparticles by changing the capping agent structures. Colloids Surf, A 450:9–14. https://doi.org/10.1016/j.colsurfa.2014.03.013

Zhou S, Li J, Gilroy KD, Tao J, Zhu C, Yang X, Sun X, Xia Y (2016) Facile synthesis of silver nanocubes with sharp corners and edges in an aqueous solution. ACS Nano 10(11):9861–9870. https://doi.org/10.1021/acsnano.6b05776

Zhu Q-L, Xu Q (2016) Immobilization of ultrafine metal nanoparticles to high-surface-area materials and their catalytic applications. Chem 1(2):220–245. https://doi.org/10.1016/j.chempr.2016.07.005

Zong R-L, Zhou J, Li Q, Du B, Li B, Fu M, Qi X-W, Li L-T, Buddhudu S (2004) Synthesis and optical properties of silver nanowire arrays embedded in anodic alumina membrane. J Phys Chem B 108(43):16713–16716. https://doi.org/10.1021/jp0474172

Zou S, Schatz GC (2006) Metal nanoparticle array waveguides: proposed structures for subwavelength devices. Phys Rev B 74(12):125111. https://doi.org/10.1103/physrevb.74.125111

Textile Fabric Processing and Their Sustainable Effluent Treatment Using Enzymes—Insights and Challenges

Debasree Kundu, M. S. Thakur, and Sanjukta Patra

Abstract Textile manufacturing and distribution sector is a promising segment in India. The functional and aesthetic value of textile is improved through de-sizing, scouring, bleaching, dying, printing, and finishing stages. During this process of textile manufacturing, both water consumption and waste generation are based on the operations. One of the key environmental issues that impede growth of this textile industry is its wastewater. At this juncture, enzyme technology exhibits enormous promise in the textile fabric processing and sustainable effluent treatment. It is a green alternative and minimizes various chemically induced hazardous effluents generated during the textile fabric processing. A wide variety of enzymes have been used in different steps of textile fabric processing which play major role in improvement of textile quality, texture, strength, market values, and treatment of produced effluents, for instance, amylase in de-sizing, pectinases in bio-scouring, cellulase in stone-washing and bio-polishing, catalase in bleaching and removal of excess hydrogen peroxide during dyeing, laccase and oxidase in lignin degradation and textile effluent treatment, etc. However, most of the enzymes cannot withstand harsh textile fabric processing. Enzyme immobilization has become the most reliable method to tailor the enzyme properties such as enhancing stability, improving catalytic activity, specificity, reusability, storage stability, etc., to make the enzyme withstand harsh industrial processes. Overall, this chapter gives an insight into the effectiveness of different enzymes in the functional finishing of textiles and its effluent treatment, nano-immobilization of the enzymes for improving robustness of these biocatalysts and addresses the current challenges in enzymatic textile fabric processing and treatment of textile effluents.

Keywords Textile processing · Textile effluent treatment · Green alternative · Enzymes · Nano-immobilization

---

D. Kundu · S. Patra (✉)
Department of Biosciences and Bioengineering, Indian Institute of Technology Guwahati, Guwahati, Assam 781039, India
e-mail: sanjukta@iitg.ac.in

M. S. Thakur
Central Food Technological Research Institute, Mysuru, Karnataka 570020, India

A. Tripathi and J. S. Melo (eds.), *Immobilization Strategies*,
Gels Horizons: From Science to Smart Materials,
https://doi.org/10.1007/978-981-15-7998-1_19

1 Introduction

The textile industry sector is one of the pillars of the national economy and has become the second largest industry after agriculture in Indian context (Raichurkar and Ramachandran 2015). It accounts for 11% of overall exports; contributes 63% to the world textile, apparel/garment market; and ranks second in textile manufacturing worldwide. Presently, the Indian textile market size has touched around US$ 108 billion and is likely to increase to US$ 141 billion by 2021 (Raichurkar and Ramachandran 2015). Based on the manufacturing/processing operations, the textile sector is further categorized into dry and wet fabric industry. While dry fabric industries discharge solid wastes, liquid wastes are emanated from wet fabric industries. The present chapter will focus on the wet fabric category.

Conventional textile manufacturing process operations undergo numerous chemical and non-chemical treatments process as products are treated with various agents from sizing to finishing processes. From pre-treatment of textile to its finishing, the chemicals can be categorized as per the application into (i) pre-treatment chemicals; (ii) textile dyeing chemicals; (iii) printing chemicals; (iv) finishing chemicals; and (v) antistatic agents.

The textile industry processes consume huge quantity of chemicals and water that are discharged into the environment as effluents. During the course of the movement of the effluents, the diverse chemicals ranging from acids, alkalis, dyes, oils, inorganics, organics, and polymers percolate into soil thereby contaminating soil, underground water, and agriculture which is detrimental to the environment (Banat et al. 1996). Thus, enzymes as green alternative for varied textile manufacturing and processing purpose are gaining acceptance due to their potential to substitute toxic materials as reflected in environmental footprint analysis (Yachmenev et al. 2002). Enzymes having required parameter can be employed in many textile mill operation steps. Engineered enzymes and extremozymes can be better alternatives as they can resist the extremes of temperature, pH, denaturing agents applied during pre- and post-treatment and also find its application in the effluent treatment (Tzanov et al. 2003a, b).

Nonetheless, there are limitations in the applicability of enzymes such as (i) lack of resistance to process environment; (ii) less thermodynamic stability (and thus inhibition) at high concentrations of reaction materials; (iii) lack of operational stability; (iv) difficulty in recovery; and (v) reusability of the enzyme that restricts its widespread industrial use (Guisan 2006; Hernandez and Fernandez-Lafuente 2011). To circumvent these limitations, enzymes are immobilized through various matrices/carriers that significantly improve robustness, specificity/activity, storage stability, reusability, and continuous production favorable for industrial applications. Enhanced activity of immobilized enzymes results from the multipoint attachment and microenvironment of the immobilization carrier (Cao 2005). Of various approaches and matrices used for immobilization, nanoscale carriers with large surface areas provide a flexible technology that offer loading of more enzyme and decrease internal as well as external mass transfer resistance for substrates.

Moreover, nanoscale material possesses unique properties including confinement properties, solution behavior, and interfacial effects that can profoundly increase the performances of enzymes (Misson et al. 2014; Khan et al. 2019).

The present chapter offers an inclusive overview on diverse wet processing approaches in textile production, focuses on enzyme performance as green biocatalysts to replace the chemical hazards, and highlights the use of nano-carriers as effective immobilization process for textile processing and its holistic effluent treatment. The challenges and suggestions for future outlook are also delineated. Overall, this chapter converges upon the effectiveness of immobilized enzymes as emerging tools for textile fabric processing and their sustainable effluent treatment.

2 Textile Manufacturing Process and the Role of Enzymes

Enzymes are touted as an integral component of the textile manufacturing and processing and play a key role as an alternative route. The use of enzymes in the textile processes and products is aiming swift momentum worldwide as a result of reduced processing duration, less energy input, cost competitiveness, non-toxicity, and 'green' features with escalating requirements for diminishing pollution in textile industrial method (Li et al. 2012; Choi et al. 2015). Among the different classes, enzymes employed in fabric operations are mainly categorized into the class of (i) hydrolase (amylase, cellulase, cutinase, protease, pectinase, lipase) required for de-sizing, scouring, finishing, denim finishing, anti-felting of wool, etc.; and (ii) oxidoreductase (laccase, peroxidase, ligninase, catalase) finding its application in bleaching, dye decolorization, wool finishing, and others (Mojsov 2011; Chen et al. 2013). Role of different enzymes in textile processes and products are detailed in the proceeding sections and reflected in Table 1.

Table 1 Application of enzymes in textile processing steps

Textile processing step	Type of fabric	Key enzymes used
De-sizing	Cotton, cotton/polyester blend, nylon/cotton blend, silk	Amylases, lipases
Scouring	Cotton, cotton/polyester blend, wool, nylon/cotton blend, silk	Pectinases, cellulases, proteases, lipases
Bleaching	Cotton, cotton/polyester blend, wool, nylon/cotton blend, silk	Laccases, glucose oxidases, Arylesterases
Mercerization	Cotton, cotton/polyester blend, wool, nylon/cotton blend, silk	Cellulases, pectinases, lipases, proteases
Dyeing and printing	Cotton, cotton/polyester blend, wool, nylon/cotton blend, silk	Proteases, amylases, lipases, diasterases
Bio-polishing/Finishing and Bio-washing	Cotton, cotton/polyester blend, wool finishing, nylon/cotton blend, silk, polyester	Cellulases, proteases, lipases

2.1 De-sizing

In textile industry, the de-sizing process is one in which the 'size' is removed from fabric prior to completing the pre-treatment process. Yarns contain starch as a removable protective layer along with additives. Through de-sizing, this surface coating is removed so that dyes and chemicals can penetrate in succeeding steps of wet processing. In the conventional process, mild concentrations of inorganic acids or, oxidizing agents are used for de-sizing purpose. The classical enzymatic de-sizing process helps in degrading starch contents on cotton fabrics with advantages of (i) increase in yarn strength; (ii) assistance in the interweaving process; and (iii) protection of yarn against abrasion. Enzyme-based de-sizing is considered as an excellent, quick, and efficient starch removal process that works without damaging the fabric (Horvathova et al. 2001).

Amylase finds application in the de-sizing process and is classified based upon the source of origin. The α-amylases are known to randomly cleave 1,4-glucosidic linkage. The β-amylases follow a systematic hydrolytic process liberating maltose. α-Amylases are more suitable for the process than β-amylases as the later are heat sensitive (Hanson and Gilbert 1974). Increase in de-sizing temperature helps in enhanced starch removal in less time. For instance, the thermophile *Bacillus licheniformis* α-amylase functions efficiently at high temperature to remove starch with improved absorbance (Saravanan et al. 2011). Further, incorporation of salts (enzyme stabilizer) or ultrasonic exposure increases the enzyme effectiveness (Chand et al. 2012; Wang et al. 2012; Hao et al. 2013). To accomplish efficient de-sizing at high temperature, research has been accorded towards how mesophilic α-amylase is affected by treatment with additives of various kinds (Fu et al. 2015). Madhu and Chakraborty have reported the application of amylases, its textile de-sizing, and bio-catalytic performance (Madhu and Chakraborty 2017). The application of other enzymes in association with amylases removes the natural impurities from cellulosic materials, thereby improving the process. To cite, integration of lipase facilitates removal of the hydrophobic part of size that are difficult to remove in scouring. The entire enzyme-based process is thus, a benign substitute to the chemical assisted textile fabric processing and the environment.

2.2 Scouring

The presence of non-cellulosic impurities like pectin, dirt, oils, waxes, gums proteins, natural colorants, etc., in the cellulosic fibers result in poor wet-ability and filthy appearance that limits the dyeing and finishing properties of the fabric. Scouring is thus done to eliminate these natural and additive impurities having hydrophobic moieties. Alkaline scouring using sodium hydroxide is the conventional process to remove such contaminations. It requires excessive alkali usage that needs repeated rinses that in turn affects the fabric carbohydrates (Buschle-Diller et al. 1998). Thus, bio-scouring is gaining wide acceptance that uses a vast spectrum of enzymes and their cocktail like pectinases, cellulases, proteases, and lipases (Li and Hardin 1997). Bio-scouring is environment-friendly, energy-saving substitute where the natural soft properties of the cotton fibers are preserved.

Pectin present in cotton functions as a binding molecule where it bridges cellulosic and non-cellulosic parts. Thus, removal of pectin helps in eliminating the non-cellulosic contaminants. Pectins, rich in galacturonic acid, are hydrolyzed by pectinases into simpler molecules. Depending upon the mode of action, they are categorized into polygalacturonases, pectinesterases and pectin lyases (Presa and Tavcer 2007). Pectinases are broadly active at pH 5–9 and has optimum activity at temperature between 40–60 °C which necessitates a high temperature rinsing after bio-scouring (Choe et al. 2004). Addition of surfactants along-with mechanical agitation and ultrasound reduces the concentration of required pectinases and the process time. Solbak et al. (2005) found a novel pectinase through gene mutagenesis, that even in low concentration, functioned well at high temperature and alkaline pH (Solbak et al. 2005). Despite lots of research been undertaken on application of pectinases alone as well as with cocktail of different enzymes for bio-souring, there is lack of commercialization on an industrial scale. More research efforts in this direction is awaited.

2.3 Bleaching

After scouring, the next process in textile industry is bleaching that improves the wet-based finishing. Bleaching destroys the natural pigments present in the fibers that are responsible for the grayness of the fabric. Conventionally, whitening is achieved with oxidizing agents containing chlorine. However, since last few decades, hydrogen peroxide is utilized like a highly dosed bleaching molecule that leads to reduce the polymerization degree resulting in deterioration of quality of fibers and incurs huge water loss for removing the bleaching agent from textiles (Basto et al. 2007). The exploitability of enzyme-based bleaching systems thus stands advantageous over the conventional process as it (i) limits the water consumption; (ii) reduces the usage of chemicals; (iii) creates mild reaction environment; and (iv) offers reusability of de-sizing waste baths (Chatha et al. 2017). Enzymatic bleaching is practiced using laccase/mediator systems, glucose oxidases and arylesterase.

The use of laccase (benzenediol: oxygen oxidoreductases, EC 1.10. 3.2) in textile industry is growing very rapidly replacing the traditional textile functioning where chemical, energy and water is highly consumed. Laccases, for example, bleach fabrics and acts as surface modifier as well as to decolorize textile effluents (Madhu and Chakraborty 2017). Laccases are non-specific multi-copper-containing oxidoreductase enzymes. They oxidize phenols and aromatic amines and catalyze electron reduction of molecular oxygen to water. Initially for the first time, Tzanov et al. (2003a, b) reported use of low concentration laccases to enhance the bleaching effect on cotton fabrics in reduced time span that enables the suitability of this process for continuous operations (Tzanov et al. 2003a, b). Further, it was found that a combined ultrasound-laccase treatment with a treatment of low-intensity ultrasound energy (7 W) facilitated more bleaching efficacy of laccase on textile materials (Basto et al. 2007). Consequently, the application of laccase is an eco-friendly alternative in textile processing and is gaining acceptance.

Glucose oxidase (β-D-glucose: oxygen 1-oxidoreductase, EC 1.1.3.4) has also been considered as a plausible method that produces hydrogen peroxide (H_2O_2) for bio-bleaching. It is a flavoenzyme whose biological function is to catalyze the oxidation of glucose to glucolactone. This glucolactone then spontaneously yields gluconic acid and H_2O_2 as products. The generated hydrogen peroxide acts as bleaching agent for disintegration of any coloring composites in cotton and other cellulosic fabrics. The gluconic acid produced acts as metal chelator, thus, eliminating the addition of any stabilizers (Tzanov et al. 2002). It was found that with increased supply of aeration than mechanical agitation, production of hydrogen peroxide increased considerably (Tzanov et al. 2001). However, during enzymatic bleaching with glucose oxidase, discoloration of cotton fabrics was observed due to the presence of residual glucose post reaction (Shin et al. 2004). This problem of discoloration can be eliminated through the use of high concentration of glucose oxidase for long time of incubation (Saravanan et al. 2010). This will also help in re-using the de-sizing waste baths as source of glucose resulting in decrease in water consumption and wastewater pollution. Glucose oxidase generates peroxides at slightly acidic to neutral conditions under low temperatures while the bleaching effect of H_2O_2 requires a high temperature of 80–90 °C and alkaline pH (Tzanov et al. 2001; Anis et al. 2009; Ramadan 2009; Farooq et al. 2013). In order to accomplish bleaching with enzymatically produced H_2O_2 at low temperature in neutral media, the H_2O_2 is converted into reactive peracids using peracid precursors. The formation of the resultant in situ generated peracids is dependent on concentrations of peroxide, pH, temperature, and activators used in the bleaching bath (Tavčer 2012). Arylesterase (EC 3.1.1.2) based bleaching system is available commercially for bleaching purpose that functions enzymatically with arylesterases and hydrogen peroxide. It is well documented that during the perhydrolysis of propylene glycol diacetate by arylesterases, propylene glycol and peracetic acid are generated as bleaching agents in situ (Auterinen et al. 2012). Further, a combination of laccase and glucose oxidase is found to have more significant effect as on one side it removes lignin, and on the other side, it effectively bleaches linen (Ren and Buschle-Diller 2007).

2.4 *Mercerizing*

Post bleaching, the next step in textile processing is mercerizing. Here, the textile is subjected to caustic solution treatment resulting in swelling of the fibers. This further improves the luster, physical and dyeing properties. Enzymes such as cellulases, pectinase, lipases, and proteases are employed for enzymatic mercerization processes

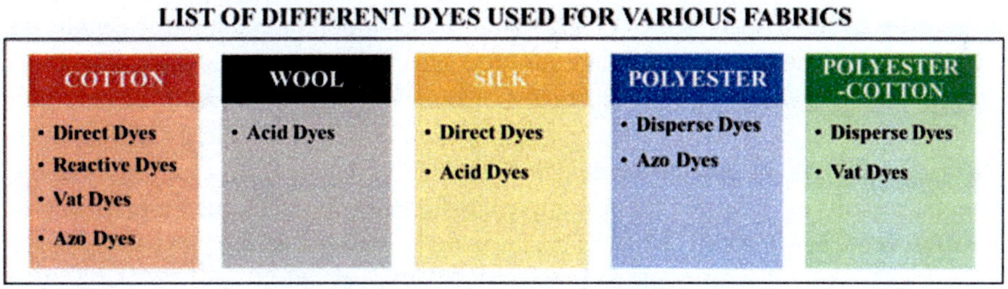

Fig. 1 Different types of dyes used for various kinds of fabrics

(Hartzell and Hsieh 1998; Buchert et al. 2000). Besides, the effect of a mixture of cellulases and hemicellulases for bleaching of wool and cotton had also been investigated (Heine and Höcker 1995).

2.5 *Dyeing and Printing*

Dyeing is a method for coloring of fabric/yarn. Chromophore groups in the dyes like azo (NN), carbonyl (CO), nitro (NO), quinoid groups; and auxochrome groups like amine, carboxyl, sulfonate, and hydroxyl are responsible for the color (Waring and Hallas 2013). Among them, azo and anthraquinone groups are the utmost significant ones. It was found that a combination of enzymes (protease, amylase, diasterase, lipase) with tannic acid enhanced the dye-ability (Vankar et al. 2007). It was observed that natural colorants from *Punica granatum*, *Rheum emodi*, and *Terminalia arjuna* when applied with diasterase, lipase, and protease, respectively, exhibited the finest results (Vankar et al. 2007; Chatha et al. 2017). The different types of dyes for dyeing various kinds of fabrics are depicted in Fig. 1 (Waring and Hallas 2013).

2.6 *Bio-polishing and Bio-washing*

Bio-polishing is a finishing process carried out before, during, or, after dyeing. It enhances fabric quality; reduces the bent towards development of lint ball, and provides a smooth surface with less fuzz (Ibrahim et al. 2011). The enzyme cellulase is exploited for bio-polishing. Cellulase is from the cluster of enzymes which catalyzes the hydrolysis of cellulose by degrading β-(1–4) glycosidic linkages (Madhu and Chakraborty 2017). There are three major classes of cellulolytic systems working in a synergistic mode. They are (i) endoglucanases or, endocellulases randomly cleaving internal bonds at amorphous sites that create new chain ends; (ii) exoglucanases or cellobiohydrolases that cleaves two to four units from the ends of the exposed chains produced by endocellulase, resulting in tetrasaccharides or, disaccharides such as cellobiose; and (iii) β-glucosidases which finally convert these exocellulase product into individual monosaccharides of glucose. The commercially available cellulases for bio-polishing is a combination of endogluconases, exoglucanases and cellobiases (Cavaco-Paulo 1998). It is reported that mechanical action and surfactant addition helps in improved enzymatic hydrolysis of cotton (Traore and Buschle-Diller 1999; Esfandiari et al. 2014). Researchers also initiated the recovery and recycling of the desorbed cellulases post-treatment using ultra-filtration and obtained a recovery of 62% (Azevedo et al. 2002). Bio-treatment using cellulases and proteases improve softness and dye-ability of blended fabrics (Ibrahim et al. 2008).

Currently, cellulase aided bio-washing is an eco-friendly alternative that imparts desired look to denims. Conventionally, denim garments are washed with pumice stones. However, it has various drawbacks like (i) processed garments contain residual pumice; (ii) physical damage to garments; and (iii) destruction of machines (Madhu and Chakraborty 2017). There are different types of cellulases having their own characteristics that are used either singly or, in combination to obtain the desired appearance and quality. The optimum activity of cellulases is obtained in a temperature range of 30–60 °C. Depending on the application, cellulases are categorized into acid (pH 4.5–5.5), neutral (pH 6.6–7) or, alkali (pH 9–10) cellulase. Cellulase attacks primarily on the surface of the cellulose fiber by removing the indigo present in the surface layer of the fiber. However, the re-deposition of released indigo and a reddening of the dyes is the major hurdle. The main reason of this back-staining is the high affinity between indigo and cellulase enzyme as well as the cellulases to cellulose of cotton. Acid cellulases are more attracted toward indigo for which neutral and endoglucanase rich cellulase are the preferred choice (Araujo et al. 2008). Thus, enzyme mediated bio-washing finds its way as a green alternative to the conventional chemical process

2.7 *Application of Other Enzymes in Textile Processing*

2.7.1 Proteases

Proteases are known to hydrolyze peptide bonds of proteins. In textile industry, proteases are used for degumming of silk as well as for carbonization of wool. In silk, proteases are known to remove the gum layer of sericin to provide more softness and luster (Doshi and Shelke 2001). This process is known as degumming. The traditional method of degumming using soap and alkaline treatment is both energy intensive and results in harsh texture. Enzymatic degumming does not attack fibroin and leaves the finished silk smoother. Further, natural wool is known to shrink and is known as felting. Felting of wool results from the scale-like surface structure of the wool. Conventionally, chlorine and polymers are used to control this which in turn increases the pollution of the generated wastewater. Thus, in enzymatic treatment, proteases degrade the cuticle and imparts anti-felting property to the wool. However, it is reported that through this treatment, fiber cuticles are severely damaged resulting in decrease in weight and strength (Vílchez et al. 2010). Transglutaminase is thus used to recover the wool and silk damages. A commercial protease (Esperase) covalently linked to Eudragit S-100 was found to be effective for wool shrink-resist finishing with less fiber damage (Soares et al. 2011).

2.7.2 Other Hydrolases

Cutinases, lipases and esterases finds wide application in textile industry. Cutinase can hydrolyze the non-cellulose component like cuticle layer of cotton that improves the fabric dyeing properties and the moisture regain. Further, addition of nonionic surfactants reduces the enzyme penetration time whereas, salts serves as activators and increases the enzyme activity (Lee and Song 2010). Similarly, lipases were found to be effective in modification of polyesters. Generally, the wettability of hydrophobic fibers is increased through alkaline hydrolysis that results in the decrease of the strength in fiber apart from the environmental load. Lipases are thereby found to be a greener alternative and have been reported to be effective on sulfonated and microdenier polyester fabrics (Doshi and Shelke 2001). Esterases were previously reported to have reduced potential than cutinases or, lipases. However, a new esterase from *Thermobifida halotolerans* was reported to function in a similar effective way like that of cutinases (Ribitsch et al. 2012).

2.7.3 Catalases

Currently, in textile industry, peroxide bleaching is used over conventional chlorine bleaching. Traces of peroxides on fabric lead to uneven dyeing. Thus, after bleaching, the residual H_2O_2 is neutralized by traditional reducing agents at high temperature which is cost intensive. Catalases are used as an attractive alternative that catalyzes the breakdown of H_2O_2 into water and oxygen (Doshi and Shelke 2001). In comparison with the traditional clean-up methods, the enzymatic process results in cleaner wastewater or, reduced water consumption leading to reduction of energy and time.

3 Role of Enzymes in Treatment of Textile Effluent

Globally, the textile effluent discharges about 2,80,000 tons of textile dyes per year, mainly the recalcitrant azo dyes in the environment (Jin et al. 2007). A cursory scrutiny revealed that around 10–15% of the dyes employed in the dyeing are unable to attach with the textile fabrics, thereby, percolating into the biosphere (Asad et al. 2007). Azo dyes are xenobiotics with electron withdrawing moieties that impart electron deficiency in the dye molecules rendering them difficult to degrade (Singh et al. 2014). The release of these colored dyes into ecosystem poses detrimental effect as most of the azo dyes and their degradation derivatives are toxic and mutagenic to living organisms (Weisburger 2002; Xu et al. 2007). The recalcitrant nature of the

Table 2 Indian textile industry standards for aqueous effluent discharge

Parameters	Standard limit
pH	6.9
Temperature	Below 42 °C
BOD	30 ppm
COD	250 ppm
TDS	2000 ppm
Sulfide	2 ppm
Chloride	500 ppm
Calcium	75 ppm
Magnesium	50 ppm

modern synthetic dyes has necessitated the requirement to achieve rigorous environmental regulations and guidelines (Singh et al. 2015). The Indian textile industry standards for aqueous effluent discharge are depicted in Table 2 (Holkar et al. 2016). The standards are established by Central Pollution Control Board (CPCB) which is dependent on the locality and environmental safety necessities (Holkar et al. 2016).

In recent times, due to scarcity of water, the recovery and reuse of wastewater has received considerable attention. Thus, it is imperative to find solutions that will cater to the need of producing reusable water, eradicate toxicity, mineralize aromatic compounds, recover the dyes/salts, and lessen the presence of sludge (Holkar et al. 2016). This warrants the search for cost-effective strategies for the treatment of the textile effluent.

The possible avenues available comprises of physical, chemical, biological, and hybrid methods. Physical process of treatment of the effluent mainly includes filtration, precipitation, flocculation, adsorption, etc., whereas, the chemical processes involves advanced chemical oxidation, chemical reduction, and electrolysis. Biological methods encompass anaerobic, aerobic, and hybrid processes (Georgiou et al. 2003). However, the various physical, chemical, and biological methods have their own merits, demerits, and limitations along-with huge operational cost. In physical process, rather than degradation, the pollutants are transferred from one form to another. The chemical methods generally lead to formation of concentrated sludge or, toxic end products. Biological treatment involving the use of microbial cells is dependent on the availability of nutrients as well as optimum growth conditions (Chhabra et al. 2015). Microbial growth is not supported to a large extent in textile effluent as it is nutrient deficient. Microbial systems undergo aerobic, anaerobic, or sequential aerobic-anaerobic pathways to either attack the azo bonds or, convert it to amines; and in the process, more toxic and mutagenic molecules are produced. At this juncture, the use of enzyme-based technologies has been advocated as an imperative alternative for sustainable treatment of textile effluent.

Predominantly, the dyes are classified into azo, anthraquinone, phthalocyanine, and triphenylmethane. Azo dyes are organic compounds bearing the azo functional group (–N=N–). They represent the most commonly used class of dyes in the textile industry (Mojsov et al. 2016). The dyes are generally degraded to form aromatic amines that are mutagenic and poses toxicity to the ecosystem. Enzymes from white rot fungi *Phanaerochaete* and *Trametes* play important role in dye degradation (Asgher et al. 2008). The main lignin modifying enzymes from the white rot fungus includes lignin peroxidase, manganese peroxidase and laccase that are capable of oxidative free radical cleavage of the azo bond. There are other enzymes like azo reductase, peroxidase, glyoxal oxidase, aryl-alcohol oxidase, oxalate decarboxylase and formate dehydrogenase found in white rot fungus that facilitates dye degradation (Wesenberg et al. 2003). The main reasons behind the enzymatic process being more holistic and viable are (i) biodegradability; (ii) acts at a broad spectrum of substrate concentrations; (iii) active in a broad range of pH, temperature and salinity; and (iv) easy to handle. These properties of the biocatalysts along-with the possibility of designing enzymes with desired characteristics through protein engineering advocate the potential application of this process in the treatment of effluents.

Lignin peroxidase degrades azo dyes by oxidation of the phenolic group to produce a radical at the carbon bearing the azo linkage. The phenolic carbon is then attacked by a water molecule resulting in its break down to produce phenyldiazene which is easily oxidized by a one-electron reaction to generate nitrogen (Mojsov et al. 2016). Lignin peroxidase produced by different white rot fungi have varying molecular mass (37–50 kDa); veratryl alcohol and 2-chloro-1,4-dimethoxybenzene are found to act as good redox mediators for oxidative catalysis of recalcitrants; performs best at pH of 2–5 and temperature of 35–55 °C, respectively (Christian et al. 2005; Asgher et al. 2007).

Manganese peroxidase is a specific enzyme that oxidizes Mn^{2+} to Mn^{3+} ions which are highly reactive. The optimum activities of manganese peroxidase are found at pH and temperature of 4–7 and 40–60 °C, respectively. It is also reported that Mn^{2+} chelator anions improves the enzyme activity and chelated Mn^{3+} acts as a low molecular weight diffusible redox mediator that is highly reactive (Cheng et al. 2007; Asgher et al. 2008).

Laccase enzymes have broad substrate specificity involving the removal of an H^+ atom from hydroxyl and amino groups of the ortho- and para-substituted mono and polyphenolic substrates, and aromatic amines (Mojsov et al. 2016). The optimum pH and temperature depends upon the strain and varies in the range of pH 2–10 and 40–65 °C, respectively (Lu et al. 2005; Ullrich et al. 2005; Murugesan et al. 2006; Quaratino et al. 2007). The activity of white rot fungus laccase increases with the addition of copper, cadmium, nickel, molybdenum, and manganese metal ions and decreases in the presence of silver, mercury, lead, zinc, sodium chloride, and hydrogen peroxide (Asgher et al. 2008).

Azoreductase is a reducing enzyme and through reductive cleavage route, the enzyme degrades the azo dyes to produce colorless amines (Sarkar et al. 2017). It requires low molecular weight reducing equivalent such as FADH or, NADH as the electron donor in the form of a redox reaction. The enzyme is categorized into three types on the basis of coenzyme used; using NADH only, using NADPH only, or using both. Azoreductase can be either cytoplasmic or, membrane-bound. Although cytoplasmic azoreductase is not easily diffusible through a cell membrane and many varieties of dye are not readily degraded, enzymes play important role due to substrate specificity. Azoreductases catalyze the reductive cleavage of azo bonds (–N=N–) to give colorless aromatic amines. The intermediate formed is further degraded by aerobic or, micro-aerophilic process. Under anaerobic condition, redox mediator is used as an electron shuttle by cell membrane-bound azoreductase. This redox mediator-dependent mechanism is however, different for membrane-bound and cytoplasmic azoreductase (Sarkar et al. 2017). The enzyme is oxygen-sensitive and functions best under anaerobic conditions.

The innovative use of various enzymes is expanding rapidly in all areas of textile processing due to their non-toxic and eco-friendly nature. Their versatility and efficiency even in mild reaction conditions gives them the advantage over the conventional physico-chemical treatment methods. List of some microbial enzymes responsible for azo dye degradation are illustrated in Table 3. Nevertheless, the main constraint in the practical applicability of enzymes is that it is not economical. However, as enzymes are the potent catalyst applied in bio-catalysis and are not consumed in reactions, reusability of these enzymes can be the finest cost-effective alternative. Thus, in the present scenario, immobilization of enzymes is a favorable path reflecting an efficient and competent green process.

4 Immobilization of Enzymes: Role of Nano-carriers and Its Advances

The field of enzymatic bio-catalysis alone is exciting and in renewed progress, propelled by both an enhanced inventory of rational enzyme/proton engineering tools and green chemistry momentum (Bommarius and Paye 2013; Liese and Hilterhaus 2013; Sheldon and van Pelt 2013; Ortiz et al. 2019). Enzymes, catalysts with a biological origin, possess a multitude of fascinating properties. So far, they are the proven natural catalysts with high stereo-, chemo-, and regio-selectivity, effective in their functionality under very mild environmental settings (for example, at low pressure and temperature) in aqueous media (Cipolatti et al. 2014). However, even with their biocompatibility and advantageous features, the field application of enzymes is still hampered as the issue of thermodynamic and kinetic stability still remains to be cleared. More often than not, they require to be stabilized to thrive at variety of factors like pH, temperature, salt type and concentration, solvents/co-solvents, as well as reactor conditions such as shear and surface forces. These drawbacks of stability, recovery, and recyclability can generally be surmounted by enzyme immobilization, making them industrially and commercially feasible so as to improve enzyme stability, increase volume specific biocatalyst loading and simplify biocatalyst recycling and downstream processing.

In general, approaches of enzyme immobilization can be divided into three types, binding to a support (carrier)/noncovalent adsorption, entrapment (encapsulation) and cross-linking (covalent bonding), each with their own advantages and demerits.

Table 3 List of some microbial enzymes responsible for azo dye degradation

Enzyme	Producer organism	Type of dye degraded	References
Laccase	*Geobacillus stearothermophilus*	Indigo carmine Congo red Remazole Brilliant Blue R (RBBR), Brilliant Green	Mehta et al. (2016)
Laccase	*Micrococcus luteus*	Azo dyes	Kanagaraj et al. (2015)
Laccase	*Ganoderma* sp.	Indigo Carmine Remazol Brilliant Blue R (RBBR) Bromophenol Blue, Crystal Violet Malachite Green Congo Red Direct Blue 15 Direct Red 23	Teerapatsakul et al. (2017)
Laccase and veratryl alcohol oxidase	*Providencia rettgeri* strain HSL1 and *Pseudomonas* sp. SUK1	C.I. Reactive Blue 172 (RB 172)	Lade et al. (2015)
Laccase, NADH-DCIP reductase, azoreductase	*Aeromonas* sp. DH-6	Methyl Orange (MO)	Du et al. (2015)
Lignin peroxidase, laccase, tyrosinase, azoreductase, and DCIP reductase	*Providencia* sp. SRS82	Acid Black 210 triazodye	Agrawal et al. (2014)
Peroxidase	*Pseudomonas putida* MET94 and *Bacillus subtilis*	Azo dyes Phenolics, Methoxylated aromatics	Santos et al. (2014)
Manganese peroxidase	*Trametes pubescens* strain i8	Remazol Brilliant Violet 5R (RBV5R) Direct Red 5B (DR5B) Methyl Green Acid Blue Indigo Carmine	Rekik et al. (2019)
Azoreductase (membrane-bound)	*Shewanella* sp. Strain IFN4	Reactive Black 5 Acid Red 88 Direct Red 81 Acid Yellow 19, Disperse Orange 3	Imran et al. (2016)

(continued)

Table 3 (continued)

Enzyme	Producer organism	Type of dye degraded	References
Azoreductase	*Pseudomonas entomophila* BS1	Reactive Black 5	Khan and Abdul (2015)
Azoreductase	*Enterobacter* sp. SXCR	Sulfonated azo dye (Congo red)	Prasad and Aikat (2014)
Azoreductase and NADH-DCIP reductase	*Alishewanella* sp. strain KMK6	Reactive Blue 59 Golden Yellow HER	Kolekar et al. (2013)
Superoxide dismutase and catalase	*Lysinibacillus* sp.	Sulfonated azo dye Reactive Orange 16 (RO16)	Bedekar et al. (2014)

Irrespective of the approach, the important functionality of immobilization is optimization of catalytic activities of immobilized enzyme. This is usually realized through improved thermal and environmental stability, mainly the prevention of denaturation resulting in deactivation. Immobilization also serves to insolubilize the enzymes making it easier for recovery with potential for reuse. The link by physical surface adsorption is the simplest one and drives immobilizing enzymes on solid supports through low energy interactions, such as hydrogen bonds, van der Waals or, hydrophobic effects, among others. Often, these physical interactions alone are too weak to keep the enzyme from desorbing from the support carrier as observed in rigorous industrial settings (for example, high reactant and product concentrations; and high ionic strength). The issue of leaching or, leakage and excessive denaturing is reduced through encapsulation wherein the physical inclusion or, confinement of the guest enzyme into a host support matrix (such as, organic or inorganic polymer matrices/hollow fiber membrane/microcapsule) leads to favorable microenvironment. Carrierless macroparticles may also be formed by a covalent bond involving amino acid residues (–NH_2, –CO_2, –SH) of the enzyme to a water-insoluble support matrix, or, by crosslinking with the matrix as seen in the recent upsurge on the use of cross-linked enzyme crystals (CLECs) and cross-linked enzyme aggregates (CLEAs). This strong and irreversible strategy offers distinct advantages. Large surface area supports matrices with big pore diameters where substrate and product can diffuse freely with much lesser enzyme leaching, greater operating stability, low production costs as additional (and often expensive) carrier is obviated.

The first report on immobilization of enzyme dates back in 1916 (Robinson 2015). Since then, approaches and strategies for enzyme immobilization are reviewed and published in over numerous publications (Cao 2006). Nowadays, nanoparticles are considered to be one of the most promising support materials for enzyme immobilization as they provide large surface areas to load more enzyme, reduces mass transfer resistance for substrates, reduces protein unfolding, possesses unique properties such as mobility, confinement effects, solution behavior and interfacial properties that profoundly increases the performance of enzymes (Khan et al. 2019). Thus, there are many investigations depicting the immobilization of enzymes on different types of nanoparticles (metal nanoparticles, metal oxide nanoparticles, magnetic nanoparticles, porous and polymeric nanoparticles, etc.) (Ahmad and Sardar 2015). Several materials such as multi-walled carbon nanotubes (MWCNT), graphene oxide (GO), titanium dioxide (TiO_2), polymers, biopolymers (agarose, chitosan), inorganics (silica, kaolinite, etc.), metal-organic frameworks (MOFs) have been utilized to immobilize the enzymes so as to trade-off three contending interests like (i) retaining high enzymatic activity; (ii) maintaining good long-term stability against physicochemical factors including temperature, dehydration, organic solvents, and or aggressive pH; and (iii) facilitating a regulation or, reversible switching of enzyme activity. Utilization of nanoparticles as enzyme carriers enables to overcome both the need to engineer enzymes and to reduce the negative effects of enzyme immobilization on micro/macro surfaces, partly due to high surface area to volume ratios that augment catalysis; surface chemistry well-suited for bioconjugation/biofunctionalization; and length scales that integrate well with and accordingly influence biological processes such as cell uptake/metabolism and gene expression (Ansari and Husain 2012).

Numerous examples can be traced on the use of nanoparticles/nanomaterials to immobilize enzymes for the applications in textile industry, particularly for the wastewater treatments (Ding et al. 2015). In the context of dye decolorization, several attempts have been reported. For instance, ginger peroxidase on amino-functionalized silica-coated TiO_2 nanocomposite for acid yellow 42 (Ali et al. 2017), laccase through chitosan nanoparticles on glass beads for congo red (Sadighi and Faramarzi 2013), laccase on Fe_3O_4/SiO_2 nanoparticles for Procion Red MX-5B (Wang et al. 2013), laccase on poly(p-phenylenediamine)/Fe_3O_4 nanocomposite for reactive blue 19 dye (Liu et al. 2016), laccase on functionalized multi-walled carbon nanotube membranes for Reactive Black 5, laccase on MOF-based nano-biocatalyst for degrading organic pollutant (Acid Blue 92: AB92) (Othman et al. 2016; Mahmoodi and Abdi 2019), immobilization of HRP on nanocomposite support for the removal of Acid Blue 113 and Acid Black 10 BX (Jun et al. 2019), phenolic compounds such as p-chlorophenol and Bisphenol A (Lu et al. 2017), etc. The nanoparticles/nanomaterials are the key components of the future that will provide efficient and biocompatible environment for enzyme immobilization.

5 Challenges in Application of Immobilized Enzymes and a Paradigm Shift Towards In Situ Immobilization

Although protein editing/engineering is used for enhancing enzyme properties, immobilization through covalent attachment or, encapsulation to carrier matrices constitutes a crucial approach for stability and process performance. In textile industry, where enzymes act on macromolecular substrates, the immobilization of enzymes poses problems such as insolubility, increased size of enzymes, loss of specific activity. Immobilization of enzymes helps in increasing the enzymatic stability and reusability while retaining the catalytic activity. However, selection of the most cost-effective method is of utmost importance. Thus, it would be beneficial to reduce the production steps of enzyme immobilization. The in situ immobilization is a new emergent approach that enables production of active enzymes and carrier materials at one go, simultaneously, by bioengineering microbes (Rehm et al. 2016). It is an emerging strategy that put forth a cost-effective method by eliminating the immobilization step; eradicates the application of rough conditions that imposes negative impact on enzyme activity; and simultaneously, simplifies the purification steps (Rehm et al. 2017). In situ enzyme immobilization paves the way for economic production of soft materials like immobilized biocatalysts for treatment of textile effluent. The in situ immobilization approaches use molecules such as protein, lipid, and polymer; magnetosomes; membraneous structures (vesicles); and insoluble inclusion body. As attachment of enzyme to carrier occurs in situ, a high level of functionality could be retained opening new possibilities of establishing multi-enzyme cascade reaction (Rehm et al. 2017). The in situ immobilization strategies is eventually broadening the horizon of immobilization techniques and emerging as one of the most promising approach for immobilizing enzymes.

6 Concluding Remarks and Outlook

The development of enzyme-based process in textile industry is considered as a green alternative and till now they are established as a dependable tool in textile processing. To cater to the need of large scale industrial processes, optimization or modulation of the properties of enzymes needs to be taken care of. Enzyme immobilization offers and enables efficient enzyme recovery as well as reusability, improved enzyme performance, enhanced storage and operational stability. However, the cost intensiveness of the process limits its large scale commercialization. Herein, in situ immobilization approaches are promising as a cost-competitive single-step option to produce bulk quantities of enzymes. It enables enzyme engineering so as to self-assemble into nano-/micro-sized supramolecular structures inside the cell thus, obviating the requirement of building a matrix separately for enzyme encapsulation or, attachment. The enzymes produced are already immobilized and so, this useful process paves way for cost-competitive enzymes production. Thus, further investigations are warranted to exhibit the potential and realization of in situ immobilization strategies for textile industry processing and its sustainable effluent treatment.

Acknowledgements The authors are grateful to Indian Institute of Technology Guwahati, Assam, India for the infrastructural facilities.

Conflict of Interest No conflict of interest to declare.

References

Agrawal S, Tipre D, Patel B et al (2014) Optimization of triazo Acid Black 210 dye degradation by *Providencia* sp. SRS82 and elucidation of degradation pathway. Process Biochem 49:110–119. https://doi.org/10.1016/j.procbio.2013.10.006

Ahmad R, Sardar M (2015) Enzyme immobilization: an overview on nanoparticles as immobilization matrix. Biochem Anal Biochem 4:1–8. https://doi.org/10.4172/2161-1009.1000178

Ali M, Husain Q, Alam N, Ahmad M (2017) Enhanced catalytic activity and stability of ginger peroxidase immobilized on amino-functionalized silica-coated titanium dioxide nanocomposite: a cost-effective tool for bioremediation. Wat Air Soil Poll 228:22. https://doi.org/10.1007/s11270-016-3205-4

Aniş P, Davulcu A, Eren HA (2009) Enzymatic pre-treatment of cotton. Part 2: peroxide generation in desizing liquor and bleaching. FIBRES Text East Eur 17:87–90

Ansari SA, Husain Q (2012) Potential applications of enzymes immobilized on/in nano materials: a review. Biotechnol Adv 30:512–523. https://doi.org/10.1016/j.biotechadv.2011.09.005

Araujo R, Casal M, Cavaco-Paulo A (2008) Application of enzymes for textiles fibers processing. Biocatal Biotechnol 26:332–349. https://doi.org/10.1080/10242420802390457

Asad S, Amoozegar MA, Pourbabaee A, Sarbolouki MN, Dastgheib SMM (2007) Decolorization of textile azo dyes by newly isolated halophilic and halotolerant bacteria. Bioresour Technol 98:2082–2088. https://doi.org/10.1016/j.biortech.2006.08.020

Asgher M, Asad MJ, Bhatti HN, Legge RL (2007) Hyperactivation and thermostabilization of *Phanerochaete chrysosporium* lignin peroxidase by immobilization in xerogels. World J Microbiol Biotechnol 23:525–531. https://doi.org/10.1007/s11274-006-9255-9

Asgher M, Bhatti HN, Ashraf M, Legge RL (2008) Recent developments in biodegradation of industrial pollutants by white rot fungi and their enzyme system. Biodegradation 19:771–783. https://doi.org/10.1007/s10532-008-9185-3

Auterinen AL, Prozzo B, Redling E, Vermeersch L, Yoon MY (2012) U.S. Patent Application No. 13/063,140

Azevedo H, Bishop D, Cavaco-Paulo A (2002) Possibilities for recycling cellulases after use in cotton processing. Appl Biochem Biotechnol 101:61–75. https://doi.org/10.1385/ABAB:101:1:61

Banat IM, Nigam P, Singh D, Marchant R (1996) Microbial decolorization of textile-dye containing effluents: a review. Bioresour Technol 58:217–227. https://doi.org/10.1016/S0960-8524(96)00113-7

Basto C, Tzanov T, Cavaco-Paulo A (2007) Combined ultrasound-laccase assisted bleaching of cotton. Ultrason Sonochem 14:350–354. https://doi.org/10.1016/j.ultsonch.2006.07.006

Bedekar PA, Saratale RG, Saratale GD et al (2014) Oxidative stress response in dye degrading bacterium *Lysinibacillus* sp. RGS exposed to Reactive Orange 16, degradation of RO16 and evaluation of toxicity. Environ Sci Pollut Res 21:11075–11085. https://doi.org/10.1007/s11356-014-3041-2

Bommarius AS, Paye MF (2013) Stabilizing biocatalysts. Chem Soc Rev 42:6534–6565. https://doi.org/10.1039/C3CS60137D

Buchert J, Pere J, Puolakka A, Nousiainen P (2000) Scouring of cotton with pectinases, proteases, and lipases. Text Chem Colorist Am Dyest Report 32:48–52

Buschle-Diller G, El Mogahzy Y, Inglesby MK, Zeronian SH (1998) Effects of scouring with enzymes, organic solvents, and caustic soda on the properties of hydrogen peroxide bleached cotton yarn. Text Res J 68:920–929. https://doi.org/10.1177/004051759806801207

Cao L (2005) Immobilised enzymes: science or art? Curr Opin Chem Biol 9:217–226. https://doi.org/10.1016/j.cbpa.2005.02.014

Cao L (2006) Carrier-bound immobilized enzymes: principles, application and design. Wiley, Hoboken

Cavaco-Paulo A (1998) Mechanism of cellulase action in textile processes. Carbohydr Polym 37:273–277. https://doi.org/10.1016/S0144-8617(98)00070-8

Chand N, Nateri AS, Sajedi RH, Mahdavi A, Rassa M (2012) Enzymatic desizing of cotton fabric using a Ca^{2+}independent α-amylase with acidic pH profile. J Mol Catal B: Enzymatic 83:46–50. https://doi.org/10.1016/j.molcatb.2012.07.003

Chatha SAS, Asgher M, Iqbal HMN (2017) Enzyme-based solutions for textile processing and dye contaminant biodegradation—a review. Environ Sci Pollut Res 24:14005–14018. https://doi.org/10.1007/s11356-017-8998-1

Chen S, Su L, Chen J, Wu J (2013) Cutinase: characteristics, preparation, and application. Biotechnol Adv 31:1754–1767. https://doi.org/10.1016/j.biotechadv.2013.09.005

Cheng XB, Rong JIA, Ping-Sheng LI, Qin ZHU, Shi-Qian TU, Wen-Zhong TANG (2007) Studies on the properties and co-immobilization of manganese peroxidase. Chin J Biotechnol 23:90–96. https://doi.org/10.1016/S1872-2075(07)60006-5

Chhabra M, Mishra S, Sreekrishnan TR (2015) Combination of chemical and enzymatic treatment for efficient decolorization/degradation of textile effluent: High operational stability of the continuous process. Biochem Eng J 93:17–24. https://doi.org/10.1016/j.bej.2014.09.007

Choe EK, Nam CW, Kook SR, Chung C, Cavaco-Paulo A (2004) Implementation of batchwise bioscouring of cotton knits. Biocatal Biotransfor 22:375–382. https://doi.org/10.1080/10242420400024540

Choi JM, Han SS, Kim HS (2015) Industrial applications of enzyme biocatalysis: current status and future aspect. Biotechnol Adv 33:1443–1454. https://doi.org/10.1016/j.biotechadv.2015.02.014

Christian V, Shrivastava R, Shukla D, Modi H, Rajiv B, Vyas M (2005) Mediator role of veratryl alcohol in the lignin peroxidase-catalyzed oxidative decolorization of Remazol Brilliant Blue R. Enzyme Microbial Technol 36:426–431. https://doi.org/10.1016/j.enzmictec.2004.06.007

Cipolatti EP, Silva MJA, Klein M et al (2014) Current status and trends in enzymatic nanoimmobilization. J Mol Catal B: Enzymatic 99:56–67. https://doi.org/10.1016/j.molcatb.2013.10.019

Ding S, Cargill AA, Medintz IL, Claussen JC (2015) Increasing the activity of immobilized enzymes with nanoparticle conjugation. Curr Opin Biotechnol 34:242–250. https://doi.org/10.1016/j.copbio.2015.04.005

Doshi R, Shelke V (2001) Enzymes in textile industry-An environment-friendly approach. Indian J Fibre Text Res 26:202–205

Du LN, Li G, Zhao YH et al (2015) Efficient metabolism of the azo dye methyl orange by *Aeromonas* sp. strain DH-6: characteristics and partial mechanism. Int Biodeterior Biodegrad 105:66–72. https://doi.org/10.1016/j.ibiod.2015.08.019

Esfandiari A, Firouzi-Pouyaei E, Aghaei-Meibodi P (2014) Effect of enzymatic and mechanical treatment on combined desizing and bio-polishing of cotton fabrics. J Text Inst 105:1193–1202. https://doi.org/10.1080/00405000.2014.880222

Farooq A, Ali S, Abbas N, Fatima GA, Ashraf MA (2013) Comparative performance evaluation of conventional bleaching and enzymatic bleaching with glucose oxidase on knitted cotton fabric. J Clean Prod 42:167–171. https://doi.org/10.1016/j.jclepro.2012.10.021

Fu KL, Wang DB, Li Y, Lu DN (2015) Effect of additives on mesophilic α-amylase and its application in the desizing of cotton fabrics. J Text I106(13):22–1327

Georgiou D, Aivazidis A, Hatiras J, Gimouhopoulos K (2003) Treatment of cotton textile wastewater using lime and ferrous sulfate. Water Res 37:2248–2250. https://doi.org/10.1016/S0043-1354(02)00481-5

Guisan JM (2006) Immobilization of enzymes as the 21st century begins. In: Guisan JM (ed) Immobilization of enzymes and cells, 2nd edn. Humana Press Inc., New Jersey (NJ), p 113

Hanson MA, Gilbert RD (1974) A new look at desizing with enzymes. Text Chem Color 6:28–31

Hao L, Wang R, Fang K, Liu J (2013) Ultrasonic effect on the desizing efficiency of α-amylase on starch-sized cotton fabrics. Carbohydr Polym 96:474–480. https://doi.org/10.1016/j.carbpol.2013.04.003

Hartzell MM, Hsieh YL (1998) Enzymatic scouring to improve cotton fabric wettability. Text Res J 68:233–241. https://doi.org/10.1177/004051759806800401

Heine E, Höcker H (1995) Enzyme treatments for wool and cotton. Rev Prog Color Relat Top 25:57–70. https://doi.org/10.1111/j.1478-4408.1995.tb00104.x

Hernandez K, Fernandez-Lafuente R (2011) Control of protein immobilization: coupling immobilization and side-directed mutagenesis to improve biocatalyst or biosensor performance. Enzyme Microb Technol 48:107–122. https://doi.org/10.1016/j.enzmictec.2010.10.003

Holkar CR, Jadhav AJ, Pinjari DV, Mahamuni NM, Pandit AB (2016) A critical review on textile wastewater treatments: possible approaches. J Environ Manag 182:351–366. https://doi.org/10.1016/j.jenvman.2016.07.090

Horvathova V, Janecek S, Sturdik E (2001) Amylolytic enzymes: molecular aspects of their properties. Gen Physiol Biophys 20:7–32

Ibrahim NA, Allam EA, El-Hossamy MB, El-Zairy WM (2008) Enzymatic modification of cotton/wool and viscose/wool blended fabrics. J Nat Fibers 5:154–169. https://doi.org/10.1080/15440470801929648

Ibrahim NA, El-Badry K, Eid BM, Hassan TM (2011) A new approach for biofinishing of cellulose-containing fabrics using acid cellulases. Carbohydr Polym 83:116–121. https://doi.org/10.1016/j.carbpol.2010.07.025

Imran M, Negm F, Hussain S et al (2016) Characterization and purification of membrane-bound azoreductase from azo dye degrading *Shewanella* sp. strain IFN4. CLEAN–Soil Air Wat 44:1523–1530. https://doi.org/10.1002/clen.201501007

Jin XC, Liu GQ, Xu ZH, Tao WY (2007) Decolorization of a dye industry effluent by *Aspergillus fumigatus* XC6. Appl Microbiol Biotechnol 74:239–243. https://doi.org/10.1007/s00253-006-0658-1

Jun LY, Yon LS, Mubarak et al (2019) An overview of immobilized enzyme technologies for dye, phenolic removal from wastewater. J Environ Chem Eng 14:102961. https://doi.org/10.1016/j.jece.2019.102961

Kanagaraj J, Senthilvelan T, Panda RC (2015) Degradation of azo dyes by laccase: biological method to reduce pollution load in dye wastewater. Clean Technol Environ Policy 17:1443–1456. https://doi.org/10.1007/s10098-014-0869-6

Khan S, Abdul M (2015) Degradation of reactive black 5 dye by a newly isolated bacterium *Pseudomonas entomophila* BS1. Can J Microbiol 62:220–232. https://doi.org/10.1139/cjm-2015-0552

Khan MF, Kundu D, Hazra C, Patra S (2019) A strategic approach of enzyme engineering by attribute ranking and enzyme immobilization on zinc oxide nanoparticles to attain thermostability in mesophilic *Bacillus subtilis* lipase for detergent formulation. Inter J Biol Macromol 136:66–82. https://doi.org/10.1016/j.ijbiomac.2019.06.042

Kolekar YM, Konde PD, Markad VL, Kulkarni SV, Chaudhari AU, Kodam KM (2013) Effective bioremoval and detoxification of textile dye mixture by *Alishewanella* sp. KMK6. Appl Microbiol Biotechnol 97:881–889. https://doi.org/10.1007/s00253-011-3698-0

Lade H, Kadam A, Paul D et al (2015) Biodegradation and detoxification of textile azo dyes by bacterial consortium under sequential microaerophilic/aerobic processes. EXCLI J Exp Clin Sci 14:158–174. https://doi.org/10.17179/excli2014-642

Lee SH, Song WS (2010) Surface modification of polyester fabrics by enzyme treatment. Fibers Polym 11:54–59. https://doi.org/10.1007/s12221-010-0054-4

Li Y, Hardin ZR (1997) Enzymatic scouring of cotton-effects on structure and properties. Text Chem Colorist 29:71–76

Li S, Yang X, Yang S et al (2012) Technology prospecting on enzymes: application, marketing and engineering. Comput Struct Biotechnol J 2:1–11. https://doi.org/10.5936/csbj.201209017

Liese A, Hilterhaus L (2013) Evaluation of immobilized enzymes for industrial applications. Chem Scoc Rev 42:6236–6249. https://doi.org/10.1039/C3CS35511J

Liu Y, Yan M, Geng Y, Huang J (2016) Laccase immobilization on poly (p-phenylenediamine)/Fe_3O_4 nanocomposite for reactive blue 19 dye removal. Appl Sci 6:232. https://doi.org/10.3390/app6080232

Lu R, Shen XL, Xia LM (2005) Studies on laccase production by Coriolus versicolor and enzymatic decoloration of dye. Linchan Huaxue Yu Gongye/Chem Ind Forest Prod 25:73–76

Lu YM, Yang Q, Wang LM et al (2017) Enhanced activity of immobilized horseradish peroxidase by carbon nanospheres for phenols removal. CLEAN–Soil Air Wat 45:1600077. https://doi.org/10.1002/clen.201600077

Madhu A, Chakraborty JN (2017) Developments in application of enzymes for textile processing. J Clean Prod 145:114–133. https://doi.org/10.1016/j.jclepro.2017.01.013

Mahmoodi NM, Abdi J (2019) Metal-organic framework as a platform of the enzyme to prepare novel environmentally friendly nanobiocatalyst for degrading pollutant in water. J Ind Eng Chem 80:606–613. https://doi.org/10.1016/j.jiec.2019.08.036

Mehta R, Singhal P, Singh H et al (2016) Insight into thermophiles and their wide-spectrum applications. 3 Biotech 6:1–9. https://doi.org/10.1007/s13205-016-0368-z

Misson M, Zhang H, Jin B (2014) Nanobiocatalyst advancements and bioprocessing applications. J R Soc Interface 12:20140891. https://doi.org/10.1098/rsif.2014.0891

Mojsov K (2011) Application of enzymes in the textile industry: a review. In: Proceedings of II international congress on engineering, ecology and materials in the processing industry, pp 230–239

Mojsov KD, Andronikov D, Janevski A, Kuzelov A, Gaber S (2016) The application of enzymes for the removal of dyes from textile effluents. Adv Technol 5:81–86

Murugesan K, Arulmani M, Nam I-H, Kim Y-M, Chang Y-S, Kalaichelvan PT (2006) Purification and characterization of laccase produced by a white rot fungus *Pleurotus sajor-caju* under submerged culture condition and its potential in decolorization of azo dyes. Appl Microbiol Biotechnol 72:939–946. https://doi.org/10.1007/s00253-006-0403-9

Ortiz C, Ferreira ML, Barbosa O et al (2019) Novozym 435: the "perfect" lipase immobilized biocatalyst? Cat Sci Technol 9:2380–2420. https://doi.org/10.1039/C9CY00415G

Othman AM, González-Domínguez E, Sanromán Á, Correa-Duarte M, Moldes D (2016) Immobilization of laccase on functionalized multiwalled carbon nanotube membranes and application for dye decolorization. RSC Adv 6:114690–114697. https://doi.org/10.1039/C6RA18283F

Prasad SS, Aikat K (2014) Study of bio-degradation and bio-decolourization of azo dye by *Enterobacter* sp. SXCR Environ Technol 35:956–965. https://doi.org/10.1080/09593330.2013.856957

Presa P, Tavcer PF (2007) Pectinases as agents for bioscouring. Tekstilec 50:16–34

Quaratino D, Federici F, Petruccioli M, Fenice M, D'Annibale A (2007) Production, purification and partial characterisation of a novel laccase from the white-rot *fungus Panus tigrinus* CBS577.79. Antonie Van Leeuwenhoek Int J Genet Mol Microbiol 91:57–69. https://doi.org/10.1007/s10482-006-9096-4

Raichurkar P, Ramachandran M (2015) Recent trends and developments in textile industry in India. Int J Text Eng Proc 1:47–50

Ramadan AA (2009) Characterization of biobleaching of cotton/linen fabrics. J Tex App Technol Manag 6:1–12

Rehm FB, Chen S, Rehm BH (2016) Enzyme engineering for in situ immobilization. Molecules 21:1370. https://doi.org/10.3390/molecules21101370

Rehm FB, Chen S, Rehm BH (2017) Bioengineering toward direct production of immobilized enzymes: a paradigm shift in biocatalyst design. Bioengineered 9:6–11. https://doi.org/10.1080/21655979.2017.1325040

Rekik H, Jaouadi NZ, Bouacem K et al (2019) Physical and enzymatic properties of a new manganese peroxidase from the white-rot fungus *Trametes pubescens* strain i8 for lignin biodegradation and textile-dyes biodecolorization. Int J Biol Macromol 125:514–525. https://doi.org/10.1016/j.ijbiomac.2018.12.053

Ren X, Buschle-Diller G (2007) Oxidoreductases for modification of linen fibers. Colloids Surfaces A Physicochem Eng Aspects 299:15–21. https://doi.org/10.1016/j.colsurfa.2006.11.011

Ribitsch D, Herrero Acero E, Greimel K et al (2012) A new esterase from *Thermobifida halotolerans* hydrolyses polyethylene terephthalate (PET) and polylactic acid (PLA). Polymers 4:617–629. https://doi.org/10.3390/polym4010617

Robinson PK (2015) Enzymes: principles and biotechnological applications. Essays Biochem 59:1–41. https://doi.org/10.1042/bse0590001

Sadighi A, Faramarzi MA (2013) Congo red decolorization by immobilized laccase through chitosan nanoparticles on the glass beads. J Taiwan Inst Chem Eng 44:156–162. https://doi.org/10.1016/j.jtice.2012.09.012

Santos A, Mendes S, Brissos V et al (2014) New dye-decolorizing peroxidases from *Bacillus subtilis* and *Pseudomonas putida* MET94: towards biotechnological applications. Appl Microbiol Biotechnol 98:2053–2065. https://doi.org/10.1007/s00253-013-5041-4

Saravanan D, Vasanthi NS, Raja KS, Das A, Ramachandran T (2010) Bleaching of cotton fabrics using hydrogen peroxide produced by glucose oxidase. Indian J Fibre Text Res 35:281–283

Saravanan D, Prakash AA, Jagadeeshwaran D, Nalankilli G, Ramachandran T, Prabakaran C (2011) Optimization of thermophile *Bacillus licheniformis*-amylase desizing of cotton fabrics. Ind J Fib Tex Res 36:253–258

Sarkar S, Banerjee A, Halder U, Biswas R, Bandopadhyay R (2017) Degradation of synthetic azo dyes of textile industry: a sustainable approach using microbial enzymes. Water Conserv Sci Eng 2:121–131. https://doi.org/10.1007/s41101-017-0031-5

Sheldon RA, van Pelt S (2013) Enzyme immobilisation in biocatalysis: why, what and how. Chem Scoc Rev 42:6223–6235. https://doi.org/10.1039/C3CS60075K

Shin Y, Hwang S, Ahn IS (2004) Enzymatic bleaching of desized cotton fabrics with hydrogen peroxide produced by glucose oxidase. J Ind Eng Chem 10:577–581

Singh RP, Singh PK, Singh RL (2014) Bacterial decolorization of textile azo dye acid orange by *Staphylococcus hominis* RMLRT03. Toxicol Int 21:160–166. https://doi.org/10.4103/0971-6580.139797

Singh RL, Singh PK, Singh RP (2015) Enzymatic decolorization and degradation of azo dyes—a review. Int Biodeterio Biod 104:21–31. https://doi.org/10.1016/j.ibiod.2015.04.027

Soares JC, Moreira PR, Queiroga AC, Morgado J, Malcata FX, Pintado ME (2011) Application of immobilized enzyme technologies for the textile industry: a review. Biocatal Biotransform 29:223–237. https://doi.org/10.3109/10242422.2011.635301

Solbak AI, Richardson TH, McCann RT et al (2005) Discovery of pectin-degrading enzymes and directed evolution of a novel pectate lyase for processing cotton fabric. J Biol Chem 280:9431–9438. https://doi.org/10.1074/jbc.M411838200

Tavčer PF (2012) Low-temperature bleaching of cotton induced by glucose oxidase enzymes and hydrogen peroxide activators. Biocatal Biotransfor 30:20–26. https://doi.org/10.3109/10242422.2012.644437

Teerapatsakul C, Parra R, Keshavarz T, Chitradon L (2017) Repeated batch for dye degradation in an airlift bioreactor by laccase entrapped in copper alginate. Int Biodeterior Biodegradation 120:52–57. https://doi.org/10.1016/j.ibiod.2017.02.001

Traore MK, Buschle-Diller G (1999) Influence of wetting agents and agitation on enzymatic hydrolysis of cotton. Text Chem Colorist Am Dyest Report 31:51–56

Tzanov T, Calafell M, Guebitz GM, Cavaco-Paulo A (2001) Bio-preparation of cotton fabrics. Enzyme Microbiol Technol 29:357–362. https://doi.org/10.1016/S0141-0229(01)00388-X

Tzanov T, Costa SA, Gübitz GM, Cavaco-Paulo A (2002) Hydrogen peroxide generation with immobilized glucose oxidase for textile bleaching. J Biotechnol 93:87–94. https://doi.org/10.1016/S0168-1656(01)00386-8

Tzanov T, Andreaus J, Gübitz GM, Cavaco-Paulo A (2003a) Protein interactions in enzymatic processes in textiles. Electron J Biotechnol 6:17–23. https://doi.org/10.4067/S0717-34582003000300013

Tzanov T, Basto C, Gübitz GM, Cavaco-Paulo A (2003b) Laccases to improve the whiteness in a conventional bleaching of cotton. Macromol Mater Eng 288:807–810. https://doi.org/10.1002/mame.200300100

Ullrich R, Le MH, Nguyen LD, Hofrichter M (2005) Laccase from the medicinal mushroom *Agaricus blazei*: production, purification and characterization. Appl Microbiol Biotechnol 67:357–363. https://doi.org/10.1007/s00253-004-1861-6

Vankar PS, Shanker R, Verma A (2007) Enzymatic natural dyeing of cotton and silk fabrics without metal mordants. J Clean Prod 15:1441–1450. https://doi.org/10.1016/j.jclepro.2006.05.004

Vílchez S, Jovančić P, Erra P (2010) Influence of chitosan on the effects of proteases on wool fibers. Fibers Polym 11:28–35. https://doi.org/10.1007/s12221-010-0028-6

Wang WM, Yu B, Zhong CJ (2012) Use of ultrasonic energy in the enzymatic desizing of cotton fabric. J Clean Prod 33:179–182. https://doi.org/10.1016/j.jclepro.2012.04.010

Wang H, Zhang W, Zhao J, Xu L, Zhou C, Chang L, Wang L (2013) Rapid decolorization of phenolic azo dyes by immobilized laccase with Fe_3O_4/SiO_2 nanoparticles as support. Ind Eng Chem Res 52:4401–4407. https://doi.org/10.1021/ie302627c

Waring DR, Hallas G (eds) (2013) The chemistry and application of dyes. Springer Science & Business Media

Weisburger JH (2002) Comments on the history and importance of aromatic and heterocyclic amines in public health. Mut Res 506:9–20. https://doi.org/10.1016/S0027-5107(02)00147-1

Wesenberg D, Kyriakides I, Agathos SN (2003) White-rot fungi and their enzymes for the treatment of industrial dye effluents. Biotechnol Adv 22:161–187. https://doi.org/10.1016/j.biotechadv.2003.08.011

Xu H, Heinze TM, Chen S, Cerniglia CE, Chen H (2007) Anaerobic metabolism of 1-amino-2-naphthol-based azo dyes (Sudan dyes) by human intestinal microflora. Appl Environ Microbiol 73:7759–7762. https://doi.org/10.1128/AEM.01410-07

Yachmenev VG, Bertoniere NR, Blanchard EJ (2002) Intensification of the bio-processing of cotton textiles by combined enzyme/ultrasound treatment. J Chem Technol Biotechnol 77:559–567. https://doi.org/10.1002/jctb.579